STATISTICS with APPLICATIONS in BIOLOGY and GEOLOGY

STATISTICS with APPLICATIONS in BIOLOGY and GEOLOGY

Preben Blæsild
Jørgen Granfeldt

A CRC Press Company
Boca Raton London New York Washington, D.C.

Library of Congress Cataloging-in-Publication Data

Blæsild, Preben.
Statistics with applications in biology and geology / Preben Blæsild and Jørgen Granfeldt.
p. cm.
Includes bibliographical references (p.).
ISBN 1-58488-309-X (alk. paper)
1. Biometry. 2. Geology—Statistical methods. I. Granfeldt, Jørgen. II. Title.

QH323.5 .B58 2002
570′.1′5195—dc21 2002034862

Direct all inquiries to CRC Press LLC, 2000 N.W. Corporate Blvd., Boca Raton, Florida 33431.

International Standard Book Number 1-58488-309-X
Library of Congress Card Number 2002034862
Printed in the United States of America 1 2 3 4 5 6 7 8 9 0
Printed on acid-free paper

Preface

The book gives a sound basis for analyzing data using fundamental parametric statistical models. It evolved out of an introductory course in statistics for students of biology and geology at the University of Aarhus, Denmark. The course has had this form since 1993. Prior to the introductory course in statistics, students of biology and geology take a course in basic probability theory and a course in mathematics.

Examples and exercises in the book are motivated by applications in biology and geology.

The fundamental models are one sample, two samples, one- and two-way analysis of variance, and linear regression for normal data and similar models for binomial, multinomial, and Poisson data. In addition, all those models are united in a chapter on generalized linear models. Moreover, the statistical analysis of directional data is treated. The emphasis is on parametric statistical models, but a single chapter is devoted to the most important nonparametric tests. Correlation and the connection between correlation and regression are explained based on the bivariate normal distribution.

Behind the book are fairly strong views on statistical analysis in general and on introductory courses in statistics, primarily aimed at users from other fields in particular.

The focus is on analyzing data to answer questions that arise in a scientific context but with a firm basis in models, the likelihood method, and numeracy.

The basis of any statistical analysis is a statistical model. The statistical model is a concise specification of the assumptions underlying an analysis, and it is the basis of the communication of the results. It is an important requirement of the model that it can be used to answer the scientific questions that are of interest. Thus the parameters of the model must have an interpretation in the scientific context, and interesting scientific hypotheses must correspond to restrictions on the parameters of the model. Restricting the parameters of a model leads to a new model that is a submodel of the original model. A particular scientific question is answered by testing whether the data are consistent with the reduction to a submodel that is simpler because it has fewer parameters. With this approach the readers quickly learn to correctly analyze fairly complicated problems. In addition, the analysis becomes transparent and understandable, not only for those who make the analysis but also for people who are neither specialists in the subject nor professional statisticians.

The likelihood method is used both to present statistics as a coherent methodology and also to prepare the reader for more complicated models where the likelihood method is inevitable. This is demonstrated in the chapter where the generalized linear models are presented based on the paradigm of the statistical models and the likelihood method.

Numeracy is of paramount importance in statistics. For this purpose a statistics course today must use a statistical package; SAS has been chosen as the statistical package that is used to illustrate all the statistical tools described. But we firmly believe in learning by doing, and in this context this means that for the simplest models the calculations must be made both by hand on a pocket calculator as well as on a computer using the statistical package. Here the simplest models include the models for one, two, and k normal samples, as well as linear regression. More complicated models, such as two-way analysis of variance and comparison of two or more regression lines, will rely on a statistical package for the heavy calculations, but the test of the reductions from a model to a submodel will be computed by hand using relevant information in the output from the two models.

SAS programs and data used in examples and exercises are available from the website
`http://www.imf.au.dk/biogeostatistics`

When using a statistical package the models are very important. First, the models need to be specified in the language of the package so if one is not clear about the model, there is no way a statistical analysis can be performed. Secondly, output from statistical packages contain a lot of potentially useful information but again it depends on the model and its relevant submodels to determine what parts of the output will be useful.

A one-semester course in biostatistics or geostatistics usually covers Chapters 1 through 5, 7, a little more than half of Chapter 8 and parts of Chapter 10. A course starts with Chapter 2 on graphical methods and continues with the models for normally distributed data that are covered in Chapters 3, 4, and 5. Next the models for multinomial data in Chapter 7 and Poisson data in Chapter 8 are treated and, finally, the course concludes with Chapter 10 on directional data. Chapter 1 begins with two examples of statistical analyses and continues with a general introduction to the main aspects of the analysis of a parametric statistical model. More mathematical details are considered in Chapter 11. Chapter 6 is concerned with correlation, and the connection between correlation and linear regression is discussed based on the bivariate normal distribution. Finally, some important nonparametric tests are considered in Chapter 12.

It is possible to read the chapters in a different order, starting with discrete data in Chapter 7 and Chapter 8, but it should be noted that the basic statistical concepts of null hypothesis, test, significance probability (p-value), significance level, and so on, are presented in connection with a single normal sample in Section 3.1.

Without data that originates from a real problem, a textbook and a course in statistics for users becomes fairly uninteresting. We would like to thank colleagues and students who have made data and their history available for the examples and the exercises of this book.

We are grateful to Michael Sørensen for his patience and the help we have received over the years with the computer system.

Finally, we would like to thank Lars Madsen, the local LaTeX expert, for his enthusiasm, and his prompt and efficient help with all kinds of LaTeX problems.

Preben Blæsild and Jørgen Granfeldt

Contents

CHAPTER 1

Statistical Analysis

This chapter describes the most important ingredients in statistical analysis. This is done in two ways. We first present two examples of a statistical analysis without assuming any knowledge of statistics. Next we present basic concepts and ideas as clearly as possible. These concepts are discussed in great detail in connection with concrete examples in later chapters and, furthermore, in the course of the exposition they will be applied and illustrated over and over again.

Section 1.2 is concerned with data from observational studies or scientific experiments. We have chosen to focus on three main ingredients or activities in a statistical analysis, which are considered in Sections 1.3–1.5.

(i) model specification
(ii) model checking
(iii) statistical inference

Throughout the book the statistical inference is based on *the likelihood function* that will be introduced in Section 3.1.4 on page 70. A theoretical discussion starts on page 481 in Chapter 11. The final section in this chapter, Section 1.6, contains some concluding remarks.

1.1 Two Examples of a Statistical Analysis

Example 1.1

Based on morphological characters, the cod has been divided into a number of geographical races or subspecies. In Danish waters two subspecies are distinguished: the Baltic cod (*Gadus morhua calarias*) living in the eastern Baltic and the Atlantic cod (*Gadus morhua morhua*) inhabiting the western Baltic and the other Danish waters. The difference between the subspecies may reflect divergence of reproductively isolated populations, but alternatively it may just be a direct consequence of the difference in environment in the two areas. The aim of the following genetic investigation was to resolve this question.

At three locations, Fehmarn Belt, Bornholm, and the Åland Islands, marked by the numbers 1, 2, and 3 on the map in Figure 1.1, a number of cods were caught and their hemoglobin types were determined by electrophoresis. In cods the hemoglobin types are determined by a gene with two allelic types A and a resulting in the genotypes AA, Aa, and aa. The results from Sick (1965) are reviewed in Table 1.1.

Table 1.1 *The genotypes observed in the three samples*

	AA	Aa	aa
Fehmarn Belt	27	30	12
Bornholm	14	20	52
Åland Islands	0	5	75

If there is random mating and no selection in a population, then the population genotypes are

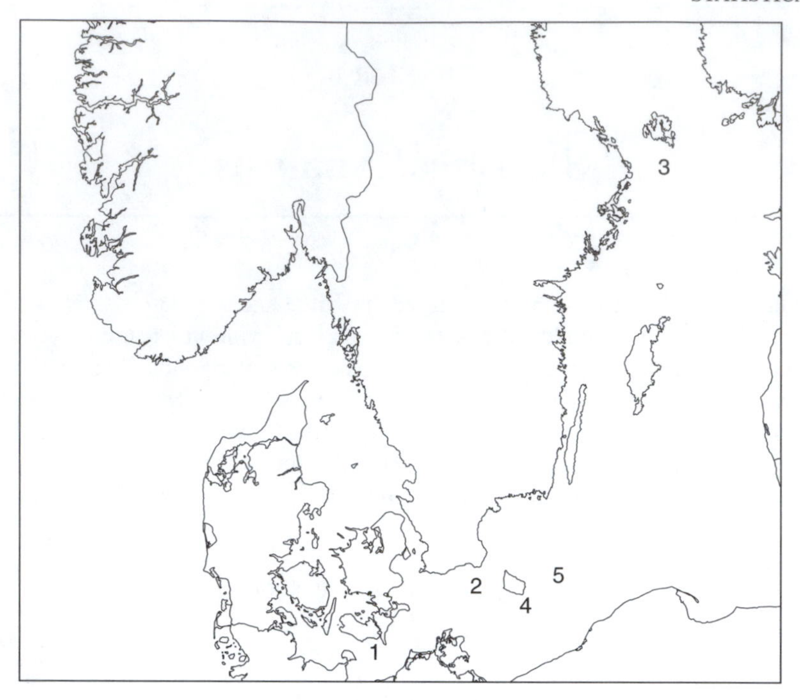

Figure 1.1 *The fishing grounds in Fehmarn Belt and near Bornholm and the Åland Islands.*

expected to occur in Hardy-Weinberg proportions, i.e., the genotypes are in the ratio $p^2 : 2p(1-p) : (1-p)^2$, where p is the frequency* of the allele A in the population. If the Baltic cod and the Atlantic cod are reproductively isolated populations and if both subspecies are represented at the location, then the assumption of random mating is not fulfilled and we cannot expect to find the cod population in Hardy-Weinberg proportions. It is therefore of interest to see if the hemoglobin data are consistent with the assumption that the genotypic frequencies at the three locations are in Hardy-Weinberg proportions.

First, we formulate a model for the data or the data collection. A *statistical model* gives the probability of all possible configurations of genotypes in the sample as a function of the genotype frequencies in the population. If we restrict attention to one location, the probability is

$$p(x_{AA}, x_{Aa}, x_{aa}) = \binom{n}{x_{AA}\ x_{Aa}\ x_{aa}} \pi_{AA}^{x_{AA}} \pi_{Aa}^{x_{Aa}} \pi_{aa}^{x_{aa}}. \tag{1.1}$$

Here x_{AA}, x_{Aa}, and x_{aa} denote the counts of the cods in the sample with genotypes AA, Aa, and aa, and therefore their sum is equal to the total number of cods in the sample, n, i.e., $x_{AA} + x_{Aa} + x_{aa} = n$. Similarly, π_{AA}, π_{Aa}, and π_{aa} denote the frequencies of the AA, Aa, and aa genotypes in the population, and their sum is equal to 1. Finally, the first factor on the right-hand side of (1.1) is the number of such samples, i.e., samples with n cods of which x_{AA}, x_{Aa}, and x_{aa} are of type AA, Aa, and aa, respectively. The probability distribution (1.1) in the model is the multinomial distribution,

* In genetics the "frequency of an allele A" is the proportion of genes in a population that are of allele type A and therefore a number between 0 and 1. This is in contrast to the meaning in statistics where a "frequency" is an integer. Although it would be more appropriate to use the term "relative frequency" in this context, we adopt the terminology from genetics in examples from this scientific area. In the same way "genotypic frequencies" means "relative genotypic frequencies" in genetics.

which will be treated in Chapter 7. The model holds under a number of assumptions that seem to be satisfied in this case.

Next, we formulate the *hypothesis* of Hardy-Weinberg proportions as the restriction on the genotype frequencies in the model specified by

$$\pi_{AA} = p^2 \qquad \pi_{Aa} = 2p(1-p) \qquad \pi_{aa} = (1-p)^2, \tag{1.2}$$

where p is the frequency of the allele A. We will refer to this hypothesis as the hypothesis of Hardy-Weinberg proportions.

As the next step we *estimate* the allele frequency p on the basis of the observed counts. We use the method of maximum likelihood, which gives as an estimate the value of the parameter that makes the observations most probable. Here the estimate $\hat{p}$ of p is the number of A genes in the sample divided by the total number of genes in the sample. In other words, the estimate of the frequency of the allele A in the population is the frequency of the gene A in the sample. For the sample in Fehmarn Belt, for instance, the estimate is

$$\hat{p} = \frac{2x_{AA} + x_{Aa}}{2n} = \frac{54+30}{138} = 0.609.$$

We can now calculate the *expected counts*[†] of the genotypes under the hypothesis of Hardy-Weinberg proportions. The expected counts are the means of the genotype counts assuming the multinomial model and the hypothesis of Hardy-Weinberg proportions. The estimated allele frequencies are given in Table 1.2 in addition to the expected counts of the genotypes (with an accuracy of one decimal place) under the hypothesis.

Comparing the expected counts in Table 1.2 with the samples in Table 1.1, it is striking that the observed and expected counts are very similar for the Fehmarn Belt and the Åland Islands samples, but far apart for the Bornholm sample. We will use the discrepancy between the observed counts and the expected counts under the hypothesis to decide whether we will reject the hypothesis. The larger the discrepancy, the smaller the faith we have in the hypothesis. We will add the discrepancies over the genotypes, and theoretical arguments support to use as *test statistic* either the function of the observations,

$$T_1(\mathbf{x}) = \sum \frac{(observed - expected)^2}{expected},$$

or the function,

$$T_2(\mathbf{x}) = 2\sum observed \ln\left(\frac{observed}{expected}\right),$$

where ln denotes the natural logarithm, and where the summation extends over all genotypes. The two functions are mathematically very similar and both measure the discrepancy between observed and expected values. Actually, $T_2(\mathbf{x})$ is the log likelihood ratio test statistic, which will be used extensively in the book, and $T_1(\mathbf{x})$ is a classical test statistic that can be seen as an approximation to the log likelihood ratio test statistic. We will use T_1 in the calculations here. The values of T_1 are 0.53 for the Fehmarn Belt sample, 15.32 for the Bornholm sample, and 0.08 for the sample near the Åland Islands.

The test statistic has the largest value for the sample near Bornholm in agreement with our visual inspection of Tables 1.1 and 1.2. We do expect, however, that the value of the test statistic will vary even if the hypothesis is true, and values of the statistic that occur often will evidently not make us reject the hypothesis. We therefore calculate the probability of obtaining a larger value of the test statistic than the one we have observed in the experiment. The probability is calculated under the assumption that the hypothesis is true. Since large values of the test statistic are critical for the hypothesis, we are calculating the probability of observing a value of the test statistic that is

[†] In the statistical terminology, "expected frequencies" is the usual term. We use the term "expected counts" here to avoid confusion with the special usage of frequency in genetics.

Table 1.2 *The estimated frequencies of the allele A and the expected counts of the genotypes under the hypothesis of Hardy-Weinberg proportions*

		AA	*Aa*	*aa*
Fehmarn Belt	$\hat{p} = 0.609$	25.6	32.9	10.6
Bornholm	$\hat{p} = 0.279$	6.7	34.6	44.7
Åland Islands	$\hat{p} = 0.0313$	0.1	4.8	75.1

more critical or just as critical for the hypothesis as the observation we have in the experiment. This probability is called the *significance probability*. The test statistics and the significance probabilities for the three samples are given in Table 1.3.

Table 1.3 *The test statistic and the significance probabilities for the hypothesis of Hardy-Weinberg proportions*

	Test statistic	*Significance probability*
Fehmarn Belt	0.53	0.47
Bornholm	15.32	0.00009
Åland Islands	0.08	0.77

We see that for the Fehmarn Belt sample we will obtain a more critical value of the test statistic with a probability of 0.47, or, approximately, once every two times the experiment is repeated. For the sample near the Åland Islands we will see a more critical value of the test statistic with a probability of 0.77, or, approximately three out of four times the experiment is repeated. Thus the deviations between the observed and the expected counts we can see for those two samples comparing Tables 1.1 and 1.2 are indeed small in the sense that very often we will see larger deviations even if the hypothesis is true. Therefore we conclude that the samples from Fehmarn Belt and the Åland Islands are consistent with the hypothesis of Hardy-Weinberg proportions.

On the other hand, the deviation between the observed and the expected counts for the sample near Bornholm is large; for if the hypothesis is true, we will observe deviations as large or larger less than once in every 1000 repetitions of the experiment. Although rare events may occur, it is nevertheless strange that we should observe this rare event, so we decide to reject the hypothesis of Hardy-Weinberg proportions for the population near Bornholm and seek another explanation.

A possible explanation of the fact that the population near Bornholm does not satisfy the Hardy-Weinberg law is that the population is a mixture of the two cod populations in that a proportion α of the population are Baltic cods and the rest are Atlantic cods. Similarly, the reason that the populations in the Fehmarn Belt and near the Åland Islands seem to be in Hardy-Weinberg proportions might be that they each consist of either Atlantic cods or Baltic cods, exclusively.

To examine this hypothesis, we let

$$\boldsymbol{\pi}_W = (p_W^2, 2p_W(1-p_W), (1-p_W)^2),$$

where $p_W = 0.609$ is the estimated frequency of the gene A in Fehmarn Belt (West population), and, similarly, we let

$$\boldsymbol{\pi}_E = (p_E^2, 2p_E(1-p_E), (1-p_E)^2),$$

where $p_E = 0.0313$ is the estimated frequency of the gene A near the Åland Islands (East population). The hypothesis may then be formulated as

$$H_0\colon \boldsymbol{\pi}_{\text{Bornholm}} = (1-\alpha)\boldsymbol{\pi}_W + \alpha\boldsymbol{\pi}_E, \tag{1.3}$$

where $\boldsymbol{\pi}_{\text{Bornholm}}$ is the vector of probabilities of the genotypes *AA*, *Aa*, and *aa* near Bornholm. We will refer to the hypothesis in (1.3) as the *mixture hypothesis*.

In order to estimate α, the proportion of Baltic cods in the population at the location of the catch, we use maximum likelihood. In this case there is no expression for the estimate in a closed form, and we have to use numerical methods to determine the estimate of α, which we denote by $\hat{\alpha}$. The maximum likelihood estimate $\hat{\alpha}$ of α and the expected counts under H_0 are as given in Table 1.4.

Table 1.4 *The estimated proportion of Baltic cods and the estimated counts near Bornholm under the mixture hypothesis*

	$\hat{\alpha}$	*AA*	*Aa*	*aa*
Bornholm	0.574	13.6	20.4	51.9

The agreement with the observed counts in Table 1.1 seems fine, but we need to use the test statistic $T_1(\mathbf{x})$ and its significance probability to judge the agreement. Here $T_1(\mathbf{x}) = 0.019$ and the significance probability is 0.89. Thus we can expect to observe a more critical value of the test statistic 9 times out of 10 if the hypothesis is true. This does not give any reason to doubt the hypothesis that the population near Bornholm is a mixture of Atlantic cods and of Baltic cods.

Not being able to reject a hypothesis is not necessarily a strong support of the hypothesis. So in order to further substantiate the hypothesis of a mixed population near Bornholm, cods were caught at two more places near Bornholm indicated by the numbers 4 and 5 in Figure 1.1.

The samples at the three localities near Bornholm are given in Table 1.5.

Table 1.5 *The genotypes observed in the three samples near Bornholm*

Locality	*AA*	*Aa*	*aa*	*Total*
2	14	20	52	86
4	12	12	54	78
5	2	4	66	72

Samples 4 and 5 deviate from the hypothesis of Hardy-Weinberg proportions. The sigificance probabilities are below 0.0001 for both samples. The expected counts, the estimated proportion of Baltic cods, the test statistics, and the significance probabilities for the mixture hypothesis in the three samples near Bornholm are given in Table 1.6. For all samples the hypothesis that the population near Bornholm is a mixture of Atlantic and Baltic cods cannot be rejected. The estimates of the proportion of Baltic cods increase as the position of the fishing ground becomes more easterly, and that is supporting the mixture hypothesis. Estimates are subject to error so to further substantiate the mixture hypothesis we test the hypothesis that the proportions of Baltic cods are the same at locations 2, 4, and 5. If this hypothesis of homogeneity is rejected it is confirmed that the proportion of Baltic cods does increase as we move to the East.

The expected counts under the assumption of homogeneity are given in Table 1.7. These do not agree very well with those in Table 1.6. The hypothesis of a common α is tested comparing the expected counts in Tables 1.6 and 1.7 by means of the test statistics $T_1(\mathbf{x})$ or $T_2(\mathbf{x})$, but with the slight modification, however, that the expected counts of Table 1.6 are used as observed values in the test statistics and the expected values of Table 1.7 are used as expected values. The value of the T_1 statistic is 18.35 and this corresponds to a significance probability of 0.0001. The hypothesis of a common α is thus rejected.

Table 1.6 *Expected counts under the mixture hypothesis, the estimated proportion of Baltic cods, the test statistics, and the significance probabilities in the three samples near Bornholm*

Locality	*AA*	*Aa*	*aa*	$\hat{\alpha}$	*Test statistic*	*Significance probability*
2	13.6	20.4	51.9	0.574	0.019	0.89
4	9.5	15.3	53.3	0.675	1.39	0.24
5	1.4	5.9	64.7	0.949	0.86	0.35

Table 1.7 *Expected counts in the sample near Bornholm under the mixture hypothesis and assuming a common* α *for all three locations*

Locality	*AA*	*Aa*	*aa*
2	8.8	15.0	62.2
4	8.0	13.6	56.4
5	7.4	12.6	52.1

The conclusion of this investigation is that with respect to reproduction, the Atlantic cods and the Baltic cods form two isolated breeding populations, and the cods caught near Bornholm is a mixture of individuals from the two populations.

The data in this example are considered Exercise 7.10. □

Example 1.2
The data for this example are from a study of the sand lizard *Lacerta agilis* conducted by Mark Goldsmith from the Institute of Biological Sciences, Aarhus University.

Eggs from the sand lizard were placed in incubators under carefully controlled conditions and a range of variables were recorded for the hatchlings, including the incubation time, which is the variable studied in this example.

The eggs were obtained from 30 pregnant sand lizards that were caught during the breeding season in 1999 and brought to the University of Aarhus. Here the sand lizards were caged separately in plastic boxes while their eggs were collected, and they were released on the point of capture after oviposition.

Mark Goldsmith wanted to study the effect of temperature and humidity during incubation on the incubation duration. It was decided to investigate all 20 combinations of 5 levels of temperature and 4 levels of humidity. The incubation duration was recorded in hours, but is given here in days to one decimal point.

Five wooden incubators with 4 shelves were equipped with thermostatically controlled heating devices and ventilators. The incubators were set at the temperatures 19, 22, 25, 28, and 31 °C, which were the 5 levels chosen for the temperature. The eggs were incubated individually in transparant plastic boxes fully buried in 500 g of sand with one of the 4 water contents 0.0075, 0.01, 0.02, or 0.1 $gH_2O/gsand$. We will refer to the water content as the humidity. A total of 160 eggs from 30 clutches were randomly assigned to one of the 20 treatments in such a way that each treatment was applied to 8 eggs. Not all eggs hatched and only 119 incubation durations were recorded. The data are shown in Table 1.8.

Plots of the incubation duration against temperature for each humidity are shown in Figure 1.2. The plots show that the incubation duration decreases as the temperature increases, and they also show that the decrease in incubation duration for a 3 degree increase in temperature diminishes as

Table 1.8 *Incubation duration in days by temperature and humidity in the incubators*

	Temperature				
Humidity	19 °C	22 °C	25 °C	28 °C	31 °C
0.0075	92.9	59.2	40.4	32.8	25.5
	93.9	58.0	40.9	31.8	26.0
	83.9	56.8	38.5	31.8	26.2
		52.4	39.1	33.5	
		55.8	39.6	29.6	
		55.3	40.4	31.6	
		55.9	40.8		
0.01	96.7	55.4	42.9	31.4	25.8
	100.0	56.8	41.1	31.1	26.9
	102.3	58.1	39.8	30.3	27.0
	85.2	54.2	40.6	30.3	25.6
		55.6	38.0	31.9	26.6
		56.1	38.0	31.3	26.0
			40.0	30.6	
			39.8	31.8	
0.02	88.9	53.9	39.9	33.0	26.8
	93.7	55.9	40.0	31.4	28.3
	97.7	58.3	41.9	35.1	25.7
		55.8	40.1	32.9	29.8
		54.9	39.9	30.0	25.3
		55.3	40.6	32.8	26.8
			38.8	30.5	
			40.0		
0.1	91.9	57.0	40.8	32.2	26.0
	89.1	56.0	40.2	30.7	26.4
	93.9	55.4	43.7	31.0	27.0
		55.3	41.5	32.1	27.9
		56.3	41.0	30.7	26.8
		58.3	41.1	29.9	26.5
			38.8	32.1	26.7
			38.8		

the temperature gets higher. We think of this as a systematic effect. The plots also reveal variation in the incubation duration: eggs that have received the same treatment, at least as far as the humidity and the temperature are concerned, do not necessarily have the same incubation duration. We think and speak of this as a random effect or a random variation.

We use probability distributions to model both the random variation and the systematic effects in the data. The probability distributions model the random variation and the parameters of the probability distributions are used to model the systematic effects and the the size of the random variation. The normal distribution plays a prominent role in models for these kinds of data, and the model for the data in Table 1.8 which we would like to use is that the observations are independent and

$$X_{ijk} \sim N(\mu_{ij}, \sigma^2), \qquad i = 1, \ldots, 2,\ j = 1, 2,\ k = 1, \ldots, 5, \tag{1.4}$$

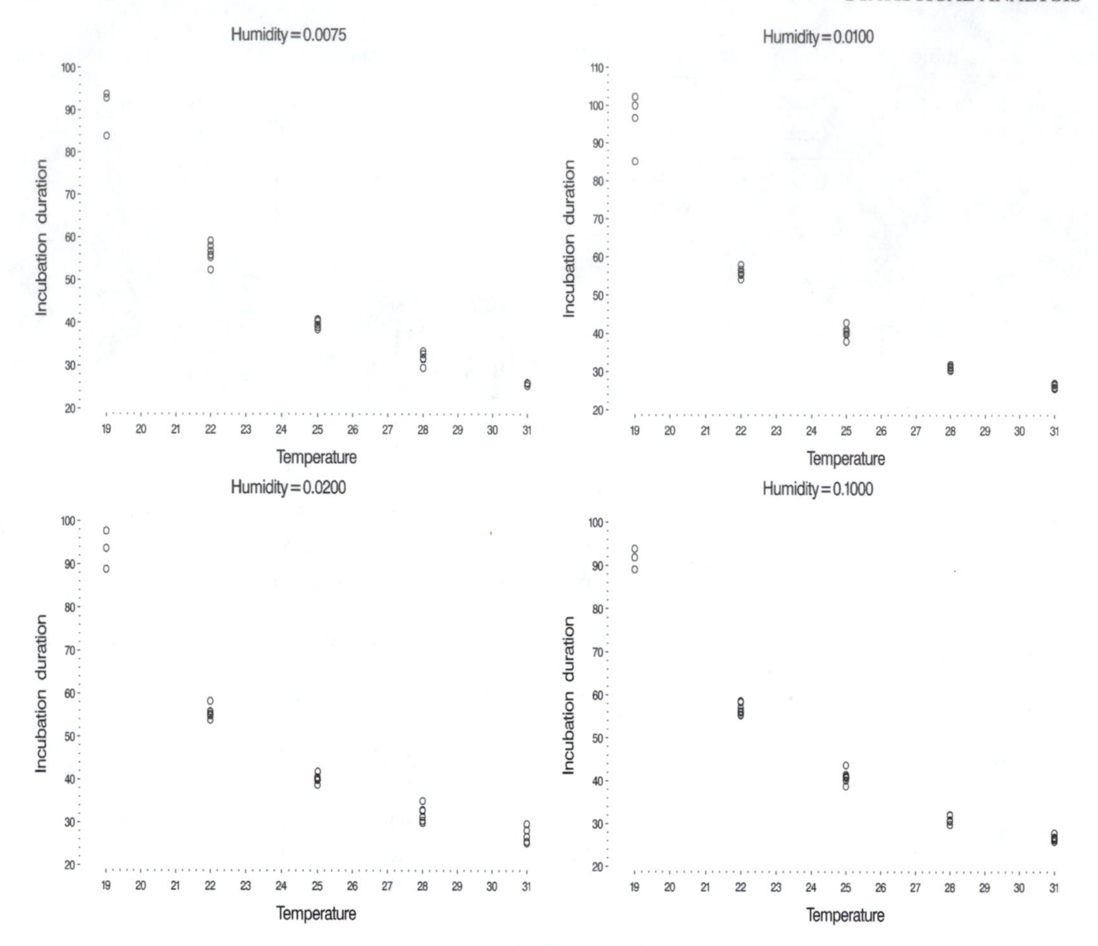

Figure 1.2 *Incubation duration plotted against temperature for each humidity.*

where i index the humidities, j index the temperatures, and k index the incubation durations for eggs that have been exposed to the same humidity and temperature. The model asserts that the incubation durations are realizations of random variables whose means, μ_{ij}, depend on the ith level of temperature and on the jth level of humidity, and whose variance, σ^2, is the same for all observations. In this model, the normal distribution describes the shape of the random variation and the variance describes the size of the random variation, whereas the means describe the systematic effects in the data.

One aspect of this model is not satisfactory for these data, and that is the assumption of constant variance. In Figure 1.2 it is clear that the variation is larger when the incubation times are larger. Rather than making the model more complicated by allowing the variance to depend on the levels of temperature, we use a transformation of the data for which the assumption of a constant variance is reasonable. The larger variation for higher incubation times indicates that a logarithmic transformation might be suitable.

In Figure 1.3 the *logarithm of the incubation duration*, ln(id), is plotted against temperature for each humidity, and a smooth curve is superimposed. Figure 1.3 has the same features as Figure 1.2, except for the increasing variation for large observations. For this reason we will adopt the model in (1.4) for the logarithm of the incubation duration rather than for the incubation duration itself.

The statistical analysis proceeds with further model checking and model reduction. Data are, for example, consistent with the assumption of a constant variance and also with the hypothesis that the

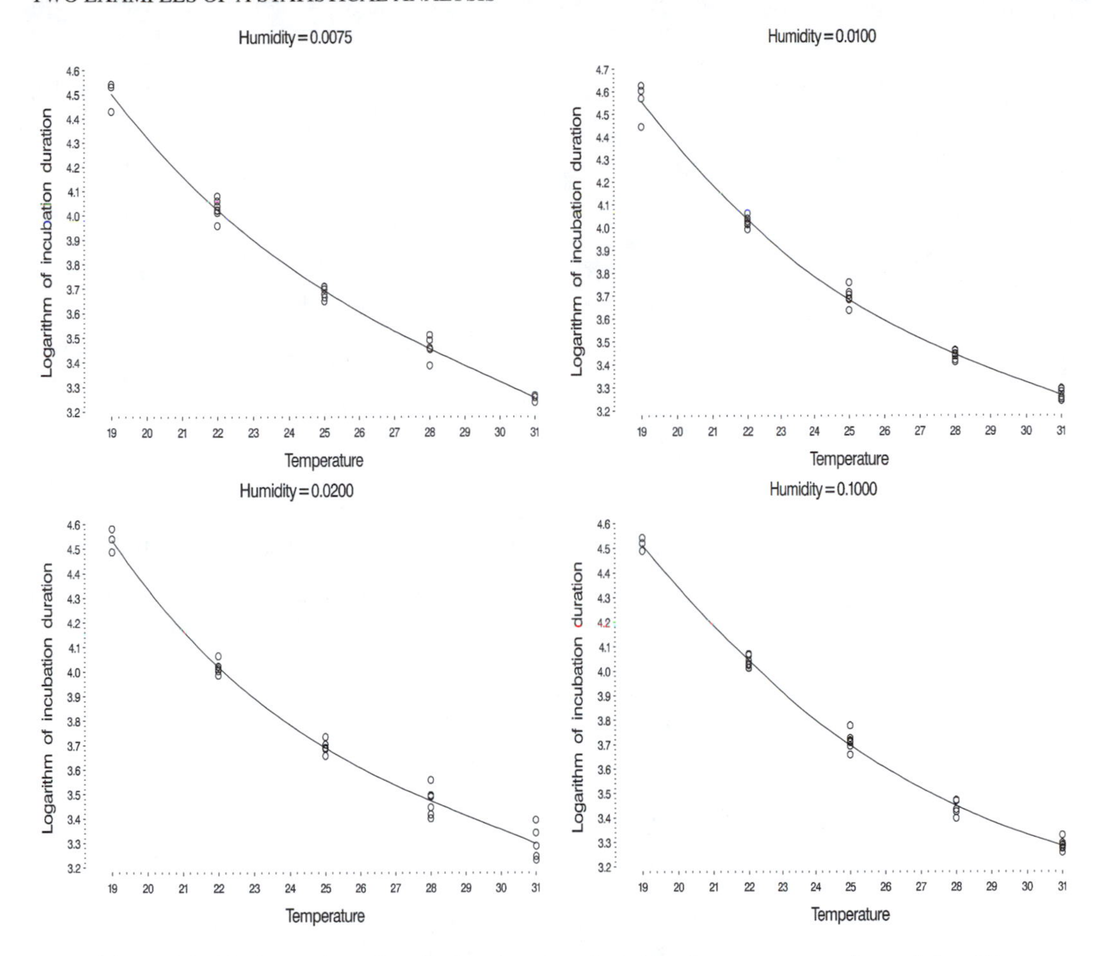

Figure 1.3 *The logarithm of incubation duration plotted against temperature for each humidity.*

incubation duration does not depend on the humidity. We will not go into details but refer the reader to Exercise 4.6.

The conclusion of the analysis is illustrated in Figure 1.4. All incubation durations are plotted againt temperature in the same plot because humidity did not have a significant effect on incubation duration. Superimposed on the plot are five curves. The middle curve is the estimated median of the incubation durations. For a given temperature this curve gives median incubation duration, which means that 50% of all incubation durations at that temperature will be below that value and 50% will be above that value.

The remaining four curves are used to express the uncertainty of the estimation. The inner curves are 95% confidence limits of the estimated median. The 95% confidence curves are close to the estimated median, in particular in the higher end of the range of temperatures. In the lower end, which corresponds to higher incubation durations, the 95% confidence curves are farther from the median curve. This is consistent with the larger variation we noted for incubation durations in the lower end of the range of observations. The outer curves are 95% prediction limits for a new observation. Considering a fixed temperature the interval between the 95% prediction limits is one answer to the question, "what will the incubation time be for this egg with this incubation temperature?" Rather than giving a single value as the answer, the stochastic nature of the observations makes it natural to give an interval where the observation will lie with an asserted confidence, here 95%. □

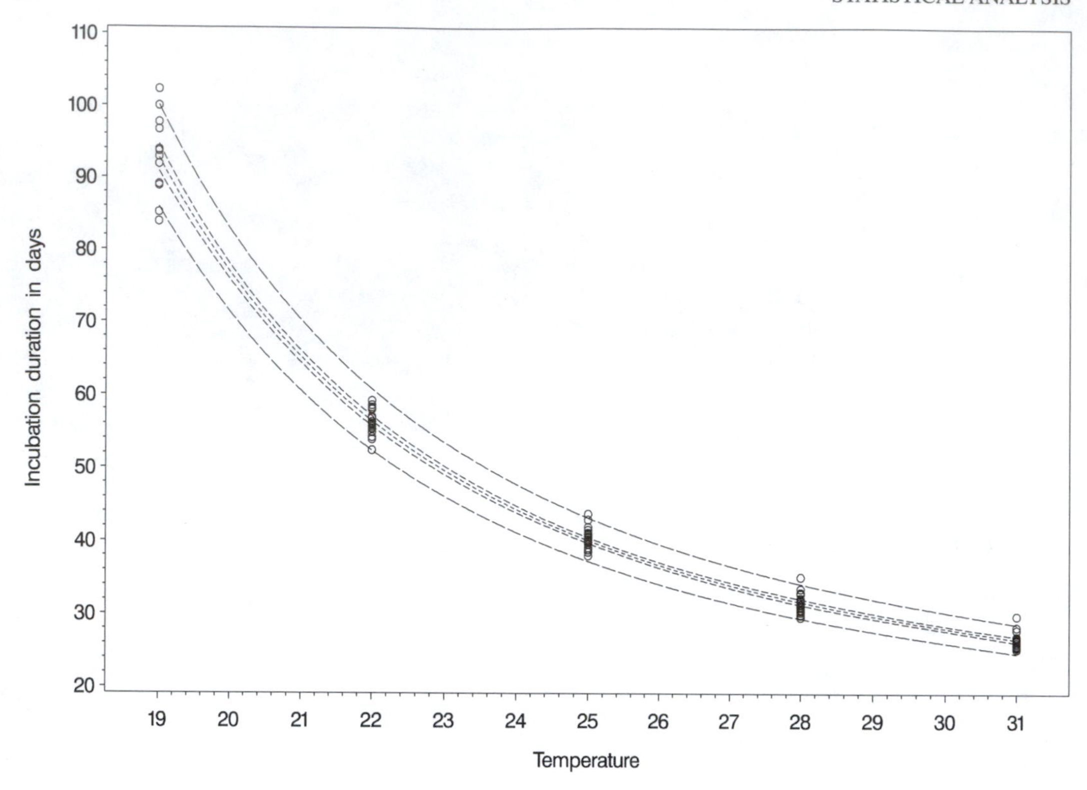

Figure 1.4 *Incubation duration plotted against temperature.*

1.2 Data

The starting point for a statistical analysis is a *data set,* which is the result of an *observational study* or an *experiment* carried out with the purpose of gaining insight into a particular *scientific context*. Geological data can, for instance, be determination of a particular mineral in collected pieces of rocks, the grain size distribution of a sand sample determined by sifting or determined by sedimentation times, measurements of water level, wind speed and wind direction, recordings of precipitation by day for a number of locations, statements concerning the number of oil-containing drillings, etc. As examples of biological data, we may think of registrations of the geno- or phenotype distribution in a subpopulation, for instance, blood groups in humans or in cods, measurements of length, width, and weight by individuals, measurements of biochemical oxygen consumption, concentration of heavy metals, and other variables in connection with investigations of pollution, measurements of pH-values and contents of various chemical compounds measured at different places within an area.

The *data set* is the set of all the data that are collected for a particular problem, but usually the data set is structured to facilitate the retrieval and handling of information. We think of the data set as a table or a matrix with the row representing the smallest item in the investigation for which we have data and the columns containing the recorded information for the items. Naturally, the recordings are made in the same order for all items, so each column will contain the same type of information. We will refer to the columns as *variables,* and we will label the variables with names that reflect the type of information they contain, such as height, weight, or pH.

If the data comprises n items and p variables, the corresponding data set is a table or a matrix with n row and p columns. The p values in the ith row contain the information concerning the ith item and, similarly, the n values in the jth column contain the information in the jth variable.

There are at least two different ways to classify the variables in a data set.

One classification is based on how the values of a variable are expressed. A variable is *numerical* or *quantitative* if its elements may be expressed numerically. If a numerical variable in principle may take all values in an interval of the real axis $\mathbb{R}$, it is called a *continuous variable*. In contrast, a *discrete numerical variable* can only take values in a subset of $\mathbb{Z}$, the set of integers. Variables with values that can only be expressed qualitatively in terms of labelled categories are called *qualitative variables* or *nominal variables*. If, in addition, the categories may be ordered, such variables are referred to as *ordinal variables*.

A second classification of the variables in a data set is related to the way the variables are used in the analysis. Some of the variables in the data set are of primary interest for the problem that is being investigated, and they are obviously to be considered as realizations of random variables, i.e., if the study is repeated under precisely the same circumstances, the values of such variables would not necessarily be the same as before. No special names are used for such a variable, but in special contexts it is sometimes referred to as *a dependent variable* or *a response variable*.

Other variables contain additional information on the collection of the data, for example, variables with information on where, when, or how the data has been recorded. In the analysis of the study, the values of these variables are considered as fixed. The terminology concerning such *deterministic* variables is very extensive. It depends on the analysis of the data and it includes, *independent variable*, *covariable* (or *covariate*), *factor*, *explanatory variable*, etc.

A data set is only relevant in connection with statistical analysis if it has random components, and we assume from now on that this is the case.

Sometimes a data set is referred to simply as data and, with an abuse of terminology, variables representing the randomness are sometimes referred to as data, in particular, if the data set consists of one variable only.

1.3 Model Specification

To make the discussion easier, we assume by way of introduction that the data set only has one variable with values $x_1, \ldots, x_n$, which are considered as realizations of random variables. We use the notation $\mathbf{x} = (x_1, \ldots, x_n)^*$, where $*$ indicates transposition, for the corresponding column vector in the data set. We refer to an element x_i of $\mathbf{x}$ as an *individual observation* or simply as an *observation*, and we will speak of the *observations in* $\mathbf{x}$ but also sometimes speak of the *observation* $\mathbf{x}$ in the singular because we think of $\mathbf{x}$ as the observation of the corresponding variable. Occasionally we will also refer to $\mathbf{x}$ as the *outcome*.

The observation $\mathbf{x}$ represents the *randomness* of the data set. Despite the fact that the outcome of the observational study or the experiment cannot be stated in advance, very often there is regularity on a higher level that may be recognized precisely in the situation where the study is repeated a number of times. Consequently, a basic element in the description of an observational study is a *probabilistic model*.

A probabilistic model consists of three components: (1) the *sample space*, $\mathcal{X}$, which is all possible values (outcomes) obtainable by the study; (2) the *system of events*, $\mathcal{A}$, which comprises all the events we will consider; and (3) the *probability measure*, P, which gives the probability of all events in $\mathcal{A}$.

The randomness in an observational study is described by the system of events that includes all events in which we are interested and the probability measure, which specifies the probability of the events. We describe the random feature of a data set by considering $\mathbf{x}$ as a realization of a random vector $\mathbf{X}$. This random vector may be thought of as the identity mapping on the sample space $\mathcal{X}$ and its distribution as given by the probability measure P.

We restrict ourselves to consider *discrete* and *continuous* random vectors only. If $\mathbb{N}_0$ denotes the set of nonnegative integers and $\mathbb{R}_+$ the set of positive real numbers, we will mainly consider the sample spaces $\mathbb{N}_0^n$ and $\mathbb{Z}^n$ corresponding to discrete data and the sample spaces $\mathbb{R}_+^n$ and $\mathbb{R}^n$ corresponding to continuous data. The system of events will include all sets consisting of one point, all intervals and all sets that can be formed from these sets using the usual set operations, such as union of sets, intersection of sets, and complement of sets. The probability measures on these systems of events can be represented either by their *distribution function* F or by their *probability density function* f.

A *statistical model* is a parameterized set of probabilistic models. In the models considered in this book, the sample spaces and the systems of events are identical for all the probabilistic models, and therefore one may think of a statistical model as a probabilistic model where the probability measure has been replaced with a *parameterized class* of probability measures, $\mathcal{P} = \{P_{\boldsymbol{\omega}} \mid \boldsymbol{\omega} \in \Omega\}$, i.e., a class of probability measures indexed by a vector $\boldsymbol{\omega} = (\omega_1, \ldots, \omega_k)^*$ varying in a subset Ω of $\mathbb{R}^k$. Alternatively, the class of probability measures can be represented as a parameterized class of distribution functions, $\mathcal{F} = \{F_{\boldsymbol{\omega}} \mid \boldsymbol{\omega} \in \Omega\}$, or as a parameterized class of probability density functions $\{f(\cdot;\boldsymbol{\omega}) \mid \boldsymbol{\omega} \in \Omega\}$. The vector $\boldsymbol{\omega}$ is called a *parameter* and Ω is referred to as the *parameter space* (*parameter set*). The parameter $\boldsymbol{\omega}$ should be chosen in such a way that it is relevant for the scientific problem that was the occasion of the observational study. That is, the parameter should be chosen in such a way that statements concerning the scientific problem can be formulated in terms of $\boldsymbol{\omega}$.

In most cases information concerning the observational circumstances has to be taken into account when a statistical models is specified. However, since this information in the data set is represented by means of deterministic variables, we suppress the influence of such variables in the notation introduced above.

All the statistical models considered in this book are of the form

$$(\mathcal{X}, \mathcal{A}; \mathcal{P}) = (\mathcal{X}, \mathcal{A}; \{P_{\boldsymbol{\omega}} \mid \boldsymbol{\omega} \in \Omega\}).$$

Our preferred representation of the probability measures is in terms of probability density functions, and we refer to the function,

$$\begin{array}{rcl} \mathcal{X} \times \Omega & \to & \mathbb{R} \\ (\mathbf{x}, \boldsymbol{\omega}) & \to & f(\mathbf{x}; \boldsymbol{\omega}), \end{array} \tag{1.5}$$

as the *model function*. The model function is the probability density function considered as a function of both the outcome $\mathbf{x}$ and the parameter $\boldsymbol{\omega}$.

We have now introduced the terminology and the notation we will use when discussing statistical models. The *model specification* we understand as the process where one identifies the components of the statistical model: the sample space, the system of events, and the class of distributions. It is usually not difficult to determine the sample space and the system of events. The essential part of the work is related to identification of the parameterized class of distributions in the statistical model. This also means that when referring to the models one often omits mention of the whole triplet $(\mathcal{X}, \mathcal{A}; \{P_{\boldsymbol{\omega}} \mid \boldsymbol{\omega} \in \Omega\})$, but focuses on the distributions $\{P_{\boldsymbol{\omega}} \mid \boldsymbol{\omega} \in \Omega\}$. As a matter of fact, one sometimes specifies the parameter set Ω only, the sample space, the system of events, and the class of distributions being implicit.

For data sets in which the randomness is represented by more than one variable, the model specification is similar to that discussed above with the exception of a more extensive sample space $\mathcal{X}$. If, for instance, the random part of the data includes c continuous variables and d discrete variables with values in $\mathbb{N}_0$, the sample space is $\mathcal{X} = (\mathbb{R}^c \times \mathbb{N}_0^d)^n$. As above, in the rest of this chapter we will use $\mathbf{x}$ as a short notation for all the variables in the data set with values that are realizations of random variables and, similarly, we refer to $\mathbf{x}$ as the observation or the outcome.

When identifying a class of distributions, one uses common and specific knowledge about the circumstances under which the observational study or the experiment has been conducted and from time to time also experiences from statistical analysis of similar data sets. Usually, the preliminary graphical procedures that are mentioned in Chapter 2 are very useful in connection with specification of the model. This stage in a statistical analysis often requires a considerable amount of insight into the scientific problem and thus close cooperation between the expert (for example a biologist or a geologist) and the statistician is necessary.

1.4 Model Checking

This stage in a statistical analysis is concerned with the appropriateness of the proposed statistical model. We investigate whether the observation $\mathbf{x}$ conflicts with one or several consequences of the model. If this is the case, the model is rejected and a new model has to be specified; if not, one is ready for the next stage in the analysis, statistical inference. It is difficult to give a general description of this stage in a statistical analysis since the methods depend partly on the model and partly on the considered aspects of the model.

It has to be emphasized that model checking is not limited to the preliminary stages of a statistical analysis. In many models, for instance, regression models, the most important part of model checking is performed after the estimation of the parameters in the model.

As it will appear from nearly all the following chapters, graphical investigations as well as numerical investigations has a part to play in the check of a model.

1.5 Statistical Inference

The purpose of a statistical analysis is to gain insight into the scientific problem that originated the collection of data. By the model specification the parameter $\boldsymbol{\omega}$ was chosen in such a way that it represents those aspects of the scientific problem that are of particular interest. Statistical inference is concerned with the question of formulating and evaluating statements about the parameter $\boldsymbol{\omega}$ – and, consequently, about the scientific problem – on the basis of the observation $\mathbf{x}$, the outcome of the observational study or the experiment. The purpose of these statements is to indicate to what extent the different parameter values $\boldsymbol{\omega}$, or to be more precise, the corresponding distributions functions $F_{\boldsymbol{\omega}}$ (or probability density functions $f(\cdot;\boldsymbol{\omega})$) give a reasonable description of the observation $\mathbf{x}$. *Estimation* of the parameters and *testing hypotheses* about the parameters are important activities in statistical inference. A concept connected to both estimation and statistical test is that of a *confidence region*.

Estimation

Estimation is the process of qualified guessing of the value of the parameter based on the value of the observations. When we estimate, we are assuming that the model is correct in the sense that the data variable $\mathbf{x}$ is an observation of the random vector $\mathbf{X}$ with distribution $P_{\boldsymbol{\omega}}$ and guided by $\mathbf{x}$, we choose one of the parameters in Ω as our estimate of the parameter.

What has been described here can be phrased in mathematical terms as a mapping from the sample space to the parameter space,

$$\begin{aligned} \tilde{\boldsymbol{\omega}}: \quad & \mathcal{X} \to \Omega \\ & \mathbf{x} \to \tilde{\boldsymbol{\omega}}(\mathbf{x}), \end{aligned} \tag{1.6}$$

which assigns the parameter $\tilde{\boldsymbol{\omega}}(\mathbf{x})$ to the observation $\mathbf{x}$ (see Figure 1.5). We call $\tilde{\boldsymbol{\omega}}(\mathbf{x})$ the *estimate* of $\boldsymbol{\omega}$ and the mapping is called an *estimator*. We will often suppress the observation $\mathbf{x}$ and simply write $\tilde{\boldsymbol{\omega}}$ for the estimate. It is important to distinguish between the estimate of the parameter and the value of the parameter, and to facilitate this distinction we use the notation $\tilde{\boldsymbol{\omega}} \to \boldsymbol{\omega}$, which is read "$\tilde{\boldsymbol{\omega}}$ is an estimate of $\boldsymbol{\omega}$" or "$\tilde{\boldsymbol{\omega}}$ estimates $\boldsymbol{\omega}$."

Observe that the estimator as a mapping from the sample space is a random vector, and we will write $\tilde{\boldsymbol{\omega}}(\mathbf{X})$ when we want to focus on that aspect of $\tilde{\boldsymbol{\omega}}$.

We will not be satisfied with an arbitrary mapping from the sample space to the parameter space as an estimator. Estimators are judged from their properties as random vectors. We want estimators to pick the estimate close to the true value with a high probability.

So far the description of estimation has been fairly general and considered a vector parameter $\boldsymbol{\omega}$. Often the components of $\boldsymbol{\omega}$ are considered separately and let ω be one such parameter. It is often required that the estimator of ω picks the correct value on the average, which in mathematical terms is formulated as

$$E_{\boldsymbol{\omega}}\tilde{\omega}(\mathbf{X}) = \omega,$$

i.e., the mean of $\tilde{\omega}(\mathbf{X})$ calculated using the probability measure $P_{\boldsymbol{\omega}}$ is ω. An estimator with this property is said to be *unbiased* and an estimator without this property is said to be *biased*. This property does not ensure that an estimate is close to ω. The variance of a random variable is a measure of its variability. Generally, unbiased estimators with the smallest possible variance are preferred.

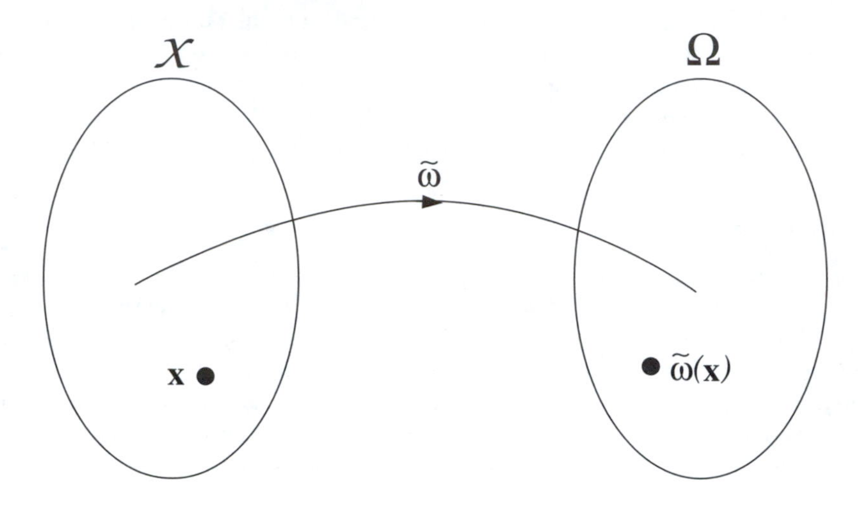

Figure 1.5 *Illustration of an estimator* $\tilde{\boldsymbol{\omega}}$.

Hypotheses and Tests

We require of our models that the parameter can be interpreted in the scientific context and indeed the fundamental questions that we investigate are often succinctly formulated as restrictions on the parameter of the model. For example, that two components of the parameter are equal or that a component of the parameter is equal to zero. A restriction on the parameter of a model is called a hypothesis or a null hypothesis. The model and the hypothesis together specify a new model that is simpler in the sense that its parameter space has fewer parameters. In this sense a null hypothesis is always a reduction from one model to a simpler one.

Very often the hypothesis is formulated mathematically as a condition on the parameter space Ω of the model and here we formulate the hypothesis as

$$H_0\colon \boldsymbol{\omega} \in \Omega_0, \tag{1.7}$$

where Ω_0 is a subset of Ω, i.e., $\Omega_0 \subset \Omega_0$.

If Ω_0 has one element $\boldsymbol{\omega}_0$ only, the hypothesis is called a *simple hypothesis* or a *point hypothesis*. Otherwise the hypothesis is *composite*.

The decision on whether to reject H_0 is based on a *test statistic* or a *testor* T, which is a function of the observation $\mathbf{x}$ with the property that a certain type of departure or certain types of departures from the hypothesis may be expressed in terms of T such that, for instance, the larger the value of T is, the stronger the evidence against H_0 is. Thus, such a test statistic T implicitly gives an ordering of the sample space $\mathcal{X}$ and we say that the observation $\mathbf{x}_1$ is more critical or just as critical for the hypothesis as the observation $\mathbf{x}_2$ if

$$T(\mathbf{x}_1) \geq T(\mathbf{x}_2).$$

Suppose for a while that the hypothesis is simple, i.e. H_0: $\boldsymbol{\omega} = \boldsymbol{\omega}_0$, and let $p_{obs}(\mathbf{x})$ denote the probability calculated under the hypothesis of outcomes $\mathbf{y}$ that are more or just as critical as the observed outcome $\mathbf{x}$, i.e.,

$$p_{obs}(\mathbf{x}) = P_{\boldsymbol{\omega}_0}(\{\mathbf{y} \,|\, T(\mathbf{y}) \geq T(\mathbf{x})\}). \tag{1.8}$$

This probability is referred to as either the *significance probability*, the *observed significance level,* or the *p-value*, and $p_{obs}(\mathbf{x})$ is used to decide whether H_0 is rejected. If $p_{obs}(\mathbf{x})$ is small, the hypothesis is *rejected* since the probability of outcomes that are more or just as critical as the observed outcome $\mathbf{x}$ is small, so judged by T the outcome $\mathbf{x}$ is a significant deviation from the hypothesis. In contrast, a large value of $p_{obs}(\mathbf{x})$ indicates that judged by T the observed outcome $\mathbf{x}$ does not deviate significantly from what is expected under the hypothesis, which therefore is *not rejected.*

The final decision on whether the hypothesis is rejected is made by comparing $p_{obs}(\mathbf{x})$ with the *significance level* α ($\in (0,1)$). We reject H_0 if

$$p_{obs}(\mathbf{x}) \leq \alpha.$$

In examples and exercises throughout the book, we will use the significance level 5% that is traditional in statistics.

A test with significance level α partitions the sample space $\mathcal{X}$ into two disjoint subsets,

$$R_\alpha = \{\mathbf{x} \in \mathcal{X} \mid p_{obs}(\mathbf{x}) \leq \alpha\}, \tag{1.9}$$

which is called the *region of rejection* (or the *critical region*) and its complementary set,

$$A_\alpha = \{\mathbf{x} \in \mathcal{X} \mid p_{obs}(\mathbf{x}) > \alpha\}, \tag{1.10}$$

the *region of no rejection*, see Figure 1.6. The reason for using the letter A in connection with this set is that in the literature, the set is often referred to as the *region of acceptance* reflecting the fact that a hypothesis is often said to be "accepted" if it is "not rejected." We prefer the latter phrase since it better reflects what has happened, namely, that the hypothesis was not rejected because we did not find significant deviations from the hypothesis.

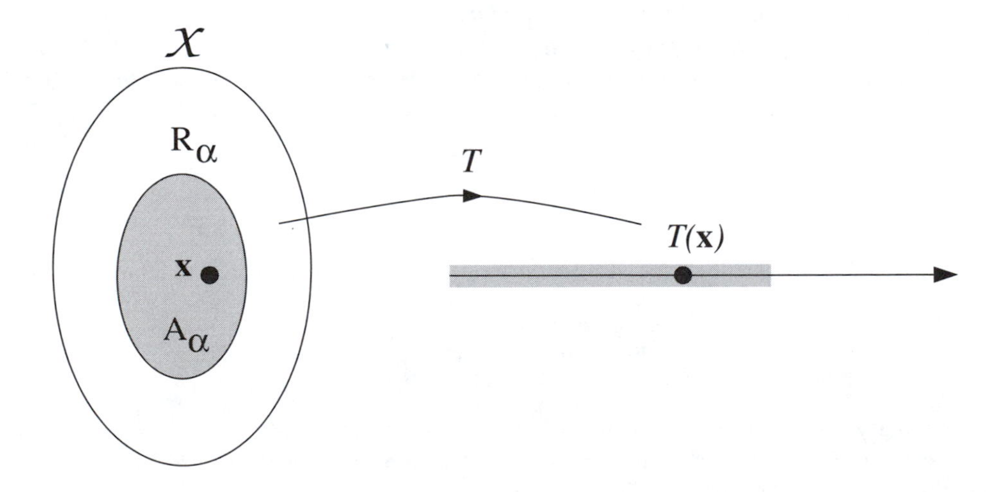

Figure 1.6 *Illustration of a testor T of the hypothesis H_0 and the induced partition of the sample space.*

If the hypothesis H_0 is a composite hypothesis, i.e., if the subset specifying the hypothesis has more than one element, the significance probability is defined as

$$p_{obs}(\mathbf{x}) = \sup_{\omega \in \Omega_0} P_\omega(\{\mathbf{y} \in \mathcal{X} \mid T(\mathbf{y}) \geq T(\mathbf{x})\}), \tag{1.11}$$

i.e., as the largest of the probabilities under H_0 of the outcomes $\mathbf{y}$ that are more or just as critical as the observed outcome $\mathbf{x}$ as judged by T. The corresponding regions of rejection and no rejection are defined as in (1.9) and (1.10).

Confidence regions

A concept closely connected to a test of a simple hypothesis is that of confidence regions. Let $A_\alpha(\omega_0)$ denote the acceptance region for a test with significance level α of the simple hypothesis H_0: $\omega = \omega_0$, i.e.,

$$A_\alpha(\omega_0) = \{\mathbf{x} \in \mathcal{X} \mid H_0\text{: } \omega = \omega_0 \text{ is not rejected on the basis of } \mathbf{x}\}$$

and

$$P_{\omega_0}(\mathbf{X} \in A_\alpha(\omega_0)) = 1 - \alpha.$$

The $(1-\alpha)$ confidence regions for the parameter ω corresponding to the observation $\mathbf{x}$ is the subset of the parameter space defined as

$$C_{1-\alpha}(\mathbf{x}) = \{\omega_0 \in \Omega \mid \mathbf{x} \in A_a(\omega_0)\}, \tag{1.12}$$

i.e., as the set of parameters values ω_0 for which the simple hypothesis H_0: $\omega = \omega_0$ is not rejected. If the confidence region is large, many values of the parameter give a reasonable description of the observation $\mathbf{x}$ and therefore $\mathbf{x}$ contains limited information on ω. In contrast, if the confidence region is small, only relatively few values of the parameter give a reasonable description of the observation $\mathbf{x}$ and therefore $\mathbf{x}$ contains much information on ω.

Construction of Estimators and Tests

Via heuristic arguments it is often possible to find estimators and testors in simple, concrete situations. However, it is of course of value to have a general methodology, based on simple principles, which also gives good estimators and testors in more complicated situations. The methodology we consider in the following is based on the *likelihood function* that is introduced in Section 3.1.4 on page 70 in a simple situation. The corresponding quantities are called the *maximum likelihood estimator* and the *likelihood ratio test statistic*, respectively. This methodology is used throughout the book, and Chapter 11 contains a theoretical discussion of statistical inference based on the likelihood function.

1.6 Concluding Remarks

As mentioned in the introduction, we consider the main ingredients in a statistical analysis to be

(i) model specification
(ii) model checking
(iii) statistical inference.

It is customary that the analysis traverses one or several cyclic phases; for if, during the model checking or the statistical inference, unsatisfactory features of the model are discovered, we must return to the model specification in order to revise the model.

As described in Section 1.3, the emphasis is on statistical models in which the distributions are a parametrized family of distributions and where the statistical inference is based on the likelihood function. Nevertheless, the final chapter is about nonparametric statistics, because a nonparametric test sometimes is useful. We often meet the view that nonparametric statistical inference does not involve assumptions and thus is safe to use. This is a serious misunderstanding. Very often a non-parametric test is derived under strict assumptions about independence, identical distributions, and sometimes even symmetrical distributions. Thus, these assumptions are common for the parametric statistics we present here and for the nonparametric statistics, and it is only a little step further to specify a parametric statistical model. In the end the additional work of finding a valid parametric statistical model is profitable, since it gives occasions for specifying more detailed mathematical models, which as a rule is the motivation behind most experimental work in science, including biology and geology.

Earlier in this chapter we mentioned that it is often part of a statistical analysis to investigate whether a simpler model than the one that was specified as a starting point gives a satisfactory description of the observations, and we outlined how to use statistical tests in order to evaluate this. Here it is very important to be aware of the fact that one is never able to prove that a hypothesis is true by means of a statistical test. One can only falsify in the sense that it is possible to convince oneself that the observations conflict with a simpler model being tested. With a limited amount of data one runs the risk of not being able to reject the reduction to a simpler model, which in reality is false. Here *design of experiments* is introduced. This discipline is concerned with how, taking the resources into account, one can plan experiments, including collection of data, in order to get sufficient information to make convincing conclusions in the relevant scientific context. Because of the size of this book, we are not able to consider this aspect of a statistical analysis in great detail. However, Chapter 5 contains a few examples of considerations of this kind.

CHAPTER 2

Preliminary Investigations

The first point in a statistical analysis is to specify a useful statistical model for the actual data. This is sometimes difficult but it is always crucial. Once the model has been chosen, the ensuing statistical inference will often be given. Several issues have to be considered, such as, the way the data has been collected, information about the scientific problem the data should elucidate and experiences, which can be personal or obtained by studies of literature, concerning analyses of similar problems. When specifying the model the statistician uses information obtained by summarizing the data in different ways, such as tabulations and/or graphical representations. Furthermore, it is important that the proposed statistical model on the one hand is sufficiently simple from a mathematical point of view and on the other hand sufficiently structured to give relevant information about the scientific problem under investigation. Finally, the model should of course give a reasonable description of the data in order to be used in further analysis. In particular, the model must not be contradicted by the data.

It will become evident from this book that *graphical investigations* are relevant in nearly all stages of a statistical analysis. In this chapter we discuss some *graphical procedures* that are relevant in the *preliminary stage* of a statistical analysis, where the main emphasis is on summarizing the data and on finding a suitable class of probability distributions by which it is possible to describe the random variation in the data.

We only consider data sets with variables that are either *discrete* or *continuous*, i.e., either variables obtained by *counting* or variables obtained by *measuring*. For such variables the relevant classes of distributions consist of discrete distributions and continuous distributions, respectively. Furthermore, in this chapter we consider mostly data sets in which the randomness is represented by a single variable, which sometimes is referred to as data.

We distinguish between *grouped variables* and *ungrouped variables.* If a data set consists of n items and if the values, $x_1, x_2, \ldots, x_n$, of all n observations of a variable are known, we say that the variable is *ungrouped*. Now and then the sample space $\mathcal{X}$ of the observations is partitioned into m disjoint subsets, $A_1, \ldots, A_m$, and instead of recording the n observations, $x_1, x_2, \ldots, x_n$, one records $a_1, a_2, \ldots, a_m$, where a_j is the number of observations belonging to the subset, A_j, $j = 1, 2, \ldots, m$. In that case we refer to $a_1, a_2, \ldots, a_m$ as observations of a *grouped* variable. Strictly speaking, in the terminology concerning data sets introduced in Chapter 1, at least two variables enter the specification of a grouped variable, one variable with the values, $a_1, a_2, \ldots, a_m$, and at least one variable characterizing the subsets, A_j, $j = 1, 2, \ldots, m$, which usually are chosen as intervals on the real axis.

In this chapter we restrict ourselves to consider *models for one sample*. By a *sample* of *size n*, we understand n independent observations, $x_1, x_2, \ldots, x_n$, from the same distribution. We give various simple numerical and graphical methods, which summarize the data and give some indications of the form of the common distribution of the observations. In Section 2.1 we consider *dot diagrams* and *bar charts* and in Section 2.2 we discuss *histograms*. The *fractile diagram* introduced in Section 2.3 is a graphical procedure for continuous variables by which it may be decided if a proposed model (distribution) gives a reasonable description of the observations. Section 2.4 contains a detailed discussion of fractile diagrams in the case where the proposed model is based on the normal distribution. In Section 2.5 we offer some remarks concerning *transformation of variables* and Section 2.6 contains some concluding remarks. Many statistical packages contain graphical procedures for representing and summarizing the variables in a data set. In an annex to this chapter, we illustrate some of the procedures in *SAS*.

Despite the simplicity of the methods mentioned in this chapter, these methods are very important since they give us guidance in specifying a statistical model and in checking its validity. All following calculations and conclusions are determined by the statistical model. If the model is seriously wrong, the conclusions drawn on the basis of it will probably be wrong, too.

2.1 Dot Diagrams and Bar Charts

The *dot diagram* is a graphical procedure that summarizes the observations of a variable and may give a first impression of the underlying distribution of the variable. The dot diagram is constructed by plotting the observations in a coordinate system in the following way. For each observation a dot is marked above the point on the first coordinate axis that corresponds to the value of the observation. The dot diagram also provides a possibility of arranging the observations according to size when the data set is not recorded in electronic form.

Example 2.1
The following data set originates from an investigation of children suffering from asthma in the county of Funen. The investigation was carried out by professor Bent Juhl, Aarhus University Hospital, in the period from 1 December 1968 until 3 March 1969. 14 different measurements on each child were performed, including measurements of height. At the time of investigation, the 247 girls in the study were between 10 and 12 years of age. The measurements of height (in cm) of these girls are shown in Table 2.1. The variable is ungrouped since the height was measured and recorded in cm. Notice, however, that if the heights originally were measured in mm and recorded in cm, the variable would have been grouped. In this situation the underlying distribution is continuous. The corresponding dot diagram is shown in Figure 2.1. □

Table 2.1 *Height (in cm) of 247 girls at the age of 10 to 12 years suffering from asthma*

139	128	139	125	132	137	146	129	146	150	141	161	143
131	128	134	132	136	137	137	129	140	140	143	148	148
149	132	144	147	137	142	127	127	126	135	136	144	130
132	141	126	135	129	132	130	139	139	134	132	134	127
138	134	127	133	134	126	140	133	142	130	143	140	140
143	150	144	144	128	135	131	135	138	131	135	148	134
132	137	113	150	155	155	155	161	142	142	146	140	141
146	140	139	137	146	142	130	145	149	156	149	155	152
144	139	157	144	149	161	150	144	141	138	140	141	141
147	142	146	156	140	144	145	137	126	134	144	159	134
134	144	130	126	131	130	133	125	122	145	140	132	139
139	128	146	137	139	138	145	133	139	133	139	151	150
138	142	151	140	142	144	136	139	135	141	132	139	140
144	142	127	147	151	141	138	142	147	153	148	144	138
139	124	127	122	123	133	133	136	134	140	137	132	133
132	128	128	136	122	122	123	123	128	145	152	152	156
149	160	148	149	159	145	156	149	153	154	144	153	144
134	140	135	149	136	145	143	139	143	138	137	140	137
144	147	151	166	147	144	159	156	147	154	150	162	159

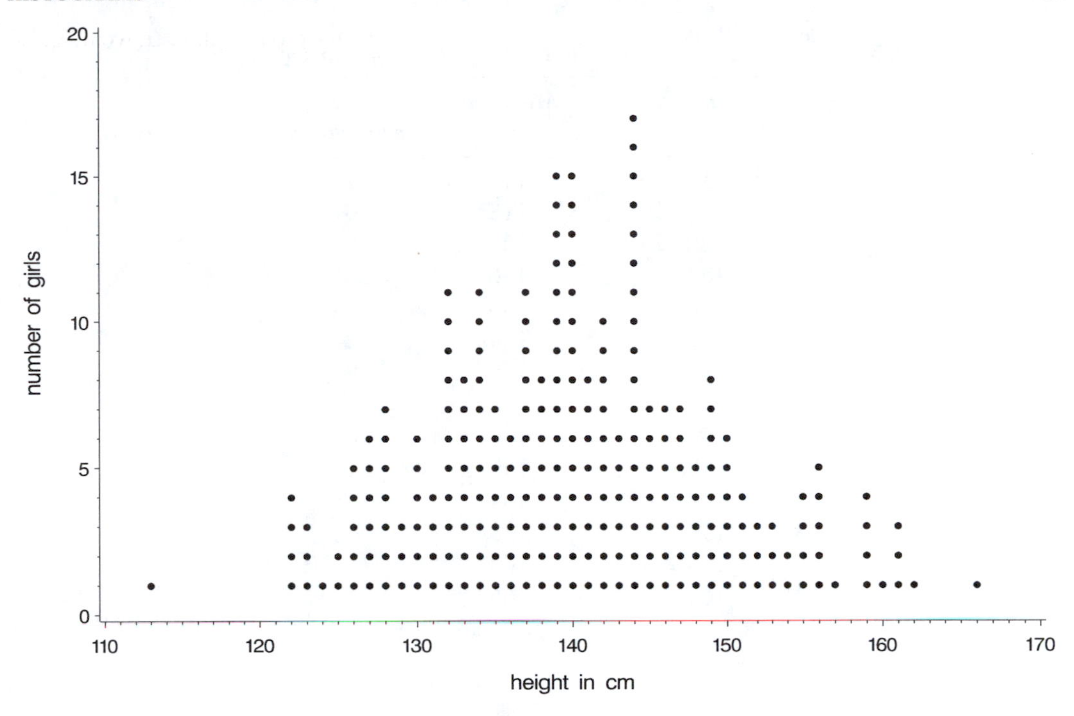

Figure 2.1 *The dot diagram of the heights in Table 2.1.*

The *bar chart* is used to represent observations grouped into intervals. For each interval a bar is put up over the midpoint of the interval, and the height of the bar represents the number of observations in the interval (or the relative number of observations in the interval, i.e., the number of observations in the interval divided by the total number of observations).

Example 2.1 (Continued)
Table 2.2 shows the result of grouping the variable in Table 2.1 into 4-cm intervals of length. The corresponding bar chart is shown in Figure 2.2. □

2.2 Histograms

If the variable is continuous, a *histogram* may be used in order to get an impression of the appearance of the probability density function of the underlying distribution. Histograms may therefore be a great help when choosing the class of distributions in the statistical model.

A histogram is constructed in the following way. The n observations, $x_1, x_2, \dots, x_n$, are grouped into a number of intervals, m. Let $t_1, t_2, \dots, t_m$ and $\Delta t_1, \Delta t_2, \dots, \Delta t_m$ denote the midpoints and the length of the intervals, respectively. If a_j denotes the number of observations in the jth interval and $h_j = a_j/n$ the *relative frequency* of observations in the jth interval, $j = 1, 2, \dots, m$, the histogram is the step function h defined by

$$h(t) = \frac{h_j}{\Delta t_j} \quad \text{if } t \in \left] t_j - \frac{\Delta t_j}{2},\, t_j + \frac{\Delta t_j}{2} \right]. \tag{2.1}$$

Notice that in a histogram the relative frequency h_j is represented as the area of a rectangle with sides of length Δt_j and $h_j/\Delta t_j$. Consequently, the total area under the step function h is 1.

If the m intervals are of the same length, i.e., if $\Delta t_1 = \Delta t_2 = \cdots = \Delta t_m = \Delta t$, one often considers,

Table 2.2 *The observations in Table 2.1 grouped into intervals of length 4 cm*

Interval	*Midpoint*	*Number of observations*
]112,116]	114	1
]116,120]	118	0
]120,124]	122	8
]124,128]	126	20
]128,132]	130	24
]132,136]	134	32
]136,140]	138	49
]140,144]	142	41
]144,148]	146	26
]148,152]	150	21
]152,156]	154	14
]156,160]	158	6
]160,164]	162	4
]164,168]	166	1

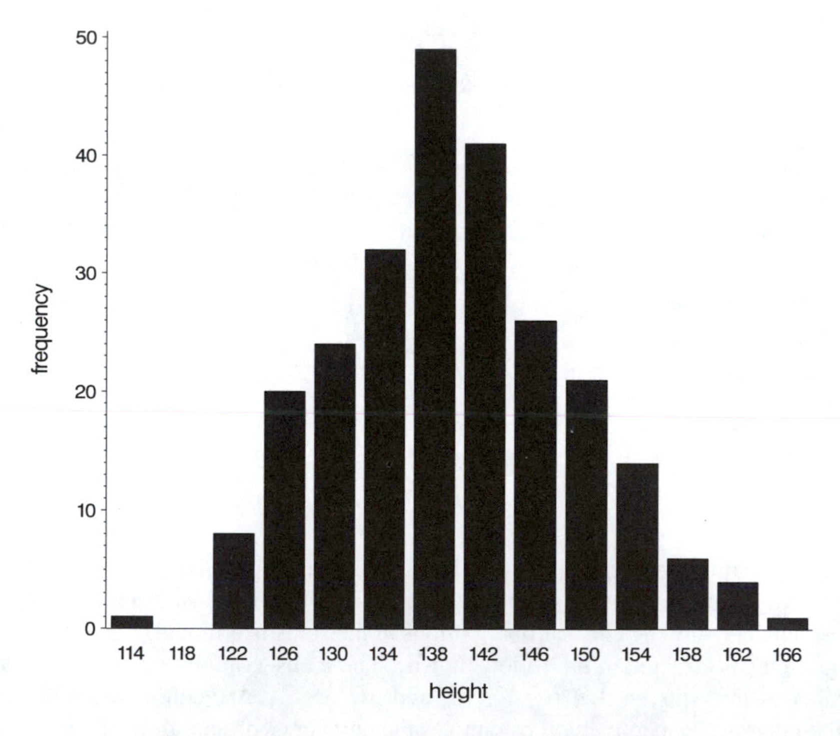

Figure 2.2 *The bar chart for the heights in Table 2.2.*

because of computational convenience, the function,

$$\tilde{h}(t) = n\Delta t h(t) = a_j \quad \text{if } t \in \,]\, t_j - \frac{\Delta t}{2},\, t_j + \frac{\Delta t}{2}\,], \tag{2.2}$$

instead of h. A plot of this function is also referred to as a histogram. Note that the area under $\tilde{h}$ is $n\Delta t$.

Figures 2.3 through 2.8 show different histograms for the variable in Table 2.1. In each histogram the interval length is constant, but the interval length varies from figure to figure. As it is seen from these figures, it is important to choose a suitable number of intervals in the histogram in order to get a good impression of the probability density function of the underlying distribution that generated the observations. Too many intervals give an irregular impression of the function and too few give an impression that is too rough. Most statistical computer packages are able to make histograms and the default value of the number of intervals m is often set to be $\sqrt{n}$. For the variable in Table 2.1, $\sqrt{n} \approx 16$, the number of intervals in Figure 2.5 and Figure 2.6 are 19 and 16, respectively. The figures indicate that the probability density function of the underlying distribution for the variable in Table 2.1 has the same bell-shaped form as the probability density function of the normal distribution, see Figure 3.14 on page 161; the figures suggest a statistical model based on the normal distribution for the heights in Table 2.1.

2.3 Fractile Diagrams

By means of a *fractile diagram* one can assess the assumption that given a sample, $x_1, x_2, \ldots, x_n$, the common distribution function of the corresponding independent and identically distributed random variables, $X_1, X_2, \ldots, X_n$, is $F(x) = P(X_i \leq x)$, or in other words, the assumption that the sample, $x_1, x_2, \ldots, x_n$, can be considered as a sample from the distribution with distribution function F. This assumption may be based on information from dot diagrams, bar charts, or histograms, or it may be based on experience obtained by analyzing similar data sets. In this section we offer a general discussion of fractile diagrams, before we consider the special case in Section 2.4 where F is the distribution function of a normal distribution.

Before we get down to the fractile diagrams, we introduce two data sets that will be used in illustrations of the methods in this section and the section that follows.

Example 2.2

In this example we consider 16 repeated measurements of the propagation velocity of sound in one piece of rock. Velocities are given in units of meters per second and are referred to as *p-wave velocities*. The manufacturer of the measuring device quantifies the measuring uncertainty in terms of a nominal standard deviation of 115 m/s. It is assumed that there are no systematic errors, i.e., the mean error is zero. The 16 measurements are:

4283	4349	4350	4394
4426	4427	4440	4464
4530	4554	4556	4600
4630	4639	4669	4784

The data have been collected in order to evaluate the hypothesis that the p-wave velocity is 4500 m/s. For computational convenience we subtract 4500 from each of the observations and obtain:

−217	−151	−150	−106
−74	−73	−60	−36
30	54	56	100
130	139	169	284

□

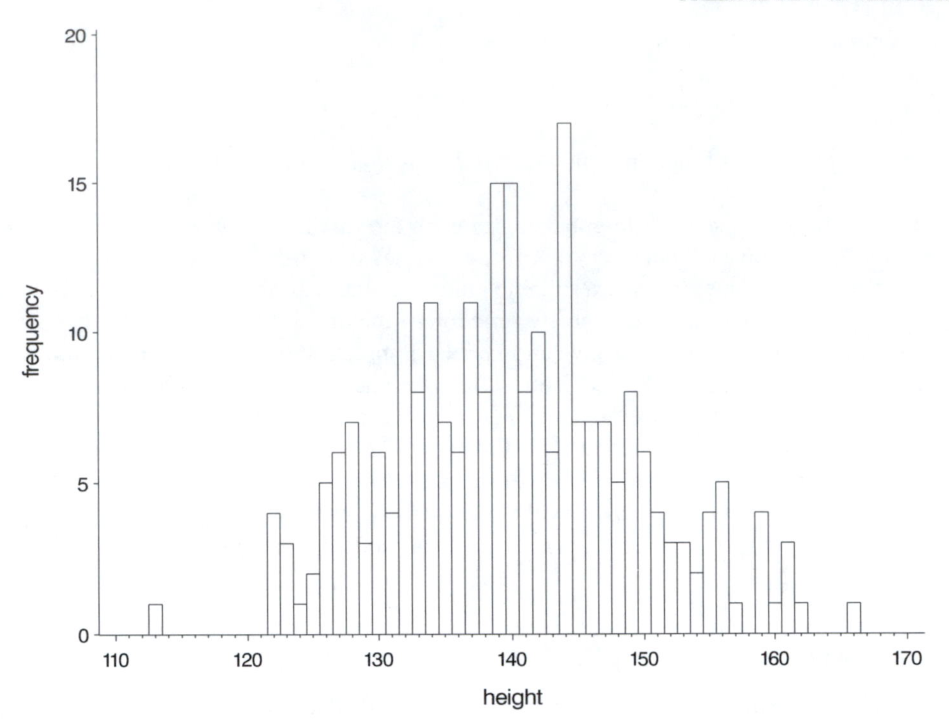

Figure 2.3 *Histogram for heights (in cm) of 247 girls. Interval length 1 cm.*

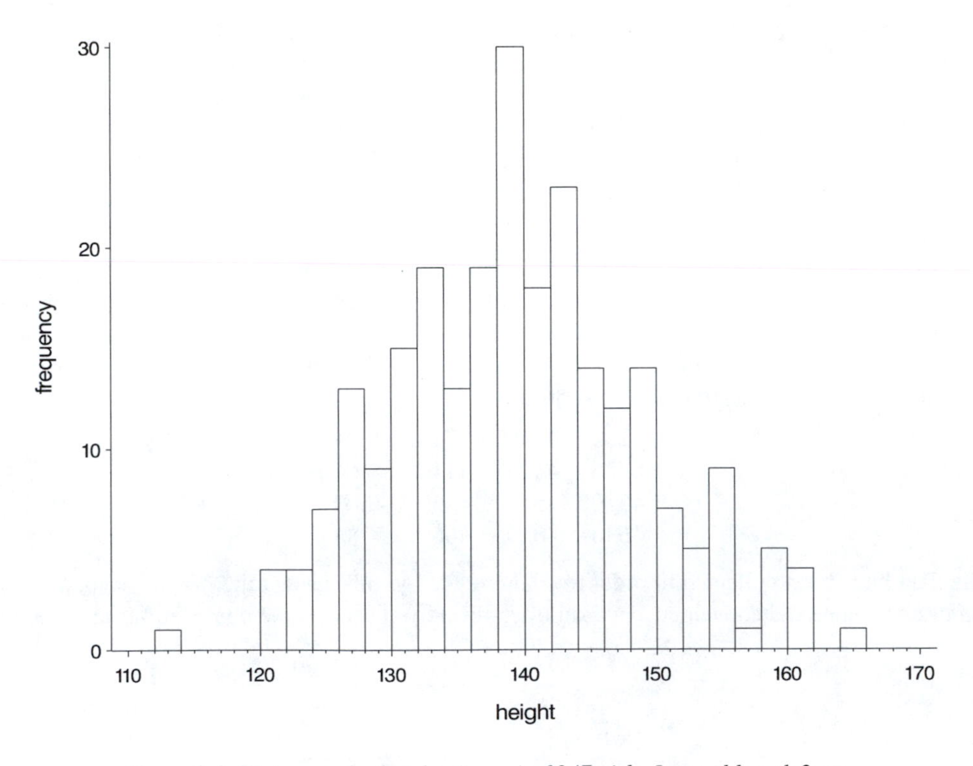

Figure 2.4 *Histogram for heights (in cm) of 247 girls. Interval length 2 cm.*

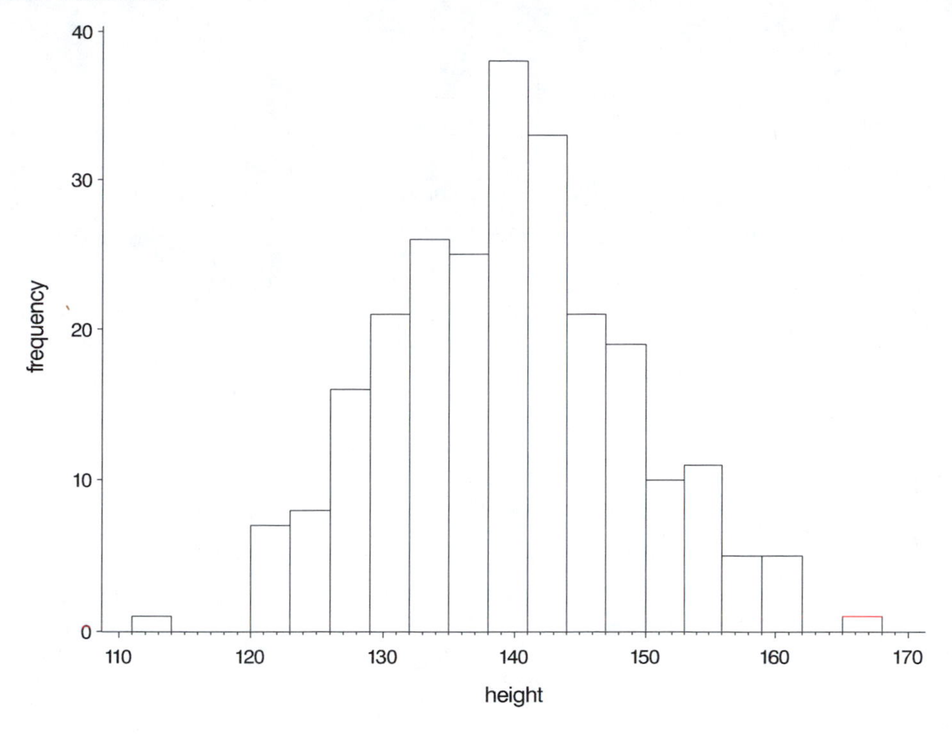

Figure 2.5 *Histogram for heights (in cm) of 247 girls. Interval length 3 cm.*

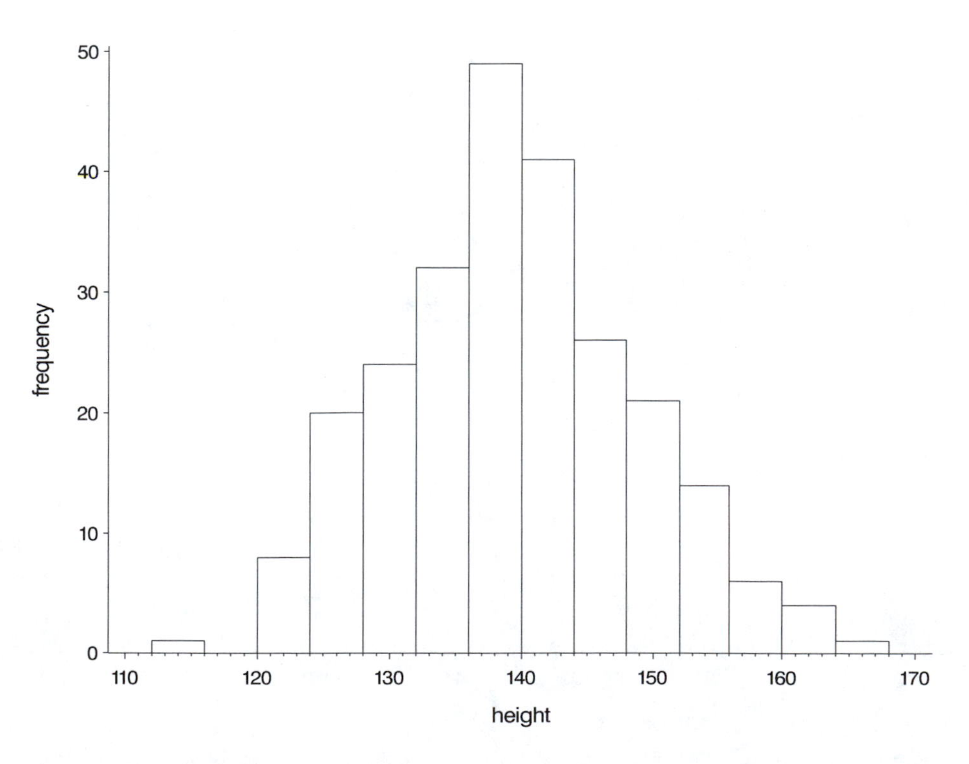

Figure 2.6 *Histogram for heights (in cm) of 247 girls. Interval length 4 cm.*

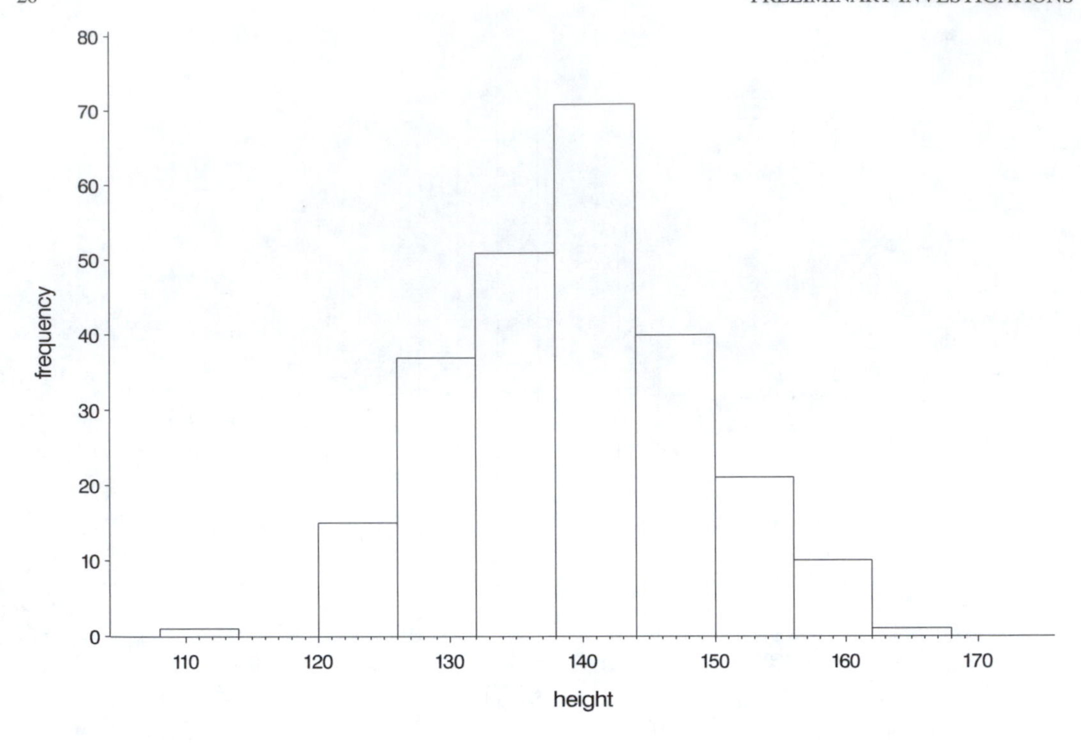

Figure 2.7 *Histogram for heights (in cm) of 247 girls. Interval length 6 cm.*

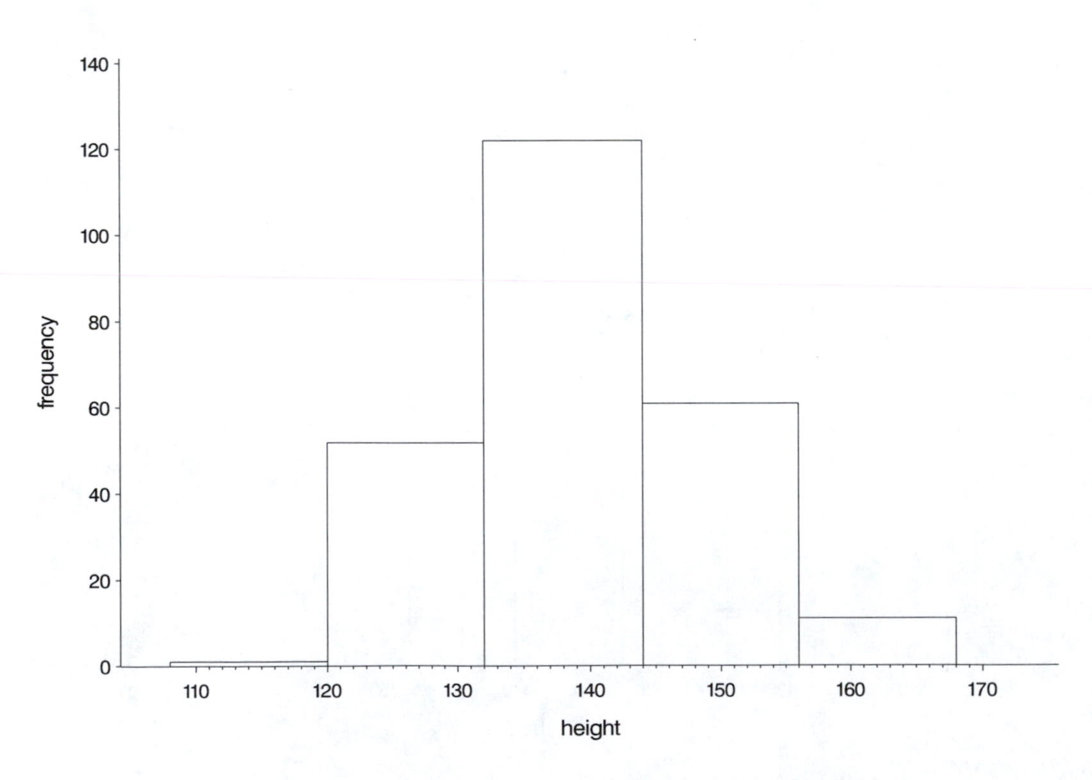

Figure 2.8 *Histogram for heights (in cm) of 247 girls. Interval length 12 cm.*

Example 2.3
In Anderson (1935) one may find the following measurements of the length (in cm) of the petal on 25 plants of the species *Iris versicolor:*

3.0	3.3	3.5	3.7	3.8
3.9	4.0	4.0	4.1	4.1
4.2	4.2	4.2	4.3	4.4
4.4	4.4	4.5	4.5	4.5
4.6	4.7	4.8	5.0	5.1

□

We begin the discussion of fractile diagrams by giving the definition of the *p-fractile* of a distribution function F. If, as usual, $F(x-)$ denotes the limiting value from the left of F at the point x, the definition is:

Definition 2.1 Let F be the distribution function of a random variable X. For each $p \in [0,1]$ the p-fractile *of* F is the set x_p given by

$$x_p = \{x \in \mathbb{R} \mid F(x-) \leq p \leq F(x)\}.$$

In particular, the fractile $x_{0.50}$ is the *median* of F, while $x_{0.25}$ and $x_{0.75}$ are referred to as the *lower* and *upper quartile* of F, respectively. ▲

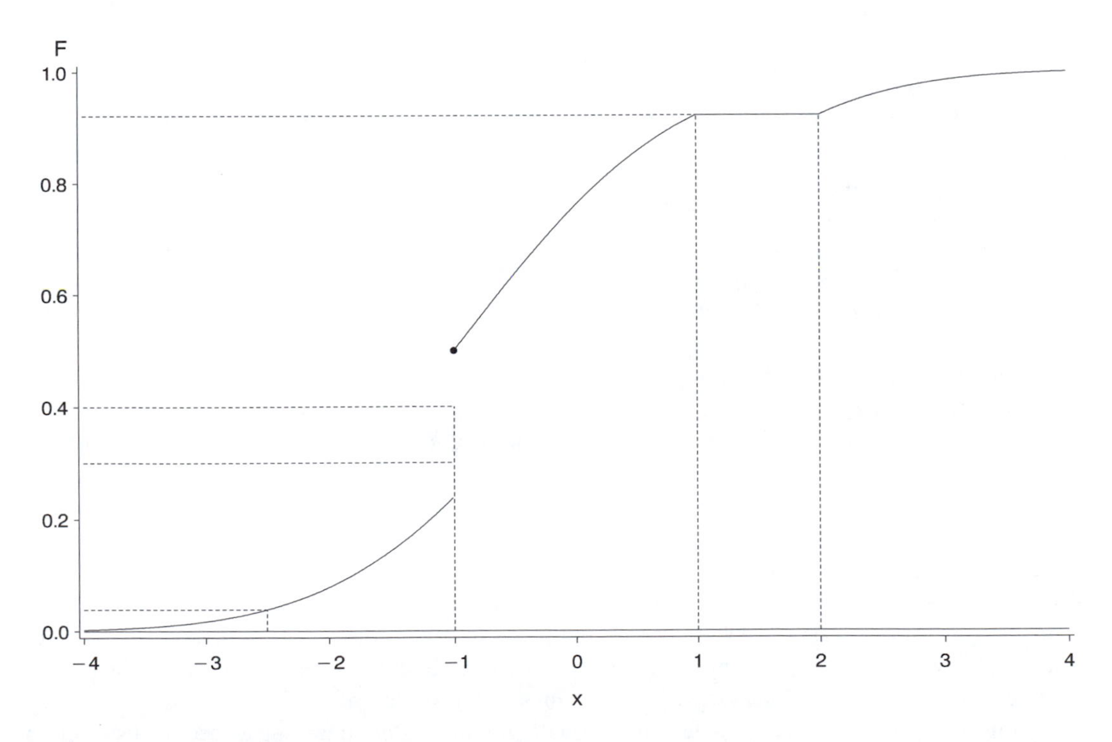

Figure 2.9 *Selected fractiles of the distribution function F: $x_{0.05} = -2.5$, $x_{0.3} = x_{0.4} = -1.0$ and $x_{0.9214} = [1,2]$.*

Note that if F is continuous and strictly increasing, then the p-fractile x_p consists of one point only, namely $F^{-1}(p)$,

$$x_p = F^{-1}(p), \quad p \in]0,1[.$$

The following theorem applies to fractiles:

Theorem 2.1

(a) Let F denote the distribution function of a random variable X. Then F is characterized by its fractiles, that is, by the set of sets, $\{ x_p \mid p \in [0,1] \}$, in the sense that by knowing F, one can find the set x_p for all $p \in [0,1]$ and *vice versa*.

(b) Let Y denote a random variable with distribution function F_Y and let $\alpha \in \mathbb{R}$ and $\beta > 0$. The distribution function F_X of the random variable, $X = \alpha + \beta Y$, is then

$$F_X(x) = F_Y(\frac{x-\alpha}{\beta}), \quad x \in \mathbb{R}.$$

Furthermore, if y_p and x_p denote the p-fractile of Y and X, respectively, then for every $p \in [0,1]$,

$$y_p = \frac{x_p - \alpha}{\beta} = \{\frac{x-\alpha}{\beta} \mid x \in x_p\}.$$

♦

In a fractile diagram a comparison is made between the theoretical distribution function F and the *empirical distribution function* F_n of a sample, $x_1, x_2, \ldots, x_n$. The value of the empirical distribution function of a sample evaluated at x is the number of observations in the sample that are less than or equal to x divided by the number n of observations in the sample:

$$F_n(x) = \frac{\#\{i \mid x_i \leq x\}}{n}, \quad x \in \mathbb{R}, \tag{2.3}$$

where # denotes the number of elements of a set.

Note that the empirical distribution function is a step function whose jumps are multiples of $1/n$ and, furthermore, note that for every $x \in \mathbb{R}$, the number $F_n(x)$ is just the relative frequency of the number of observations in the sample that are less than or equal to x. Using probability theory it may be shown that $F_n(x)$ converges to $F(x)$ when the number of observations tends to infinity. Thus for large values of n, $F_n(x)$ approximates $F(x) = P(X \leq x)$, i.e., $F_n(x)$ approximates the probability that the random variable X is less than or equal to x. The larger n is the better is the approximation of $F(x)$ by $F_n(x)$. Examples of empirical distribution functions may be seen in Figure 2.10 and Figure 2.11.

The assumption that F is the common distribution function of the random variables, $X_1, X_2, \ldots, X_n$, may be evaluated by comparing the graphs of F and F_n. However, this comparison is seldom made because it is difficult to compare two curved graphs, in particular when the graphs are close to 0 and 1.

Instead, one compares the *fractiles* of F and F_n. The idea behind this method is that a distribution function is characterized by its fractiles, see Theorem 2.1. In a *fractile diagram* one plots the p-fractile of the theoretical distribution function F against the p-fractile of the empirical distribution function F_n for a suitable selection of p-values. In order to use F as the distribution in the statistical model for the sample, $x_1, x_2, \ldots, x_n$, the points in the fractile diagram should vary around the identity line. In this way, the comparison of the two functions, F_n and F, is reduced to an evaluation of whether points in a coordinate system exhibit a suitable linear pattern.

When specifying a statistical model, one is usually not interested in finding a specific distribution F describing the random variation in the observations but rather in finding a family of distributions containing a distribution that may describe the variation. Quite often this family of distributions is a *location-scale family* $\mathcal{F}$ in the following sense. Suppose Y is a random variable with distribution function F_Y in $\mathcal{F}$. For all $\mu \in \mathbb{R}$ and for all $\sigma > 0$, $X = \sigma Y + \mu$ is a random variable with distribution function,

$$F_X(x) = F_Y(\frac{x-\mu}{\sigma}). \tag{2.4}$$

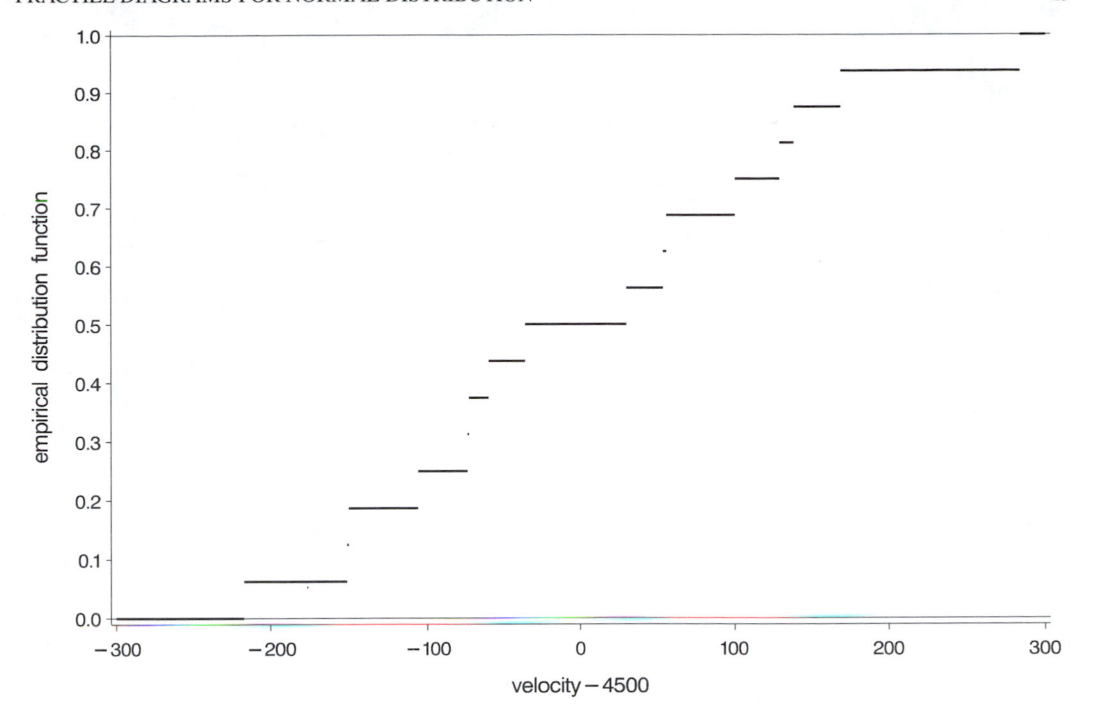

Figure 2.10 *The empirical distribution function of the data in Example 2.2.*

The class of distributions,

$$\mathcal{F} = \{F_Y(\frac{x-\mu}{\sigma}) \mid \mu \in \mathbb{R},\ \sigma > 0\},$$

is called the *position-scale family generated by* F_Y. It consists of all the distributions that may be brought out by changing the reference point and the scale of the measurements.

If x_p and y_p denote the p-fractile of F_X and F_Y, respectively, it follows from (2.4) that

$$y_p = \frac{x_p - \mu}{\sigma}, \qquad \text{for } p \in \,]0,1[. \tag{2.5}$$

It is seen from (2.5) that if we want to assess whether the distribution function F_X gives a satisfactory description of the observations, $x_1, x_2, \ldots, x_n$, we can plot the p-fractiles of F_Y against the p-fractiles of F_n and assess whether the points vary around a straight line $(y = (x-\mu)/\sigma)$.

2.4 Fractile Diagrams for Normal Distribution

Suppose U is a random variable distributed according to the standard normal distribution, $U \sim N(0,1)$ for short, and let $\mu \in \mathbb{R}$ and $\sigma > 0$. From formula (3.81) on page 161, it then follows that $X = \sigma U + \mu \sim N(\mu, \sigma^2)$, which shows that the family of all normal distributions, $\mathcal{N} = \{N(\mu,\sigma^2) \mid \mu \in \mathbb{R},\ \sigma > 0\}$, is the location-scale family generated by the standard normal distribution $N(0,1)$. Consequently, it follows from the discussion in Section 2.3 that the applicability of a statistical model based on the normal distribution for the observations, $x_1, x_2, \ldots, x_n$, may be evaluated by plotting the p-fractiles u_p of the standard normal distribution against the p-fractiles of the empirical distribution function F_n. We now give a detailed description of this method for assessing the assumption that the sample, $x_1, x_2, \ldots, x_n$, may be described by a statistical model based on the normal distribution.

First we introduce the necessary notation. Suppose that there are m *different values* in the sample, $x_1, x_2, \ldots, x_n$, and let $y_1, y_2, \ldots, y_m$ denote the different values ordered according to size,

$$y_1 < y_2 < \ldots < y_m.$$

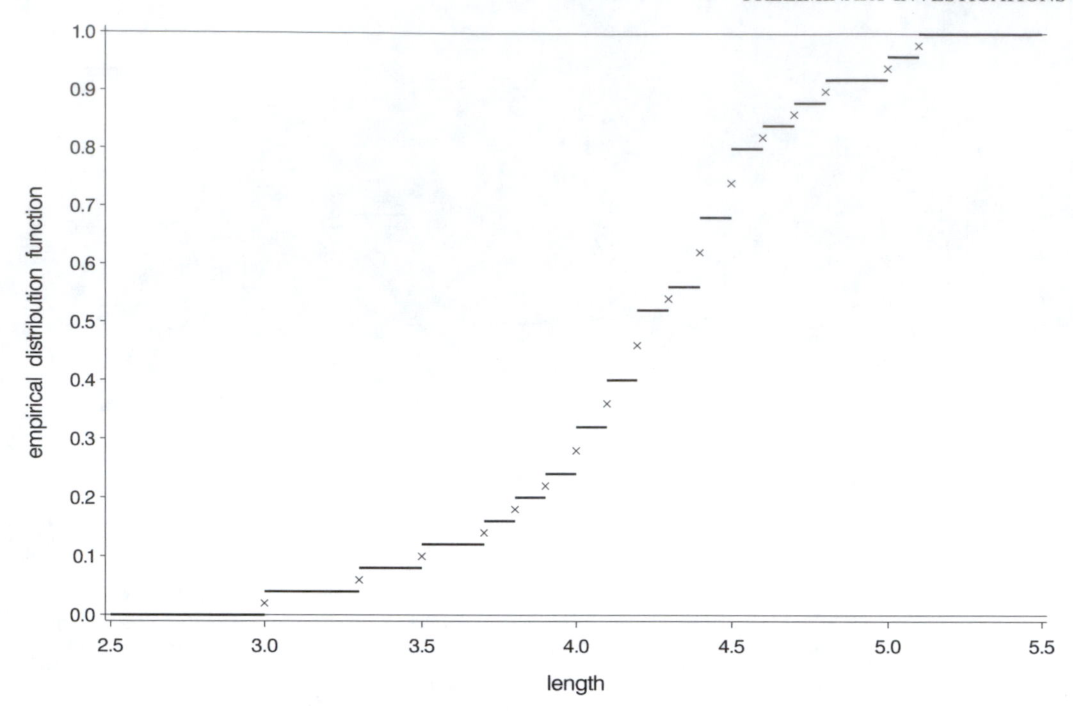

Figure 2.11 *The empirical distribution function of the data in Example 2.3. In the figure* $\times$ *indicates the p-values used in the fractile diagram.*

For $j = 1,2,\ldots,m$ we let a_j denote the *number* of observations in the sample that are equal to y_j and, furthermore, we let k_j denote the *cumulative number*, i.e., $k_j = a_1 + \cdots + a_j$ and k_j is the number of observations in the sample that are less than or equal to y_j. Finally, we let $k_0 = 0$.

For an ungrouped variable with values, $x_1, x_2, \ldots, x_n$, the empirical distribution function F_n is completely known and with the notation introduced above, y_j is p-fractile for all values of p in the interval $[k_{j-1}/n, k_j/n]$. In the fractile diagram we choose the p-value p_j corresponding to y_j as the midpoint of the interval of values for which y_j is a p-fractile, and we choose $p_j = (k_{j-1} + k_j)/(2n)$, see Figure 2.11.

In the fractile diagram we plot the points,

$$(y_j, u_{p_j}), \qquad j = 1,\ 2, \ldots, m.$$

The necessary calculations for the fractile diagram are shown in Table 2.3.

Sometimes a fractile diagram is referred to as a *fractile plot* or a *QQ-plot*.

When the data are not available in electronic form, the fractile diagram may be plotted on *probability paper,* see Figure 2.12. Probability paper exists in different manufactures. A common feature for all of them is that they have a linear first axis for marking the observations and two second axes. One of the second axes is linear and is used for marking the fractiles; the other second axis is a percentage axis. The percentage axis is nonlinear since the percentage $100p$ is marked opposite u_p. The whole point with the two second axes is that reference to a table to find u_p is built into the connection between the two axes and the fractile diagram can be made by plotting the points,

$$(y_j, 100p_j), \qquad j = 1,2,\ldots,m,$$

using the percentage axis. Thus, by using probability paper one avoids calculation of the numbers in the last column of Table 2.3.

Certain types of probability paper use an axis displaced parallel to the linear second axis since the

Table 2.3 *Template for calculations of the fractile diagram for an ungrouped variable*

Observation y	*Number* a	*Cumulative number* k	*Probability in %* p *in %*	p *Fractile* u_p
y_1	a_1	$k_1 = a_1$	$p_1 = 100 \times k_1/(2n)$	$\Phi^{-1}(p_1)$
y_2	a_2	$k_2 = a_1 + a_2$	$p_2 = 100 \times (k_1 + k_2)/(2n)$	$\Phi^{-1}(p_2)$
y_3	a_3	$k_3 = a_1 + a_2 + a_3$	$p_3 = 100 \times (k_2 + k_3)/(2n)$	$\Phi^{-1}(p_3)$
$\vdots$	$\vdots$	$\vdots$	$\vdots$	$\vdots$
y_j	a_j	$k_j = a_1 + \cdots + a_j$	$p_j = 100 \times (k_{j-1} + k_j)/(2n)$	$\Phi^{-1}(p_j)$
$\vdots$	$\vdots$	$\vdots$	$\vdots$	$\vdots$
y_m	a_m	$k_m = a_1 + \cdots + a_m$	$p_m = 100 \times (k_{m-1} + k_m)/(2n)$	$\Phi^{-1}(p_m)$

Table 2.4 *Calculation of the fractile diagram for the data in Example 2.2*

Observation y	*Number* a	*Cumulative number* k	*Probability in %* p *in %*	p *Fractile* u_p
−217	1	1	3.125	−1.863
−151	1	2	9.375	−1.318
−150	1	3	15.625	−1.010
−106	1	4	21.875	−0.776
−74	1	5	28.125	−0.579
−73	1	6	34.375	−0.402
−60	1	7	40.625	−0.237
−36	1	8	46.875	−0.078
30	1	9	53.125	0.078
54	1	10	59.375	0.237
56	1	11	65.625	0.402
100	1	12	71.875	0.579
130	1	13	78.125	0.776
139	1	14	84.375	1.010
169	1	15	90.625	1.318
284	1	16	96.875	1.863

so-called *probits* are used. Probit $prob(p)$ corresponding to a p-value is defined as $prob(p) = u_p + 5$. A fractile diagram plotted on this type of probability paper is often called a *probit diagram*. An example of a probit diagram may be found in Figure 2.12.

For the data in Example 2.2, the template for calculations in Table 2.3 is filled in with Table 2.4 as a result. The corresponding probit diagram is shown in Figure 2.12. Table 2.5 contains the calculations for the data in Example 2.3 and the fractile diagram is shown in Figure 2.13. Finally, the fractile diagram for the variable height in Example 2.1 is seen in Figure 2.14.

When evaluating whether the points in a fractile (probit) diagram exhibit a sufficiently linear pattern to support the assumption that the observations may be considered as a sample from the normal distribution, the following three facts must be taken into account:

1. The variability of the points in a fractile diagram around a straight line decreases when the sample size n increases.

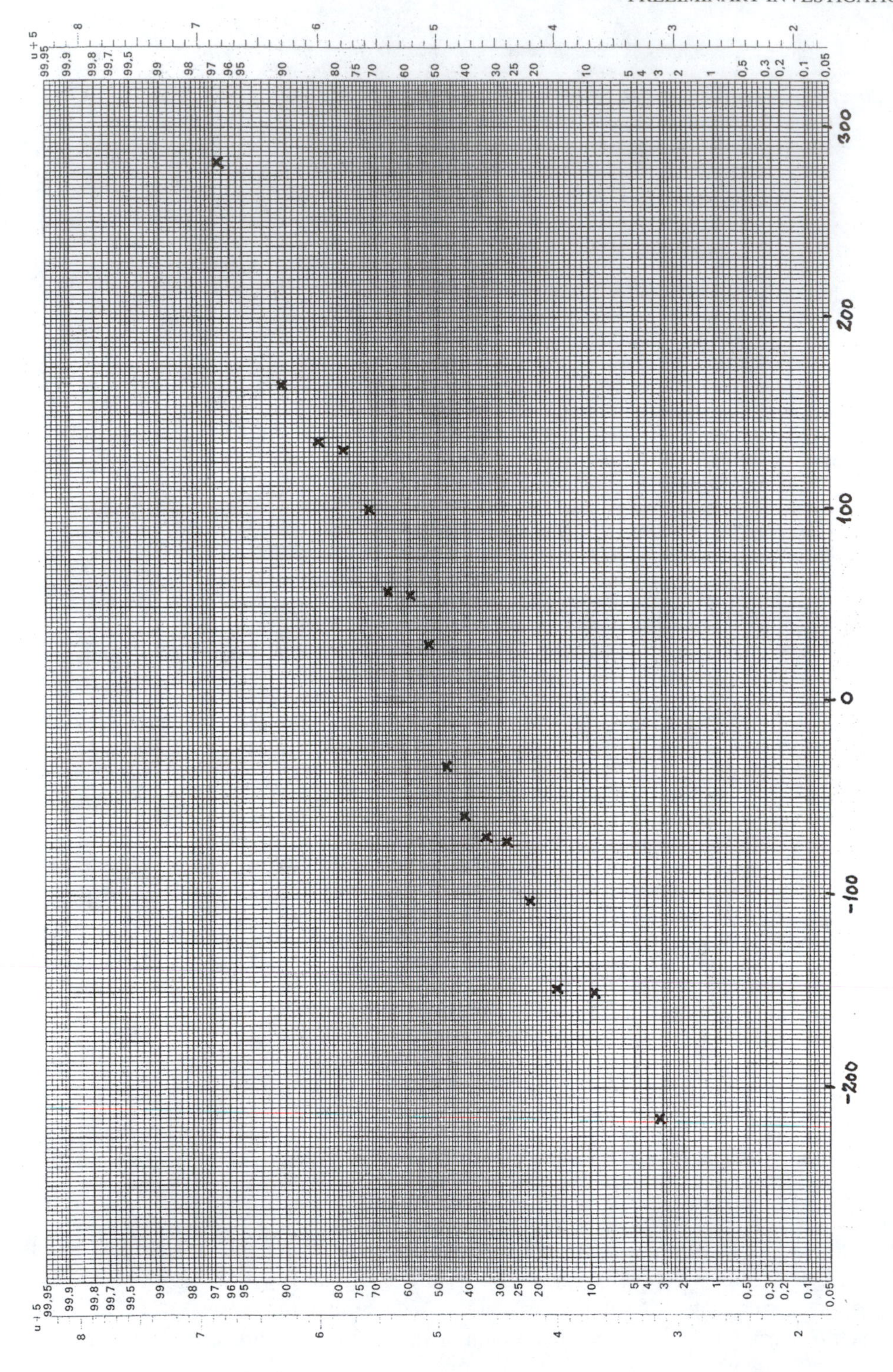

Figure 2.12 *The probit diagram for the data in Example 2.2.*

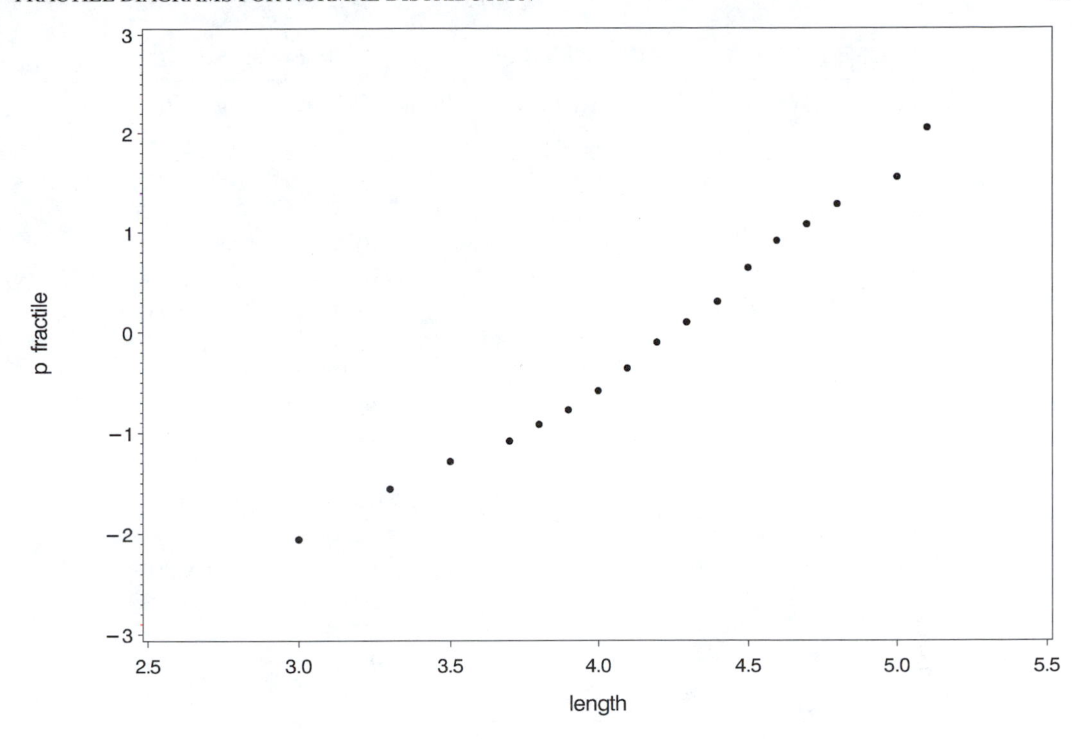

Figure 2.13 *The fractile diagram for the data in Example 2.3.*

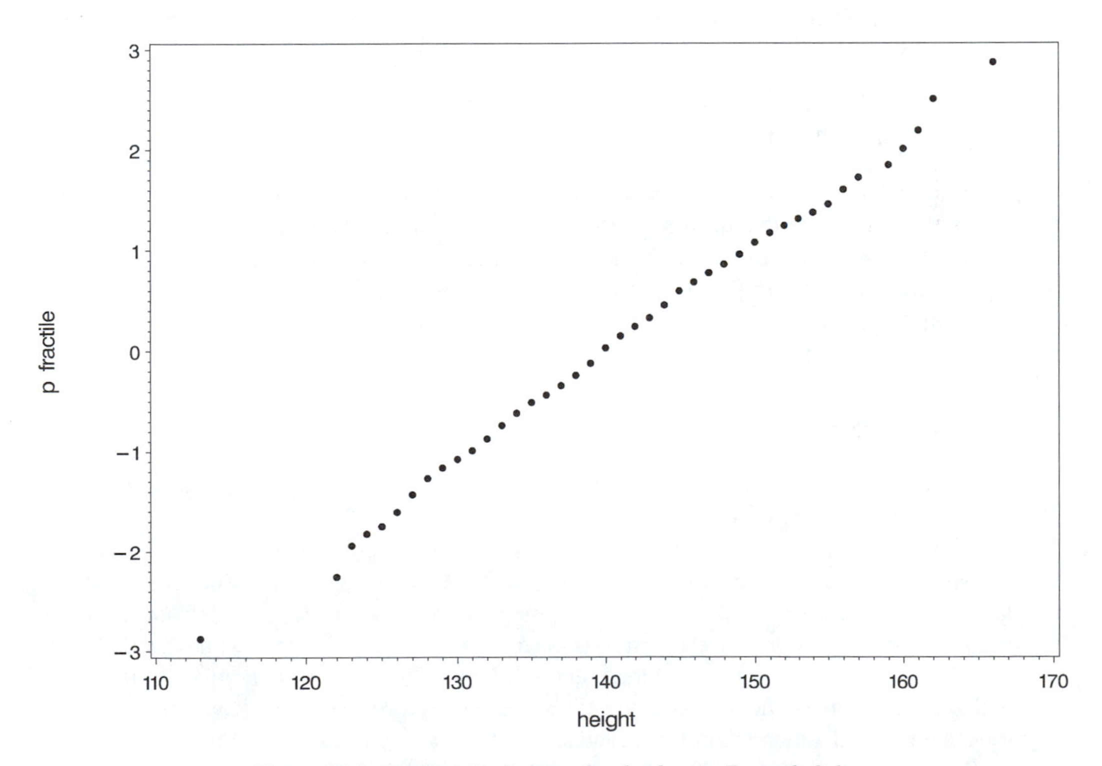

Figure 2.14 *The fractile diagram for the data in Example 2.1.*

Table 2.5 *Calculation of the fractile diagram for the data in Example 2.3*

Observation y	*Number* a	*Cumulative number* k	*Probability in %* p *in %*	p *Fractile* u_p
3.0	1	1	2.0	−2.054
3.3	1	2	6.0	−1.555
3.5	1	3	10.0	−1.282
3.7	1	4	14.0	−1.080
3.8	1	5	18.0	−0.915
3.9	1	6	22.0	−0.772
4.0	2	8	28.0	−0.583
4.1	2	10	36.0	−0.358
4.2	3	13	46.0	−0.100
4.3	1	14	54.0	0.100
4.4	3	17	62.0	0.305
4.5	3	20	74.0	0.643
4.6	1	21	82.0	0.915
4.7	1	22	86.0	1.080
4.8	1	23	90.0	1.282
5.0	1	24	94.0	1.555
5.1	1	25	98.0	2.054

2. The points in a fractile diagram are correlated and as a consequence they have a tendency to twist around the line.

3. The variation of the points around the line is largest in the ends of the fractile diagram and, consequently, the most attention should be paid to points in the middle of the diagram.

A statistical model based on the normal distribution is first and foremost *rejected* if the points in the diagram show a pattern that *deviates systematically* from a linear pattern.

The evaluation of fractile or probit diagrams is primarily a matter of experience, which may be obtained by studying diagrams for normally distributed samples generated by numerical simulation. The figures in Appendix A starting on page 527 show fractile diagrams for a number of different samples of this type with varying sample size n. The figures indicate that for small values of n, the variation of the points around the line may be rather large.

Comparing Figure 2.12, Figure 2.13, and Figure 2.14 with the figures in Appendix A (with the relevant sample size), it may be concluded that there is no evidence against using a statistical model based on the normal distribution for the heights in Example 2.1, the p-wave velocities in Example 2.2, and the petal lengths in Example 2.3.

If the fractile (probit) diagram provides no reasons for rejecting a statistical model based on the normal distribution, *crude* estimates of the mean μ and the variance σ^2 of the normal distribution may be obtained by drawing a line through the points in the diagram. It follows from (2.5) that μ may be estimated as the value on the first axis corresponding to the value 0 on the second axis in a fractile diagram (or to the value 5 on the outer second axis in a probit diagram). Similarly, (2.5) implies that an estimate of the standard deviation σ is the reciprocal of the slope of the line in the diagram. (In a probit diagram the outer second axis should be used when determining the slope.)

If the data set consists of one sample only and if the sample size is small, $n < 10$, it is seen from the figures in Appendix A that the fractile diagram is of limited value when the assumption of normality is evaluated. If the data set consists of several samples, the diagrams may be valuable

Table 2.6 *Template for calculation of the fractile diagram for a grouped variable*

Right endpoint y	*Number* a	*Cumulative number* k	*Probability in %* p *in %*	p *Fractile* u_p
y_1	a_1	$k_1 = a_1$	$p_1 = 100 \times k_1/n$	$\Phi^{-1}(p_1)$
y_2	a_2	$k_2 = a_1 + a_2$	$p_2 = 100 \times k_2/n$	$\Phi^{-1}(p_2)$
y_3	a_3	$k_3 = a_1 + a_2 + a_3$	$p_3 = 100 \times k_3/n$	$\Phi^{-1}(p_3)$
$\vdots$	$\vdots$	$\vdots$	$\vdots$	$\vdots$
y_j	a_j	$k_j = a_1 + \cdots + a_j$	$p_j = 100 \times k_j/n$	$\Phi^{-1}(p_j)$
$\vdots$	$\vdots$	$\vdots$	$\vdots$	$\vdots$
y_{m-1}	a_{m-1}	$k_{m-1} = a_1 + \cdots + a_{m-1}$	$p_{m-1} = 100 \times k_{m-1}/n$	$\Phi^{-1}(p_{m-1})$

even for sample sizes less than 10 by indicating deviations from a linear pattern that reproduce from sample to sample.

With some reluctance we conclude this section with a brief discussion of fractile diagrams for *grouped* variables. Our reluctance is due to the fact that continuous and grouped variables seldom occur in reality. Most frequently a grouped variable originates from an ungrouped one and the motivation for grouping the variable is nearly always practical, for instance, reducing the amount of work in a manual processing of the data, saving space in journals and textbooks; the grouping is very seldom a result of scientific considerations. The procedure of grouping a variable is illustrated by the variable in Table 2.1 and Table 2.2. The variable in Table 2.2 represents a summary of the variable in Table 2.1, which clearly is not as informative as the original variable itself. Statistical procedures should use *all information* in the data and not only a part hereof. However, variables in literature, especially older literature, are often grouped and this is the reason for discussing fractile diagrams for such variables here.

Suppose that the number of observations in the m intervals, $]y_0, y_1],]y_1, y_2], \ldots,]y_{m-1}, y_m]$, are a_1, $a_2, \ldots, a_m$, respectively. For $j = 1, 2, \ldots, m$ we let $k_j = a_1 + a_2 + \cdots + a_j$ denote the cumulative number of observations. Note that $k_m = n$, the total number of observations.

For a grouped variable we only know the value of the empirical distribution function F_n at the right interval endpoints, $y_1, y_2, \ldots, y_m$. Letting $p_j = F_n(y_j) = k_j/n, j = 1, 2, \ldots, m$, we have, in particular, that $p_m = F_n(y_m) = 1$ and since the corresponding u fractile is $u_1 = \infty$, we only plot the following $m-1$ points in the fractile diagram:

$$(y_j, u_{p_j}), \qquad j = 1, 2, \ldots, m-1.$$

Similarly, in a probit diagram the following points are plotted:

$$(y_j, 100p_j), \qquad j = 1, 2, \ldots, m-1.$$

The necessary calculations of the fractile diagram for a grouped variable are shown in Table 2.6.

Example 2.4

Figure 2.15 shows the fractile diagram for a data set, which originates from geochemical prospecting in Finnmark. For 300 specimens the total content of the elements Cu, Zn, and Pb was measured in ppm (parts per million). The measurements from Saxov (1978) given in intervals of length corresponding to 1 ppm are shown in Table 2.7.

From the fractile diagram in Figure 2.15, it may be concluded that the normal distribution gives an acceptable description of the data in Table 2.7. □

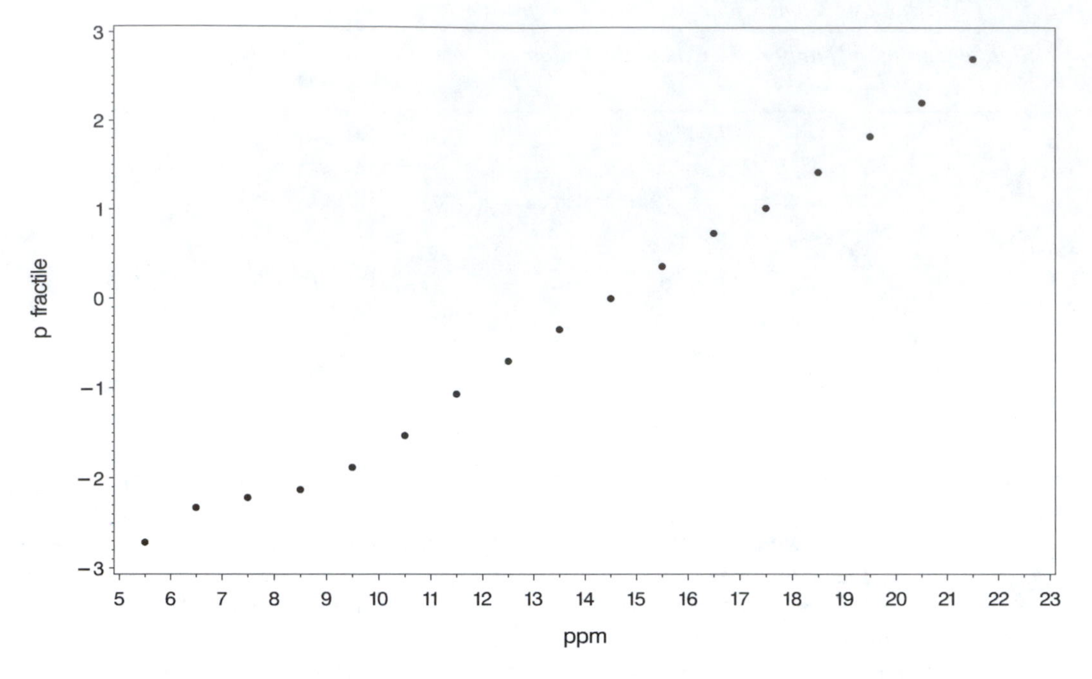

Figure 2.15 *Fractile diagram for the data in Table 2.7.*

Table 2.7 *300 Grouped measurements of the total content of Cu, Zn, and Pb (in ppm) and the calculation of the corresponding fractile diagram*

Right endpoint y	*Number* a	*Cumulative number* k	*Probability in %* p *in %*	*p Fractile* u_p
5.5	1	1	0.333	−2.713
6.5	2	3	1.000	−2.326
7.5	1	4	1.333	−2.216
8.5	1	5	1.667	−2.128
9.5	4	9	3.000	−1.881
10.5	10	19	6.333	−1.527
11.5	24	43	14.333	−1.065
12.5	30	73	24.333	−0.696
13.5	37	110	36.667	−0.341
14.5	41	151	50.333	0.008
15.5	42	193	64.333	0.367
16.5	38	231	77.000	0.739
17.5	23	254	84.667	1.022
18.5	23	277	92.333	1.428
19.5	13	290	96.667	1.833
20.5	6	296	98.667	2.216
21.5	3	299	99.667	2.713

2.5 Transformation

When it is not possible to find a reasonably simple statistical model, i.e., distribution, for the original variable, $x_1, x_2, \ldots, x_n$, it may sometimes be achieved for a *transformed* version of the variable, that is, for $y_1, y_2, \ldots, y_n$, where $y_i = h(x_i), i = 1, 2, \ldots, n$, and h is a known function that is one-to-one (injective). All transformations met in practice are strictly monotone over the range of the observations. Quite often the scientist has a special reason for proposing a special transformation. The most frequently used transformations are $h(x) = \ln(x)$, $h(x) = \sqrt{x}$, and $h(x) = 1/x$.

When searching for a transformation one has to take into account that all the most frequently applied transformations are *locally linear*. Thus, if the observations are all in a narrow interval, the evaluation of the fractile diagram will be the same for both the transformed variable and the untransformed (original) variable. A rule-of-thumb for application of the logarithmic transformation is that the observations have to stretch over a decade, and preferably two, in order that the transformation has a visible effect. There should be at least a factor 10 and preferably a factor 100 between the smallest and the largest observation.

Even though a preliminary investigation using fractile (probit) diagrams is negative in the sense that a statistical model based on the normal distribution is not reasonable for the original observations, $x_1, x_2, \ldots, x_n$, the diagrams may provide information on how to transform the observations in order to obtain a model based on the normal distribution. This fact is illustrated in the following example.

Example 2.5
The measurements in this example are taken from two experiments where the reaction time of flies exposed to a particular nerve gas was recorded. The reaction time of a fly is defined as the time that passes from the moment it comes into contact with the gas until the moment where it is unable to stand on its feet. In the first experiment the flies were exposed to the gas in 30 s and in the second experiment in 60 s. The measurements of the reaction time from Hald (1952) may be seen in Table 2.8.

Table 2.8 *Reaction times of 31 flies exposed to nerve gas*

	Reaction time in minutes							
Contact time	3	5	5	7	9	9	10	12
30 seconds	20	24	24	34	43	46	58	140
Contact time	2	5	5	7	8	9	14	18
60 seconds	24	26	26	34	37	42	90	

From Hald, A., *Statistical Theory with Engineering Applications*, John Wiley & Sons, New York, 1960. With permission.

From Figure 2.16 it is evident that the points in the corresponding fractile diagrams deviate systematically from straight lines. Actually, the curves resemble the graph of the logarithmic function indicating that the observations after a logarithmic transformation may be normally distributed. The fractile diagrams for the logarithmically transformed variable may be obtained from the fractile diagrams for the original variable by using a logarithmic transformation on the first axis. This is done in Figure 2.17, which shows that the transformed variable may be considered as normally distributed. Moreover, Figure 2.17 indicates that the contact time does not seem to have any influence on the reaction time since the points in the diagrams seem to vary around the same line. □

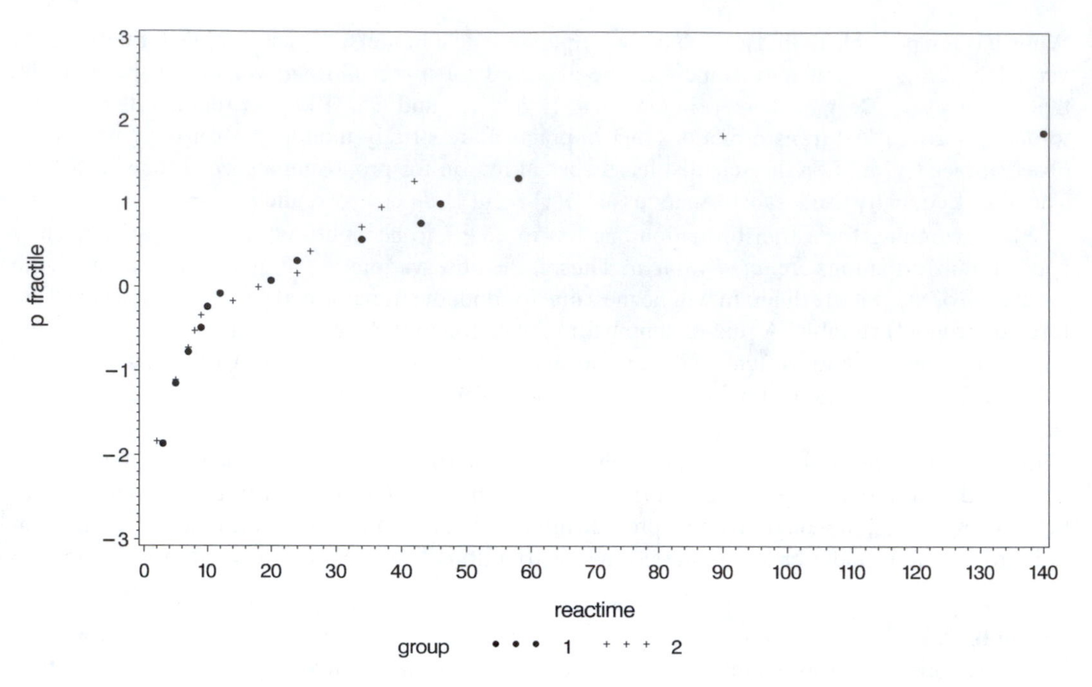

Figure 2.16 *Fractile diagrams of the reaction times in Example 2.5.*

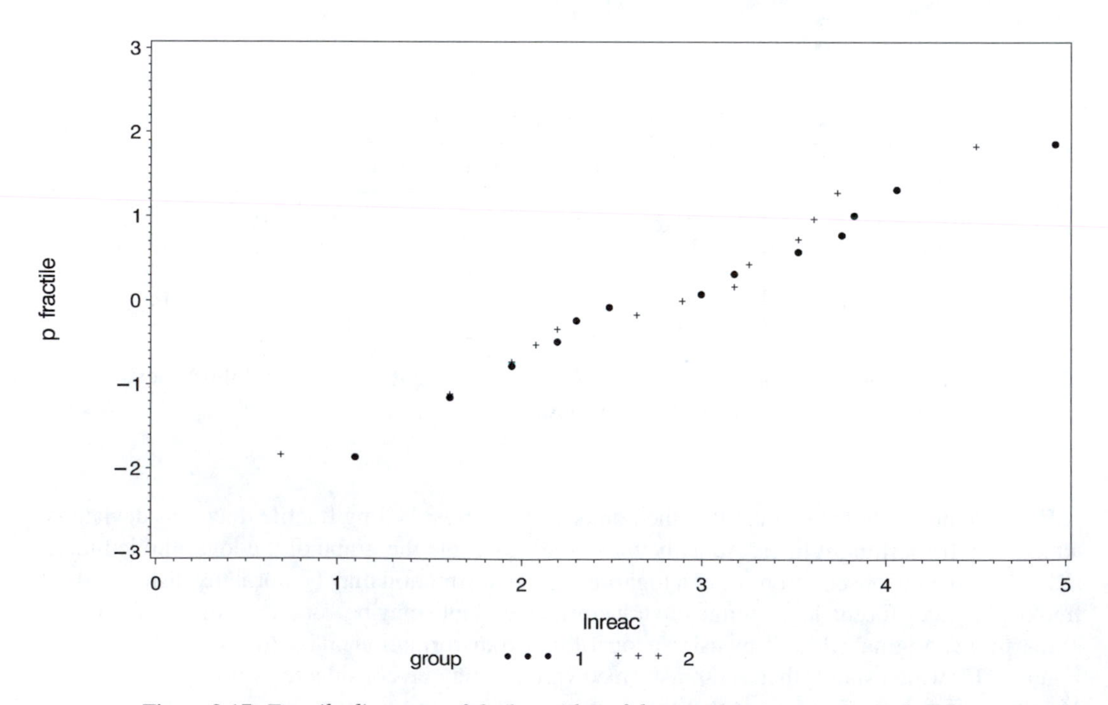

Figure 2.17 *Fractile diagrams of the logarithm of the reaction times in Example 2.5.*

2.6 Concluding Remarks

Despite the simplicity of the graphical methods mentioned in this chapter, the methods are used over and over again when a statistical model is specified and its validity evaluated. The statistical model is the basis of statistical inference whose aim is to formulate statements about the scientific problem that originated the collection of data. It is obvious that statements based on a wrong model is without any value.

The calculations in all the graphical procedures mentioned here are easily done on a pocket calculator. If the data set is large, the calculations may be both tedious and time-consuming, so statistical computer packages may be of great help. Most statistical computer packages have standard procedures for making dot diagrams, bar charts, histograms, and fractile diagrams. As pointed out in Section 2.3 and Section 2.4, there is no reason for grouping an ungrouped variable before making the fractile diagram. Statistical calculations should use *all relevant information* and not only parts hereof. In computers one may store and perform calculations on even very large data sets, and the difference between efficient and less efficient calculations is mainly a question of a negligible increase in computing time.

Annex to Chapter 2

Calculations in SAS

Most chapters in this book have an annex with this title where we give SAS programs for the calculations we have made. In this case the annex is mostly concerned with making plots in SAS.

The plots in this chapter are generated using five procedures. *PROC GPLOT*, *PROC GCHART*, and *PROC CAPABILITY* make plots in high-resolution graphics, whereas the sister procedures *PROC PLOT* and *PROC CHART* give low-resolution graphics in the *OUTPUT* window.

PROC GPLOT is simply a procedure for plotting two or several variables against each other but by controlling the axes and the plot symbols, including whether the points are to be connected and if so with which type of lines, one may obtain figures with a high level of information. *PROC GPLOT* has been used to produce most of the illustrations in the book.

PROC GCHART makes histograms (bar charts), pie charts, and other types of diagrams.

The procedures *PROC GPLOT* and *PROC GCHART* are found in the *GRAPHICS* module in SAS.

PROC CAPABILITY is a procedure in the *QUALITY* module in SAS. Capability analysis is concerned with documenting the ability of industrial productive processes to produce goods that conform with specifications. An element of this is to precisely identify a model for the distribution of properties by the produced goods. Consequently, *PROC CAPABILITY* provides opportunities for making histograms, empirical distribution functions, and fractile diagrams, using a few commands only.

PROC CAPABILITY is a relatively new product and many of the figures in this chapter could have been made using this procedure, as illustrated in the following. We also show how the figures are made using *PROC GPLOT*. This line of action has the advantage that the important *PROC GPLOT* is introduced and, furthermore, it gives us an opportunity to point out that SAS is also a programming language.

As a rare exception we show how most of the figures in this chapter may be produced using the graphical procedures in SAS. The address `http://www.imf.au.dk/biogeostatistics` contains files with information about how the figures in the examples of the remaining chapters may be produced.

In the beginning of all programs producing high-resolution graphics, one has to specify certain options concerning the procedures, the so-called `GOPTIONS`. We use the following options that have been chosen in order to facilitate the insertion of plots on the *PostScript* files into the manuscript.

```
FILENAME GSASFILE 'c:\biogeostatistics\chapter2\examples\plot.ps';
GOPTIONS DEV=win CTEXT=black GACCESS=gsasfile GSFMODE=replace;
GOPTIONS TARGETDEVICE=PS300A4 ROTATE=landscape
         HORIGIN=2 cm VORIGIN=2 cm
         CBACK=white GUNIT=pct HTITLE=4 HTEXT=2.7
         FTEXT=swiss CTEXT=black CSYMBOL=black;
OPTIONS NODATE PAGESIZE=45 PAGENO=1 LS=80;
```

In SAS two kinds of histograms can be made by means of *PROC GCHART*, which we illustrate using the observations in Table 2.1.

Example 2.1 (Continued)

PROC CHART can be used to make bar charts. When the intervals all are of the same length, the bar charts can be viewed as rough histograms. The bar charts are produced using simply the statements HBAR, which produces a chart with horizontal bars, and VBAR, which produces a chart with vertical bars.

The MIDPOINT option makes it possible to control the intervals. Choosing the midpoints as 115, 119, ..., and so on in steps of four units to 167 gives the same number of observations in the intervals as in Table 2.2. The reason for this is that SAS uses intervals that are closed on the left and open on the right, whereas in the construction of histograms we use intervals that are open on the left and closed on the right.

The following program begins with a data step where the data set ex2_1 is created and then a bar chart in the *OUTPUT* window is made by *PROC CHART*.

```
TITLE1 'Example 2.1';

DATA ex2_1;
INPUT height@@;
DATALINES;
139 128 139 125 132 137 146 129 146 150 141 161 143
131 128 134 132 136 137 137 129 140 140 143 148 148
149 132 144 147 137 142 127 127 126 135 136 144 130
132 141 126 135 129 132 130 139 139 134 132 134 127
138 134 127 133 134 126 140 133 142 130 143 140 140
143 150 144 144 128 135 131 135 138 131 135 148 134
132 137 113 150 155 155 155 161 142 142 146 140 141
146 140 139 137 146 142 130 145 149 156 149 155 152
144 139 157 144 149 161 150 144 141 138 140 141 141
147 142 146 156 140 144 145 137 126 134 144 159 134
134 144 130 126 131 130 133 125 122 145 140 132 139
139 128 146 137 139 138 145 133 139 133 139 151 150
138 142 151 140 142 144 136 139 135 141 132 139 140
144 142 127 147 151 141 138 142 147 153 148 144 138
139 124 127 122 123 133 133 136 134 140 137 132 133
132 128 128 136 122 122 123 123 128 145 152 152 156
149 160 148 149 159 145 156 149 153 154 144 153 144
134 140 135 149 136 145 143 139 143 138 137 140 137
144 147 151 166 147 144 159 156 147 154 150 162 159
;
RUN;

PROC CHART DATA=ex2_1;
    HBAR height/MIDPOINTS = 115 TO 167 BY 4;
RUN;
```

The output from *PROC CHART* is shown on the upper part of the next page. The bar chart on the left should be compared to Figure 2.2 and the table on the right to Table *2.2*.

In the following program VBAR has been substituted for HBAR.

```
PROC CHART DATA=ex2_1;
    VBAR height/MIDPOINTS = 115 TO 167 BY 4;
RUN;
```

It gives the output on the lower part of the next page. This should be compared to the bar chart in Figure 2.2 and the histogram in Figure 2.6. Note that the table with the number of observations in the intervals is not produced with the VBAR statement.

```
                         Example 2.1                                  1

 height                             Cum.               Cum.
Midpoint                     Freq   Freq   Percent   Percent
          |
   115    |*                    1      1      0.40      0.40
          |
   119    |                     0      1      0.00      0.40
          |
   123    |****                 8      9      3.24      3.64
          |
   127    |**********          20     29      8.10     11.74
          |
   131    |************        24     53      9.72     21.46
          |
   135    |****************    32     85     12.96     34.41
          |
   139    |*************************  49  134  19.84   54.25
          |
   143    |*********************  41   175     16.60     70.85
          |
   147    |*************       26    201     10.53     81.38
          |
   151    |***********         21    222      8.50     89.88
          |
   155    |*******             14    236      5.67     95.55
          |
   159    |***                  6    242      2.43     97.98
          |
   163    |**                   4    246      1.62     99.60
          |
   167    |*                    1    247      0.40    100.00
          |
          -----+----+----+----+----+
              10   20   30   40   50

                  Frequency
```

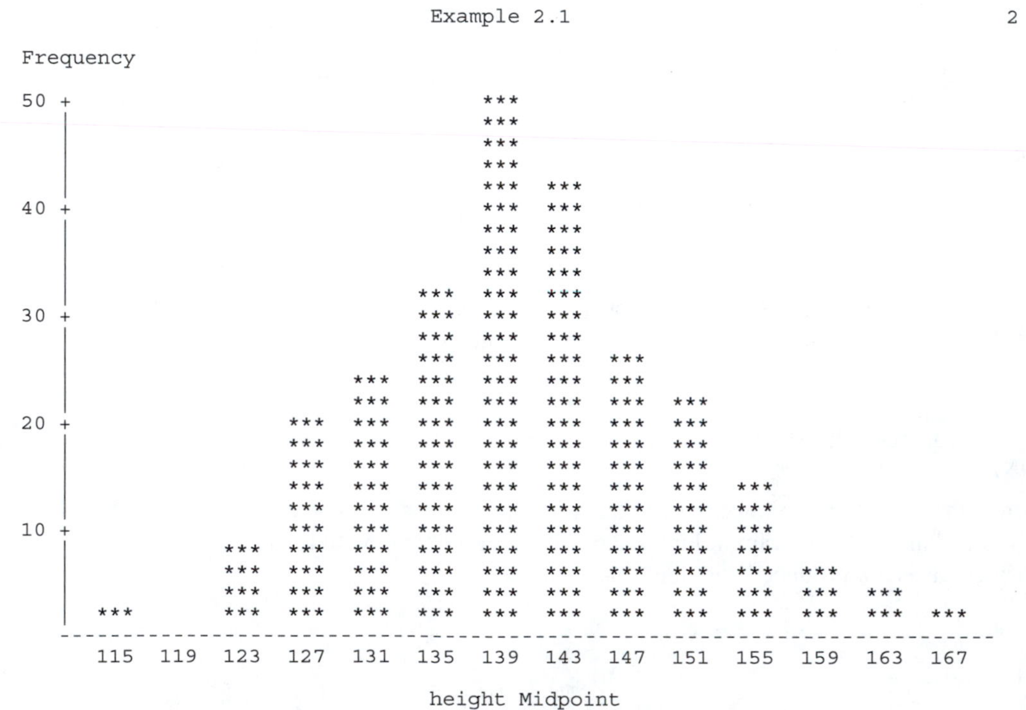

Figure 2.1 (Dot diagram of heights in Table 2.1)
The plot is made by means of the following program:

```
PROC SORT DATA=ex2_1;
   BY height;
RUN;

PROC SUMMARY DATA=ex2_1;
   BY height;
   VAR height;
   OUTPUT OUT=table N=freq;
RUN;

DATA dot;
SET table(KEEP=height freq);
   DO i=1 TO freq;
      dot=i;
      OUTPUT;
   END;
RUN;

SYMBOL1 I=none C=black V=dot H=2;
AXIS1 LABEL=(H=3 'height in cm') VALUE=(H=2.7)
      WIDTH=2 ORDER=110 TO 170 BY 10;
AXIS2 LABEL=(H=3 ANGLE=90 'number of girls') VALUE=(H=2.7)
      WIDTH=2 ORDER=0 TO 20 BY 5 MINOR=(N=4);
PROC GPLOT DATA=dot;
     PLOT dot*height/HAXIS=axis1
                     VAXIS=axis2
                     NOFRAME;
RUN;
QUIT;
```

First the data is sorted in ascending order using *PROC SORT*. By means of *PROC SUMMARY*, a data set `table` is constructed with variables `height` and `freq` of primary interest. The variable `height` in `table` records the different values of `height` in the original data set and the variable `freq` records for each value of `height` the number of girls with that height. Plotting `freq` against `height` would give the topmost dot over each observed value of `height`.

The data set `dot` is an expansion of the data set `table` to give `freq` dots above each observed value of `height`.

The data set is finally used as input to *PROC GPLOT* and `dot` is plotted against `height`. Before this procedure is called the symbols (`SYMBOL`) and axes (`AXIS`) used in the plot are defined.

Figure 2.2 (The bar chart for the heights in Table 2.2)
The plot is made using *PROC GCHART* as shown in the program segment below. Since SAS uses intervals half-open on the right and we are using intervals half-open on the left, we correct for this difference by subtracting 1 from all the observations. This is done in the data set `stick`, which is used as input to *PROC GCHART*. We stress that in other applications of *PROC GCHART* there is no need for this correction. It is only in order to reproduce Figure 2.2 that the corrections have been made here.

```
DATA stick;
SET ex2_1;
   height=height-1;
RUN;

AXIS1 LABEL=(H=3 ANGLE=90 'frequency') VALUE=(H=2.7)
```

```
      WIDTH=2 ORDER=0 TO 50 BY 10;
AXIS2 LABEL=(H=3 'height') VALUE=(H=2.7) WIDTH=2;

PROC GCHART DATA=stick;
     VBAR height/MIDPOINTS = 114 TO 166 BY 4
                 MAXIS=axis2
                 RAXIS=axis1
                 NOFRAME;
RUN;
```

Figure 2.6 (Histogram for the heights of 247 girls; interval length 4 cm)
As mentioned earlier the SAS procedure *PROC CHART* is using intervals half-open on the right. However, the option RTINCLUDE in the procedure *PROC CAPABILITY* ensures that intervals half-open on the left are considered. Consequently, the plot can be made by this procedure as shown in the program segment:

```
AXIS1 LABEL=(H=3 'height') VALUE=(H=2.7)
      WIDTH=2 ORDER=110 TO 170 BY 10;
AXIS2 LABEL=(H=3 ANGLE=90 'frequency') VALUE=(H=2.7)
      WIDTH=2 ORDER=0 TO 50 BY 10;
PROC CAPABILITY data=ex2_1 GRAPHICS;
     HISTOGRAM height/MIDPOINTS=114 TO 166 BY 4
                      RTINCLUDE VSCALE=count
                      HAXIS=axis1 VAXIS=axis2
                      HMINOR=9
                      NOFRAME;
RUN;
```

Figure 2.14 (Fractile diagram for the heights in Example 2.1)
The calculation and the plot of the fractile diagram for an *ungrouped* variable can be made by means of the macro normgraph. This macro has been produced in order to make fractile diagrams for ungrouped variables precisely in the way described in this book. The option graphics=yes implies that the plot is made in high-resolution graphics. As default, the plot is made in low-resolution graphics in the *OUTPUT* window; see the continuation of Example 2.3 on page 49. The fractile diagram can be made by means of the following program:

```
AXIS2 LABEL=(H=3 'height') VALUE=(H=2.7)
      WIDTH=2 ORDER=110 TO 170 BY 10;
/* AXIS2 controls the data axis in normgraph */

%normgraph(data=ex2_1,var=height,graphics=yes);
```

Since the call of the macro contains graphics=yes, the fractile diagram is plotted by means of *PROC GPLOT*. □

Example 2.2 (Continued)
The program segment

```
TITLE1 'Example 2.2';

DATA ex2_2;
INPUT p_wave@@;
pw_4500=p_wave - 4500;
DATALINES;
4283 4349 4350 4394
4426 4427 4440 4464
4530 4554 4556 4600
4630 4639 4669 4784
;
```

```
RUN;

PROC SORT DATA=ex2_2;
   BY pw_4500;
RUN;

PROC SUMMARY DATA=ex2_2;
   BY pw_4500;
   VAR pw_4500;
   OUTPUT OUT=table N=freq;
RUN;
```

reads the data and produces a SAS data set `table` with two variables `pw_4500` (the different values of the p-wave velocity −4500) and `freq` (the number of times the different values are observed).

Figure 2.10 (Empirical distribution function of the data in Example 2.2)
By means of the data set `table` the empirical distribution function F_n is calculated in the data set `emp_dist` as shown below. Notice that the cumulative numbers are calculated using `cum+freq`. The data set `end` contains the coordinates of F_n at the left and right endpoints of the interval where we want to plot F_n, in this case the interval $[-300, 300]$. The values are added to the data set `emp_dist`.

```
DATA emp_dist;
SET tabel(KEEP=pw_4500 freq);
cum+freq;
F_n=cum/16;
RUN;

DATA end;
INPUT pw_4500 F_n;
DATALINES;
-300 0
 300 1
;
RUN;

DATA emp_dist;
SET emp_dist end;
RUN;
```

The empirical distribution function F_n is plotted using the segment:

```
SYMBOL1 I=steps V=none C=black L=1 WIDTH=7;
AXIS1 LABEL=(H=3 'velocity-4500')  WIDTH=2 VALUE=(H=2.7)
      MINOR=none ORDER=-300 TO 300 BY 100;
AXIS2 LABEL=(H=3 ANGLE=90 'empirical distribution function')
      VALUE=(H=2.7) WIDTH=2 MINOR=none ORDER=0 TO 1 BY 0.1;
PROC GPLOT DATA=emp_dist;
     PLOT F_n*pw_4500/HAXIS=axis1
                      VAXIS=axis2
                      VREF=0 1
                      NOFRAME;
RUN;
QUIT;
```

The `I=steps` option in the `SYMBOL1` statement implies that F_n is plotted as a step function, and the `L=1` option requests that a full line is used. Finally, the `WIDTH=7` option determines the line width; the default is `WIDTH=1` so the step function is plotted with a line that is thicker than the rest of the plot.

The option `VREF=0 1` specifies horizontal lines intersecting the vertical axis at 0 and 1.

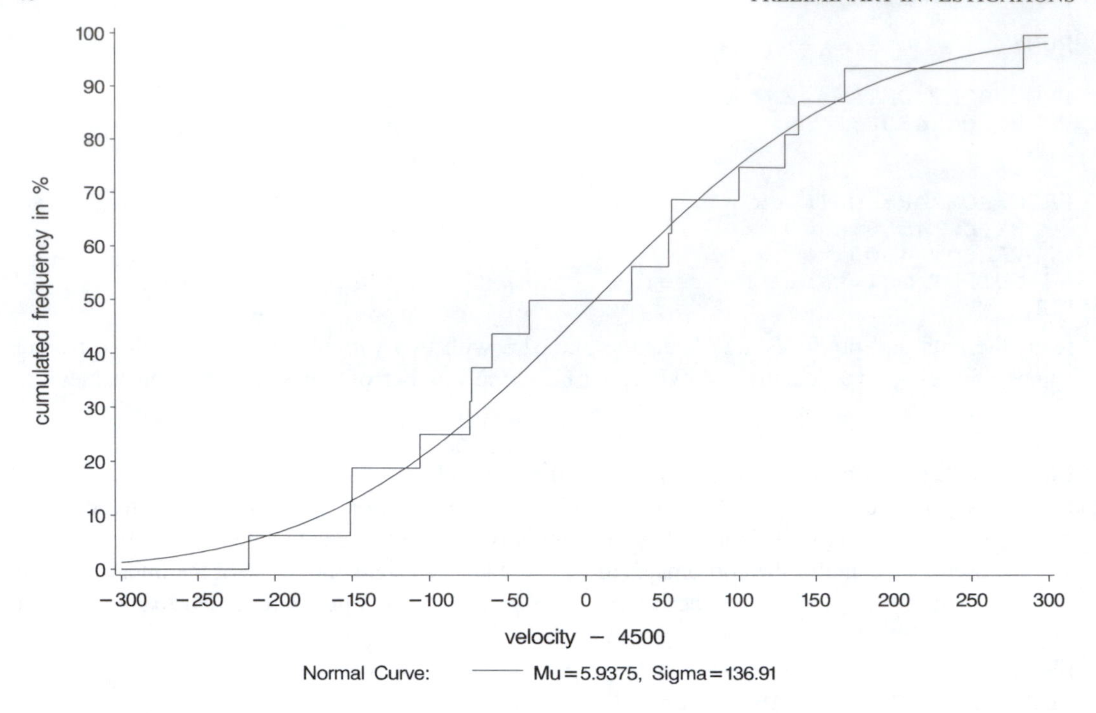

Figure 2.18 *The empirical distribution function and the distribution function of the normal distribution with estimated mean value and standard deviation for p-wave velocities in Example 2.2.*

PROC CAPABILITY has facilities for plotting the empirical distribution function as well as the fractile diagram. We illustrate this by showing how figures corresponding to Figure 2.10 and Figure 2.12 may be obtained. The result is a graph of the empirical distribution function similar to that in Figure 2.10 with the commands:

```
PROC CAPABILITY DATA=ex2_2;
   CDFPLOT pw_4500/;
RUN;
```

The `CDFPLOT` statement has several options to enhance the plot. An important one is the possibility to overlay the empirical distribution function with a theoretical distribution function from one of several families of distributions frequently met in practice.

It is also useful that the `AXIS` statements can be used to control the axes of the graphs in the same way as in the `PLOT` statement in *PROC GPLOT*. The graph in Figure 2.18 is generated by the commands:

```
AXIS1 LABEL=(H=3 ANGLE=90 'cumulated frequency in %') VALUE=(H=2.7)
      WIDTH=2 ORDER=0 TO 100 BY 10;
AXIS2 LABEL=(H=3 'velocity - 4500') VALUE=(H=2.7) WIDTH=2.0
      ORDER=-300 TO 300 BY 50;

PROC CAPABILITY DATA=ex2_2;
   CDFPLOT pw_4500/NORMAL (MU=EST SIGMA=EST)
                   HAXIS=axis2
                   VAXIS=axis1
                   NOFRAME;
RUN;
```

Here the option

```
NORMAL (MU=EST SIGMA=EST)
```

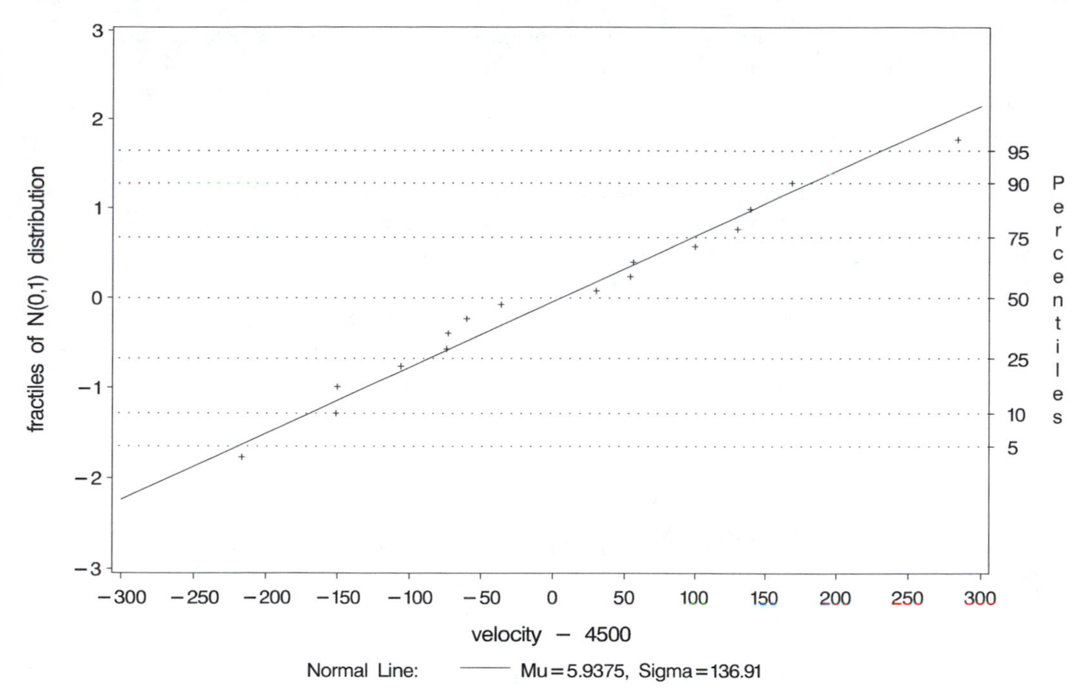

Figure 2.19 *The fractile diagram of the p-wave velocities in Example 2.2.*

requests a plot of the distribution function of the normal distribution with mean and standard deviation estimated from the sample. If a comparison with a specific normal distribution is desired, its mean and standard deviation must be specified instead of EST in the option.

A fractile diagram corresponding to (but not precisely similar to) Figure 2.12 may be constructed using the commands:

```
PROC CAPABILITY DATA=ex2_2;
   QQPLOT pw_4500/ ;
RUN;
```

There is a possibility of plotting a straight reference line corresponding to a theoretical normal distribution and, furthermore, a possibility of having both a linear fractile axis and a percentage axis just as on probability paper. These facilities are used in the following program segment producing Figure 2.19.

```
AXIS1 LABEL=(H=3 ANGLE=90 'fractiles of N(0,1) distribution')
      VALUE=(H=2.7) WIDTH=2 ORDER=-3 TO 3 BY 1 ;
AXIS2 LABEL=(H=3 'velocity - 4500') VALUE=(H=2.7) WIDTH=2.0
      ORDER=-300 TO 300 BY 50;

PROC CAPABILITY DATA=ex2_2;
   QQPLOT pw_4500/NORMAL (MU=EST SIGMA=EST)
                  ROTATE
                  HAXIS=axis2
                  VAXIS=axis1
                  PCTLAXIS ( LABEL='Percentiles' GRID LGRID=35 );
RUN;
```

Here the option

```
NORMAL (MU=EST SIGMA=EST)
```

requests a line corresponding to the normal distribution with mean and standard deviation estimated from the sample. If a comparison with a specific normal distribution is desired, its mean and standard deviation must be specified instead of `EST` in the option.

The percentage axis is obtained with the option

```
PCTLAXIS ( LABEL='Percentiles' GRID LGRID=35 );
```

which also determines the type of horizontal line between the fractile axis and the percentage axis.

The command `ROTATE` ensures that the data axis becomes the first axis in the fractile diagram. In American literature the data axis is often the second axis.

We noted above that the fractile diagram in Figure 2.19 is not identical to the fractile diagram in Figure 2.12. A comparison between the two figures shows that more points have been plotted in Figure 2.19 than in Figure 2.12. `QQPLOT` in *PROC CAPABILITY* plots a point for all n observations, whereas we only plot points corresponding to the m different values in a data set.

Moreover, even though all observations in the data set were different, Figure 2.12 and Figure 2.19 would not have been identical. If

$$x_{(1)} \leq \cdots \leq x_{(i)} \leq \cdots \leq x_{(n)}$$

denotes the ordered sample, *PROC CAPABILITY*'s `QQPLOT` plots the points,

$$\left(x_{(i)}, \Phi^{-1}\left(\frac{i-3/8}{n+1/2}\right)\right), \qquad i = 1, \ldots, n, \tag{2.6}$$

whereas the recommendation in Table 2.3, when all the observations are different, is to plot the points,

$$\left(x_{(i)}, \Phi^{-1}\left(\frac{i-1/2}{n}\right)\right), \qquad i = 1, \ldots, n. \tag{2.7}$$

The reason behind the choice in *PROC CAPABILITY*'s `QQPLOT` is an investigation in a thesis from 1958 by the Swedish statistician Gunnar Blom: If one is going to estimate a straight line through the points in a fractile diagram using the methods from linear regression, which will be discussed in Section 3.3, and hereby estimate σ as the reciprocal slope, the points in (2.6) gives an estimate $\tilde{\sigma}_{\text{qqplot}}$ of σ, which practically has mean σ, whereas the points in (2.7) gives an estimate $\tilde{\sigma}_{\text{normgraph}}$, which does not have mean σ. However, measured in terms of expected squared deviation, the two estimates are in fact equally good, since

$$E(\tilde{\sigma}_{\text{qqplot}} - \sigma)^2 \doteq E(\tilde{\sigma}_{\text{normgraph}} - \sigma)^2$$

and close to the minimal value of the expected square deviation of an estimate.

The following question emerges: Should one use *PROC CAPABILITY*'s `QQPLOT` or the fractiles from Table 2.3 as implemented in the macro `normgraph`?

The answer is that one can safely use whichever method is the most convenient. Both methods produce plots well-suited for evaluating if a sample may be considered to be normally distributed. An advantage of *PROC CAPABILITY*'s `QQPLOT` is that it can plot straight lines in the fractile diagram corresponding to a normal distribution with parameters that are either estimated or have fixed values. An advantage of `normgraph` is the possibility of having the fractile diagrams for several samples plotted in the same graph, cf. Figure 3.2 on page 87 and Figure 3.3 on page 97.

Even though it was an argument for the choice of the points in *PROC CAPABILITY*'s `QQPLOT` that an estimate of σ should have good properties, one should never estimate the parameters of the normal distribution on the basis of the straight line in the fractile diagram. The estimates have to be found as described in Section 3.1 and these are also the estimates given under the diagram in Figure 2.19. □

Example 2.3 (Continued)

In the program below the data set `ex2_3` is created in the data step and `normgraph` is used to give a fractile diagram in low resolution in the *OUTPUT* window as well as the table used to plot the fractile diagram. The output corresponds to Table 2.5 and Figure 2.13.

```
DATA ex2_3;
INPUT length@@;
DATALINES;
3.0 3.3 3.5 3.7 3.8
3.9 4.0 4.0 4.1 4.1
4.2 4.2 4.2 4.3 4.4
4.4 4.4 4.5 4.5 4.5
4.6 4.7 4.8 5.0 5.1
;
RUN;
%normgraph(data=ex2_3,var=length,graphics=no);
```

Output:

Example 2.3 1

length	frequency	cumulated frequency	p	p fractile
3.0	1	1	0.020	-2.05
3.3	1	2	0.060	-1.55
3.5	1	3	0.100	-1.28
3.7	1	4	0.140	-1.08
3.8	1	5	0.180	-.915
3.9	1	6	0.220	-.772
4.0	2	8	0.280	-.583
4.1	2	10	0.360	-.358
4.2	3	13	0.460	-.100
4.3	1	14	0.540	0.100
4.4	3	17	0.620	0.305
4.5	3	20	0.740	0.643
4.6	1	21	0.820	0.915
4.7	1	22	0.860	1.080
4.8	1	23	0.900	1.282
5.0	1	24	0.940	1.555
5.1	1	25	0.980	2.054

Example 2.3 2

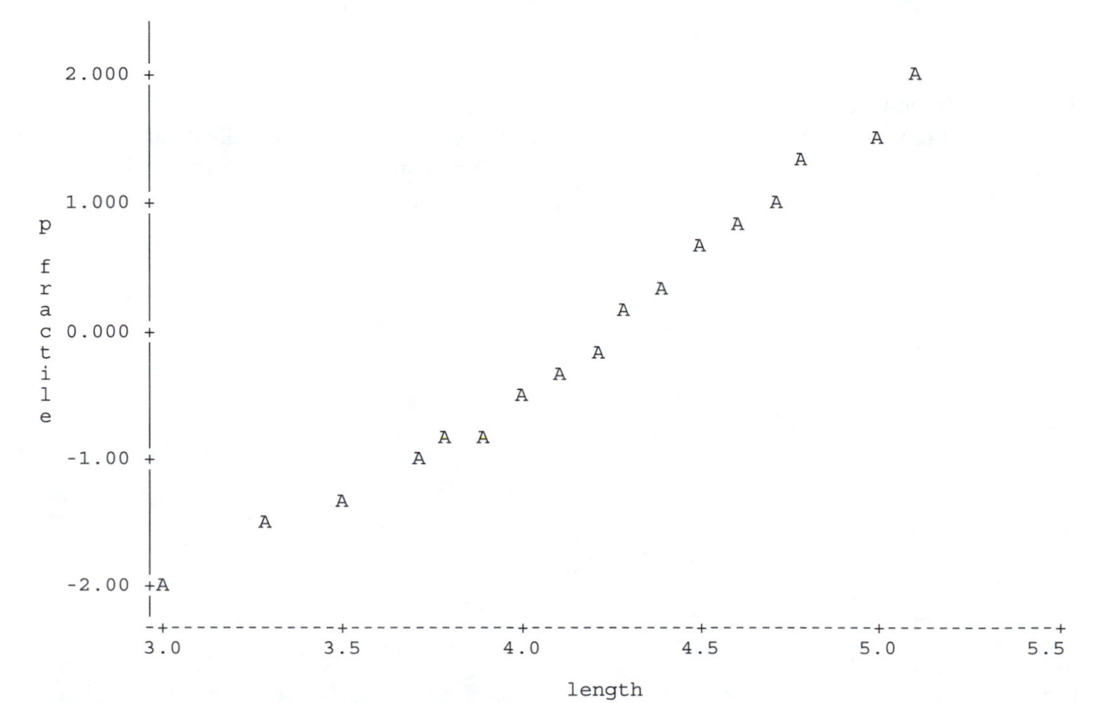

Figure 2.11 (Empirical distribution function of the data in Example 2.3)
The calculations are based on the data set `ex2_3` that was created above. The empirical distribution function F_n is calculated in a SAS data set `emp_dist` as in Example 2.2 above. The p-values that are used in the fractile diagram are calculated in the data step below and stored in the variable `fractile`.

```
DATA emp_dist;
SET emp_dist;
fractile=(F_n+lag(F_n))/2;
IF length=5.5 THEN fractile=.;
RUN;
```

and the plot in Figure 2.11 is the result of:

```
SYMBOL1 I=steps V=none C=black L=1 WIDTH=7;
SYMBOL2 I=none V=x C=black H=2;
AXIS1 LABEL=(H=3 'length') WIDTH=2 VALUE=(h=2.7)
      MINOR=none ORDER=2.5 TO 5.5 BY 0.5;
AXIS2 LABEL=(h=3 ANGLE=90 'empirical distribution function')
      VALUE=(h=2.7) WIDTH=2 MINOR=none ORDER=0 TO 1 BY 0.1;
PROC GPLOT DATA=emp_dist;
     PLOT (F_n fractile)*length/OVERLAY
                                HAXIS=axis1
                                VAXIS=axis2
                                VREF=0 VREF=1
                                NOFRAME;
RUN;
QUIT;
```

□

Example 2.4 (Continued)
The variable in this example is grouped and therefore we cannot use the SAS macro `normgraph` for calculating and plotting the points in the fractile diagram.

Figure 2.15 (Fractile diagram for the data in Example 2.4)
The SAS program below performs the necessary calculations and plots the fractile diagram for the grouped variable. We use the SAS function `PROBIT`, which is the inverse Φ^{-1} of the distribution function Φ of the $N(0,1)$ distribution (the u-distribution), to compute the fractiles.

```
DATA ex2_4;
INPUT ppm number;
DATALINES;
  5      1
  6      2
  7      1
  8      1
  9      4
 10     10
 11     24
 12     30
 13     37
 14     41
 15     42
 16     38
 17     23
 18     23
 19     13
```

```
 20        6
 21        3
 22        1
;
RUN;

DATA table;
SET ex2_4;
ppm=ppm+0.5;
cum+number;
F_n=cum/300;
ufrac=PROBIT(F_n);
LABEL ufrac='p fractile';
RUN;

SYMBOL1 I=none V=dot C=black  H=2;
AXIS1 LABEL=(H=3 ANGLE=90 'p fractile') VALUE=(H=2.7) WIDTH=2
      ORDER=-3 TO 3 BY 1;
AXIS2 LABEL=(H=3 'ppm') VALUE=(H=2.7) WIDTH=2
      ORDER=5 TO 23 BY 1 MINOR=(NUMBER=1);
PROC GPLOT DATA=table;
     PLOT ufrac*ppm/VAXIS=axis1
                    HAXIS=axis2
                    FRAME;
RUN;
QUIT;
```

□

Example 2.5 (Continued)
Quite often, as in this example, there are two or more samples of the same variable measured under varying experimental conditions. When one wishes to evaluate whether several samples may be considered as normally distributed, it is an advantage to plot the fractile diagrams in the same figure. In this way it is possible to investigate whether the points in the diagrams deviate systematically from a linear pattern. If this is not the case, one may get a first impression of whether the samples have the same variance by comparing the slopes of the lines in the figure. If this is the case and, furthermore, if the points in the diagrams vary around the same straight line, then the samples are possibly from the same normal distribution, i.e., the samples have possibly the same mean and the same variance.

Fractile diagrams for several samples may be plotted in the same figure by means of the SAS macro normgraph.

Figure 2.16 and Figure 2.17 (Fractile plots for the two samples in Example 2.5)
The plots in these figures are the result of

```
DATA reac;
INPUT group reactime ;
DATALINES;
1   3 1   5 1   5 1   7 1   9 1    9 1 10 1 12 1 20 1 24
1 24 1 34 1 43 1 46 1 58 1 140
2   2 2   5 2   5 2   7 2   8 2    9 2 14 2 18 2 24
2 26 2 26 2 34 2 37 2 42 2   90
;
RUN;

DATA reac;
SET reac;
lnreac= LOG(reactime);
```

```
RUN;
AXIS2 LABEL=(H=3 'reactime') VALUE=(H=2.7)
      ORDER=0 TO 140 BY 10 WIDTH=2;
/* AXIS2 controls the appearence of the data axis in %normgraph */

%normgraph(data=reac,var=reactime,bygroup=group,graphics=yes);

AXIS2 LABEL=(H=3 'lnreac') VALUE=(H=2.7)
      ORDER=0 TO 5 BY 1 WIDTH=2;

%normgraph(data=reac,var=lnreac,bygroup=group,graphics=yes);
```

If `AXIS2` has not been defined earlier in the program the result will be as in Figure 2.16.

The macro `normgraph` gives the user possibility of defining his own first axis in the plot. This is done by defining `AXIS2` as shown in the second call of the macro in the program segment above. If one uses the same values as here of `LABEL`, `VALUE` and `WIDTH` one gets a first axis with the same appearance as the second axis. (If `AXIS2` is used in connection with *PROC GPLOT* later in the program it may be necessary to redefine `AXIS2`.) □

CHAPTER 3

Normal Data

The models in this chapter have the common feature that the observations are assumed to be normally distributed. The models differ in the complexity of the structure of the means and the variances of the observations. The different models will be treated and referred to under the headings:

One sample
Two or more samples
Linear regression

The models are on the one hand relatively simple, but on the other hand sufficiently flexible to be used to solve a wide range of problems in many areas, ranging from engineering and industrial production over all the natural sciences, including biology and geology, to medical and social science.

Although the models are flexible and widely used, they will not be the only models for normal data that a biologist or a geologist may need in his or her professional career. However, the concepts and routines that are developed in connection with the models of this chapter will make it possible to understand and handle far more complicated models.

The simplest model is that for one normal sample, which will be treated first. This model will be used to introduce important statistical concepts, such as *estimation*, *test*, *significance level*, *significance probability* (*p-value*), and *confidence interval.*

Section 3.4 on page 160 reviews some of the most important properties of the normal distribution and related distributions (t, χ^2, and F) that appear in the analysis of normally distributed data.

3.1 One Sample

Introduction

A single sample forms the simplest example of a statistical model. A *sample* of n observations is n independent observations from the same distribution. We shall denote the observations by $x_1,\ldots,x_n$, and we consider them as realizations of independent random variables, $X_1,\ldots,X_n$, with a common distribution F. If the common distribution is normal, we shall speak of normally distributed data or, simply, *normal data* or a *normal sample*. The questions that motivate the collection of the data must be translated into characteristics of F, and it will typically be in terms of mean and variance.

Chapter 2 contained several examples of samples.

In Example 2.1 the observations are the heights of 10- to 12-year-old girls and the normal distribution is a model for the distribution of heights of 10- to 12-year-old girls with asthma in the county of Funen, Denmark.

In Example 2.2 the normal distribution is a model for the variation in the seismic p-wave velocity at a particular site.

In Example 2.3 the normal distribution is a model for the population of petal lengths of the species *Iris versicolor*.

In Example 2.4 the normal distribution is a model for the variation in the measured contents of three metals in 300 samples. There is insufficient information to decide whether and to what extent the variation is due to the measurement method, or whether it reflects the variation of the contents of the metals in the samples.

Example 2.5 involves two samples, one for each contact time. Initially, it seems reasonable to believe that the reaction times for the flies will depend on the length of time the flies have been exposed to the poison. It is, in fact, an essential part of the experiment to investigate this. Here the normal distribution is a model for the logarithm of the reaction times. It is obvious that the mean of this distribution reflects the length of the reaction time. But the flies react very differently, so the variance is needed to quantify this in a fairly simple way.

In all cases the models have been checked in Chapter 2 using fractile diagrams.

In the treatment of a single normal sample, one traditionally distinguishes between whether *the variance is known* or whether *the variance is unknown*. The case of a known variance is the simplest one and it will be used to introduce the important statistical concepts.

The variance is rarely known in biological or geological data. In laboratories some measurement methods may be so well established that the variance may be known, and the same may be true of some processes in industrial production. But in any case it is wise to take nothing for granted and check that the assumptions that justify a statistical analysis are not violated.

The analysis of one normal sample will be introduced in connection with Example 3.1.

Example 3.1

Keys for a keyboard are injection molded. A critical dimension is the diameter of a circular peg under the keys that is used to attach the key to the keyboard. It will be referred as the diameter of the key.

Table 3.1 *40 Diameters of keys in μm. 5160 μm has been subtracted from all diameters*

54	36	32	42	37
50	38	31	50	37
57	48	31	53	37
50	44	31	49	42
53	46	40	43	43
38	46	38	50	40
26	44	41	38	45
37	32	38	36	43

The diameter must not be too small, for then the keys will be too loose and the diameter must not be too large; the keys will not move smoothly and in the worst case, they will not even fit into the keyboard. All diameters must lie between the *lower specification limit LSL* and the *upper specification limit USL,* which are 5170 μm and 5230 μm, respectively (μm denotes a micrometer, i.e., 10^{-6} meter or 10^{-3} millimeter). Ideally, all diameters should be 5200 μm and this value is referred to as the *target value* or simply the *target*.

Engineering questions are whether the production is on target and how frequently keys with diameters outside the tolerance limits are produced. Keys with diameters outside the specification limits do not conform to specifications and are called nonconforming keys.

Table 3.1 shows the diameters in μm of 40 keys that have been sampled from the production during three hours on the same morning. From all diameters 5160 μm is subtracted and with this origin for the computations, the *LSL* is 10 μm and the *USL* is 70 μm while the target value is 40 μm.

We will use the model of a normally distributed sample. Experience has shown that the diameters of the keys can be considered to be normally distributed. Nevertheless, the assumption is checked with the fractile plot shown in Figure 3.1. The points do not deviate systematically from a straight line and hence the fractile plot does not lead us to doubt the assumption of a normal distribution.

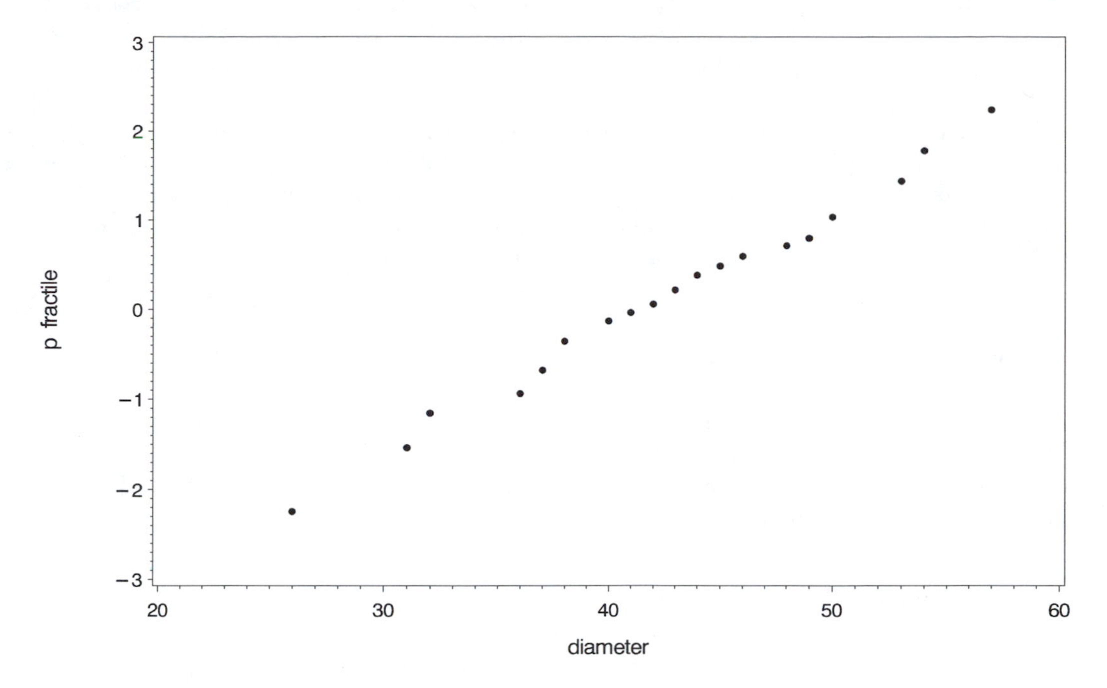

Figure 3.1 *Fractile diagram of the 40 diameters in Table 3.1.*

Thus the model specifies that the $n = 40$ observations in Table 3.1, i.e., the diameters $-5160\,\mu$m, $x_1, \ldots, x_n$ are realizations of independent, identically normally distributed random variables, $X_1, \ldots,$ X_n, with unknown mean μ and *known* variance σ_0^2. Experience with the production has led to the belief that the standard deviation σ_0 is 10 μm and we will use that assumption for convenience. We will, however, later check the assumption that the standard deviation is 10 μm.

We will briefly write the model as

$$X_i \sim N(\mu, \sigma_0^2), \qquad i = 1, \ldots, n,$$

subsuming the independence of the observations. (Hopefully, it does not cause any misunderstandings that μ is used both as an abbreviation of 10^{-6} or *micro* and as the symbol for the mean.)

Known Variance

We will first look at what we can say about the mean μ based on the observations. We say that we *estimate* μ. Traditionally, one uses the average of the observations,

$$\bar{x}_{\cdot} = \frac{1}{n}\sum_{i=1}^{n} x_i = 41.65\mu\text{m}. \tag{3.1}$$

Notice that $\bar{x}_{\cdot}$ is a realization of the normally distributed random variable, cf. (3.82) on page 161,

$$\bar{X}_{\cdot} = \frac{1}{n}\sum_{i=1}^{n} X_i \sim N(\mu, \frac{\sigma_0^2}{n}), \tag{3.2}$$

which has the correct mean μ and a variance σ_0^2/n, which decreases with the number of observations. It is those two properties that justify the usage of the average to estimate the mean μ.

We will use the notation,

$$\bar{x}_{\cdot} \sim\sim N(\mu, \frac{\sigma_0^2}{n}), \tag{3.3}$$

as a concatenation of (3.1) and (3.2). The formula (3.3) is expressed verbally as "$\bar{x}.$ is a realization of a random variable $\bar{X}.$, which is distributed as $N(\mu, \sigma_0^2/n)$". The first $\sim$ is a reminder that the estimate is a realization of a random variable and the second $\sim$ has its usual interpretation as "is distributed as".

The notation $\sim\sim$ is in no way standard, but it is convenient because in one line it conveys both information that an estimate, which is calculated from a set of data, is subject to variation and also the precise form of that variation. It is important to keep the distinction between the theoretical but unknown parameter and its numerical estimate, which is calculated from observations. For example, we distiguish between the mean μ of a normal distribution and the average $\bar{x}.$ of the observations in a sample. In order to keep this distinction, we will use the notations, $\bar{x}. \rightarrow \mu$ or $\mu \leftarrow \bar{x}.$, which are read as "$\bar{x}.$ is an estimate of μ" or "μ is estimated by $\bar{x}.$".

In this example we have $41.65\ \mu\text{m} \rightarrow \mu$.

The engineering question is whether the production is on the target value of 5200 μm and in the model for the data in Table 3.1 this is formulated as the question, "Is the mean μ equal to 40 μm?" In the statistical tradition the question we ask is formulated as a null hypothesis, and we answer the question by *testing the null hypothesis*, i.e., by checking if the hypothesis is consistent with the data. In this case the null hypothesis is

$$H_0\colon \mu = \mu_0 = 40\mu m.$$

As a starting point, we consider the difference between the estimate of the mean $\bar{x}.$ and the mean μ under the null hypothesis $\bar{x}. - 40\ \mu\text{m} = 1.65\ \mu\text{m}$. The greater this difference is, in absolute value, the more critical it is for the null hypothesis. But whether 1.65 μm is considered to be large or not depends on the variance of the measurements or more precisely of the average. We therefore normalize the difference, dividing it by the square root of its standard deviation and arrive at the *test statistic,*

$$u(\mathbf{x}) = u(x_1, \ldots, x_n) = \frac{\bar{x}. - \mu_0}{\sqrt{\sigma_0^2/n}} = \frac{41.65 - 40}{\sqrt{100/40}} = 1.044. \tag{3.4}$$

The test statistic $u(\mathbf{x})$ is a realization of the random variable,

$$u(\mathbf{X}) = u(X_1, \ldots, X_n) = \frac{\bar{X}. - \mu_0}{\sqrt{\sigma_0^2/n}} = \frac{\bar{X}. - 40}{\sqrt{100/40}}, \tag{3.5}$$

which is $N(0,1)$-distributed under the null hypothesis. This follows from $\bar{X}.$ being normally distributed according to (3.2) and from a property (3.83) of the normal distribution. Those values of the test statistic $u(\mathbf{x})$ that would have been more critical for H_0 are those values that are smaller than -1.044 and larger than 1.044. It is not possible offhand to decide whether 1.044 is such a large value that one would doubt the null hypothesis. Therefore, a final transformation is performed, calculating *the significance probability* or, simply, *the p-value* $p_{obs}(\mathbf{x})$, which is *the probability under the null hypothesis of a more critical value of the test statistic than the one observed.* If Φ denotes the distribution function of the standard normal distribution, the significance probability is

$$\begin{aligned} p_{obs}(\mathbf{x}) &= \Phi(-1.044) + (1 - \Phi(1.044)) \\ &= 2(1 - \Phi(1.044)) \\ &= 0.296. \end{aligned} \tag{3.6}$$

In this calculation we use the symmetry of the standard normal distribution around 0, see (3.76). The interpretation of this p-value is that if the null hypothesis is true, one would, in 3 out of 10 cases where one repeated the experiment, obtain a value of the test statistic that was more critical for the hypothesis than that obtained in the experiment. This is not a rare event by any standards and so it does not cause us to doubt the null hypothesis. The null hypotheses will be rejected if the p-values are below a limit that is called the *significance level* and is denoted by α. Conventional significance levels are 0.05 or 0.01 and in this book the significance level of 0.05 is used throughout.

The whole procedure of formulating the null hypothesis, computing the test statistic and the significance probability, and finally making a decision by comparing the p-value with the significance level is called a *significance test*.

The reasoning behind the significance test is treated later on page 63 in a separate section for easy reference under the heading, *The Logic of Significance Tests.*

Unknown Variance

If the variance is unknown, it is still possible to test H_0: $\mu = \mu_0 = 40\ \mu\text{m}$. The starting point is (as before) the difference $\bar{x}. - \mu_0 = 41.65\ \mu\text{m} - 40\ \mu\text{m} = 1.65\ \mu\text{m}$ between the estimate of the mean, $\bar{x}.$, and the mean, μ_0, under the null hypothesis. We would have liked to evaluate this difference taking the standard deviation of the average, $\sqrt{\sigma^2/n}$, into account, but that is not possible because σ^2 is unknown. It seems that the only way out of this problem is to use an estimate of σ^2 when σ^2 itself is not available. The estimate of σ^2 is denoted by s^2 and it is

$$s^2 = \frac{1}{n-1}\sum_{i=1}^{n}(x_i - \bar{x}.)^2 = 53.31. \tag{3.7}$$

We will see later how we arrive at this estimate and how it is computed. Here we will only mention that s^2 is a realization of the random variable,

$$s^2(\mathbf{X}) = \frac{1}{n-1}\sum_{i=1}^{n}(X_i - \bar{X}.)^2, \tag{3.8}$$

which is $\sigma^2\chi^2(n-1)/(n-1)$ distributed, cf. (3.95). In particular, the mean is

$$E\,s^2(\mathbf{X}) = \sigma^2$$

and the variance is

$$Var\,s^2(\mathbf{X}) = \frac{2\sigma^4}{(n-1)},$$

and those properties make s^2 a good estimate of σ^2. It has the right mean and its variance decreases with the sample size. It was exactly the same properties that made the average a good estimate of the mean.

Using s^2 instead of the unknown variance σ^2, we obtain the test statistic,

$$t(\mathbf{x}) = t(x_1,\ldots,x_n) = \frac{\bar{x}. - \mu_0}{\sqrt{s^2/n}} = \frac{41.65 - 40}{\sqrt{53.31/40}} = 1.429,$$

which is a realization of the random variable,

$$t(\mathbf{X}) = t(X_1,\ldots,X_n) = \frac{\bar{X}. - \mu_0}{\sqrt{s^2(\mathbf{X})/n}}. \tag{3.9}$$

The random variable $t(\mathbf{X})$ has a t-distribution with $n-1$ degrees of freedom, see (3.97), i.e., 39 degrees of freedom in the present example.

The t-distribution is available in tables, on some pocket calculators with statistical functions, and in statistical packages, so the significance probabilities, $p_{obs}(\mathbf{x})$, can be computed. The values of the test statistic that are more critical for the null hypothesis than the observed value of 1.429 are values smaller than -1.429 and larger than 1.429.

The significance probability is

$$\begin{aligned} p_{obs}(\mathbf{x}) &= F_{t(n-1)}(-1.429) + (1 - F_{t(n-1)}(1.429)) \\ &= 2(1 - F_{t(n-1)}(1.429)) \\ &= 0.161. \end{aligned}$$

Here $F_{t(n-1)}$ denotes the distribution function of the t-distribution with $n-1$ degrees of freedom. The calculation of the p-values has used the fact that the t-distributions, like the standard normal distribution $N(0,1)$, are symmetric around 0, i.e., $F_{t(n-1)}(-t) = 1 - F_{t(n-1)}(t)$ for all t, see (3.98).

The significance probability is larger than 0.05, so the test does not reject the null hypothesis. We reach the same conclusion as when the variance was considered to be known and equal to 100 μm^2.

This conclusion is that the data are consistent with the hypothesis that the mean is 40 μm or, in other words, that the process is on target.

Test of the Hypothesis H_0: $\sigma^2 = \sigma_0^2$

As mentioned on page 55, experience with the production process was that the standard deviation σ_0 was 10 μm. It is always wise to check the assumptions if possible. Sometimes assumptions and common knowledge are merely wishful thinking. So we formulate the null hypothesis,

$$H_0\colon \sigma^2 = \sigma_0^2 = 100(\mu m)^2.$$

Like the previous test, this test starts with the estimate of the parameter in question. Here it is the estimate of the variance $s^2 = 53.31$ and as the test statistic, one chooses the ratio between the estimate and σ_0^2, i.e.,

$$\frac{s^2}{\sigma_0^2} = \frac{53.31}{100} = 0.5331.$$

The test statistic is a realization of the random variable,

$$\frac{s^2(\mathbf{X})}{\sigma_0^2}. \tag{3.10}$$

$s^2(\mathbf{X})$ is distributed as $\sigma^2\chi^2(n-1)/(n-1)$ so *under the null hypothesis* the test statistic is $\chi^2(n-1)/(n-1)$- distributed. These distributions are available in tables, statistical pocket calculators, and in statistical software packages.

If the test statistic is close to 1, it corresponds to good agreement between the estimate and σ_0^2. In contrast, small values of the test statistic are critical for H_0; for it corresponds to the estimate of the variance being smaller than σ_0^2 indicating that the variance is smaller than σ_0^2. By the same argument, large values of the test statistic are critical H_0; for they indicate that the estimate of the variance is larger than σ_0^2, thus indicating that the variance is greater than σ_0^2. As with previous tests, the decisions are based on the significance probabilities. In this example, values of the test statistic smaller that the observed ratio of 0.5331 are more critical for the hypothesis. The probability that a $\chi^2(39)/39$-distributed random variable is smaller than 0.5331 is

$$F_{\chi^2(39)/39}(0.5331) = 0.00743.$$

The $\chi^2(f)/f$-distribution does not have the symmetry of the normal distribution, so it not so easy to point out the value above 1 that is exactly as critical for the null hypothesis as 0.5331. Therefore the significance probability is defined simply as twice the probability of obtaining a smaller value of the test statistic than 0.5331, i.e., as

$$p_{obs}(\mathbf{x}) = 2F_{\chi^2(39)/39}(0.5331) = 0.01486.$$

This corresponds to deciding that the values greater than the $1 - 0.00743 = 0.99257$ fractile of the

$\chi^2(39)/39$-distribution will be at least as critical for H_0 as the observed 0.5331. Incidentally, the 0.99257 fractile of the $\chi^2(39)/39$-distribution is equal to 1.6347.

The significance probability is below 0.05 and the null hypothesis is rejected. The estimate of the variance is smaller than $\sigma_0^2 = 100\ \mu\text{m}^2$ and this indicates that the variance is smaller than σ_0^2. The estimate of the variance is 53.31 μm^2, which can be expressed briefly as 53.31 $\mu\text{m}^2 \rightarrow \sigma^2$ or, equivalently, as 7.3014 $\mu\text{m} \rightarrow \sigma$.

It will have no dramatic consequences that the null hypothesis is rejected; for the smaller the variance the better the product, so it is just a relief that the production apparently runs extremely well on this particular morning.

If the observed values of the test statistic $s^2(\mathbf{X})/\sigma_0^2$ had been larger than 1, for example, equal to 1.6347, the significance probability would be calculated as

$$p_{obs}(\mathbf{x}) = 2\left[1 - F_{\chi^2(39)/39}(1.6347)\right] = 0.01486.$$

We frequently use the test statistic in (3.10) multiplied by $n-1$,

$$\frac{(n-1)s^2}{\sigma_0^2}, \tag{3.11}$$

to test H_0: $\sigma^2 = \sigma_0^2$. This statistic has a $\chi^2(n-1)$-distribution under the null hypothesis. Again, small and large values of the test statistic are critical for H_0 and if the test statistic is smaller than $n-1$, the lower tail of the distribution is used in the computation of the p-value. Otherwise, the upper tail is used. Thus the formula for computing the p-value is

$$p_{obs}(\mathbf{x}) = \begin{cases} 2F_{\chi^2(n-1)}\left(\dfrac{(n-1)s^2}{\sigma_0^2}\right) & \text{if } \dfrac{(n-1)s^2}{\sigma_0^2} \leq n-1, \\[2ex] 2\left(1 - F_{\chi^2(n-1)}\left(\dfrac{(n-1)s^2}{\sigma_0^2}\right)\right) & \text{if } \dfrac{(n-1)s^2}{\sigma_0^2} \geq n-1. \end{cases} \tag{3.12}$$

The reason for preferring $(n-1)s^2/\sigma_0^2$ to s^2/σ_0^2 is that tables of the χ^2-distribution are more detailed and readily available than tables of the $\chi^2(f)/f$-distribution and it is the χ^2-distribution that is available on pocket calculators and in statistical packages. Whether the test statistic in (3.10) or that in (3.11) is used, it is actually the same test that is performed; the significance probability is the same and hence the conclusion will be the same.

The Fraction of Nonconforming Keys

The fraction of keys with diameters outside the specification limits is a very important quantity that has serious implications for the quality of the keyboards that are produced, so let us look at the fraction of nonconforming keys under two different assumptions about the variance of the process or, in other words, under two different models I and II.

I. $\mu = 40\ \mu\text{m}$ and $\sigma^2 = 100\ \mu\text{m}^2$

This model has been questioned by the data as far as the variance is concerned. But 100 μm^2 was, after all, the variance that the production engineers believed that the process had. We first compute the fraction of keys within the tolerance limits.

$$
\begin{aligned}
P_{\mu,\sigma^2}[10\,\mu\text{m} \le X \le 70\,\mu\text{m}] &= P_{\mu,\sigma^2}\left[\frac{10\,\mu\text{m}-\mu}{\sigma} \le \frac{X-\mu}{\sigma} \le \frac{70\,\mu\text{m}-\mu}{\sigma}\right] \\
&= P_{\mu,\sigma^2}\left[\frac{10-40}{10} \le \frac{X-\mu}{\sigma} \le \frac{70-40}{10}\right] \\
&= P_{\mu,\sigma^2}\left[-3 \le \frac{X-\mu}{\sigma} \le 3\right] \\
&= \Phi(3) - (1-\Phi(3)) \\
&= 2\Phi(3) - 1 \\
&= 0.997300.
\end{aligned}
$$

The fraction of keys with too large or too small a diameter is $1 - 0.997300 = 0.002700$ under model I.

II. $\mu = 40\ \mu$m and $\sigma^2 = 53.31\ \mu\text{m}^2$

In this model the variance is assumed to be equal to estimated variance in the sample. Again, we first calculate the fraction of keys inside the tolerance interval.

$$
\begin{aligned}
P_{\mu,\sigma^2}[10\,\mu\text{m} \le X \le 70\,\mu\text{m}] &= P_{\mu,\sigma^2}\left[\frac{10\,\mu\text{m}-\mu}{\sigma} \le \frac{X-\mu}{\sigma} \le \frac{70\,\mu\text{m}-\mu}{\sigma}\right] \\
&= P_{\mu,\sigma^2}\left[\frac{10-40}{7.3014} \le \frac{X-\mu}{\sigma} \le \frac{70-40}{7.3014}\right] \\
&= P_{\mu,\sigma^2}\left[-4.109 \le \frac{X-\mu}{\sigma} \le 4.109\right] \\
&= \Phi(4.109) - (1-\Phi(4.109)) \\
&= 2\Phi(4.109) - 1 \\
&= 0.999960.
\end{aligned}
$$

Under model II, one expects that the fraction of keys with too small or too large a diameter is $1 - 0.999960 = 0.000040$. The fraction of keys with too small or too large a diameter is 2700 per 10^6 keys under model I, whereas it is only 40 per 10^6 keys under model II. It is a dramatic difference. If one were too choose between models I and II based on the evidence of the data, one would definitely choose model II. But how certain can we be that the variance is $53.31\ \mu\text{m}^2$? An important tool for answering this kind of question is the *confidence interval.*

Incidentally, we should not accept having to choose between models I and II. A more careful and reasonable summary of our analysis once we have the confidence intervals at our disposal will be along the following lines. The data do not support the aspect of model I that the variance is $100\ \mu\text{m}^2$ and it seems to be smaller than $100\ \mu\text{m}^2$; in fact, the estimate is $53.31\ \mu\text{m}^2$ and we are willing to specify an interval where we are confident that the variance is. This interval will be the confidence interval and “confident” will be made precise.

Confidence Intervals

Confidence Interval for the Variance

Let σ^2 denote the unknown value of the variance. Then $s^2(\mathbf{X})/\sigma^2$ is $\chi^2(f)/f$-distributed and the probability of an observation between the $\alpha/2$ and the $1-\alpha/2$ fractiles is $1-\alpha$,

$$1-\alpha = P\left[\frac{\chi^2_{\alpha/2}(f)}{f} \le \frac{s^2(\mathbf{X})}{\sigma^2} \le \frac{\chi^2_{1-\alpha/2}(f)}{f}\right] \tag{3.13}$$

$$= P\left[\frac{fs^2(\mathbf{X})}{\chi^2_{1-\alpha/2}(f)} \le \sigma^2 \le \frac{fs^2(\mathbf{X})}{\chi^2_{\alpha/2}(f)}\right].$$

In other words, the inequality,

$$\frac{fs^2}{\chi^2_{1-\alpha/2}(f)} \le \sigma^2 \le \frac{fs^2}{\chi^2_{\alpha/2}(f)}, \tag{3.14}$$

holds with a probability of $1-\alpha$. The interval,

$$\left[\frac{fs^2}{\chi^2_{1-\alpha/2}(f)}, \frac{fs^2}{\chi^2_{\alpha/2}(f)}\right], \tag{3.15}$$

is called a $(1-\alpha)$ *confidence interval for* σ^2, and $1-\alpha$ is denoted the *confidence coefficient*. The interpretation of the confidence interval and the confidence coefficient $1-\alpha$ is that either the parameter σ^2 belongs to the interval or an event has occurred with a probability smaller than α. The smaller α is the greater is the confidence that σ^2 belongs to the confidence interval, hence its name. It is exactly the same reasoning that is behind the rejection of a null hypothesis when the significance probability is small, see Section 3.1.1, *The Logic of Significance Tests,* on page 63.

The justification of the confidence interval and the confidence coefficient is sometimes given as a frequency interpretation where we imagine that we repeat the experiment a large number of times. In each repetition we compute the confidence interval according to (3.15). Those confidence intervals will be different because s^2 will vary according to its $\chi^2(f)/f$-distribution, and some of them will cover the variance σ^2 and some will not. But the fraction of confidence intervals that cover σ^2 will become arbitrarily close to $1-\alpha$ as the number of repetitions increases. This interpretation of the confidence intervals is referred to as the *frequency interpretation* and the confidence coefficient is interpreted as the *coverage probability*.

We stress that it is s^2, which is random, and not σ^2. We have not been able to find fixed limits between which σ^2 will lie with the probability $1-\alpha$. It is the variation in s^2 between samples that is the basis of the probability statement about the interval in (3.14). The choice of the term *confidence* may be seen as an attempt to stress that σ^2 has not suddenly become a random variable with a distribution.

The $(1-\alpha)$ confidence interval that was constructed here was based on the test statistic, s^2/σ^2, and therefore it also has the property that it consists of those values of σ^2 that would not be rejected as a null hypothesis if it were tested using the test statistic, s^2/σ^2, and using the significance level α.

The limits of the confidence interval are called the *confidence limits* and the confidence interval is the interval between the *lower confidence limit* and the *upper confidence limit.*

It is often preferable to report the standard deviation at the end of an investigation because it is easier to interpret, see Subsection 3.1.3 on page 70. A $(1-\alpha)$ confidence interval for the standard deviation σ can be obtained from the confidence interval for σ^2. The confidence limits for σ are simply the square roots of the confidence limits for σ^2. Thus the $(1-\alpha)$ confidence interval for the standard deviation σ is

$$\left[\sqrt{\frac{fs^2}{\chi^2_{1-\alpha/2}(f)}}, \sqrt{\frac{fs^2}{\chi^2_{\alpha/2}(f)}}\right]. \tag{3.16}$$

In the actual example the 95% confidence interval for the variance is

$$\begin{aligned}\left[\frac{fs^2}{\chi^2_{1-\alpha/2}(f)}, \frac{fs^2}{\chi^2_{\alpha/2}(f)}\right] &= \left[\frac{fs^2}{\chi^2_{0.975}(f)}, \frac{fs^2}{\chi^2_{0.025}(f)}\right] \\ &= \left[\frac{39 \times 53.31}{58.1201}\mu m^2, \frac{39 \times 53.31}{23.6543}\mu m^2\right] \\ &= \left[35.77\,\mu m^2, 87.89\,\mu m^2\right]. \end{aligned} \tag{3.17}$$

This interval does not contain the value 100 μm^2 because that value was rejected as a null hypothesis above.

The 95% confidence interval for the standard deviation based on (3.16) and obtained simply by taking square roots in (3.17) is

$$[5.981\,\mu\text{m}, 9.375\,\mu\text{m}].$$

Confidence Interval for the Mean

Confidence intervals for the mean μ are constructed in a similar way. They are based on the test for a fixed value of μ. This test is a u-test if the variance is known and a t-test if the variance is unknown and, consequently, the confidence intervals for the mean depend on whether the variance is known or not.

Variance known: $\sigma^2 = \sigma_0^2$.

Let μ denote the unknown value of the mean. Now

$$u(\mathbf{X}) = \frac{\bar{X}. - \mu}{\sqrt{\sigma_0^2/n}}$$

has a $N(0,1)$ distribution and therefore the probability that it takes values between the $\alpha/2$ and $1-\alpha/2$ fractiles of the $N(0,1)$-distribution is $1-\alpha$, thus

$$\begin{aligned}1-\alpha &= P\left[u_{\alpha/2} \le \frac{\bar{X}. - \mu}{\sqrt{\sigma_0^2/n}} \le u_{1-\alpha/2}\right] \\ &= P\left[\bar{X}. - \sqrt{\frac{\sigma_0^2}{n}}u_{1-\alpha/2} \le \mu \le \bar{X}. + \sqrt{\frac{\sigma_0^2}{n}}u_{1-\alpha/2}\right]. \end{aligned} \tag{3.18}$$

Here $u_{\alpha/2}$ and $u_{1-\alpha/2}$ are the $\alpha/2$ and $1-\alpha/2$ fractiles of the $N(0,1)$-distribution, and the last step has used the fact that $u_{\alpha/2} = -u_{1-\alpha/2}$. In other words, the inequality,

$$\bar{x}. - \sqrt{\frac{\sigma_0^2}{n}}u_{1-\alpha/2} \le \mu \le \bar{x}. + \sqrt{\frac{\sigma_0^2}{n}}u_{1-\alpha/2}, \tag{3.19}$$

is true with the probability $1-\alpha$. The interval,

$$\left[\bar{x}. - \sqrt{\frac{\sigma_0^2}{n}}u_{1-\alpha/2}, \bar{x}. + \sqrt{\frac{\sigma_0^2}{n}}u_{1-\alpha/2}\right], \tag{3.20}$$

is a $(1-\alpha)$ *confidence interval* for the mean. If $u_{0.975} = 1.960$ is used in (3.20) a 0.95 or a 95% confidence interval is obtained.

Variance unknown

The derivation is very similar to the case with the variance known. As usual μ denotes the unknown value of the mean. Because

$$t(\mathbf{X}) = \frac{\bar{X}. - \mu}{\sqrt{s^2(\mathbf{X})/n}}$$

has a t-distribution with $f = n-1$ degrees of freedom, the probability that it takes values between the $\alpha/2$ and $1-\alpha/2$ fractiles of the $t(f)$-distribution is $1-\alpha$,

$$\begin{aligned} 1-\alpha &= P\left[t_{\alpha/2}(f) \leq \frac{\bar{X}. - \mu}{\sqrt{s^2(\mathbf{X})/n}} \leq t_{1-\alpha/2}(f)\right] \quad (3.21)\\ &= P\left[\bar{X}. - \sqrt{\frac{s^2(\mathbf{X})}{n}}t_{1-\alpha/2}(f) \leq \mu \leq \bar{X}. + \sqrt{\frac{s^2(\mathbf{X})}{n}}t_{1-\alpha/2}(f)\right]. \end{aligned}$$

Here $t_{\alpha/2}(f)$ and $t_{1-\alpha/2}(f)$ are the $\alpha/2$ and the $1-\alpha/2$ fractiles of the t-distribution with f degrees of freedom and the last step has used $t_{\alpha/2}(f) = -t_{1-\alpha/2}(f)$. Thus the inequality,

$$\bar{x}. - \sqrt{\frac{s^2}{n}}t_{1-\alpha/2}(f) \leq \mu \leq \bar{x}. + \sqrt{\frac{s^2}{n}}t_{1-\alpha/2}(f), \quad (3.22)$$

holds with probability $1-\alpha$. The interval,

$$\left[\bar{x}. - \sqrt{\frac{s^2}{n}}t_{1-\alpha/2}(f),\ \bar{x}. + \sqrt{\frac{s^2}{n}}t_{1-\alpha/2}(f)\right], \quad (3.23)$$

is a $(1-\alpha)$ *confidence interval* for the mean.

In the present example, the 95% confidence interval for the mean is

$$\begin{aligned} &\left[\bar{x}. - \sqrt{\frac{s^2}{n}}t_{1-\alpha/2}(f),\ \bar{x}. + \sqrt{\frac{s^2}{n}}t_{1-\alpha/2}(f)\right] \\ &= \left[41.65 - \sqrt{\frac{53.31}{40}}2.023,\ 41.65 + \sqrt{\frac{53.31}{40}}2.023\right] \mu\text{m} \\ &= [41.65 - 2.34,\ 41.65 + 2.34]\ \mu\text{m} \\ &= [39.31,\ 43.99]\ \mu\text{m}. \end{aligned}$$

This concludes Example 3.1. □

3.1.1 The Logic of Significance Tests

The proposition behind the conclusions based on significance tests and the basis of the faith we have in confidence intervals is simply: *An event with a small probability does not occur*. "Small" in this context is defined in terms of the *significance level*, which is usually denoted by α. We choose to disregard the possibility that events with probabilities below the significance level may occur.

The risk that a person will be injured in a plane crash is smaller than 10^{-6} per flying hour, and most people consider this risk to be negligible. It is probabilities of the same order of magnitude that are considered acceptable when the risk of accidents at atomic power plants and other large industrial sites are considered.

In statistics we work with significance levels of the order of magnitude 10^{-2}. As previously mentioned, we will use $\alpha = 5 \times 10^{-2} = 0.05$ everywhere in this book, and we will almost exclusively express it as a percentage. We will always report the significance probability together with the conclusion, because it gives the reader the possibility to draw his or her own conclusion should they prefer a different significance level.

The reasoning behind the rejection of a null hypothesis because a small significance probability has been obtained is:

Either the null hypothesis is false *or* an event with a small probability has occurred. Events with small probabilities do not occur, *therefore* the null hypothesis is false.

Thus it is an entirely classical logical reasoning:

Either A *or* B. B is false, *therefore* A.

What is new and perhaps surprising is the bold formulation of the proposition: *Events with small probabilities do not occur.*

When the significance probability is smaller than the significance level a convincing conclusion can be made from the significance test. This conclusion may be formulated along the lines, "The data shows a significant deviation from the null hypothesis which is rejected."

In contrast, no conclusions are supported by the significance test when the significance probability is greater than the significance level. All that can be said is, "There is no evidence in the data that, etc."

3.1.2 Calculations

In the treatment of a single sample, one must calculate the average $\bar{x}.$ as an estimate of the mean and the sample variance s^2 as an estimate of the variance, whereas fractiles that are needed to compute significance probabilities and confidence intervals must be found in tables, on a pocket calculator, or using a statistical package on a computer. The formulas for $\bar{x}.$ and s^2 are

$$\bar{x}. = \frac{1}{n}\sum_{i=1}^{n} x_i = \frac{1}{n}S,$$

where the symbol S is used to denote the sum of the observations and

$$s^2 = \frac{1}{n-1}\sum_{i=1}^{n}(x_i - \bar{x}.)^2 = \frac{1}{n-1}SSD. \tag{3.24}$$

Here SSD is an abbreviation of *S*um of *S*quares of *D*eviations, $\sum_{i=1}^{n}(x_i - \bar{x}.)^2$. The formula (3.24) for s^2 is not suitable for calculations on a pocket calculator, but SSD can be rewritten as

$$\begin{aligned} SSD &= \sum_{i=1}^{n}(x_i - \bar{x}.)^2 \qquad (3.25) \\ &= \sum_{i=1}^{n}(x_i^2 + \bar{x}.^2 - 2x_i\bar{x}.) \\ &= USS + n\bar{x}.^2 - 2\bar{x}.S \\ &= USS - \frac{S^2}{n}, \end{aligned}$$

where USS is short for *U*ncorrected *S*um of *S*quares, $\sum_{i=1}^{n} x_i^2$. Now s^2 can be calculated from the formula,

$$s^2 = \frac{1}{n-1}(USS - \frac{S^2}{n}), \tag{3.26}$$

and, finally, s as the square root of s^2. Note the distinction between the capitalized S, which denotes the sum of the observations, and the lower case s, which denotes the sample standard deviation.

Calculation of S and USS will be referred to as the *standard calculations*.

Example 3.1 (Continued)
The computations in Example 3.1 are based on the following results of the standard calculations:

$$S = 1666\,\mu\text{m}, \qquad USS = 71468\,\mu\text{m}^2.$$

□

Example 2.1 (Continued)
The example is concerned with the heights of 247 10- to 12-year-old girls with a diagnosis of asthma. The model is one normal sample with unknown mean μ and unknown variance σ^2. There are no meaningful hypotheses to be tested about either the mean or the variance, and we conclude this example by giving the estimates and the confidence intervals for the parameters.

Standard calculations give the results,

$$S = 34613\text{ cm}, \qquad USS = 4871559\text{ cm}^2,$$

and, noting that the sample size is $n = 247$, this gives the estimates of the mean, the variance, and the standard deviation

$$\bar{x}_{\bullet} = 140.13\text{ cm}, \quad s^2 = 85.8317\text{ cm}^2 \quad \text{and} \quad s = 9.3\text{ cm}.$$

The 95% confidence intervals for mean, variance, and standard deviation are as follows:

$$\begin{aligned}
&\left[\bar{x}_{\bullet} - \sqrt{\frac{s^2}{n}}t_{1-\alpha/2}(f),\ \bar{x}_{\bullet} + \sqrt{\frac{s^2}{n}}t_{1-\alpha/2}(f)\right] \\
&= \left[140.13 - \sqrt{\frac{85.8317}{247}}1.9697,\ 140.13 + \sqrt{\frac{85.8317}{247}}1.9697\right]\text{ cm} \\
&= [140.13 - 1.16,\ 140.13 + 1.16]\text{ cm} \\
&= [138.97,\ 141.29]\text{ cm},
\end{aligned}$$

$$\begin{aligned}
&\left[\frac{fs^2}{\chi^2_{1-\alpha/2}(f)},\ \frac{fs^2}{\chi^2_{\alpha/2}(f)}\right] \\
&= \left[\frac{fs^2}{\chi^2_{0.975}(f)},\ \frac{fs^2}{\chi^2_{0.025}(f)}\right] \\
&= \left[\frac{246 \times 85.83167}{291.34},\ \frac{246 \times 85.83167}{204.45}\right]\text{ cm}^2 \\
&= [72.4741,\ 103.2751]\text{ cm}^2,
\end{aligned}$$

and

$$[8.5,\ 10.2]\text{ cm}.$$

□

Example 2.2 (Continued)
The example is concerned with the description of 16 repeated measurements of p-wave velocity in one piece of rock. The sample has $n = 16$ measurements. The model is one normal sample with unknown mean μ and unknown variance σ^2.

The p-wave velocity is recorded by measurement equipment and the manufacturer of the equipment maintains that the standard deviation σ is 115 m/s. Therefore we will first test the hypothesis, $H_{0\sigma^2}$: $\sigma^2 = 115^2$ m^2/s^2. Data has been collected to investigate if it can be assumed that the mean,

μ, of the p-wave velocity is 4500 m/s, so the next hypothesis to be tested is $H_{0\mu}$: $\mu = 4500$ m/s. But whether $H_{0\mu}$ will be tested with a u-test or a t-test depends on the outcome of the test of $H_{0\sigma^2}$.

To facilitate computations, 4500 m/s is subtracted from all observations. For the resulting data the hypothesis of the mean is $H_{0\mu}$: $\mu = 0$.

Standard calculations give

$$S = 95 \text{ m/s} \qquad \text{and} \qquad USS = 281717 \text{ m}^2/\text{s}^2,$$

and the sample size is $n = 16$. This gives the following estimates of the mean, the variance, and the standard deviation,

$$\bar{x}. = 5.938 \text{ m/s}, \qquad s^2 = 18743.53 \text{ m}^2/\text{s}^2 \qquad \text{and} \qquad s = 136.91 \text{ m/s}.$$

The test statistic for $H_{0\sigma^2}$ is

$$\frac{fs^2}{115^2} = \frac{15 \times 18743.53}{115^2} = 21.259.$$

Under the null hypothesis $H_{0\sigma^2}$, the test statistic has a χ^2-distribution with 15 degrees of freedom and therefore the probability of an observation of the test statistic that is larger than that observed is

$$P\left[\chi^2(15) > 21.259\right] = 1 - F_{\chi^2(15)}(21.259) = 0.129,$$

and the significance probability is obtained as twice this value,

$$p_{obs}(\mathbf{x}) = 2 \times 0.129 = 0.258.$$

This does not lead to rejection of $H_{0\sigma^2}$.

The 95% confidence interval of the variance σ^2 is found to be

$$\left[\frac{15 \times 18743.53}{27.488}, \frac{15 \times 18743.53}{6.262}\right] \text{ m}^2/\text{s}^2 = [10228.2, 44898.27] \text{ m}^2/\text{s}^2,$$

and the 95% confidence interval of the standard deviation σ becomes

$$[101.1, 211.9] \text{ m/s}.$$

The test of $H_{0\mu}$: $\mu = 0$ can be tested with the u-test statistic,

$$u(\mathbf{x}) = \frac{\bar{x}. - 0}{\sqrt{\sigma_0^2/n}} = \frac{5.938 - 0}{115/\sqrt{16}} = 0.207,$$

which should be evaluated in an $N(0,1)$-distribution. The p-value is

$$\begin{aligned} p_{obs}(\mathbf{x}) &= 2(1 - \Phi(0.207)) \\ &= 0.836, \end{aligned}$$

which is larger than $\alpha = 0.05$ and the conclusion is that the experiment does not contradict the assumption that the mean, μ, of the p-wave velocity is 4500 m/s.

The 95% confidence interval for ($\mu - 4500$ m/s) is

$$\begin{aligned} &\left[\bar{x}. - \frac{\sigma_0}{\sqrt{n}} u_{1-\alpha/2}, \bar{x}. + \frac{\sigma_0}{\sqrt{n}} u_{1-\alpha/2}\right] \\ &= \left[5.938 - \frac{115}{\sqrt{16}} 1.960, 5.938 + \frac{115}{\sqrt{16}} 1.960\right] \text{ m/s} \\ &= [-50.412, 62.288] \text{ m/s}, \end{aligned}$$

and the 95% confidence interval for the mean μ is

$$[4449.6, 4562.9] \text{ m/s}.$$

If one does not wish to consider the hypothesis of a standard deviation equal to 115 m/s before testing the hypothesis of the mean, due to a general lack of faith in manufacturers, for example, one can choose to test $H_{0\mu}$ with a t-test. Then

$$t(x) = \frac{\bar{x}. - 0}{\sqrt{s^2/n}} = \frac{5.938 - 0}{\sqrt{18743.53/16}} = 0.173,$$

which is to be evaluated in a t-distribution with 15 degrees of freedom. The significance probability is

$$\begin{aligned} p_{obs}(\mathbf{x}) &= 2(1 - F_{t(15)}(0.173)) \\ &= 0.865, \end{aligned}$$

which does not give any reason to question $H_{0\mu}$.

The 95% confidence interval for $\mu - 4500$ m/s based on the t-distribution is

$$\begin{aligned} &\left[\bar{x}. - \sqrt{\frac{s^2}{n}} t_{1-\alpha/2}(f),\ \bar{x}. + \sqrt{\frac{s^2}{n}} t_{1-\alpha/2}(f)\right] \\ &= \left[5.938 - \sqrt{\frac{18743.53}{16}} 2.131,\ 5938 + \sqrt{\frac{18743.53}{16}} 2.131\right] \text{ m/s} \\ &= [-66.999,\ 78.875] \text{ m/s}, \end{aligned}$$

and the 95% confidence interval for μ based on the t-distribution is

$$[4433.0,\ 4578.8] \text{ m/s}.$$

□

Example 2.3 (Continued)

Here the purpose is to describe the distribution of the petal length of the species *Iris versicolor*. The model is one normal sample with unknown mean μ and unknown variance σ^2. There are no reasonable hypotheses to be tested about either the mean or the variance, and we conclude this example by giving the estimates and the confidence intervals for the parameters. The standard calculations give the result,

$$S = 105.2 \text{ cm} \quad \text{and} \quad USS = 448.64 \text{ cm}^2.$$

Noting that $n = 25$, this gives the estimates of the mean, the variance, and the standard deviation as

$$\bar{x}. = 4.208 \text{ cm}, \quad s^2 = 0.2483 \text{ cm}^2 \text{ and } \quad s = 0.498 \text{ cm}.$$

The 95% confidence interval for the mean is

$$\begin{aligned} &\left[\bar{x}. - \sqrt{\frac{s^2}{n}} t_{1-\alpha/2}(f),\ \bar{x}. + \sqrt{\frac{s^2}{n}} t_{1-\alpha/2}(f)\right] \\ &= \left[4.208 - \sqrt{\frac{0.2483}{25}} 2.064,\ 4.208 + \sqrt{\frac{0.2483}{25}} 2.064\right] \text{ cm} \\ &= [4.208 - 0.206,\ 4.208 + 0.206] \text{ cm} \\ &= [4.00,\ 4.41] \text{ cm}, \end{aligned}$$

the 95% confidence interval for the variance σ^2 is

$$\begin{aligned}
&\left[\frac{fs^2}{\chi^2_{1-\alpha/2}(f)}, \frac{fs^2}{\chi^2_{\alpha/2}(f)}\right] \\
&= \left[\frac{fs^2}{\chi^2_{0.975}(f)}, \frac{fs^2}{\chi^2_{0.025}(f)}\right] \\
&= \left[\frac{24 \times 0.2483}{39.364}, \frac{24 \times 0.2483}{12.401}\right] \text{ cm}^2 \\
&= [0.1514,\ 0.4805] \text{ cm}^2,
\end{aligned}$$

and the 95% confidence interval for the standard deviation σ is

$$[0.389 \text{ cm}, 0.693 \text{ cm}].$$

□

Example 2.4 (Continued)

Here the normal sample with unknown mean μ and unknown variance σ^2 is a model of the contents of Zn, Cu, and Pb in 300 samples from a location in Finnmark. There is no information about hypotheses to be tested. But the mean and the variance are important parameters, so therefore their estimates and their confidence intervals will be given.

Standard calculations give

$$S = 4342 \text{ ppm} \quad \text{and} \quad USS = 65224 \text{ ppm}^2,$$

and with the sample size $n = 300$, this gives the estimates,

$$\bar{x}. = 14.4733 \text{ ppm} \quad s^2 = 7.9625 \text{ ppm}^2 \quad \text{and} \quad s = 2.8218 \text{ ppm}.$$

The 95% confidence interval for the mean is

$$\begin{aligned}
&\left[\bar{x}. - \sqrt{\frac{s^2}{n}} t_{1-\alpha/2}(f),\ \bar{x}. + \sqrt{\frac{s^2}{n}} t_{1-\alpha/2}(f)\right] \\
&= \left[14.4733 - \sqrt{\frac{7.9625}{300}} 1.968,\ 14.4733 + \sqrt{\frac{7.9625}{300}} 1.968\right] \text{ ppm} \\
&= [14.4733 - 0.3206,\ 14.4733 + 0.3206] \text{ ppm} \\
&= [14.1527,\ 14.7939] \text{ ppm},
\end{aligned}$$

and the 95% confidence interval for the variance is

$$\begin{aligned}
&\left[\frac{fs^2}{\chi^2_{1-\alpha/2}(f)}, \frac{fs^2}{\chi^2_{\alpha/2}(f)}\right] \\
&= \left[\frac{fs^2}{\chi^2_{0.975}(f)}, \frac{fs^2}{\chi^2_{0.025}(f)}\right] \\
&= \left[\frac{299 \times 7.9625}{348.79}, \frac{299 \times 7.9625}{252.99}\right] \text{ ppm}^2 \\
&= [6.826,\ 9.411] \text{ ppm}^2.
\end{aligned}$$

The 95% confidence interval for the standard deviation σ is

$$[2.61 \text{ ppm}, 3.07 \text{ ppm}].$$

□

Example 2.5 (Continued)
This example has two independent samples, one for each contact time, and the model check revealed that it was the logarithm of the reaction time that could be considered to be normally distributed. Therefore the *calculations are performed on the logarithm of the reaction times*. Using three decimal places for the logarithm, the standard calculations give:

Contact time	*S*	*USS*
30 seconds	44.915	142.626517
60 seconds	40.731	124.933851

This gives the estimates of the parameters in the table below.

Contact time	*Mean* μ	*Variance* σ^2	*Standard deviation* σ
30 seconds	2.8072	1.102779	1.050133
60 seconds	2.7154	1.023778	1.011819

The 95% confidence intervals are given in the following table.

		95% Confidence intervals	
Contact time	*Mean*	*Variance*	*Standard deviation*
30 seconds	[2.2476, 3.3668]	[0.6018, 2.6416]	[0.7758, 1.6253]
60 seconds	[2.1551, 3.2757]	[0.5489, 2.5469]	[0.7409, 1.5959]

Note that the estimated means reported here are the means of the *logarithm* of the reaction times measured in minutes. This result is not readily interpretable and it should be transformed back to the original scale of the measurements and be interpreted there.

For contact time of 30 s one obtains

$$e^{\mu_{30}} \leftarrow e^{\bar{x}_{\bullet 30}} \text{ minutes} = 16.56 \text{ minutes},$$

and the 95% confidence interval for the transformed mean is

$$[9.47, 28.98] \text{ minutes}.$$

For contact time of 60 s the results is

$$e^{\mu_{60}} \leftarrow e^{\bar{x}_{\bullet 60}} \text{ minutes} = 15.11 \text{ minutes},$$

with 95% confidence interval,

$$[8.63, 26.46] \text{ minutes}.$$

Now $e^{\mu_{30}}$ and $e^{\mu_{60}}$ are not the means in the distributions of reaction times. The means are preserved only under linear transformations. But the mean of a normal distribution is also the 50% fractile or the *median* and that property is preserved by a monotone transformation. Therefore $e^{\bar{x}_{\bullet 30}}$ and $e^{\bar{x}_{\bullet 60}}$ are estimates of the 50% fractile in the distribution of contact times. In other words they are estimates of the time it takes until 50% of the flies have reacted.

So far we have not answered the question if reaction times depend on whether the contact times are 30 seconds or 60 seconds. There is a great similarity between the confidence intervals of the means of the logarithms of the reaction times and between the medians of the reaction times, so most people will presumably feel that there is no statistically significant difference between the two samples. Incidentally, this is also the impression from the fractile plot in Figure 2.17. This turns out to be true when we use this example in Section 3.2.1 to develop methods for comparing variances and means of two normal samples.

Section 3.2 as a whole is devoted to developing methods for comparing two or more samples. □

3.1.3 Variance or Standard Deviation

In the previous examples we have reported the estimate of the standard deviation σ and its confidence interval. But, in addition, we have reported the estimate and the confidence interval for the variance σ^2. This seems to be superfluous bearing in mind that the standard deviation is in one-to-one correspondence with the variance, and even more so when we think about the fact that it is the standard deviation that is easy to interpret for normal data.

The standard deviation gives concise information about the spread of the observations around the mean. It is useful to remember that 68.27%, 95.45%, and 99.73% of the observations, respectively, in an $N(\mu,\sigma^2)$ distribution fall within intervals of lengths 2σ, 4σ, and 6σ centered on the mean μ.

There are at least two reasons why the variance plays an important role in the analysis. First, certain rules are easier to formulate in terms of the variance than in terms of the standard deviation. For example, *the variance of a sum of independent random variables is the sum of the variances of the individual random variables*. Secondly, one has chosen to tabulate the χ^2-distributions and the estimate s^2 of σ^2 is a realization of a random variable that has a $\sigma^2\chi^2(n-1)/(n-1)$-distribution. Therefore tests of hypotheses of the standard deviation are formulated as hypotheses about the variance and tested using the χ^2-distribution. Similarly, the precision of the estimate of the standard deviation in terms of a confidence interval for σ is first made as a confidence interval for σ^2 by means of the χ^2-distribution using the formula (3.15) and it is then transformed into a confidence interval for the standard deviation σ.

To summarize, the variance is more important than the standard deviation during the analysis of the data and in the final stage of the analysis, their roles are interchanged and the standard deviation becomes the more important parameter.

In this connection it is suitable to mention the term *standard error*. The standard error is the standard deviation of an estimator. The term is used frequently by statistical packages where estimates are always followed by their *estimated* standard errors.

3.1.4 The Likelihood Method

We will use the model of a single normal sample to introduce the likelihood method. We shall find the same estimates of the parameters and the same tests of the hypotheses that have already been derived in the treatment of Example 3.1 using intuition and common sense.

The intention is to show how the likelihood method gives sensible results in a simple situation. Hopefully, this gives faith and confidence in the results that are obtained with the likelihood method in models that are so complicated that intuition and common sense offer no guidance.

It should be noted, however, that the foundations of the likelihood method are much stronger than a collection of examples where it can be seen to work well. Mathematical statisticians have proved quite generally that the method possesses so many desirable properties that it is safe to state that it is one of the most important tools in modern statistics. The likelihood method is treated in more detail in Chapter 11.

For simplicity the case with σ^2 *known* will be treated first and both the maximum likelihood estimate of μ and the likelihood ratio test of H_0: $\mu=\mu_0$ will be derived. Next the case with σ^2 *unknown* will be treated and estimates of μ and σ^2 are derived. The sections ends with a very brief derivation of the likelihood ratio test of H_0: $\mu=\mu_0$.

Variance Known: $\sigma^2=\sigma_0^2$

Model: The observations, $x_1,\ldots,x_n$, are realizations of the independent and identically distributed random variable, $X_1,\ldots,X_n$, $X_i\sim N(\mu,\sigma_0^2), i=1,\ldots,n$.

The likelihood function is the joint density function of the observations but considered as a function of the unknown mean μ. The joint density function is

$$\begin{aligned} p(\mathbf{x};\mu) &= f_{\mathbf{X}}(\mathbf{x}) \\ &= \prod_{i=1}^{n} \frac{1}{\sqrt{2\pi\sigma_0^2}} e^{-\frac{1}{2\sigma_0^2}(x_i-\mu)^2} \\ &= (\frac{1}{2\pi\sigma_0^2})^{\frac{n}{2}} e^{-\frac{1}{2\sigma_0^2}\sum_{i=1}^{n}(x_i-\mu)^2}. \end{aligned}$$

and therefore the likelihood function that traditionally is denoted by L is

$$L(\mu) = (\frac{1}{2\pi\sigma_0^2})^{\frac{n}{2}} e^{-\frac{1}{2\sigma_0^2}\sum_{i=1}^{n}(x_i-\mu)^2}.$$

The interpretation of the likelihood function is that it is the probability of the observations as a function of the parameters. (Strictly speaking, this is not true for continuous data because it is not until the density function is multiplied by a volume element $\prod_{i=1}^{n} dx_i$ that

$$f_{\mathbf{X}}(\mathbf{x}) \prod_{i=1}^{n} dx_i$$

has the interpretation of the probability of an observation in a small volume around $\mathbf{x}$. But this volume element has no influence on the applications of the likelihood function and therefore it is omitted. When maximizing the likelihood function to find estimates of the parameters, the volume element is a constant and when forming the likelihood ratio statistics, the volume elements cancel out.)

Often the mathematical treatment of L is facilitated by considering the natural logarithm of the likelihood function. It is called the *log likelihood function* and it is denoted by l. The log likelihood function is

$$l(\mu) = -\frac{n}{2}\ln(2\pi\sigma_0^2) - \frac{1}{2\sigma_0^2}\sum_{i=1}^{n}(x_i-\mu)^2.$$

Estimation of the Mean μ

The *maximum likelihood estimate* is the value of the parameter that maximizes the likelihood function. The justification is that it is this value of the parameter that assigns the highest probability to the observations. In precisely this sense it is this value of the parameter that gives the best explanation of the observations. If the log likelihood function is differentiated once with respect to μ and equated to 0 the *likelihood equation* is obtained:

$$0 = \frac{\partial l}{\partial \mu}(\mu) = \frac{1}{\sigma_0^2}\sum_{i=1}^{n}(x_i-\mu).$$

Solving the likelihood equation with respect to μ gives the solution,

$$\hat{\mu} = \bar{x}_{\bullet} = \frac{1}{n}\sum_{i=1}^{n} x_i,$$

which maximizes l. The logarithm is an increasing function so the value that maximizes the log likelihood function is also maximizing the likelihood function. The maximum likelihood estimate of the mean μ is the average of the observations. As explained in connection with Example 3.1, this is an intuitively reasonable estimate. It is a realization of the random variable,

$$\bar{X}_{\bullet} = \frac{1}{n}\sum_{i=1}^{n} X_i \sim N(\mu, \frac{\sigma_0^2}{n}),$$

which has the right mean μ and a variance σ_0^2/n, which decreases with the number of observations.

Test of the Hypothesis H_0: $\mu = \mu_0$

The likelihood ratio test statistic $Q(\mathbf{x})$ is the ratio of the maximum of the likelihood function under the null hypothesis H_0 and the maximum of the likelihood function without the restriction specified by H_0, i.e.,

$$Q(\mathbf{x}) = \frac{\max\limits_{\mu \in H_0} L(\mu)}{\max\limits_{\mu \in \mathbb{R}} L(\mu)} = \frac{L(\mu_0)}{L(\bar{x}_\cdot)}.$$

The likelihood ratio test statistic is the ratio of the maximal probability of the observation under the hypothesis or the prospective model to the maximal probability of the observation under the model. If the maximal probability of an observation under a model is viewed as the ability of the model to account for the observation $\mathbf{x}$, then $Q(\mathbf{x})$ can be seen as the ratio of the abilities of the two models to explain the observation. If $Q(\mathbf{x})$ is small, it means that it is more difficult to explain the observation under the hypothesis or the prospective model than under the model. Consequently, if $Q(\mathbf{x})$ is small, it will be difficult to change from the existing model to the prospective model on the evidence of the observation $\mathbf{x}$. In other words, small values of $Q(\mathbf{x})$ are critical for the hypothesis or the prospective model.

The values that are more critical to H_0 than the actual observation $\mathbf{x}$ are those values $\mathbf{y}$ that have a value of the likelihood ratio statistic that is smaller than $Q(\mathbf{x})$, i.e. $\{\mathbf{y} \mid Q(\mathbf{y}) \leq Q(\mathbf{x})\}$. For technical reasons it is convenient to look at the logarithm of the likelihood ratio statistic, $\ln Q$, which is referred to as the *log likelihood ratio test statistic*, i.e.,

$$\begin{aligned}
\ln Q(\mathbf{x}) &= l(\mu_0) - l(\bar{x}_\cdot) \\
&= -\frac{1}{2\sigma_0^2}[\sum_{i=1}^{n}(x_i - \mu_0)^2 - \sum_{i=1}^{n}(x_i - \bar{x}_\cdot)^2] \\
&= -\frac{n(\bar{x}_\cdot - \mu_0)^2}{2\sigma_0^2} \\
&= -\frac{1}{2}u^2(\mathbf{x}),
\end{aligned}$$

where $u(\mathbf{x})$ is the test statistic in (3.4) that was derived in Example 3.1. The observations that are more critical for H_0 than the observation $\mathbf{x}$ are

$$\begin{aligned}
\{\mathbf{y} \mid Q(\mathbf{y}) \leq Q(\mathbf{x})\} &= \{\mathbf{y} \mid -2\ln Q(\mathbf{y}) \geq -2\ln Q(\mathbf{x})\} \\
&= \{\mathbf{y} \mid u^2(\mathbf{y}) \geq u^2(\mathbf{x})\} \\
&= \{\mathbf{y} \mid \mid u(\mathbf{y}) \mid \geq \mid u(\mathbf{x}) \mid\},
\end{aligned}$$

and it turns out that the likelihood ratio test for H_0 is the same as the test based on (3.4):

$$u(\mathbf{x}) = u(x_1, \ldots, x_n) = \frac{\bar{x}_\cdot - \mu_0}{\sqrt{\sigma_0^2/n}}.$$

Variance Unknown

Estimation of μ and σ^2

The likelihood function is now a function of both unknown parameters μ and σ^2, i.e.,

$$L(\mu, \sigma^2) = (\frac{1}{2\pi\sigma^2})^{\frac{n}{2}} e^{-\frac{1}{2\sigma^2}\sum_{i=1}^{n}(x_i - \mu)^2},$$

and the log likelihood function is

$$l(\mu,\sigma^2) = -\frac{n}{2}\ln(2\pi\sigma^2) - \frac{1}{2\sigma^2}\sum_{i=1}^{n}(x_i-\mu)^2.$$

From the treatment of the case with σ^2 known, it follows that for σ^2 fixed the log likelihood function considered as a function of μ is maximized for $\mu = \bar{x}.$, i.e.,

$$l(\mu,\sigma^2) \le l(\bar{x}.,\sigma^2) = -\frac{n}{2}\ln(2\pi\sigma^2) - \frac{1}{2\sigma^2}\sum_{i=1}^{n}(x_i-\bar{x}.)^2,$$

so in order to find the maximum likelihood estimate of (μ,σ^2) all that is needed is to find the value $\hat{\sigma}^2$ of σ^2 that maximize $l(\bar{x}.,\sigma^2)$ as a function of σ^2. Differentiating with respect to σ^2 gives

$$\frac{\partial l(\bar{x}.,\sigma^2)}{\partial\sigma^2} = -\frac{n}{2\sigma^2} + \frac{1}{2(\sigma^2)^2}\sum_{i=1}^{n}(x_i-\bar{x}.)^2,$$

and equating to 0 and solving with respect to σ^2 gives the solution,

$$\hat{\sigma}^2 = \frac{1}{n}\sum_{i=1}^{n}(x_i-\bar{x}.)^2,$$

which maximize $l(\bar{x}.,\sigma^2)$ with respect to σ^2. Thus

$$l(\mu,\sigma^2) \le l(\bar{x}.,\sigma^2) \le l(\bar{x}.,\hat{\sigma}^2)$$

and the maximum likelihood estimate for (μ,σ^2) is

$$(\hat{\mu},\hat{\sigma}^2) = (\bar{x}.,\frac{1}{n}\sum_{i=1}^{n}(x_i-\bar{x}.)^2).$$

Here $\bar{x}.$ is the well-known average, while $\hat{\sigma}^2$ except for the factor $n/(n-1)$ is equal to the estimate s^2 that was used in Example 3.1, i.e.,

$$s^2 = \frac{1}{n-1}\sum_{i=1}^{n}(x_i-\bar{x}.)^2 = \frac{n}{n-1}\hat{\sigma}^2.$$

Although the difference between s^2 and $\hat{\sigma}^2$ becomes negligible when the sample size increases, we always prefer to use s^2 as an estimate of σ^2. One reason for this is that $s^2(\mathbf{X})$ in contrast to $\hat{\sigma}^2(\mathbf{X})$ has the right mean σ^2. But it is much more important that $s^2(\mathbf{X})$ has a $\sigma^2\chi^2(f)/f$-distribution with $f = n-1$ degrees of freedom. Tables of the fractiles of this class of distributions has been available since the early days of statistics. Moreover, the distributions of the ratio between independent $\sigma^2\chi^2(f)/f$-distributed estimates of the variance, possibly with different degrees of freedom, also form a well-known and important class of distributions, the F-distributions, which have a prominent part to play later in this chapter.

Test of H_0: $\mu = \mu_0$

We will briefly write down the likelihood ratio tested for H_0 and point out that it is equivalent to the t-test that was used in Example 3.1. The likelihood ratio test statistic is

$$Q(\mathbf{x}) = \frac{\max_{\sigma^2\in\mathbb{R}_+} L(\mu_0,\sigma^2)}{\max_{(\mu,\sigma^2)\in\mathbb{R}\times\mathbb{R}_+} L(\mu,\sigma^2)} = \frac{L(\mu_0,\frac{1}{n}\sum_{i=1}^{n}(x_i-\mu_0)^2)}{L(\bar{x}.,\hat{\sigma}^2)},$$

where the numerator is

$$L(\mu_0,\frac{1}{n}\sum_{i=1}^{n}(x_i-\mu_0)^2) = \frac{1}{(2\pi\frac{1}{n}\sum_{i=1}^{n}(x_i-\mu_0)^2)^{\frac{n}{2}}}e^{-\frac{n}{2}},$$

and the denominator is

$$L(\bar{x}_{\cdot}, \frac{1}{n}\sum_{i=1}^{n}(x_i - \bar{x}_{\cdot})^2) = \frac{1}{(2\pi \frac{1}{n}\sum_{i=1}^{n}(x_i - \bar{x}_{\cdot})^2)^{\frac{n}{2}}} e^{-\frac{n}{2}}.$$

$Q(\mathbf{x})$ can be rewritten as follows:

$$Q(\mathbf{x}) = \left[\frac{\sum_{i=1}^{n}(x_i - \bar{x}_{\cdot})^2}{\sum_{i=1}^{n}(x_i - \mu_0)^2}\right]^{\frac{n}{2}} = \left[\frac{\sum_{i=1}^{n}(x_i - \bar{x}_{\cdot})^2}{\sum_{i=1}^{n}(x_i - \bar{x}_{\cdot})^2 + n(\bar{x}_{\cdot} - \mu_0)^2}\right]^{\frac{n}{2}}$$

$$= \left[\frac{1}{1 + n\frac{(\bar{x}_{\cdot} - \mu_0)^2}{\sum_{i=1}^{n}(x_i - \bar{x}_{\cdot})^2}}\right]^{\frac{n}{2}} = \left[\frac{1}{1 + \frac{t^2(x)}{n-1}}\right]^{\frac{n}{2}}. \qquad (3.27)$$

Small values of $Q(\mathbf{x})$ are critical for H_0, which is equivalent to large values of $t^2(\mathbf{x})$ being critical where $t(\mathbf{x})$ is the test statistic that was used in Example 3.1,

$$t(\mathbf{x}) = t(x_1, \ldots, x_n) = \frac{\bar{x}_{\cdot} - \mu_0}{\sqrt{s^2/n}}.$$

Annex to Section 3.1

Calculations in SAS

SAS programs for the examples can be found at the address
`http://www.imf.au.dk/biogeostatistics/`

Computations for a single normal sample are conveniently made with *PROC MEANS*.

Example 2.4 (Continued)
The following program creates the data set, and the computations needed for the statistical analysis are made by *PROC MEANS*:

```
TITLE1 'Example 2.4';

DATA ex2_4;
INPUT ppm number;
DATALINES;
   5       1
   6       2
   7       1
   8       1
   9       4
  10      10
  11      24
  12      30
  13      37
  14      41
  15      42
  16      38
  17      23
  18      23
  19      13
  20       6
  21       3
  22       1
 ;
RUN;

PROC MEANS DATA=ex2_4 ALPHA=0.05 VARDEF=DF FW=8 MAXDEC=4
     N SUM USS MEAN CSS VAR STD STDERR CLM ;
VAR ppm;
FREQ number;
RUN;
```

The two statements `FW=8` and `MAXDEC=4` control the output of computed quantities. `FW`, field width, is the maximal number of digits including the decimal point that will be printed and `MAXDEC` is the maximal number of decimal places.

The `VAR ppm;` statement requests computations on the variable `ppm`. Data in Example 2.4 are grouped and therefore SAS needs to be told which variable contains the information on how many times each value of `ppm` have been observed. This is done with the `FREQ number;` statement because the variable `number` in the present data set contains this information. When observations are not grouped the `FREQ number;` statement must be excluded from the program.

The options `ALPHA=0.05` requests that the confidence coefficient shall be 95%. The confidence interval for the mean is requested with the statistic keyword `CLM`.

Output listing from *PROC MEANS*:

```
                              Example 2.4                                   1

                          The MEANS Procedure

                       Analysis Variable : ppm

                                                                        Std
      N         Sum         USS        Mean         CSS   Variance    Std Dev      Error
-----------------------------------------------------------------------------------------
    300    4342.000    65224.00     14.4733    2380.787     7.9625     2.8218     0.1629
-----------------------------------------------------------------------------------------

                       Analysis Variable : ppm

                         Lower 95%        Upper 95%
                       CL for Mean      CL for Mean
                       ----------------------------
                           14.1527          14.7939
                       ----------------------------
```

A comparison with the computations in Example 2.4 on page 68 shows good agreement. Note, however, that neither the confidence interval for the variance nor for the standard deviation is available from *PROC MEANS*. □

Example 3.1 (Continued)
The following program creates the data set for Example 3.1, and all the computations that are necessary for the analysis are performed by *PROC MEANS*:

```
TITLE1 'Example 3.1';

DATA diam;
INPUT  diameter@@;
diam_40=diameter-40;
DATALINES;
54 50 57 50 53 38 26 37 36 38
48 44 46 46 44 32 32 31 31 31
40 38 41 38 42 50 53 49 43 50
38 36 37 37 37 42 43 40 45 43
;
RUN;

PROC MEANS ALPHA=0.05 DATA=diam VARDEF=DF FW=10 MAXDEC=4
     N MEAN STDERR CSS USS VAR SUM CLM T PRT;
VAR diameter diam_40;
RUN;
```

In the data step the variable `diam_40` is made by subtracting 40 from all the values of `diameter`. *PROC MEANS* can compute the t-test for mean 0 (requested by the statistic keyword `T`) and also the p-value (requested by the statistic keyword `PRT`). We are interested in the test for the mean equal to 40 μm for the variable `diameter`, but this is the same as the test for the mean equal to 0 μm for the variable `diam_40`.

```
                              Example 3.1                                   1

                          The MEANS Procedure

Variable      N          Mean     Std Error           CSS           USS      Variance
--------------------------------------------------------------------------------------
diameter     40       41.6500        1.1545     2079.1000    71468.0000       53.3103
diam_40      40        1.6500        1.1545     2079.1000     2188.0000       53.3103
--------------------------------------------------------------------------------------
```

```
                                  Lower 95%         Upper 95%
Variable              Sum      CL for Mean       CL for Mean     t Value      Pr > |t|
----------------------------------------------------------------------------------------
diameter        1666.0000          39.3149           43.9851       36.08        <.0001
diam_40           66.0000          -0.6851            3.9851        1.43        0.1609
----------------------------------------------------------------------------------------
```

□

The quantity `Std Error` in the output is particularly useful for calculating confidence intervals and t-test for the mean. Combined with the estimate of μ in `Mean` and the $1-\alpha/2$ fractile of the $t(f)$-distribution, $t_{1-\alpha/2}(f)$, the limits of the $(1-\alpha)$ confidence interval for μ are obtained as

$$\texttt{Mean} \pm \texttt{Std Error} \times t_{1-\alpha/2}(f)$$

with $f = n-1$.

Furthermore, `Std Error` is the denominator in the t-test of $H_{0\mu}$: $\mu = \mu_0$, so the t-test is easily calculated for any μ_0 as

$$\frac{\texttt{Mean} - \mu_0}{\texttt{Std Error}}$$

without calculating the adjusted variable `diameter` $-\mu_0$.

Note that neither the confidence interval for the variance nor the confidence interval for the standard deviation are supplied by *PROC MEANS*.

Main Points in Section 3.1

The main points in the analysis of one sample from a normal distribution are summarized in this section. Results that use the t- and χ^2-distributions depend on the degrees of freedom, f, of the estimate of the variance.

For one sample the degrees of freedom are $f = n-1$, where n is the number of observations, but some of the results, particularly those concerning the variance and the standard deviation, are generally applicable so we use f to denote the degrees of freedom.

Model:
The observations, $x_1,\dots,x_n$, are realizations of independent and identically distributed random variables, $X_1,\dots,X_n$, $X_i \sim N(\mu,\sigma^2), i = 1,\dots,n$.

Model check:
Fractile plot if the number of observations is sufficiently large ($n > 10$).

Estimation:

$$\mu \leftarrow \bar{x}. \sim\sim N(\mu, \frac{\sigma^2}{n}),$$

$$\sigma^2 \leftarrow s^2 \sim\sim \sigma^2\chi^2(f)/f.$$

Test of H_0: $\mu = \mu_0$:
I. Variance known: $\sigma^2 = \sigma_0^2$.
Test statistic:

$$u(\mathbf{x}) = \frac{\bar{x}. - \mu_0}{\sqrt{\sigma_0^2/n}} \sim\sim N(0,1).$$

Significance probability:

$$p_{obs}(\mathbf{x}) = 2\,[1 - \Phi(|\, u(\mathbf{x})\,|)]\,.$$

$(1-\alpha)$ confidence interval for μ:

$$\{\mu \,|\, \bar{x}. - \sqrt{\frac{\sigma_0^2}{n}}u_{1-\alpha/2} \le \mu \le \bar{x}. + \sqrt{\frac{\sigma_0^2}{n}}u_{1-\alpha/2}\},$$

where $u_{1-\alpha/2}$ is the $1-\alpha/2$ fractile in the $N(0,1)$-distribution.

II. Variance unknown:
Test statistic:

$$t(\mathbf{x}) = \frac{\bar{x}. - \mu_0}{\sqrt{s^2/n}} \sim\sim t(f).$$

Significance probability:

$$p_{obs}(\mathbf{x}) = 2\left[1 - F_{t(f)}(|\, t(\mathbf{x})\,|)\right].$$

$(1-\alpha)$ confidence interval for μ:

$$\{\mu \,|\, \bar{x}. - \sqrt{\frac{s^2}{n}}t_{1-\alpha/2}(f) \le \mu \le \bar{x}. + \sqrt{\frac{s^2}{n}}t_{1-\alpha/2}(f)\},$$

where $t_{1-\alpha/2}(f)$ is the $1-\alpha/2$ fractile in the t-distribution with f degrees of freedom.

Test of H_0: $\sigma^2 = \sigma_0^2$:
Test statistic:

$$\frac{fs^2}{\sigma_0^2}.$$

Significance probability:

$$p_{obs}(\mathbf{x}) = \begin{cases} 2F_{\chi^2(f)}(\frac{fs^2}{\sigma_0^2}) & \text{if } \frac{fs^2}{\sigma_0^2} \leq f, \\ 2(1 - F_{\chi^2(f)}(\frac{fs^2}{\sigma_0^2})) & \text{if } \frac{fs^2}{\sigma_0^2} \geq f. \end{cases}$$

$(1-\alpha)$ confidence interval for σ^2:

$$\left\{\sigma^2 \mid \frac{fs^2}{\chi^2_{1-\alpha/2}(f)} \leq \sigma^2 \leq \frac{fs^2}{\chi^2_{\alpha/2}(f)}\right\},$$

where $\chi^2_{1-\alpha/2}(f)$ and $\chi^2_{\alpha/2}(f)$ are the $1-\alpha/2$ and $\alpha/2$ fractiles of the χ^2-distribution of f degrees of freedom.

$(1-\alpha)$ confidence interval for σ:

$$\left\{\sigma \mid \sqrt{\frac{fs^2}{\chi^2_{1-\alpha/2}(f)}} \leq \sigma \leq \sqrt{\frac{fs^2}{\chi^2_{\alpha/2}(f)}}\right\},$$

where $\chi^2_{1-\alpha/2}(f)$ and $\chi^2_{\alpha/2}(f)$ are the $1-\alpha/2$ and $\alpha/2$ fractiles of the χ^2-distribution with f degrees of freedom.

3.2 Two or More Samples

For two or more (k) normal samples or groups, it is the hypothesis of a common variance, $H_{0\sigma^2}$: $\sigma_1^2 = \cdots = \sigma_k^2$, and the hypothesis of identical means, $H_{0\mu}$: $\mu_1 = \cdots = \mu_k$, that are of interest.

In the treatment of two or more samples we shall distinguish between exactly two samples and more than two samples. For more than two samples ($k > 2$), we will only test the hypothesis of a common mean under the assumption of a common variance, i.e. $H_{0\sigma^2}$ has been tested first and no evidence of a conflict with the data has been detected. For exactly two samples ($k = 2$) the hypothesis of a common variance, $H_{0\sigma^2}$: $\sigma_1^2 = \sigma_2^2$, is tested first and depending on the outcome of this test, it is decided which test to use when testing the hypothesis of a common mean, $H_{0\mu}$: $\mu_1 = \mu_2$.

The model for k normal samples is that the observations,

$$
\begin{array}{c}
x_{11}, \ldots, x_{1j}, \ldots, x_{1n_1} \\
\cdots \\
x_{i1}, \ldots, x_{ij}, \ldots, x_{in_i} \\
\cdots \\
x_{k1}, \ldots, x_{kj}, \ldots, x_{kn_k},
\end{array}
$$

are realizations of independent normally distributed random variables,

$$X_{ij} \sim N(\mu_i, \sigma_i^2), \qquad j = 1, \ldots, n_i, \; i = 1, \ldots, k.$$

This model will briefly be referred to as the model for k normal samples and in this section occasionally as M_0. We will use i to index the samples and j to index the observations within samples. The number of observations in the ith sample is denoted by n_i, and there is no requirement that the number of observations in each sample must be the same. In each sample estimates of the mean and the variance are computed. The estimate of the mean in the ith sample is the average of the observations in the ith sample and it is denoted by $\bar{x}_{i\cdot}$. The estimate of the variance in the ith sample is the sample variance s^2 computed in the ith sample and it is denoted by $s_{(i)}^2$. The seemingly superfluous parentheses around the sample index on the sample variances $s_{(i)}^2$ are introduced in order to distinguish them from the quantities s_1^2 and s_2^2, which will be introduced later in this chapter in connection with a sequence of increasingly more restrictive hypotheses of the means, see page 134. A similar notation will be used for the degrees of freedom and the SSDs. Thus $f_{(i)} = n_i - 1$ denotes the degrees of freedom of $s_{(i)}^2$, and $s_{(i)}^2 = SSD_{(i)}/f_{(i)}$, where $SSD_{(i)}$ is the SSD from the ith sample. A summary of the notation in connection with k samples can be seen in the template for calculations in k samples on page 116.

In this section we have used the notation $\bar{x}_{i\cdot}$ to denote the average of the ith sample. It is a commonly used convention that for indexed variables the sum of the variable over one or more indices will be denoted by the same symbol as the indexed variable but replacing the summation indices with a $\cdot$. Thus $x_{i\cdot}$ denotes the sum of the observations x_{ij} in the ith sample. Similarly, $x_{\cdot\cdot}$ denotes the sum of all observations and $n_{\cdot}$ denotes the total number of observations. Often a sum is used to calculate an average. A bar, $^-$, over the name of the variable indicates an average of *all* the observations included in the sum. Thus $\bar{x}_{i\cdot}$ denotes the average of the n_i observations in the ith sample and $\bar{x}_{\cdot\cdot}$ denotes the average of all $n_{\cdot}$ observations.

3.2.1 Two Samples

The treatment of two normal samples consists of investigating whether the variances can be assumed to be identical, i.e., testing the hypothesis $H_{0\sigma^2}$: $\sigma_1^2 = \sigma_2^2$, and then investigating whether the means can be assumed to be equal, i.e., testing the hypothesis $H_{0\mu}$: $\mu_1 = \mu_2$. The treatment concludes with the computation of the relevant confidence intervals and which confidence intervals will be relevant depends on the fate of the hypotheses $H_{0\sigma^2}$ and $H_{0\mu}$. The way the hypothesis of a common mean,

$H_{0\mu}$: $\mu_1 = \mu_2$, is tested depends on whether a common variance of the two samples is assumed or not. The two cases are treated separately in subsections.

Two Samples with a Common Variance

The treatment of two samples with a common variance is based on Example 2.5.

Example 2.5 (Continued)
Recall that the observations are the natural logarithm of the reaction time of flies to a nerve poison and that flies in the first group ($i = 1$) have been exposed to the poison for 30 seconds and that flies in the second group ($i = 2$) have been exposed to the poison for 60 seconds.

First the hypothesis of a common variance in the two samples is tested:
$H_{0\sigma^2}$: $\sigma_1^2 = \sigma_2^2$.

The starting point is the estimates of the variances in the two samples,

$$s_{(1)}^2 = 1.102779 \rightarrow \sigma_1^2$$
$$s_{(2)}^2 = 1.023778 \rightarrow \sigma_2^2,$$

and as the test statistic we use the ratio of the estimates of the variances,

$$F = \frac{s_{(1)}^2}{s_{(2)}^2} = \frac{1.102779}{1.023778} = 1.077.$$

Both large values and small values of the F-statistic are critical for $H_{0\sigma^2}$; for such values reflect very different estimates of the variances in the two samples and conceivably different variances. Values of the test statistic in the neighborhood of 1 are not critical. In this example the probability of a larger value of the test statistic than the one observed is

$$P[F > 1.077] = 1 - F_{F(f_{(1)},f_{(2)})}(1.077) = 0.447,$$

where $F(f_{(1)}, f_{(2)})$ is the F-distribution with $f_{(1)} = n_1 - 1$ degrees of freedom in the numerator and $f_{(2)} = n_2 - 1$ degrees of freedom in the denominator. In this example $f_{(1)} = 15$ and $f_{(2)} = 14$.

Small values of the F-statistic are also critical for $H_{0\sigma^2}$ and therefore the significance probability is chosen to be two times this probability, i.e.,

$$p_{obs}(\mathbf{x}) = 2P[F > 1.077] = 0.894.$$

This does not contradict $H_{0\sigma^2}$ and we will adopt the model of two independent samples with a common variance.

The calculation of the significance probability was based on the F-distribution. The F-distribution with $f_{(1)}$ degrees of freedom in the numerator and $f_{(2)}$ degrees of freedom in the denominator is the distribution of the ratio between a $\chi^2(f_{(1)})/f_{(1)}$-distributed random variable and a $\chi^2(f_{(2)})/f_{(2)}$-distributed random variable that are independent, cf. Section 3.4.4 on page 166.

Now $s_{(1)}^2$ and $s_{(2)}^2$ are realizations of a $\sigma_1^2\chi^2(f_{(1)})/f_{(1)}$ and a $\sigma_2^2\chi^2(f_{(2)})/f_{(2)}$-distributed random variable, respectively, and exactly under the hypothesis $H_{0\sigma^2}$: $\sigma_1^2 = \sigma_2^2$ the ratio σ_1^2/σ_2^2 is equal to 1 and then the ratio $F = s_{(1)}^2/s_{(2)}^2$ will be a realization of a $F(f_{(1)}, f_{(2)})$-distributed random variable.

It is instructive to see how different the two estimates of variances can be without rejecting the hypothesis of a common variance. The 97.5% fractile in the $F(15,14)$-distribution is 2.95 so an observation of the ratio of 2.95 of the variances would lead to a significance probability of exactly 5%, and this means that the estimated variances would have to differ by roughly a factor of 3 or more before the hypothesis $H_{0\sigma^2}$: $\sigma_1^2 = \sigma_2^2$ will be rejected with the present sample sizes.

As mentioned above, the failure to reject $H_{0\sigma^2}$ leads to the model of two independent samples with a common variance that we will write as

$$M_1\colon X_{ij} \sim N(\mu_i, \sigma^2), \qquad j = 1, \ldots, n_i,\ i = 1, 2.$$

This is an abbreviation of the more precise statement:
Data x_{ij}, $j=1,\ldots,n_i$, $i=1,2$, are realizations of independent random variables X_{ij}, where

$$X_{ij} \sim N(\mu_i, \sigma^2), \qquad j=1,\ldots,n_i,\ i=1,2.$$

When we specify the models in the following we will only write down the distributions of the random variables and subsume the independence of the random variables.

We will not derive the estimates of the parameters in any detail but only mention that the maximum likelihood estimates of μ_1 and μ_2 are the averages of the two samples, i.e., $\bar{x}_{1\bullet}$ and $\bar{x}_{2\bullet}$, respectively, whereas the maximum likelihood estimate of σ^2 is

$$\hat{\sigma}^2 = \frac{SSD_1}{n_\bullet} = \frac{SSD_{(1)}+SSD_{(2)}}{n_1+n_2}.$$

As for distributions $SSD_{(1)} \sim \sigma^2\chi^2(f_{(1)})$ and $SSD_{(2)} \sim \sigma^2\chi^2(f_{(2)})$ and $SSD_{(1)}$ and $SSD_{(2)}$ are independent and therefore $SSD_1 \sim \sigma^2\chi^2(f_{(1)}+f_{(2)})$, cf. (3.91) on page 163. In order to get an estimate with mean σ^2, we will use the estimate,

$$s_1^2 = \frac{SSD_1}{f_{(1)}+f_{(2)}} = \frac{SSD_1}{f_1}, \tag{3.28}$$

where $f_1 = f_{(1)}+f_{(2)}$ denotes the degrees of freedom of s_1^2. The estimates of the parameters in M_1 and their distribution can be summarized in the following way:

$$\mu_1 \leftarrow \bar{x}_{1\bullet} \sim\sim N(\mu_1, \frac{\sigma^2}{n_1}),$$

$$\mu_2 \leftarrow \bar{x}_{2\bullet} \sim\sim N(\mu_2, \frac{\sigma^2}{n_2}),$$

and

$$\sigma^2 \leftarrow s_1^2 = \frac{f_{(1)}s_{(1)}^2 + f_{(2)}s_{(2)}^2}{f_{(1)}+f_{(2)}} = \frac{SSD_{(1)}+SSD_{(2)}}{f_{(1)}+f_{(2)}} \sim\sim \sigma^2\chi^2(f_{(1)}+f_{(2)})/(f_{(1)}+f_{(2)}).$$

Recall that the notation,

$$\bar{x}_{1\bullet} \sim\sim N(\mu_1, \frac{\sigma^2}{n_1}),$$

must be read as "$\bar{x}_{1\bullet}$ is a realization of a random variable $\bar{X}_{1\bullet}$ which is distributed as $N(\mu_1, \sigma^2/n_1)$". The first $\sim$ is a reminder that the estimate is a realization of a random variable and the second $\sim$ has its usual interpretation as "is distributed as".

We have now described the marginal distributions of the estimates and with the marginal distributions being independent the joint distribution of the estimates has been described. It is beyond the scope of this book to prove the independence of the distributions of the estimates.

Note that the formula $s_1^2 = (f_{(1)}s_{(1)}^2 + f_{(2)}s_{(2)}^2)/(f_{(1)}+f_{(2)})$ shows that s_1^2 is a weighted average of $s_{(1)}^2$ and $s_{(2)}^2$ with *weights* $f_{(1)}$ and $f_{(2)}$. Therefore, that estimate of the variances that has the larger number of degrees of freedom will have the greater influence on s_1^2.

Returning to the example we summarize the estimates in the table below.

Contact time	*Mean* μ	*Variance* σ^2
30 seconds	2.8072	1.064638
60 seconds	2.7154	

We will now test the hypothesis of a common mean in the two samples:
$H_{0\mu}$: $\mu_1 = \mu_2$.

The test is based on a comparison of the estimates of the means. The estimates and their distributions are

$$\bar{x}_{1\cdot} \sim\sim N(\mu_1, \frac{\sigma^2}{n_1}) \text{ and } \bar{x}_{2\cdot} \sim\sim N(\mu_2, \frac{\sigma^2}{n_2})$$

and because the distributions of $\bar{x}_{1\cdot}$ and $\bar{x}_{2\cdot}$ are independent one has that

$$\bar{x}_{1\cdot} - \bar{x}_{2\cdot} \sim\sim N\left(\mu_1 - \mu_2, \sigma^2\left(\frac{1}{n_1} + \frac{1}{n_2}\right)\right).$$

Under the null hypothesis the distribution is

$$\bar{x}_{1\cdot} - \bar{x}_{2\cdot} \sim\sim N\left(0, \sigma^2\left(\frac{1}{n_1} + \frac{1}{n_2}\right)\right),$$

and therefore

$$\frac{\bar{x}_{1\cdot} - \bar{x}_{2\cdot}}{\sqrt{\sigma^2\left(\frac{1}{n_1} + \frac{1}{n_2}\right)}} \sim\sim N(0,1). \tag{3.29}$$

The estimate of the variance, s_1^2, is substituted for σ^2 because the variance is unknown and one obtains the test statistic,

$$t(\mathbf{x}) = \frac{\bar{x}_{1\cdot} - \bar{x}_{2\cdot}}{\sqrt{s_1^2\left(\frac{1}{n_1} + \frac{1}{n_2}\right)}} \sim\sim t(f_1). \tag{3.30}$$

It is explained in Section 3.4.3 on page 164 that $t(\mathbf{x})$ is a realization of a t-distributed random variable with f_1 degrees of freedom because $s_1^2 \sim\sim \sigma^2\chi^2(f_1)/f_1$ and s_1^2 is independent of the distribution of (3.29). The observed value of the test statistic $t(\mathbf{x})$ is

$$t(\mathbf{x}) = \frac{2.8072 - 2.7154}{\sqrt{1.064638\left(\frac{1}{16} + \frac{1}{15}\right)}} = \frac{0.0918}{0.3708} = 0.2475. \tag{3.31}$$

The values of $t(\mathbf{x})$, which are more critical for the null hypothesis than the observed value of 0.2475, are $t < -0.2475$ and $t > 0.2475$. The significance probability is

$$p_{obs}(\mathbf{x}) = 2\left[1 - F_{t(29)}(0.2475)\right] = 0.81,$$

where $F_{t(29)}$ denotes the distribution function of the t-distribution with 29 degrees of freedom. Now $p_{obs}(\mathbf{x}) > 0.05$ and it is concluded that the data do not contradict $H_{0\mu}$, and the model for the data is now

$$M_2\colon X_{ij} \sim N(\mu, \sigma^2), \qquad j = 1, \ldots, n_i,\ i = 1, 2.$$

This is the well-known model of one normal sample, in spite of a slightly more complicated indexing of the individual observations than was necessary when the model was first treated in Section 3.1. The estimates of the parameters are

$$\mu \leftarrow \bar{x}_{\cdot\cdot} = 2.762677 \sim\sim N(\mu, \frac{\sigma^2}{n_\cdot})$$

$$\sigma^2 \leftarrow s^2 = 1.031464 \sim\sim \sigma^2\chi^2(n_\cdot - 1)/(n_\cdot - 1),$$

where $n_\cdot = n_1 + n_2$.

Note that when the standard calculations have been performed for each of the two samples, i.e., S_1, S_2, USS_1, and USS_2 have been calculated, one obtains S and USS for the combined sample as $S = S_1 + S_2$ and $USS = USS_1 + USS_2$. Standard calculations for the two samples are given in the table on page 69 and one finds

$$S = 85.646, \qquad USS = 267.560368,$$

which has been used to compute the estimates of μ and σ^2 above.

95% confidence intervals for μ, σ^2 and σ are given in the table below.

Contact time	*95% Confidence intervals* *Mean* μ	*Variance* σ^2	*Standard deviation* σ
30 and 60 seconds	[2.3903, 3.1353]	[0.6587, 1.8429]	[0.8116,1.3575]

The confidence intervals here are shorter than the confidence intervals for the same parameter based on each of the two samples, cf. page 69. This is due to the to the larger sample size in the combined sample ($n = 31$) than in each of the samples ($n_1 = 16$ and $n_2 = 15$).

We have been analyzing a model for the *logarithm* for reaction times measured in minutes. If we transform back to the original scale of measurements one finds that the 95% confidence interval for e^{μ} is

$$[10.91, \ 22.99] \text{ minutes.}$$

In order to interpret this result recall that e^{μ} is the median, i.e., the 50% fractile, in the distribution of the reaction times.

The conclusion of the experiment can now be stated briefly: With the present sample sizes, it was not possible to prove a difference in reaction times between contact times of 30 seconds or 60 seconds, and the median in the distribution of reaction times was estimated to be $e^{\bar{x}_{\cdot\cdot}} = 15.8$ minutes with a 95% confidence interval ranging from 11 to 23 minutes. □

Confidence Intervals in Two Samples with a Common Variance

In Example 2.5 the conclusion was that the t-test did not contradict the hypothesis of a common mean, $H_{0\mu}$: $\mu_1 = \mu_2$, and the analysis ended with a $(1-\alpha)$ confidence interval for the common mean μ and for the common variance σ^2. If $H_{0\mu}$ had been rejected, the final model would have been M_1, two samples with different means and a common variance. In that case it might be of interest to report the confidence intervals for the means μ_1 and μ_2, which are given by the formulas,

$$\{\mu_1 \mid \bar{x}_{1\cdot} - \sqrt{\frac{s_1^2}{n_1}} t_{1-\alpha/2}(f_1) \le \mu_1 \le \bar{x}_{1\cdot} + \sqrt{\frac{s_1^2}{n_1}} t_{1-\alpha/2}(f_1)\},$$

$$\{\mu_2 \mid \bar{x}_{2\cdot} - \sqrt{\frac{s_1^2}{n_2}} t_{1-\alpha/2}(f_1) \le \mu_2 \le \bar{x}_{2\cdot} + \sqrt{\frac{s_1^2}{n_2}} t_{1-\alpha/2}(f_1)\},$$

where it should be noted that in agreement with the final model M_1, the estimate of the common variance s_1^2 is used with its degrees of freedom f_1. But it will often be far more interesting to calculate the confidence interval for the difference of the means of the two samples, i.e., for $\mu_1 - \mu_2$. Very often a two-sample experiment is planned with the purpose of demonstrating a difference in the means and the width of the confidence interval for $\mu_1 - \mu_2$ indicates the precision with which the difference is determined. The confidence interval is based on the t-test in (3.30) and the $(1-\alpha)$ confidence interval is

$$\left\{\bar{x}_{1\cdot} - \bar{x}_{2\cdot} - \sqrt{s_1^2(\frac{1}{n_1} + \frac{1}{n_2})}\, t_{1-\alpha/2}(f_1) \le \mu_1 - \mu_2 \le \bar{x}_{1\cdot} - \bar{x}_{2\cdot} + \sqrt{s_1^2(\frac{1}{n_1} + \frac{1}{n_2})}\, t_{1-\alpha/2}(f_1)\right\}. \tag{3.32}$$

Even when the test of $H_{0\mu}$ does not reject the hypothesis and the conclusion therefore is that the data do not contradict the hypothesis of a common mean, $H_{0\mu}$, it will be interesting to calculate the confidence interval for $\mu_1 - \mu_2$, which is given by the formula (3.32). The reason is the quite general one that when a statistical test of a hypothesis is not statistically significant, one has not proved that the hypothesis is true. It has been demonstrated only that as far as this particular test is concerned, the data do not contradict the hypothesis. The explanation might well be that the experiment was deficient in some way. In the present case the explanation might be that one had conducted an experiment with too large a variance, samples that were too small, or both. The width of the confidence interval for $\mu_1 - \mu_2$ can be used to evaluate the quality of the experiment. An experiment that results in a narrow confidence interval will seem more convincing than an experiment with a wide confidence interval. It will depend on the context of the experiment whether a confidence interval will be considered to be wide or narrow.

Example 2.5 (Continued)
The 95% confidence interval for the difference in means $\mu_1 - \mu_2$ is

$$\bar{x}_{1\cdot} - \bar{x}_{2\cdot} \pm \sqrt{s_1^2(1/n_1 + 1/n_2)}t_{0.975}(29) = 0.0918 \pm 0.3708 \times 2.0452 = [-0.6666, 0.8502]\,. \quad (3.33)$$

The width of this interval is not easy to interpret because it is for the difference of the means of the logarithm of the reaction times. Fortunately, it can be transformed into a 95% confidence interval for $\exp(\mu_1 - \mu_2) = \exp\mu_1 / \exp\mu_2$, which is the ratio of the median in the distribution of reaction times for flies that have been exposed to the poison for 30 seconds and the median in the distribution of reaction times for flies that have been exposed to the poison for 60 seconds. This confidence interval is

$$[\exp(-0.6666), \exp(0.8502)] = [0.51, 2.34].$$

The median is the same as the 50% fractile, so we can conclude that the 95% confidence interval for the ratio between the time it takes for half of the flies to die after the exposure to the poison for 30 or for 60 seconds, respectively, is from 0.51 to 2.34. This is hardly a precision that can be called impressive. □

Confidence intervals for σ^2 are computed as described for one sample. It is based on the distribution of the estimate of σ^2, $s_1^2 \sim\sim \sigma^2\chi^2(f_1)/f_1$, and the $(1-\alpha)$ confidence interval is

$$\left\{\sigma^2 \mid \frac{f_1 s_1^2}{\chi^2_{1-\alpha/2}(f_1)} \leq \sigma^2 \leq \frac{f_1 s_1^2}{\chi^2_{\alpha/2}(f_1)}\right\},$$

and it can be transformed into a $(1-\alpha)$ confidence interval for σ

$$\left\{\sigma \mid \sqrt{\frac{f_1 s_1^2}{\chi^2_{1-\alpha/2}(f_1)}} \leq \sigma \leq \sqrt{\frac{f_1 s_1^2}{\chi^2_{\alpha/2}(f_1)}}\right\}.$$

Two Samples with Different Variances

We saw in the previous section the F-test for the hypothesis of a common variance, $H_{0\sigma^2}$: $\sigma_1^2 = \sigma_2^2$, and the t-test for the hypothesis of a common mean, $H_{0\mu}$: $\mu_1 = \mu_2$, *assuming a common variance* or, in other words, assuming that $H_{0\sigma^2}$ was not rejected. This was illustrated with the data from Example 2.5. Even in cases where the hypothesis of a common variance is not tenable, it may be of interest to test the hypothesis of a common mean, $H_{0\mu}$: $\mu_1 = \mu_2$. This may be done using an approximate t-test with degrees of freedom that depend not only on the number of observations in each sample but also on the estimates of the variances. We will introduce the test in connection with Example 3.2.

Example 3.2

In Garrison Bay in the state of Washington, one of the recreational activities is clam digging. In order to gain information on how this affects the population of each species, a periodic survey by stratified random sampling of the abundance and size distributions of the standing stock was performed (Gallucci, 1985). In a study section five strata were chosen; two below the tide line and three between the tide line and the high water mark. Each stratum is 100 m long by 20 feet wide. Each sample unit is a square measuring 0.375 m on each side. The data in this example are from only two sample units; one from the stratum just above the tide line and one from the stratum just below the tide line. Moreover, only the length of the littleneck clam, *Protothaca staminea*, is used here. The data are shown in Table 3.2. The length of a clam is recorded in tenths of a millimeter.

Table 3.2 *Length in tenths of a mm of Protothaca staminea*

Stratum 1 just below the tide line	*Stratum 2 just above the tide line*
420	305
449	330
452	389
455	393
459	402
468	410
471	420
472	465
475	474
479	486
481	508
485	
487	
505	
512	
517	
530	

From Gallucci, V.F., The Garrison Bay Project, Stock Assessment and Dynamics of the Littleneck Clam, *Protothaca staminea*, in Andrews, D.F. and Herzberg, A.M., *Data. A Collection of Problems from Many Fields for the Student and Research Worker*, Springer-Verlag, New-York, 1985. With permission.

The fractile plots of the lengths of the clams in each stratum can be seen in Figure 3.2. The plots do not question the assumption of a normal distribution of the length of the clams in each stratum and we begin the analysis with the model M_0: two normal samples. It is important to note that the slopes of the fractile plots are very different and we expect that the hypothesis of a common variance will be rejected.

A summary of the standard calculations is given in the table below.

	n	S	USS
Stratum 1	17	8117	3887839
Stratum 2	11	4582	1948960

This gives the following estimate of the parameters.

	Mean μ	*Variance* σ^2
Stratum 1	477.4706	763.1397
Stratum 2	416.5455	4034.873

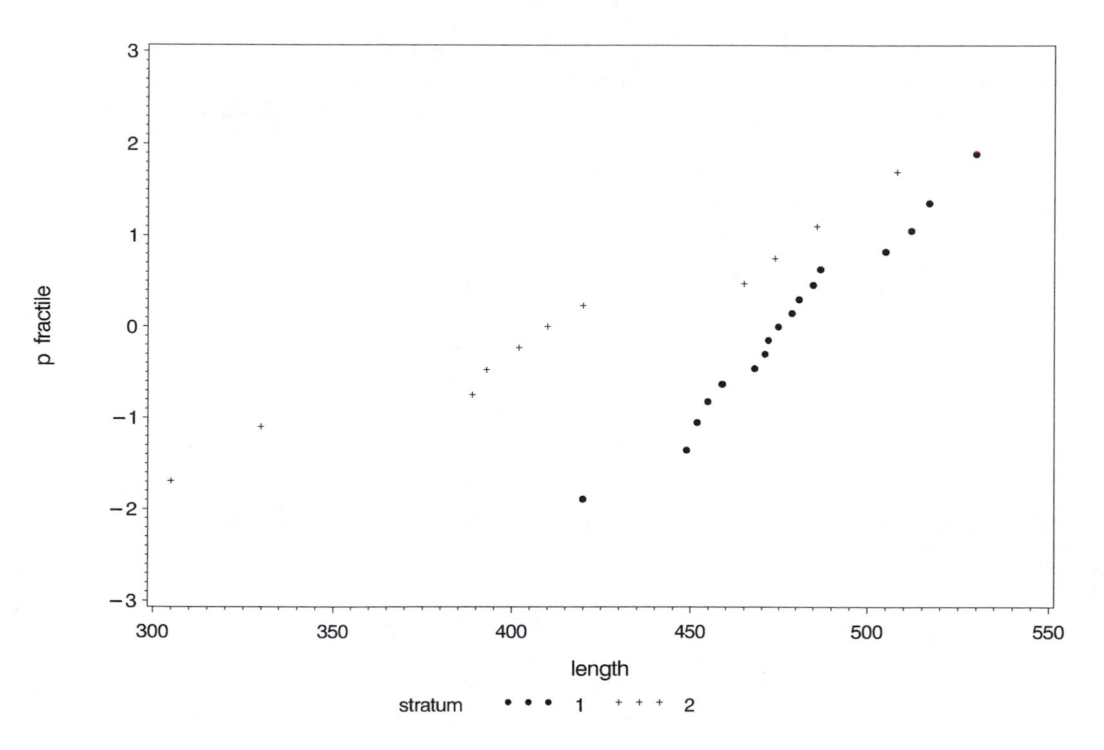

Figure 3.2 *Fractile plots for the two samples in Table 3.2.*

As always the hypothesis of a common variance in the two samples is tested first: $H_{0\sigma^2}\colon \sigma_1^2 = \sigma_2^2$.
The F-test statistic is

$$F = \frac{s^2_{(1)}}{s^2_{(2)}} = \frac{763.1397}{4034.873} = 0.1891.$$

The test statistic is to be evaluated in an F-distribution with 16 degrees of freedom in the numerator and 10 degrees of freedom in the denominator. Here,

$$P[F < 0.1891] = F_{F(16,10)}(0.1891) = 0.001672,$$

and the significance probability is twice that value, i.e.,

$$p_{obs}(\mathbf{x}) = 2P[F < 0.1891] = 0.0033.$$

The significance probability is below 0.05 and $H_{0\sigma^2}$ is rejected. It cannot be assumed that the two samples have a common variance. The estimate of the variance is larger for stratum 2 above the tide line.

Even if $\sigma_1^2 \neq \sigma_2^2$, it is still of interest to test the hypothesis of a common mean $H_{0\mu}$: $\mu_1 = \mu_2$. Just as in the intuitive derivation of the t-test in the case of a common variance, the starting point is the estimated difference in means normalized with the square root of its variance. Under $H_{0\mu}$,

$$\frac{\bar{x}_{1\cdot} - \bar{x}_{2\cdot}}{\sqrt{\sigma_1^2/n_1 + \sigma_2^2/n_2}} \sim\sim N(0,1), \tag{3.34}$$

but since the variances are unknown they are replaced by their estimates in (3.34). This gives the test statistic,

$$t(\mathbf{x}) = \frac{\bar{x}_{1\cdot} - \bar{x}_{2\cdot}}{\sqrt{s_{(1)}^2/n_1 + s_{(2)}^2/n_2}}.$$

This statistic is not exactly t-distributed but its distribution can be well approximated by a t-distribution with degrees of freedom $\bar{f}$, which is calculated from the estimates of the variances in the two samples and the degrees of freedom of each of the two samples. It can be proved that

$$\min\{f_{(1)}, f_{(2)}\} \leq \bar{f} \leq f_{(1)} + f_{(2)}.$$

The formula for computation of $\bar{f}$ is

$$\bar{f} = \frac{\left(\frac{s_{(1)}^2}{n_1} + \frac{s_{(2)}^2}{n_2}\right)^2}{\frac{\left(\frac{s_{(1)}^2}{n_1}\right)^2}{f_{(1)}} + \frac{\left(\frac{s_{(2)}^2}{n_2}\right)^2}{f_{(2)}}}. \tag{3.35}$$

The reasoning behind the formulas is that the distribution of $s_{(1)}^2/n_1 + s_{(2)}^2/n_2$ is assumed to be approximated by a $(\sigma_1^2/n_1 + \sigma_2^2/n_2)\chi^2(\bar{f})/\bar{f}$-distribution with the same variance as the distribution of $s_{(1)}^2/n_1 + s_{(2)}^2/n_2$. The variance of a $\chi^2(f)/f$-distribution is equal to $2/f$ and using this result repeatedly we find that

$$Var\left[\frac{s_{(1)}^2}{n_1} + \frac{s_{(2)}^2}{n_2}\right] = \frac{2\sigma_1^4}{n_1^2 f_{(1)}} + \frac{2\sigma_2^4}{n_2^2 f_{(2)}}, \tag{3.36}$$

and

$$Var\left[\left(\frac{\sigma_1^2}{n_1} + \frac{\sigma_2^2}{n_2}\right)\chi^2(\bar{f})/\bar{f}\right] = \left(\frac{\sigma_1^2}{n_1} + \frac{\sigma_2^2}{n_2}\right)^2 2/\bar{f}. \tag{3.37}$$

Equating the right-hand sides in (3.36) and (3.37) gives the formula for $\bar{f}$.

$$\bar{f}^{-1} = \frac{1}{\left(\frac{\sigma_1^2}{n_1} + \frac{\sigma_2^2}{n_2}\right)^2}\left(\frac{\sigma_1^4}{n_1^2 f_{(1)}} + \frac{\sigma_2^4}{n_2^2 f_{(2)}}\right)$$

$$= \frac{\frac{\left(\frac{\sigma_1^2}{n_1}\right)^2}{f_{(1)}} + \frac{\left(\frac{\sigma_2^2}{n_2}\right)^2}{f_{(2)}}}{\left(\frac{\sigma_1^2}{n_1} + \frac{\sigma_2^2}{n_2}\right)^2}. \tag{3.38}$$

When $s_{(1)}^2$ and $s_{(2)}^2$ are substituted for σ_1^2 and σ_2^2 in (3.38) formula (3.35) for computation of $\bar{f}$ is obtained.

Returning to the actual example we find that

$$t(\mathbf{x}) = \frac{477.4706 - 416.5455}{\sqrt{763.1397/17 + 4034.873/11}} = \frac{60.9251}{20.2903} = 3.003. \tag{3.39}$$

The significance probability is found using formula (3.35) with degrees of freedom $\bar{f} = 12.48$. Tables of the t-distribution and most pocket calculators only have integer degrees of freedom in which case $\bar{f}$ is rounded to 12 and the significance probability is calculated as

$$p_{obs}(\mathbf{x}) = 2\left[1 - F_{t(12)}(3.003)\right] = 0.01.$$

The significance probability is smaller than 0.05 and $H_{0\mu}$ is rejected. Thus both the hypothesis of a common variance and the hypothesis of a common mean for the two strata are rejected. In stratum 1, just below the tide line, the clams are larger and more homogeneous. □

Confidence Intervals in Two Samples with Different Variances

Confidence intervals for means and variances in each of the two samples are computed as described for one normal sample, see *Main Points in Section 3.1* on page 78. An approximative confidence interval for $\mu_1 - \mu_2$ can based on the approximating $t(\bar{f})$-distribution of

$$t(\mathbf{x}) = \frac{\bar{x}_{1\cdot} - \bar{x}_{2\cdot}}{\sqrt{s^2_{(1)}/n_1 + s^2_{(2)}/n_2}},$$

and this gives the $(1-\alpha)$ confidence interval,

$$\left\{\bar{x}_{1\cdot} - \bar{x}_{2\cdot} - \sqrt{\frac{s^2_{(1)}}{n_1} + \frac{s^2_{(2)}}{n_2}}\, t_{1-\alpha/2}(\bar{f}) \leq \mu_1 - \mu_2 \leq \bar{x}_{1\cdot} - \bar{x}_{2\cdot} + \sqrt{\frac{s^2_{(1)}}{n_1} + \frac{s^2_{(2)}}{n_2}}\, t_{1-\alpha/2}(\bar{f})\right\}. \tag{3.40}$$

Example 3.2 (Continued)
The 95% confidence limits for $\mu_1 - \mu_2$ are

$$\bar{x}_{1\cdot} - \bar{x}_{2\cdot} \pm \sqrt{\frac{s^2_{(1)}}{n_1} + \frac{s^2_{(2)}}{n_2}}\, t_{0.975}(12.48) = 60.9251 \pm 20.2903 \times 2.1600 \tag{3.41}$$

$$= [17.10, 104.75]. \tag{3.42}$$

Recalling that the lengths of the clams have been recorded in tenths of an mm, we can conclude that the length of the confidence interval for the difference in the means in the two strata is approximately 9 mm and although the clams in stratum 1 are significantly larger than those in stratum 2, it cannot be ruled out that the difference in means is as small as 1.71 mm.

In (3.41) the fractile $t_{0.975}(12.48) = 2.1600$ has been given accurately, although the degrees of freedom are not an integer. Tables and some pocket calculators can only give t-fractiles for integer degrees of freedom and in that case one can use the nearest integer degrees of freedom. Using 12 degrees of freedom one finds the 97.5% fractile $t_{0.975}(12) = 2.1788$ and the confidence interval, [16.72, 105.13], which is slightly longer than (3.42). □

The F-test of $H_{0\sigma^2}$: $\sigma_1^2 = \sigma_2^2$ and Statistical Tables

Tables of the F-distribution are often limited to the upper tail of the distribution, i.e., to fractiles greater than 0.5. This is sufficient because the F-distributions has the property,

$$F_{F(f_1,f_2)}(x) = 1 - F_{F(f_2,f_1)}\left(\frac{1}{x}\right), \quad x > 0. \tag{3.43}$$

The identity follows because if $X \sim F(f_1, f_2)$, then $1/X \sim F(f_2, f_1)$, so for any $x > 0$,

$$F_{F(f_1,f_2)}(x) = P[X \leq x] = P\left[\frac{1}{X} \geq \frac{1}{x}\right] = 1 - P\left[\frac{1}{X} \leq \frac{1}{x}\right] = 1 - F_{F(f_2,f_1)}(\frac{1}{x}).$$

The fact that the calculation of p-values previously relied on tables has influenced the way the F-test for the identity of two variances is presented even in statistical packages of today.

In Example 3.2 the F-test for $H_{0\sigma^2}\colon \sigma_1^2 = \sigma_2^2$ is given as

$$F = \frac{s_{(1)}^2}{s_{(2)}^2} = \frac{763.1397}{4034.873} = 0.1891,$$

and the significance probability is

$$p_{obs}(\mathbf{x}) = 2F_{F(16,10)}(0.1891) = 0.0033.$$

Few, if any, tables of the F-distribution can be used to look up the value of 0.1891 directly, but using (3.43) the significance probability can be calculated as

$$p_{obs}(\mathbf{x}) = 2F_{F(16,10)}(0.1891) = 2\left[1 - F_{F(10,16)}(\frac{1}{0.1891})\right] = 2\left[1 - F_{F(10,16)}(5.29)\right] = 0.0033,$$

where 5.29 belongs to the upper tail of the $F(10,16)$ distribution. If instead of F its reciprocal had been calculated,

$$\frac{1}{F} = \frac{s_{(2)}^2}{s_{(1)}^2} = \frac{4034.873}{763.1397} = 5.29,$$

it should have been evaluated in an $F(10,16)$ distribution and the significance probability would be calculated as

$$p_{obs}(\mathbf{x}) = 2\left[1 - F_{F(10,16)}(5.29)\right] = 0.0033.$$

It follows that regardless of whether the test statistic F or its reciprocal F^{-1} is calculated, the same significance probability is obtained, and its calculation is always based on the larger of F or F^{-1}. This has led to the practice of using the larger of $s_{(1)}^2$ and $s_{(2)}^2$ in the numerator of the F-test statistic so a value greater than 1 is obtained.

The test procedure can briefly be described as follows:

Set $s_{numerator}^2 = \max\{s_{(1)}^2, s_{(2)}^2\}$, and let $f_{numerator}$ denote the degrees of freedom of $s_{numerator}^2$, and, similarly, set $s_{denominator}^2 = \min\{s_{(1)}^2, s_{(2)}^2\}$ and let $f_{denominator}$ denote the degrees of freedom of $s_{denominator}^2$.

Then the F-test statistic of the hypothesis $H_{0\sigma^2}\colon \sigma_1^2 = \sigma_2^2$ is

$$F = \frac{s_{numerator}^2}{s_{denominator}^2}.$$

The significance probability is calcutated as

$$p_{obs}(\mathbf{x}) = 2\left[1 - F_{F(f_{numerator}, f_{denominator})}(F)\right],$$

where $F(f_{numerator}, f_{denominator})$ is the F-distribution with $f_{numerator}$ degrees of freedom in the numerator and $f_{denominator}$ degrees of freedom in the denominator.

Annex to Subsection 3.2.1

Calculations in SAS

The relevant SAS procedure to make the computations for a comparison of two normal samples is *PROC TTEST*.

Example 2.5 (Continued)
The following program creates the data set, and the computations for the analysis of two normal samples are performed by *PROC TTEST*. This includes the tests of the hypotheses, $H_{0\sigma^2}$: $\sigma_1^2 = \sigma_2^2$, and $H_{0\mu}$: $\mu_1 = \mu_2$, as well as several potentially useful confidence intervals.

```
TITLE1 'Example 2.5';

DATA react;
INPUT group reactime @@;
lnreact=LOG(reactime);
DATALINES;
1  3 1  5 1  5 1  7 1  9 1   9 1 10 1 12 1 20 1 24
1 24 1 34 1 43 1 46 1 58 1 140
2  2 2  5 2  5 2  7 2  8 2   9 2 14 2 18 2 24
2 26 2 26 2 34 2 37 2 42 2  90
;
RUN;

PROC TTEST DATA=react;
CLASS group;
VAR lnreact;
RUN;
```

The output from *PROC TTEST* consists of three parts that are referred to by SAS as `Statistics`, `T-Tests` and `Equality of Variances`, respectively.

```
                            The TTEST Procedure

                                Statistics

                             Lower CL          Upper CL  Lower CL
Variable  group          N      Mean    Mean      Mean   Std Dev   Std Dev

lnreact            1    16    2.2477  2.8072    3.3668    0.7757    1.0501
lnreact            2    15    2.1553  2.7155    3.2758    0.7407    1.0117
lnreact   Diff (1-2)          -0.667  0.0917    0.8501    0.8217    1.0317

                               Statistics

                          Upper CL
    Variable  group        Std Dev    Std Err    Minimum    Maximum

    lnreact            1    1.6252     0.2625     1.0986     4.9416
    lnreact            2    1.5955     0.2612     0.6931     4.4998
    lnreact   Diff (1-2)     1.387     0.3708


                                 T-Tests

   Variable    Method          Variances     DF    t Value    Pr > |t|

   lnreact     Pooled          Equal         29       0.25      0.8065
   lnreact     Satterthwaite   Unequal       29       0.25      0.8062
```

```
                    Equality of Variances

Variable    Method      Num DF    Den DF    F Value    Pr > F

lnreact     Folded F        15        14       1.08    0.8940
```

We will comment on the output from the bottom and start with `Equality of Variances`. This is where the F-test of the hypothesis of a common variance, $H_{0\sigma^2}$: $\sigma_1^2 = \sigma_2^2$, is found, including the degrees of freedom for the F-distribution and the significance probability $p_{obs}(\mathbf{x})$. The values are recognized from page 81.

In `T-Tests` the t-tests of the hypothesis of a common mean, $H_{0\mu}$: $\mu_1 = \mu_2$, is found. If the variances are assumed to be equal, the top one that is marked `Variances Equal` must be used. Otherwise the test in the second line that is marked `Variances Unequal` must be used.

The `Statistics` part of the output basically consists of three rows, although they may be split into two or more parts due to limitations of the line size for the output listing. Each row contains an estimate of a mean with its lower and upper 95% confidence limits and an estimate of a standard deviation and its lower and upper 95% confidence limits. One can distinguish between the three rows by the content of the column, which is labeled `group` in this example. In general the column is labeled by the name of the variable in the `CLASS` statement. Whether they are of any use depends on the conclusions based on the F-test of $H_{0\sigma^2}$ and the t-test of $H_{0\mu}$.

In this example the data neither rejects $H_{0\sigma^2}$ nor $H_{0\mu}$ and then it is the lines that are marked `Dif (1-2)` in the `group` column that are of interest.

The difference $\bar{x}_{1\cdot} - \bar{x}_{2\cdot}$, which is an estimate of $\mu_1 - \mu_2$, is found in the `Mean` column with the upper and lower limits of the 95% confidence interval for $\mu_1 - \mu_2$ on either side. One recognizes the results from page 85 although there are small deviations in the last decimal place due to rounding errors. In the computations on page 85 we have only used the observations with three decimal places, whereas all decimal places are used in the calculation by SAS.

Furthermore, `Std Dev` gives an estimate of the standard error of an individual observation. In the `Dif (1-2)` row this estimate is s_1, i.e., the square root of s_1^2 in (3.28). Note that this is the estimate in the model M_1 and not the updated estimate of variance after the reduction to the model M_2. In practice it will be satisfactory to use the estimate of σ that is given by SAS even in cases where the reduction to M_2 has not been contradicted by the data. The 95% confidence limits for σ that are given along with the estimate are calculated from the formula that was derived in the section on one normal sample and given on page 79. In the present application with degrees of freedom equal to f_1.

The rows that in this output are marked with `1` and `2` in the `group` column contain estimates of the mean and the standard deviation and the 95% confidence limits in each of the two samples, respectively. In general the two rows are labeled by the levels of the `CLASS` variable. This part of the output is of interest if either $H_{0\sigma^2}$ is rejected, in which case statements about the standard errors of the two samples may be of interest, or if $H_{0\mu}$ is rejected, which may make statements on the means of the two samples of interest. Everything in these rows is based on the formulas for one normal sample that can be found in *Main Points in Section 3.1* on pages 78 and 79 and applied here with degrees of freedom equal to $f_{(1)}$ and $f_{(2)}$, respectively. □

Example 3.2 (Continued)

The program below creates the data set, and the computations for the analysis of the two samples are performed by *PROC TTEST*.

```
TITLE1 'Example 3.2';

DATA clam;
INPUT length stratum@@;
DATALINES;
```

```
420 1 449 1 452 1 455 1 459 1 468 1
471 1 472 1 475 1 479 1 481 1 485 1
487 1 505 1 512 1 517 1 530 1
305 2 330 2 389 2 393 2 402 2 410 2
420 2 465 2 474 2 486 2 508 2
;
RUN;

PROC TTEST DATA=clam;
CLASS stratum;
RUN;
```

Output listing from *PROC TTEST:*

```
                                   Example 3.2                                  1

                               The TTEST Procedure

                                   Statistics

                                     Lower CL              Upper CL  Lower CL
Variable   stratum              N        Mean      Mean       Mean   Std Dev   Std Dev

length                  1      17     463.27   477.47     491.67    20.574    27.625
length                  2      11     373.87   416.55     459.22    44.383    63.521
length     Diff (1-2)                 25.163   60.925     96.687    35.408    44.961

                                   Statistics

                              Upper CL
     Variable   stratum        Std Dev     Std Err     Minimum     Maximum

     length                1    42.043         6.7         420         530
     length                2    111.47      19.152         305         508
     length     Diff (1-2)      61.616      17.398


                                     T-Tests

   Variable     Method           Variances        DF     t Value     Pr > |t|

   length       Pooled           Equal            26        3.50       0.0017
   length       Satterthwaite    Unequal        12.5        3.00       0.0106


                              Equality of Variances

       Variable     Method         Num DF     Den DF     F Value     Pr > F

       length       Folded F           10         16        5.29     0.0033
```

The F-test of $H_{0\sigma^2}$: $\sigma_1^2 = \sigma_2^2$ and its significance probability is found under `Equality of Variances`. The significance probability is the one given on page 87, but *PROC TTEST* reports the F-test with the larger of $s^2_{(1)}$ and $s^2_{(2)}$ in the numerator as explained on page 90. The hypothesis of a common variance is rejected.

Therefore the t-test of $H_{0\mu}$, which is marked with `Variances Unequal` in the bottom line of `T-Tests`, must be used. We recognize the value of the t-test in (3.39) and its significance probability, given on page 89.

In this case both $H_{0\sigma^2}$ and $H_{0\mu}$ are rejected. Therefore the focus is on the first two rows of the `Statistics` part of the output. Here estimates and confidence limits for the mean and the standard deviation in each of the two samples are found.

We end this section with a warning. It is of course relevant and interesting to report a confidence interval for $\mu_1 - \mu_2$ as it has been done in (3.41) based on the formula (3.40). This confidence interval is nowhere to be found in the SAS output. The confidence interval that is given for $\mu_1 - \mu_2$ in the `Mean` column in the `Dif (1-2)` row has been computed according to the formula (3.32), which assumes a common variance. □

Main Points in Subsection 3.2.1

This section summarizes the main points in Section 3.2.1 about $k = 2$ normal samples.
Model:
The model M_0 for two normal samples states that the observations,

$$x_{11}, \ldots, x_{1j}, \ldots, x_{1n_1}$$

$$x_{21}, \ldots, x_{2j}, \ldots, x_{2n_2},$$

are realizations of independent, normally distributed random variables

$$X_{ij} \sim N(\mu_i, \sigma_i^2).$$

Model checking: Fractile plots in the same graph of the two samples if the number of observations is sufficiently large ($n_i > 10$), but at least dot plots of the two samples in the same graph in order to evaluate differences and similarities.
Estimation:
Separate estimation in the two samples:

$$\mu_1 \leftarrow \bar{x}_{1\bullet} \sim\sim N(\mu_1, \frac{\sigma_1^2}{n_1}), \qquad \sigma_1^2 \leftarrow s_{(1)}^2 \sim\sim \sigma_1^2 \chi^2(f_{(1)})/f_{(1)},$$

$$\mu_2 \leftarrow \bar{x}_{2\bullet} \sim\sim N(\mu_2, \frac{\sigma_2^2}{n_2}), \qquad \sigma_2^2 \leftarrow s_{(2)}^2 \sim\sim \sigma_2^2 \chi^2(f_{(2)})/f_{(2)}.$$

Calculations: Even for two samples it will be advantageous to use the template for calculations in k normal samples on page 116.
Test of $H_{0\sigma^2}$: $\sigma_1^2 = \sigma_2^2$:
Set $s_{numerator}^2 = \max\{s_{(1)}^2, s_{(2)}^2\}$ and let $f_{numerator}$ be the degrees of freedom of $s_{numerator}^2$, and, similarly, set $s_{denominator}^2 = \min\{s_{(1)}^2, s_{(2)}^2\}$ and let $f_{denominator}$ be the degrees of freedom of $s_{denominator}^2$.
Test statistic:

$$F = \frac{s_{numerator}^2}{s_{denominator}^2}.$$

Significance probability:

$$p_{obs}(\mathbf{x}) = 2\left[1 - F_{F(f_{numerator}, f_{denominator})}(F)\right],$$

where $F(f_{numerator}, f_{demominator})$ denotes the F-distribution with $f_{numerator}$ degrees of freedom in the numerator and $f_{demominator}$ degrees of freedom in the denominator.
I. Common variance $\sigma_1^2 = \sigma_2^2 = \sigma^2$.
Estimation:

$$\mu_1 \leftarrow \bar{x}_{1\bullet} \sim\sim N(\mu_1, \frac{\sigma^2}{n_1}), \qquad \mu_2 \leftarrow \bar{x}_{2\bullet} \sim\sim N(\mu_2, \frac{\sigma^2}{n_2}),$$

$$\sigma^2 \leftarrow s_1^2 = \frac{f_{(1)} s_{(1)}^2 + f_{(2)} s_{(2)}^2}{f_{(1)} + f_{(2)}} = \frac{SSD_{(1)} + SSD_{(2)}}{f_{(1)} + f_{(2)}} \sim\sim \sigma^2 \chi^2(f_1)/f_1$$

with $f_1 = f_{(1)} + f_{(2)} = n_\bullet - 2$.
Test of $H_{0\mu}$: $\mu_1 = \mu_2$:
Test statistic:

$$t(\mathbf{x}) = \frac{\bar{x}_{1\bullet} - \bar{x}_{2\bullet}}{\sqrt{s_1^2 \left(\frac{1}{n_1} + \frac{1}{n_2}\right)}} \sim\sim t(f_1),$$

where the degrees of freedom f_1 are $n_1 + n_2 - 2$.
Significance probability:

$$p_{obs}(\mathbf{x}) = 2\left[1 - F_{t(f_1)}(|\, t(\mathbf{x})\,|)\right],$$

where $F_{t(f_1)}$ is distribution function of the t-distribution with $f_1 = n_1 + n_2 - 2$ degrees of freedom.
$(1-\alpha)$ confidence interval for the difference in means $\mu_1 - \mu_2$:

$$\left\{\bar{x}_{1\cdot} - \bar{x}_{2\cdot} - \sqrt{s_1^2(\frac{1}{n_1} + \frac{1}{n_2})}\, t_{1-\alpha/2}(f_1) \leq \mu_1 - \mu_2 \leq \bar{x}_{1\cdot} - \bar{x}_{2\cdot} + \sqrt{s_1^2(\frac{1}{n_1} + \frac{1}{n_2})}\, t_{1-\alpha/2}(f_1)\right\},$$

where $t_{1-\alpha/2}(f_1)$ is the $1-\alpha/2$ fractile in the $t(f_1)$-distribution with $f_1 = n_1 + n_2 - 2$.
II. Different variances $\sigma_1^2 \neq \sigma_2^2$.
Test of $H_{0\mu}$: $\mu_1 = \mu_2$:
Test statistic:

$$t(\mathbf{x}) = \frac{\bar{x}_{1\cdot} - \bar{x}_{2\cdot}}{\sqrt{s_{(1)}^2/n_1 + s_{(2)}^2/n_2}} \sim\sim t(\bar{f}),$$

where $t(\bar{f})$ is the t-distribution with $\bar{f}$ degrees of freedom and $\bar{f}$ is computed from the formula

$$\bar{f} = \frac{\left(\frac{s_{(1)}^2}{n_1} + \frac{s_{(2)}^2}{n_2}\right)^2}{\frac{\left(\frac{s_{(1)}^2}{n_1}\right)^2}{f_{(1)}} + \frac{\left(\frac{s_{(2)}^2}{n_2}\right)^2}{f_{(2)}}}.$$

Significance probability:

$$p_{obs}(\mathbf{x}) = 2\left[1 - F_{t(\bar{f})}(|\, t(\mathbf{x})\,|)\right].$$

$(1-\alpha)$ confidence interval for the difference in means $\mu_1 - \mu_2$:

$$\left\{\bar{x}_{1\cdot} - \bar{x}_{2\cdot} - \sqrt{\frac{s_{(1)}^2}{n_1} + \frac{s_{(2)}^2}{n_2}}\, t_{1-\alpha/2}(\bar{f}) \leq \mu_1 - \mu_2 \leq \bar{x}_{1\cdot} - \bar{x}_{2\cdot} + \sqrt{\frac{s_{(1)}^2}{n_1} + \frac{s_{(2)}^2}{n_2}}\, t_{1-\alpha/2}(\bar{f})\right\},$$

where $t_{1-\alpha/2}(\bar{f})$ is the $1-\alpha/2$ fractile in the t-distribution with $\bar{f}$ degrees of freedom.

3.2.2 *More than Two Samples*

The analysis of more than two normal samples will be introduced in connection with Example 3.3. The number of samples will be denoted by k and we will speak of k normal samples.

Example 3.3
A plant physiologist wanted to examine the influence, if any, of mechanical stress on the growth of soya bean plants. 52 seedlings were potted with one seedling in each pot. The 52 seedlings were randomly divided into four samples with 13 seedlings in each sample. The seedlings in two of the samples were stressed by being shaken for twenty minutes daily, while the seedling in the other two samples were not shaken (and thus not exposed to stress). The two samples that received the same exposure to stress were grown under different levels of light. Thus the four samples of plants were allocated to one of four treatments that were defined by the levels of the two basic treatments, stress and light. The samples were randomly allocated to treatments and the allocation can be seen from the table below.

Sample 1: Low level of light, no stress

Sample 2: Low level of light, stress

Sample 3: High level of light, no stress

Sample 4: High level of light, stress

After 16 days the plants were harvested and the total area of the leaves for each plant was measured. The measurements sorted by size within each sample are recorded in Table 3.3.

Table 3.3 *Sample 1: Low level of light, no stress; Sample 2: Low level of light, stress; Sample 3: High level of light, no stress; Sample 4: High level light, stress*

Total area in cm^2 *Sample*			
1	2	3	4
200	163	268	201
215	182	271	216
225	188	273	241
229	195	282	257
230	202	285	259
232	205	299	267
241	212	309	269
253	214	310	282
256	215	314	283
264	230	320	291
268	235	337	291
288	255	340	312
288	272	345	326

From Samuels, M.L. and Witmer, J.A., *Statistics for the Life Sciences*, Prentice-Hall, New York, 1999. With permission.

We wish to see if the data can be described by the model for $k = 4$ normal samples. In Figure 3.3 the fractile plots for the four samples are displayed in the same graph to facilitate the comparison of slopes and positions of the fractile plots. The assumption of four normal samples is not questioned by the plots. The slopes of the fractile plots do not differ very much between the samples, thus indicating that the estimated variances will be very similar. In contrast the fractile plots have very different locations along the horizontal axis indicating that the treatments may influence the mean total area of the leaves.

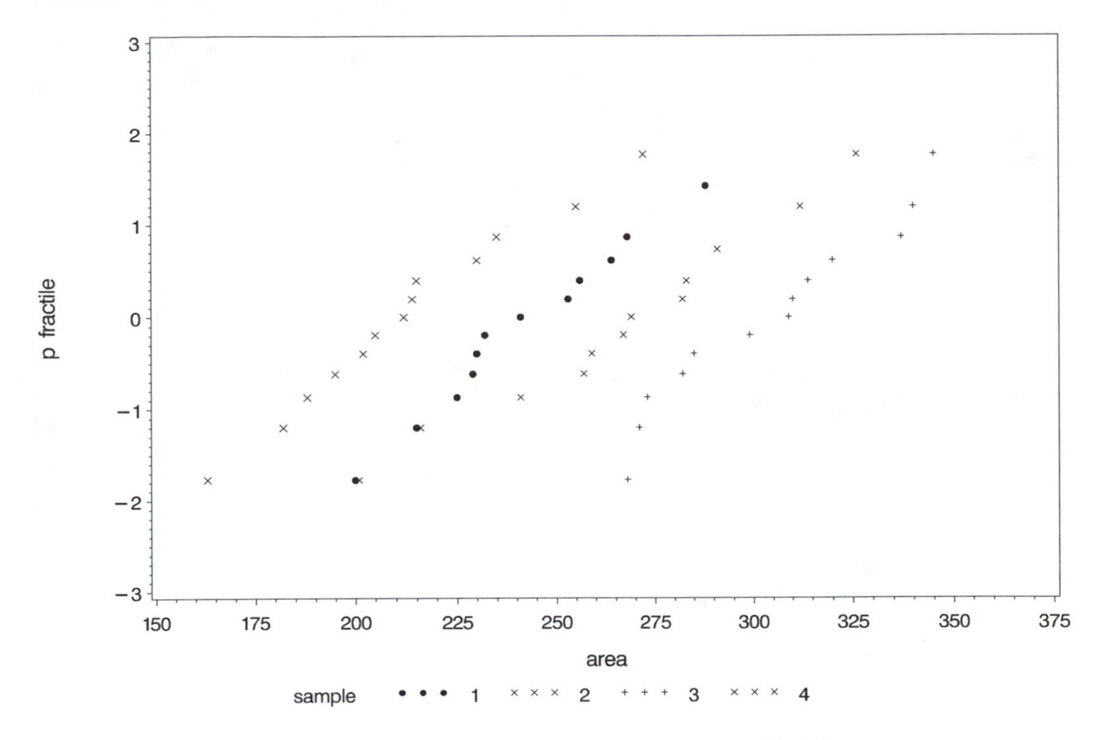

Figure 3.3 *Fractile plots for the four samples in Table 3.3.*

A simple *scatter plot* of the variables area and sample as given in Figure 3.4 is often informative. Here the scatter plot essentially gives the same information as the normal fractile plots in Figure 3.3, except that the information about the distributions is less detailed in the scatter plot. The ranges of the samples, i.e., the difference between the largest and the smallest observation, express the variability of the samples. More precisely, for a normal sample the mean of the range is proportional to the standard deviation and the factor of proportionality is a function of the sample size. A useful rule of thumb is that for normal samples of size 10, 30, 100, and 500, the mean of the range is approximately 3, 4, 5, and 6 times the standard deviation.

In this case where the sample sizes are identical, the ranges of the samples can be taken as expressions of the standard deviations in the distributions behind the samples. The ranges are very similar giving us no reason to think that the standard deviations are different. The scatter plot also gives the impression that the samples have different means. The observations tend to be higher in sample 3 and in sample 4, which were grown at the high level of light, and there seems to be a tendency that among plants that were exposed to the same level of light the total area of leaves is larger for plants that were not stressed.

An elaboration of the simple scatter plot is the *box plot* or the *box-and-whiskers plot* that is shown in Figure 3.5. The name of the plot is derived from the box that is drawn between the upper and lower quartiles. The whiskers are the lines between the smallest observation and the lower quartile and between the largest observation and the upper quartile. In addition the sample median is shown as a horizontal line in the box and the sample mean is shown as a dot. The box plot is superior to the simple scatter plot in displaying deviations from symmetry.

In the present case there are no deviations from symmetry that are common for all four samples.

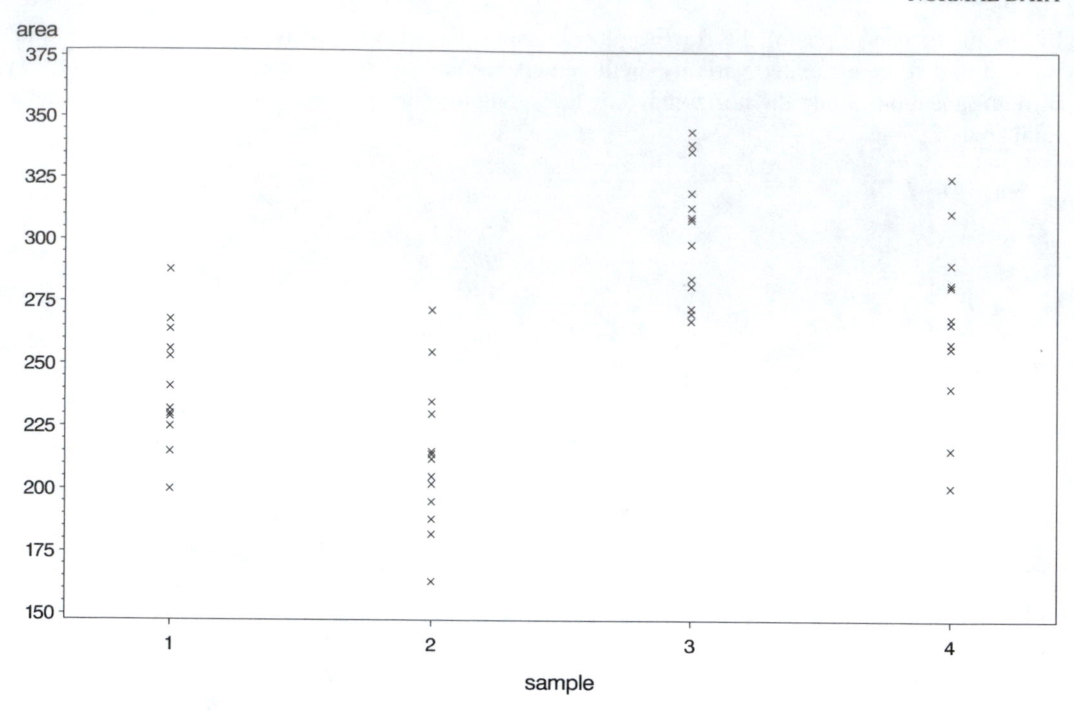

Figure 3.4 *Scatter plot of the variables area and sample in Table 3.3.*

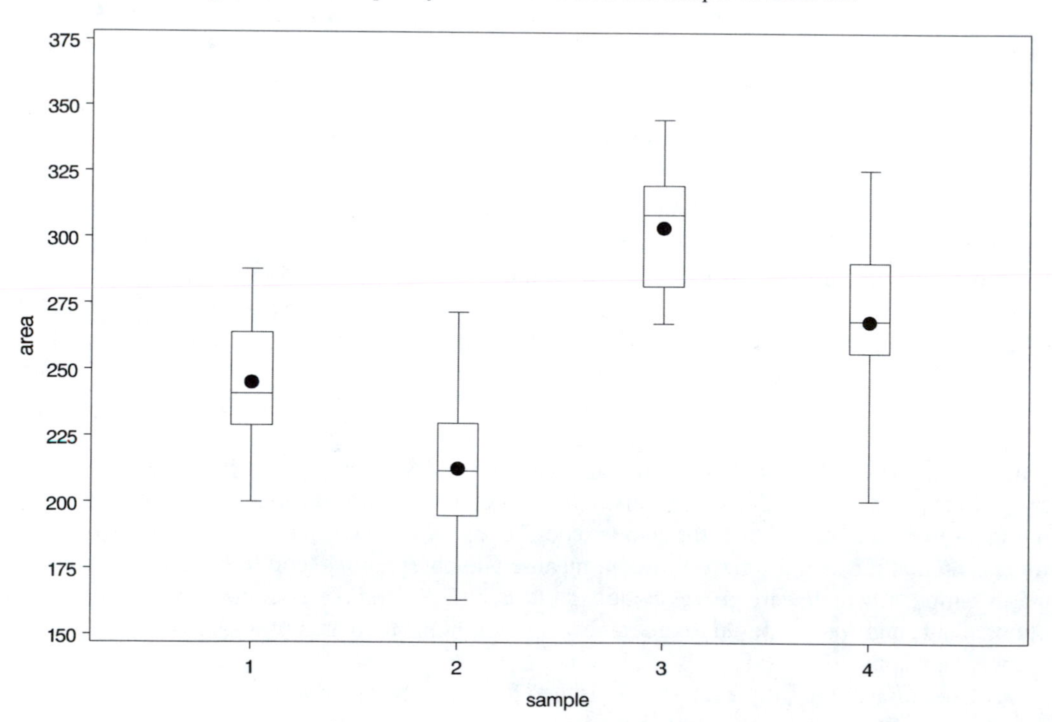

Figure 3.5 *Box-and-whiskers plot of the variables area and sample in Table 3.3.*

On the contrary, the two whiskers for each sample are approximately of the same length and the median line divides the box approximately in the middle. Moreover, there is no systematic pattern where the sample mean for all or the majority of the samples is either above or below the sample median.

We will now see if the first impressions can be confirmed by a more formal analysis. □

After the graphical analysis we have the model: four independent normal samples with unspecified means and variances. We write this model as

$$M_0\colon X_{ij} \sim N(\mu_i, \sigma_i^2), \qquad j = 1, \ldots, n_i,\ i = 1, \ldots, k,$$

and as usual the independence of the observations is subsumed. The estimates of the parameters in M_0 are easily calculated using the template for calculations in Table 3.5. This has been done for the data in Example 3.3 in Table 3.4 where the estimates of the means are found in column **9** and the estimates of the variances are found in column **8**.

We will be interested in the hypothesis $H_{0\sigma^2}$ that the variances of the k samples are identical and the hypothesis $H_{0\mu}$ that the means of the k samples are identical.

In Example 3.3 the sample sizes, n_i, are 13 for all samples, but the methods that we develop to test those hypotheses do not require the sample sizes to be identical.

Calculations

When the calculations are performed using a pocket calculator, it is convenient to organize the computations as shown in Table 3.4.

The table contains basic numerical quantities that are needed in the analysis of k samples, as will be demonstrated in the analysis of Example 3.3.

Table 3.4 *Basic calculations for the data in Table 3.3*

1	**2**	**3**	**4**	**5**	**6**	**7**	**8**	**9**
i	n_i	S_i	USS_i	$\frac{S_i^2}{n_i}$	$SSD_{(i)}$	$f_{(i)}$	$s_{(i)}^2$	$\bar{x}_{i\cdot}$
1	13	3189	791049	782286.23	8762.77	12	730.231	245.3
2	13	2768	599990	589371.08	10618.92	12	884.910	212.9
3	13	3953	1210715	1202016.08	8698.92	12	724.910	304.1
4	13	3495	954513	939617.31	14895.69	12	1241.308	268.8
Σ	52	13405	3556267	3513290.70	42976.30	48	895.340	257.8

The bottom row of the table contains the sum of the quantities in the individual samples, except for the last two entries in column **8** (s_1^2) and in column **9** ($\bar{x}_{\cdot\cdot}$) and this is indicated with the extra horizontal line.

The template for calculations is given in Table 3.5 with formulas instead of numbers. Recall that $SSD_{(i)} = USS_i - \frac{S_i^2}{n_i}$, $f_{(i)} = n_i - 1$, $s_{(i)}^2 = \frac{SSD_{(i)}}{f_{(i)}}$ and $\bar{x}_{i\cdot} = \frac{S_i}{n_i}$, so the relevant quantities for the k samples in columns **6** to **9** are easily calculated from the basic quantities of columns **2**, **3**, and **4**. It follows immediately from the notation that each of the quantities $n_\cdot$, $S_\cdot$, $USS_\cdot$ and $\Sigma_{i=1}^k \frac{S_i^2}{n_i}$ in the bottom row is the sum of the k numbers in the column above, and the same applies to $SSD_1 = \Sigma_{i=1}^k SSD_{(i)}$ and $f_1 = \Sigma_{i=1}^k f_{(i)}$. In contrast, $s_1^2 = \frac{SSD_1}{n_1}$ and $\bar{x}_{\cdot\cdot} = \frac{S_\cdot}{n_\cdot}$ and we use a double line above

Table 3.5 *Template for calculations in k samples*

1	**2**	**3**	**4**	**5**	**6**	**7**	**8**	**9**
Sample	*Sample size*	*Sum*	*Sum of squares*			*Degrees of freedom*	*Estimate of variance*	*Average*
i	n_i	S_i	USS_i	$\frac{S_i^2}{n_i}$	$SSD_{(i)}$	$f_{(i)}$	$s^2_{(i)}$	$\bar{x}_{i\cdot}$
1	n_1	S_1	USS_1	$\frac{S_1^2}{n_1}$	$SSD_{(1)}$	$f_{(1)}$	$s^2_{(1)}$	$\bar{x}_{1\cdot}$
$\vdots$	$\vdots$	$\vdots$	$\vdots$	$\vdots$	$\vdots$	$\vdots$	$\vdots$	$\vdots$
i	n_i	S_i	USS_i	$\frac{S_i^2}{n_i}$	$SSD_{(i)}$	$f_{(i)}$	$s^2_{(i)}$	$\bar{x}_{i\cdot}$
$\vdots$	$\vdots$	$\vdots$	$\vdots$	$\vdots$	$\vdots$	$\vdots$	$\vdots$	$\vdots$
k	n_k	S_k	USS_k	$\frac{S_k^2}{n_k}$	$SSD_{(k)}$	$f_{(k)}$	$s^2_{(k)}$	$\bar{x}_{k\cdot}$
Total	$n.$	$S.$	$USS.$	$\sum_{i=1}^{k}\frac{S_i^2}{n_i}$	SSD_1	f_1	s_1^2	$\bar{x}_{\cdot\cdot}$

s_1^2 and $\bar{x}_{\cdot\cdot}$ to indicate that those two quantities are calculated in a different way from the rest of the quantities of the bottom row. Finally, note that $SSD_2 = \sum_{i=1}^{k}\frac{S_i^2}{n_i} - \frac{S_\cdot^2}{n_\cdot}$ can be calculated from the results in columns **5**, **2**, and **3** in the bottom row.

Testing the Hypothesis of a Common Variance

We first test the hypothesis of a common variances in the k samples:

$$H_{0\sigma^2} : \sigma_1^2 = \cdots = \sigma_k^2 = \sigma^2.$$

If $H_{0\sigma^2}$ is not rejected, the model is reduced to

$$M_1\colon X_{ij} \sim N(\mu_i, \sigma^2), \qquad j = 1, \ldots, n_i,\ i = 1, \ldots, k.$$

When $k > 2$ it is not easy to find a test statistic based solely on intuition and common sense. But the likelihood method can be used. The starting point is the estimates, $s^2_{(1)}, \ldots, s^2_{(k)}$, of the variances, $\sigma_1^2, \ldots, \sigma_k^2$. The distribution of the estimates is

$$s^2_{(i)} \sim\sim \sigma_i^2 \chi^2(f_{(i)})/f_{(i)}, \qquad i = 1, \ldots, k,$$

and the estimates are independent. It is easy then to write the joint density function of $s^2_{(1)}, \ldots, s^2_{(k)}$ as the product of the individual densities and when considered as a function of $\sigma_1^2, \ldots, \sigma_k^2$, this is the likelihood function. The likelihood function may then be used to give an estimate of σ^2 in the model M_1 and the likelihood ratio test for $H_{0\sigma^2}$. We will not consider the derivation in great detail but mention in passing that the density of the $\sigma^2\chi^2(f)/f$-distribution is given in (3.88) on page 163.

We find that the maximum likelihood estimate of the common variance σ^2 based on this likelihood function is

$$s_1^2 = \frac{SSD_1}{f_1} = \frac{SSD_{(1)} + \cdots + SSD_{(k)}}{f_{(1)} + \cdots + f_{(k)}} = \frac{f_{(1)}s_{(1)}^2 + \cdots + f_{(k)}s_{(k)}^2}{f_{(1)} + \cdots + f_{(k)}},$$

and that the likelihood ratio test statistic is

$$-2\ln Q(\mathbf{x}) = f_1 \ln s_1^2 - \sum_{i=1}^{k} f_{(i)} \ln s_{(i)}^2.$$

As noted in Section 11.3, *Approximative Likelihood Theory*, the distribution of $-2\ln Q$ is approximately χ^2-distributed with $k-1$ degrees of freedom because it tests the reduction from a model with k parameters $(\sigma_1^2, \ldots, \sigma_k^2)$ to a model with 1 parameter (σ^2). M.S. Bartlett (1937) showed that the approximation to the χ^2-distribution was improved if the $-2\ln Q$ statistic was divided by the quantity,

$$C = 1 + \frac{1}{3(k-1)}\left[\left(\sum_{i=1}^{k} \frac{1}{f_{(i)}}\right) - \frac{1}{f_1}\right].$$

The test statistic

$$Ba = \frac{-2\ln Q(\mathbf{x})}{C} \tag{3.44}$$

and its evaluation in the $\chi^2(k-1)$-distribution is called *Bartlett's test*. The approximation of the distribution of Ba with a $\chi^2(k-1)$-distribution is satisfactory if $f_{(i)} \geq 2$, $i = 1, \ldots, k$, i.e., if each sample has at least three observations.

Q is a likelihood ratio test statistic and therefore small values of Q are critical for the null hypothesis and hence *large* values of $-2\ln Q$ are critical for the null hypothesis. The significance probability is calculated as

$$p_{obs}(\mathbf{x}) = 1 - F_{\chi^2(k-1)}(Ba),$$

where $F_{\chi^2(k-1)}$ is the distribution function of the χ^2-distribution with $k-1$ degrees of freedom.

The likelihood ratio statistic $Q(\mathbf{x})$ takes values between 0 and 1 only, $0 < Q(\mathbf{x}) \leq 1$ so $-2\ln Q(\mathbf{x}) \geq 0$. Consequently, if a calculation by hand gives a negative value of $-2\ln Q(\mathbf{x})$, there is only one explanation: **miscalculation!**

All the quantities needed to calculate Bartlett's test are available from the template for calculations in Table 3.5.

If Bartlett's test does not reject $H_{0\sigma^2}$, further analysis will be based on the model,

$$M_1\colon X_{ij} \sim N(\mu_i, \sigma^2), \qquad j = 1, \ldots, n_i,\ i = 1, \ldots, k.$$

The estimates of the parameters in the model are

$$\mu_i \leftarrow \bar{x}_{i\cdot} \sim\sim N(\mu_i, \frac{\sigma^2}{n_i}), \qquad i = 1, \ldots, k,$$

and

$$\sigma^2 \leftarrow s_1^2 \sim\sim \sigma^2\chi^2(f_1)/f_1,$$

and the distributions of the estimates are independent.

The estimates of the means are the same as in model M_0 and they are found in column **9** in the template for calculations in Table 3.5 and the estimate of the variance, s_1^2, is found in the last row of column **8**.

Example 3.3 (Continued)
In the actual example we find from the basic calculations in Table 3.4 on page 99 that

$$-2\ln Q = 1.1999$$

and

$$C = 1.03472$$

and, finally, that

$$Ba = \frac{-2\ln Q}{C} = \frac{1.1999}{1.03472} = 1.160.$$

The significance probability is

$$p_{obs}(\mathbf{x}) = 1 - F_{\chi^2(3)}(1.160) = 0.763.$$

The significance probability is greater than 0.05 so Bartlett's test does not reject $H_{0\sigma^2}$ and one chooses to work with the model,

$$M_1 \colon X_{ij} \sim N(\mu_i, \sigma^2), \qquad j = 1, \ldots, n_i,\ i = 1, \ldots, 4.$$

□

Testing the Hypothesis of a Common Mean

Next, we consider the test of the hypothesis of a common mean:

$$H_{0\mu} : \mu_1 = \cdots = \mu_k = \mu.$$

If $H_{0\mu}$ is not rejected, one will consider the model,

$$M_2 \colon X_{ij} \sim N(\mu, \sigma^2), \qquad j = 1, \ldots, n_i,\ i = 1, \ldots, k,$$

as the model for the data and this is the model for one normal sample.

We will use the likelihood ratio test and mention without proof that the likelihood ratio test for $H_{0\mu}$ is

$$\begin{aligned} Q(\mathbf{x}) &= \left[\frac{1}{1 + \frac{SSD_2}{SSD_1}} \right]^{\frac{n}{2}} \\ &= \left[\frac{1}{1 + \frac{k-1}{n-k} \frac{s_2^2}{s_1^2}} \right]^{\frac{n}{2}} \\ &= \left[\frac{1}{1 + \frac{k-1}{n-k} F(\mathbf{x})} \right]^{\frac{n}{2}}. \end{aligned}$$

Here

$$SSD_2 = \sum_{i=1}^{k} n_i (\bar{x}_{i\bullet} - \bar{x}_{\bullet\bullet})^2$$

and

$$s_2^2 = \frac{SSD_2}{k-1},$$

and the F-statistic is the ratio between s_2^2 and s_1^2

$$F(\mathbf{x}) = \frac{s_2^2}{s_1^2}. \tag{3.45}$$

We want to express Q in terms of F because the distribution of F under $H_{0\mu}$ is an F-distribution with $k-1$ degrees of freedom in the numerator and $n_{\cdot}-k$ degrees of freedom in the denominator. This makes it possible to compute exact p-values using the F-distribution, and one does not have to rely on large sample results about $-2\ln Q$ being *approximately* χ^2-distributed.

Note that it is always small values of the likelihood ratio statistic Q that are critical for $H_{0\mu}$ and that small values of Q correspond to large values of F. Thus large values of F are critical. The significance probability is calculated as

$$p_{obs}(\mathbf{x}) = 1 - F_{F(k-1,n_{\cdot}-k)}(F(\mathbf{x})),$$

where $F_{F(k-1,n_{\cdot}-k)}$ is the distribution function of the F-distribution with $k-1$ degrees of freedom in the numerator and $f_1 = n_{\cdot} - k$ degrees of freedom in the denominator.

In the template for calculations in k samples, which is given on page 116, f_1 is found in the bottom row of column **7** and s_1^2 is found in the bottom row of column **8**. SSD_2 can be rewritten as follows to give a formula that is suitable for calculations by hand:

$$\begin{aligned} SSD_2 &= \sum_{i=1}^{k} n_i(\bar{x}_{i\cdot} - \bar{x}_{\cdot\cdot})^2 \\ &= \sum_{i=1}^{k} n_i(\bar{x}_{i\cdot}^2 - 2\bar{x}_{i\cdot}\bar{x}_{\cdot\cdot} + \bar{x}_{\cdot\cdot}^2) \\ &= \left(\sum_{i=1}^{k} \frac{x_{i\cdot}^2}{n_i}\right) - 2x_{\cdot\cdot}\bar{x}_{\cdot\cdot} + n_{\cdot}\bar{x}_{\cdot\cdot}^2 \\ &= \left(\sum_{i=1}^{k} \frac{x_{i\cdot}^2}{n_i}\right) - \frac{x_{\cdot\cdot}^2}{n_{\cdot}} \\ &= \left(\sum_{i=1}^{k} \frac{S_i^2}{n_i}\right) - \frac{S_{\cdot}^2}{n_{\cdot}}, \end{aligned}$$

where both S_i and $x_{i\cdot}$ denote the sum of the observations in the ith sample.

The quantities $\Sigma S_i^2/n_i$, $S_{\cdot}$ and $n_{\cdot}$ are found in the bottom row in the columns **5**, **3**, and **2**, respectively, in the template for calculations on page 116.

Example 3.3 (Continued)
All basic numerical quantities needed for Example 3.3 are given in Table 3.4. Using the formula above SSD_2 is calculated as

$$\begin{aligned} SSD_2 &= 3513290.70 - \frac{13405^2}{52} \\ &= 57636.365, \end{aligned}$$

and this in turn gives

$$s_2^2 = \frac{SSD_2}{k-1} = \frac{57636.365}{3} = 19212.122.$$

Now the test statistic F is computed as

$$F = \frac{s_2^2}{s_1^2} = \frac{19212.122}{895.340} = 21.46,$$

and finally F is evaluated in an $F(3,48)$ distribution and large values of F being critical, the significance probability is

$$p_{obs}(\mathbf{x}) = 1 - F_{F(3,48)}(21.46) = 0.000000006.$$

$H_{0\mu}$ or, equivalently, the reduction to the model M_2 is rejected because $p_{obs}(\mathbf{x}) < 0.05$. The model for the data is still M_1: four normal samples with different means and a common variance.

The estimates of the means can be found in the last column of the template for calculations and the estimate of σ^2 in the bottom row of column **8** in Table 3.4 and the estimation can be summarized as

$$\mu_1 \leftarrow 245.3 \sim\sim N(\mu_1, \frac{\sigma^2}{13}),$$
$$\mu_2 \leftarrow 212.9 \sim\sim N(\mu_2, \frac{\sigma^2}{13}),$$
$$\mu_3 \leftarrow 304.1 \sim\sim N(\mu_3, \frac{\sigma^2}{13}),$$
$$\mu_4 \leftarrow 268.8 \sim\sim N(\mu_4, \frac{\sigma^2}{13}),$$

and

$$\sigma^2 \leftarrow 895.340 \sim\sim \sigma^2\chi^2(f_1)/f_1.$$

Only a weak conclusion can be drawn from this analysis. All that can be said is that the data provide evidence that not all four means are equal. But we cannot without further analysis say whether all four means are significantly different or a single mean is different from the remaining three. □

Confidence Intervals

Confidence intervals for the parameters in the model M_1 can be calculated as described for one normal sample. For the variance the $(1-\alpha)$ confidence interval is

$$\{\sigma^2 \mid \frac{f_1 s_1^2}{\chi^2_{1-\alpha/2}(f_1)} \leq \sigma^2 \leq \frac{f_1 s_1^2}{\chi^2_{\alpha/2}(f_1)}\},$$

where $\chi^2_{1-\alpha/2}(f_1)$ and $\chi^2_{\alpha/2}(f_1)$ are $1-\alpha/2$ and $\alpha/2$ fractiles, respectively, of the χ^2-distribution with f_1 degrees of freedom.

For the mean in the ith sample, μ_i, the $(1-\alpha)$ confidence interval is

$$\{\mu_i \mid \bar{x}_{i\cdot} - \sqrt{\frac{s_1^2}{n_i}}t_{1-\alpha/2}(f_1) \leq \mu_i \leq \bar{x}_{i\cdot} + \sqrt{\frac{s_1^2}{n_i}}t_{1-\alpha/2}(f_1)\}, \tag{3.46}$$

where $t_{1-\alpha/2}(f_1)$ is the $1-\alpha/2$ fractile in the t-distribution with f_1 degrees of freedom. Note that in the calculations of the confidence intervals for the means in the model M_1, we use the estimate of variance in M_1, which is s_1^2.

Confidence intervals for differences of means in the samples can be computed using the formula (3.32), which for $\mu_i - \mu_j$ becomes

$$\left\{\bar{x}_{i\cdot} - \bar{x}_{j\cdot} - \sqrt{s_1^2(\frac{1}{n_i}+\frac{1}{n_j})}\, t_{1-\alpha/2}(f_1) \leq \mu_i - \mu_j \leq \bar{x}_{i\cdot} - \bar{x}_{j\cdot} + \sqrt{s_1^2(\frac{1}{n_i}+\frac{1}{n_j})}\, t_{1-\alpha/2}(f_1)\right\}. \tag{3.47}$$

It is possible to calculate this confidence interval for all differences between pairs of means and when the reduction to the model M_2 has been rejected, i.e., the hypothesis of a common mean has been rejected; it is interesting to do so in an attempt to interpret which means are different. Often, the focus will be on the confidence intervals that do not contain 0, and then the procedure will be equivalent to performing all possible two-sample t-tests for equality of pair of means in the k samples, using s_1^2 as the estimate of variance in all tests. We emphasize that the consideration of all two-sample t-tests is only justified if the F-test has rejected the reduction to M_2 for reasons to be explained below.

If the hypothesis of a common mean has not been rejeced, it may be even more tempting for the experimenter to perform all possible two-sample t-tests between the k samples or calculate all the confidence intervals for differences between pairs of means in an effort to find at least some significant features in the data. But this is not advisable. Note that $k(k-1)/2$ test are being made and that this number increases quickly with k. Also note that the significance level is the probability of rejecting the null hypothesis even if it is true and the probability of making more than one wrong conclusion increases with the number of tests being made. In this context the significance level is said to control the probability of a wrong conclusion in each comparison and it is referred to as the *comparisonwise error rate*. The other type of error to be concerned with is to make one or more wrong conclusion in the whole experiment, i.e., in one or more of the comparisons, and this is called the *experimentwise error rate*. The problems with comparisonwise error versus experimentwise error and the many procedures that have been developed to control the experimentwise error are known as *multiple comparisons.* It is beyond the scope of this book to discuss the multiple comparisons in detail, but in Example 3.3 below we use the simplest multiple comparison method, known as *Fisher's least significant difference method*.

Fisher's least significant difference method consists of two steps. First, perform the F-test for the hypothesis of a common mean. If the hypothesis is not rejected, nothing further is done; we do not try to find pairs of means that we may consider significantly different. If the hypothesis is rejected, we go on to consider all confidence intervals of differences of pairs of means (3.47). The idea in Fisher's least significant difference method is to let the initial F-test for the hypothesis of a common mean control the experimentwise error rate.

Example 3.3 (Continued)

The confidence intervals of the $4 \times 3/2 = 6$ different pairs of means in Example 3.3 are given in the following table and we have observed whether the means are significantly different. Except for μ_4 and μ_1, where 0 is in the 95% confidence interval of $\mu_4 - \mu_1$, all pairs of means are significantly different. We conclude that treatments 1 (low level of light, no stress) and 4 (high level of light, stress) may have the same effect on foliage but all other pairs of treatments have different effects.

$\mu_i - \mu_j$	$\bar{x}_{i\cdot} - \bar{x}_{j\cdot}$	*Lower 95% confidence limit*	*Upper 95% confidence limit*	*Significant*
$\mu_3 - \mu_4$	35.23	11.63	58.83	Yes
$\mu_3 - \mu_1$	58.77	35.17	82.37	Yes
$\mu_3 - \mu_2$	91.15	67.56	114.75	Yes
$\mu_4 - \mu_1$	23.54	−0.06	47.14	No
$\mu_4 - \mu_2$	55.92	32.33	79.52	Yes
$\mu_1 - \mu_2$	32.38	8.79	55.98	Yes

In this example all samples have the same size n and therefore the confidence intervals of differences between two means (3.47) all have the same length. In this situation, half the length of the confidence interval is called the *least significant difference*, abbreviated *LSD*, and it is

$$LSD = \sqrt{s_1^2(\frac{1}{n} + \frac{1}{n})}\, t_{1-\alpha/2}(f_1).$$

The explanation for the name is that the t-test rejects the hypothesis H: $\mu_i = \mu_j$ if and only if

$$|t| = \frac{|\bar{x}_{i\cdot} - \bar{x}_{j\cdot}|}{\sqrt{s_1^2(\frac{1}{n} + \frac{1}{n})}} > t_{1-\alpha/2}(f_1),$$

which is equivalent to

$$|\bar{x}_{i\cdot} - \bar{x}_{j\cdot}| > LSD.$$

Thus the search for pairs of means that are significantly different simply amounts to finding the pairs of samples for which the absolute value of the differences in estimated means exceeds the *LSD*.

In Example 3.3 $LSD = 23.60$ and only $|\bar{x}_4. - \bar{x}_1.| = \bar{x}_4. - \bar{x}_1. = 23.54$ fails to exceed LSD. □

In Example 3.3 one could not reduce to the model M_2: one sample. But when it happens, confidence intervals for the mean and the variance are calculated using the methods of Section 3.1, cf. *Main Points in Section 3.1* on page 78.

So far the analysis has not answered the questions about the influence of light and stress on the foliage of soya bean plants. Those questions will be considered and answered in Section 4.3.

Two or More Samples as Analysis of Variance

In the derivation of the test of the hypothesis of a common mean, $H_{0\mu}$, the likelihood method was used and we used our general knowledge of this method to find not only the test statistic but also those values of the test statistic that were critical for the hypothesis.

We will now give an alternative derivation of the test because it gives an introduction to a general method, which is called *analysis of variance*, and because it gives an understanding of the way the output is organized in output listings from statistical packages.

The derivation relies on a partitioning of the total variation,

$$SSD_{0\infty} = \sum_{i=1}^{k} \sum_{j=1}^{n_i} (x_{ij} - \bar{x}..)^2,$$

into components SSD_1 and SSD_2.

The partitioning is derived as follows from expanding the square of a sum of two terms and observing that the sum of twice the product of the terms is equal to 0.

$$\begin{aligned} SSD_{0\infty} &= \sum_{i=1}^{k} \sum_{j=1}^{n_i} (x_{ij} - \bar{x}..)^2 \\ &= \sum_{i=1}^{k} \sum_{j=1}^{n_i} (x_{ij} - \bar{x}_i. + \bar{x}_i. - x..)^2 \\ &= \sum_{i=1}^{k} \sum_{j=1}^{n_i} (x_{ij} - \bar{x}_i.)^2 + \sum_{i=1}^{k} n_i(\bar{x}_i. - \bar{x}..)^2 \\ &\quad + 2\sum_{i=1}^{k} \sum_{j=1}^{n_i} (x_{ij} - \bar{x}_i.)(\bar{x}_i. - \bar{x}..) \\ &= \sum_{i=1}^{k} \sum_{j=1}^{n_i} (x_{ij} - \bar{x}_i.)^2 + \sum_{i=1}^{k} n_i(\bar{x}_i. - \bar{x}..)^2 \\ &= SSD_1 + SSD_2. \end{aligned}$$

Here the sum of twice the product of the terms is equal to 0 because of the definition of $\bar{x}_i.$:

$$2\sum_{i=1}^{k} \sum_{j=1}^{n_i} (x_{ij} - \bar{x}_i.)(\bar{x}_i. - \bar{x}..) = 2\sum_{i=1}^{k} (\bar{x}_i. - \bar{x}..) \sum_{j=1}^{n_i} (x_{ij} - \bar{x}_i.) = 0.$$

$SSD_{0\infty}$ is a measure of the variation of the observations around the total average of all the observations, $\bar{x}..$, and $SSD_{0\infty}/(n. - 1)$ is the estimate of the variance, σ^2, in the model M_2. SSD_1 summarize the variation of the observations around the sample averages, $\bar{x}_i.$. Finally, SSD_2 measures the variation of the sample averages, $\bar{x}_i.$, around the total average, $\bar{x}..$.

The three types of variation can be visualized in Figure 3.6 where the averages in the $k = 4$ samples have been indicated with a dot, •, and $\bar{x}..$ has been indicated by a horizontal line.

SSD_2 is sometimes referred to as that part of the total variation that is *explained by the model*. Here the model is M_1: four independent normal samples with different means and a common variance. If we did not know that the data could be divided into four samples, all of the variation would have been unexplained and $SSD_{0\infty}$ would be the basis for the estimation of the variance. In the model M_1, however, it is only the variation of the observations around the sample means that is unexplained and which will be used to estimate the variance via SSD_1.

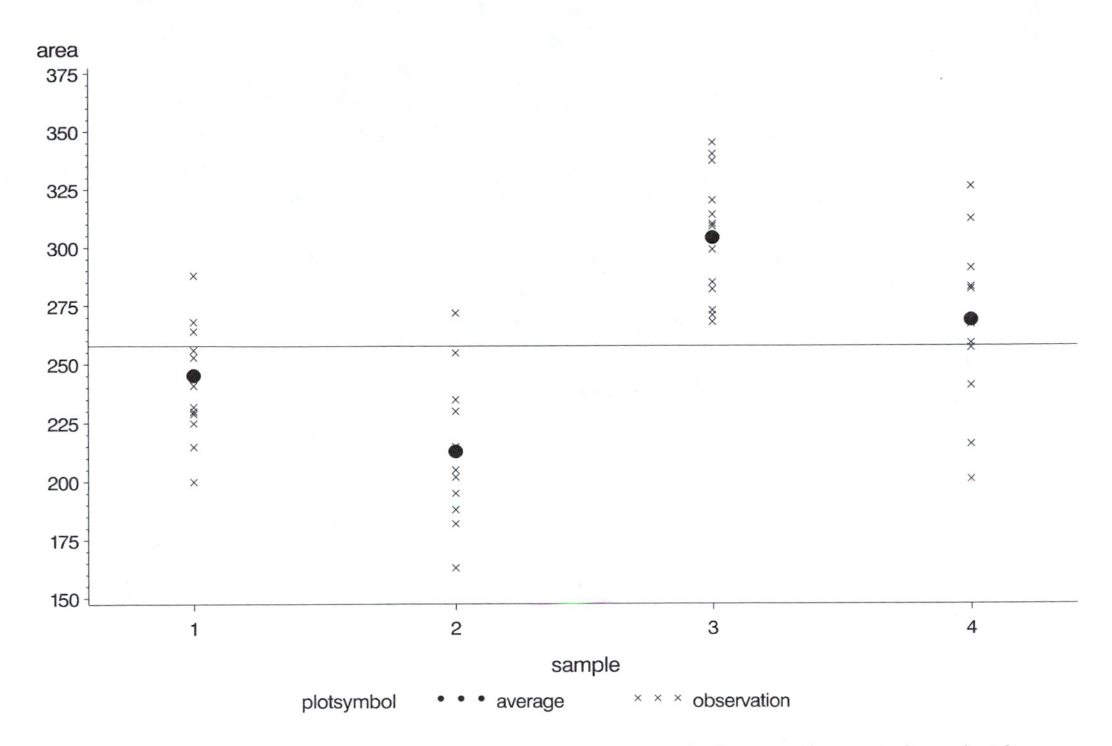

Figure 3.6 *The horizontal line indicates* $\bar{x}_{\cdot\cdot}$, *and the averages in the four samples are indicated with a* $\bullet$.

The test of the reduction from M_1 to the simpler model M_2 or, in other words, the test of the hypothesis $H_{0\mu}$ is an evaluation of whether SSD_2 is greater than expected under M_2, i.e., the model that all observations are realizations of independent normally distributed random variables with the same mean and variance.

It can be proved that SSD_1 and SSD_2 are independent in the model M_1 and that $SSD_1 \sim\sim \sigma^2\chi^2(f_1)$, whereas $SSD_2 \sim\sim \sigma^2\chi^2(k-1)$ only if M_2 is true. In this way we see that it is sensible to base the test of $H_{0\mu}$ on the test statistic,

$$F = \frac{s_2^2}{s_1^2},$$

that F under $H_{0\mu}$ has an $F(k-1, f_1)$-distribution, and that large values of F are critical for $H_{0\mu}$, which we already know from the connection with the likelihood ratio test.

We show in Annex to Section 3.2.2 the first of many examples of how statistical packages organize the output after a division of the total variation into a contribution that is explained by the model and accordingly is denoted by `MODEL` and the unexplained variation that is denoted by `ERROR`.

In spite of its name, *analysis of variance*, it is a method for testing a simpler structure of the means of a data set. In defense of the name, note that the method relies on an evaluation of how much the estimates of variance change if a current model is to be replaced by a simpler model.

Testing the identity of the means of k samples as described here is known as *one-way analysis of variance*.

Differences and Similarities in the Treatment of Two and More than Two Samples

To test the hypothesis of a common mean, $H_{0\mu}$, we have used a t-test when we had two samples and an F-test when we had more than two samples. The F-test can also be used in the case of two samples and it is equivalent to the t-test; for it can be shown that $t^2 = F$ and the significance probabilities are identical regardless of which of the two tests are applied in the two-sample case.

The special case of two normal samples is treated in great detail due to the possibility of finding confidence intervals for the difference in means, $\mu_1 - \mu_2$, which may be very interesting regardless of whether $H_{0\mu}$ is rejected or not. As seen above, confidence intervals for differences of means based on the t-test are of interest also for k samples if the hypothesis of a common mean is rejected by the F-test in (3.45).

We have also used different tests for a common variance, $H_{0\sigma^2}$, in the two cases. Technically, it is possible to compute Bartlett's test even in the case of two samples, but this is not recommended. The F-test based on the ratio between the estimates of variance in the two samples is much quicker to calculate, and it is furthermore an *exact* test where the significance probabilities can be calculated exactly. In contrast, an approximation is used when the significance probabilities are calculated using Bartlett's test.

Annex to Subsection 3.2.2

Calculations in SAS

We have made the macro bartlett to fill in the template for calculations in k normal samples and to calculate Bartlett's test. Although the macro is less useful since Bartlett's test has become available in *PROC GLM,* it is convenient to have an easy way to check calculations made by hand.

The SAS macros made for this book as well as SAS programs can be found at the address http://www.imf.au.dk/biogeostatistics/

How the bartlett macro is used appears from the example below.

The test of $H_{0\mu}$ is executed by the SAS procedure *PROC GLM,* which is a very powerful and general procedure for the computations in all of the models for normal data that are treated in this book. In more recent versions of SAS, Bartlett's test can also be computed by *PROC GLM* and this will also be demonstrated.

Example 3.3 (Continued)
The data set stress is created in the data step. Next, the bartlett macro is used to fill in the template for calculations and to calculate Bartlett's test of $H_{0\sigma^2}$. Finally, the F-test of $H_{0\mu}$ is calculated with *PROC GLM*.

```
TITLE1 'Example 3.3';

DATA stress;
INPUT sample area@@;
DATALINES;
1 200 1 215 1 225 1 229 1 230 1 232 1 241
1 253 1 256 1 264 1 268 1 288 1 288
2 235 2 188 2 195 2 205 2 212 2 214 2 182
2 215 2 272 2 163 2 230 2 255 2 202
3 314 3 320 3 310 3 340 3 299 3 268 3 345
3 271 3 285 3 309 3 337 3 282 3 273
4 283 4 312 4 291 4 259 4 216 4 201 4 267
4 326 4 241 4 291 4 269 4 282 4 257
;

%bartlett(data=stress,group=sample,var=area);

PROC GLM DATA=stress;
CLASS sample;
MODEL area=sample/SS1;
MEANS sample/HOVTEST=BARTLETT T CLM CLDIFF;
LSMEANS sample/ PDIFF;
RUN;
QUIT;
```

In the call of the macro bartlett the data to be used is specified by the command data= , the variable that divides the data set into samples is identified with the command group= and, finally, the variable (x) to be used in the calculations is specified with var= .

In *PROC GLM* it is specified by the statement DATA=stress that the data set to be used in the computations is stress. The specification of the model consists of two statements. The statement CLASS sample; specifies that the variable sample defines the subdivision of the data set into samples and in the model statement MODEL area=sample/SS1; it is MODEL area=sample that specifies that the model for area is k samples, which are defined by the different values of `sample`. The purpose of the option `/SS1` to the model statement is to limit the amount of output from *PROC GLM*.

The output from the macro `bartlett` is

```
                              Example 3.3                              1

                           Macro "BARTLETT"

                             Data set: STRESS
                       Response variable: AREA
                          Group variable: GROUP


                       Calculations in k samples:

                                                                 Sample      Sample
i     ni       Si            USSi          Si2/ni         SSDi      fi  Variance      Mean

1     13  3189.000   791049.0000   782286.2308   8762.7692   12   730.23077  245.3077
2     13  2768.000   599990.0000   589371.0769  10618.9231   12   884.91026  212.9231
3     13  3953.000  1210715.0000  1202016.0769   8698.9231   12   724.91026  304.0769
4     13  3495.000   954513.0000   939617.3077  14895.6923   12  1241.30769  268.8462
-----------------------------------------------------------------========================
      52 13405.000  3556267.0000  3513290.6923  42976.3077   48   895.33974


                            Bartlett's test:

                  C          -2lnQ          Ba        Pr > ChiSq

             1.03472       1.19997       1.15970       0.76268
```

The output from the `bartlett` macro should be easy to interpret. The order of the columns is the same as in the template for calculations in k normal samples on page 116 in *Main Points in Section 3.2.2* and the contents can be compared to Table 3.4 on page 99.

Furthermore, Bartlett's test can also be computed by *PROC GLM*. It is requested by the option `HOVTEST=BARTLETT` to the `MEANS` statement

```
MEANS sample/HOVTEST=BARTLETT;
```

in *PROC GLM* and the output is

```
                              Example 3.3                              4

                           The GLM Procedure

          Bartlett's Test for Homogeneity of area Variance

            Source          DF     Chi-Square     Pr > ChiSq

            sample           3         1.1597         0.7627
```

The two statements

```
CLASS sample;
MODEL area=sample/SS1;
```

are the core of *PROC GLM* code for the analysis of k samples. In the `CLASS` statement the variable `sample` is identified as a variable that can be used to group the observations into classes. Observations with the same value of `sample` are considered to belong to the same class and observations with different values of `sample` are allocated to different classes. When `sample` is used on the right-hand side of the `MODEL` equation, `area=sample`, this means that the model for `area` will be a number of samples defined by the `CLASS` variable `sample`.

The two statements give the central part of the output listing where both the F-test for the hypothesis of a common mean and the estimate of the variance in model M_1 are found.

```
                    Example 3.3                                  2

                 The GLM Procedure

              Class Level Information

          Class          Levels    Values

          sample              4    1 2 3 4


           Number of observations    52
                    Example 3.3                                  3

                 The GLM Procedure

Dependent Variable: area


                                       Sum of
 Source                      DF       Squares    Mean Square   F Value   Pr > F

 Model                        3    57636.3654     19212.1218     21.46   <.0001

 Error                       48    42976.3077       895.3397

 Corrected Total             51   100612.6731


               R-Square     Coeff Var      Root MSE     area Mean

               0.572854      11.60728      29.92223      257.7885


 Source                      DF     Type I SS    Mean Square   F Value   Pr > F

 sample                       3   57636.36538    19212.12179     21.46   <.0001
```

The output is most easily explained by repeating the most essential part, the *analysis of variance table*, with formulas instead of numbers:

The GLM Procedure

Dependent Variable: area

Source	DF	Sum of Squares	Mean Square	F Value	Pr > F
Model	$k-1$	SSD_2	s_2^2	s_2^2/s_1^2	$p_{obs}(x)$
Error	$n_{\cdot}-k$	SSD_1	s_1^2		
Corrected Total	$n_{\cdot}-1$	$SSD_{0\infty}$			

PROC GLM can supply the confidence intervals for the means in the model M_1, cf. (3.46). In the statement

```
MEANS sample/HOVTEST=BARTLETT T CLM CLDIFF;
```

it is the option `T CLM` that requests the confidence intervals for the means. For these data all samples have the same number of observations and a least significant difference exists. It is explained on page 105.

```
                    Example 3.3                                  5

                 The GLM Procedure

          t Confidence Intervals for area

     Alpha                                     0.05
     Error Degrees of Freedom                    48
     Error Mean Square                     895.3397
     Critical Value of t                    2.01063
     Half Width of Confidence Interval     16.68612
```

```
sample          N           Mean      95% Confidence Limits

3              13        304.077      287.391       320.763
4              13        268.846      252.160       285.532
1              13        245.308      228.622       261.994
2              13        212.923      196.237       229.609
```

PROC GLM can also give the confidence intervals for the differences between means in the model M_1, cf. (3.47). In the statement

```
MEANS sample/HOVTEST=BARTLETT T CLM CLDIFF;
```

it is the option `CLDIFF` that requests the confidence intervals for the differences in means.

```
                          Example 3.3                              5

                       The GLM Procedure

                     t Tests (LSD) for area

  NOTE: This test controls the Type I comparisonwise error rate, not the
                    experimentwise error rate.


                 Alpha                                0.05
                 Error Degrees of Freedom               48
                 Error Mean Square                895.3397
                 Critical Value of t               2.01063
                 Least Significant Difference       23.598


     Comparisons significant at the 0.05 level are indicated by ***.


                           Difference
             sample          Between       95% Confidence
           Comparison         Means            Limits

           3     - 4          35.23       11.63     58.83   ***
           3     - 1          58.77       35.17     82.37   ***
           3     - 2          91.15       67.56    114.75   ***
           4     - 3         -35.23      -58.83    -11.63   ***
           4     - 1          23.54       -0.06     47.14
           4     - 2          55.92       32.33     79.52   ***
           1     - 3         -58.77      -82.37    -35.17   ***
           1     - 4         -23.54      -47.14      0.06
           1     - 2          32.38        8.79     55.98   ***
           2     - 3         -91.15     -114.75    -67.56   ***
           2     - 4         -55.92      -79.52    -32.33   ***
           2     - 1         -32.38      -55.98     -8.79   ***
```

The two sample t-tests for all pairs of means in the k samples can be obtained using the option `PDIFF` to the `LSMEANS` statement:

```
LSMEANS sample/ PDIFF;
```

The output listing is:

```
                          Example 3.3                              6

                       The GLM Procedure
                      Least Squares Means

                                                LSMEAN
               sample         area LSMEAN       Number

               1               245.307692            1
               2               212.923077            2
               3               304.076923            3
               4               268.846154            4


                 Least Squares Means for effect sample
                  Pr > |t| for H0: LSMean(i)=LSMean(j)
```

Dependent Variable: area

i/j	1	2	3	4
1		0.0082	<.0001	0.0506
2	0.0082		<.0001	<.0001
3	<.0001	<.0001		0.0043
4	0.0506	<.0001	0.0043	

NOTE: To ensure overall protection level, only probabilities associated with
pre-planned comparisons should be used.

□

Main Points in Section 3.2

This is a summary of the main points in Section 3.2 about more than two normal samples. The main points for two samples are on page 94.

Model:

The model M_0 for k normal samples states that data,

$$\begin{array}{c} x_{11}, \ldots, x_{1j}, \ldots, x_{1n_1} \\ \cdots \\ x_{i1}, \ldots, x_{ij}, \ldots, x_{in_i} \\ \cdots \\ x_{k1}, \ldots, x_{kj}, \ldots, x_{kn_k}, \end{array}$$

are realizations of independent, normally distributed random variables

$$X_{ij} \sim N(\mu_i, \sigma_i^2).$$

Model checking:

Fractile plots in the same graph for those samples where the sample size is sufficiently large ($n_i > 10$), but at least a scatter plot of the observations and the samples in order to evaluate differences and similarities between samples.

Estimation:

The template for calculations on page 116 is useful for numerical computations by hand.

The estimates of the parameters in the model M_0 are

$$\mu_i \leftarrow \bar{x}_{i\bullet} \sim\sim N(\mu_i, \frac{\sigma_i^2}{n_i}), \qquad i = 1, \ldots, k,$$

and

$$\sigma_i^2 \leftarrow s_{(i)}^2 \sim\sim \sigma_i^2 \chi^2(f_{(i)})/f_{(i)}, \qquad i = 1, \ldots, k,$$

and the distributions of the estimates are independent.

The estimates of the parameters in the model M_1 are

$$\mu_i \leftarrow \bar{x}_{i\bullet} \sim\sim N(\mu_i, \frac{\sigma^2}{n_i}), \qquad i = 1, \ldots, k,$$

and

$$\sigma^2 \leftarrow s_1^2 \sim\sim \sigma^2 \chi^2(f_1)/f_1$$

and the distributions of the estimates are independent.

Test of $H_{0\sigma^2}$: $\sigma_1^2 = \cdots = \sigma_k^2 = \sigma^2$:

Test statistic:

$$Ba = \frac{-2 \ln Q(\mathbf{x})}{C},$$

where

$$-2 \ln Q(\mathbf{x}) = f_1 \ln s_1^2 - \sum_{i=1}^{k} f_{(i)} \ln s_{(i)}^2$$

and

$$C = 1 + \frac{1}{3(k-1)} \left[\left(\sum_{i=1}^{k} \frac{1}{f_{(i)}} \right) - \frac{1}{f_1} \right].$$

Here

$$s_1^2 = \frac{\sum_{i=1}^{k} f_{(i)} s_{(i)}^2}{\sum_{i=1}^{k} f_{(i)}} = \frac{\sum_{i=1}^{k} SSD_{(i)}}{\sum_{i=1}^{k} f_{(i)}}$$

and

$$f_1 = \sum_{i=1}^{k} f_{(i)} = n_{\bullet} - k.$$

Significance probability, provided that $f_{(i)} \geq 2$, $i = 1, \ldots, k$,

$$p_{obs}(\mathbf{x}) = 1 - F_{\chi^2(k-1)}(Ba),$$

where $F_{\chi^2(k-1)}$ is the distribution function of the χ^2-distribution with $k-1$ degrees of freedom.
Test of $H_{0\mu}$: $\mu_1 = \cdots = \mu_k$:
Test statistic:

$$F = \frac{s_2^2}{s_1^2}.$$

Significance probability:

$$p_{obs}(\mathbf{x}) = 1 - F_{F(k-1, n_{\bullet}-k)}(F),$$

where $F_{F(k-1, n_{\bullet}-k)}$ is the distribution function of the F-distribution with $k-1$ degrees of freedom in the numerator and $n_{\bullet} - k$ degrees of freedom in the denominator.

Here

$$s_2^2 = \frac{SSD_2}{k-1},$$

and the formula for calculating SSD_2 by hand is

$$SSD_2 = \left(\sum_{i=1}^{k} \frac{S_i^2}{n_i} \right) - \frac{S_{\bullet}^2}{n_{\bullet}}.$$

The quantities needed for the calculation of SSD_2 are found at the bottom of the columns **5**, **2**, and **3** in the template for calculations on page 116.

Template for calculations in k normal samples

1 *Sample*	2 *Sample size*	3 *Sum*	4 *Sum of squares*	5	6	7 *Degrees of freedom*	8 *Estimate of variance*	9 *Average*
i	n_i	S_i	USS_i	$\frac{S_i^2}{n_i}$	$SSD_{(i)}$	$f_{(i)}$	$s^2_{(i)}$	$\bar{x}_{i\cdot}$
1	n_1	S_1	USS_1	$\frac{S_1^2}{n_1}$	$SSD_{(1)}$	$f_{(1)}$	$s^2_{(1)}$	$\bar{x}_{1\cdot}$
$\vdots$	$\vdots$	$\vdots$	$\vdots$	$\vdots$	$\vdots$	$\vdots$	$\vdots$	$\vdots$
i	n_i	S_i	USS_i	$\frac{S_i^2}{n_i}$	$SSD_{(i)}$	$f_{(i)}$	$s^2_{(i)}$	$\bar{x}_{i\cdot}$
$\vdots$	$\vdots$	$\vdots$	$\vdots$	$\vdots$	$\vdots$	$\vdots$	$\vdots$	$\vdots$
k	n_k	S_k	USS_k	$\frac{S_k^2}{n_k}$	$SSD_{(k)}$	$f_{(k)}$	$s^2_{(k)}$	$\bar{x}_{k\cdot}$
Total	$n_\cdot$	$S_\cdot$	$USS_\cdot$	$\sum_{i=1}^{k} \frac{S_i^2}{n_i}$	SSD_1	f_1	s_1^2	$\bar{x}_{\cdot\cdot}$

Recall that $SSD_{(i)} = USS_i - \frac{S_i^2}{n_i}$, $f_{(i)} = n_i - 1$, $s^2_{(i)} = \frac{SSD_{(i)}}{f_{(i)}}$ and $\bar{x}_{i\cdot} = \frac{S_i}{n_i}$, so the relevant quantities for the k samples in columns **6** to **9** are easily calculated from the basic quantities of columns **2**, **3**, and **4**. It follows immediately from the standard notation that each of the quantities $n_\cdot$, $S_\cdot$, $USS_\cdot$ and $\sum_{i=1}^{k} \frac{S_i^2}{n_i}$ in the bottom row is the sum of the k numbers in the column above, and the same applies to $SSD_1 = \sum_{i=1}^{k} SSD_{(i)}$ and $f_1 = \sum_{i=1}^{k} f_{(i)}$. In contrast, $s_1^2 = \frac{SSD_1}{n_1}$ and $\bar{x}_{\cdot\cdot} = \frac{S_\cdot}{n_\cdot}$ and we use a double line above s_1^2 and $\bar{x}_{\cdot\cdot}$ to indicate that they are not the sum of the numbers of the columns above. Finally, note that $SSD_2 = \sum_{i=1}^{k} \frac{S_i^2}{n_i} - \frac{S_\cdot^2}{n_\cdot}$ can easily be calculated from the results in columns **5**, **2**, and **3** in the bottom row.

3.3 Linear Regression

The introduction of linear regression is based on two examples. We begin with Example 3.4.

Example 3.4
In Table 3.6 age and systolic blood pressure have been recorded for 28 males, 13 of which are journalists and 15 university teachers. It is wellknown that the blood pressure of humans increases with age and making a plot of the data with age along the horizontal axis and blood pressure along the vertical axis one notes that this increase is also visible in the present data, cf. Figure 3.7.

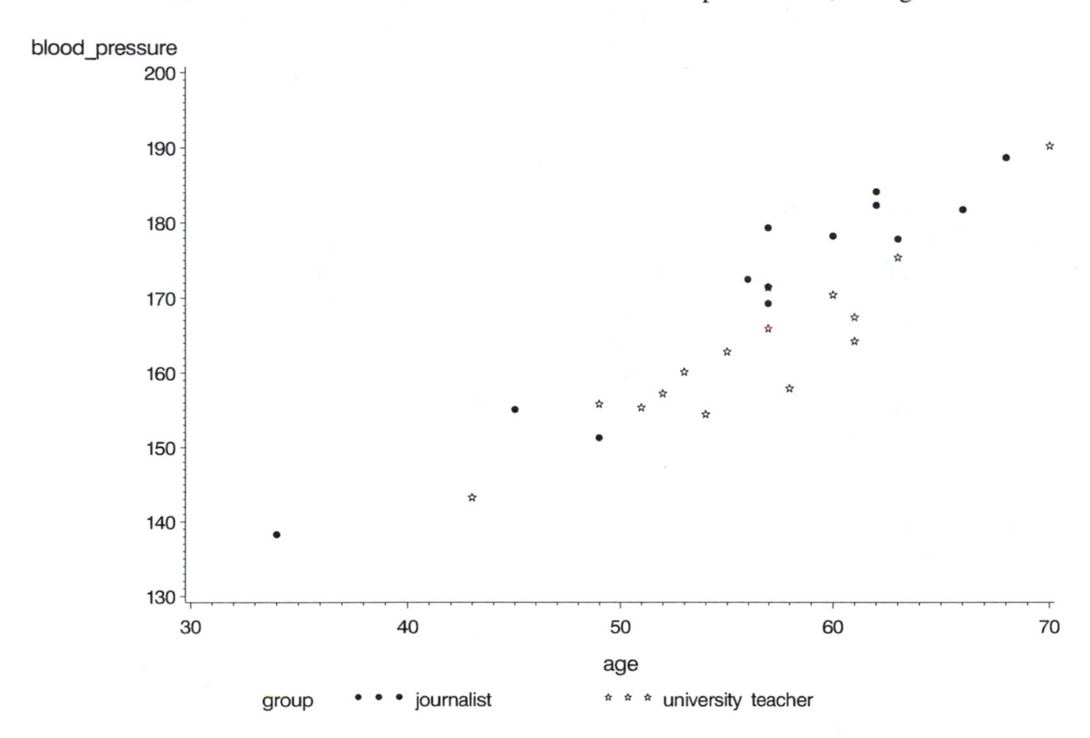

Figure 3.7 *Scatter plots of systolic blood pressure plotted against age for 13 journalists and 15 university teachers.*

We will introduce a model, the linear regression model, which can be used to describe how one variable depends linearly on another variable and use it in the present example to describe how blood pressure depends linearly on age. We shall consider the two professions separately in this chapter, but in Chapter 4 we shall compare the blood pressure for the two professions.

We will consider age as a known or deterministic quantity, which we will denote by t. In contrast, blood pressure is considered to be a realization of a random variable. In this example the blood pressures are considered to be observations that are drawn from the population of blood pressures of males of the same age and profession. So far the observations that are realizations of random variables have been denoted by x and we will keep this tradition. Data for which the linear regression model may be used consist of pairs of observations, $(t_i, x_i), i = 1, \ldots, n$.

Figure 3.8 is similar to Figure 3.7 but shows, in addition, straight lines through the scatter of points for each profession. The lines in Figure 3.8 are actually regression lines fitted by the method of linear regression but lines fitted by eye will do just as well for this initial discussion of whether the regression model fits the data.

For both professions the scatter of points lie closely around the straight line and therefore one decides to work with the linear regression model for each profession:

$$x_i \sim\sim N(\alpha + \beta t_i, \sigma^2), \qquad i = 1, \ldots, n.$$

Table 3.6 *Systolic blood pressure and age for 13 journalists and 15 university teachers*

Systolic blood pressure in mm Hg			
Journalists		*University teachers*	
Age	*Blood pressure*	*Age*	*Blood pressure*
68	188.7	55	162.9
62	184.2	49	155.9
49	151.3	60	170.5
62	182.4	54	154.5
34	138.3	58	158.0
66	181.8	51	155.4
45	155.1	43	143.3
60	178.3	52	157.3
57	179.4	53	160.2
57	171.5	61	164.3
63	177.9	61	167.5
56	172.5	57	171.5
57	169.3	57	166.0
		70	190.3
		63	175.5

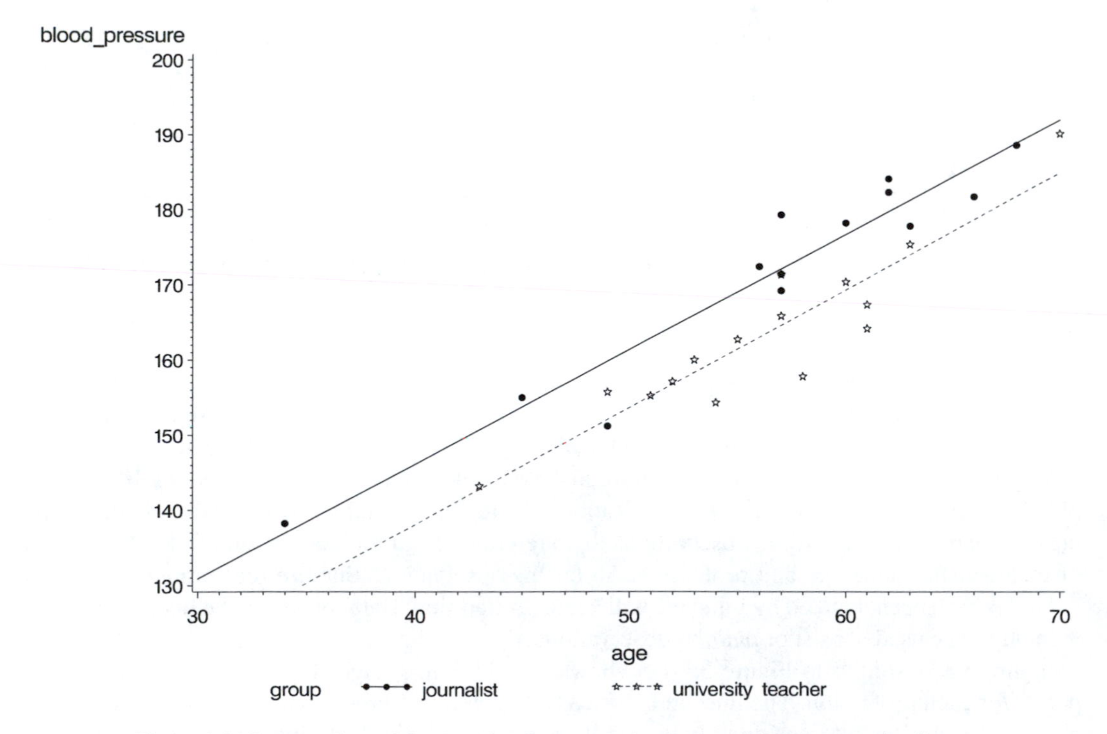

Figure 3.8 *Scatter plots of the data for the two professions with fitted regression lines.*

We now leave Example 3.4 for a moment to consider the linear regression model in general. □

3.3.1 The Linear Regression Model

The model of linear regression corresponding to the data pairs $(t_i, x_i), i = 1, \ldots, n$ is

$$M_2\colon X_i \sim N(\alpha + \beta t_i, \sigma^2), \qquad i = 1, \ldots, n, \tag{3.48}$$

and the X_i's are independent. The model is referred to as a *linear regression* of x on t. The line $x(t) = \alpha + \beta t$ is called the *regression line*. The parameter α is the intercept of the vertical axis corresponding to $t = 0$ and β is the slope of the regression line. The parameter α is called the *intercept* or sometimes the *position* and β is called the *slope* or the *regression coefficient*. The variable x is called the *dependent* variable and t is referred to as the *independent* variable. This manner of speech is inspired from considering x as a function of t and independence in this context should not be confused with independence of random variables. Alternatively, x will be called the *response* and t will be called the *explanatory* variable.

Two points are needed to determine a straight line and therefore the regression model cannot be considered unless at least two different values of the explanatory variable are available.

Model Checking

The regression model (3.48) has three characteristic features. First, the mean of an observation is a linear function of the explanatory variable, t. Secondly, the variance is the same for all observations. Thirdly, the observations are normally distributed. Standard methods of checking the model focus on those three features.

The assumption of linearity is checked with a scatter plot. Here we try to decide if the points scatter around a straight line. If the scatter of points has a tendency to curve, for example, in the shape of a banana, it may be an indication that the means of the responses are not a linear function of t. If the observations have a constant variance, the points will lie in an area of approximately constant width around the line. In judging this, one must bear in mind that the spread around the line is a function of the variance but also of the number of observations. In cases with more observations in the center of the range of t than at the ends, a scatter of points that looks like an ellipse may be entirely consistent with the assumption of constant variance. A typical and frequently encountered deviation from the hypothesis of constant variance is that the variance increases with t or decreases with t, so that the observations spread out more for large or for small values of t.

Later in Section 3.3.8 we will see how, under special circumstances, both the hypothesis of the mean being a linear function of t and the hypothesis of a constant variance may be tested. But often a graphical check of the model is the only possibility.

The third feature of the model is the normal distribution. If the data are very extensive and contain several pairs of observations with the same value of the explanatory variable, normal fractile diagrams may be used to check the assumption of normality. Otherwise, the check is based on the *residuals* in the linear regression model. In general, a residual is what is left unexplained by the model and in all models based on the normal distribution the residuals are the difference between the observations and the estimated means of the observations in the model. For the regression model the residuals are

$$r_i = x_i - \hat{\alpha} - \hat{\beta} t_i, \qquad i = 1, \ldots, n, \tag{3.49}$$

where $\hat{\alpha}$ and $\hat{\beta}$ denote the estimates of α and β to be derived in Section 3.3.2. We will refer to the residuals in (3.49) as the *raw residuals*. It can be shown that the raw residuals do not have a common variance, so we prefer to standardize the residuals to have a common variance before they are used for any type of plots. This leads to the *scaled residuals* that are defined in (3.59). The assumption of a normal distribution can be checked making a probit diagram based on the scaled residuals.

This is not a very sensitive check and the residuals are much more useful in other aspects of model checking.

A plot of the residuals as a function of the estimated mean is a useful check of the assumption a constant variance, because when the variance is not constant, it is often an increasing function of the means of the observations. With only one explanatory variable this plot is equivalent to the scatter plot of the observations as a function of the explanatory variable as in Figure 3.7 and it is not shown here. When more than one variable are considered for inclusion in the model as an explanatory variable, it is useful to make scatter plots of the residuals against variables that are considered for inclusion in the model and also against variables that are already included in the model. If the model is satisfactory the residuals will be distributed evenly around zero. Deviations from that pattern for a variable is an indication that it should be included in the model if it is not already one of the explanatory variables, and if it is an explanatory variable a function of it may need to be included as well. In the latter case the scatter plot will indicate what kind of function will be needed. If the points scatter around a parabola, for example, the squared explanatory variable should be included.

In addition to checking aspects of the model, residuals are also useful when checking for errors in the observations. Observing an atypically large residual may be an indication that the observation is in error in some way and the protocol should be consulted for a possible explanation. The easiest explanation to deal with is if the observation was not recorded correctly.

Residuals are so important in statistical analysis that many variations have been suggested, and some of them are explained in Section 3.3.5.

It is beyond the scope of the this book to consider regression models with several explanatory variables, usually referred to as multiple regression, but after the careful treatment of the simple regression model in this section and the linear normal models in the next chapter, the reader should be able to use a computer package to fit multiple regression models and to test reductions in multiple regression models.

Example 3.4 (Continued)
In order to illustrate the application of residuals in model checking, we have fitted the regression of blood pressure on age without taking the professions into account and computed the scaled residuals (3.59). The estimated regression line and the data are shown in Figure 4.6 on page 192. Figure 3.9 shows a probit diagram of the scaled residuals and this probit diagram is consistent with the assumption of normally distributed observations. In Figure 3.10 the scaled residuals are plotted against profession. If the model were correct, the residuals should be distributed symmetrically around 0 regardless of profession, but this is clearly not the case and Figure 3.10 is a clear indication that the profession should be included in the model. □

Interpretation of the Parameters

The variance σ^2 is an important parameter in the regression model that quantifies the variation around the regression line. Usually, however, the parameters in the regression line are at the center of interest. The interpretation of the parameters of the regression line is fairly obvious and it is reflected in the names used for the parameters: slope and intercept. The slope β is the change in the mean of the response for each unit increase of the explanatory variable. The intercept α is the mean of the response at the value zero of the explanatory variable. Although the interpretation of β is seems simple, care should be taken to consider the collection of the data before the interpretation of the slope is formulated. The data in Example 3.4 consist of the recording of the blood pressure once for a number of men so the interpretation of β is that it is the difference in mean blood pressure between individuals of a certain age and individuals who are one year younger. This may not be the same as the expected change in blood pressure for an individual for each year as he ages.

Similarly, care should be taken to interpret the intercept α as the mean of the response at the value

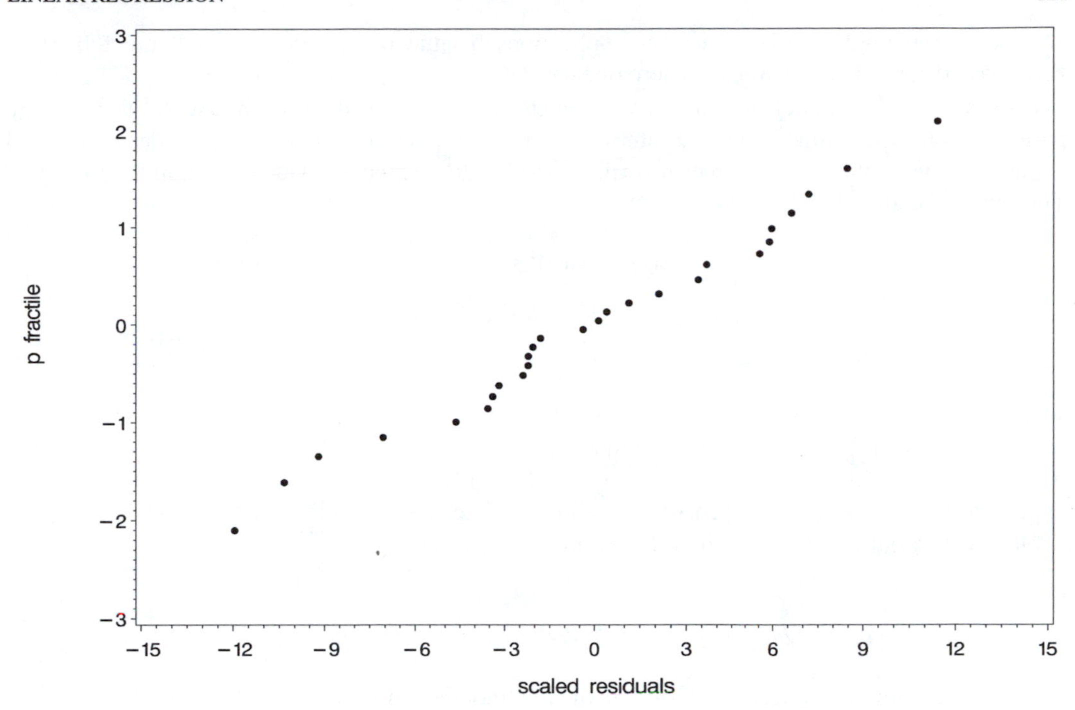

Figure 3.9 *Probit diagram of scaled residuals from a linear regression of blood pressure on age ignoring the professions. This plot does not contradict the assumption of normally distributed errors.*

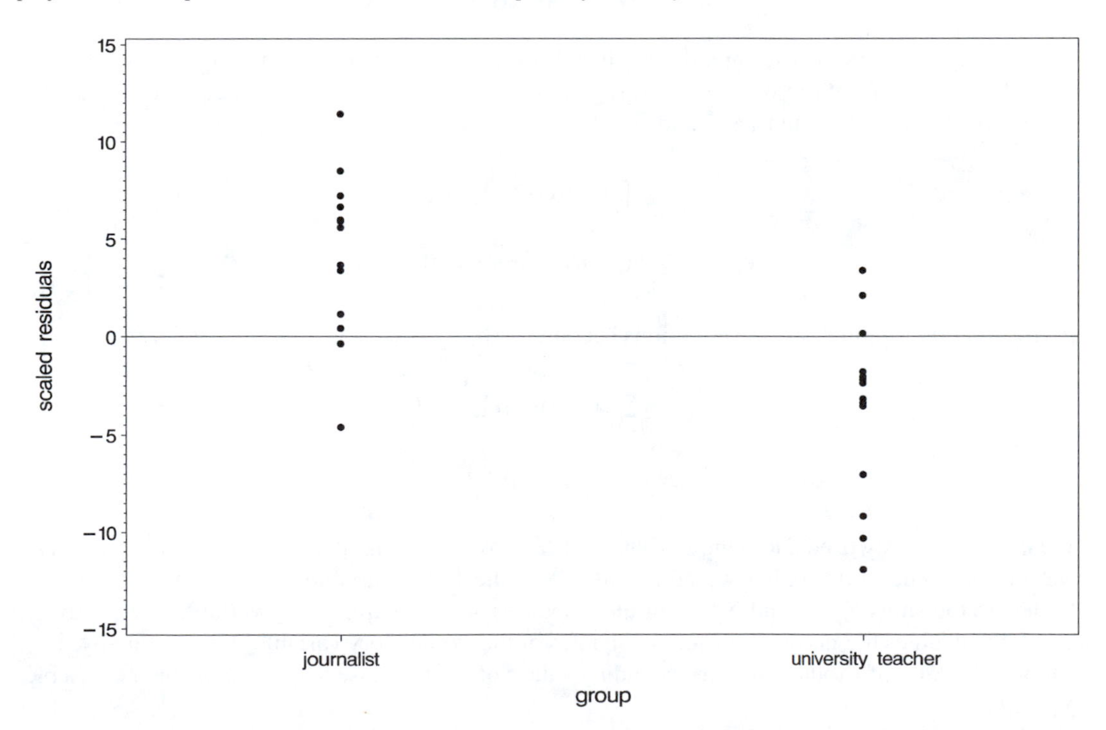

Figure 3.10 *Scaled residuals from a linear regression of blood pressure on age plotted against profession. The profession was not included in the model but the residuals tend to be positive for journalists and negative for university teachers and this indicates that profession should be included in the model.*

zero of the explanatory variable. In some applications this may be a meaningless statement, because zero may be far outside the region where one has data on the explanatory variable.

Unless the slope β is clearly the only parameter of interest, it does seem a waste to have one parameter out of two that cannot be interpreted. Therfore one may choose to consider a modified explanatory variable $t - t_0$, which is the original explanatory variable t with a constant t_0 subtracted. The regression line then looks like

$$\begin{aligned} x(t) &= \alpha + \beta t \\ &= \alpha + \beta t_0 + \beta(t - t_0) \\ &= \alpha_{t_0} + \beta(t - t_0), \end{aligned}$$

so when the explanatory variable $t - t_0$ is used the intercept is $\alpha_{t_0} = \alpha + \beta t_0$, which is the mean of responses with the value t_0 of the original explanatory variable. Here t_0 is chosen in the region where the regression model is assumed to be valid. In Example 3.4 one choice for the modified explanatory variable might be "years of age above 50," i.e., $t - 50$, and the intercept when using this explanatory variable is the mean blood pressure at 50 years of age.

3.3.2 Estimation in the Linear Regression Model

The estimation in the model M_2 is based on the likelihood method. The log likelihood function is

$$l(\alpha, \beta, \sigma^2) = -\frac{n}{2}\ln(2\pi\sigma^2) - \frac{1}{2\sigma^2}\sum_{i=1}^{n}\{x_i - (\alpha + \beta t_i)\}^2.$$

First, σ^2 is considered as fixed and the log likelihood function is maximized as a function of α and β. Differentiating l with respect to α and β gives the following two likelihood equations that need to be solved to find the estimates $\hat{\alpha}$ and $\hat{\beta}$:

$$\begin{aligned} &\sum_{i=1}^{n}\{x_i - (\alpha + \beta t_i)\} = 0 \\ &\sum_{i=1}^{n}\{x_i - (\alpha + \beta t_i)\}t_i = 0. \end{aligned}$$

Performing the summations, the equations become

$$\begin{aligned} \sum_{i=1}^{n} x_i - n\alpha - \beta\sum_{i=1}^{n} t_i &= 0 \\ \sum_{i=1}^{n} x_i t_i - \alpha\sum_{i=1}^{n} t_i - \beta\sum_{i=1}^{n} t_i^2 &= 0. \end{aligned}$$

We have previously used S to denote a sum and USS to denote an uncorrected sum of squares and we will continue to do so, but we add a suffix to indicate the variable in question. Thus, S_x and S_t denote the sums $\sum_{i=1}^{n} x_i$ and $\sum_{i=1}^{n} t_i$ of the responses and the explanatory variables, respectively, and USS_t denotes the uncorrected sum of squares of the explanatory variable, $\sum_{i=1}^{n} t_i^2$. Finally, SP_{xt}, denotes the *S*um of *P*roducts of corresponding values of the response and the explanatory variable, $\sum_{i=1}^{n} x_i t_i$.

With this notation the equations above can be written as

$$S_x - n\alpha - \beta S_t = 0 \tag{3.50}$$

$$SP_{xt} - \alpha S_t - \beta USS_t = 0. \tag{3.51}$$

If equation (3.50) is multiplied by S_t/n and subtracted from equation (3.51), one obtains

$$\frac{S_xS_t}{n} - \alpha S_t - \beta\frac{S_t^2}{n} = 0$$

$$SP_{xt} - \frac{S_xS_t}{n} - \beta(USS_t - \frac{S_t^2}{n}) = 0. \tag{3.52}$$

Equation (3.52) can now be solved with respect to β to give the solution,

$$\hat{\beta} = \frac{SP_{xt} - \frac{S_xS_t}{n}}{USS_t - \frac{S_t^2}{n}} = \frac{SPD_{xt}}{SSD_t}, \tag{3.53}$$

where $SSD_t > 0$ because at least two pairs of observations have different values of the explanatory variable. If $\hat{\beta}$ is substituted for β in equation (3.50), which is then solved with respect to α, the result is

$$\hat{\alpha} = \frac{S_x}{n} - \hat{\beta}\frac{S_t}{n} = \bar{x}. - \hat{\beta}\bar{t}.. \tag{3.54}$$

Note that neither $\hat{\alpha}$ nor $\hat{\beta}$ depends on σ^2 so the maximum likelihood estimate, $\hat{\sigma}^2$, of σ^2 is obtained by finding the value of σ^2 that maximizes

$$l(\hat{\alpha},\hat{\beta},\sigma^2) = -\frac{n}{2}\ln(2\pi\sigma^2) - \frac{1}{2\sigma^2}\sum_{i=1}^{n}\{x_i - (\hat{\alpha} + \hat{\beta}t_i)\}^2$$

with respect to σ^2. This gives

$$\hat{\sigma}^2 = \frac{1}{n}\sum_{i=1}^{n}\{x_i - (\hat{\alpha} + \hat{\beta}t_i)\}^2.$$

It is beyond the scope of this book to prove results about the distribution of $\hat{\sigma}^2$ but we will certainly use them. Now

$$n\hat{\sigma}^2 = SSD_{02} = \sum_{i=1}^{n}\{x_i - (\hat{\alpha} + \hat{\beta}t_i)\}^2 \sim\sim \sigma^2\chi^2(n-2),$$

and therefore

$$s_{02}^2 = \frac{1}{n-2}\sum_{i=1}^{n}\{x_i - (\hat{\alpha} + \hat{\beta}t_i)\}^2 \tag{3.55}$$

is preferred as an estimate of σ^2, because it has the nice properties that it has the correct mean σ^2, and that it is a realization of a $\sigma^2\chi^2(f_{02})/f_{02}$-distributed random variable with $f_{02} = n-2$. Note that *the degrees of freedom of the estimate of variance are equal to the number of observations minus the number of unknown parameters in the mean*. This is a general rule that will be used frequently in this and the following chapter.

3.3.3 Computations

In the expression for $\hat{\beta}$ in (3.53), the notation SPD_{xt} is used for $SP_{xt} - S_xS_t/n$. SPD is an abbreviation of *S*um of *P*roducts of *D*eviations and that is exactly what SPD_{xt} is, for

$$\begin{aligned} SPD_{xt} &= \sum_{i=1}^{n}(x_i - \bar{x}.)(t_i - \bar{t}.) \\ &= \sum_{i=1}^{n}x_it_i - \sum_{i=1}^{n}x_i\bar{t}. - \sum_{i=1}^{n}\bar{x}.t_i + n\bar{x}.\bar{t}. \\ &= SP_{xt} - S_x\frac{S_t}{n} - \frac{S_x}{n}S_t + n\frac{S_x}{n}\frac{S_t}{n} \\ &= SP_{xt} - \frac{S_xS_t}{n}. \end{aligned}$$

The formulas (3.53) and (3.54) show how $\hat{\beta}$ and $\hat{\alpha}$ can be calculated by hand. To find a formula for calculating SSD_{02} by hand, note first that

$$\hat{\alpha} + \hat{\beta} t_i = \bar{x}. - \hat{\beta}\bar{t}. + \hat{\beta} t_i = \bar{x}. + \hat{\beta}(t_i - \bar{t}.),$$

so SSD_{02} can be rewritten in the following way:

$$\begin{aligned}
SSD_{02} &= \sum_{i=1}^{n} \{x_i - (\hat{\alpha} + \hat{\beta} t_i)\}^2 \\
&= \sum_{i=1}^{n} \{x_i - \bar{x}. - \hat{\beta}(t_i - \bar{t}.)\}^2 \\
&= \sum_{i=1}^{n} \{(x_i - \bar{x}.)^2 + \hat{\beta}^2 (t_i - \bar{t}.)^2 - 2\hat{\beta}(x_i - \bar{x}.)(t_i - \bar{t}.)\} \\
&= SSD_x + \hat{\beta}^2 SSD_t - 2\hat{\beta} SPD_{xt} \\
&= SSD_x - \frac{SPD_{xt}^2}{SSD_t}. \qquad (3.56)
\end{aligned}$$

Going through the computation formulas (3.53), (3.54), and (3.56), it should be noted that once the five basic sums S_x, S_t, USS_x, USS_t, and SP_{xt} are available, it is easy to calculate $\hat{\beta}$, $\hat{\alpha}$, and s_{02}^2.

The formulas are summarized in the template for calculations in Table 3.7.

Example 3.4 (Continued)

In Table 3.8 and Table 3.9 the templates for calculations in linear regression are filled in with the data for the two professions. The estimates of the parameters in the regressions models are found in Table 3.8 and Table 3.9 and in addition $\hat{\alpha}_{50}$ is calculated as $\hat{\alpha}_{50} = \hat{\alpha} + \hat{\beta} \times 50$.

Journalists:					University teachers:				
$\hat{\beta}$	=	1.529568	$\rightarrow$	β	$\hat{\beta}$	=	1.562384	$\rightarrow$	β
$\hat{\alpha}$	=	84.995220	$\rightarrow$	α	$\hat{\alpha}$	=	75.629862	$\rightarrow$	α
$\hat{\alpha}_{50}$	=	161.473620	$\rightarrow$	α_{50}	$\hat{\alpha}_{50}$	=	153.749062	$\rightarrow$	α_{50}
s_{02}^2	=	18.14949	$\rightarrow$	σ^2	s_{02}^2	=	19.44953	$\rightarrow$	σ^2

The regression coefficient, β, is the change in the mean of the response, x, for a unit change in the explanatory variable, t. Here β is estimated to be 1.529568 and 1.562384 mm Hg year^{-1} for the two professions. Although this is a clear statement, it is likely that it will nevertheless be misunderstood, for β is not the mean increase in blood pressure for an individual for each year as he ages. It is the difference in mean blood pressure between individuals of a certain age and individuals who are one year younger. If we wanted the first interpretation, we would have to make a different data collection in which the same individuals were followed over time. For the present data we have only one observation on each individual.

The parameter α is the intercept of the vertical axis corresponding to $t = 0$. If the range of the explanatory variable does not contain 0, it may be meaningless to interpret α except as a parameter that is necessary to fix the position of the regression line. In this example α should be interpreted as the blood pressure at birth for a future journalist or university teacher. Clearly, this is a very dubious interpretation. For these data, 0 is far outside the range of the explanatory variable and the regression model may not have any justification outside the range of the explanatory variable.

Using the modified explanatory variable *age*$-$50 the intercept is mean blood pressure at 50 years of age and the estimate is found as $\hat{\alpha}_{50} = \hat{\alpha} + \hat{\beta} \times 50$. The estimates have been printed above with an excessive number of decimal places compared to the precision with which they have been determined in the experiment. But before we can decide how well the estimates have been determined in the experiment we need to know their distribution. □

Table 3.7 *Template for calculations in linear regression*

	x	t
n	n	
S	S_x	S_t
USS	USS_x	USS_t
SP	SP_{xt}	
SSD	$USS_x - \frac{S_x^2}{n}$	$USS_t - \frac{S_t^2}{n}$
SPD	$SP_{xt} - \frac{S_x S_t}{n}$	
$\hat{\beta}$	$\frac{SPD_{xt}}{SSD_t}$	
$\hat{\alpha}$	$\frac{1}{n}\left[S_x - \hat{\beta} S_t\right]$	
SSD_{02}	$SSD_x - \frac{SPD_{xt}^2}{SSD_t}$	
s_{02}^2	$\frac{1}{n-2} SSD_{02}$	

3.3.4 Distribution of Parameter Estimates

The distributions of the interesting estimates in the model are given below:

$$\hat{\alpha} \sim\sim N\left(\alpha, \sigma^2\left(\frac{1}{n} + \frac{\bar{t}_{\cdot}^2}{SSD_t}\right)\right)$$

$$\hat{\beta} \sim\sim N\left(\beta, \frac{\sigma^2}{SSD_t}\right)$$

$$\hat{\alpha} + \hat{\beta} t \sim\sim N\left(\alpha + \beta t, \sigma^2\left(\frac{1}{n} + \frac{(t - \bar{t}_{\cdot})^2}{SSD_t}\right)\right)$$

$$s_{02}^2 \sim\sim \sigma^2 \chi^2(f_{02})/f_{02},$$

where $f_{02} = n - 2$. Furthermore, it is important to know that the distribution of $(\hat{\alpha}, \hat{\beta})$ is independent of the distribution of s_{02}^2, and that the distributions of $\hat{\beta}$ and $\bar{x}_{\cdot}$ are independent.

It is beyond the scope of this book to prove that $s_{02}^2 \sim\sim \sigma^2\chi^2(f_{02})/f_{02}$, that the distributions of s_{02}^2 and $(\hat{\alpha}, \hat{\beta})$ are independent, and that the distributions of $\hat{\beta}$ and $\bar{x}_{\cdot}$ are independent. But using

Table 3.8 *The template for calculations with data from 13 journalists*

	Journalists x	t
n	13	
S	2230.7	736
USS	385387.57	42702
SP	127872.1	
SSD	2616.6092	1033.0769
SPD	1580.161538	
$\hat{\beta}$	1.529568	
$\hat{\alpha}$	84.995220	
SSD_{02}	199.644416	
s_{02}^2	18.14949	

Table 3.9 *The template for calculations with data from 15 university teachers*

	University teachers x	t
n	15	
S	2453.1	844
USS	402870.43	48078
SP	138947.9	
SSD	1690.4560	588.9333
SPD	920.1400	
$\hat{\beta}$	1.562384	
$\hat{\alpha}$	75.629862	
SSD_{02}	252.843932	
s_{02}^2	19.44953	

those results and the fact that the distribution of a sum of independent normally distributed random variables is again normally distributed, elementary calculations with means and variances show that the estimates $\hat{\alpha}$, $\hat{\beta}$ and $\hat{\alpha}+\hat{\beta}t$ have the distributions given above.

Note first that

$$SPD_{xt} = \sum_{i=1}^{n}(x_i - \bar{x}.)(t_i - \bar{t}.) = \sum_{i=1}^{n} x_i(t_i - \bar{t}.),$$

so that

$$
\begin{aligned}
SPD_{xt} &\sim\sim N\left(\sum_{i=1}^{n}(\alpha+\beta t_i)(t_i-\bar{t}.),\ \sum_{i=1}^{n}\sigma^2(t_i-\bar{t}.)^2\right)\\
&= N\left(\beta\sum_{i=1}^{n}t_i(t_i-\bar{t}.),\ \sigma^2\sum_{i=1}^{n}(t_i-\bar{t}.)^2\right)\\
&= N\left(\beta\, SSD_t, \sigma^2 SSD_t\right).
\end{aligned}
$$

Now $\hat{\beta} = SPD_{xt}/SSD_t$ and therefore

$$\hat{\beta} \sim\sim N(\beta, \frac{\sigma^2}{SSD_t}).$$

In order to find the distribution of $\hat{\alpha} = \bar{x}. - \hat{\beta}\bar{t}.$ we note that

$$\bar{x}. \sim\sim N(\alpha + \beta\bar{t}., \frac{\sigma^2}{n})$$

and that

$$\hat{\beta}\bar{t}. \sim\sim N(\beta\bar{t}., \frac{\sigma^2\bar{t}^2_.}{SSD_t}).$$

Because the distributions of $\hat{\beta}$ and $\bar{x}.$ are independent, we obtain the distribution of the estimate of the intercept,

$$\hat{\alpha} = \bar{x}. - \hat{\beta}\bar{t}. \sim\sim N\left(\alpha, \sigma^2(\frac{1}{n} + \frac{\bar{t}^2_.}{SSD_t})\right).$$

Since $\hat{\alpha} + \hat{\beta}t = \bar{x}. + \hat{\beta}(t - \bar{t}.)$, we obtain, again because of the independence of $\hat{\beta}$ and $\bar{x}.$, that the distribution of the estimated regression line at t is

$$\hat{\alpha} + \hat{\beta}t = \bar{x}. + \hat{\beta}(t - \bar{t}.) \sim\sim N\left(\alpha + \beta t, \sigma^2(\frac{1}{n} + \frac{(t - \bar{t}.)^2}{SSD_t})\right).$$

The variances of the estimates are interesting.

The variance of $\hat{\beta}$ is σ^2/SSD_t so the larger the SSD_t is, the smaller the variance is, and the better the estimate of β is. The SSD_t is a measure of the variation or the range of the explanatory variable. Broadly speaking, the larger the range of the explanatory variable the greater SSD_t is and the better the estimate of β is. This should be kept in mind in the planning of an experiment that involves a linear regression, if the experimenter is free to choose the values of the explanatory value t.

The variance of $\hat{\alpha}$ supplements the previous warnings that it may be meaningless to try to interpret α when the range of the explanatory variables is far from 0. In those cases the factor $\bar{t}^2_.$ in the expression of the variance of $\hat{\alpha}$ will make the variance so large that it will be obvious that the estimate of the intercept is virtually useless.

The variance of $\hat{\alpha} + \hat{\beta}t$ depends on the explanatory variable t through $(t - \bar{t}.)^2$. The variance attains its minimum value for $t = \bar{t}.$, and at that point the variance is σ^2/n, agreeing with the fact that $\hat{\alpha} + \hat{\beta}\bar{t}. = \bar{x}.$.

The regression line $\hat{\alpha} + \hat{\beta}t$ is most precisely determined, i.e., it has the smallest variance, for values of t in the middle of the range of the explanatory variable.

3.3.5 Residuals

The application of residuals was briefly mentioned in the section on *Model Checking* on page 119 and illustrated using Example 3.3.1 on page 120. Several slightly different definitions of a residual have been suggested and are now being offered as options in statistical packages. In this section the connection between the residuals is explained and we recommend concentrating on at most two of the residuals. This section may be omitted at a first reading.

Residuals are in general used to

- check the distribution of the observations
- check the model for the means
- check for errors in the observations

The residuals are a very important tool for these purposes so a closer look at the residuals and their properties is called for. We will consider the residuals in the linear regression model, but the results hold for a large class of models of great practical importance, and this will be explained at the end of this section.

The raw residuals are in general defined as

$$r_i = x_i - \hat{\mu}_i, \qquad i = 1, \ldots, n, \tag{3.57}$$

where $\hat{\mu}_i$ is the estimate of the mean of x_i and in the linear regression model it is

$$\hat{\mu}_i = \hat{\alpha} + \hat{\beta} t_i, \qquad i = 1, \ldots, n.$$

In the rest of this section, the index i runs through the sample and all quantities indexed by i are defined for all observations in the sample.

The raw residuals as defined in (3.57) are not independent and they are not identically distributed. It may be proved that

$$r_i \sim\sim N(0, \sigma^2(1 - h_i)),$$

where h_i is denoted as the *leverage* and in the linear regression model it is

$$h_i = \frac{1}{n} + \frac{(t_i - \bar{t}.)^2}{SSD_t}. \tag{3.58}$$

This may be used to define the *scaled residual* for the ith observation as

$$r_i^{\dagger} = \frac{r_i}{\sqrt{(1 - h_i)}}. \tag{3.59}$$

The scaled residuals are normally distributed with a common variance,

$$r_i^{\dagger} \sim\sim N(0, \sigma^2),$$

but they are still not independent. The distribution of the scaled residuals depends on the unknown σ^2 and this has led to the introduction of the *studentized scaled residuals,*

$$r_i^{\ddagger} = \frac{r_i}{\sqrt{s^2(1 - h_i)}}, \tag{3.60}$$

where s^2 denotes the estimate of the variance in the model. The term *studentized* is used in general for the result of the operation where the influence of an unknown variance σ^2 is eliminated by division with the square root of an *independent* estimate of σ^2, which has a $\sigma^2\chi^2(f)/f$-distribution in order to obtain a quantity which has a t-distribution. It is a slight problem here that the raw residuals r_i are not independent of s^2 so the studentized scaled residuals are $r_i^{\ddagger}$ are only approximately $t(f)$-distributed.

Moreover, neither the scaled residuals nor the studentized scaled residuals are independent.

The names used for the residuals have not been standardized in the literature. The residual in (3.60) has been referred to both as a "standardized residual" and as "studentized residual" in the literature.

In particular when checking for potential errors in the observations, the *deletion residuals* are useful and intuitively appealing. The concern is that an outlying observation may influence the estimation so much that the residual becomes small so it will not reveal the outlier. Therefore the mean of the ith observation is estimated without using x_i. The raw deletion residuals that corresponds to the raw residuals in (3.57) may be expressed as

$$r_{(i)} = x_i - \hat{\mu}_{(i)} \tag{3.61}$$

and in the linear regression model it is

$$\hat{\mu}_{(i)} = \hat{\alpha}_{(i)} + t_i \hat{\beta}_{(i)}.$$

Here the index "(i)" on $\hat{\mu}_{(i)}$, $\hat{\alpha}_{(i)}$ and $\hat{\beta}_{(i)}$ is used to remind the reader that the ith observation has not been used in the estimation. This intuitively reasonable idea has the further advantage that the distribution of the studentized scaled deletion residuals to be defined below in (3.64) has an exact t-distribution in contrast to the residuals defined in (3.60).

Note first that x_i and $\hat{\mu}_{(i)}$ are independent because they are based on independent subsets of the n independent observations so

$$r_{(i)} = x_i - \hat{\mu}_{(i)} \sim\sim N(0, \sigma^2(1 + h_{(i)})).$$

Here $h_{(i)}$ comes from the variance of $\hat{\mu}_{(i)}$, which is $\sigma^2 h_{(i)}$. The expression for $h_{(i)}$ is very similar to the expression for h_i, and in the linear regression model it is

$$h_{(i)} = Var(\hat{\alpha}_{(i)} + t_i \hat{\beta}_{(i)}) = \frac{1}{n-1} + \frac{(t_i - \bar{t}_{\bullet(i)})^2}{SSD_{t(i)}},$$

where the index "(i)" on $\bar{t}_{\bullet(i)}$ and $SSD_{t(i)}$ indicates that t_i has not been used in the calculations of those quantities. It is remarkable that it can be proved that $h_{(i)}$ can be expressed very easily in terms of h_i as

$$h_{(i)} = \frac{h_i}{1-h_i}. \tag{3.62}$$

The scaled version of the deletion residual $r_{(i)}$ is defined as

$$r^{\dagger}_{(i)} = \frac{x_i - \hat{\mu}_{(i)}}{\sqrt{1 + h_{(i)}}}. \tag{3.63}$$

Finally, the studentized scaled version of $r_{(i)}$ is defined as

$$r^{\ddagger}_{(i)} = \frac{x_i - \hat{\mu}_{(i)}}{\sqrt{s^2_{(i)}(1 + h_{(i)})}}, \tag{3.64}$$

where $s^2_{(i)}$ is the estimate of the variance based on the data without the ith observation. Therefore $s^2_{(i)} \sim\sim \sigma^2 \chi(n-1-2)/(n-1-2)$ and $s^2_{(i)}$ is independent of $r_{(i)} = x_i - \hat{\alpha}_{(i)} - t_i \hat{\beta}_{(i)}$, so $r^{\ddagger}_{(i)}$ is exactly t-distributed:

$$r^{\ddagger}_{(i)} \sim\sim t(n-1-2).$$

There is a very simple relation between the raw residuals r_i in (3.49) and the raw deletion residuals $r_{(i)}$ in (3.61), which is

$$r_{(i)} = \frac{r_i}{1-h_i}. \tag{3.65}$$

This relation together with (3.62) shows that the scaled residuals $r^{\dagger}_i$ in (3.59) and the scaled deletion residuals $r^{\dagger}_{(i)}$ in (3.63) are identical; for

$$r^{\dagger}_{(i)} = \frac{r_{(i)}}{\sqrt{1 + \frac{h_i}{1-h_i}}} = r_{(i)}\sqrt{1-h_i} = \frac{r_i}{\sqrt{1-h_i}} = r^{\dagger}_i, \tag{3.66}$$

and we will refer to them simply as *scaled residuals*.

There is also a fairly simple relation between the studentized scaled residuals $r_i^\ddagger$ in (3.60) and the studentized scaled deletion residuals $r_{(i)}^\ddagger$ in (3.64), which is

$$r_{(i)}^\ddagger = \frac{r_i^\ddagger}{\left(\frac{n-2-(r_i^\ddagger)^2}{n-1-2}\right)^{1/2}}. \quad (3.67)$$

The relations (3.65), (3.66), and (3.67) follow from elementary but rather long calculations. It is interesting and useful that the various deletion residuals may be computed using quantities that are available after the estimation in the regression model using all the observations, so it is not necessary to estimate in the regression models n times without the ith observation.

Having noticed that the scaled residuals in (3.59) and the scaled deletion residuals in (3.63) are identical, five different residuals have been defined. We recommend using at most two of those residuals, namely either the scaled residuals, $r_i^\dagger$, or the studentized scaled deletion residuals, $r_{(i)}^\ddagger$, and it is a matter of personal preferences which of the two will be preferred. The normally distributed scaled residuals are slightly easier to use in a routine fashion because it is easier to make fractile plots based on the normal distribution than plots based on the t-distribution.

The assumption of a normal distribution may be checked with fractile plots for the residuals. The fractile plots may be based either on the normal distribution for the scaled residuals or the $t(n-1-2)$-distribution for the studentized scaled residuals. In addition, the residuals are used in plots as described in the section on model checking on page 119.

In the beginning of this section, it was mentioned that the results for the residuals apply to a large class of models of great practical importance. This class of models is the linear normal models, where the means of the observations depend on d unknown parameters. This class of models is considered in Chapter 4 and all the models of Chapter 3 are examples of linear normal models. All the definitions of residuals above apply to the linear normal model, and so does the relation (3.65) between the raw residuals and the raw deletion residuals, and the identity between the scaled residuals and the scaled deletion residuals in (3.66). As for distributions, the scaled residuals are $N(0,\sigma^2)$-distributed and the studentized deletion residuals are $t(n-1-d)$-distributed. The degrees of freedom of the t-distribution reflects that the general model has d unknown parameters in the mean, whereas the linear regression has 2 unknown parameters in the mean. There is also a general version of the relation between the studentized scaled residuals and the studentized scaled deletion residuals in (3.67), which is

$$r_{(i)}^\ddagger = \frac{r_i^\ddagger}{\left(\frac{n-d-(r_i^\ddagger)^2}{n-1-d}\right)^{1/2}}.$$

The recommendation to use either the scaled residuals or the studentized scaled deletion residuals apply quite generally.

It may seem a waste of time to introduce six residuals and notice that two of them are identical and end up recommending the use of at most two of the five residuals. But every statistical package supplies many of these residuals, and a lot of time may be saved by realizing once and for all that in most situations, it makes no difference which residuals are used in the analysis and that the logical choice is either the scaled residuals or the studentized scaled deletion residuals.

3.3.6 Confidence Intervals for the Parameters in the Linear Regression Model

The formulas for the confidence intervals for the parameters α, β and $\alpha+\beta t$ have the general form of confidence intervals for parameters whose estimates are normally distributed with a variance $k\sigma^2$, where k is a known constant and σ^2 is unknown but an estimate s^2 of σ^2 is available, with a $\sigma^2\chi^2(f)/f$-distribution, independent of the distribution of the parameter estimate. If ξ denotes this parameter and the estimate is denoted by $\hat{\xi}$, the $(1-\alpha)$ confidence interval for ξ is given by the formula,

$$\left[\hat{\xi} - \sqrt{s^2 k}\, t_{1-\alpha/2}(f),\ \hat{\xi} + \sqrt{s^2 k}\, t_{1-\alpha/2}(f)\right].$$

Note that $\sqrt{s^2 k}$ is an estimate of the standard deviation of $\hat{\xi}$ and that statistical packages always give an estimate of the standard deviation of the estimate, often under the heading `Standard Error`, together with the value of the estimate.

The $(1-\alpha)$ confidence interval for the intercept α is

$$\left[\hat{\alpha} - \sqrt{s_{02}^2\left(\frac{1}{n} + \frac{\bar{t}_{\cdot}^2}{SSD_t}\right)}\, t_{1-\alpha/2}(f_{02}),\ \hat{\alpha} + \sqrt{s_{02}^2\left(\frac{1}{n} + \frac{\bar{t}_{\cdot}^2}{SSD_t}\right)}\, t_{1-\alpha/2}(f_{02})\right].$$

In actual calculations by hand or on a pocket calculator of the confidence interval for α, it is useful to notice that

$$\frac{1}{n} + \frac{\bar{t}_{\cdot}^2}{SSD_t} = \frac{USS_t}{nSSD_t},$$

and that n, USS_t and SSD_t are available in the template for computation in linear regression.

The $(1-\alpha)$ confidence interval for the slope β is

$$\left[\hat{\beta} - \sqrt{\frac{s_{02}^2}{SSD_t}}\, t_{1-\alpha/2}(f_{02}),\ \hat{\beta} + \sqrt{\frac{s_{02}^2}{SSD_t}}\, t_{1-\alpha/2}(f_{02})\right].$$

The $(1-\alpha)$ confidence interval for the value $\alpha + \beta t$ of the regression line at t is

$$\left[\hat{\alpha} + \hat{\beta}t - \sqrt{s_{02}^2\left(\frac{1}{n} + \frac{(t-\bar{t}_{\cdot})^2}{SSD_t}\right)}\, t_{1-\alpha/2}(f_{02}),\ \hat{\alpha} + \hat{\beta}t + \sqrt{s_{02}^2\left(\frac{1}{n} + \frac{(t-\bar{t}_{\cdot})^2}{SSD_t}\right)}\, t_{1-\alpha/2}(f_{02})\right].$$

$\alpha + \beta t$ is the mean of observations that has the value t of the explanatory variable. Occasionally the purpose of a linear regression model is to give estimates of the mean of observations for different values of the explanatory variable t. In those cases the $(1-\alpha)$ confidence interval for $\alpha + \beta t$ becomes interesting.

The 95% confidence interval for the regression line is plotted in Figure 3.11 based on the data for journalists in Example 3.4.

The $(1-\alpha)$ confidence interval for the variance σ^2 is

$$\left[\frac{s_{02}^2}{\chi^2_{1-\alpha/2}(f_{02})/f_{02}},\ \frac{s_{02}^2}{\chi^2_{\alpha/2}(f_{02})/f_{02}}\right],$$

where $\chi^2_{1-\alpha/2}(f_{02})$ and $\chi^2_{\alpha/2}(f_{02})$ are $1-\alpha/2$ and $\alpha/2$ fractiles for the χ^2-distribution with $f_{02} = n-2$ degrees of freedom. The derivation of the confidence interval for σ^2 is given on page 61 in connection with a single normal sample. This confidence interval may be transformed into the $(1-\alpha)$ confidence interval for the standard deviation σ:

$$\left[\sqrt{\frac{s_{02}^2}{\chi^2_{1-\alpha/2}(f_{02})/f_{02}}},\ \sqrt{\frac{s_{02}^2}{\chi^2_{\alpha/2}(f_{02})/f_{02}}}\right].$$

3.3.7 Prediction Interval for a New Observation

In some applications the regession model is used to predict a new observation corresponding to a fixed value of the explanatory variable. The prediction, if we were only allowed one guess, would be the estimated mean, but that is not satisfactory; for we want an answer that expresses the variability similar to what the confidence interval does for estimation.

Let Y denote the new observation corresponding to the value t of the explanatory variable. Y is independent of the estimated regression line $\hat{\alpha}+\hat{\beta}t$, so

$$Y-\hat{\alpha}-\hat{\beta}t \sim\sim N\left(0, \sigma^2(1+\frac{1}{n}+\frac{(t-\bar{t}.)^2}{SSD_t})\right)$$

and because s_{02}^2 is independent of $Y-\hat{\alpha}-\hat{\beta}t$ we find that

$$\frac{Y-\hat{\alpha}-\hat{\beta}t}{\sqrt{s_{02}^2\left(1+\frac{1}{n}+\frac{(t-\bar{t}.)^2}{SSD_t}\right)}} \sim\sim t(f_{02}),$$

and therefore the inequalities,

$$-t_{1-\alpha/2}(f_{02}) < \frac{Y-\hat{\alpha}-\hat{\beta}t}{\sqrt{s_{02}^2\left(1+\frac{1}{n}+\frac{(t-\bar{t}.)^2}{SSD_t}\right)}} < t_{1-\alpha/2}(f_{02})$$

hold with a probability of $1-\alpha$. The inequalities are solved for Y to give the $(1-\alpha)$ prediction interval,

$$\hat{\mu}(t)-\sqrt{s_{02}^2\left(1+\frac{1}{n}+\frac{(t-\bar{t}.)^2}{SSD_t}\right)}t_{1-\alpha/2}(f_{02}) < Y < \hat{\mu}(t)+\sqrt{s_{02}^2\left(1+\frac{1}{n}+\frac{(t-\bar{t}.)^2}{SSD_t}\right)}t_{1-\alpha/2}(f_{02}),$$

where $\hat{\mu}(t)=\hat{\alpha}+\hat{\beta}t$. The arguments leading to the prediction interval are completely analogous to the ones used to find a confidence interval and occasionally the prediction interval is referred to as a confidence interval of an individual observation. Note that the $(1-\alpha)$ prediction interval is wider than the $(1-\alpha)$ confidence interval for the regression line because it accounts for the variation in the observation as well as the variation of the estimated regression line.

Example 3.4 (Continued)
For journalists the 95% confidence interval for β is found to be

$$\begin{aligned}&\left[\hat{\beta}-\sqrt{\frac{s_{02}^2}{SSD_t}}\,t_{1-\alpha/2}(f_{02}),\ \hat{\beta}+\sqrt{\frac{s_{02}^2}{SSD_t}}\,t_{1-\alpha/2}(f_{02})\right]\\ &=[1.529568-0.132546\times 2.201,\ 1.529568+0.132546\times 2.201]\ \text{mmHg} \qquad (3.68)\\ &=[1.237834,\ 1.821302]\ \text{mmHg},\end{aligned}$$

and for university teachers the 95% confidence interval for β is found to be

$$\begin{aligned}&\left[\hat{\beta}-\sqrt{\frac{s_{02}^2}{SSD_t}}\,t_{1-\alpha/2}(f_{02}),\ \hat{\beta}+\sqrt{\frac{s_{02}^2}{SSD_t}}\,t_{1-\alpha/2}(f_{02})\right]\\ &=[1.562384-0.181728\times 2.160,\ 1.562384+0.181728\times 2.160]\ \text{mmHg} \qquad (3.69)\\ &=[1.169852,\ 1.954916]\ \text{mmHg}.\end{aligned}$$

In light of these confidence intervals, it will be appropriate to give the estimates of the slopes for the two professions as 1.53 mm Hg and 1.56 mm Hg, respectively.

The $(1-\alpha)$ confidence interval for the regression line at t for journalists is plottet for a range of values of t in Figure 3.11 together with the $(1-\alpha)$ prediction interval for a new observation. □

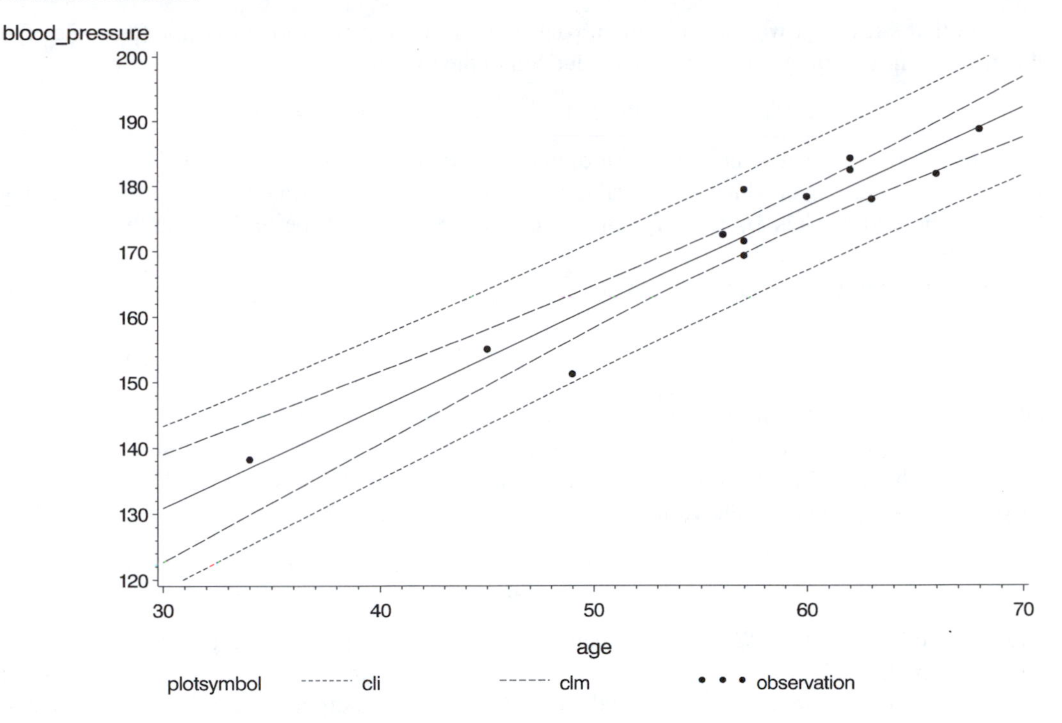

Figure 3.11 *The 95% confidence interval for the regression line for journalists (clm) and the 95% prediction interval (cli) for a new observation plotted with the estimated regression line and the observations.*

3.3.8 *Test of the Hypothesis of Linear Regression*

We stated on page 119 that under special circumstances it would be possible to test both the hypothesis of a constant variance and the hypothesis of linear regression. This is the case if several pairs of observations have the same value of the explanatory variable, so that the data can be divided into k samples after a sorting by the explanatory variable:

Response:	Explanatory variable:	
$x_{11},\dots,x_{1j},\dots,x_{1n_1}$	$t_{1j}=t_1,$	$j=1,\dots,n_1$
$\dots$	$\dots$	$\dots$
$x_{i1},\dots,x_{ij},\dots,x_{in_i}$	$t_{ij}=t_i,$	$j=1,\dots,n_i$
$\dots$	$\dots$	$\dots$
$x_{k1},\dots,x_{kj},\dots x_{kn_k}$	$t_{kj}=t_k,$	$j=1,\dots,n_k$

The n_i observations in sample i has the same value of the explanatory variable and that value is denoted by t_i.

When the data has this structure, the starting point for the analysis will be the basic model,

$$M_0\colon X_{ij}\sim N(\mu_i,\sigma_i^2), \qquad j=1,\dots,n_i,\ i=1,\dots,k,$$

which may be checked using k normal plots, one for each sample, if the number of observations in each sample is sufficiently large, $n_i\geq 10$ say. If the checks do not indicate that M_0 is invalid, one considers the reduction from M_0 to the model,

$$M_1\colon X_{ij}\sim N(\mu_i,\sigma^2), \qquad j=1,\dots,n_i,\ i=1,\dots,k.$$

This reduction is equivalent to the hypothesis, $H_{01}=H_{0\sigma^2}\colon \sigma_1^2=\cdots=\sigma_k^2=\sigma^2$, which can be tested with Bartlett's test that was introduced on page 101. If the reduction to M_1 is not contradicted by the data, the next step may be to test the reduction to the model,

$$M_2\colon X_{ij} \sim N(\alpha + \beta t_i, \sigma^2), \qquad j = 1,\ldots,n_i,\ i = 1,\ldots,k.$$

This is equivalent to testing the hypothesis of linear regression $H_{02}\colon \mu_i = \alpha + \beta t_i$. This test is a typical analysis of variance test, which is an evaluation of how much the estimates of the variance change between the two models M_1 and M_2 compared to the estimate of the variance under the current model M_1.

The estimate of the variance in the model M_1 is

$$s_1^2 = \frac{1}{f_1} SSD_1$$

and $f_1 = n - k$. Furthermore, it is wellknown that f_1, SSD_1 and s_1^2 may be found both in the bottom row in the template for calculations on page 116, and in the ERROR line in the output from *PROC GLM* when the model for k samples has been specified in *PROC GLM*, cf. page 111.

Similarly, the estimate of the variance in the regression model M_2 is

$$s_{02}^2 = \frac{1}{f_{02}} SSD_{02},$$

and it is wellknown that SSD_{02} and s_{02}^2 can be found in the template for calculation in linear regression, see Table 3.7, page 125. It will be pointed out in the Annex to Section 3.3, page 144, that f_{02}, SSD_{02}, and s_{02}^2 can also be found in the line ERROR in the output from *PROC GLM* when the regression model has been specified.

When these quantities are available, the test of the reduction from M_1 to M_2 can be calculated as

$$F(\mathbf{x}) = \frac{\dfrac{SSD_{02} - SSD_1}{f_{02} - f_1}}{s_1^2} = \frac{s_2^2}{s_1^2}.$$

The test statistic is evaluated in an F-distribution with $f_2 = f_{02} - f_1 = (n-2) - (n-k) = k-2$ degrees of freedom in the numerator and $f_1 = n - k$ degrees of freedom in the denominator. Large values only of the test statistic are critical for H_{02} so the p-value is computed as

$$p_{obs}(\mathbf{x}) = 1 - F_{F(k-2,n-k)}(F(\mathbf{x})),$$

where $F_{F(k-2,n-k)}$ is the distribution function of the F-distribution with $k-2$ degrees of freedom in the numerator and $f_1 = n - k$ degrees of freedom in the denominator. Note that here n denotes the total number of observations. This number has previously been denoted by $n.$ in connection with the model of k samples.

3.3.9 Notation in Connection with a Sequence of Models

The test for the hypothesis of linear regression, H_{02}, had the form $F(\mathbf{x}) = s_2^2/s_1^2$. Exactly the same expression was used for the test of the hypothesis of a common mean, $H_{0\mu}\colon \mu_1 = \cdots = \mu_k = \mu$, in the comparison of k samples in (3.45) on page 102. While s_1^2 denotes the same quantity in the two tests s_2^2 *does not*. This apparent confusion is the result of a carefully chosen notation. We consider a sequence of models,

$$M_1 \to M_2 \to \cdots \to M_{i-1} \to M_i \to \cdots,$$

which consists of gradually simpler specifications of the means of the observations. It is assumed that all observations have the same variance. The hypothesis that describes the reduction from one model to the next,

$$M_{i-1} \to M_i,$$

is denoted by H_{0i}. The estimate of the variance in the model M_i is denoted by s_{0i}^2 and it has the form,

$$s_{0i}^2 = \frac{1}{f_{0i}} SSD_{0i},$$

where SSD_{0i} is the sum of squared deviations of the observations from their estimated means in the model M_i. f_{0i} denotes the degrees of freedom of s_{0i}^2 and SSD_{0i}. The test statistic of H_{0i} is

$$F(\mathbf{x}) = \frac{\dfrac{SSD_{0i} - SSD_{0i-1}}{f_{0i} - f_{0i-1}}}{s_{0i-1}^2} = \frac{s_i^2}{s_{0i-1}^2},$$

and the p-value is calculated using the F-distribution with $f_i = f_{0i} - f_{0i-1}$ degrees of freedom in the numerator and f_{0i-1} degrees of freedom in the denominator as

$$p_{obs}(\mathbf{x}) = 1 - F_{F(f_i, f_{0i-1})}(F(\mathbf{x})).$$

The interpretation of the degrees of freedom for the numerator is that it is the difference in the number of parameters in the means of the two models M_{i-1} and M_i. The degrees of freedom of the denominator is the number of observations minus the number of parameters needed to specify the mean in the model M_{i-1}.

On two occasions we have failed to live up to those general rules for the notation. The first is that the estimate of the variance in the model M_1 is denoted by s_1^2 instead of by s_{01}^2. The second is that in the comparison of k samples we used the slightly more informative designation $H_{0\mu}$ for the hypothesis of identical means in the k samples instead of H_{02}.

We noted in the beginning of this section that on two different occasions we had used the symbol s_2^2 for the numerator of an F-test. This is exactly the explanation; for it was in two different connections or, to be precise, in two different sequences of hypotheses, so misunderstandings should be ruled out.

We conclude this section by admitting that there seem to be some exceptions to the general rule for constructing tests of hypotheses of the means. The exceptions are tests of hypotheses that a parameter has a fixed value. A hypothesis like that may be of interest and it is tested with a t-test that is based on the estimate of the parameter in question. We will give the test for a fixed value of the intercept, $\alpha = \alpha_0$, and the test for a fixed value of the slope, $\beta = \beta_0$, in Subsection 3.3.10 on page 138.

Incidentally, a t-test of the kind just mentioned is no exception to the general rule for constructing an F-test for a model reduction, because the squared t-test is equivalent to an F-test that is constructed using the general rule. The t-test is usually preferred to its equivalent F-test because of the connection between the t-tests and confidence intervals of the parameters.

We will use Example 3.5 to illustrate the test of the hypothesis of linear regression and the test of a fixed value of the intercept.

Example 3.5

If sand with grains of approximately the same size is placed in a sieve and we start to rotate the sieve, we will notice that the amount of sand that comes through the holes in the sieve will gradually decrease. This observation is consistent with the following simple rule: the mass of sand that passes through the sieve per unit time is proportional to the mass of sand still remaining in the sieve. If $x(t)$ denotes the mass of sand in the sieve at time t, this may be formulated as

$$\frac{dx(t)}{dt} = cx(t).$$

Here time t is measured from the start of the rotations. This differential equation has the solution,

$$x(t) = x(0)e^{ct}, \tag{3.70}$$

where c is a constant, which in the present application must be negative because the mass of sand in the sieve decreases with time. $x(0)$ is the mass of sand in the sieve at time 0, i.e., before the rotations start. Taking the logarithm on both sides of (3.70) gives

$$\ln x(t) = \ln x(0) + ct \tag{3.71}$$

and we have found that the simple rule we formulated above has led to a simple relationship between the amount of sand in the sieve at time t and the time t elapsed since the start of the rotations: the logarithm of the mass of sand in the sieve at time t is a linear function of time. The latter formulation is much easier to check than the initial formulation and a small experiment has been performed to do so.

Table 3.10 gives the results of 12 siftings of the same amount of sand with a mass 60 g. In four siftings the sieve has been rotated 170 times (the sieve rotates with a constant angular velocity and makes 43 rotations per minute), in four other siftings 530 rotations have been performed, and in the last four 880 rotations. The mass left in the sieve when it stops has been recorded in grams. We will refer to this mass as the residual mass in the text and in tables, but simply as "mass" on plots. Table 3.10 gives the logarithm of the residual mass and the corresponding number of rotations.

It is important that the sieve rotates with a constant angular velocity, so that the number of rotations is proportional to time and the model in (3.71) holds regardless of whether t denotes the elapsed time or the number of rotations. In the following, the number of rotations is the explanatory variable, which is denoted by t.

Table 3.10 *Data from the sifting experiment*

Number of rotations	*Logarithm of residual mass in* g			
170	3.89	3.95	3.85	3.91
530	3.61	3.58	3.66	3.71
880	3.30	3.26	3.30	3.34

In Figure 3.12 the logarithm of the residual mass has been plotted as a function of the number of rotations. The plot confirms the considerations that led to the regression model. There is a clear linear relationship between the logarithm of the residual mass and the number of rotations. Furthermore, the plot does not contradict the assumption of a common variance of the observations, nor does it contradict the assumption of a normal distribution of the observations. In other words, the plot in Figure 3.12 does not contradict the regression model M_2. The structure of the present data admits the testing of two of the assumptions of the regression model, so we will follow the general principle of assuming as little as possible and check the assumptions as far as possible. We therefore start by assuming only the basic model,

$$M_0\colon X_{ij} \sim N(\mu_i, \sigma_i^2), \qquad j = 1, \ldots, 4,\ i = 1, \ldots, 3,$$

and based on that model test the hypothesis of a common variance, $H_{01} = H_{0\sigma^2}\colon \sigma_1^2 = \cdots = \sigma_3^2 = \sigma^2$, or, in other words, test the reduction to the model,

$$M_1\colon X_{ij} \sim N(\mu_i, \sigma^2), \qquad j = 1, \ldots, 4,\ i = 1, \ldots, 3.$$

$H_{0\sigma^2}$ is tested with Bartlett's test. The computations are performed in SAS with the `bartlett` macro and the output is displayed in Table 3.11. See page 149 for computation of Bartlett's test using the `HOVTEST=BA` option in *PROC GLM*.

Bartlett's test statistic is Ba = 0.82107. Only large values are critical for the hypothesis and after evaluation in the $\chi^2(2)$-distribution we find the significance probability $p_{obs}(\mathbf{x}) = 0.66330$. Now $p_{obs}(\mathbf{x}) > 0.05$ and H_{01} is not rejected and the model is now M_1. The estimate of the variance in

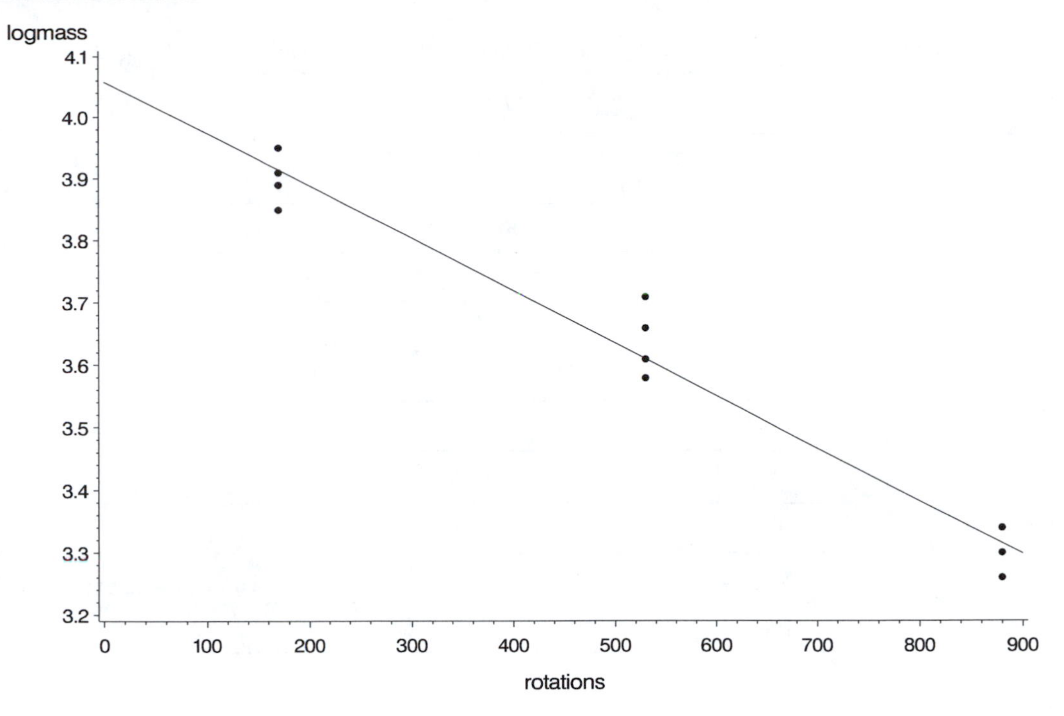

Figure 3.12 *Logarithm of residual mass as a function of the number of rotations.*

Table 3.11 *The output from the macro* `bartlett` *applied to data in Table 3.10*

Calculations in k samples:

i	ni	Si	USSi	Si2/ni	SSDi	fi	Sample Variance	Sample Mean
170	4	15.600	60.8452	60.8400	0.0052	3	0.00173	3.9000
530	4	14.560	53.0082	52.9984	0.0098	3	0.00327	3.6400
880	4	13.200	43.5632	43.5600	0.0032	3	0.00107	3.3000
	12	43.360	157.4166	157.3984	0.0182	9	0.00202	

Bartlett's test:

C	-2lnQ	Ba	Pr > ChiSq
1.14815	0.94271	0.82107	0.66330

the model M_1 is $s_1^2 = 0.00202$ and that estimate is the denominator in the test of H_{02}: $\mu_i = \alpha + \beta t_i$ which, if it is not rejected, leads to the model,

$$M_2: X_{ij} \sim N(\alpha + \beta t_i, \sigma^2), \qquad j = 1, \ldots, 4,\ i = 1, \ldots, 3.$$

In order to compute the numerator of the test we need to know the estimate of the variance in M_2 and therefore we fill in the template for calculations for linear regression. The result is given in Table 3.12.

The numerator in the F-statistic can now be calculated. First SSD_2 as

$$SSD_2 = SSD_{02} - SSD_1 = 0.0234 - 0.0182 = 0.0052,$$

Table 3.12 *The template for calculations for linear regression with data from Table 3.10*

	Sifting of sand	
	x	t
n	12	
S	43.36	6320
USS	157.4166	4336800
SP	21984.8	
SSD	0.742467	1008266.67
SPD	−851.4667	
$\hat{\beta}$	−0.0008445	
$\hat{\alpha}$	4.0581	
SSD_{02}	0.023416	
s^2_{02}	0.002342	

and next the degrees of freedom, f_2, of SSD_2 as

$$f_2 = n - 2 - (n - k) = k - 2 = 3 - 2 = 1,$$

and finally the test statistic,

$$F(\mathbf{x}) = \frac{\frac{SSD_2}{f_2}}{s_1^2} = \frac{0.0052}{0.00202} = 2.57.$$

The significance probability is

$$p_{obs}(\mathbf{x}) = 1 - F_{F(1,9)}(2.57) = 0.14.$$

The significance probability is greater than 0.05 and the reduction to the regression model M_2 is not rejected.

Estimates of the parameters in the model can be found in Table 3.12 and confidence intervals for the parameters can be computed as explained in Section 3.3.6 on page 130 if one does not plan to reduce the model further by testing hypotheses about fixed values of α or β or both.

This would in fact be a sensible thing to do. As mentioned on page 136, the initial mass, $x(0)$, was 60 g, so if the model holds, the data should not contradict the assertion that the regression line intercepts the vertical axis at $\alpha_0 = \ln(60) = 4.0943$. It will therefore be interesting to test the hypothesis, H^*_{03}: $\alpha = \alpha_0 = 4.0943$, and we will do so on page 141. But first we will consider in general terms the models where one or both regression parameters are known. □

3.3.10 Hypotheses about Regression Parameters

Here we consider the hypotheses with known slope and/or known intercept and the models they lead to. The connection between the models and the hypotheses that connect them can be displayed graphically:

$$\begin{array}{ccccc} & & M_3\colon X_i \sim N(\alpha+\beta_0 t_i,\sigma^2) & & \\ & H_{03}\colon \beta=\beta_0 \nearrow & & \searrow H_{04}\colon \alpha=\alpha_0 & \\ M_2\colon X_i \sim N(\alpha+\beta t_i,\sigma^2) & & & & M_4\colon X_i \sim N(\alpha_0+\beta_0 t_i,\sigma^2) \\ & H^*_{03}\colon \alpha=\alpha_0 \searrow & & \nearrow H^*_{04}\colon \beta=\beta_0 & \\ & & M^*_3\colon X_i \sim N(\alpha_0+\beta t_i,\sigma^2) & & \end{array}$$

We will give the tests of the four hypotheses and the estimates of the parameters in the models M_3, M_3^*, and M_4 without proof. Although H^*_{03} and H_{04} both are the hypothesis of a fixed value of the intercept, $\alpha = \alpha_0$, they are very different. They represent the reduction from two different models to two different models and therefore it should be no big surprise that the tests are different. When H_{04} is tested it is based on model M_3, where the slope has a fixed value and in that model the estimate of the intercept will have a much smaller variance than the estimate of α in M_2. Similar comments apply to the hypotheses H_{03} and H^*_{04} that both specify a fixed slope $\beta = \beta_0$.

Test of H_{03}: $\beta = \beta_0$.

Test statistic:

$$t(\mathbf{x}) = \frac{\hat{\beta} - \beta_0}{\sqrt{s^2_{02}/SSD_t}} \sim\sim t(n-2).$$

Significance probability:

$$p_{obs}(\mathbf{x}) = 2\left[1 - F_{t(n-2)}(|\, t(\mathbf{x})\,|)\right].$$

Estimates of the parameters in M_3:

$$\alpha \leftarrow \hat{\alpha}_{M_3} \quad = \quad \bar{x}. - \beta_0 \bar{t}. \sim\sim N(\alpha, \frac{\sigma^2}{n})$$

$$\begin{aligned} \sigma^2 \leftarrow s^2_{03} &= \frac{1}{n-1}\sum_{i=1}^{n}\{x_i - (\hat{\alpha}_{M_3} + \beta_0 t_i)\}^2 \\ &= \frac{1}{n-1}\left[SSD_{02} + (\hat{\beta} - \beta_0)^2 SSD_t\right] \\ &\sim\sim \sigma^2\chi^2(n-1)/(n-1). \end{aligned}$$

Test of H^*_{03}: $\alpha = \alpha_0$:

Test statistic:

$$t(\mathbf{x}) = \frac{\hat{\alpha} - \alpha_0}{\sqrt{s^2_{02}\left(\frac{1}{n} + \frac{\bar{t}^2_{\cdot}}{SSD_t}\right)}} \sim\sim t(n-2).$$

Significance probability:

$$p_{obs}(\mathbf{x}) = 2\left[1 - F_{t(n-2)}(|\, t(\mathbf{x})\,|)\right].$$

Estimates of the parameters in M^*_3:

$$\beta \leftarrow \hat{\beta}_{M^*_3} = \frac{\sum_{i=1}^{n} t_i(x_i - \alpha_0)}{\sum_{i=1}^{n} t_i^2} = \frac{SP_{xt} - \alpha_0 S_t}{USS_t} \sim\sim N(\beta, \frac{\sigma^2}{USS_t}),$$

$$\begin{aligned}\sigma^2 \leftarrow s_{03}^{*2} &= \frac{1}{n-1}\sum_{i=1}^{n}\{x_i - (\alpha_0 + \hat{\beta}_{M_3^*} t_i)\}^2 \\ &= \frac{1}{n-1}\left[USS_x + n\alpha_0^2 - 2\alpha_0 S_x - \hat{\beta}_{M_3^*}^2 USS_t\right] \\ &\sim\sim \sigma^2\chi^2(n-1)/(n-1).\end{aligned}$$

Test of H_{04}^*: $\beta = \beta_0$:
Test statistic:

$$\begin{aligned}t(\mathbf{x}) &= \frac{\hat{\beta}_{M_3^*} - \beta_0}{\sqrt{s_{03}^{*2}/USS_t}} \\ &= \frac{SP_{xt} - \alpha_0 S_t - \beta_0 USS_t}{\sqrt{s_{03}^{*2} USS_t}} \\ &\sim\sim t(n-1).\end{aligned}$$

Significance probability:

$$p_{obs}(\mathbf{x}) = 2\left[1 - F_{t(n-1)}(|\,t(\mathbf{x})\,|)\right].$$

Estimate of the parameter in M_4:

$$\begin{aligned}\sigma^2 \leftarrow s_{04}^2 &= \frac{1}{n}\sum_{i=1}^{n}\{x_i - (\alpha_0 + \beta_0 t_i)\}^2 \\ &= \frac{1}{n}\left[USS_x + n\alpha_0^2 + \beta_0^2 USS_t - 2\alpha_0 S_x - 2\beta_0 SP_{xt} + 2\alpha_0\beta_0 S_t\right] \\ &\sim\sim \sigma^2\chi^2(n)/(n).\end{aligned}$$

Test of H_{04}: $\alpha = \alpha_0$:
Test statistic:

$$\begin{aligned}t(\mathbf{x}) &= \frac{\hat{\alpha}_{M_3} - \alpha_0}{\sqrt{s_{03}^2/n}} \\ &= \frac{S_x - \beta_0 S_t - \alpha_0 n}{\sqrt{s_{03}^2 n}} \\ &\sim\sim t(n-1).\end{aligned}$$

Significance probability:

$$p_{obs}(\mathbf{x}) = 2\left[1 - F_{t(n-1)}(|\,t(\mathbf{x})\,|)\right].$$

Estimate of the parameter in M_4:

$$\begin{aligned}\sigma^2 \leftarrow s_{04}^2 &= \frac{1}{n}\sum_{i=1}^{n}\{x_i - (\alpha_0 + \beta_0 t_i)\}^2 \\ &= \frac{1}{n}\left[USS_x + n\alpha_0^2 + \beta_0^2 USS_t - 2\alpha_0 S_x - 2\beta_0 SP_{xt} + 2\alpha_0\beta_0 S_t\right] \\ &\sim\sim \sigma^2\chi^2(n)/(n).\end{aligned}$$

Example 3.4 (Continued)
In the example with the regression of blood pressure on age it is meaningful to test the hypothesis H_{03}: $\beta = 0$, which states that blood pressure does not depend on age. For journalists one finds from the template for calculations in Table 3.8 on page 126 the quantities that are needed to compute the

t-test:

$$t(\mathbf{x}) = \frac{\hat{\beta} - \beta_0}{\sqrt{s_{02}^2/SSD_t}} = \frac{1.529568 - 0}{\sqrt{18.14949/1033.0769}} = 11.54 \sim\sim t(11).$$

The significance probability is 0.00000017.

Correspondingly one finds from Table 3.9 on page 126 the test statistic for university teachers:

$$t(\mathbf{x}) = \frac{\hat{\beta} - \beta_0}{\sqrt{s_{02}^2/SSD_t}} = \frac{1.562384 - 0}{\sqrt{19.44953/588.9333}} = 8.60 \sim\sim t(13).$$

The significance probability is 0.000001.

For both professions the hypothesis that blood pressure does not depend on age is rejected. Hypotheses about a fixed intercept are not meaningful in this example. □

Example 3.5 (Continued)
Some simple considerations on page 136 led to the model,

$$\ln x(t) = \ln x(0) + ct, \tag{3.72}$$

where $x(t)$ is the logarithm of residual mass at time t. For each measurement the starting point was an initial amount of sand of 60 g. Thus $\ln x(0) = \ln 60 = 4.0943$ and if the model holds, the regression line must have an intercept of $\alpha_0 = 4.0943$. It is therefore interesting to test the hypothesis H_{03}^*: $\alpha = \alpha_0 = 4.0943$.

The numbers to be used in the test statistic are found in Table 3.12 on page 138, and the test statistic becomes:

$$\begin{aligned} t(\mathbf{x}) &= \frac{\hat{\alpha} - \alpha_0}{\sqrt{s_{02}^2\left(\frac{1}{n} + \frac{\bar{t}_{\cdot}^2}{SSD_t}\right)}} \\ &= \frac{4.0581 - 4.0943}{\sqrt{0.002342\left(\frac{1}{12} + \frac{(6320/12)^2}{1008266.67}\right)}} \\ &= -1.249 \sim\sim t(10). \end{aligned}$$

The significance probability is 0.24 and one can conclude that the data do not question the model,

$$M_3^*: X_{ij} \sim N(\alpha_0 + \beta t_i, \sigma^2).$$

M_3^* will be the final model, which will be the basis of confidence intervals for the parameters. Note that if the data had rejected H_{03}^*, it would not simply mean that M_2 had remained the model for the data; for we know that the mass of sand at time 0 is 60 g and if the model had failed to account for that fact, it would seriously have questioned the model of linear regression.

The estimate of β in M_3^* is found to be

$$\beta \leftarrow \beta_{M_3^*} = \frac{SP_{xt} - \alpha_0 S_t}{USS_t} = \frac{21984.8 - 4.0943 \times 6320}{4336800} = -0.00089725 \sim\sim N(\beta, \frac{\sigma^2}{USS_t}).$$

The numerical value of the estimate is not very different from the value in the model M_2. The important distinction is the distribution of the estimates of β in the two models or, more precisely, the different variances of the estimates in the two models. In M_2 the variance of the estimate of β is equal to σ^2/SSD_t, whereas the variance of the estimate of β in M_3^* is σ^2/USS_t. For the actual data the variance under M_3^* is less than one fourth of the variance under M_2.

The estimate of σ^2 under M_3^* is

$$\begin{aligned}
s_{03}^{*2} &= \frac{1}{n-1}\left[USS_x + n\alpha_0^2 - 2\alpha_0 S_x - \hat{\beta}_{M_3^*}^2 USS_t\right] \\
&= \frac{1}{11}\left[157.4166 + 12\times 4.0943^2 - 2\times 4.0943\times 43.36 - 0.00089725^2\times 4336800\right] \\
&= 0.0024582 \\
&\sim\sim \sigma^2\chi^2(11)/11
\end{aligned}$$

and it is of the same order of magnitude as the estimate under M_2. □

Example 3.5 (Continued)

The simple idea that the amount of sand filtering through the sieve at a given point in time is proportional to the amount of sand in the sieve and some mathematics lead to consideration of the linear regression of the *logarithm* of the residual mass on the number of rotations. Is anything wrong with the more direct approach of considering simply the regression of the residual mass on the number of rotations, i.e., the residual mass without transformations of any kind? The residual mass is plotted as a function of the number of rotations in Figure 3.13 together with the regression line. The dashed line is the regression line in the regression of the logarithm of residual mass transformed back to the original scale of measurements.

We use computations made with SAS and printed in *Annex to Section 3.3* on pages 152 to 154.

Note first that neither the hypothesis of a common variance nor the hypothesis of linearity is rejected for the residual mass, which is no surprise in view of Figure 3.13.

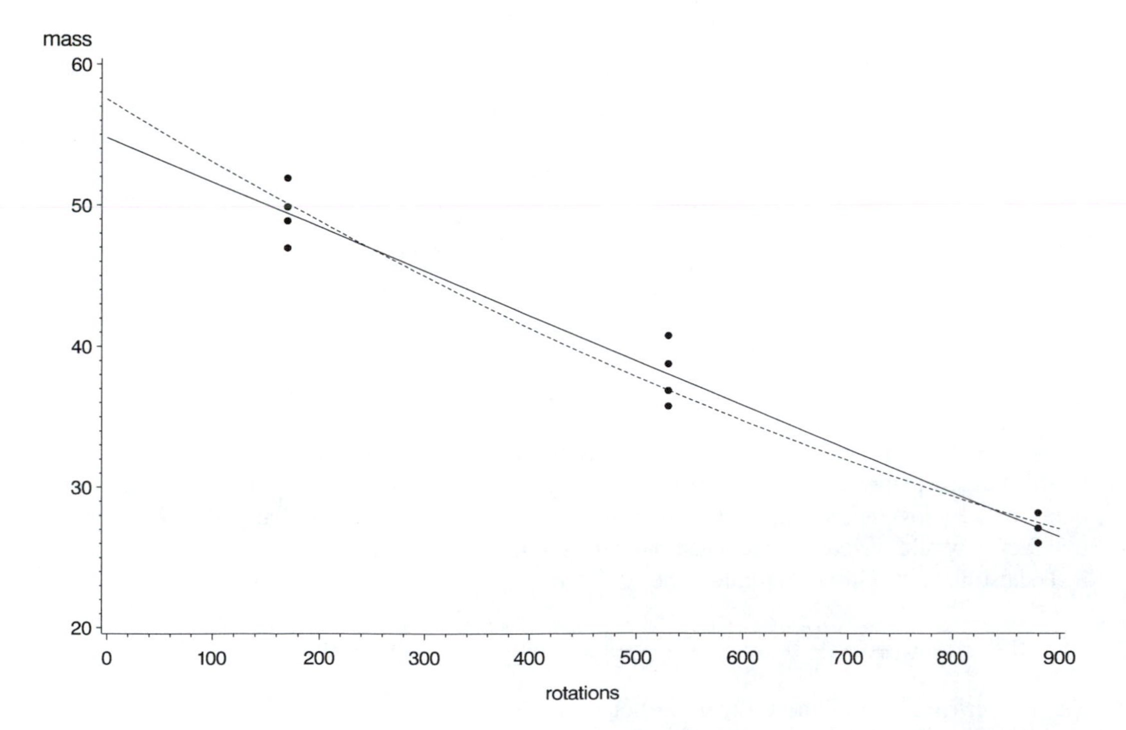

Figure 3.13 *Residual mass and the expected residual mass as a function of the number of rotations. The dashed line is the expected residual mass in the regression model for the logarithm of the residual mass but transformed* (exp) *to the original scale of measurements. The full line is the residual mass from the regression model for untransformed residual mass.*

Next, we note that if the purpose of the modeling is to predict the residual mass after a fixed number of rotations, the two models will work equally well at least for a number of rotations in the vicinity of $\bar{t}_\cdot$. As an example let us look at the confidence intervals of the median of the residual mass after $\bar{t}_\cdot = 526.7$ rotations.

Using first the regression model for the logarithm of the residual mass, we find the confidence interval for the mean of the logarithm of the residual mass to be

$$3.6133 \pm 0.031126 = [3.5822,\ 3.6445],$$

which in turn can be transformed (exp) to a 95% confidence interval for the *median* of the residual mass after $\bar{t}_\cdot = 526.7$ rotations:

$$[36.0,\ 38.3]\ \text{g}.$$

The estimate of the median is 37.1 g.

For the untransformed mass we find the confidence interval for the mean to be

$$[38.23 \pm 1.10]\ \text{g} = [37.1, 39.3]\ \text{g}.$$

Now the mean in a normal distribution is also the median, so we have two confidence intervals for the median based on different models and they are not very different.

We have in both cases used the model M_2 where the intercepts and the slopes of the regression lines are unspecified, i.e., they are not forced to intersect the vertical axis in $\alpha_0 = 4.0943$.

It is no big surprise that the two models give similar results in an interval around $\bar{t}_\cdot$ for a linear function is a good approximation to the function,

$$x(t) = e^{4.0581} e^{-0.0008445t},$$

in an interval around $\bar{t}_\cdot$.

So far the two models have done equally well. But we have a slight preference for the regression model for the logarithm of the residual mass; for this model can explain the fact that the initial mass is 60 g as we saw on page 141 and the regression model for the residual mass is not consistent with that fact. If the regression line of the residual mass on Figure 3.13 is extended to intersect the vertical axis, it does not look as if the regression line can be forced to intersect in 60 g and, indeed, the test of $H^*_{03}\colon \alpha = \alpha_0 = 60$ rejects that hypothesis.

Using the output listing on page 3.3.10 we find that the estimate of the intercept is 54.79317641 with a standard error of 1.02738680, so the t-test of $H^*_{03}\colon \alpha = \alpha_0 = 60$ is

$$t = \frac{54.7931764 - 60}{1.0273868} = -5.068,$$

which gives a significance probability below 0.001. □

Annex to Section 3.3

Calculations in SAS

Example 3.4 (Continued)
The following program first creates the data set and then the calculation of a linear regression for each of the two professions is performed by *PROC GLM*. The calculations are restricted to one of the two professions with the `WHERE` statement, which selects observations from a SAS data set that meet certain conditions.

The usage of *PROC FORMAT* to define a format and the assignment of that format to the variable `group` in the format statement in the data step has no influence on the computations. The purpose is to have the names of the groups appear in the graphs and in the output listings and still be able to use numbers to denote the groups in the coding of the data.

The statements in *PROC GLM* that produce a linear regression are similar to the statements that produce a one-way analysis of variance but with the important exception that in a linear regression the variable that contains the explanatory variable is **not** included in a `CLASS` statement.

```
TITLE1 'Example 3.4';

PROC FORMAT;
   VALUE $grpfmt    '1'='journalist' '2'='university teacher';
RUN;

DATA blood_pressure;
INPUT group$ blood_pressure age@@;
FORMAT group $grpfmt.;
DATALINES;
1 188.7 68 1 184.2 62 1 151.3 49 1 182.4 62 1 138.3 34
1 181.8 66 1 155.1 45 1 178.3 60 1 179.4 57 1 171.5 57
1 177.9 63 1 172.5 56 1 169.3 57
2 162.9 55 2 155.9 49 2 170.5 60 2 154.5 54 2 158.0 58
2 155.4 51 2 143.3 43 2 157.3 52 2 160.2 53 2 164.3 61
2 167.5 61 2 171.5 57 2 166.0 57 2 190.3 70 2 175.5 63
;
RUN;

TITLE2 'journalist';
PROC GLM DATA=blood_pressure(WHERE=(group='1'));
   MODEL blood_pressure=age/SS1;
RUN;

TITLE2 'university teacher';
PROC GLM DATA=blood_pressure(WHERE=(group='2'));
   MODEL blood_pressure=age/SS1;
RUN; QUIT;
TITLE;
```

Output listing from *PROC GLM*:

On page 145 it is indicated in the output listing concerning journalists on SAS page 2, where the important quantities f_{02}, SSD_{02}, s^2_{02}, $\hat{\alpha}$, and $\hat{\beta}$ are found.

Furthermore, we note that the F-test that is given in the analysis of variance table in the `Model`-line is the test of H_{03}: $\beta = 0$. The t-test for the same hypothesis is given in the line `age` in the `Parameter` part of the output at the bottom of SAS page 2. Finally, the t-test of H^*_{03}: $\alpha = 0$ is given in the line `Intercept`.

The information in the `Parameter` part of the output can also be used to compute confidence intervals of the regression parameters. The formula for the limits of the $(1-\alpha)$ confidence interval is

$$\texttt{Estimate} \pm \texttt{Standard Error} \times t_{1-\alpha/2}(f),$$

where `Estimate` and its `Standard Error` is found in the same line at the bottom of the page. The $t_{1-\alpha/2}(f)$-fractile can be obtained once it is noted that the degrees of freedom, f, are found under `DF` in the `Error` line in the analysis of variance table. It is shown on page 166 how the t-fractile can be calculated using the SAS function `TINV`.

Taking the output listing for journalists on page 145 as an example the limits of the 95% confidence interval of the slope β for journalists are found as

$$\texttt{1.52956813} \pm \texttt{0.13254581} \times 2.201$$

where $2.201 = t_{0.975}(11)$, cf. formula (3.68).

The SAS statements:

```
TITLE2 'journalist';
PROC GLM DATA=blood_pressure(WHERE=(group='1'));
   MODEL blood_pressure=age/SS1;
RUN;
```

give the output listing:

```
                                example 3.4                              1
                                 journalist

                             The GLM Procedure

                        Number of observations    13

                                example 3.4                              2
                                 journalist

                             The GLM Procedure

Dependent Variable: blood_pressure
                                             Sum of
Source                        DF            Squares    Mean Square    F Value     Pr > F

Model                          1          2416.9647      2416.9647     133.17     <.0001

Error                   f02 =11        199.6445=SSD02 18.1495=s02^2

Corrected Total               12          2616.6092


          R-Square      Coeff Var       Root MSE      blodtryk Mean

          0.923701       2.482759         4.2602             171.59


Source                         DF       Type I SS    Mean Square   F Value     Pr > F

age                             1       2416.9647      2416.9647    133.17     <.0001


                                                Standard
Parameter             Estimate                     Error     t Value     Pr > |t|

Intercept          84.99521966=α̂             7.59658557       11.19       <.0001
age                 1.52956813=β̂             0.13254581       11.54       <.0001
```

In this case the output listing is edited. Some of the decimal places in the output listing have been removed to make room for f_{02}, SSD_{02}, s_{02}^2, $\hat{\alpha}$ and $\hat{\beta}$.

The SAS statements:

```
TITLE2 'university teacher';
PROC GLM DATA=blood_pressure(WHERE=(group='2'));
   MODEL blood_pressure=age/SS1;
RUN;
```

give the output listing:

```
                          Example 3.4                                 3
                       university teacher

                       The GLM Procedure

                  Number of observations    15
                          Example 3.4                                 4
                       university teacher

                       The GLM Procedure

Dependent Variable: blood_pressure

                                         Sum of
 Source                      DF         Squares     Mean Square    F Value    Pr > F

 Model                        1     1437.611987     1437.611987      73.91    <.0001

 Error                       13      252.844013       19.449539

 Corrected Total             14     1690.456000

        R-Square     Coeff Var      Root MSE    blood_pressure Mean

        0.850429      2.696688      4.410163               163.5400

 Source                      DF       Type I SS     Mean Square    F Value    Pr > F

 age                          1     1437.611987     1437.611987      73.91    <.0001

 Source                      DF     Type III SS     Mean Square    F Value    Pr > F

 age                          1     1437.611987     1437.611987      73.91    <.0001

                                            Standard
     Parameter          Estimate              Error    t Value    Pr > |t|

     Intercept       75.62986190        10.28843126       7.35      <.0001
     age              1.56238397         0.18172789       8.60      <.0001
```

□

Residuals in PROC GLM

Four of the residuals defineded in the section on residuals on page 127 are calculated by *PROC GLM*. They are the raw residuals in (3.49), the studentized scaled residuals in (3.60), the raw deletion residuals in (3.61), and, finally, the studentized scaled deletion residuals in (3.64). They can be written to a SAS data set using the keywords `RESIDUAL` (or `R`), `STUDENT`, `PRESS` and `RSTUDENT`, respectively, in the `OUTPUT` statement.

The scaled residuals are not directly available, but the leverage is available with the keyword `H`, so the scaled residuals can be computed from the formula (3.59).

Example 3.4 (Continued)

The residuals in a regression of blood pressure on age ignoring the profession are obtained with the following SAS program. In the data step that follows, the scaled residuals are computed and the residuals are labeled as a reminder of the meaning of the keywords but also in order to enhance plots and output.

The data set `out_resid` was used to make the plots in Figure 3.9 and Figure 3.10 .

```
PROC GLM DATA=blood_pressure;
MODEL blood_pressure=age;
OUTPUT OUT=out_resid H=leverage R=raw_residual STUDENT=s_s_r
                            PRESS=raw_del_r   RSTUDENT=s_s_del_r;
RUN;

DATA out_resid;
SET out_resid;
scaled_r=raw_residual/sqrt(1-leverage);
LABEL
leverage    ='leverage'
raw_residual='raw residuals'
s_s_r       ='studentized scaled residuals'
raw_del_r   ='raw deletion residuals'
s_s_del_r   ='studentized scaled deletion residuals'
scaled_r    ='scaled residuals'
;
RUN;
```

□

Example 3.5 (Continued)

The SAS program below delivers all the necessary results for the analysis of the logarithm of the residual mass as well as the residual mass.

```
TITLE1 'Example 3.5';

DATA sifting;
INPUT rotations logmass @@;
DATALINES;
170 3.89 170 3.95 170 3.85 170 3.91
530 3.61 530 3.58 530 3.66 530 3.71
880 3.30 880 3.26 880 3.30 880 3.34
;
RUN;
DATA untrans;
SET sifting;
mass=exp(logmass);
RUN;

TITLE3 'logarithm of residual mass';
%bartlett(data=sifting,group=rotations,var=logmass);
RUN;

TITLE2 'one-way analysis of variance';
PROC GLM DATA=sifting;
   CLASS rotations;
   MODEL logmass = rotations / SS1;
   MEANS rotations/HOVTEST=BARTLETT;
RUN;

TITLE2 'regression analysis';
PROC GLM DATA=sifting;
   MODEL logmass = rotations / SS1;
   OUTPUT OUT=temp P=predlmas;
RUN;

TITLE2;
TITLE3 'untransformed residual mass';
%bartlett(data=untrans,group=rotations,var=mass);

TITLE2 'one-way analysis of variance';
TITLE3 'untransformed residual mass';
PROC GLM DATA=untrans;
   CLASS rotations;
   MODEL mass = rotations / SS1;
   MEANS rotations/HOVTEST=BARTLETT;
RUN;

TITLE2 'regression analysis';
TITLE3 'untransformed residual mass';
PROC GLM DATA=untrans;
   MODEL mass = rotations / SS1;
   OUTPUT OUT=temput P=predmas;
RUN; QUIT;
TITLE;
```

The output listings of the program are displayed on this and the following pages. For easy reference the SAS statements are given along with their output. Programs and listings on pages 149 to 151 concern the logarithm of the residual mass and programs and listings on pages 152 to 154 concern the residual mass.

The SAS statements:

```
TITLE3 'logarithm of residual mass';
%bartlett(data=sifting,group=rotations,var=logmass);
RUN;
```

give the output listing:

```
                              Example 3.5                              1

                       logarithm of residual mass

                            Macro "BARTLETT"

                               Data set: SIFTING
                       Response variable: LOGMASS
                          Group variable: ROTATIONS


                        Calculations in k samples:

                                                            Sample     Sample
i     ni     Si        USSi       Si2/ni      SSDi     fi  Variance     Mean

170    4   15.600    60.8452     60.8400    0.0052   3    0.00173    3.9000
530    4   14.560    53.0082     52.9984    0.0098   3    0.00327    3.6400
880    4   13.200    43.5632     43.5600    0.0032   3    0.00107    3.3000
-----------------------------------------------------======================
      12   43.360   157.4166    157.3984    0.0182   9    0.00202


                             Bartlett's test:

                    C         -2lnQ        Ba       Pr > ChiSq

                 1.14815     0.94271    0.82107     0.66330
```

Almost the same result is obtained using the statement

```
MEANS rotation/HOVTEST=BARTLETT;
```

in *PROC GLM*:

```
                             Example 3.5                            4
                       one-way analysis of variance

                           The GLM Procedure

             Bartlett's Test for Homogeneity of logmass Variance

               Source            DF     Chi-Square     Pr > ChiSq

               rotations          2         0.8211         0.6633
                             Example 3.5                            5
                       one-way analysis of variance

                           The GLM Procedure

              Level of            -----------logmass-----------
              rotations      N            Mean          Std Dev

              170            4      3.90000000       0.04163332
              530            4      3.64000000       0.05715476
              880            4      3.30000000       0.03265986
```

It is worth noting that the standard deviations given in the `Std Dev` column above are the estimated standard deviations in the individual samples without the assumption of a common variance.

The SAS statements:

```
TITLE2 'one-way analysis of variance';
PROC GLM DATA=sifting;
   CLASS rotation;
   MODEL logmass = rotation / SS1;
/*   MEANS rotation/HOVTEST=BARTLETT; */
RUN;
```

give the output listing:

```
                      Example 3.5                              2
                 one-way analysis of variance

                      The GLM Procedure

                   Class Level Information

            Class           Levels     Values

            rotations            3     170 530 880


                 Number of observations    12
                      Example 3.5                              3
                 one-way analysis of variance

                      The GLM Procedure

Dependent Variable: logmass

                                        Sum of
 Source                      DF        Squares    Mean Square   F Value   Pr > F

 Model                        2     0.72426667     0.36213333    179.08   <.0001

 Error                        9     0.01820000     0.00202222

 Corrected Total             11     0.74246667


           R-Square     Coeff Var      Root MSE    logmass Mean

           0.975487      1.244533      0.044969        3.613333


 Source                      DF      Type I SS    Mean Square   F Value   Pr > F

 rotations                    2     0.72426667     0.36213333    179.08   <.0001
```

Note that the statement

```
MEANS rotation/HOVTEST=BARTLETT;
```

has been made a comment. The result of the statement has been shown on the previous page.

The SAS statements:

```
TITLE2 'regression analysis';
PROC GLM DATA=sifting;
   MODEL logmass = rotation / SS1;
RUN;
```

give the output listing:

```
                          Example 3.5                                    6
                      regression analysis

                       The GLM Procedure

                  Number of observations      12
                          Example 3.5                                    7
                      regression analysis

                       The GLM Procedure

Dependent Variable: logmass

                                     Sum of
 Source                   DF        Squares    Mean Square  F Value  Pr > F

 Model                     1     0.71905133     0.71905133   307.09  <.0001

 Error                    10     0.02341534     0.00234153

 Corrected Total          11     0.74246667


           R-Square     Coeff Var      Root MSE    logmass Mean

           0.968463      1.339190      0.048389        3.613333


 Source                   DF      Type I SS    Mean Square  F Value  Pr > F

 rotations                 1     0.71905133     0.71905133   307.09  <.0001


                                          Standard
     Parameter         Estimate              Error    t Value    Pr > |t|

     Intercept      4.058095742         0.02897054     140.08      <.0001
     rotations     -0.000844486         0.00004819     -17.52      <.0001
```

The SAS commands:

```
TITLE3 'untransformed residual mass';
%bartlett(data=untrans,group=rotations,var=mass);
```

give the output listing:

```
                              Example 3.5                              8

                      untransformed residual mass

                           Macro "BARTLETT"

                           Data set: SIFTING
                      Response variable: MASS
                         Group variable: ROTATIONS


                      Calculations in k samples:

                                                                Sample      Sample
i     ni     Si        USSi        Si2/ni       SSDi     fi   Variance      Mean

170    4   197.700   9783.8300   9771.3225    12.5075    3    4.16917    49.4250
530    4   152.700   5843.8300   5829.3225    14.5075    3    4.83583    38.1750
880    4   108.400   2940.0600   2937.6400     2.4200    3    0.80667    27.1000
-----------------------------------------------------------=========================
      12   458.800  18567.7200  18538.2850    29.4350    9    3.27056

                            Bartlett's test:

                   C        -2lnQ        Ba      Pr > ChiSq

                1.14815    2.29786    2.00137     0.36763
```

The SAS commands:

```
TITLE2 'one-way analysis of variance';
TITLE3 'untransformed residual mass';
PROC GLM DATA=untrans;
   CLASS rotation;
   MODEL mass = rotation / SS1;
/* MEANS rotation/HOVTEST=BARTLETT; */
RUN;
```

give the output listing:

```
                          Example 3.5                                  9
                   one-way analysis of variance
                   untransformed residual mass

                       The GLM Procedure

                    Class Level Information

              Class          Levels    Values

              rotations           3    170 530 880


                   Number of observations    12
                          Example 3.5                                 10
                   one-way analysis of variance
                   untransformed residual mass

                       The GLM Procedure

Dependent Variable: mass

                                      Sum of
 Source                      DF      Squares   Mean Square  F Value   Pr > F

 Model                        2   996.831667    498.415833   152.39   <.0001

 Error                        9    29.435000      3.270556

 Corrected Total             11  1026.266667


           R-Square     Coeff Var      Root MSE     mass Mean

           0.971318      4.730081      1.808468      38.23333


 Source                      DF     Type I SS   Mean Square  F Value   Pr > F

 rotations                    2   996.8316667   498.4158333   152.39   <.0001
```

The SAS commands:

```
TITLE2 'regression analysis';
TITLE3 'untransformed residual mass';
PROC GLM DATA=untrans;
   MODEL mass = rotation / SS1;
RUN;
```

give the output listing:

```
                              Example 3.5                              13
                          regression analysis
                      untransformed residual mass

                            The GLM Procedure

                      Number of observations    12
                              Example 3.5                              14
                          regression analysis
                      untransformed residual mass

                            The GLM Procedure

Dependent Variable: mass

                                        Sum of
 Source                      DF        Squares    Mean Square    F Value    Pr > F

 Model                        1     996.818706     996.818706     338.50    <.0001

 Error                       10      29.447961       2.944796

 Corrected Total             11    1026.266667


              R-Square     Coeff Var      Root MSE     mass Mean

              0.971306      4.488337      1.716041      38.23333


 Source                      DF      Type I SS    Mean Square    F Value    Pr > F

 rotations                    1    996.8187058    996.8187058     338.50    <.0001


                                               Standard
       Parameter            Estimate              Error    t Value    Pr > |t|

       Intercept         54.79317641         1.02738680      53.33      <.0001
       rotations         -0.03144274         0.00170899     -18.40      <.0001
```

□

Main Points in Section 3.3

Data consist of pairs of observations of the explanatory variable t and the response x, (t_i, x_i), $i = 1, \dots, n$.

Model:

The model M_2 of linear regression specifies that the observations, x_i, $i = 1, \dots, n$, are realizations of independent random variables,

$$X_i \sim N(\alpha + \beta t_i, \sigma^2),\ 1, \dots, n,$$

which in the shorthand version is written as

$$M_2\colon X_i \sim N(\alpha + \beta t_i, \sigma^2),\ i = 1, \dots, n.$$

Model checking:

At least a scatter plot of the pairs of observations, (t_i, x_i), $i = 1, \dots, n$. If several pairs of observations have the same value of the explanatory variable t_i, normal plots, test of homogeneity of variance, and test of the linear regression may be useful.

Test of linear regression:

Test statistic:

$$F(\mathbf{x}) = \frac{\dfrac{SSD_{02} - SSD_1}{f_{02} - f_1}}{s_1^2} = \frac{s_2^2}{s_1^2} \sim\sim F(k-2, n-k).$$

See page 134 for an explanation of how the quantities of the test statistic are calculated or found in SAS output listings.

p–value:

$$p_{obs}(\mathbf{x}) = 1 - F_{F(k-2,n-k)}(F(\mathbf{x})).$$

Estimation:

Formulas for the estimates appear from the template of calculations in linear regression on page 159.

$$\hat{\alpha} \sim\sim N\left(\alpha, \sigma^2\left(\frac{1}{n} + \frac{\bar{t}_\cdot^2}{SSD_t}\right)\right),$$

$$\hat{\beta} \sim\sim N\left(\beta, \frac{\sigma^2}{SSD_t}\right),$$

$$\hat{\alpha} + \hat{\beta} t \sim\sim N\left(\alpha + \beta t, \sigma^2\left(\frac{1}{n} + \frac{(t - \bar{t}_\cdot)^2}{SSD_t}\right)\right),$$

$$s_{02}^2 \sim\sim \sigma^2 \chi^2(f_{02})/f_{02},$$

where $f_{02} = n - 2$ (number of observations minus the number of unknown parameters in the mean).

Confidence intervals:

$(1-\alpha)$ confidence interval for the intercept α:

$$\left[\hat{\alpha} - \sqrt{s_{02}^2\left(\frac{1}{n} + \frac{\bar{t}_\cdot^2}{SSD_t}\right)}\, t_{1-\alpha/2}(f_{02}),\ \hat{\alpha} + \sqrt{s_{02}^2\left(\frac{1}{n} + \frac{\bar{t}_\cdot^2}{SSD_t}\right)}\, t_{1-\alpha/2}(f_{02})\right].$$

$(1-\alpha)$ confidence interval for the slope β:

$$\left[\hat{\beta} - \sqrt{\frac{s_{02}^2}{SSD_t}}\, t_{1-\alpha/2}(f_{02}),\ \hat{\beta} + \sqrt{\frac{s_{02}^2}{SSD_t}}\, t_{1-\alpha/2}(f_{02})\right].$$

$(1-\alpha)$ confidence interval for the value $\alpha+\beta t$ of the regression line at t:

$$\left[\hat{\alpha}+\hat{\beta}t-\sqrt{s_{02}^2\left(\frac{1}{n}+\frac{(t-\bar{t}.)^2}{SSD_t}\right)}\,t_{1-\alpha/2}(f_{02}),\ \hat{\alpha}+\hat{\beta}t+\sqrt{s_{02}^2\left(\frac{1}{n}+\frac{(t-\bar{t}.)^2}{SSD_t}\right)}\,t_{1-\alpha/2}(f_{02})\right].$$

$(1-\alpha)$ confidence interval for the variance σ^2:

$$\left[\frac{s_{02}^2}{\chi^2_{1-\alpha/2}(f_{02})/f_{02}},\ \frac{s_{02}^2}{\chi^2_{\alpha/2}(f_{02})/f_{02}}\right].$$

$(1-\alpha)$ confidence interval for the standard deviation σ:

$$\left[\sqrt{\frac{s_{02}^2}{\chi^2_{1-\alpha/2}(f_{02})/f_{02}}},\ \sqrt{\frac{s_{02}^2}{\chi^2_{\alpha/2}(f_{02})/f_{02}}}\right].$$

Prediction interval:
$(1-\alpha)$ prediction interval for a new observation:

$$\left[\hat{\mu}(t)-\sqrt{s_{02}^2\left(1+\frac{1}{n}+\frac{(t-\bar{t}.)^2}{SSD_t}\right)}\,t_{1-\alpha/2}(f_{02}),\ \hat{\mu}(t)+\sqrt{s_{02}^2\left(1+\frac{1}{n}+\frac{(t-\bar{t}.)^2}{SSD_t}\right)}\,t_{1-\alpha/2}(f_{02})\right],$$

where $\hat{\mu}(t)=\hat{\alpha}+\hat{\beta}t$.

Submodels of the regression model or hypotheses about the regression parameters

We consider models with known slope and/or known intercept. The connection between the models and the hypotheses that connect them can be displayed graphically:

$$\begin{array}{ccccc} & & M_3\colon X_i\sim N(\alpha+\beta_0 t_i,\sigma^2) & & \\ & H_{03}\colon\beta=\beta_0\nearrow & & \searrow H_{04}\colon\alpha=\alpha_0 & \\ M_2\colon X_i\sim N(\alpha+\beta t_i,\sigma^2) & & & & M_4\colon X_i\sim N(\alpha_0+\beta_0 t_i,\sigma^2) \\ & H_{03}^*\colon\alpha=\alpha_0\searrow & & \nearrow H_{04}^*\colon\beta=\beta_0 & \\ & & M_3^*\colon X_i\sim N(\alpha_0+\beta t_i,\sigma^2) & & \end{array}$$

Test of H_{03}: $\beta=\beta_0$:
Test statistic:

$$t(\mathbf{x})=\frac{\hat{\beta}-\beta_0}{\sqrt{s_{02}^2/SSD_t}}\sim\sim t(n-2).$$

Significance probability:

$$p_{obs}(\mathbf{x})=2\left[1-F_{t(n-2)}(|\,t(\mathbf{x})\,|)\right].$$

Estimates of the parameters in M_3:

$$\alpha\leftarrow\hat{\alpha}_{M_3}=\bar{x}.-\beta_0\bar{t}.\sim\sim N(\alpha,\frac{\sigma^2}{n}),$$

$$\begin{aligned}\sigma^2\leftarrow s_{03}^2&=\frac{1}{n-1}\sum_{i=1}^{n}\{x_i-(\hat{\alpha}_{M_3}+\beta_0 t_i)\}^2\\&=\frac{1}{n-1}\left[SSD_{02}+(\hat{\beta}-\beta_0)^2 SSD_t\right]\\&\sim\sim\sigma^2\chi^2(n-1)/(n-1).\end{aligned}$$

Test of H_{03}^*: $\alpha=\alpha_0$:

Test statistic:

$$t(\mathbf{x}) = \frac{\hat{\alpha} - \alpha_0}{\sqrt{s_{02}^2(\frac{1}{n} + \frac{\bar{t}_\cdot^2}{SSD_t})}} \sim\sim t(n-2).$$

Significance probability:

$$p_{obs}(\mathbf{x}) = 2\left[1 - F_{t(n-2)}(|\,t(\mathbf{x})\,|)\right].$$

Estimates of the parameters in M_3^*:

$$\beta \leftarrow \hat{\beta}_{M_3^*} = \frac{\sum_{i=1}^{n} t_i(x_i - \alpha_0)}{\sum_{i=1}^{n} t_i^2} = \frac{SP_{xt} - \alpha_0 S_t}{USS_t} \sim\sim N(\beta, \frac{\sigma^2}{USS_t}),$$

$$\begin{aligned}
\sigma^2 \leftarrow s_{03}^{*2} &= \frac{1}{n-1}\sum_{i=1}^{n}\{x_i - (\alpha_0 + \hat{\beta}_{M_3^*} t_i)\}^2 \\
&= \frac{1}{n-1}\left[USS_x + n\alpha_0^2 - 2\alpha_0 S_x - \hat{\beta}_{M_3^*}^2 USS_t\right] \\
&\sim\sim \sigma^2\chi^2(n-1)/(n-1).
\end{aligned}$$

Test of H_{04}^*: $\beta = \beta_0$:
Test statistic:

$$\begin{aligned}
t(\mathbf{x}) &= \frac{\hat{\beta}_{M_3^*} - \beta_0}{\sqrt{s_{03}^{*2}/USS_t}} \\
&= \frac{SP_{xt} - \alpha_0 S_t - \beta_0 USS_t}{\sqrt{s_{03}^{*2} USS_t}} \\
&\sim\sim t(n-1).
\end{aligned}$$

Significance probability:

$$p_{obs}(\mathbf{x}) = 2\left[1 - F_{t(n-1)}(|\,t(\mathbf{x})\,|)\right].$$

Estimate of the parameter in M_4:

$$\begin{aligned}
\sigma^2 \leftarrow s_{04}^2 &= \frac{1}{n}\sum_{i=1}^{n}\{x_i - (\alpha_0 + \beta_0 t_i)\}^2 \\
&= \frac{1}{n}\left[USS_x + n\alpha_0^2 + \beta_0^2 USS_t - 2\alpha_0 S_x - 2\beta_0 SP_{xt} + 2\alpha_0\beta_0 S_t\right] \\
&\sim\sim \sigma^2\chi^2(n)/(n).
\end{aligned}$$

Test of H_{04}: $\alpha = \alpha_0$:
Test statistic:

$$\begin{aligned}
t(\mathbf{x}) &= \frac{\hat{\alpha}_{M_3} - \alpha_0}{\sqrt{s_{03}^2/n}} \\
&= \frac{S_x - \beta_0 S_t - \alpha_0 n}{\sqrt{s_{03}^2 n}} \\
&\sim\sim t(n-1).
\end{aligned}$$

Significance probability:

$$p_{obs}(\mathbf{x}) = 2\left[1 - F_{t(n-1)}(|\,t(\mathbf{x})\,|)\right].$$

Estimate of the parameter in M_4:

$$\begin{aligned}\sigma^2 \leftarrow s_{04}^2 &= \frac{1}{n}\sum_{i=1}^{n}\{x_i - (\alpha_0 + \beta_0 t_i)\}^2 \\ &= \frac{1}{n}\left[USS_x + n\alpha_0^2 + \beta_0^2 USS_t - 2\alpha_0 S_x - 2\beta_0 SP_{xt} + 2\alpha_0\beta_0 S_t\right] \\ &\sim\sim \sigma^2\chi^2(n)/(n).\end{aligned}$$

Template for calculations in linear regression

	x	t
n	n	
S	S_x	S_t
USS	USS_x	USS_t
SP	SP_{xt}	
SSD	$USS_x - \frac{S_x^2}{n}$	$USS_t - \frac{S_t^2}{n}$
SPD	$SP_{xt} - \frac{S_x S_t}{n}$	
$\hat{\beta}$	$\frac{SPD_{xt}}{SSD_t}$	
$\hat{\alpha}$	$\frac{1}{n}\left[S_x - \hat{\beta} S_t\right]$	
SSD_{02}	$SSD_x - \frac{SPD_{xt}^2}{SSD_t}$	
s_{02}^2	$\frac{1}{n-2} SSD_{02}$	

Supplement to Chapter 3

In statistical inference in models based on the normal distribution the conclusions are often drawn by means of either the χ^2-distribution, the t-distribution, or the F-distribution. Section 3.4 below may be considered as a small catalogue of the most important properties, from a statistical point of view, of the normal distribution and of these three distributions. For each of the distributions the definition and the most basic properties are reviewed and for the χ^2-distribution, the t-distribution, and the F-distribution, the relation to the normal distribution is discussed under the heading *Distributional results*. Furthermore, the use of SAS to calculate distribution functions and fractiles is illustrated under the heading *Calculations in SAS*. Finally, the beta distribution is mentioned briefly.

3.4 Normal Distribution and Related Distributions

3.4.1 Normal Distribution

Definition

A continuous random variable X has a normal distribution with mean μ ($\in \mathbb{R}$) and variance $\sigma^2 (> 0)$ if the probability density function of X is

$$f_X(x) = \frac{1}{\sqrt{2\pi\sigma^2}} e^{-\frac{(x-\mu)^2}{2\sigma^2}}, \quad x \in \mathbb{R}. \tag{3.73}$$

The distribution is denoted by $N(\mu, \sigma^2)$, and if X has the probability density function (3.73), we write $X \sim N(\mu, \sigma^2)$.

The distribution $N(0,1)$ is referred to as the *standard normal distribution* or as the *u-distribution*. Traditionally its probability density function is denoted by φ and its distribution function by Φ, i.e.,

$$\varphi(x) = \frac{1}{\sqrt{2\pi}} e^{-\frac{x^2}{2}}, \quad x \in \mathbb{R} \tag{3.74}$$

and

$$\Phi(x) = \int_{-\infty}^{x} \frac{1}{\sqrt{2\pi}} e^{-\frac{z^2}{2}} dz, \quad x \in \mathbb{R}. \tag{3.75}$$

The probability density function of the standard normal distribution is symmetric around 0:

$$\varphi(-x) = \varphi(x), \quad x \in \mathbb{R}$$

and this is reflected in the distribution function by the fact that

$$\Phi(-x) = 1 - \Phi(x), \quad x \in \mathbb{R}. \tag{3.76}$$

If $X \sim N(\mu, \sigma^2)$, the probability density function of X may be expressed by the probability density function of the standard normal distribution,

$$f_X(x) = \frac{1}{\sigma}\varphi\left(\frac{x-\mu}{\sigma}\right), \tag{3.77}$$

and the distribution function of X may be expressed by the distribution function of the standard normal distribution,

$$F_X(x) = \Phi\left(\frac{x-\mu}{\sigma}\right). \tag{3.78}$$

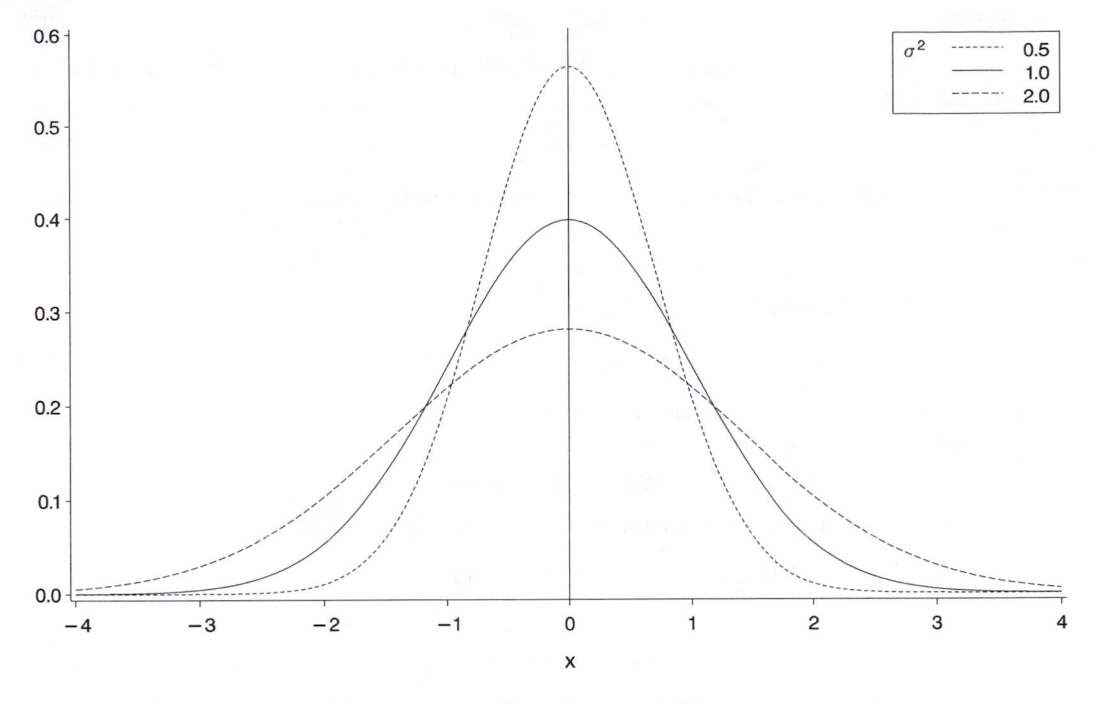

Figure 3.14 *The probability density function of* $N(0, \sigma^2)$ *for* $\sigma^2 = 0.5$, *1.0, and 2.0.*

Mean and Variance

If $X \sim N(\mu, \sigma^2)$, it follows that

$$EX = \mu \tag{3.79}$$

and

$$VarX = \sigma^2. \tag{3.80}$$

Distributional Results

Let $X_1, \ldots, X_n$ be independent random variables such that $X_i \sim N(\mu_i, \sigma_i^2)$, $i = 1, \ldots, n$. If Y is an affine function of the X's, i.e., Y is of the form,

$$Y = c_0 + c_1 X_1 + \cdots + c_n X_n,$$

where $c_0, \ldots, c_n$ are constants, we have

$$Y \sim N(c_0 + c_1\mu_1 + \cdots + c_n\mu_n, c_1^2\sigma_1^2 + \cdots + c_n^2\sigma_n^2). \tag{3.81}$$

If, in addition, the X's are identically distributed, i.e., $X_i \sim N(\mu, \sigma^2)$, we have in particular that

$$X. = \sum_{i=1}^{n} X_i \sim N(n\mu, n\sigma^2) \quad \text{and} \quad \bar{X}. = \frac{1}{n}\sum_{i=1}^{n} X_i \sim N(\mu, \frac{\sigma^2}{n}). \tag{3.82}$$

Finally, notice that (3.81) implies that

$$X \sim N(\mu, \sigma^2) \quad \Leftrightarrow \quad \frac{X - \mu}{\sigma} \sim N(0, 1). \tag{3.83}$$

Calculations in SAS

If u_p and x_p denotes the p-fractile of the $N(0,1)$ distribution and of the $N(\mu,\sigma^2)$ distribution, respectively, we get

$$u_p = \frac{x_p - \mu}{\sigma}. \tag{3.84}$$

The fractiles u_p of the standard normal distribution are related to the so-called *probits* in the following way:

$$probit(p) = u_p + 5, \quad p \in]0,1[. \tag{3.85}$$

Since $u_p = \Phi^{-1}(p)$, formula (3.85) is equivalent to

$$\Phi^{-1}(p) = probit(p) - 5. \tag{3.86}$$

In SAS the function `PROBNORM` calculates the values of the distribution function Φ of the standard normal distribution since

$$\texttt{PROBNORM(x)} = \Phi(x).$$

Thus, for instance, we have with an accuracy of three decimal places that

$$\Phi(2.57) = \texttt{PROBNORM(2.57)} = \texttt{0.995}$$

and

$$\Phi(-1.96) = \texttt{PROBNORM(-1.96)} = \texttt{0.025}.$$

In view of (3.85) it is somewhat strange that SAS uses the name `PROBIT` for the function calculating the values of the function Φ^{-1}, i.e.,

$$\texttt{PROBIT(p)} = \Phi^{-1}(p) = u_p.$$

We have, for instance, that

$$\Phi^{-1}(0.005) = \texttt{PROBIT(0.005)} = \texttt{-2.576}$$

and

$$\Phi^{-1}(0.975) = \texttt{PROBIT(0.975)} = \texttt{1.960}.$$

3.4.2 χ^2-distribution

Definition

The *χ^2-distribution with f degrees of freedom,* denoted $\chi^2(f)$, is a particular case of the *gamma distribution* $\Gamma(\alpha,\lambda)$, which is a continuous distribution on $]0,\infty[$ with probability density function,

$$\gamma(x;\alpha,\lambda) = \frac{\lambda^\alpha}{\Gamma(\alpha)} x^{\alpha-1} e^{-\lambda x}, \quad x \in]0,\infty[. \tag{3.87}$$

Here $\alpha > 0$, $\lambda > 0$ and Γ denotes the gamma function,

$$\Gamma(\alpha) = \int_0^\infty x^{\alpha-1} e^{-x} dx, \quad \alpha > 0.$$

More precisely, the $\chi^2(f)$-distribution is the gamma distribution $\Gamma(f/2,1/2)$. Most applications of the χ^2-distribution in this book is in connection with confidence intervals for variances in normal models and calculation of p-values based on likelihood ratio statistics and therefore the fractiles of the $\chi^2(f)$-distribution are of primary interest.

If Y is a random variable such that $Y/\sigma^2 \sim \chi^2(f)$, we often say that Y has a $\sigma^2\chi^2(f)$-distribution,

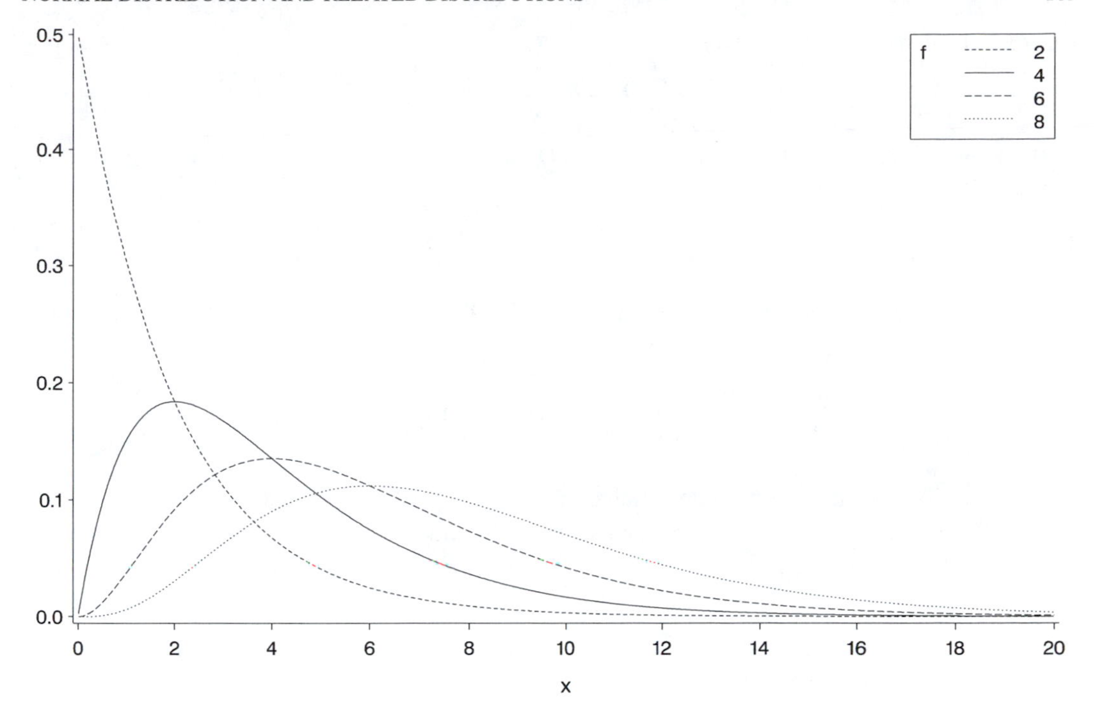

Figure 3.15 *The probability density function of* $\chi^2(f)$ *for* $f = 2, 4, 6$, *and 8.*

and if Z is a random variable such that $fZ/\sigma^2 \sim \chi^2(f)$, we say that Z is $\sigma^2\chi^2(f)/f$-distributed. For completeness we give the density of the $\sigma^2\chi^2(f)/f$-distribution,

$$f_{\sigma^2\chi^2(f)/f}(x;\sigma^2,f) = \frac{(f/2)^{f/2}}{\Gamma(f/2)(\sigma^2)^{f/2}}x^{\frac{f}{2}-1}\exp(-\frac{xf}{2\sigma^2}), \quad x \in]0,\infty[. \tag{3.88}$$

Mean and Variance

If $X \sim \chi^2(f)$, then we get

$$EX = f \tag{3.89}$$

and

$$VarX = 2f. \tag{3.90}$$

Distributional Results

If X_1 and X_2 are independent random variables, we have

$$X_i \sim \chi^2(f_i), \quad i = 1,2 \quad \Rightarrow \quad X_1 + X_2 \sim \chi^2(f_1 + f_2). \tag{3.91}$$

The fundamental connection between the normal distribution and the χ^2-distribution is the result:

$$U \sim N(0,1) \quad \Rightarrow \quad U^2 \sim \chi^2(1). \tag{3.92}$$

Combining the results (3.83), (3.91) and (3.92) we find that if $X_1,\ldots,X_n$ are independent and $N(\mu,\sigma^2)$ distributed, then

$$\sum_{i=1}^{n} \frac{(X_i - \mu)^2}{\sigma^2} \sim \chi^2(n)$$

or, equivalently,

$$\sum_{i=1}^{n}(X_i-\mu)^2 \sim \sigma^2\chi^2(n). \tag{3.93}$$

If the mean μ is replaced by the average $\bar{X}. = (X_1+\cdots+X_n)/n$ of the Xs, it may be shown that

$$\sum_{i=1}^{n}(X_i-\bar{X}.)^2 \sim \sigma^2\chi^2(n-1) \tag{3.94}$$

and, furthermore, that the random variables $\bar{X}.$ and $\sum_{i=1}^{n}(X_i-\bar{X}.)^2$ are stochastically independent. It follows from (3.94) that

$$s^2(\mathbf{X}) = \frac{1}{n-1}\sum_{i=1}^{n}(X_i-\bar{X}.)^2 \sim \sigma^2\chi^2(n-1)/(n-1), \tag{3.95}$$

and, in addition, that the two random variables $\bar{X}.$ and $s^2(\mathbf{X})$, which in statistics are used as estimators of the mean μ and the variance σ^2, are stochastically independent.

Calculations in SAS

The distribution function and the fractiles of the distributions $\chi^2(f)$ and $\chi^2(f)/f$ may be calculated in SAS using the functions `PROBCHI` and `CINV` since

$$\texttt{PROBCHI(x,f)} = F_{\chi^2(f)}(x)$$

and

$$\texttt{CINV(p,f)} = F^{-1}_{\chi^2(f)}(p).$$

As illustrations we have

$$F_{\chi^2(3)}(7.81) = \texttt{PROBCHI(7.81,3)} = \texttt{0.95},$$

$$F^{-1}_{\chi^2(8)}(0.60) = \texttt{CINV(0.60,8)} = \texttt{8.35},$$

$$F_{\chi^2(5)/5}(0.1662) = \texttt{PROBCHI(5*0.1662,5)} = \texttt{0.025}$$

and

$$F^{-1}_{\chi^2(12)/12}(0.95) = \texttt{CINV(0.95,12)/12} = \texttt{1.7522}.$$

3.4.3 t-distribution

Definition

If U and Z are independent random variables such that $U \sim N(0,1)$ and $Z \sim \chi^2(f)/f$, then the quantity,

$$t = \frac{U}{\sqrt{Z}}, \tag{3.96}$$

is *t-distributed with f degrees of freedom* and we write $t \sim t(f)$. Symbolically, the definition of the t-distribution may be reviewed as

$$t(f) = \frac{N(0,1)}{\sqrt{\chi^2(f)/f}}$$

if we remember that numerator and denominator symbolize *independent* random variables.

The distribution is sometimes called the *Student distribution* or *Student's t*-distribution.

As for the χ^2-distribution it is the fractiles of the $t(f)$-distribution that are of primary interest in connection with inference in models based on the normal distribution and not the probability density function itself. This function is

$$f_{t(f)}(x) = \frac{1}{\sqrt{f}B(1/2, f/2)}(1 + f^{-1}x^2)^{-(f+1)/2}, \quad x \in \mathbb{R},$$

where B denotes the beta function defined in formula (3.106) below.

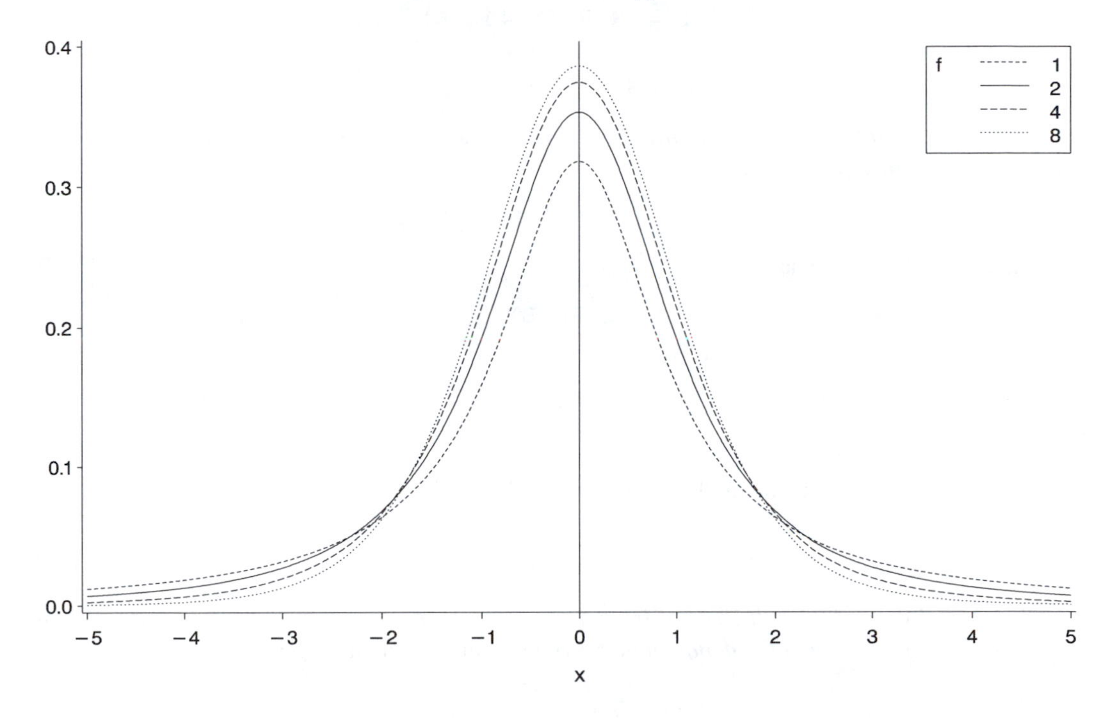

Figure 3.16 *The probability density function of $t(f)$ for $f = 1, 2, 4$, and 8.*

The $t(f)$-distribution converges in law to the standard normal distribution $N(0,1)$ as $f \to \infty$.

Distributional Results

Suppose that $X_1, \ldots, X_n$ are independent and identically $N(\mu, \sigma^2)$-distributed and let $\bar{X}_{\bullet}$ and $s^2(\mathbf{X})$ denote the empirical mean and variance, respectively. It follows from (3.82), (3.95) and (3.96) together with the independence of $\bar{X}_{\bullet}$ and $s^2(\mathbf{X})$ that

$$t = \frac{\bar{X}_{\bullet} - \mu}{\sqrt{s^2(\mathbf{X})/n}} \sim t(n-1). \tag{3.97}$$

Calculations in SAS

The probability density function of the $t(f)$-distribution is symmetric around 0, and this implies that

$$F_{t(f)}(-x) = 1 - F_{t(f)}(x), \quad x \in \mathbb{R}, \tag{3.98}$$

where $F_{t(f)}$ denotes the distribution function of the $t(f)$-distribution. If $t_p(f)$ denotes the p-fractile of the $t(f)$-distribution, formula (3.98) implies that

$$t_{1-p}(f) = -t_p(f), \quad p \in]0,1[. \tag{3.99}$$

In SAS the distribution function and the fractiles of the $t(f)$-distribution may be calculated using the functions PROBT and TINV, respectively, since

$$\texttt{PROBT(x,f)} = F_{t(f)}(x)$$

and

$$\texttt{TINV(p,f)} = F_{t(f)}^{-1}(p) = t_p(f).$$

For instance, we have

$$F_{t(6)}(1.440) = \texttt{PROBT(1.440,6)} = \texttt{0.90}$$

and

$$t_{0.975}(17) = F_{t(17)}^{-1}(0.975) = \texttt{TINV(0.975,17)} = \texttt{2.110}.$$

Finally, note that probabilities of the form $P(|t(f)| \geq x)$, where $t(f)$ symbolizes a $t(f)$ distributed random variable may be calculated as

$$P(|t(f)| \geq x) = \texttt{2*(1-PROBT(x,f))}$$

such that, for instance, we have

$$P(|t(10)| \geq 1.372) = \texttt{2*(1-PROBT(1.372,10))} = \texttt{0.20}.$$

3.4.4 *F*-distribution

Definition

Let Z_1 and Z_2 be independent random variables such that $Z_i \sim \chi^2(f_i)/f_i$, $i = 1,2$. The random variable,

$$F = \frac{Z_1}{Z_2}, \qquad (3.100)$$

is *F-distributed with* (f_1, f_2) *degrees of freedom*, or *with* f_1 *degrees of freedom in the numerator and* f_2 *degrees of freedom in the denominator*. Symbolically, the definition is

$$F(f_1, f_2) = \frac{\chi^2(f_1)/f_1}{\chi^2(f_2)/f_2},$$

where the numerator and the denominator symbolize *independent* random variables.

Again it is the fractiles of the distribution that are of primary interest in statistical inference. The probability density function of the $F(f_1, f_2)$ distribution is

$$f_{F(f_1,f_2)}(x) = \frac{f_1^{f_1/2} f_2^{f_2/2}}{B(f_1/2, f_2/2)} x^{f_1/2-1}(f_2 + f_1 x)^{-(f_1+f_2)/2}, \quad x > 0,$$

where B is the beta function given in formula (3.106).

Distributional Results

In statistics the distribution emerges in analysis of variance where one wishes to compare two empirical variances in a model based on the normal distribution. Assume, for instance, that $X_1, \ldots, X_n$ and $Y_1, \ldots, Y_m$ are independent random variables such that $X_i \sim N(\mu_X, \sigma^2)$, $i = 1, \ldots, n$, and $Y_j \sim N(\mu_Y, \sigma^2)$, $j = 1, \ldots, m$. Note that the variance is assumed to be the same for all the random variables. From (3.95) we find for the empirical variances $s^2(\mathbf{X})$ and $s^2(\mathbf{Y})$ that

$$s^2(\mathbf{X}) = \frac{1}{n-1} \sum_{i=1}^{n} (X_i - \bar{X}.)^2 \sim \sigma^2 \chi^2(n-1)/(n-1)$$

and

$$s^2(\mathbf{Y}) = \frac{1}{m-1} \sum_{j=1}^{m} (Y_j - \bar{Y}.)^2 \sim \sigma^2 \chi^2(m-1)/(m-1).$$

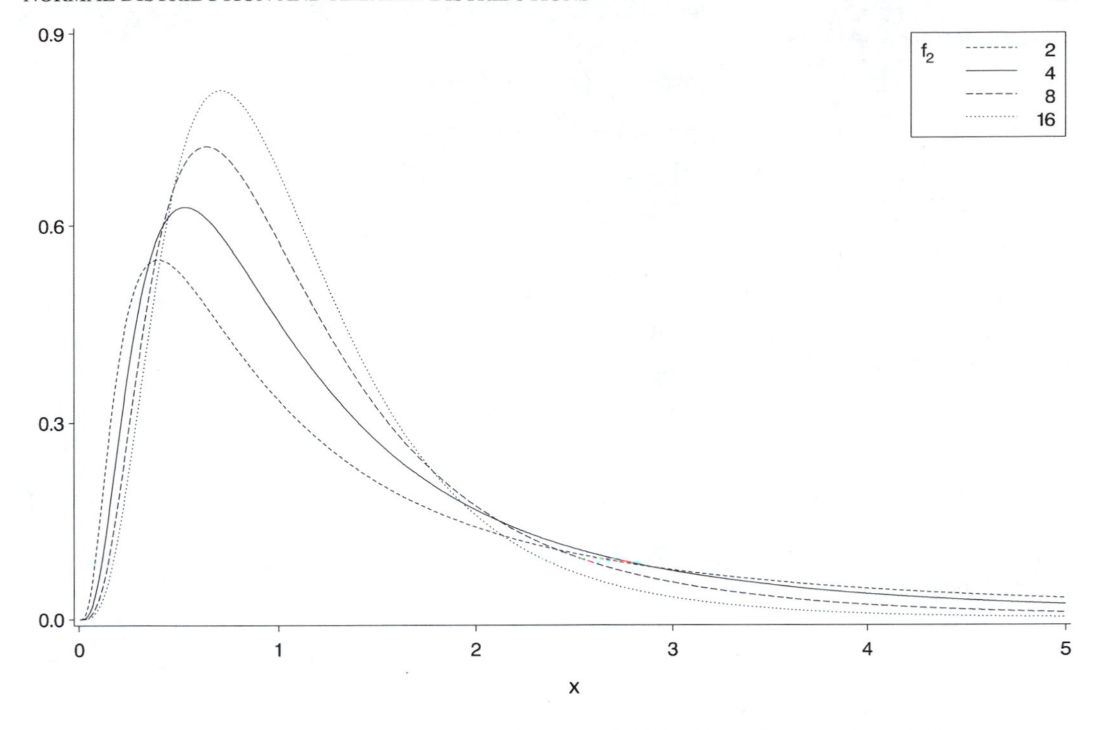

Figure 3.17 *The probability density function of* $F(10, f_2)$ *for* $f_2 = 2$, *4, 8, and 16.*

Since $s^2(\mathbf{X})$ and $s^2(\mathbf{Y})$ are stochastically independent, formula (3.100) implies that

$$F = \frac{s^2(\mathbf{X})}{s^2(\mathbf{Y})} \sim F(n-1, m-1).$$

Furthermore, the formulas (3.92), (3.96) and (3.100) imply that

$$t \sim t(f) \quad \Rightarrow \quad t^2 \sim F(1, f). \tag{3.101}$$

Finally, note that from (3.100) we have

$$X \sim F(f_1, f_2) \quad \Rightarrow \quad \frac{1}{X} \sim F(f_2, f_1),$$

which may be used to prove

$$F_{F(f_1,f_2)}(x) = 1 - F_{F(f_2,f_1)}(\frac{1}{x}), \quad x > 0. \tag{3.102}$$

This equation is useful when significance probabilities must be found using a table of the F-distribution. These tables are often restricted to the upper tail of the distributions, i.e., to probabilities greater than 0.5.

Calculations in SAS

In SAS the distribution function $F_{F(f_1,f_2)}$ and the p-fractiles $F_p(f_1, f_2)$ of the $F(f_1, f_2)$-distribution is calculated by means of the functions PROBF and INVF, respectively:

$$\texttt{PROBF(x,f}_1\texttt{,f}_2\texttt{)} = F_{F(f_1,f_2)}(x)$$

and

$$\texttt{INVF(p,f}_1\texttt{,f}_2\texttt{)} = F_p(f_1, f_2).$$

Thus, for instance, we have

$$F^{-1}_{F(9,15)}(0.95) = F_{0.95}(9,15) = \texttt{INVF(0.95,9,15)} = \texttt{2.59}$$

and

$$F_{F(13,6)}(7.66) = \texttt{PROBF(7.66,13,6)} = \texttt{0.99}.$$

3.4.5 Beta Distribution

The beta distribution is related to the gamma distribution in the following way. If X_1 and X_2 are independent random variables such that $X_i \sim \Gamma(\alpha_i, \lambda)$, $i = 1, 2$, then the random variables $X_1 + X_2$ and $X_1/(X_1 + X_2)$ are independent and, furthermore,

$$X_1 + X_2 \sim \Gamma(\alpha_1 + \alpha_2, \lambda) \tag{3.103}$$

and

$$\frac{X_1}{X_1 + X_2} \sim B(\alpha_1, \alpha_2), \tag{3.104}$$

where $B(\alpha_1, \alpha_2)$ denotes the *beta distribution* with *parameters* $\alpha_1 > 0$ and $\alpha_2 > 0$ whose probability density function is

$$\beta(x; \alpha_1, \alpha_2) = \frac{1}{B(\alpha_1, \alpha_2)} x^{\alpha_1 - 1}(1 - x)^{\alpha_2 - 1}, \quad x \in [0, 1]. \tag{3.105}$$

Here $B(\alpha_1, \alpha_2)$ is the beta function given by

$$B(\alpha_1, \alpha_2) = \int_0^1 x^{\alpha_1 - 1}(1 - x)^{\alpha_2 - 1} dx. \tag{3.106}$$

If X is a random variable with this probability density function, we write $X \sim B(\alpha_1, \alpha_2)$. (Thus we use the symbol $B(\alpha_1, \alpha_2)$ in two senses, namely as the symbol of the beta function and also as the symbol of the beta distribution.) The beta function can be expressed by the gamma function since

$$B(\alpha_1, \alpha_2) = \frac{\Gamma(\alpha_1)\Gamma(\alpha_2)}{\Gamma(\alpha_1 + \alpha_2)}. \tag{3.107}$$

Distributional Results

The beta distribution is related to the F-distribution by simple monotone transformations. One relation is the following:

$$X \sim F(f_1, f_2) \quad \Rightarrow \quad \frac{f_1 X}{f_2 + f_1 X} \sim B(f_1/2, f_2/2), \tag{3.108}$$

with its reverse

$$X \sim B(f_1/2, f_2/2) \quad \Rightarrow \quad \frac{f_2 X}{f_1(1 - X)} \sim F(f_1, f_2). \tag{3.109}$$

3.5 Multivariate Normal Distributions

Definition

Let $\mathbf{U} = (U_1, \ldots, U_m)^*$, where $U_1, \ldots, U_m$ are independent and $N(0, 1)$-distributed. Suppose that k is an integer, let $\boldsymbol{\mu} = (\mu_1, \ldots, \mu_k)^* \in \mathbb{R}^k$ and let B be a $k \times m$ matrix. Then the distribution of the k- dimensional random vector $\mathbf{X} = \boldsymbol{\mu} + B\mathbf{U}$ is the *k-dimensional normal distribution* with *vector of means* $\boldsymbol{\mu}$ and *covariance matrix* $\Sigma = BB^*$, where B^* is the transpose of B. The distribution is denoted $N_k(\boldsymbol{\mu}, \Sigma)$.

Vector of Means and Covariance Matrix

If $\mathbf{X} = (X_1, \ldots, X_k)^* \sim N_k(\boldsymbol{\mu}, \Sigma)$, then $\boldsymbol{\mu} = (\mu_1, \ldots, \mu_k)^*$ and $\Sigma = \{\sigma_{ij}\}$ are the vector of means and the covariance matrix of $\mathbf{X}$, i.e.,

$$\mu_i = EX_i, \quad i = 1, \ldots, k, \tag{3.110}$$

$$\sigma_{ij} = Cov(X_i, X_j) = E\{(X_i - \mu_i)(X_j - \mu_j)\}, \quad i \neq j = 1, \ldots, k, \tag{3.111}$$

and

$$\sigma_{ii} = VarX_i = E\{(X_i - \mu_i)^2\}, \quad i = 1, \ldots, k. \tag{3.112}$$

Distributional Results

The distributional theory related to the k-dimensional normal distributions is very comprehensive, but we restrict ourselves to mention two results that are used in this book.

(i) Suppose that $\mathbf{X} = (X_1, \ldots, X_k)^* \sim N_k(\boldsymbol{\mu}, \Sigma)$. Then the ith component X_i of $\mathbf{X}$ is normally distributed, more precisely,

$$X_i \sim N(\mu_i, \sigma_{ii}), \tag{3.113}$$

i.e., the mean of the ith component X_i of $\mathbf{X}$ is the ith component μ_i of the vector of means $\boldsymbol{\mu}$ and the variance of X_i is the ith diagonal element σ_{ii} of the covariance matrix Σ.

(ii) If $\mathbf{X} \sim N_k(\boldsymbol{\mu}, \Sigma)$, where Σ has an inverse Σ^{-1}, then,

$$(\mathbf{X} - \boldsymbol{\mu})^* \Sigma^{-1} (\mathbf{X} - \boldsymbol{\mu}) \sim \chi^2(k). \tag{3.114}$$

Exercises for Chapter 3

SAS programs for the exercises can be found at the address
`http://www.imf.au.dk/biogeostatistics/`

Exercise 3.1 In online monitoring of production of keys for a keyboard, keys were sampled. The weight in milligrams of a sample of 15 keys is given in the table below:

1215	1181	1195	1189	1185
1206	1192	1203	1213	1156
1164	1177	1167	1206	1193

(1) Show that the observations may be assumed to be normally distributed.

The following questions may be answered assuming that the observations are normally distributed. Standard calculations for the 15 observations give the following results:

$$S = 17842 \qquad USS = 21226970$$

(2) The target value for the production is a mean of 1200 mg for the keys. Do the observations give us any reason to believe that the process has drifted from the target?

(3) Compute a 95% confidence interval for the mean of the weight of the keys.

(4) Compute a 95% confidence interval for the standard deviation of the weight of the keys.

(5) Can the standard deviation be assumed to be 20 milligrams?

Exercise 3.2 Two measurement methods used to determine the content of CaO in pieces of rock are being compared. For this purpose the content of CaO in nine pieces of rock has been determined by both methods.

Method 1	*Method* 2	*Difference*
10.4	10.6	−0.2
9.9	9.7	0.2
9.1	8.8	0.3
9.6	8.9	0.7
8.5	8.4	0.1
7.4	6.8	0.6
8.1	7.9	0.2
6.6	6.3	0.3
7.2	6.6	0.6

(1) Show that it may be assumed that the difference between the determinations of the content of CaO in the same piece of rock by the two methods is normally distributed.

(2) Do the two measurements methods give the same determinations of the CaO content?

(3) Compute a 95% confidence interval for the mean of the differences.

The table below contains the standard calculations for the data in this exercise.

	n	S	USS
Method 1	9	76.8	669.16
Method 2	9	74.0	625.56
Difference	9	2.8	1.52

Exercise 3.3 It is often argued that when measurements of a physical quantity are repeated twice by the same person, there is a tendency that the first result influences the second one. In order to investigate this, 20 test samples were randomly divided into 2 groups of 10 samples. In the tables below, the test samples are numbered from 1 to 20 to indicate the measurements from the different samples. The test samples contained slightly varying quantities of a chemical substance B.

A laboratory assistant received the first 10 samples and was asked to determine the content of B in each sample twice. The results were:

Measurement	S_1	S_2	S_3	S_4	S_5	S_6	S_7	S_8	S_9	S_{10}
1	5.25	6.10	5.70	5.76	5.00	6.04	5.49	5.33	4.84	5.89
2	5.17	6.19	5.61	5.67	4.94	6.02	5.55	5.42	4.94	5.81

Next, each test sample in the second group was divided into 2 portions, and the resulting 20 portions were given to the laboratory assistant in random order and in such a way that he could not identify the samples that were subsamples of the same test sample. The results were as follows:

Measurement	S_{11}	S_{12}	S_{13}	S_{14}	S_{15}	S_{16}	S_{17}	S_{18}	S_{19}	S_{20}
3	4.79	5.61	5.33	5.86	5.42	4.91	5.83	5.85	5.25	5.01
4	4.80	5.55	5.03	6.11	5.32	5.04	6.02	5.77	5.39	5.22

(1) What is the conclusion of the experiment regarding the allegation?

(2) What is the standard error of the measurement method?

Exercise 3.4 In an attempt to quantify the pollution, the content of SO_2 (in ppm) in the air has been measured in a rural area and in an urban area. Analyze the data.

Rural area	3.3	0.6	0.8	2.1	0.2	1.2	0.5	0.1			
Urban area	24.7	2.5	15.0	3.2	5.0	14.1	5.0	22.3	2.0	17.3	10.1

Exercise 3.5 (Prediction intervals)

When we want to make statements about a new observation based on data and a model for those data, it is called *prediction*. We will not attempt to predict the exact value of the new observation but rather an interval that will contain the new observation with an asserted probability, such as, for example, a 95% *prediction interval*.

Consider a normal sample,

$$X_i \sim N(\mu, \sigma^2), \qquad i = 1, \ldots, n,$$

and let Y denote a random variable, which is independent of $X_1, \ldots, X_n$ and $N(\mu, \sigma^2)$-distributed.

(1) Show that

$$\frac{Y - \bar{X}.}{s(\mathbf{X})\sqrt{1+1/n}} \sim t(n-1),$$

and that the inequality,

$$\bar{X}. - t_{1-\alpha/2}(n-1)s(\mathbf{X})\sqrt{1+1/n} < Y < \bar{X}. + t_{1-\alpha/2}(n-1)s(\mathbf{X})\sqrt{1+1/n}, \tag{3.115}$$

holds with probability $1-\alpha$. The interval (3.115) with $\bar{X}.$ and $s(\mathbf{X})$ replaced by the observed values $\bar{x}.$ and s from the sample is a $(1-\alpha)$ prediction interval.

(2) Compare the $(1-\alpha)$ confidence interval for the mean to the $(1-\alpha)$ prediction interval for a new observation.

(3) Consider the linear regression model of Section 3.3 and give a $(1-\alpha)$ prediction interval for a new observation that is independent of the original sample and with the value t of the explanatory variable.

Exercise 3.6 The data in this exercise is part of a larger investigation of caries among school children and consist of the observed number of DMF teeth (Damaged, Missing, or Filled) of 12-year-old boys. The purpose of the investigation is to shed light on the assumption that a high natural content of fluoride in the water will diminish the number of DMF teeth. The investigation was conducted in the 1960s before fluoride was allowed to be added to toothpaste in Denmark. Four waterworks with different amount of natural fluoride in the water were selected and all the 12-year-old boys whose homes were supplied by one of the four waterworks were examined.

The data are shown in Table 3.13 divided into the four groups defined by the waterworks. The average concentration of fluoride in the water from the four waterworks is shown in the bottom row of the table as ppm F (mg fluoride per kg). In Vejen, for example, it was found that 3 12-year-old boys had 4 DMF teeth, 4 had 6 DMF teeth, and so on. The square root of the number of DMF teeth is also given in the table. Theoretical considerations suggest that the square root of the number of DMF teeth may be normally distributed.

(1) Use normal plots to see if the square root of the number of DMF teeth among 12-year-old boys in each of the four groups can be assumed to be normally distributed.

Based on the normal plots in (1) we conclude that we can consider the square root of the number of DMF teeth to be normally distributed. The square root of the number of DMF teeth will be considered further in the rest of the exercise. In the two tables below the results of some standard calculations are given.

(2) Investigate whether the variance of the square root of the number of DMF teeth can be assumed to be the same for the four waterworks.

(3) Investigate whether the square root of the number of DMF teeth can be assumed to depend linearly on the fluoride content of the drinking water.

(4) Investigate whether the square root of the number of DMF teeth can be assumed to be independent of the fluoride content of the drinking water.

The number of observations, the sums and the uncorrected sums of squares of the square root of the number of DMF teeth in the four samples are given in the following table.

i	n	S	USS
1	39	127.91	434.8519
2	122	379.42	1233.2858
3	40	90.84	219.0978
4	45	90.05	193.0717
Sum	246	688.22	2080.3072

Table 3.13 *Number of DMF teeth among 12-year-old boys living in communities serviced by four waterworks*

Number of DMF teeth	*Square root of the number of DMF teeth*	*Vejen*	*Slagelse*	*Næstved new waterwork*	*Næstved old waterwork*
0	0.00				1
1	1.00				1
2	1.41		1	6	8
3	1.73		2	5	3
4	2.00	3	8	6	13
5	2.24		8	3	7
6	2.45	4	7	7	8
7	2.65	2	8	4	2
8	2.83	1	13	3	
9	3.00	2	7	4	2
10	3.16	5	14	1	
11	3.32	2	10		
12	3.46	5	8		
13	3.61	4	12	1	
14	3.74	4	4		
15	3.87		7		
16	4.00	2	7		
17	4.12	3	2		
18	4.24	2	1		
19	4.36		1		
20	4.47				
21	4.58		2		
Fluoride content in drinking water in ppm F		0.05	0.34	1.20	1.90

Standard calculations for a linear regression analysis are given in the table below.

	Square root of the number of DMF teeth x	*Fluoride content in drinking water* t
n	246	
S	688.22	176.93
USS	2080.3072	234.2507
SP	415.5013	

Exercise 3.7 The data for this exercise are from a study of the Noble fir seed chalcid *Megastimus pinus* conducted by Trine Iversen from the Institute of Biological Sciences, Aarhus University.

In the spring, the Noble fir seed chalcid oviposits in the seeds of the noble fir. Each cone of the noble fir contains 500–600 seeds, and each seed can at most contain one egg. The egg develops to a larva which eats everything in the seed. The larva remains in the seed until the fall when a fully developed seed chalcid emerges from the seed.

One spring a number of cones were collected. In order to get an idea of the number of infected seeds in each cone, 100 seeds from each cone were X-rayed and the number of seeds containing a larva were counted. This number divided by 100 is called the relative frequency of infection, p_{inf}, of the cone.

Theoretical considerations indicate that the transformed relative frequency of infection

$$x_{inf} = \sin^{-1}(\sqrt{p_{inf}})$$

is approximately normally distributed with a constant variance. This will be partially checked in the sequel.

One of the purposes of the study is to investigate whether there is a connection between physical features of the cone and the relative frequency of infection. Here we restrict attention to the diameter that was measured in millimeter to an accuracy of 5 millimeters. Due to the poor accuracy, only four different diameters were recorded: 40 mm, 45 mm, 50 mm, or 55 mm. The diameters and the transformed relative frequency of infection are given in the following table.

Diameter	*Transformed relative frequency of infection*								
40	0.60	0.61	0.72	0.82	0.82	0.84	0.87	0.87	0.89
	0.89	0.95	0.96	1.00	1.02	1.07			
45	0.50	0.66	0.76	0.76	0.81	0.82	0.82	0.82	0.86
	0.87	0.88	0.88	0.89	0.93	0.94	0.97	0.97	0.97
50	0.54	0.54	0.62	0.67	0.67	0.69	0.69	0.73	0.74
	0.76	0.77	0.78	0.79	0.79	0.80	0.82	0.84	0.85
	0.86	0.91	0.95	0.95					
55	0.42	0.45	0.52	0.59	0.59	0.60	0.61	0.62	0.63
	0.72	0.95							

(1) Find and check a model that can be used to answer the following questions.

In the two tables below the results of some standard calculations are given.

(2) Show that it may be assumed that the variance of x_{inf} does not depend on the diameter.

(3) Investigate whether it can be assumed that the mean of x_{inf} depend linearly of the diameter.

(4) Give estimates and confidence intervals in the final model.

Diameter	*Number of observations*	$SSD_{(i)}$	$f_{(i)}$	*Sample variance*
40	15	0.26264	14	0.01876
45	18	0.23869	17	0.01404
50	22	0.27033	21	0.01287
55	11	0.19889	10	0.01989
Total	66	0.97055	62	0.01565

	x_{inf}	*Diameter*
S	51.50	3115
USS	41.6492	148725
SP	2403.65	

Exercise 3.8 Data for this exercise is part of a larger investigation of the occurrence of fungi in the soil of the moor, and in particular a group of fungi which grows on the roots of the heather and is called *Mykorrhiza*. The investigation was performed by Marianne Johansson and Torben Riis-Nielsen, Botanical Institute, University of Copenhagen.

On an area of the moor five locations are chosen at random and a cylinder of soil is taken with an earth auger. The roots are washed, the fungi in the roots are quantified under a microscope, and the percentage of root cells with fungi is calculated.

In order to investigate a possible variation over the year, this harvest of roots is repeated four times during a year. Harvest 1 is taken in October 1992, harvest 2 in February 1993, harvest 3 in May 1993, and harvest 4 in August 1993.

The soil on the moor consists of two characteristic layers, moor and bleached sand. The whole procedure is repeated twice; once for roots in the moor layer and once for roots in the layer of bleached sand. Table 3.14 contains data from the moor layer and Table 3.15 contains data from the layer of bleached sand.

Table 3.14 *Data for the moor layer*

Mykorrhiza percentage in the moor layer			
Harvest 1	*Harvest* 2	*Harvest* 3	*Harvest* 4
37.00	42.00	46.75	38.00
44.50	31.00	43.50	47.50
47.25	43.25	37.50	28.25
47.50	46.75	50.50	35.50
45.75	37.25	47.50	39.75

Table 3.15 *Data for the layer of bleached sand*

Mykorrhiza percentage in the layer of bleached sand			
Harvest 1	*Harvest* 2	*Harvest* 3	*Harvest* 4
43.75	52.00	47.50	34.25
40.50	47.00	45.00	51.25
49.75	38.75	38.50	38.25
39.00	41.00	51.25	39.50
39.50	41.50	55.00	41.50

A closer examination shows that it can be assumed that the Mykorrhiza percentages from the same harvest and the same layer are independent observations from the same normal distribution.

In answering the questions the standard calculations given in Tables 3.16 and 3.17 may be useful. In the first two questions only the data from the moor layer are considered.

(1) Show that it can be assumed that the variance of the Mykorrhiza percentage does not depend on the time of the harvest.

(2) Show that it can be assumed that the mean of the Mykorrhiza percentage does depend on the time of the harvest.

An analogous analysis for the layer of bleached sand leads to the same model; neither the variance nor the mean of the Mykorrhiza percentage depend on the time of the harvest.

Table 3.16 *Standard calculations for the moor layer*

```
                         Data set: MOOR
                Response variable: MYK_PCT  Mykorrhiza percentage
                   Group variable: TIME

                Calculations in k samples:

                                                                Sample       Sample
i      ni      Si           USSi         Si2/ni        SSDi      fi  Variance       Mean

1       5   222.000     9931.1250     9856.8000     74.3250    4   18.58125    44.4000
2       5   200.250     8168.6875     8020.0125    148.6750    4   37.16875    40.0500
3       5   225.750    10290.5625    10192.6125     97.9500    4   24.48750    45.1500
4       5   189.000     7338.6250     7144.2000    194.4250    4   48.60625    37.8000
---------------------------------------------------------------========================
       20   837.000    35729.0000    35213.6250    515.3750   16   32.21094
```

Table 3.17 *Standard calculations for the layer of bleached sand*

```
                         Data set: BLEACHED_SAND
                Response variable: MYK_PCT  Mykorrhiza percentage
                   Group variable: TIME

                Calculations in k samples:

                                                                Sample       Sample
i      ni      Si           USSi         Si2/ni        SSDi      fi  Variance       Mean

1       5   212.500     9110.6250     9031.2500     79.3750    4   19.84375    42.5000
2       5   220.250     9817.8125     9702.0125    115.8000    4   28.95000    44.0500
3       5   237.250    11415.0625    11257.5125    157.5500    4   39.38750    47.4500
4       5   204.750     8545.1875     8384.5125    160.6750    4   40.16875    40.9500
---------------------------------------------------------------========================
       20   874.750    38888.6875    38375.2875    513.4000   16   32.08750
```

(3) Investigate whether the variance of the Mykorrhiza percentage can be assumed to be the same for the moor layer and the layer of bleached sand.

(4) Investigate whether the mean of Mykorrhiza percentage can be assumed to be the same in the moor layer and in the layer of bleached sand.

CHAPTER 4

Linear Normal Models

In Chapters 3.1, 3.2, and 3.3, we considered a variety of different models for normally distributed data: one sample, two samples, k samples, and linear regression. Although the models are different, a difference that is further emphasized by the different scientific questions that the models are used to answer, from a mathematical point of view they are merely examples of *the linear normal model*. This is the reason why the calculations in statistical packages can be handled by the same procedure. In SAS this procedure is *PROC GLM*.

Many other models in addition to those already mentioned and treated in Chapter 3 are examples of the linear normal model. Therefore, it is worthwhile to master the linear normal model. We first treat the model in general terms in Section 4.1 showing how to estimate the parameters and test hypotheses in the model. We also point out how the important statistics are found in the output from a statistical package. Throughout the chapter we use the well-known examples from Chapter 3: one sample, two samples, k samples, and linear regression. New examples of the linear normal model will be comparison of two or more regression lines in Section 4.2 and two-way analysis of variance in Section 4.3.

Up to now we have emphasized that we should be able to make the calculations both by hand and using the output from a statistical package. In the general treatment of the linear normal model, we will rely on the statistical package to do the heavy calculations and only use the pocket calculator to make an F-test or calculate a confidence interval based on a few statistics that are available in the output from a statistical package.

4.1 The Linear Normal Model

Let $x_1,\ldots,x_n$ be observed values of a sample of independent normally distributed random variables, $X_1,\ldots,X_n$, where

$$X_i \sim N(\mu_i,\sigma^2), \qquad i=1,\ldots,n.$$

Notice that all random variables are assumed to have the same variance. This is an assumption that must be checked, if possible. We consider the observations as a vector in $\mathbb{R}^n$,

$$\mathbf{x} = (x_1,\ldots,x_n)^*,$$

and, accordingly, also the means of the random variables are considered as a vector, the *vector of means*,

$$\boldsymbol{\mu} = (\mu_1,\ldots,\mu_n)^*.$$

We will consider the vectors to be column vectors and use * to denote the transpose of a vector. The models are specified by restricting the vector of means to a linear subspace of $\mathbb{R}^n$. By the phrase *the linear model M specified by the subspace L*, we shall understand that the observations, $x_1,\ldots,x_n$, are realizations of independent random variables, $X_1,\ldots,X_n$, with $X_i \sim N(\mu_i,\sigma^2)$ and the vector of means belonging to L, i.e.,

$$\boldsymbol{\mu} = (\mu_1,\ldots,\mu_n)^* \in L.$$

We will use the shorthand notation,

$$M\colon \boldsymbol{\mu} \in L,$$

for this model and tacitly assume the independence as well as the normal distributions of the individual observations.

We have previously mentioned that the number of parameters of the vector of means of a model determines the number of degrees of freedom f in the $\chi^2(f)/f$-distribution of the estimator of the variance and hence also the degrees of freedom of the F-tests, see page 123 or page 155. A precise mathematical formulation of the number of parameters of the vector of means of a linear normal model is the dimension of the subspace L that defines the model. We shall therefore repeatedly record the dimension, d, of the vector space L. A precise version of the rule for finding the degrees of freedom of the estimate of variance in a linear normal model is: *The degrees of freedom of the estimate of variance in a linear normal model are the number of observations minus the dimension of the subspace for the vector of means.*

Example 4.1
(One normal sample) Here $\mu_i = \mu$ for all i, so

$$\boldsymbol{\mu} = (\mu_1, \ldots, \mu_n)^* = (\mu, \ldots, \mu)^* = \mu(1, \ldots, 1)^*.$$

One observes that the model for one normal sample is specified by the subspace L that is generated by the vector $\mathbf{e} = (1, \ldots, 1)^*$, i.e.,

$$L = \text{span}(\mathbf{e}),$$

and the dimension of L is $d = 1$. □

Example 4.2
(Two normal samples) Here $n = n_1 + n_2$ and $\mu_i = \eta_1$ for $i = 1, \ldots, n_1$ and $\mu_i = \eta_2$ for $i = n_1 + 1, \ldots, n_1 + n_2$. Therefore,

$$\begin{aligned} \boldsymbol{\mu} &= (\mu_1, \ldots, \mu_n)^* = (\eta_1, \ldots, \eta_1, \eta_2, \ldots, \eta_2)^* \\ &= \eta_1(1, \ldots, 1, 0, \ldots, 0)^* + \eta_2(0, \ldots, 0, 1, \ldots, 1)^* \\ &= \eta_1 \mathbf{e}_1 + \eta_2 \mathbf{e}_2, \end{aligned}$$

where the coordinates of $\mathbf{e}_1$ and $\mathbf{e}_2$ are given by

$$(\mathbf{e}_1)_j = \begin{cases} 1 & \text{for } j = 1, \ldots, n_1, \\ 0 & \text{for } j = n_1 + 1, \ldots, n_1 + n_2, \end{cases}$$

and

$$(\mathbf{e}_2)_j = \begin{cases} 0 & \text{for } j = 1, \ldots, n_1, \\ 1 & \text{for } j = n_1 + 1, \ldots, n_1 + n_2. \end{cases}$$

Thus the model for two normal samples is specified by the subspace L, which is generated by the vectors $\mathbf{e}_1$ and $\mathbf{e}_2$, i.e.,

$$L = \text{span}(\mathbf{e}_1, \mathbf{e}_2).$$

The dimension of L is $d = 2$ because $\mathbf{e}_1$ and $\mathbf{e}_2$ are linearly independent. □

Example 4.3
(k normal samples) The situation with k samples is analogous to the one with exactly two samples and one observes that the model for k samples is the linear normal model specified by the subspace,

$$L = \text{span}\{\mathbf{e}_1, \ldots, \mathbf{e}_k\},$$

where the vectors are defined by the allocation of observations to the k samples as follows:

$$(\mathbf{e}_i)_j = \begin{cases} 1 & \text{if } x_j \text{ belongs to the } i\text{th sample}, \\ 0 & \text{otherwise.} \end{cases}$$

The dimension of L is k because $\mathbf{e}_1, \ldots, \mathbf{e}_k$ are linearly independent. □

Example 4.4
(Linear regression) Here $\mu_i = \alpha + \beta t_i$ and

$$\boldsymbol{\mu} = (\alpha+\beta t_1, \ldots, \alpha+\beta t_n)^* = \alpha(1, \ldots, 1)^* + \beta(t_1, \ldots, t_n)^*.$$

It follows that $L = \text{span}(\mathbf{e}, \mathbf{t})$ where $\mathbf{e}$ is defined in Example 4.1 and

$$\mathbf{t} = (t_1, \ldots, t_n)^*.$$

The dimension of L is 2 provided that $\mathbf{t}$ is not proportional to $\mathbf{e}$, or, in other words, that the explanatory variable t has at least two different values. □

4.1.1 Estimation

The likelihood function is

$$\begin{aligned} L(\boldsymbol{\mu}, \sigma^2) &= \prod_{i=1}^{n} \frac{1}{\sqrt{2\pi\sigma^2}} e^{-\frac{1}{2\sigma^2}(x_i-\mu_i)^2} \\ &= (2\pi\sigma^2)^{-\frac{n}{2}} e^{-\frac{1}{2\sigma^2}\sum_{i=1}^{n}(x_i-\mu_i)^2} \\ &= (2\pi\sigma^2)^{-\frac{n}{2}} e^{-\frac{1}{2\sigma^2}\|\mathbf{x}-\boldsymbol{\mu}\|^2}, \end{aligned} \tag{4.1}$$

where the length of a vector $\mathbf{v} = (v_1, \ldots, v_n)^*$ is denoted by double brackets $\|\ \|$ and it is defined by

$$\|\mathbf{v}\| = \sqrt{v_1^2 + \cdots + v_n^2}.$$

We first maximize the likelihood function as a function of $\boldsymbol{\mu}$ for σ^2 fixed. Because of the minus in the exponent, maximizing $L(\boldsymbol{\mu}, \sigma^2)$ with respect to $\boldsymbol{\mu}$ corresponds to minimizing the squared length $\|\mathbf{x} - \boldsymbol{\mu}\|^2$ with respect to $\boldsymbol{\mu}$. The maximum likelihood estimate $\hat{\boldsymbol{\mu}}$ for $\boldsymbol{\mu}$ is that value of $\boldsymbol{\mu}$ in L which satisfies

$$\|\mathbf{x} - \hat{\boldsymbol{\mu}}\|^2 = \min_{\boldsymbol{\mu} \in L} \|\mathbf{x} - \boldsymbol{\mu}\|^2. \tag{4.2}$$

Thus the estimate $\hat{\boldsymbol{\mu}}$ of the vector of means is that point in L which has the shortest distance to $\mathbf{x}$ or, in other words, is closest to $\mathbf{x}$.

Mathematicians can prove that there is exactly one point in L that is closest to $\mathbf{x}$. This point is called the *projection of* $\mathbf{x}$ *onto* L. We will denote it $P(\mathbf{x})$. It can furthermore be proved that $P(\mathbf{x})$ is characterized by the fact that $\mathbf{x} - P(\mathbf{x})$ is orthogonal to any vector in L. As a consequence of this the triangle with corners at $\mathbf{0}$, $\mathbf{x}$ and $P(\mathbf{x})$ is a right-angled triangle, as indicated in Figure 4.1.

Furthermore, $P(\mathbf{x})$ can always be calculated once the vectors that span L are known. How this is done need not concern us. This is one of the heavy calculations that are left to the statistical package.

We have now found the maximum likelihood estimate of $\boldsymbol{\mu}$ and we note the important property that it does not depend on σ^2. For any σ^2 we have

$$L(\boldsymbol{\mu}, \sigma^2) \leq L(P(\mathbf{x}), \sigma^2) = \left(\frac{1}{2\pi\sigma^2}\right)^{\frac{n}{2}} e^{-\frac{1}{2\sigma^2}\|\mathbf{x}-P(\mathbf{x})\|^2}. \tag{4.3}$$

When the right-hand side of (4.3) is maximized as a function of σ^2, we obtain the solution

$$\hat{\sigma}^2 = \frac{1}{n}\|\mathbf{x} - P(\mathbf{x})\|^2. \tag{4.4}$$

For later use we note that if $\hat{\sigma}^2$ is substituted for σ^2 on the right-hand side of (4.3), one finds the

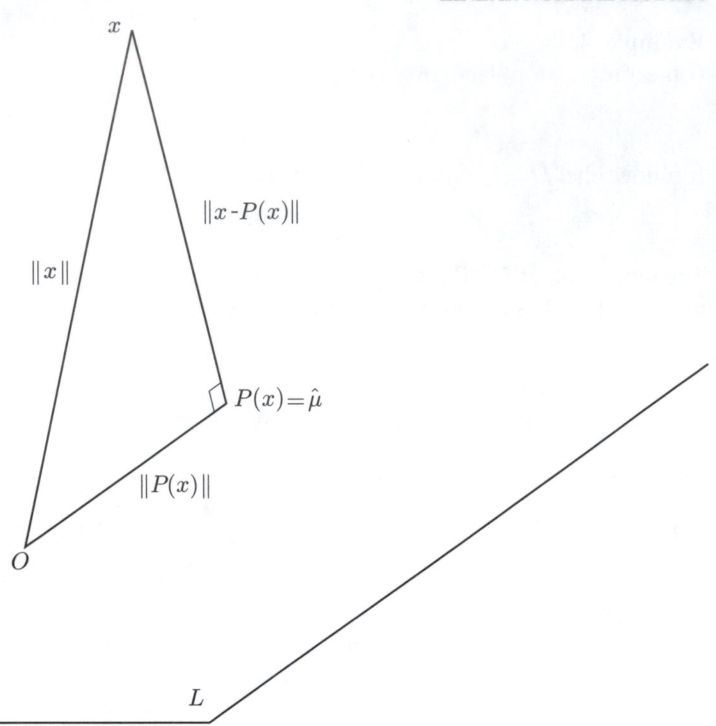

Figure 4.1 *The geometry behind the estimation in a linear normal model.*

maximum of the likelihood function,

$$L(\hat{\boldsymbol{\mu}}, \hat{\sigma}^2) = L(P(\mathbf{x}), \frac{1}{n}\|\mathbf{x} - P(\mathbf{x})\|^2) = \left(\frac{n}{2\pi\|\mathbf{x} - P(\mathbf{x})\|^2}\right)^{\frac{n}{2}} e^{-\frac{n}{2}}. \tag{4.5}$$

It can be proved that

$$\|\mathbf{x} - P(\mathbf{x})\|^2 \sim\sim \sigma^2\chi^2(n-d), \tag{4.6}$$

where d is the dimension of L, and as an estimate of σ^2 one often uses

$$s^2 = \frac{1}{n-d}\|\mathbf{x} - P(\mathbf{x})\|^2 \sim\sim \sigma^2\chi^2(n-d)/(n-d), \tag{4.7}$$

which has mean value σ^2. In accordance with the notation in Chapter 3 we shall sometimes write SSD for $\|\mathbf{x} - P(\mathbf{x})\|^2$.

4.1.2 Test

We shall now consider more than one model at the same time and therefore we will have to index the models and their characteristics, such as linear subspace, dimension, projection, and estimate of variance. For the model M_i those quantities are denoted by L_i, d_i, $P_i(\mathbf{x})$, and s_{0i}^2, respectively. Furthermore, $SSD_{0i} = \|\mathbf{x} - P_i(\mathbf{x})\|^2$.

We consider the basic model,

$$M_1\colon \boldsymbol{\mu} \in L_1,$$

and we want to test whether the reduction to the model,

$$M_2\colon \boldsymbol{\mu} \in L_2,$$

is contradicted by the data. Here M_2 is always a model that gives a simpler description of the data than M_1. In terms of the linear subspaces that define the models, this means that L_2 *is a subspace of* L_1. We write this as $L_2 \subset L_1$. We say that the model M_2 *is contained in* M_1 or M_2 *is a submodel of* M_1 if the corresponding subspaces satisfy $L_2 \subset L_1$. The geometry of the two models is illustrated in Figure 4.2.

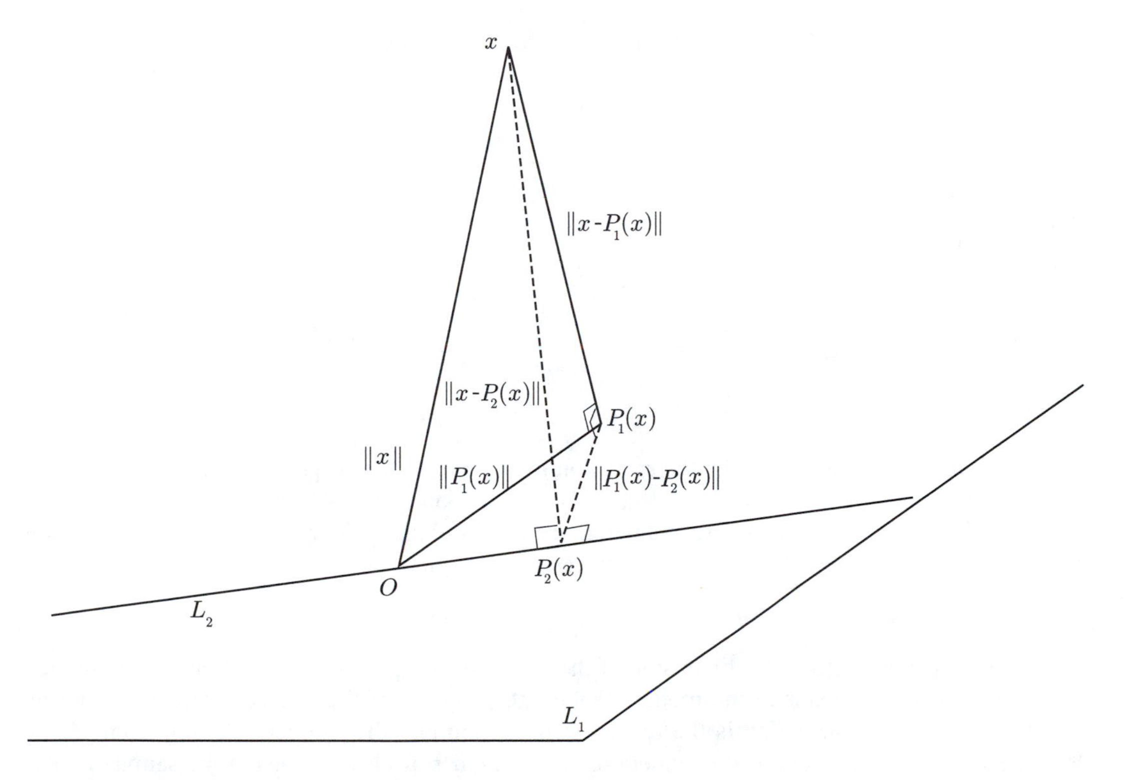

Figure 4.2 *The geometry in linear normal models specified by the subspaces $L_2 \subset L_1$.*

It is easy to derive the likelihood ratio test for the reduction from M_1 to M_2. The maximum of the likelihood function under the two models can be seen from (4.5) and one finds that

$$Q(\mathbf{x}) = \frac{L\left(P_2(\mathbf{x}), \frac{1}{n}||\mathbf{x}-P_2(\mathbf{x})||^2\right)}{L\left(P_1(\mathbf{x}), \frac{1}{n}||\mathbf{x}-P_1(\mathbf{x})||^2\right)}$$
$$= \left[\frac{||\mathbf{x}-P_1(\mathbf{x})||^2}{||\mathbf{x}-P_2(\mathbf{x})||^2}\right]^{\frac{n}{2}}.$$

Although this expression for $Q(\mathbf{x})$ is simple and easy to calculate, we shall continue the mathematical manipulations in order to express $Q(\mathbf{x})$ as a function of an F-distributed random variable.

Now $L_2 \subset L_1$ so $P_1(\mathbf{x}) - P_2(\mathbf{x}) \in L_1$ and hence $\mathbf{x} - P_1(\mathbf{x})$ is orthogonal to $P_1(\mathbf{x}) - P_2(\mathbf{x})$. The triangle with corners at $\mathbf{x}$, $P_1(\mathbf{x})$ and $P_2(\mathbf{x})$ has a right angle at $P_1(\mathbf{x})$. According to Pythagoras' theorem,

$$||\mathbf{x}-P_2(\mathbf{x})||^2 = ||\mathbf{x}-P_1(\mathbf{x})||^2 + ||P_1(\mathbf{x})-P_2(\mathbf{x})||^2, \tag{4.8}$$

and we can complete the mathematical manipulations on Q as follows:

$$Q(\mathbf{x}) = \left[\frac{\|\mathbf{x}-P_1(\mathbf{x})\|^2}{\|\mathbf{x}-P_2(\mathbf{x})\|^2}\right]^{\frac{n}{2}}$$

$$= \left[\frac{\|\mathbf{x}-P_1(\mathbf{x})\|^2}{\|\mathbf{x}-P_1(\mathbf{x})\|^2+\|P_1(\mathbf{x})-P_2(\mathbf{x})\|^2}\right]^{\frac{n}{2}}$$

$$= \left[\frac{1}{1+\frac{\|P_1(\mathbf{x})-P_2(\mathbf{x})\|^2}{\|\mathbf{x}-P_1(\mathbf{x})\|^2}}\right]^{\frac{n}{2}}$$

$$= \left[\frac{1}{1+\frac{d_1-d_2}{n-d_1}F}\right]^{\frac{n}{2}},$$

where

$$F = \frac{\|P_1(\mathbf{x})-P_2(\mathbf{x})\|^2/(d_1-d_2)}{\|\mathbf{x}-P_1(\mathbf{x})\|^2/(n-d_1)}. \tag{4.9}$$

Small values of Q are critical for the reduction to M_2 and this corresponds to large values of F. It can be proved that if the model M_2 is true, F is a realization of an F-distributed random variable with d_1-d_2 degrees of freedom in the numerator and $n-d_1$ degrees of freedom in the denominator and, consequently, the significance probability can be calculated as

$$p_{obs}(\mathbf{x}) = 1 - F_{F(d_1-d_2,\, n-d_1)}(F(\mathbf{x})).$$

As mentioned previously the dimension of the subspace is a precise formulation of the number of parameters of the means in the model, so the interpretation of the degrees of freedom of the numerator in the F-distribution is the reduction in the number of parameters going from M_1 to M_2. The degrees of freedom of the numerator is the number of observations in the sample minus the number of parameters in M_1, and this may be interpreted as the reduction in the number of parameters by going from a saturated model where each observation has its own parameter to the model M_1.

Example 4.1 (Continued)
Let the vector space L in Example 4.1 be L_1. Now $\dim L_1 = 1$ so the only possibility for a subspace that is contained in L_1 is the subspace $L_2 = \{\mathbf{0}\}$, which has dimension 0. Thus M_2 is the model where all observations have mean 0.

This reduction of the model was treated in Section 3.1 and tested using a t-test c.f. the formula (3.9) or the treatment on page 74. The squared t-statistic is precisely the F-test in (4.9) and the significance probability is the same regardless of whether the t-test or the F-test is used, see (3.101). □

Example 4.2 (Continued)
Let L_1 be the vector space L from Example 4.2. The model specified by the subspace $L_2 = \text{span}\{\mathbf{e}_1+\mathbf{e}_2\} = \text{span}\{\mathbf{e}\}$ is the model for one sample. The dimensions are $d_1 = 2$ and $d_2 = 1$ and the degrees of freedom for the F-test for the reduction from M_1 to M_2 are found to be $d_1-d_2 = 1$ for the numerator and $n-d_1 = n-2$ for the denominator.

This reduction of the model was tested with a t-test in Section 3.2.1, see page 83 or page 94. The squared t-statistic is exactly the F-test in (4.9) and the significance probability is the same whether the t-test or the F-test is used. □

Example 4.3 (Continued)
Let L_1 be the k-dimensional subspace from Example 4.3. The model for one sample corresponds to specifying the subspace $L_2 = \text{span}\{\mathbf{e}_1 + \cdots + \mathbf{e}_k\} = \text{span}\{\mathbf{e}\}$.

This was the reduction that was the subject of Section 3.2.2 and the F-test on page 102 is a special case of (4.9). On page 106 we showed in the subsection, *Two or More Samples as Analysis of Variance,* a partitioning of the total variation $SSD_{0\infty}$ in the components SSD_1 and SSD_2. That partitioning is (4.8) written for this particular case and

$$SSD_{0\infty} = \|\mathbf{x} - P_2(\mathbf{x})\|^2, \; SSD_1 = \|\mathbf{x} - P_1(\mathbf{x})\|^2 \text{ and } SSD_2 = \|P_1(\mathbf{x}) - P_2(\mathbf{x})\|^2.$$

When $k > 2$ there are several possibilities for subspaces that correspond to meaningful statistical models. One example is the subspace $L_2 = \text{span}\{\mathbf{e}_1 + \cdots + \mathbf{e}_i, \mathbf{e}_{i+1} + \cdots + \mathbf{e}_k\}$, which corresponds to the model where data is divided into two groups where the means of the observations are the same in each group but possibly different between groups. We shall consider this model again when we return to Example 3.3. □

Example 4.4 (Continued)
L_1 is the two-dimensional subspace of Example 4.4. Here we have two obvious possibilities for subspaces L_2 of L_1that correspond to meaningful statistical models.

$L_2 = \text{span}\{\mathbf{e}\}$ is the model for one sample and the test for the reduction to M_2 corresponds to testing that the slope of the regression line is 0, i.e., $\beta = 0$. The F-test in (4.9), which will have $d_1 - d_2 = 2 - 1 = 1$ degrees of freedom in the numerator and $n - d_1 = n - 2$ degrees of freedom in the denominator, is the squared t-test of $\beta = 0$ that was given on page 139 and the significance probability is the same whether the t-test or the F-test is used.

The other obvious possibility is M_2^* specified by $L_2^* = \text{span}\{\mathbf{t}\}$. The reduction to M_2^* corresponds to testing $\alpha = 0$ and also in this case the t-test on page 139 and the F-test in (4.9) are equivalent in the sense of giving the same significance probability. □

4.1.3 Calculations

The F-test for the reduction from M_1 to M_2 is according to (4.9) given by

$$F = \frac{\|P_1(\mathbf{x}) - P_2(\mathbf{x})\|^2 / (d_1 - d_2)}{\|\mathbf{x} - P_1(\mathbf{x})\|^2 / (n - d_1)}.$$

The numerator in the F-test is the estimate of the variance in M_1, and if

$$\|\mathbf{x} - P_2(\mathbf{x})\|^2 \text{ and } \|\mathbf{x} - P_1(\mathbf{x})\|^2$$

are known, one can calculate $\|P_1(\mathbf{x}) - P_2(\mathbf{x})\|^2$ using Pythagoras' theorem on the triangle with corners at $\mathbf{x}$, $P_1(\mathbf{x})$ and $P_2(\mathbf{x})$ in Figure 4.3:

$$\|P_1(\mathbf{x}) - P_2(\mathbf{x})\|^2 = \|\mathbf{x} - P_2(\mathbf{x})\|^2 - \|\mathbf{x} - P_1(\mathbf{x})\|^2. \tag{4.10}$$

The two quantities on the right-hand side of (4.10) can be found from the analysis of variance tables from the estimation of the models using a statistical package. The analysis of variance tables are very similar for all statistical packages. In Table 4.1 and Table 4.2 the contents of the analysis of variance tables in SAS for the models M_1 and M_2 are explained in the notation that is used in this section. Notice that the relevant quantities are found in the line `Error` in the column headed by `Sum of Squares`.

Similarly, the degrees of freedom, $d_1 - d_2$, of $\|P_1(\mathbf{x}) - P_2(\mathbf{x})\|^2$ can be found from the numbers in the column headed by `DF` in the line `Error` as

$$d_1 - d_2 = (n - d_2) - (n - d_1). \tag{4.11}$$

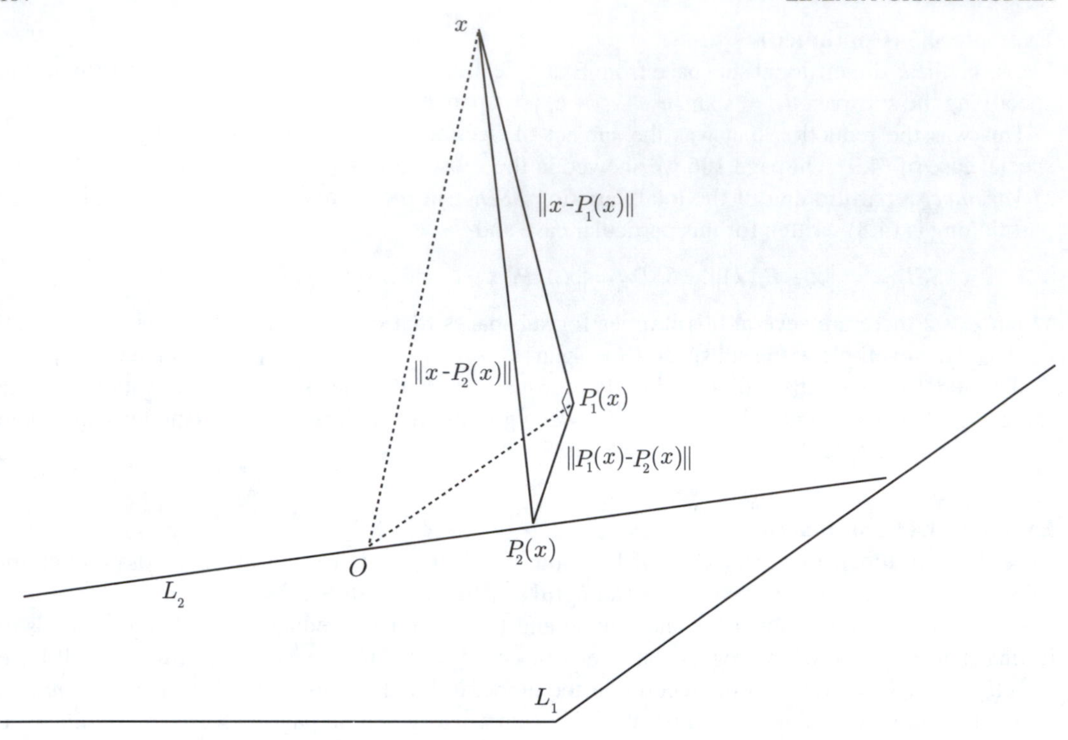

Figure 4.3 *The essential quantities in testing the reductionen from M_1 to M_2 are the squared length of the sides in the triangle with corners at $\mathbf{x}$, $P_1(\mathbf{x})$ and $P_2(\mathbf{x})$. The relation between the squared lengths of the sides of the triangle is given in (4.8) and (4.10).*

The following inelegant expression of the F-statistic explains clearly how it is calculated because exactly those quantities that can be found in the output of the statistical package are mentioned:

$$F = \frac{\dfrac{\|\mathbf{x}-P_2(\mathbf{x})\|^2 - \|\mathbf{x}-P_1(\mathbf{x})\|^2}{(n-d_2)-(n-d_1)}}{s_{01}^2}. \tag{4.12}$$

Table 4.1 *The analysis of variance table from model M_1*

The GLM Procedure

Model M_1

Dependent Variable: X

	Source	DF	Sum of Squares	Mean Square	F Value	Pr > F
	Model	$d_1 - 1$	$\|P_1(\mathbf{x}) - P_\infty(\mathbf{x})\|^2$	$s_{1\infty}^2$	$s_{1\infty}^2/s_{01}^2$	$p_{obs}(\mathbf{x})$
$\rightarrow$	Error	$n - d_1$	$\|\mathbf{x} - P_1(\mathbf{x})\|^2$	s_{01}^2		
	Corrected Total	$n - 1$	$\|\mathbf{x} - P_\infty(\mathbf{x})\|^2$			

It is now possible to explain the contents of the analysis of variance table for a particular model and the notation that has been chosen in Tables 4.1 and 4.2.

For any model that is specified it is always possible to consider the submodel corresponding to the subspace $L_\infty = \text{span}\{\mathbf{e}\}$, where $\mathbf{e}$ denotes the vector with all coordinates identical and equal to 1. The quantities that are given in the analysis of variance table are the relevant ones for testing

Table 4.2 *The analysis of variance table from model M_2*

The GLM Procedure
Model M_2

Dependent Variable: X

	Source	DF	Sum of Squares	Mean Square	F Value	Pr > F
	Model	d_2-1	$\|P_2(\mathbf{x})-P_\infty(\mathbf{x})\|^2$	$s_{2\infty}^2$	$s_{2\infty}^2/s_{02}^2$	$p_{obs}(\mathbf{x})$
$\rightarrow$	Error	$n-d_2$	$\|\mathbf{x}-P_2(\mathbf{x})\|^2$	s_{02}^2		
	Corrected Total	$n-1$	$\|\mathbf{x}-P_\infty(\mathbf{x})\|^2$			

whether the data contradicts the reduction from the specified model to M_∞. In particular, the F-test for that reduction is given in the line `Model`.

In principle it is possible to specify the model M_∞ with the subspace $L_\infty = \text{span}\{\mathbf{e}\}$ and then the only possible submodel is the one corresponding to the subspace $\{0\}$, i.e., the model that specifies that all observations have mean 0.

Example 4.5
In the section, *Test of the Hypothesis of Linear Regression,* on page 133 we gave the conditions for a linear regression model to be contained in the model for k independent samples and we showed how the hypothesis of linear regression is tested. It is a special case of what is described here in that M_1 is the model for k independent samples and M_2 is the model for linear regression. □

4.1.4 Treatment of a Sequence of Models

As mentioned previously a statistical analysis is often performed by testing a sequence of models that gradually gives a simpler description of the data. But in reality one need not consider more than two models at a time. The most recent model that was not contradicted by the data will play the role of the established model M_1 above and the next model in the sequence will play the role of M_2.

If the test does not reject the reduction to M_2, it will take M_1's role as a model and the next model in line will take M_2's role as a candidate for the new model.

4.1.5 Which Models are Linear Normal Models?

We have explained that all the models that were treated in Chapter 3 were linear normal models. Obvious questions are if there are any other linear normal models or which models are linear normal ones. The main question is which models are specified by a linear subspace for the vector of means? The answer is that all models that have the means identical in groups or depending linearly on one or more explanatory variables are models that can be specified by a linear subspace for the mean.

The only requirement is that the parameters of the mean must be unspecified. Thus a linear regression model with the slope fixed at the value of 2, for example, is not a linear model. This is not a serious limitation, however. For if a linear normal model for data $\mathbf{x}$ with the subspace, $L = span\{\mathbf{e}_1, \mathbf{e}_2, \ldots, \mathbf{e}_d\}$, is adequate and there is a need to consider the coefficient of $\mathbf{e}_1$, β_{01}, as fixed, for example, then the adjusted response $\mathbf{x} - \beta_{01}\mathbf{e}_1$ will follow a linear normal model with subspace, $\tilde{L} = span\{\mathbf{e}_2, \ldots, \mathbf{e}_d\}$.

The model in Example 3.5 was a linear regression corresponding to the subspace, $L = span\{\mathbf{e}, \mathbf{t}\}$, where $\mathbf{t}$ contained the number of rotations. In that case it was reasonable to work with the model where the intercept was $\ln 60$ corresponding to the initial mass of sand in the sieve, i.e., the coefficient to $\mathbf{e}$ is $\ln 60$. Then $\mathbf{x} - \ln 60\mathbf{e}$ is a linear normal model with the subspace $span\{\mathbf{t}\}$ or, in other words, a linear regression with intercept 0.

The user of these models need not be able to specify the vectors that generate a particular subspace nor is he or she required to be able to calculate the dimension correctly. It is only required that the user can specify the model in a statistical package and must be able to recognize when a model is included in another model such that the reduction can be tested.

It will be obvious from the examples how the models are specified in SAS.

Main Points in Section 4.1

The salient feature of linear normal models is the ease with which a reducution from one linear normal model,

$$M_1\colon \boldsymbol{\mu} \in L_1,$$

to a simpler linear normal model,

$$M_2\colon \boldsymbol{\mu} \in L_2,$$

where simpler means that L_2 is included in L_1.
Test of the reduction $M_1 \longrightarrow M_2$:
Test statistic

$$F = \frac{\dfrac{\|\mathbf{x} - P_2(\mathbf{x})\|^2 - \|\mathbf{x} - P_1(\mathbf{x})\|^2}{(n-d_2)-(n-d_1)}}{s_{01}^2} \sim\sim F(d_1 - d_2, n - d_1)$$

with significance probability

$$p_{obs}(\mathbf{x}) = 1 - F_{F(d_1-d_2,\, n-d_1)}(F(\mathbf{x})).$$

The quantities needed to calculate the F-test are found in the ERROR lines of the analysis of variance tables of the two models as indicated in the tables below.

Table 4.3 *The analysis of variance table from model* M_1

The GLM Procedure
Model M_1

Dependent Variable: X

	Source	DF	Sum of Squares	Mean Square	F Value	Pr > F
	Model	$d_1 - 1$	$\|P_1(\mathbf{x}) - P_\infty(\mathbf{x})\|^2$	$s_{1\infty}^2$	$s_{1\infty}^2/s_{01}^2$	$p_{obs}(\mathbf{x})$
→	Error	$n - d_1$	$\|\mathbf{x} - P_1(\mathbf{x})\|^2$	s_{01}^2		
	Corrected Total	$n - 1$	$\|\mathbf{x} - P_\infty(\mathbf{x})\|^2$			

Table 4.4 *The analysis of variance table from model* M_2

The GLM Procedure
Model M_2

Dependent Variable: X

	Source	DF	Sum of Squares	Mean Square	F Value	Pr > F
	Model	$d_2 - 1$	$\|P_2(\mathbf{x}) - P_\infty(\mathbf{x})\|^2$	$s_{2\infty}^2$	$s_{2\infty}^2/s_{02}^2$	$p_{obs}(\mathbf{x})$
→	Error	$n - d_2$	$\|\mathbf{x} - P_2(\mathbf{x})\|^2$	s_{02}^2		
	Corrected Total	$n - 1$	$\|\mathbf{x} - P_\infty(\mathbf{x})\|^2$			

4.2 Comparison of Regression Lines

The starting point for the comparison of two or more regression lines is the model M_0: independent random variables X_{hi} $h=1,\dots,m$, $i=1,\dots,n_h$ and $X_{hi} \sim N(\alpha_h+\beta_h t_{hi}, \sigma_h^2)$. Here (X_{hi}, t_{hi}) denote the pair of observations of the random variable X_{hi} and the explanatory variable t_{hi}. The model M_0 is the model for m independent regression models with no restrictions on either the regression parameters or the variance parameters. Note that h is used to index the m regression lines and i is used to index the observations within each regression line.

The models that we will consider are:

M_1: $X_{hi} \sim N(\alpha_h+\beta_h t_{hi}, \sigma^2)$, common variance for all m regressions lines.
M_2: $X_{hi} \sim N(\alpha+\beta_h t_{hi}, \sigma^2)$, common intercept.
M_2^*: $X_{hi} \sim N(\alpha_h+\beta t_{hi}, \sigma^2)$, common slope.
M_3: $X_{hi} \sim N(\alpha+\beta t_{hi}, \sigma^2)$, identical regression lines.

Notice that M_0 is not a linear normal model because the observations from different regression lines do not have a common variance. But starting with M_1 the rest of the models above are linear normal ones. The reduction to M_1 is tested using Bartlett's test based on the estimated variances in each of the m regression lines if $m > 2$, see page 114, or by comparing the two estimated variances using an F-test, as described on page 94, if $m = 2$.

The relationship between the four linear normal models can be described in a diagram where an arrow goes from a model to a submodel:

$$
\begin{array}{ccccc}
 & & M_2^*\colon X_{hi} \sim N(\alpha_h+\beta t_{hi}, \sigma^2) & & \\
 & H_{02}^*\colon \beta_h=\beta \quad \nearrow & & \searrow \quad H_{03}^*\colon \alpha_h=\alpha & \\
M_1\colon X_{hi} \sim N(\alpha_h+\beta_h t_{hi}, \sigma^2) & & & & M_3\colon X_{hi} \sim N(\alpha+\beta t_{hi}, \sigma^2) \\
 & H_{02}\colon \alpha_h=\alpha \quad \searrow & & \nearrow \quad H_{03}\colon \beta_h=\beta & \\
 & & M_2\colon X_{hi} \sim N(\alpha+\beta_h t_{hi}, \sigma^2) & &
\end{array}
$$

The diagram emphasizes that neither M_2 nor M_2^* are contained in the other. Depending on the scientific context one can choose first to test whether the intercepts are identical, i.e., to test the reduction from M_1 to M_2, or to test whether the slopes are identical, i.e., to test the reduction from M_1 to M_2^*.

Example 3.4 (Continued)
13 male journalists and 15 male university teachers have had their systolic blood pressure and their age recorded. For each profession the data have been described as linear regressions with systolic blood pressure as response (X) and age as explanatory variable (t). The regression models were accepted solely on the evidence of Figures 3.7 and 3.8 on pages 117 and 118. In Figure 3.8 the estimated regression lines in the model M_1 (the same estimates as in M_0) are plotted with the data.

Table 4.5 *Estimates of the parameters in model M_0, obtained from SAS pages 2 and 4 on page 196 and 197*

Group	α	β	σ^2
Journalist	84.99521966	1.52956813	18.149500
University teacher	75.62986190	1.56238397	19.449539

Testing the reduction from M_0 to M_1 is equivalent to testing the hypothesis, H_{01}: $\sigma_1^2 = \sigma_2^2$. The test for H_{01} is based on the estimates of the variances in the model M_0, see page 94.

SAS programs and SAS output listings are collected in Annex to Section 4.2, which begins on page 194. The output listings are labeled with the model names `M0`, `M1`, `M2`, and `M2*` for easy reference.

The F-test for H_{01} is

$$F = \frac{19.449539}{18.149500} = 1.07,$$

which is evaluated in an F-distribution with 13 degrees of freedom in the numerator and 11 degrees of freedom in the denominator. The 97.5% fractile in this F-distribution is 3.39 so the significance probability is greater than 5%. Thus the hypothesis H_{01} is not contradicted by the data and M_1 is the model for the data.

The estimates of the slopes and intercepts of the regression lines are the same in the models M_1 and M_0. The estimate of σ^2 in M_1 can be calculated as the weighted average of the estimates of the variance in the two groups in M_0 with the degrees of freedom as weights:

$$s_{01}^2 = \frac{11 \times 18.149500 + 13 \times 19.449539}{11 + 13} = 18.853688.$$

The estimates are displayed in Table 4.6. We display the estimates of the parameters of the models M_1, M_2, and M_2^* in Tables 4.6, 4.7, and 4.8 for a number of reasons. First, in order to see at a glance the number of parameters of the models; secondly, to be able to see how the estimates of the parameters change between the models and thirdly, in order to explain the way the SAS output displays the estimates of the parameters. It is the last purpose that explains the excessive number of decimals used in the display of the estimates. We emphasize that Tables 4.6, 4.7, and 4.8 do not provide satisfactory summaries of the estimation process because no indication of the accuracy of the estimates is given. This is done in Tables 4.9, 4.10, and 4.11 where parameter estimates as well as 95% confidence intervals are shown.

Table 4.6 *Estimates of the parameters in model M_1*

Group	α	β	σ^2
Journalist	84.99521966	1.52956813	18.853688
University teacher	75.62986190	1.56238397	

Even if all the estimates of the unknown parameters in the model M_1 have been calculated using the estimates in M_0, we shall nevertheless use SAS to estimate in M_1 for two reasons. First, the estimates of the regression parameters are presented in a special way in the SAS output and, secondly, we will need the SAS output to test whether the model can be reduced to a simpler model.

The output from M_1 is in SAS pages 5 and 6 on page 198.

The `Error` line clearly shows that the estimate of the variance in M_1 is 18.853688, which is the value displayed in Table 4.6. The estimates of the slopes and intercepts are a little harder to recognize but the place to look is at the bottom of the output listing under the heading `Parameter Estimate`. On page 198 it is at the bottom of SAS page 6 just below the middle of page 198. Only the estimates of the intercept α_2 (75.62986190) and slope β_2 (1.56238397) of university teachers are immediately recognized. In the line `age*group journalist` it is the estimate of $\beta_1 - \beta_2$ that is found to be -0.03281584 and in the line `group journalist` it is the estimate of $\alpha_1 - \alpha_2$ that is given as 9.36535776 so a little calculation is needed to find the estimate of α_1 as $(75.62986190 + 9.36535776 = 84.99521966)$ and the estimate of β_1 $(1.56238397 - 0.03281584 = 1.52956813)$. It may seem unnecessarily complicated but it is really quite useful; for the t-tests that are given along with the estimates are the tests of the hypothesis that the corresponding parameters are 0.

In the line where the estimate of $\alpha_1 - \alpha_2$ is given, we find the t-test of the hypothesis H_{02}: $\alpha_1 = \alpha_2$ or, in other words, the test for reduction from M_1 to M_2. We shall see later that this test is equivalent to the F-test for the reduction because $t^2 = F$ and the significance probabilities are identical.

Correspondingly, we find the t-test of H^*_{02}: $\beta_1 = \beta_2$, i.e., the test of the reduction from M_1 to M^*_2 in the line where the estimate of $\beta_1 - \beta_2$ is given.

The significance probabilities are 0.4697 and 0.8849, respectively, so in model M_1 the data neither contradict the hypothesis H_{02} that the intercepts are equal nor the hypothesis H^*_{02} that the slopes are identical. This means that one can *either* adopt model M_2 *or* model M^*_2.

For these data it must be scientific arguments and not statistical ones that determine the choice between the two models. We will not go into this discussion now but assume for the moment that we prefer M_2. The estimates of the parameters in M_2 are found using *PROC GLM* and they are given in Table 4.7 for easy reference. The estimates of α, the common intercept, and β_2 are found directly in the output on page 199, while the estimate of β_1 is calculated as a sum of the estimate of β_2 and the estimate of $\beta_1 - \beta_2$, 0.13048919, which is found in the line `age*group journalist`.

Table 4.7 *Estimates of the parameters in model M_2*

Group	α	β	σ^2
Journalist	81.54148011	1.58909585	18.506454
University teatcher		1.45860666	

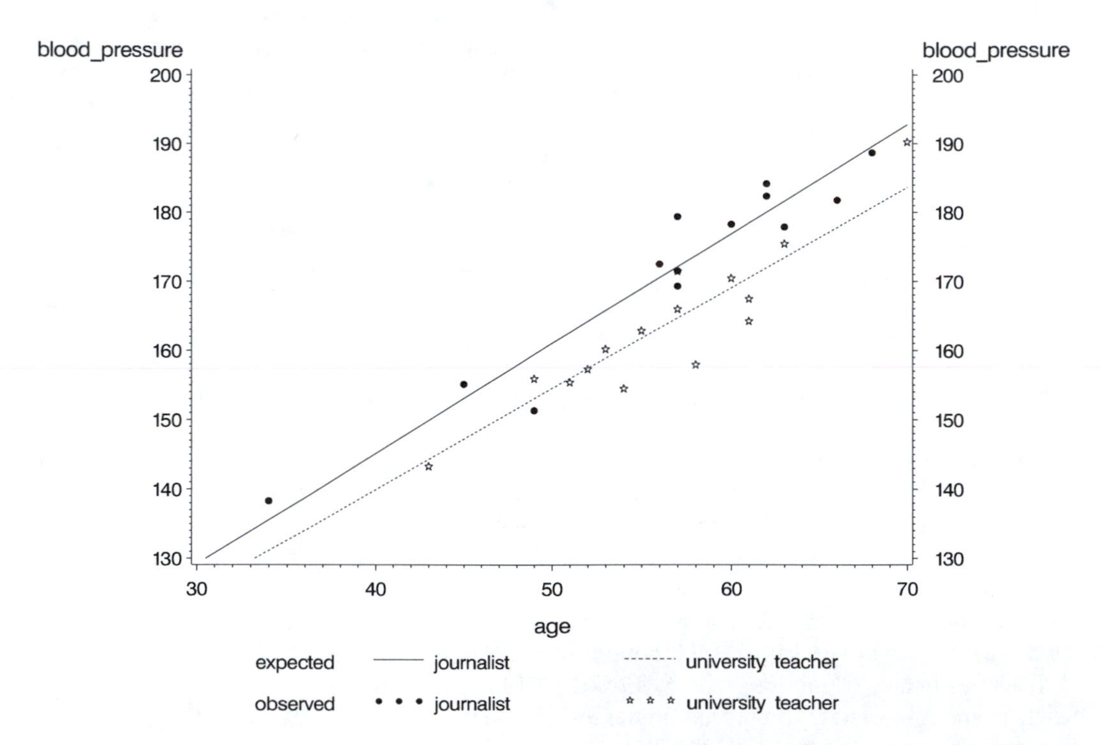

Figure 4.4 *Regression lines in model M_2: common intercept. The regression lines in model M_1 are shown in Figure 3.8.*

We shall now use formula (4.12) to calculate the F-test for the reduction from M_1 to M_2 to show that it is equivalent to the t-test that was given in the output from M_1. The quantities needed for the calculation are found in the `Error` lines in the SAS output on pages 198 and 199:

$$F = \frac{\frac{462.661342 - 452.488513}{25 - 24}}{18.853688} = 0.5396.$$

Using the $F(1,24)$-distribution the significance probability is found to be 0.4697 exactly as displayed in the `group journalist` line of the output from M_1. The mathematics behind the identical significance probabilities is the identity,

$$\sqrt{F} = \sqrt{0.5396} = 0.7346 \approx 0.73 = |t|.$$

The estimated regression lines in model M_2 are plotted in Figure 4.4.

We now consider if the model can be further reduced from M_2 to M_3, "identical regression lines." It is immediately seen from the t-test of the hypothesis, H_{03}: $\beta_1 = \beta_2$, which is given in the `age*group journalist` line in the output from M_2 on SAS page 8 on page 199, that this reduction is contradicted by the data. The t-test is 4.55 and the significance probability is less than 0.0001.

In spite of the fact that the reduction to M_3 was rejected, the estimated regression line in M_3 is shown in Figure 4.6 along with the data. The figure clearly shows that the model with identical regression lines is unreasonable; for the observations from the two professions lie almost exclusively on either side of the regression line with the observations for the journalists lying above the regression line.

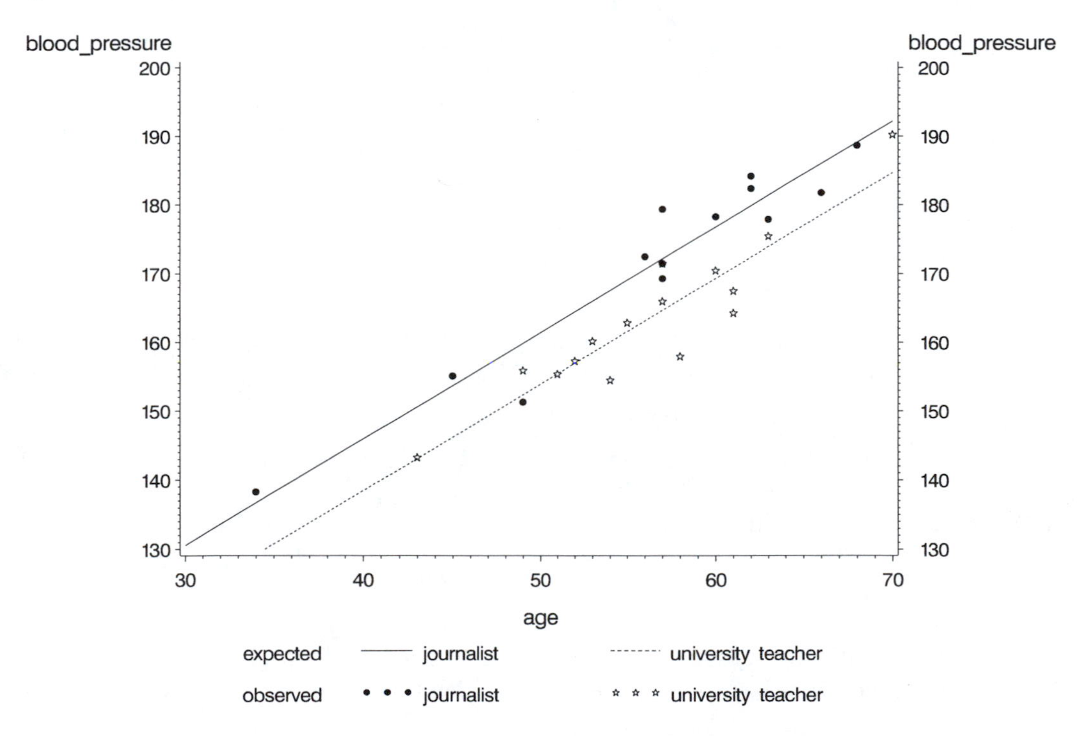

Figure 4.5 *Regression lines in model M_2^*: parallel regression lines.*

Table 4.8 *Estimates of the parameters in model M_2^**

Group	α	β	σ^2
Journalist	84.32064422	1.54148319	18.115698
University teacher	76.80587938		

The following shows the analysis if instead of M_2 we had chosen to consider M_2^* first.

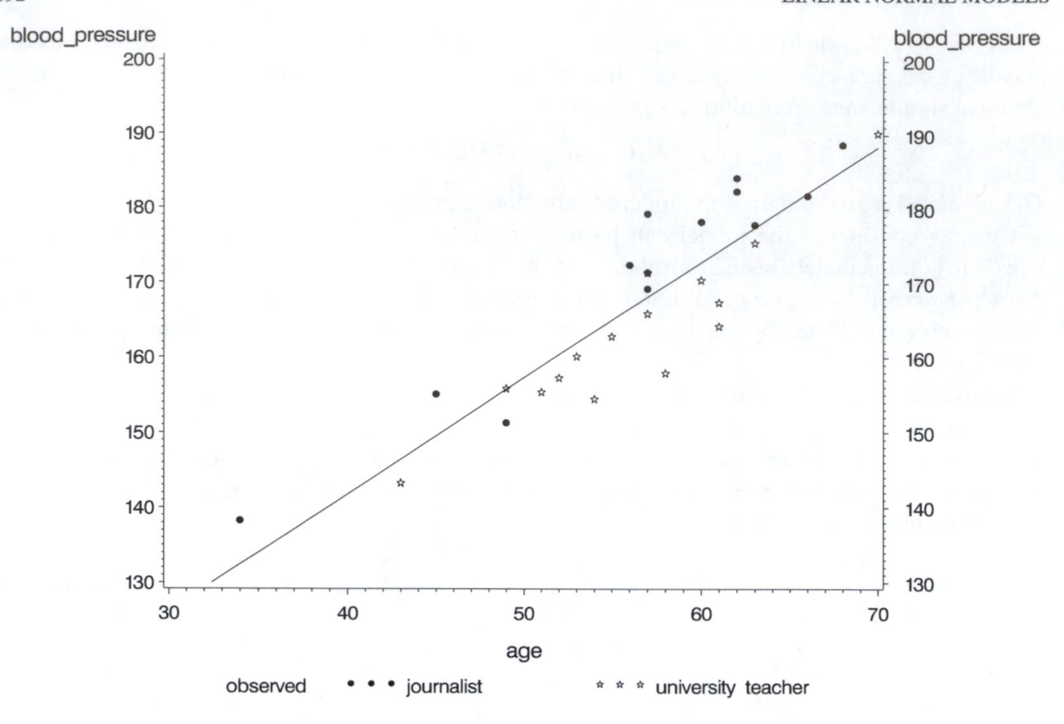

Figure 4.6 *The estimated regression line in model M_3: identical regression lines. The model is obviously wrong, as the observations from the two professions lie on either side of the regression line. Compare the residual plot in Figure 3.10.*

The estimates of the parameters in model M_2^* from the SAS output on page 200 are collected in Table 4.8. The F-test for the reduction from M_1 to M_2^* is

$$F = \frac{\frac{452.892448 - 452.488513}{25-24}}{18.853688} = \frac{0.403935}{18.853688} = 0.02142, \tag{4.13}$$

and the significance probability is found to be 0.8849 using the $F(1,24)$-distribution. We notice that

$$\sqrt{F} = \sqrt{0.02142} = 0.1464 \approx 0.15 = |t|,$$

in order to emphasize once again that the test of the reduction from M_1 to M_2^* is given in the output from M_1. It is found in the `age*group journalist` line on SAS page 6, which displays the test of parallel regressions lines or, in other words, of identical slopes of the regression lines H_{02}^*: $\beta_1 = \beta_2$.

Further reduction or simplification of the model is not possible. The output from M_2^* displays the t-test of identical intercepts, H_{03}^*: $\alpha_1 = \alpha_2$, and its value is as large as 4.66 with a significance probability less than 0.0001. For later use, however, we will calculate the F-test for the reduction from M_2^* to M_3 using the output listings from those models:

$$F = \frac{\frac{845.972201 - 452.892448}{26-25}}{18.115698} = \frac{393.079753}{18.115698} = 21.70. \tag{4.14}$$

The estimated regression lines in model M_2^* are shown in Figure 4.5 along with the data and they are very similar to the lines in Figure 3.8, which depicts the estimated regression lines in models M_0 and M_1.

The analysis of these data showed that two different models M_2 and M_2^* describe the data almost equally well. The estimates of the parameters are shown in Table 4.10 for model M_2 and in Table 4.11 for model M_2^* together with 95% confidence intervals of the parameters. The models agree that the mean systolic blood pressure of university teachers is lower than the mean of systolic blood pressure of journalists of the same age. This is exactly the impression that Figure 4.4 and Figure 4.5 give and the result of the analysis is that this impression is statistically significant.

Apart from that the two models are very different. According to M_2 the difference in mean blood pressure will increase with age, whereas according to M_2^*, the two professions have the same increase in mean blood pressure with age.

It is not possible using statistical methods on the present data to decide which of the two models are more nearly correct. But it will be easy to design a new investigation to shed light on the question. For that purpose it will be important to include many individuals between 30 and 40 years of age and many individuals between 55 and 65 years of age in the investigation. □

Table 4.9 *Estimates and 95% confidence intervals of the parameters in model* M_1

Group	α	β	σ^2
Journalist	85.0 [69.7, 100.3]	1.53 [1.25, 1.81]	18.85 [11.49, 37.31]
University teacher	75.6 [54.7, 96.5]	1.56 [1.19, 1.93]	

Table 4.10 *Estimates and 95% confidence intervals of the parameters in model* M_2

Group	α	β	σ^2
Journalist	81.5 [73.4, 89.7]	1.59 [1.37, 1.80]	18.51 [11.38, 35.26]
University teatcher		1.46 [1.23, 1.68]	

Table 4.11 *Estimates and 95% confidence intervals of the parameters in model* M_2^*

Group	α	β	σ^2
Journalist	84.3 [71.8, 96.9]	1.54 [1.32, 1.76]	18.12 [11.14, 34.52]
University teacher	76.8 [64.4, 89.3]		

In this example only two regression lines were compared, but m regression lines can be compared in almost the same way. There are only two modifications to be aware of. First, the reduction from M_0 to M_1, the hypothesis of homogeneity of variance, will be tested using Bartlett's test and, secondly, the hypotheses of common intercept of the regression lines or common slope of the regression lines can only be tested with an F-test when $m > 2$.

Annex to Section 4.2

Calculations in SAS

In this section we show how the models that are involved in the comparison of m regression lines are specified in SAS. We do this by listing the program that was used in the analysis in this section.

Finally, we explain the type I sums of squares that are listed in the output from *PROC GLM* and may be of some use.

Example 3.4 (Continued)

A few comments are given after each segment of the program.

```
TITLE1 'Example 3.4';

PROC FORMAT;
   VALUE $grpfmt    '1'='journalist' '2'='university teacher';
RUN;

DATA blood_pressure;
INPUT group$ blood_pressure age@@;
expected=group;
observed =group;
FORMAT group expected observed $grpfmt.;
DATALINES;
1 188.7 68 1 184.2 62 1 151.3 49 1 182.4 62 1 138.3 34
1 181.8 66 1 155.1 45 1 178.3 60 1 179.4 57 1 171.5 57
1 177.9 63 1 172.5 56 1 169.3 57
2 162.9 55 2 155.9 49 2 170.5 60 2 154.5 54 2 158.0 58
2 155.4 51 2 143.3 43 2 157.3 52 2 160.2 53 2 164.3 61
2 167.5 61 2 171.5 57 2 166.0 57 2 190.3 70 2 175.5 63
;
```

The commands above read the data into an SAS data set. The variable `group` defines the two professions and is originally coded as `1` for journalist and `2` for university teacher. In order to make the output and figures easier to interpret, a format is attached to the variable `group`. This is done in two steps. First a format `$grpfmt` is defined in the statement in `PROC FORMAT`. This format is attached to the variable `group` with the statement `FORMAT group $grpfmt.;` in the data step.

```
TITLE2 'comparison of regression lines';
TITLE3 'separate calculations';
TITLE4 'M0';
PROC GLM DATA=blood_pressure;
   MODEL blood_pressure = age / SS1;
   BY group;
RUN;
```

The estimation is performed independently in the two groups due to the statement `BY group`. The SAS data set must be sorted by the variable that appears in the `BY` statement. If this is not the case *PROC SORT* must be used before the procedure where the `BY` statement is used.

```
TITLE2 'comparison of regression lines';
TITLE3;
TITLE4 'M1';
PROC GLM DATA=blood_pressure;
   CLASS group;
   MODEL blood_pressure = age group age*group/SS1 SOLUTION;
RUN;
```

Because `group` is specified as a `CLASS` variable, the model with possibly different slopes in the groups is fitted with `alder*group` in the model statement. The variable name `group` also appears alone in the model statement and therefore separate slopes in the two groups are estimated.

`SOLUTION` has been added as an option in order to have the parameter estimates displayed in the listing of the output. This is necessary whenever a `CLASS` variable appears in the model.

```
TITLE2 'comparison of regression lines';
TITLE3 'common intercept';
TITLE4 'M2';
PROC GLM DATA=blood_pressure;
   CLASS group;
   MODEL blood_pressure = age age*group/
                          SS1 SOLUTION;
RUN;

TITLE2 'comparison of regression lines';
TITLE3 'common slope';
TITLE4 'M2*';
PROC GLM DATA=blood_pressure;
   CLASS group;
   MODEL blood_pressure = age group/
                          SS1 SOLUTION;
RUN;

TITLE2 'comparison of regression lines';
TITLE3 'identical regression lines';
TITLE4 'M3';
PROC GLM DATA=blood_pressure;
   MODEL blood_pressure = age/ SS1;
RUN;
```

On the following pages the output from the SAS programs has been listed.

The `TITLE` statements are used to make it easier to find the output listing from a particular model. But it is important to be able to recognize the model from the output listing alone. The model is defined from two statements in the SAS program: the `CLASS` statement and the `MODEL` statement, and in the latter it is the model formula that is important. The options after the slash (/) only serve to monitor aspects of the output. Taking M_2 as an example the statements that define the model are:

```
CLASS group;
MODEL blood_pressure = age age*group;
```

From the output listing on page 199 the model formula can be recreated:

```
blood_pressure = age age*group;
```

because the left-hand side of the model formula, the response variable, is mentioned above the analysis of variance table: `Dependent Variable: blood_pressure`. The right-hand side of the model formula can be reestablished from the `Source` section between the analysis of variance table and the parameter estimates where each term on the right-hand side of the model formula label a line of output.

Finally, the first SAS page of the of output listing for each model lists the class variables and their levels.

```
                         Example 3.4                                     1
                comparison of regression lines
                    separate calculations
                              M0

------------------------------ group=journalist ------------------------------

                        The GLM Procedure

                  Number of observations     13
                         Example 3.4                                     2
                comparison of regression lines
                    separate calculations
                              M0

------------------------------ group=journalist ------------------------------

                        The GLM Procedure

Dependent Variable: blood_pressure

                                        Sum of
 Source                      DF        Squares    Mean Square   F Value   Pr > F

 Model                        1    2416.964731    2416.964731    133.17   <.0001

 Error                       11     199.644500      18.149500

 Corrected Total             12    2616.609231

          R-Square     Coeff Var      Root MSE    blood_pressure Mean

          0.923701      2.482759      4.260223               171.5923

 Source                      DF      Type I SS    Mean Square   F Value   Pr > F

 age                          1    2416.964731    2416.964731    133.17   <.0001

                                               Standard
       Parameter          Estimate               Error    t Value    Pr > |t|

       Intercept       84.99521966          7.59658557      11.19      <.0001
       age              1.52956813          0.13254581      11.54      <.0001
```

```
                         Example 3.4                              3
                comparison of regression lines
                     separate calculations
                              M0

-------------------------- group=university teacher --------------------------

                       The GLM Procedure

              Number of observations    15
                         Example 3.4                              4
                comparison of regression lines
                     separate calculations
                              M0

-------------------------- group=university teacher --------------------------

                       The GLM Procedure

Dependent Variable: blood_pressure

                                     Sum of
 Source                     DF      Squares    Mean Square   F Value   Pr > F

 Model                       1  1437.611987    1437.611987     73.91   <.0001

 Error                      13   252.844013      19.449539

 Corrected Total            14  1690.456000

          R-Square     Coeff Var      Root MSE    blood_pressure Mean

          0.850429      2.696688      4.410163               163.5400

 Source                     DF    Type I SS    Mean Square   F Value   Pr > F

 age                         1  1437.611987    1437.611987     73.91   <.0001

                                          Standard
       Parameter         Estimate            Error    t Value    Pr > |t|

       Intercept      75.62986190      10.28843126       7.35      <.0001
       age             1.56238397       0.18172789       8.60      <.0001
```

```
                        Example 3.4                                5
                comparison of regression lines
                       common variance
                             M1
                     The GLM Procedure

                  Class Level Information

         Class         Levels    Values

         group              2    journalist university teach

                 Number of observations     28
                        Example 3.4                                6
                comparison of regression lines
                       common variance
                             M1

                     The GLM Procedure

Dependent Variable: blood_pressure
```

Source	DF	Sum of Squares	Mean Square	F Value	Pr > F
Model	3	4306.138630	1435.379543	76.13	<.0001
Error	24	452.488513	18.853688		
Corrected Total	27	4758.627143			

R-Square	Coeff Var	Root MSE	blood_pressure Mean
0.904912	2.595720	4.342083	167.2786

Source	DF	Type I SS	Mean Square	F Value	Pr > F
age	1	3912.654942	3912.654942	207.53	<.0001
group	1	393.079752	393.079752	20.85	0.0001
age*group	1	0.403936	0.403936	0.02	0.8849

Parameter		Estimate	Standard Error	t Value	Pr > |t|
Intercept		75.62986190 B	10.12960842	7.47	<.0001
age		1.56238397 B	0.17892255	8.73	<.0001
group	journalist	9.36535776 B	12.74974956	0.73	0.4697
group	university teach	0.00000000 B	.	.	.
age*group	journalist	-0.03281584 B	0.22419481	-0.15	0.8849
age*group	university teach	0.00000000 B	.	.	.

An alternative MODEL statement for the model M_1 is

```
MODEL blood_pressure = group age*group / SS1 SOLUTION NOINT ;
```

which differ in the Source and Parameter part of the output that is displayed below. The Parameter part is particularly useful if estimates and confidence intervals of the individual regression parameters are of interest.

Source	DF	Type I SS	Mean Square	F Value	Pr > F
group	2	783950.9348	391975.4674	20790.4	<.0001
age*group	2	3854.5767	1927.2884	102.22	<.0001

Parameter		Estimate	Standard Error	t Value	Pr > |t|
group	journalist	84.99521966	7.74255430	10.98	<.0001
group	university teach	75.62986190	10.12960842	7.47	<.0001
age*group	journalist	1.52956813	0.13509268	11.32	<.0001
age*group	university teach	1.56238397	0.17892255	8.73	<.0001

```
                          Example 3.4                                      7
                 comparison of regression lines
                        common intercept
                               M2

                         The GLM Procedure

                      Class Level Information

          Class          Levels    Values

          group               2    journalist university teach

                   Number of observations    28
                          Example 3.4                                      8
                 comparison of regression lines
                        common intercept
                               M2

                         The GLM Procedure

Dependent Variable: blood_pressure

                                         Sum of
 Source                      DF         Squares    Mean Square   F Value   Pr > F

 Model                        2     4295.965801    2147.982901    116.07   <.0001

 Error                       25      462.661342      18.506454

 Corrected Total             27     4758.627143

         R-Square     Coeff Var      Root MSE    blood_pressure Mean

         0.902774      2.571706      4.301913               167.2786

 Source                      DF       Type I SS    Mean Square   F Value   Pr > F

 age                          1     3912.654942    3912.654942    211.42   <.0001
 age*group                    1      383.310859     383.310859     20.71   0.0001

                                                     Standard
Parameter                           Estimate            Error    t Value    Pr > |t|

Intercept                        81.54148011       6.09450878      13.38      <.0001
age                               1.45860666 B     0.10877197      13.41      <.0001
age*group journalist              0.13048919 B     0.02867217       4.55      0.0001
age*group university teach        0.00000000 B      .                .          .
```

An alternative MODEL statement for the model M_2 is

```
MODEL blood_pressure = age*group / SS1 SOLUTION ;
```

which differ in the Source and Parameter part of the output that is displayed below. The Parameter part is particularly useful if estimates and confidence intervals of the individual regression parameters are of interest.

```
 Source                      DF       Type I SS    Mean Square   F Value   Pr > F

 age*group                    2     4295.965801    2147.982901    116.07   <.0001


                                                     Standard
 Parameter                          Estimate            Error    t Value    Pr > |t|

 Intercept                       81.54148011       6.09450878      13.38      <.0001
 age*group journalist             1.58909584       0.10708631      14.84      <.0001
 age*group university teach       1.45860666       0.10877197      13.41      <.0001
```

```
                         Example 3.4                                  9
                comparison of regression lines
                         common slope
                             M2*

                     The GLM Procedure

                  Class Level Information

       Class          Levels    Values

       group               2    journalist university teach


              Number of observations    28
                         Example 3.4                                 10
                comparison of regression lines
                         common slope
                             M2*

                     The GLM Procedure

Dependent Variable: blood_pressure

                                            Sum of
 Source                      DF          Squares     Mean Square    F Value    Pr > F

 Model                        2      4305.734695     2152.867347     118.84    <.0001

 Error                       25       452.892448       18.115698

 Corrected Total             27      4758.627143


           R-Square     Coeff Var      Root MSE    blood_pressure Mean

           0.904827      2.544411      4.256254               167.2786


 Source                      DF        Type I SS     Mean Square    F Value    Pr > F

 age                          1      3912.654942     3912.654942     215.98    <.0001
 group                        1       393.079752      393.079752      21.70    <.0001


                                                       Standard
Parameter                             Estimate            Error    t Value    Pr > |t|

Intercept                        76.80587938 B       6.04706778      12.70      <.0001
age                               1.54148319         0.10568193      14.59      <.0001
group       journalist            7.51476484 B       1.61325341       4.66      <.0001
group       university teach      0.00000000 B        .                .         .

NOTE: The X'X matrix has been found to be singular, and a generalized inverse
      was used to solve the normal equations.  Terms whose estimates are
      followed by the letter 'B' are not uniquely estimable.
```

An alternative MODEL statement for the model M_2^* is

```
MODEL blood_pressure = age group / SS1 SOLUTION NOINT ;
```

which differ in the Source and Parameter part of the output. The latter is particularly useful if estimates and confidence intervals of the individual regression parameters are of interest.

```
 Source                      DF        Type I SS     Mean Square    F Value    Pr > F

 age                          1      784235.6510     784235.6510    43290.4    <.0001
 group                        2        3569.4565       1784.7283      98.52    <.0001


                                                       Standard
 Parameter                            Estimate            Error    t Value    Pr > |t|

 age                                1.54148319       0.10568193      14.59      <.0001
 group       journalist            84.32064421       6.09856330      13.83      <.0001
 group       university teach      76.80587938       6.04706778      12.70      <.0001
```

```
                          Example 3.4                              11
                   comparison of regression lines
                     identical regression lines
                               M3

                        The GLM Procedure

                 Number of observations    28
                          Example 3.4                              12
                   comparison of regression lines
                     identical regression lines
                               M3

                        The GLM Procedure

Dependent Variable: blood_pressure

                                        Sum of
 Source                      DF        Squares     Mean Square    F Value    Pr > F

 Model                        1    3912.654942     3912.654942     120.25    <.0001

 Error                       26     845.972201       32.537392

 Corrected Total             27    4758.627143


          R-Square     Coeff Var      Root MSE    blood_pressure Mean

          0.822223      3.409974      5.704156               167.2786


 Source                      DF      Type I SS     Mean Square    F Value    Pr > F

 age                          1    3912.654942     3912.654942     120.25    <.0001


                                              Standard
       Parameter         Estimate                Error    t Value    Pr > |t|

       Intercept      79.66029930           8.06245493       9.88      <.0001
       age             1.55272887           0.14159608      10.97      <.0001
```

□

Type I Sums of Squares in PROC GLM

Example 3.4 (Continued)
The model statement used to specify model M_1 in the program on page 194 was

```
MODEL blood_pressure = age group age*group/
                       SS1 SOLUTION;
```

and the option SS1 gave the following table in the output listing on page 198:

```
Source                DF     Type I SS     Mean Square    F Value    Pr > F

age                    1   3912.654942     3912.654942     207.53    <.0001
group                  1    393.079752      393.079752      20.85    0.0001
age*group              1      0.403936        0.403936       0.02    0.8849
```

This is the table we have in mind when we speak of the type I sums of squares, although strictly speaking, the type I sums of squares are those in the column labeled Type I SS. We will refer to it as the SS1 table. An entry in the Mean Square column is the entry in the same row of the Type I SS column divided by the entry in the same row in the DF column.

The entries of the SS1 table are related to some of the F-test for model reduction that were used in the analysis of Example 3.4 in Section 4.2. For easy reference we copy the F-test for the reduction from M_1 to M_2^* that was given in (4.13),

$$F = \frac{\frac{452.892448 - 452.488513}{25-24}}{18.853688} = \frac{0.403935}{18.853688} = 0.02142, \tag{4.15}$$

and the F-test for the reduction M_2^* from to M_3 that was given in (4.14),

$$F = \frac{\frac{845.972201 - 452.892448}{26-25}}{18.115698} = \frac{393.079753}{18.115698} = 21.70. \tag{4.16}$$

The bottom row of the SS1 table gives the F-test and related quantities for the reduction from M_1 to M_2^* and in (4.15) we recognize the degrees of freedom for the numerator of the F-test as the entry in the DF column, the numerator of the F-test as the entry in the Mean Square column, apart from the effect of rounding that shows in the last digit, the F-test statistic in the F Value column, and the significance probability based on the $F(1,24)$-distribution in the Pr > F column.

The row above which is labeled group **does not** give the F-test for the reduction from M_2^* to M_3, but it still gives related quantities that may be useful; the degrees of freedom for the numerator of the F-test is the entry in the DF column and the numerator of the F-test is the entry in the Mean Square column. The F value is not the F-test in (4.16) but the only difference is that the estimated variance in M_1, $s_1^2 = 18.853688$, is used in the denominator instead of the estimated variance in M_2^*, $s_{02}^{2*} = 18.115698$. The Pr > F is computed using the appropriate $F(1,24)$-distribution.

The entries in the top row of the SS1 table are similar to the group row. The table gives relevant quantities for the numerator of the F-test for the reduction from M_3 to M_∞, the model for a single sample, but s_1^2 is used in the denominator to compute the F-value. □

What has already been described for model M_1 in Example 3.4 will now be described in general terms. We will consider the table of type I sums of squares which is derived from the model formula,

$$\texttt{MODEL x = } A_k\ A_{k-1}\ \ldots\ A_2\ A_1\texttt{;}$$

where A_i $i = 1,\ldots,k$ may be a factor, i.e., it appears in the CLASS statement of *PROC GLM*, or a variable or a product of variables and factors. A model formula with k terms defines a sequence of $k+1$ simpler models, $M_1 \to M_2 \to \cdots \to M_k \to M_{k+1}$, where M_1 is defined by the full model formula and the following models are derived from M_1 by successively removing terms from the

Table 4.12 *Explanation of Type I sums of squares. SS, DF, and MS refer to sums of squares, degrees of freedom, and mean square in the same row. In the last column the function F denotes the distribution function of the test statistic,* $F = MS/s_1^2$*, in a given row. In the row* A_i *it is an F-distribution with* $d_{i+1} - d_i$ *degrees of freedom in the numerator and* $n - d_1$ *degrees of freedom in the denominator.*

Source	DF	Type I SS	Mean Square	F Value	Pr > F
A_k	$d_{k+1} - d_k$	$\|\|P_k(\mathbf{x}) - P_{k+1}(\mathbf{x})\|\|^2$	SS/DF	MS/s_1^2	$1 - F(F)$
A_{k-1}	$d_k - d_{k-1}$	$\|\|P_{k-1}(\mathbf{x}) - P_k(\mathbf{x})\|\|^2$	SS/DF	MS/s_1^2	$1 - F(F)$
...	...	...	...	...	...
A_i	$d_{i+1} - d_i$	$\|\|P_i(\mathbf{x}) - P_{i+1}(\mathbf{x})\|\|^2$	SS/DF	MS/s_1^2	$1 - F(F)$
...	...	...	...	...	...
A_1	$d_2 - d_1$	$\|\|P_1(\mathbf{x}) - P_2(\mathbf{x})\|\|^2$	SS/DF	MS/s_1^2	$1 - F(F)$

model formula from the right, first A_1, then A_2, and so on.

```
M1:      MODEL   x = Ak Ak-1 ... A2 A1;
M2:      MODEL   x = Ak Ak-1 ... A2;

Mi:      MODEL   x = Ak Ak-1 ... Ai;

Mk:      MODEL   x = Ak;
Mk+1:    MODEL   x = ;
```

The empty model M_{k+1} is the model for one sample where all observations have the same mean. This is the model we have denoted M_∞, because it is the same model regardless of the number of terms in the original model formula.

Using the notation from linear normal models, we can explain the contents of the table of sums of squares of Type I and this has been done in in Table 4.12. Recall that n denotes the number of observations, L_i refers to the linear subspace of the model M_i and d_i is the dimension of L_i. The orthogonal projection on L_i is denoted by P_i and, finally, s_{0i}^2 denotes the estimate of the variance in M_i, except in M_1, where it has been denoted by s_1^2. Note that the sum of the type I sums of squares gives $||P_1(\mathbf{x}) - P_\infty(\mathbf{x})||^2$, which is the `MODEL Sum of Squares` in the analysis of variance table from M_1.

If a particular decreasing sequence of models is of interest, it may be possible to formulate the model in such a way in the `MODEL` statement that the interesting sequence of models corresponds to the sequence in the `SS1` table. The `Type I SS` column will usually be studied first, and it will be noticed which terms in the model give the largest contributions to the `MODEL Sum of Squares`, and then the `Pr > F` column will be consulted to see which contributions are significant.

4.3 Two-Way Analysis of Variance

To fix ideas we start by using the data of Example 3.3 for a two-way analysis of variance. In the next subsection some general features and virtues of two-way analysis are explained and the section ends with an example of a two-way analysis of variance.

Example 3.3 (Continued)

A plant physiologist wanted to examine the influence of mechanical stress on the growth of soya bean plants. 52 seedlings were available for the experiment and half of those were exposed to mechanical stress by being shaken for 30 minutes daily while the other half was not exposed to stress. In order to see if a possible effect of stress depended on the amount of light available for the plants, half the plants that were stressed and half the plants that were not stressed were grown at a low intensity of light. The rest of the plants were grown at a higher intensity of light. The response for each plant is the total area of the leaves after 16 days of growth. Data are displayed in Table 4.13 and in Figure 4.7.

Table 4.13 *Total area of leaves of 52 soya bean plants grown under four different conditions*

	Stress (low)		*Stress (high)*	
	200	215	163	182
	225	229	188	195
	230	232	202	205
Light (low)	241	253	212	214
	256	264	215	230
	268	288	235	255
	288		272	
	268	271	200	215
	273	282	225	229
	285	299	230	232
Light (high)	309	310	241	253
	314	320	256	264
	337	340	268	288
	345		288	

The data from this experiment were used in Chapter 3 as an example of a one-way analysis of variance. In the course of that analysis, it was decided that the data were consistent with the model of four independent normal samples with a common variance. The hypothesis that all observations had a common mean was also tested but that hypothesis was strongly rejected by the data.

The conclusion of the analysis so far is that the treatment did have an influence on the total area of the leaves but it remains to be described if and how the intensity of light and the level of stress influence the total area of the leaves.

This is an example of a designed experiment where a number of well-defined treatments are applied to subsets of the experimental material available. It is customary to use the term a *factor* to denote a subdivision of the experimental material into subgroups that receive different treatments and the term the *factor levels* or simply the *levels* to denote the different treatments that are applied to the subdivisions given by a factor. In this case we have the factors *light* and *stress* both with two levels. For both factors we shall denote the levels as *low* and *high* and for the factor light the level low will correspond to the lower intensity of light, whereas for the factor stress the level low corresponds to no stress. When the factors light and stress are considered together, they form a factor with four levels corresponding to the four combinations of the levels of light and stress.

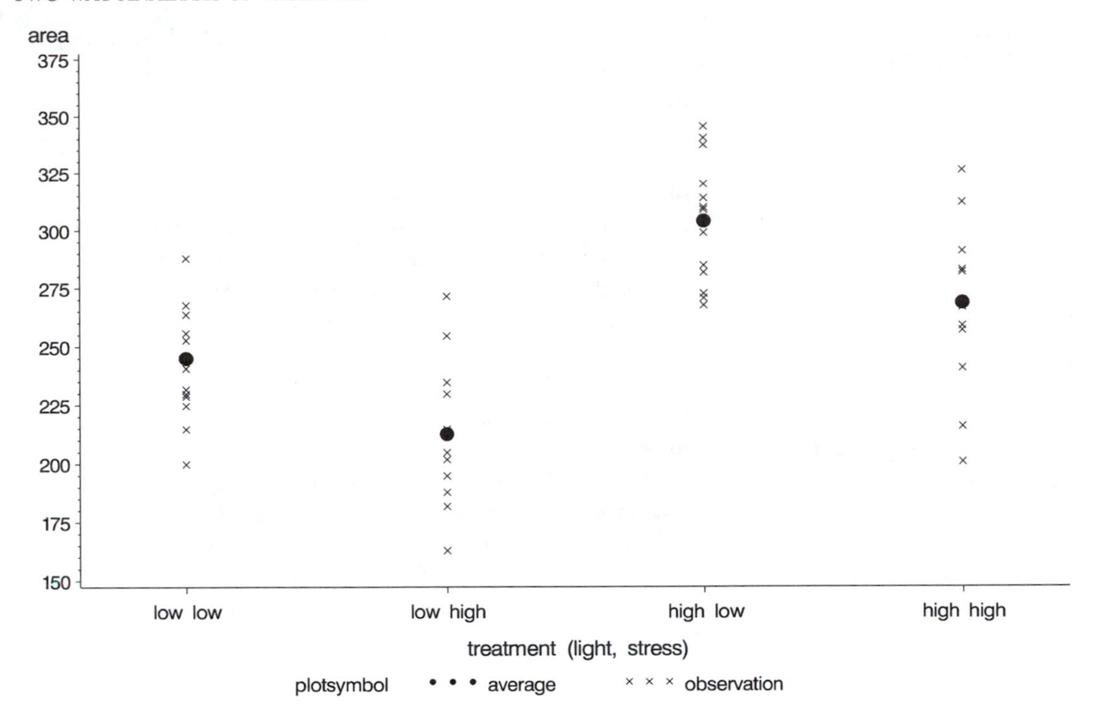

Figure 4.7 *Total area of leaves plotted against condition of growth.*

The starting point for further analysis is the model,

$$M_1: x_{ijk} \sim\sim N(\mu_{ij}, \sigma^2), \qquad i = 1,2,\ j = 1,2,\ k = 1,\ldots,13.$$

Here $i = 1,2$ index the two levels of light, $j = 1,2$ index the two levels of stress, and $k = 1,\ldots,13$ index the observations within samples, i.e., total area of leaves of plants that have been exposed to the same levels of light and stress.

Judging from Figure 4.7 the total area of the leaves tends to be lower when the plants have been exposed to stress and the total area of the leaves tends to be lower for plants grown at a low level of light than for plants grown at a high level of light. The message of the figure seems to be that stress has an effect on the total area of the leaves and this effect is to lower the total area of the leaves. Similarly, the intensity of light seems to have an effect, which is that plants grown at a low level of light develop a smaller total area of the leaves.

So far the description of the effects has only been qualitative and has been based on noting the change in location of the samples corresponding to different experimental conditions. If we want to quantify the effects, it will be done in terms of the means of the distributions. The estimated means are indicated by dots in Figure 4.7.

By the *effect of a factor* with two levels we shall understand the difference in means at the low and the high level of the factor. Obviously it may not be meaningful to speak simply of *the* effect of a factor because it may depend on the levels of other factors. In the present example it is quite conceivable that the effect of stress would be different at a low level of light from the effect of stress at a high level of light. If that were to happen, one could neither speak of the effect of stress nor of the effect of light without specifying the level of the other factor, and in that case one would say that the there was *interaction* between the factors light and stress.

Looking again at Figure 4.7, it seems that the estimate of the effect of stress at a low level of light given as the difference between the estimated means in the two leftmost samples is very close to the estimate of the effect of stress at a high level of light estimated in the same way from the two rightmost samples. Similarly, the estimates of the effect of light are very similar whether they are calculated at a low level of stress or at a high level of stress.

Thus Figure 4.7 points to the *additivity model,*

$$M_2\colon x_{ijk} \sim\sim N(\alpha_i + \beta_j, \sigma^2), \qquad i = 1,2,\ j = 1,2,\ k = 1,\ldots,13. \tag{4.17}$$

The name is derived from the fact that the mean of observations at level i of light and level j of stress is a sum of a parameter α_i that depends on the level of light and of a parameter β_j that depends on the level of stress. The interesting parameters are not α_i and β_j in themselves but rather the differences $\alpha_1 - \alpha_2$ and $\beta_1 - \beta_2$, which are the effects of light and stress, respectively; for example, the difference $\alpha_1 - \alpha_2$ is the difference in mean response between the two levels of light regardless of which level of stress is chosen to measure the difference and similarly for $\beta_1 - \beta_2$. Thus the additivity model has the property that it is meaningful to speak of the effect of the two factors without specifying at which level of the other factor the effect is being measured. When two factors have this property, we shall say that the two factors *do not interact*, or there is *no interaction* between the two factors, and the additivity model is also called the *model of no interaction*.

The mean of the total area of the leaves for the four treatments under M_2 are determined by those two differences and $\alpha_2 + \beta_2$. This illustrates that M_2 has $d_2 = 3$ parameters. We shall see later how the parameters are estimated.

In the Annex to Section 4.3, pages 216 and 217, the listings of the output from the models M_1 and M_2 are displayed. The test of the reduction from M_1 to M_2 is calculated as explained in (4.12) as

$$F = \frac{\dfrac{43002.6346 - 42976.308}{49 - 48}}{895.340} = 0.0294.$$

The significance probability is

$$p_{obs}(\mathbf{x}) = 1 - F_{F(1,48)}(0.0294) = 0.865.$$

This does not contradict the reduction of the model to M_2, which takes over as the model for the data.

The next step will be to investigate whether the model can be further reduced to either

$$M_3^r\colon x_{ijk} \sim\sim N(\alpha_i, \sigma^2), \qquad i = 1,2,\ j = 1,2,\ k = 1,\ldots,13,$$

or to

$$M_3^c\colon x_{ijk} \sim\sim N(\beta_j, \sigma^2), \qquad i = 1,2,\ j = 1,2,\ k = 1,\ldots,13.$$

In the model M_3^r the means only depend on the level of light so the test for the reduction to M_3^r is a test of the hypothesis that the level of stress has no influence on the total area of the leaves. The superscript r on M_3^r is chosen as a reminder that *the means in this model only depend on the factor of the rows.*

In the same way the means only depend on the level of stress in the model M_3^c so the test of the reduction to M_3^c is the test of the hypothesis that light has no effect on the total area of the leaves. Here the superscript c on M_3^c serves as a reminder that *the means in this model only depend on the factor of the columns.*

The models M_3^r are M_3^c cannot be compared in the sense that none of them is a submodel of the other. Both models are examples of the well-known model for two independent samples. In M_3^r the rows of Table 4.13 constitute the two samples and in M_3^c the columns of Table 4.13 constitute the two samples.

We first test the reduction to M_3^c. The output listings from the estimation in the models M_2 and M_3^c are on pages 217 and 218. The test statistic is

$$F = \frac{\dfrac{85754.1923 - 43002.6346}{50 - 49}}{877.6048} = 48.71,$$

and this gives the significance probability,

$$p_{obs}(\mathbf{x}) = 1 - F_{F(1,49)}(48.71) = 7 \times 10^{-9}.$$

Thus the data strongly reject the reduction to M_3^c and the interpretation is that the amount of light is important for the total area of the leaves. M_2 is still the model for the data.

The test of the reduction to M_3^r is based on the output listings on pages 217 and page 219.

$$F = \frac{\frac{57861.1154 - 43002.6346}{50 - 49}}{877.6048} = 16.93,$$

and the significance probability is

$$p_{obs}(\mathbf{x}) = 1 - F_{F(1,49)}(16.93) = 0.00015.$$

The reduction to M_3^r is not accepted and this is a confirmation that the amount of stress is important for the total area of the leaves.

The final model for the data is M_2, which says that both light and stress influence the total area of leaves and that the influence is additive, i.e., the influence of stress is the same for both levels of light and the influence of the amount of light is the same for both levels of stress.

An estimate of the effect of light in the model M_2 is calculated as a difference of the averages in the two groups of data corresponding to the two levels of light,

$$\hat{\alpha}_1 - \hat{\alpha}_2 = \bar{x}_{1\cdot\cdot} - \bar{x}_{2\cdot\cdot} = -57.35 \text{ cm}^2 \sim\sim N(\alpha_1 - \alpha_2, \sigma^2/13), \tag{4.18}$$

which is an estimate of the difference of the means of the total area of leaves between plants grown at the low and the high levels of light. The estimate is negative indicating that the mean total area of the leaves is smaller for plants grown at the low intensity of light.

Likewise an estimate of the effect of stress in the model M_2 is calculated as a difference of the averages in the two groups of data that correspond to the two levels of stress

$$\hat{\beta}_1 - \hat{\beta}_2 = \bar{x}_{\cdot 1\cdot} - \bar{x}_{\cdot 2\cdot} = 33.81 \text{ cm}^2 \sim\sim N(\beta_1 - \beta_2, \sigma^2/13), \tag{4.19}$$

which is an estimate of the difference of the means of the total area of leaves when the plants are grown at the low level of stress as compared to the high level of stress.

It is explained on page 217 where the estimates are found in the SAS output listing from M_2.

As usual we shall use the distribution of the estimates to calculate a 95% confidence interval for the effects to give an idea of the precision of the experiment. The standard error of the estimated effects is found in the SAS output from model M_2 on page 217.

$$\begin{array}{llcrl} \text{Light:} & \alpha_1 - \alpha_2 & \leftarrow & -57 \text{ cm}^2 & [-74 \text{ cm}^2,\ -40 \text{ cm}^2] \\ \text{Stress:} & \beta_1 - \beta_2 & \leftarrow & 34 \text{ cm}^2 & [17 \text{ cm}^2,\ 50 \text{ cm}^2] \end{array}$$

□

4.3.1 Two-Way Analysis of Variance in General

Example 3.3 had four samples, one for each combination of two factors (stress and light), each having two levels. The experiment and the subsequent analysis focused on evaluating the effects of the two factors.

In general, one can imagine rs samples corresponding to all combinations of a factor R with r levels and a factor S with s levels. The observations are denoted as x_{ijk}, where $i = 1, \ldots, r$ index the r levels of R, $j = 1, \ldots, s$ index the s levels of S, and $k = 1, \ldots, n_{ij}$ index the observations that received treatment i, j. The observations can be written as shown in Table 4.13 with r rows, s columns, and n_{ij} observations in cell i, j. Assuming as usual that all observations are independent, we can use the shorthand notation for the models of interest:

The basic model
M_1: $x_{ijk} \sim\sim N(\mu_{ij}, \sigma^2)$, $i = 1, \ldots, r,\ j = 1, \ldots, s,\ k = 1, \ldots, n_{ij}$.
The additivity model
M_2: $x_{ijk} \sim\sim N(\alpha_i + \beta_j, \sigma^2)$, $i = 1, \ldots, r,\ j = 1, \ldots, s,\ k = 1, \ldots, n_{ij}$.
The model with *row effects only*
M_3^r: $x_{ijk} \sim\sim N(\alpha_i, \sigma^2)$, $i = 1, \ldots, r,\ j = 1, \ldots, s,\ k = 1, \ldots, n_{ij}$.
The model with *column effect only*
M_3^c: $x_{ijk} \sim\sim N(\beta_j, \sigma^2)$, $i = 1, \ldots, r,\ j = 1, \ldots, s,\ k = 1, \ldots, n_{ij}$.
The model for one sample or *homogeneity*
M_∞: $x_{ijk} \sim\sim N(\mu, \sigma^2)$, $i = 1, \ldots, r,\ j = 1, \ldots, s,\ k = 1, \ldots, n_{ij}$.

The relationship between the models M_2, M_3^r, M_3^c, and M_∞ is apparent from the diagram below.

$$
\begin{array}{ccccc}
 & \nearrow & M_3^r: X_{ijk} \sim N(\alpha_i, \sigma^2) & \searrow & \\
M_2: X_{ijk} \sim N(\alpha_i + \beta_j, \sigma^2) & & & & M_\infty: X_{ijk} \sim N(\mu, \sigma^2) \\
 & \searrow & M_3^c: X_{ijk} \sim N(\beta_j, \sigma^2) & \nearrow &
\end{array}
$$

Additivity Model

We shall concentrate on the additivity model M_2, which is the only new model listed above because M_1, M_3^r, M_3^c, and M_∞ are the well-known models for rs, r, s, and 1 samples.

M_2 is a linear normal model and it can be proved that the dimension of the corresponding linear subspace has dimension $d_2 = r + s - 1$ so the F-test of the reduction from M_1 to M_2 has $d_1 - d_2 = rs - (r + s - 1) = (r - 1)(s - 1)$ degrees of freedom in the numerator and $n_{\cdot\cdot} - rs$ degrees of freedom in the denominator. Here $n_{\cdot\cdot}$ is the total number of observations.

The F-test of the reduction from M_1 to M_2 can and should be supplemented by plots. On page 205 Figure 4.7 was used to motivate the additivity model. With more than two rows and columns a plot like Figure 4.7 becomes unwieldy and, therefore, other plots are used to investigate if the additivity model is reasonable.

Under M_2 the difference in means between two different rows g and h does not depend on which column is chosen to evaluate the difference, because

$$EX_{gjk} - EX_{hjk} = (\alpha_g + \beta_j) - (\alpha_h + \beta_j) = \alpha_g - \alpha_h. \tag{4.20}$$

In the same way the difference in means between two different columns l and m does not depend on which row is chosen to measure the difference:

$$EX_{ilk} - EX_{imk} = (\alpha_i + \beta_l) - (\alpha_i + \beta_m) = \beta_l - \beta_m. \tag{4.21}$$

The means themselves are not available but the averages within cells estimate the means, $\bar{x}_{ij\cdot} \rightarrow EX_{ijk}$ so, if for each row i the points $(j, \bar{x}_{ij\cdot})$, $j = 1, \ldots, s$ are plotted and connected by straight lines, from (4.20) we have r curves with a constant vertical distance apart from random deviations.

Figure 4.10 on page 213 is an example of such a figure based on data in Table 4.14, which is the basis of Example 4.7 on page 211.

Similarly, if for each column j the points $(i, \bar{x}_{ij\cdot})$, $i = 1, \ldots, r$, are plotted and joined by straight lines we get s curves, which because of (4.21), will have a constant vertical distance apart from random deviations. Figure 4.9 on page 213 shows this drawing based on the data in Table 4.14.

Upon inspection of the plots to see whether they contradict M_2, we look for the type of deviation where the deviations increase when the averages $\bar{x}_{ij\cdot}$ increase. This could be an indication of a multiplicative structure in the means in which case one may decide to transform the data taking the logarithm of the observations to achieve the additive structure that is the defining property of M_2.

When the number of replications n_{ij} is 1, the reduction to M_2 cannot be tested using an F-test and the drawing is the only possibility whereby we can check whether the model is contradicted by the data.

Example 3.3 (Continued)
Even for tables with only two rows and columns, these drawings are very informative, as illustrated in Figure 4.8 where the data are displayed.

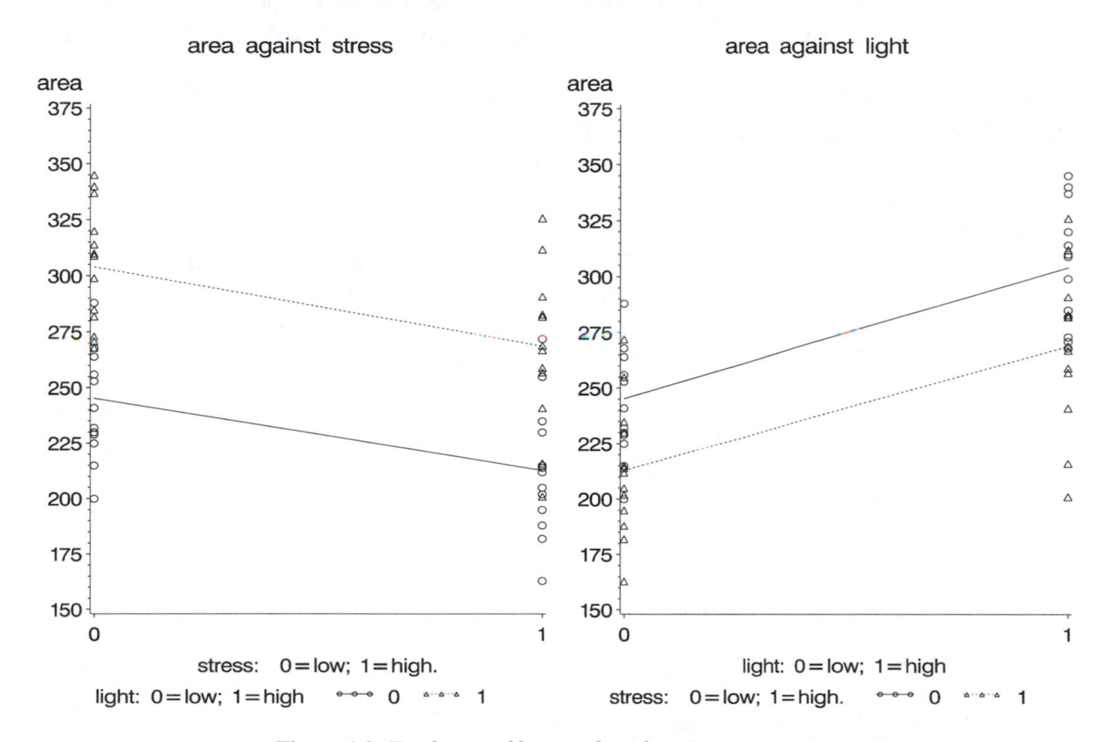

Figure 4.8 *Total area of leaves plotted against treatment.*

In this case one can hardly discern the deviation from parallelism of the two lines that connect the averages in groups with the same level of light (left) or groups with the same level of stress (right). When, as is the case, here there are more than one observation in each sample the decision of whether a deviation from parallelism contradicts M_2 must be based on the size of the variation within samples. Clearly the larger the variation is within samples the larger deviation from parallellity can be accepted without rejecting M_2. With the variation within samples observed in this example, far larger deviations from parallelism may be anticipated even when M_2 is true. □

Requirements of the Number of Observations n_{ij}

When two-way analysis of variance is treated as we do here where the estimation is left to a statistical package, there are no technical requirements of the number of observations, n_{ij}, in each cell to make the calculation simple. The reduction of the models are tested using the F-test as described in (4.12) and the package also gives the estimates $\hat{\alpha}_i - \hat{\alpha}_r$ and $\hat{\beta}_j - \hat{\beta}_s$ and their standard errors.

Some important advantages are achieved, however, by making sure that the number of observations satisfy the relation,

$$n_{ij} = \frac{n_{i\cdot}n_{\cdot j}}{n_{\cdot\cdot}}, \tag{4.22}$$

where $n_{i\cdot}$, $n_{\cdot j}$ and $n_{\cdot\cdot}$ denote the number of observations in the ith row, the number of observations in the jth column, and the total number of observations, c.f. the description of the notation on page 80. An important special case where (4.22) holds is the one where n_{ij} is the same for all combinations of i and j in which case we simply write n for n_{ij}.

One of the advantages of (4.22) is that the numerator in the F-test for the reduction from M_2 to M_3^c is the same as the numerator in the F-test for the reduction from M_3^r to M_∞. Both reductions say that the levels of R have no effect on the response, but the reduction from M_3^r to M_∞ is tested in a model where the levels of S have no effect on the response and the identity of the numerators of the two F-tests mean that the test of the effect of the levels of R *rarely* depend on whether the effect of the levels of S has been tested.

The proviso *"rarely"* is necessary because the denominator in the F-tests is s_{02}^2 and s_{03r}^2, respectively, and they will not only be numerically different but their distribution will have different degrees of freedom so the F-test will be evaluated in different F-distributions.

The situation is analogous as far as the test for the reductions from M_2 to M_3^r and from M_3^c to M_∞ is concerned.

A different but related advantage of (4.22) is that the estimate of the difference in mean response between the two different levels of R is the same in the model M_2 as in the model M_3^r so the estimates do not depend on whether or not S is included in the model.

Estimation in M_2 of $\alpha_g - \alpha_h$ and $\beta_l - \beta_m$

In the following we assume that (4.22) holds. In M_2 the average of all observations at level i of R satisfy $\bar{x}_{i\cdot\cdot} \sim\sim N(\alpha_i + \bar{\beta}_\cdot, \sigma^2/n_{i\cdot})$, where

$$\bar{\beta}_\cdot = \frac{1}{n_{\cdot\cdot}} \sum_{j=1}^{s} n_{\cdot j}\beta_j,$$

and applying this to the two levels g and h of R gives

$$\alpha_g - \alpha_h \leftarrow \bar{x}_{g\cdot\cdot} - \bar{x}_{h\cdot\cdot} \sim\sim N(\alpha_g - \alpha_h, \sigma^2(1/n_{g\cdot} + 1/n_{h\cdot})). \tag{4.23}$$

This means that the estimate of the difference between the mean of the response at level g and the mean of the response at level h of R is the intuitively obvious difference of the average of all observations at level g and the average of all observations of all observations at level h. We stress that if (4.22) does not hold, the mean of $\bar{x}_{g\cdot\cdot} - \bar{x}_{h\cdot\cdot}$ depends on the β_j's so in that case "the intuitively obvious estimate", $\bar{x}_{g\cdot\cdot} - \bar{x}_{h\cdot\cdot}$, is simply not a meaningful estimate of $\alpha_g - \alpha_h$.

Likewise, the estimate of the difference of the means of the response at levels l and m of S is

$$\beta_l - \beta_m \leftarrow \bar{x}_{\cdot l\cdot} - \bar{x}_{\cdot m\cdot} \sim\sim N(\beta_l - \beta_m, \sigma^2(1/n_{\cdot l} + 1/n_{\cdot m})), \tag{4.24}$$

when (4.22) holds.

In M_3^r, which is the model for r independent samples, it is well known that the difference of the means of the response at levels g and h of R is given by (4.23) and similarly the estimate of the difference of the means of the response at levels l and m in the model M_3^c is given by (4.24).

Example 3.3 (Continued)
The condition (4.22) is satisfied for this example with $n = n_{ij} = 13$. On page 207 the estimates of $\hat{\alpha}_1 - \hat{\alpha}_2$ and $\hat{\beta}_1 - \hat{\beta}_2$ were given as

$$\hat{\alpha}_1 - \hat{\alpha}_2 = \bar{x}_{1\cdot\cdot} - \bar{x}_{2\cdot\cdot} = -57.35 \text{ cm}^2 \sim\sim N(\alpha_1 - \alpha_2, \sigma^2/13),$$

and

$$\hat{\beta}_1 - \hat{\beta}_2 = \bar{x}_{\cdot 1\cdot} - \bar{x}_{\cdot 2\cdot} = 33.81 \text{ cm}^2 \sim\sim N(\beta_1 - \beta_2, \sigma^2/13),$$

and we note that the given distributions agree with (4.23) and (4.24). In the SAS output listing from M_2 on page 217 the estimates under `Parameter` may be found at the bottom of the page. Under

`Std Error of Estimate` the estimated standard deviation of the estimates,

$$\sqrt{s_{02}^2(1/26+1/26)} = 8.21632890,$$

is consistent with (4.23) and (4.24). □

Example 4.6

On page 218 we find numerically the same estimate $\hat{\alpha}_1 - \hat{\alpha}_2$ in the model M_3^r but `Std Error of Estimate` is now

$$\sqrt{s_{03r}^2(1/26+1/26)} = 11.48605792,$$

and that value is substantially larger than in the correct model M_2. Thus even if the estimate of the difference $\alpha_1 - \alpha_2$ has not changed, it is important to find the estimated standard error of the estimate in the right model. □

Example 4.7

The data in Table 4.14 are from six drillings in an area. Each drilling results in a core of soil, and the response that we consider is the pH-value of the soil at the top, in the middle, and at the bottom of the core (Li, 1964).

Table 4.14 *pH measured at the bottom, in the middle, and at the top of 6 cores*

Core	*Bottom*	*Middle*	*Top*	*Average*
1	7.2	7.6	7.5	7.43
2	6.7	7.1	7.2	7.00
3	7.0	7.2	7.3	7.17
4	7.0	7.4	7.5	7.30
5	7.0	7.7	7.7	7.47
6	6.9	7.7	7.6	7.40
Average	6.97	7.45	7.47	7.29

From Li, J.C.R., *Statistical Inference*, vol.1, Edwards, Ann Arbor, Mich.,1964. With Permission.

The basic factors are *core* with the six levels labelled 1 through 6 and *position* with the three levels *bottom*, *middle,* and *top*.

The six drillings were made over a fairly large area and it is neither sensible nor reasonable to consider the responses from the six drillings as replicated measurements of the same quantity. Nor is it reasonable to consider the three measurements of pH from each core as replicated measurements of the same quantity. Actually, one expects a systematic change in the level of pH through the soil and one of the objectives of the investigation is to study this change. After these considerations we start out with a model where all measurements are assumed to be independent and normally distributed with the same variance, but with possibly different means:

$$M_1\colon x_{ij} \sim\sim N(\mu_{ij}, \sigma^2), \qquad i = 1,\ldots,6,\ j = 1,2,3.$$

In contrast to the previous example, no aspect of this model can be checked because there is only one observation in each cell, i.e., $n_{ij} = 1$. A consequence of $n_{ij} = 1$ is that $L_1 = \mathbb{R}^n$ and then $P_1(\mathbf{x}) = \mathbf{x}$, $d_1 = n$ so an attempt to apply the formula (4.7) for the estimate of the variance gives $s_{01}^2 = \|\mathbf{x} - P_1(\mathbf{x})\|^2/(n-n) = 0/0$, which is undefined.

The additivity model,

$$M_2\colon x_{ij} \sim\sim N(\alpha_i + \beta_j, \sigma^2), \qquad i = 1,\ldots,6,\ j = 1,2,3,$$

gives a relatively simple description of the data and it is convenient for further analysis. But the reduction from M_1 til M_2 cannot be tested with an F-test in the usual way because the estimate of the variance in M_1 is undefined, and consequently it is not available as a denominator in the F-test for the reduction from M_1 til M_2.

The plots in Figure 4.9 and Figure 4.10 do not show any systematic deviations from the pattern expected under the additivity model M_2 that is adopted as the model for the data.

We first test the reduction to M_3^r. The output listings from the estimation in the models M_2 and M_3^r are found on page 222 and on page 223. The test statistic is

$$F = \frac{\dfrac{1.12000000 - 0.15222222}{12 - 10}}{0.01522222} = 31.79,$$

which gives the significance probability,

$$p_{obs}(\mathbf{x}) = 1 - F_{F(2,10)}(31.79) = 0.000046.$$

Thus the reduction to M_3^r is rejected, and we have confirmed that there is a difference between the pH levels at the three positions in the core.

The test for the reduction to M_3^c is calculated from the output listings on page 222 and on page 224,

$$F = \frac{\dfrac{0.64166667 - 0.15222222}{15 - 10}}{0.01522222} = 6.43,$$

which gives the significance probability,

$$p_{obs}(\mathbf{x}) = 1 - F_{F(5,10)}(6.43) = 0.0063.$$

The reduction to M_3^c is also rejected confirming the different pH levels between the 6 cores.

The final model for data is M_2 that says that pH levels depend on both the location of the drilling and of the position in the core where the pH value is measured. More precisely, the model says that the level of pH depends *additively* on the position and the location; thus the difference between means of pH in the *top*, in the *middle,* or at the *bottom* of the core is the same for all cores and, similarly, that the difference between the means of pH at the drillings is the same at the top, in the middle, and at the bottom of the core. This conclusion is supported by the plots in Figure 4.9 and Figure 4.10.

The estimates of the differences between means of pH between cores and between positions in the core can be found in the SAS output listings on page 222.

But they can also easily be obtained using the formulas (4.23) and (4.24) and exploiting the row and column averages given in the margins of Table 4.14. □

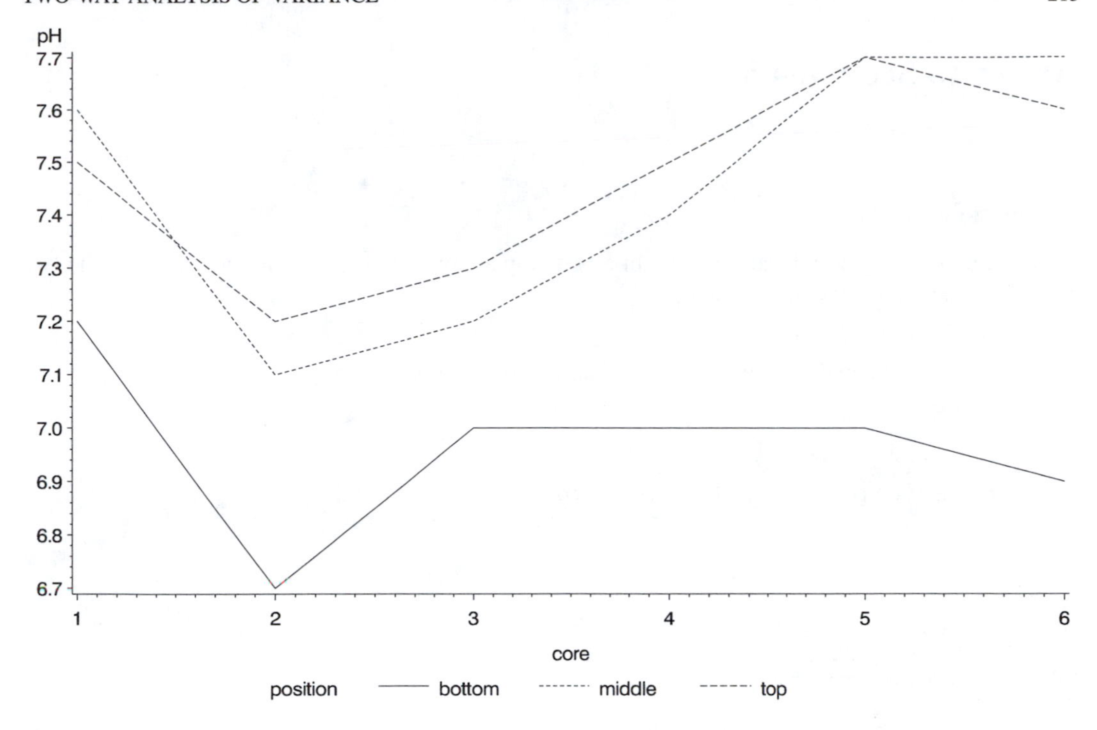

Figure 4.9 *Graphical investigation of the additivity model M_2. Under M_2 the three curves are vertical displacements of the same curve apart from random fluctuations.*

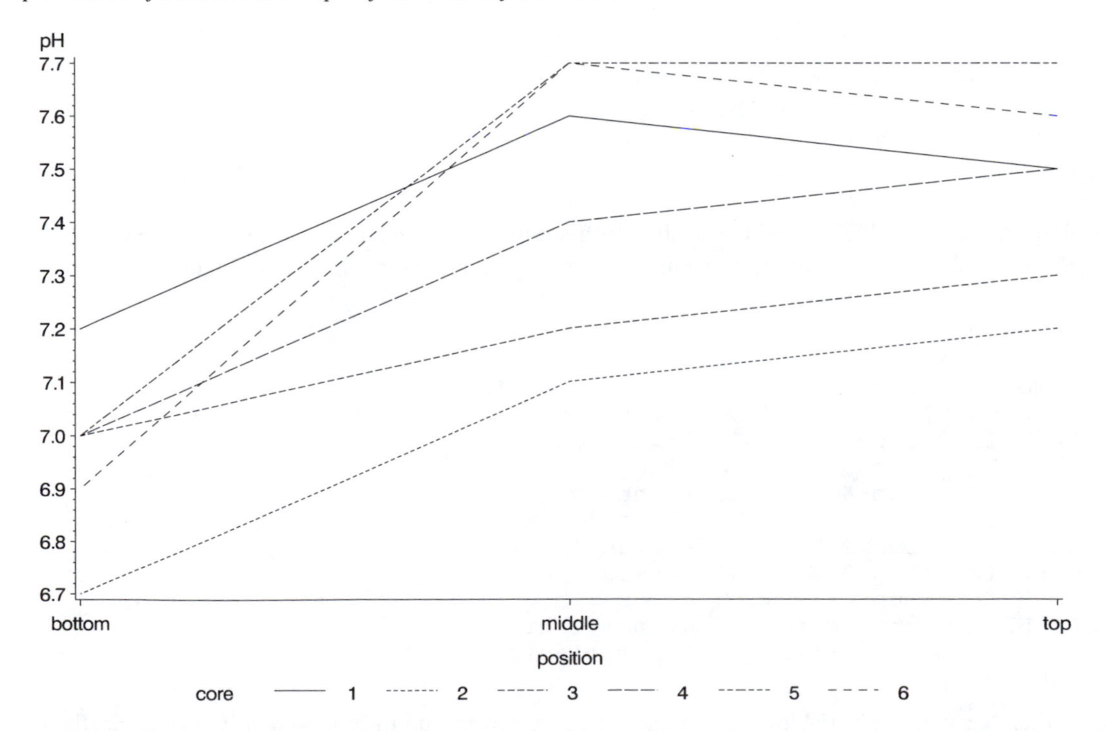

Figure 4.10 *Graphical investigation of the additivity model M_2. Under M_2 the six curves are vertical displacements of the same curve apart from random fluctuations.*

Annex to Section 4.3

Calculations in SAS

Here we show how the relevant models in a two-way analysis of variance are specified in the two examples that were used in Section 4.3.

Example 3.3 (Continued)
The program has been divided into parts and comments follow each part.

```
TITLE1 'Example 3.3';

PROC FORMAT;
    VALUE $grpfmt       '1'='low low'      '2'='low high'
                        '3'='high low'     '4'='high high'     ;
RUN;


DATA soya;
INPUT group $ area@@ ;
FORMAT group $grpfmt.;
DATALINES;
1 200 1 215 1 225 1 229 1 230 1 232 1 241
1 253 1 256 1 264 1 268 1 288 1 288
2 235 2 188 2 195 2 205 2 212 2 214 2 182
2 215 2 272 2 163 2 230 2 255 2 202
3 314 3 320 3 310 3 340 3 299 3 268 3 345
3 271 3 285 3 309 3 337 3 282 3 273
4 283 4 312 4 291 4 259 4 216 4 201 4 267
4 326 4 241 4 291 4 269 4 282 4 257
;
```

Here *PROC FORMAT* is used to supply information to the values 1, 2, 3, and 4 that are the values of the variable group. This piece of information is used in making the plot in Figure 4.7.

```
DATA stress;
SET soya;
LENGTH light  $ 4;
LENGTH stress $ 4;
IF group=1        THEN DO light='low ';        num_light=0; END;
ELSE IF group=2 THEN DO light='low ';         num_light=0; END;
ELSE IF group=3 THEN DO light='high';         num_light=1; END;
ELSE IF group=4 THEN DO light='high';         num_light=1; END;
IF group=1        THEN DO stress='low ';       num_str=0; END;
ELSE IF group=2 THEN DO stress='high';        num_str=1; END;
ELSE IF group=3 THEN DO stress='low ';        num_str=0; END;
ELSE IF group=4 THEN DO stress='high';        num_str=1; END;
LABEL num_str='stress:  0=low; 1=high.'
      num_light='light: 0=low; 1=high';
RUN;
```

In this data step the variables light and stress correspond to rows and columns, respectively, in Table 4.13. Similar numerical variables num_light and num_str are defined for use in the plots in Figure 4.8.

```
TITLE2 'M1';
PROC GLM DATA=stress ORDER=DATA;
CLASS light stress;
MODEL area=light*stress/SS1 SOLUTION NOINT;
RUN;

TITLE2 'M2';
PROC GLM DATA=stress ORDER=DATA;
CLASS light stress;
MODEL area=light stress/SS1 SOLUTION;
RUN;

TITLE2 'M3c';
PROC GLM DATA=stress ORDER=DATA;
CLASS stress;
MODEL area=stress /SS1 SOLUTION;
RUN;

TITLE2 'M3r';
PROC GLM DATA=stress ORDER=DATA;
CLASS light;
MODEL area=light/SS1 SOLUTION;
RUN;
QUIT;
```

The option `ORDER=DATA` to *PROC GLM* has been chosen to facilitate the comparison of the estimates of effects in the SAS output with those in the text. If the `ORDER=DATA` option had not been chosen, the levels of the factors would have been sorted alphabetically by SAS and the estimated effects would have had the opposite sign.

On the following pages the output listings from the program segments above are listed. It is possible to identify the model behind any listing as explained on page 195. But in addition the listings contain the model designations `M1`, `M2`, `M3c`, and `M3r`.

A few comments on the parameter estimates are given in connection with the output listings from M_1 and M_2 on pages 216 and 217.

```
                                                                      1
                                  M1

                          The GLM Procedure

                       Class Level Information

                   Class          Levels    Values

                   light               2    low high

                   stress              2    low high

                     Number of observations    52
                                                                      2
                                  M1

                          The GLM Procedure

Dependent Variable: area

                                        Sum of
 Source                      DF        Squares     Mean Square    F Value    Pr > F

 Model                        4    3513290.692      878322.673     980.99    <.0001

 Error                       48      42976.308         895.340

 Uncorrected Total           52    3556267.000


               R-Square     Coeff Var      Root MSE     area Mean

               0.572854      11.60728      29.92223      257.7885


 Source                      DF      Type I SS     Mean Square    F Value    Pr > F

 light*stress                 4    3513290.692      878322.673     980.99    <.0001


                                                   Standard
 Parameter                          Estimate          Error    t Value    Pr > |t|

 light*stress low low            245.3076923     8.29893294      29.56      <.0001
 light*stress low high           212.9230769     8.29893294      25.66      <.0001
 light*stress high low           304.0769231     8.29893294      36.64      <.0001
 light*stress high high          268.8461538     8.29893294      32.40      <.0001
```

Note that the NOINT option to the MODEL statement has been used here to obtain the estimates of the means of the four samples of Table 4.13 in the Parameter part of the output above.

If the NOINT option is dropped from the MODEL statement, the parameter estimates will be reported as shown below. The Intercept line contains the estimated mean and related quantities of observations in the cell in the bottom right-hand corner of the table that received the treatment with stress and light both on their high level. The other three nonzero numbers in the Estimate column are the estimates of the differences between the mean of observations a the treatment combination indicated in the same line in the Parameter column and the mean of the observations in the bottom right-hand corner of the table.

```
                                                     Standard
Parameter                          Estimate             Error    t Value    Pr > |t|

Intercept                      268.8461538 B       8.29893294      32.40      <.0001
light*stress low low           -23.5384615 B      11.73646352      -2.01      0.0506
light*stress low high          -55.9230769 B      11.73646352      -4.76      <.0001
light*stress high low           35.2307692 B      11.73646352       3.00      0.0043
light*stress high high           0.0000000 B         .                .          .

NOTE: The X'X matrix has been found to be singular, and a generalized inverse
      was used to solve the normal equations.  Terms whose estimates are
      followed by the letter 'B' are not uniquely estimable.
```

```
                                                                    3
                                  M2

                           The GLM Procedure

                        Class Level Information

                    Class          Levels     Values

                    light               2     low high

                    stress              2     low high

                      Number of observations     52
                                                                    4
                                  M2

                           The GLM Procedure

Dependent Variable: area

                                          Sum of
 Source                      DF          Squares     Mean Square    F Value    Pr > F

 Model                        2       57610.0385      28805.0192      32.82    <.0001

 Error                       49       43002.6346        877.6048

 Corrected Total             51      100612.6731

               R-Square     Coeff Var      Root MSE     area Mean

               0.572592      11.49175      29.62440      257.7885

 Source                      DF        Type I SS     Mean Square    F Value    Pr > F

 light                        1      42751.55769     42751.55769      48.71    <.0001
 stress                       1      14858.48077     14858.48077      16.93    0.0001

                                                 Standard
    Parameter                  Estimate             Error    t Value    Pr > |t|

    Intercept               269.5576923 B      7.11554955      37.88      <.0001
    light       low         -57.3461538 B      8.21632890      -6.98      <.0001
    light       high          0.0000000 B       .                .         .
    stress      low          33.8076923 B      8.21632890       4.11      0.0001
    stress      high          0.0000000 B       .                .         .

NOTE: The X'X matrix has been found to be singular, and a generalized inverse
      was used to solve the normal equations.  Terms whose estimates are
      followed by the letter 'B' are not uniquely estimable.
```

The `Intercept` line contains information on the estimated mean of observations in the cell in the bottom right-hand corner of the table that is the treatment with stress and light both on their high level. In the notation used in (4.17) the estimate is $\hat{\alpha}_2 + \hat{\beta}_2$.

The estimate $\hat{\alpha}_1 - \hat{\alpha}_2$ and its standard error etc. is found in the line labeled `light low` in the `Parameter` column and, similarly, the estimate $\hat{\beta}_1 - \hat{\beta}_2$ is found in the line labeled `stress low` in the `Parameter` column.

```
                                                                5
                              M3c

                       The GLM Procedure

                    Class Level Information

                 Class         Levels    Values

                 stress             2    low high

                   Number of observations    52
                                                                6
                              M3c

                       The GLM Procedure

Dependent Variable: area

                                       Sum of
 Source                      DF       Squares     Mean Square    F Value    Pr > F

 Model                        1    14858.4808      14858.4808       8.66    0.0049

 Error                       50    85754.1923       1715.0838

 Corrected Total             51   100612.6731


              R-Square     Coeff Var      Root MSE     area Mean

              0.147680      16.06494      41.41357      257.7885


 Source                      DF     Type I SS     Mean Square    F Value    Pr > F

 stress                       1   14858.48077     14858.48077       8.66    0.0049


                                               Standard
    Parameter                 Estimate            Error     t Value    Pr > |t|

    Intercept             240.8846154 B      8.12186945       29.66      <.0001
    stress    low          33.8076923 B     11.48605792        2.94      0.0049
    stress    high          0.0000000 B       .                 .         .

NOTE: The X'X matrix has been found to be singular, and a generalized inverse
      was used to solve the normal equations.  Terms whose estimates are
      followed by the letter 'B' are not uniquely estimable.
```

```
                                                                      7
                                   M3r

                             The GLM Procedure

                          Class Level Information

                      Class          Levels    Values

                      light               2    low high


                         Number of observations    52
                                                                      8
                                   M3r

                             The GLM Procedure

Dependent Variable: area

                                        Sum of
 Source                      DF        Squares    Mean Square   F Value   Pr > F

 Model                        1     42751.5577     42751.5577     36.94   <.0001

 Error                       50     57861.1154      1157.2223

 Corrected Total             51    100612.6731


               R-Square     Coeff Var      Root MSE     area Mean

               0.424912      13.19608      34.01797      257.7885


 Source                      DF      Type I SS    Mean Square   F Value   Pr > F

 light                        1    42751.55769    42751.55769     36.94   <.0001


                                                Standard
    Parameter                Estimate              Error    t Value    Pr > |t|

    Intercept             286.4615385 B       6.67147287      42.94      <.0001
    light      low        -57.3461538 B       9.43488742      -6.08      <.0001
    light      high         0.0000000 B        .                .         .

NOTE: The X'X matrix has been found to be singular, and a generalized inverse
      was used to solve the normal equations.  Terms whose estimates are
      followed by the letter 'B' are not uniquely estimable.
```

□

Example 4.6 (Continued)

The formulation of the models does not deviate from the formulation in Example 3.3, but for the sake of completeness the program is given.

```
TITLE1 'Example 4.6';

DATA ex4_6;
INPUT core pH position$7-12;
DATALINES;
1 7.5 top
2 7.2 top
3 7.3 top
4 7.5 top
5 7.7 top
6 7.6 top
1 7.6 middle
2 7.1 middle
3 7.2 middle
4 7.4 middle
5 7.7 middle
6 7.7 middle
1 7.2 bottom
2 6.7 bottom
3 7.0 bottom
4 7.0 bottom
5 7.0 bottom
6 6.9 bottom
;


TITLE2 'M1';
PROC GLM DATA=ex4_6;
   CLASS core position;
   MODEL pH = core*position / SS1 SOLUTION;
RUN;

TITLE2 'M2';
PROC GLM DATA=ex4_6;;
   CLASS core position;
   MODEL pH = core position / SS1 SOLUTION;
RUN;

TITLE2 'M3r';
PROC GLM DATA=ex4_6;;
   CLASS core;
   MODEL pH = core / SS1 SOLUTION;
RUN;

TITLE2 'M3c';
PROC GLM DATA=ex4_6;;
   CLASS position;
   MODEL pH = position / SS1 SOLUTION;
RUN;QUIT;

TITLE1; TITLE2;
```

On the following pages the output listings from the program segments above are printed. The model behind a particular output listing can be found as described on page 195, but for easy reference the output listings also contain the model designations `M1`, `M2`, `M3r`, and `M3c`.

```
                              Example 4.6                                  1
                                  M1

                           The GLM Procedure

                        Class Level Information

                 Class          Levels    Values

                 core                6    1 2 3 4 5 6

                 position            3    bottom middle top


                      Number of observations    18
                              Example 4.6                                  2
                                  M1

                           The GLM Procedure

Dependent Variable: pH

                                        Sum of
 Source                      DF        Squares     Mean Square    F Value    Pr > F

 Model                       17     1.60944444      0.09467320          .         .

 Error                        0     0.00000000          .

 Corrected Total             17     1.60944444


               R-Square     Coeff Var      Root MSE        pH Mean

               1.000000             .             .       7.294444


 Source                      DF      Type I SS     Mean Square    F Value    Pr > F

 core*position               17     1.60944444      0.09467320          .         .


                                                     Standard
Parameter                           Estimate            Error    t Value    Pr > |t|

Intercept                        7.600000000 B              .          .          .
core*position 1 bottom          -0.400000000 B              .          .          .
core*position 1 middle           0.000000000 B              .          .          .
core*position 1 top             -0.100000000 B              .          .          .
core*position 2 bottom          -0.900000000 B              .          .          .
core*position 2 middle          -0.500000000 B              .          .          .
core*position 2 top             -0.400000000 B              .          .          .
core*position 3 bottom          -0.600000000 B              .          .          .
core*position 3 middle          -0.400000000 B              .          .          .
core*position 3 top             -0.300000000 B              .          .          .
core*position 4 bottom          -0.600000000 B              .          .          .
core*position 4 middle          -0.200000000 B              .          .          .
core*position 4 top             -0.100000000 B              .          .          .
core*position 5 bottom          -0.600000000 B              .          .          .
core*position 5 middle           0.100000000 B              .          .          .
core*position 5 top              0.100000000 B              .          .          .
core*position 6 bottom          -0.700000000 B              .          .          .
core*position 6 middle           0.100000000 B              .          .          .
core*position 6 top              0.000000000 B              .          .          .

NOTE: The X'X matrix has been found to be singular, and a generalized inverse
      was used to solve the normal equations.  Terms whose estimates are
      followed by the letter 'B' are not uniquely estimable.
```

The estimate of the variance is undefined in this model as explained on page 4.7 and so are all the quantities that involve the estimate of the variance.

```
                          Example 4.6                                   3
                              M2

                        The GLM Procedure

                     Class Level Information

             Class           Levels    Values

             core                 6    1 2 3 4 5 6

             position             3    bottom middle top


                 Number of observations    18
                          Example 4.6                                   4
                              M2

                        The GLM Procedure

Dependent Variable: pH

                                        Sum of
 Source                      DF        Squares     Mean Square    F Value    Pr > F

 Model                        7     1.45722222      0.20817460      13.68    0.0002

 Error                       10     0.15222222      0.01522222

 Corrected Total             17     1.60944444


              R-Square     Coeff Var      Root MSE        pH Mean

              0.905419      1.691402      0.123378       7.294444


 Source                      DF      Type I SS     Mean Square    F Value    Pr > F

 core                         5     0.48944444      0.09788889       6.43    0.0063
 position                     2     0.96777778      0.48388889      31.79    <.0001


                                                 Standard
   Parameter                   Estimate             Error    t Value    Pr > |t|

   Intercept               7.572222222 B       0.08225225      92.06      <.0001
   core        1           0.033333333 B       0.10073802       0.33      0.7476
   core        2          -0.400000000 B       0.10073802      -3.97      0.0026
   core        3          -0.233333333 B       0.10073802      -2.32      0.0430
   core        4          -0.100000000 B       0.10073802      -0.99      0.3443
   core        5           0.066666667 B       0.10073802       0.66      0.5231
   core        6           0.000000000 B        .                .         .
   position    bottom     -0.500000000 B       0.07123254      -7.02      <.0001
   position    middle     -0.016666667 B       0.07123254      -0.23      0.8197
   position    top         0.000000000 B        .                .         .

NOTE: The X'X matrix has been found to be singular, and a generalized inverse
      was used to solve the normal equations.  Terms whose estimates are
      followed by the letter 'B' are not uniquely estimable.
```

```
                              Example 4.6                              5
                                 M3r

                          The GLM Procedure

                       Class Level Information

                  Class          Levels    Values

                  core                6    1 2 3 4 5 6


                    Number of observations     18
                              Example 4.6                              6
                                 M3r

                          The GLM Procedure

Dependent Variable: pH

                                        Sum of
 Source                      DF        Squares    Mean Square   F Value   Pr > F

 Model                        5     0.48944444     0.09788889      1.05   0.4339

 Error                       12     1.12000000     0.09333333

 Corrected Total             17     1.60944444


               R-Square     Coeff Var      Root MSE       pH Mean

               0.304108      4.188188      0.305505      7.294444


 Source                      DF      Type I SS    Mean Square   F Value   Pr > F

 core                         5     0.48944444     0.09788889      1.05   0.4339


                                               Standard
    Parameter               Estimate              Error    t Value    Pr > |t|

    Intercept            7.400000000 B       0.17638342      41.95      <.0001
    core       1         0.033333333 B       0.24944383       0.13      0.8959
    core       2        -0.400000000 B       0.24944383      -1.60      0.1348
    core       3        -0.233333333 B       0.24944383      -0.94      0.3680
    core       4        -0.100000000 B       0.24944383      -0.40      0.6955
    core       5         0.066666667 B       0.24944383       0.27      0.7938
    core       6         0.000000000 B        .                .         .

NOTE: The X'X matrix has been found to be singular, and a generalized inverse
      was used to solve the normal equations.  Terms whose estimates are
      followed by the letter 'B' are not uniquely estimable.
```

```
                         Example 4.6                              7
                            M3c

                     The GLM Procedure

                  Class Level Information

          Class          Levels    Values

          position            3    bottom middle top


                Number of observations    18
                         Example 4.6                              8
                            M3c

                     The GLM Procedure

Dependent Variable: pH

                                        Sum of
 Source                      DF        Squares     Mean Square    F Value    Pr > F

 Model                        2     0.96777778      0.48388889      11.31    0.0010

 Error                       15     0.64166667      0.04277778

 Corrected Total             17     1.60944444


              R-Square     Coeff Var      Root MSE       pH Mean

              0.601312      2.835417      0.206828      7.294444


 Source                      DF      Type I SS     Mean Square    F Value    Pr > F

 position                     2     0.96777778      0.48388889      11.31    0.0010


                                                Standard
   Parameter                     Estimate           Error    t Value    Pr > |t|

   Intercept                 7.466666667 B     0.08443713      88.43      <.0001
   position  bottom         -0.500000000 B     0.11941214      -4.19      0.0008
   position  middle         -0.016666667 B     0.11941214      -0.14      0.8909
   position  top             0.000000000 B      .                .         .

NOTE: The X'X matrix has been found to be singular, and a generalized inverse
      was used to solve the normal equations.  Terms whose estimates are
      followed by the letter 'B' are not uniquely estimable.
```

□

Exercises for Chapter 4

SAS programs for the exercises can be found at the address
http://www.imf.au.dk/biogeostatistics/
Most of the exercises in this chapter require the use of a statistical package.

Exercise 4.1 The data in this exercise are from an investigation that was undertaken by Bo Utoft from the Institute of Biological Sciences, Aarhus University.

The purpose of the investigation was to study the amount of chromium in fertilized eggs of salmon, *Salmon trutta L.* The eggs came from a fish farm and were artificially fertilized in the laboratory. The eggs were put into incubators with water containing chromium. Eight incubators containing an increasing amount of chromium from 1 μg Cr/L to 1000 μg Cr/L were used. On certain days (see the rows of Table 4.15) five eggs were taken from each incubator and the the amount of chromium in the eggs was measured exploiting a 51Cr-isotope which had been added to the water in the incubators. Using a scintillation counter the radiation from the eggs was recorded and the counts were subsequently transformed to the amount of chromium per gram of dry weight. The data are displayed in Table 4.15 to one decimal point.

The purpose of the experiment is to evaluate how the amount of chromium in the eggs depends on the exposure to chromium. In this experiment two aspects of exposure are meticulously controlled: the duration of the exposure and the concentration of chromium in the water.

A further study of the data shows that the larger the measured concentration of chromium in the eggs is the larger the variation; we therefore analyze the logarithm of the measured amount of chromium in the eggs.

Diagnostic plots shows that it is reasonable to use the basic model that the logarithm of the amount of chromium in the eggs, x_{ij}, which have been exposed to chromium for i days in the incubator with the jth concentration of chromium is a realization of a normally distributed random variable, X_{ij}, with a mean, μ_{ij}, which depends on the number of days in the water as well as on the concentration of chromium in the water, whereas the variance σ^2 may be assumed to be the same regardless of the number of days in the incubators and the concentration of chromium in the incubators. The model may be written as

$$M_1\colon X_{ij} \sim N(\mu_{ij}, \sigma^2),$$

where as usual the independence of the observations is assumed without mentioning. The diagnostic plots furthermore show that the data do not contradict the additivity model,

$$M_2\colon X_{ij} \sim N(\alpha_i + \beta_j, \sigma^2).$$

We assume that a statistical package is available for the solution of this exercise. The data are available at the address http://www.imf.au.dk/biogeostatistics/.

(1) Show that the model M_2 are not contradicted by the data.

(2) Show that the content of chromium in the eggs depend on the length of time the eggs have been in the incubators.

(3) Determine whether the amount of chromium in the eggs depends on the concentration of chromium in the water in the incubators.

(4) Compute a 95% confidence interval for the difference in means of the logarithm of the content of chromium in eggs that have been exposed to chromium for 1 day and eggs that have been exposed to chromium for 57 days.

Exercise 4.2 This exercise uses the same data as Exercise 4.1. Here we attempt to give a more detailed description of how the concentration of chromium in the eggs depends on the concentration

Table 4.15 *Amount of chromium in the eggs in µg per gram of dry weight to one decimal point. The data in each row are from eggs that have been kept in the incubators for the same number of days.*

	Amount of chromium in the water in the incubators (µg Cr/L)							
Days	1	5	10	50	100	200	500	1000
1	2.0	3.9	4.5	11.3	15.3	19.4	30.1	36.1
2	2.2	3.9	5.3	10.8	14.8	19.5	29.6	37.8
3	2.0	3.7	5.5	10.9	15.2	18.7	30.4	45.3
4	2.1	4.1	5.8	13.8	14.4	23.9	34.3	42.0
6	2.5	4.5	6.3	12.1	17.3	23.0	34.2	44.0
8	2.0	4.6	5.5	11.9	15.0	22.3	31.5	44.8
14	3.1	5.6	10.2	16.5	22.9	28.4	49.3	54.8
22	3.7	6.5	11.0	19.8	27.9	32.7	53.7	54.8
29	4.2	7.1	11.7	21.4	31.2	37.2	58.7	69.6
37	4.6	8.5	9.5	23.3	34.0	36.4	62.0	82.9
43	3.8	6.6	10.6	28.0	30.2	42.2	58.7	75.2
49	3.8	7.5	8.7	21.6	30.2	37.7	54.3	70.2
57	4.9	9.1	14.1	25.9	41.5	50.5	69.4	105.5

Table 4.16 *Corresponding values of the logarithm of the concentration of chromium in the water and the logarithm of concentration of chromium in the eggs that have been in the incubators for one day. In Exercises 4.1 and 4.2, j is used to index the concentrations of chromium in the water.*

j	*Logarithm of the concentration of chromium in the water* t_j	*Logarithm of the contents of chromium in the egg* x_j
1	0.000	0.698
2	1.609	1.352
3	2.303	1.501
4	3.912	2.422
5	4.605	2.727
6	5.298	2.965
7	6.215	3.406
8	6.908	3.587

of chromium in the water in incubators. Consider first the eggs that have been lying the water for just one day.

(1) Plot the *logarithm of the concentration of chromium in the eggs* against the *logarithm of the concentration of chromium in the water* and argue that the data does not contradict a linear regression model with the logarithm of the concentration of chromium in the eggs as the response and the logarithm of the concentration of chromium in the water as the explanatory variable.

(2) Estimate the parameters in the regression model and plot the estimated regression line in the plot in (1).

(In the estimation the following calculations may be used: $S_x = 18.658$, $S_t = 30.850$, $USS_x = 51.129352$, $USS_t = 158.817952$, and $SP_{xt} = 89.320726$.)

In the rest of the exercise we consider a model for all the data.

Table 4.17 $SSD_{(i)}$, *degrees of freedom*, $f_{(i)}$, *and estimated variances*, $s^2_{(i)}$, *in the 13 regression lines indexed by* i

i	$SSD_{(i)}$	$f_{(i)}$	$s^2_{(i)}$
1	0.04269	6	0.007116
2	0.00890	6	0.001483
3	0.02037	6	0.003394
4	0.04887	6	0.008145
5	0.00618	6	0.001030
6	0.01728	6	0.002880
7	0.07379	6	0.012299
8	0.08171	6	0.013618
9	0.04107	6	0.006844
10	0.06664	6	0.011106
11	0.10387	6	0.017311
12	0.03914	6	0.006523
13	0.03208	6	0.005347
SUM	0.58259	78	

(3) Evaluate whether it is reasonable to work with a model with 13 regression lines, one for each of the rows of Table 4.15, or, in other words, one for each group of eggs which have been exposed to chromium for a fixed number of days. More specifically, the model specifies that the logarithm of the content of chromium in the eggs, x_{ij}, after i days in the incubator with the jth concentration of chromium, is an observation of a normally distributed random variable, X_{ij}, whose mean depends linearly on the logarithm of the concentration of chromium in the incubators (t_j). Intercept, slope, and variance may for a start be assumed to depend on the number of days the eggs have been exposed to the chromium in the incubators.

The model can be expressed as

$$M_2\colon X_{ij} \sim N(\alpha_i + \beta_i t_j, \sigma_i^2),$$

where we assume that the observations are independent.

(4) Verify that it may be assumed that the variances are the same for the 13 regression lines. The information in Table 4.17 may be used to answer this question.

The following require access to a statistical package. The data are available at the address `http://www.imf.au.dk/biogeostatistics/`.

(5) Verify that it may be assumed that the 13 regression lines have identical slopes.

Exercise 4.3 The data in this exercise is part of a larger investigation performed by Bo Elberling from the Department of Earth Sciences, Aarhus University.

The purpose has been to describe the connection between the effective diffusion, D, of a core of sand as a function of the degree of cementation, C, of the sand. The effective diffusion describes how easily gasses diffuse through the core of sand and the degree of cementation is a number between 0 and 1, describing the reduction of the initial volume of the cavity between the grains of sand.

In the experiment a core of sand has been encased in a glass tube. The degree of cementation for the original core of sand is 0 and the corresponding effective diffusion is measured. Next, the degree of cementation has been changed by leading a solution of copper sulfate through the core of sand, letting part of the copper sulfate precipitate on the sand grains. Then both the effective diffusion and the degree of cementation is measured. This is being repeated a number of times and in this way a number of corresponding values of effective diffusion and degree of cementation.

It is expected that the connection between the effective diffusion and the degree of cementation can be described by an equation of the form,

$$D = \tau k(1-C)^{\beta}, \tag{4.25}$$

where k is a known constant and τ and β are unknown parameters. Taking logarithms and rewriting, we find that (4.25) is equivalent to a linear relationship between $x = \log D - \log k$ and $t = \log(1-C)$, i. e.,

$$x = \log D - \log k = \log \tau + \beta \log(1-C) = \alpha + \beta t, \tag{4.26}$$

where $\alpha = \log \tau$.

Two experiments denoted by A and B have been made. The distinction between the two experiments is that two different methods have been used to let the copper sulfate precipitate on the sand grains. The first two questions only consider experiment A where 10 corresponding values of x and t are available. These values are given in Table 4.18. In the analysis t will be considered as a deterministic variable.

Questions (1) and (2) can be solved without using a statistical package in which case the standard calculations in Table 4.19 may be useful.

(1) Make a plot of the data in Table 4.18 in order to check that the data approximately satisfy the linear relationship in (4.26) and decide on a suitable statistical model.

(2) Estimate the parameters in the regressionen of x on t.

Table 4.18 *Corresponding values of $x = \log D - \log k$ and $t = \log(1-C)$ for experiment A*

$x = \log D - \log k$	$t = \log(1-C)$
−1.452	0.000
−1.796	−0.083
−2.397	−0.174
−2.495	−0.288
−3.137	−0.367
−3.570	−0.478
−4.431	−0.616
−4.739	−0.734
−5.384	−0.892
−6.755	−1.115

Table 4.19 *Standard calculations on the data from Table 4.18*

	x	t
S	−36.156	−4.747
USS	156.599586	3.440383
SP	22.684720	

In the following questions use of a statistical package is expected. Data for experiment B are only available electronically at the address `http://www.imf.au.dk/biogeostatistics/`. In experiment B, which has 9 pairs of corresponding observations, there is a good fit to the linear relationship.

(3) Examine whether it can be assumed that the variance is the same in the two experiments.

(4) Examine whether it can be assumed that the the regression lines in the two experiments have the same intercept.

(5) Calculate a 95% confidence interval for the intercept in the final model.

Exercise 4.4 The velocity of evaporation of water from the surface of the skin (transepidermic water loss, TEWL) may be used as an indirect measure of the permeability and the barrier function of the skin. For a group of workers in the fishing industry the temperature of the skin and velocity of evaporation of water from the surface of the skin, TEWL, was measured in order to investigate the association, if any, between the two variables. The same measurements were made on a control group of persons not working in the fishing industry. In all cases the measurements were performed on the tip of the index finger.

Experience from previous analyses of larger data sets has shown that for fixed temperature of the skin, the logarithm of TEWL may be described by a normal distribution. The logarithm to base 10, $\log_{10}$, is traditionally used in this context in the literature. The data are displayed in Table 4.20.

(1) Make a plot to check that it is reasonable to use a linear regression model with skin temperature as explanatory variable and $\log_{10}$(TEWL) as response. Use different plot symbols for the two groups.

Next, we concentrate on the 22 persons in the fishing industry.

(2) Estimate the parameters in the regression model. (In calculations by hand we may use that $S_x = 34.733$, $USS_x = 55.284081$, $S_t = 503.9$, $USS_t = 11765.51$, and $SP_{xt} = 802.4084$, where x denotes $\log_{10}$(TEWL) and t denotes the skin temperature.)

For the following questions it is assumed that a statistical package is available. The data are available at the address `http://www.imf.au.dk/biogeostatistics/`.

(3) Estimate the regression of $\log_{10}$(TEWL) on skin temperature in the control group and determine whether it may be assumed that the variance is the same in the two groups.

(4) Investigate whether it may be assumed that the two regression lines have the same slope.

(5) Investigate whether it may be assumed that the two regression lines have the same intercept.

(6) Investigate whether $\log_{10}$(TEWL) depends on skin the temperature.

Exercise 4.5 Table 4.21 gives the 24-hour average of the contents of lead in the air in ng/m^3 recorded by two measurements stations in Aalborg in between 1983 and 1989. One station is located in the city center and the other in a residential area. Table 4.22 contains the natural logarithm of the measurements in Table 4.21.

(1) Make a plot of the the contents of lead as a function of the year, both for the city center and for the residential area. Make the same plot for the logarithm of the contents of lead. What do the plots show?

In the following we consider a simple description of the logarithm of the content of lead, i.e., of the data in Table 4.22, based on the linear regression of ln(lead) on year for Aalborg city center and for the residential area. Table 4.23 and Table 4.24 contain useful standard calculations.

(2) Estimate the parameters in the linear regression of ln(lead) on year for the city center and for the residential area.

(3) Show that is may be assumed that the variance is the same for the two areas.

Table 4.20 *Skin temperature and* $\log_{10}$*(TEWT) for 22 workers in the fishing industry and 17 persons in a control group*

Fishing industry Skin temperature i °C	$\log_{10}$(TEWT)	*Control group* Skin temperature i °C	$\log_{10}$(TEWT)
21.8	1.468	31.0	1.907
21.3	1.375	31.4	1.904
25.4	1.831	28.2	1.744
24.3	1.787	31.4	1,904
18.9	1.427	25.3	1.688
22.2	1.629	23.4	1.435
21.9	1.450	32.9	1.865
21.2	1.628	28.2	1.963
28.6	1.732	28.6	1.872
20.3	1.537	30.8	1.806
25.7	1.622	28.6	1.700
25.6	1.787	31.8	1.757
22.4	1.396	33.8	1.938
19.4	1.396	29.3	1.857
26.5	1.632	30.6	1.852
22.2	1.582	34.0	1.832
22.4	1.504	32.3	1.887
27.0	1.650		
29.5	1.797		
20.1	1.384		
20.3	1.504		
16.9	1.615		

Table 4.21 *24-hour averages of the contents of lead in ng/m*^{3} *recorded at two measurement stations in Aalborg*

Year	*Town center lead*	*Residential area lead*
83	865	257
84	721	201
85	671	206
86	514	189
87	381	145
88	370	130
89	343	112

Table 4.22 *Natural logarithm to a 24-hour average of the contents of lead (ng/m^3) in two measuring stations in the city of Aalborg*

Year	*Town center* log(*bly*)	*Residential area* ln(*bly*)
83	6.763	5.549
84	6.581	5.303
85	6.509	5.328
86	6.242	5.242
87	5.943	4.977
88	5.914	4.868
89	5.838	4.719

Table 4.23 *Standard calculations for the town center with the response x being* ln(*bly*) *and the explanatory variable t being the year*

S_x	USS_x	S_t	USS_t	SP_{xt}
43.790	274.754264	602	51800	3761.265

(4) May the regression coefficients be assumed to be the same for Aalborg town center and the residential area? (In answering this question the access to a statistical package is assumed.)

Exercise 4.6 The data for this exercise are from a study of the sand lizard *Lacerta agilis* conducted by Mark Goldsmith from the Institute of Biological Sciences, Aarhus University.

Eggs from the sand lizard were placed in incubators under carefully controlled conditions and a range of variables were recorded for the hatchlings, including the incubation time, which is the variable studied in this exercise.

Thirty pregnant sand lizards were caught during the breeding season in 1999 and brought to the University of Aarhus. Here the sand lizards were caged separately in plastic boxes while their eggs were collected, and they were released on the point of capture after oviposition.

Mark Goldsmith wanted to study the effect of temperature and humidity during incubation on a range of variables, including the incubation duration. It was decided to investigate all 20 combinations of 5 levels of temperature and 4 levels of humidity. The incubation duration was recorded in hours, but is given here in days to one decimal point.

Five wooden incubators with 4 shelves were equipped with thermostatically controlled heating devices and ventilators. The incubators were set at the temperatures 19, 22, 25, 28, and 31 °C, which were the 5 levels chosen for the temperature. The eggs were incubated individually in transparant plastic boxes fully buried in 500 g of sand with one of the 4 water contents 0.0075, 0.01, 0.02, or 0.1 gH_2O/gsand. We will refer to the water content as the humidity. A total of 160 eggs from 30 clutches were randomly assigned to one of the 20 treatments in such a way that each treatment was applied to 8 eggs. Not all eggs hatched and only 119 incubation durations were recorded.

For this exercise it is assumed that a statistical package is available. The data are available at the address `http://www.imf.au.dk/biogeostatistics/`.

Plots of the incubation duration against temperature for each humidity indicates that the variance of the incubation duration depends on temperature, and this is confirmed by an application of

Table 4.24 *Standard calculations for the residential area with the response x being* ln(*bly*) *and the explanatory variable t being the year*

S_x	USS_x	S_t	USS_t	SP_{xt}
35.986	185.516272	602	51800	3091.085

Bartlett's test. The variance seems to be larger for larger values of the incubation duration. Therefore the *logarithm of the incubation duration*, ln(id), is considered.

(1) Check that it is reasonable to consider ln(id) in a two-way analysis of variance with the factors temperature and humidity.

(2) Consider the effect of the factors temperature and humidity on the ln(id). Show that it can be assumed that there is no interaction between the factors temperature and humidity.

(3) Show that it can be assumed that the humidity does not have an effect on ln(id).

(4) Show that it can be assumed that temperature has an effect on ln(id).

(5) Describe the mean of the incubation duration as a function of the temperature.

Exercise 4.7 The data in Table 4.26 are from an experiment on the effect of compounds of nitrate and phosphate used as fertilizers on potatoes. Six different compounds corresponding to different combinations of the amount of phosphate (0, 1, or 2 units) and the amount of nitrate (0 or 1 units) were used in the experiment. Each of the 6 fertilizers were used on 8 plots, so 48 plots were used in the experiment. The yield in kg for each of the 48 plots is given in Table 4.26.

(1) Show that the data do not contradict a statistical model for the yields with six independent normally distributed samples corresponding to the six fertilizers.

(2) Show that it can be assumed that the variance does not depend on the fertilizers.

(3) Show that it can be assumed that phosphate and nitrate have an additive effect on the yield.

(4) Investigate whether nitrate has an effect on the yield.

(5) Investigate whether phosphate has an effect on the yield.

(6) Estimate a 95% confidence interval for the difference in the mean yield between plots receiving a fertilizer without nitrate and plots receiving a fertilizer with one unit of nitrate.

Standard calculations for the 6 samples are given in Table 4.27. In the following pages SAS programs and SAS program listings are given that can be used to answer the questions above.

Table 4.25 *Incubation duration in days by temperature and humidity in the incubators*

	Temperature				
Humidity	19 °C	22 °C	25 °C	28 °C	31 °C
0.0075	92.9	59.2	40.4	32.8	25.5
	93.9	58.0	40.9	31.8	26.0
	83.9	56.8	38.5	31.8	26.2
		52.4	39.1	33.5	
		55.8	39.6	29.6	
		55.3	40.4	31.6	
		55.9	40.8		
0.01	96.7	55.4	42.9	31.4	25.8
	100.0	56.8	41.1	31.1	26.9
	102.3	58.1	39.8	30.3	27.0
	85.2	54.2	40.6	30.3	25.6
		55.6	38.0	31.9	26.6
		56.1	38.0	31.3	26.0
			40.0	30.6	
			39.8	31.8	
0.02	88.9	53.9	39.9	33.0	26.8
	93.7	55.9	40.0	31.4	28.3
	97.7	58.3	41.9	35.1	25.7
		55.8	40.1	32.9	29.8
		54.9	39.9	30.0	25.3
		55.3	40.6	32.8	26.8
			38.8	30.5	
			40.0		
0.1	91.9	57.0	40.8	32.2	26.0
	89.1	56.0	40.2	30.7	26.4
	93.9	55.4	43.7	31.0	27.0
		55.3	41.5	32.1	27.9
		56.3	41.0	30.7	26.8
		58.3	41.1	29.9	26.5
			38.8	32.1	26.7
			38.8		

Table 4.26 *The yield in kg of potatoes for each of the plots with the amount of nitrate and phosphate in the fertilizer applied to the plot*

		Nitrate Number of units							
Phosphate		0				1			
Number of units	0	195	141	192	161	210	210	165	222
		216	130	148	189	224	181	191	218
	1	264	245	209	193	317	244	240	251
		205	166	200	221	296	209	289	279
	2	253	238	243	221	324	286	322	312
		233	251	198	251	309	250	279	263

Table 4.27 *Standard calculations for the data in Table 4.26*

```
                              Data set: POTATOES
                     Response variable: YIELD
                        Group variable: FERTILIZER


                     Calculations in k samples:

                                                                   Sample       Sample
i     ni      Si          USSi          Si2/ni         SSDi     fi  Variance      Mean

A      8  1372.000  241872.0000  235298.0000  6574.0000   7   939.14286  171.5000
B      8  1703.000  369073.0000  362526.1250  6546.8750   7   935.26786  212.8750
C      8  1888.000  448038.0000  445568.0000  2470.0000   7   352.85714  236.0000
D      8  1621.000  331651.0000  328455.1250  3195.8750   7   456.55357  202.6250
E      8  2125.000  573285.0000  564453.1250  8831.8750   7 1261.69643  265.6250
F      8  2345.000  692791.0000  687378.1250  5412.8750   7   773.26786  293.1250
-----------------------------------------------------------------======================
      48 11054.000 2656710.0000 2623678.5000 33031.5000  42   786.46429
```

```
PROC GLM DATA=potatoes;
CLASS phosphate nitrate;
MODEL yield = phosphate nitrate phosphate*nitrate/ SS1;
RUN;
```

This program gives the following output listing:

```
                              The GLM Procedure

Dependent Variable: yield

                                       Sum of
 Source                    DF         Squares     Mean Square    F Value    Pr > F

 Model                      5      78034.4167      15606.8833      19.84    <.0001

 Error                     42      33031.5000        786.4643

 Corrected Total           47     111065.9167


              R-Square     Coeff Var      Root MSE    yield Mean

              0.702596      12.17759      28.04397      230.2917


 Source                    DF       Type I SS     Mean Square    F Value    Pr > F

 phosphate                  2     49976.04167     24988.02083      31.77    <.0001
 nitrate                    1     26508.00000     26508.00000      33.71    <.0001
 phosphate*nitrate          2      1550.37500       775.18750       0.99    0.3817
```

```
PROC GLM DATA=potatoes;
CLASS phosphate nitrate;
MODEL yield = phosphate nitrate/ SS1 SOLUTION;
RUN;
```

This program gives the following output listing:

```
                          The GLM Procedure

Dependent Variable: yield

                                          Sum of
 Source                      DF          Squares     Mean Square    F Value    Pr > F

 Model                        3       76484.0417      25494.6806      32.44    <.0001

 Error                       44       34581.8750        785.9517

 Corrected Total             47      111065.9167

              R-Square     Coeff Var      Root MSE    yield Mean

              0.688636      12.17362      28.03483      230.2917

 Source                      DF        Type I SS     Mean Square    F Value    Pr > F

 phosphate                    2      49976.04167     24988.02083      31.79    <.0001
 nitrate                      1      26508.00000     26508.00000      33.73    <.0001

                                                Standard
     Parameter               Estimate              Error    t Value    Pr > |t|

     Intercept            241.0625000 B       8.09295838      29.79      <.0001
     phosphate 0          -77.5000000 B       9.91180927      -7.82      <.0001
     phosphate 1          -25.3125000 B       9.91180927      -2.55      0.0142
     phosphate 2            0.0000000 B        .                .         .
     nitrate   0           47.0000000 B       8.09295838       5.81      <.0001
     nitrate   1            0.0000000 B        .                .         .

NOTE: The X'X matrix has been found to be singular, and a generalized inverse
      was used to solve the normal equations.  Terms whose estimates are
      followed by the letter 'B' are not uniquely estimable.
```

```
PROC GLM DATA=potatoes;
CLASS phosphate;
MODEL yield = phosphate/ SS1 SOLUTION;
RUN;
```

This program gives the following output listing:

```
                              The GLM Procedure

Dependent Variable: yield

                                            Sum of
 Source                       DF          Squares     Mean Square    F Value    Pr > F

 Model                         2       49976.0417      24988.0208      18.41    <.0001

 Error                        45       61089.8750       1357.5528

 Corrected Total              47      111065.9167


              R-Square     Coeff Var      Root MSE    yield Mean

              0.449967      15.99927      36.84498      230.2917


 Source                       DF        Type I SS     Mean Square    F Value    Pr > F

 phosphate                     2      49976.04167     24988.02083      18.41    <.0001


                                                  Standard
     Parameter               Estimate                Error    t Value    Pr > |t|

     Intercept           264.5625000 B          9.21124577      28.72      <.0001
     phosphate 0         -77.5000000 B         13.02666869      -5.95      <.0001
     phosphate 1         -25.3125000 B         13.02666869      -1.94      0.0583
     phosphate 2           0.0000000 B           .                .         .

NOTE: The X'X matrix has been found to be singular, and a generalized inverse
      was used to solve the normal equations.  Terms whose estimates are
      followed by the letter 'B' are not uniquely estimable.
```

```
PROC GLM DATA=potatoes;
CLASS nitrate;
MODEL yield = nitrate/ SS1 SOLUTION;
RUN;
```

This program gives the following output listing:

```
                              The GLM Procedure

Dependent Variable: yield

                                        Sum of
 Source                      DF        Squares    Mean Square   F Value   Pr > F

 Model                        1     26508.0000     26508.0000     14.42   0.0004

 Error                       46     84557.9167      1838.2156

 Corrected Total             47    111065.9167


              R-Square     Coeff Var      Root MSE    yield Mean

              0.238669      18.61744      42.87442      230.2917


 Source                      DF      Type I SS    Mean Square   F Value   Pr > F

 nitrate                      1    26508.00000    26508.00000     14.42   0.0004


                                               Standard
     Parameter             Estimate               Error    t Value    Pr > |t|

     Intercept          206.7916667 B      8.75170360      23.63      <.0001
     nitrate    0        47.0000000 B     12.37677792       3.80      0.0004
     nitrate    1         0.0000000 B       .                .         .

NOTE: The X'X matrix has been found to be singular, and a generalized inverse
      was used to solve the normal equations.  Terms whose estimates are
      followed by the letter 'B' are not uniquely estimable.
```

Exercise 4.8 In order to make 100 concrete flagstones, a manufacturer adds 10 kg of an additive A in order to reduce the curing time. The manufacturer has become aware of an additive B which is cheaper than A and is supposed to have the same effect. In order to investigate whether B or a mixture of A and B can possibly replace A in the production, the following experiment is performed.

From a production of 100 concrete flagstones without the additives A and B, 10 flagstones are chosen at random 5 days after the production and their compressive strengths are measured in mega pascal. This procedure is repeated 3 times with productions where 10 kg of A, 10 kg of B, and 10 kg of both A and B, have been added to the production.

A large compressive strength on a certain day after the production is interpreted as a short curing time, and, similarly, a small compressive strength is interpreted as a long curing time.

Table 4.28 *Compressive strength in mega pascal tabulated according to the addition of A and B*

	B not added		10 kg *of B added*	
	2.23	2.99	3.61	3.31
A	2.34	2.43	2.87	5.68
not	1.87	2.95	3.27	5.51
added	2.99	3.41	3.34	6.57
	1.82	2.89	5.67	5.04
	5.06	3.37	8.62	8.35
10 kg	4.37	3.90	5.94	3.51
of A	4.29	5.98	7.57	8.43
added	2.08	3.48	5.95	5.90
	3.73	3.54	7.73	9.85

For this exercise it is assumed that a statistical package is available. The data are available at the address `http://www.imf.au.dk/biogeostatistics/`.

(1) Analyze the data in Table 4.28 in order to elucidate whether the additives A and B have an effect on the compressive strength and whether A and B can be assumed to have the same effect on the compressive strength.

This analysis may be perfomed in the following steps.

(2) Show that it may be assumed that the *logarithm of the compressive strength* in Table 4.28 may be assumed to be four normal samples with the same variance. It is instructive to perform the model checking both for the compressive strength and for the logarithm of the compressive strength, and note differences and similarities.

(3) Investigate whether A and B have an additive effect on the logarithm of the compressive strength, and whether each of them separately has an effect on the compressive strength.

(4) Investigate whether A and B have the same effect on the logarithm of the compressive strength?

While (2) above addresses the question of finding a model and therefore is a necessary prerequisite for further analysis, (3) and (4) could be interchanged, i.e., (5) and (4) could be considered instead of (3) and (4).

(5) Can A and B be assumed to have the same effect on the compressive strength when they are added alone?

(6) Can A and B be assumed to have an additive effect, when they are both added?

In the questions (2), (3), (5), and (6) we have considered the models:

M_1: Four normal samples.

M_2^*: Three normal samples that are defined by the amount of additive ignoring the type of additive.

M_2: The additivity model in two-way analysis of variance.

M_3: The regressionen of the logarithm of the compressive strength on the amount of additive (0, 10, or 20 kg).

(7) Explain the relationship between the four models.

CHAPTER 5

An Introduction to the Power of Tests and Design of Experiments

When a statistical investigation is undertaken it is often with the purpose of confirming a conjecture or a theory. This is most effectively done if one can obtain data, a model for the data, and a null hypothesis which must be rejected if the conjecture or the theory is to be confirmed.

If the data do not reject the hypothesis, all that can be done is to make confidence intervals of the relevant parameters. When those confidence intervals are interpreted one of the following two situations will have occurred. One may find that the confidence intervals are fairly narrow, which indicates that a good experiment has been performed and the results may even have an impact on the perception of the conjecture or the theory. Alternatively, one may find that the confidence intervals are wide, thus indicating that there is little information in the experiment or, putting it more strongly, that the experimental effort has been wasted. If the interpretation based on such experiments is that the conjecture or the theory is false in situations where in fact they are true, a halfhearted experimental effort is not merely a waste of time but it is also damaging.

In this chapter we introduce some considerations that should be taken into account when planning an experiment. The considerations focus on the variance of the individual observations and the number of observations.

In Section 5.1 the *power* of a test is defined in connection with the two-sample t-test, and we offer a method to decide how large the samples must be in order to obtain a given power of the two-sample t-test. To fix ideas we consider in Example 5.1 a scientific question that is identical to the one in Example 3.3.

In Section 5.2 we introduce the *blocking* concept as a method to reduce the variance and hence increase the power. This leads to the *paired t-test.*

A control plot that checks the assumptions behind the paired t-test is explained in Section 5.3 and, finally, the connection between the paired t-test and two-way analysis of variance is investigated.

5.1 The Power of Tests

The following example treats the same problem as Example 3.3, although in a slightly simpler form with only one factor with two levels.

Example 5.1
A plant physiologist examined the influence of mechanical stress on the growth of soya bean plants. 26 seedling were potted, one in each pot. The 26 seedlings were randomly allocated to one of two groups with 13 seedlings in each group. The seedlings in one group were exposed to stress by being shaken for 20 minutes daily, while the seedlings in the other group were not stressed. In this experiment we have one factor, *stress,* with the two levels, *stress* and *no stress*.

After 16 days of growth the plants were harvested, and the total area of the leaves of each plant were measured. The result is shown in Table 5.1.

The data in Table 5.1 has been plotted in Figure 5.1 as two normal fractile plots. The fractile plots do not contradict the assumption that the total area of the leaves is normally distributed in each group. Furthermore, the fractile plots are consistent with the assumption that the variances are identical in the two treatment groups because the slopes of the fractile plots are not very different.

Table 5.1 *The data of the experiment in Example 5.1*

Total area of leaves in cm^2			
Group			
No stress		*Stress*	
205	258	193	219
220	261	168	217
230	269	187	240
234	273	210	235
235	293	207	277
237	294	200	261
246		220	

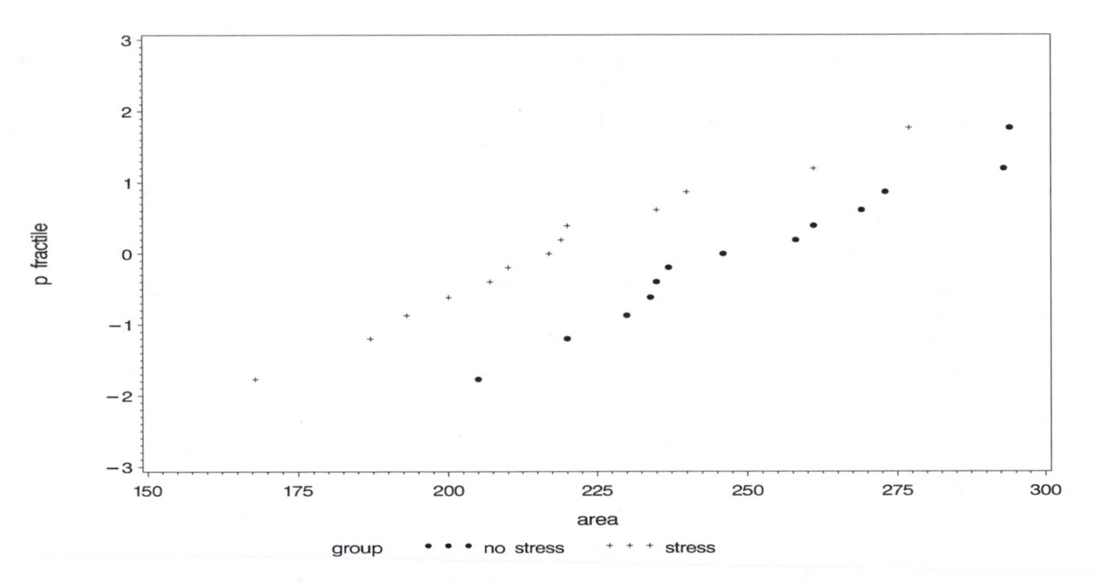

Figure 5.1 *Normal fractile plots for the two samples in Table 5.1.*

The analysis proceeds by adopting the model that the observations are independent and normally distributed with a mean μ_1 for the no-stress group and μ_2 for the stress group and a common variance σ^2. Subsuming the independence of the observations the model can be briefly stated:

$$\begin{array}{lll} \text{No stress:} & x_{11},\ldots,x_{1n_1} & \sim\sim \; N(\mu_1,\sigma^2) \\ \text{Stress:} & x_{21},\ldots,x_{2n_2} & \sim\sim \; N(\mu_2,\sigma^2) \end{array} \tag{5.1}$$

Here x_{1i} denotes the total area of the leaves for the ith plant in the no-stress group, and n_1 and n_2 denote the numbers of plants, the replications, in each group. In this example $n_1 = n_2 = 13$.

The normal fractile plots in Figure 5.1 show that the data do not contradict the assumption of a normal distribution and that the assumption of a common variance is tenable. But the assumption of a common variance can also be checked using the F-test for the identity of variances. The F-test gives a significance probability of 0.75, so there is no reason to doubt the assumption of a common variance in the two samples.

The hypothesis of interest is the null hypothesis,

$$H_{0\mu}\colon \mu_1 = \mu_2,$$

which specifies that the mean total area of the leaves is the same whether the plants are stressed or not. The hypothesis is tested by a t-test:

$$t = \frac{\bar{x}_{1\cdot} - \bar{x}_{2\cdot}}{\sqrt{s_1^2(1/n_1 + 1/n_2)}} = \frac{250.38 - 218.00}{\sqrt{814.7115(1/13 + 1/13)}} = \frac{32.38}{11.1955} = 2.89 \tag{5.2}$$

where $\bar{x}_{1\cdot}$ and $\bar{x}_{2\cdot}$ are the averages in the no-stress and the stress groups, respectively, and s_1^2 is the estimate of the common variance σ^2, see *Main Points in Section 3.2.1,* page 94, for a summary of the treatment of two normal samples.

Numerically large values of t are critical for the hypothesis. The significance probability, $p_{obs}(\mathbf{x})$, which is the probability – if the hypothesis is true – of observing a t, which is more critical for $H_{0\mu}$ than the observed value of $t = 2.89$. In this case t is to be evaluated in a t-distribution with 24 degrees of freedom and this gives the significance probability,

$$p_{obs}(\mathbf{x}) = 2\left[1 - F_{t(24)}\left(|t|\right)\right] = 0.0080.$$

The interpretation of this probability is that if the hypothesis is true, we would experience in only 8 out 1000 repetitions of the experiment a t-value that was more extreme than the one obtained in this experiment. This is such a rare event that most people would choose the explanation that the null hypothesis is wrong. Often one works with a conventional limit for how small a significance probability can get before rejecting the null hypothesis. This limit is the *significance level,* which is denoted by α. The significance level $\alpha = 5\%$ has been used throughout this book, but a significance level of $\alpha = 1\%$ is also used.

The conclusion of the experiment is that mechanical stress does have an influence on the growth of soya bean plants. The estimated mean in the no-stress group ($\bar{x}_{1\cdot} = 250.38$) is larger than the estimated mean in the stress group ($\bar{x}_{2\cdot} = 218.00$), so the effect of mechanical stress is to lower the total area of the leaves.

An estimate of the effect of stress, i.e., the difference in means of the total area of leaves in the two groups, is

$$\bar{x}_{1\cdot} - \bar{x}_{2\cdot} = 32.38 \text{ cm}^2 \rightarrow \mu_1 - \mu_2,$$

and a 95% confidence interval for $\mu_1 - \mu_2$ is

$$\begin{aligned} \bar{x}_{1\cdot} - \bar{x}_{2\cdot} \pm t_{1-\alpha/2}(f_1)\sqrt{s_1^2(1/n_1 + 1/n_2)} &= 32.38 \text{ cm}^2 \pm 2.064 \times 11.1955 \text{ cm}^2 \\ &= [9.27, 55.49] \text{ cm}^2. \end{aligned} \tag{5.3}$$

Even if the experiment showed a clearly significant effect of stress, we are not able to estimate the effect with great precision since the 95% confidence interval has a length of 46 cm^2. □

In this experiment the effect of stress was established because $H_{0\mu}$ was rejected. Was this a stroke of luck or the result of a carefully planned experiment? We shall try to answer this question and at the same time introduce some considerations that could and should be made before an experiment is conducted.

The estimate of the difference of means in the two groups is the difference of averages $\bar{x}_{1\cdot} - \bar{x}_{2\cdot}$. The variance of the estimate, i.e., the variance of the corresponding random variable, is

$$Var(\bar{X}_{1\cdot} - \bar{X}_{2\cdot}) = \sigma^2\left(\frac{1}{n_1} + \frac{1}{n_2}\right).$$

The smaller the variance is the larger the precision of the estimate is and the better the design of the experiment is. The variance of the estimate consists of two factors; σ^2, which is the variance of a single measurement, and $1/n_1 + 1/n_2$, which is determined by the size of the experiment. A consideration of both factors is necessary in the planning of the experiment.

First, we consider the second factor,

$$\frac{1}{n_1} + \frac{1}{n_2} = \frac{n_1 + n_2}{n_1 n_2} = \frac{n_.}{n_1(n_. - n_1)}, \tag{5.4}$$

where $n_. = n_1 + n_2$ denotes the total number of observations in the experiment.

Often considerations of time and money will dictate an upper limit to $n_.$. If the number of observations $n_.$ is fixed and one is free to control the number of observations in each group, the expression on the right-hand side of (5.4) considered as a function of n_1 shows that the factor is smallest when $n_1 = n_./2$ and the smallest possible value is $4/n_.$. This can be formulated as the rule:

If the total number of observations is fixed, the smallest variance of the difference in averages is obtained by having the same number of observations in each group.

This rule should not give the impression, however, that the number observations in the two groups must always be the same. In some cases the number of observations in one group may be limited, e.g., because it is a more expensive or time-consuming treatment. If n_1 is fixed, for example, it appears from the expression on the-left hand side of (5.4) that the variance of the difference $\bar{x}_{1.} - \bar{x}_{2.}$ will decrease if n_2 is increased.

The importance of the sample sizes n_1 and n_2 and the variance σ^2 for the experiment will now be illustrated in connection with the important concept of the *power of a statistical test.* Although power is a property of a test, it will be clear below that the quantities that determine the power are characteristics of the experiment, in particular the sample sizes, which in many cases can be controlled by the experimenter.

The power of a test is the probability of rejecting the null hypothesis when it is false. For the two-sample t that we consider here, the power is the probability of rejecting the hypothesis of equal means, $H_{0\mu}$: $\mu_1 = \mu_2$, when the means are different. The power depends on the variance of a single observation, σ^2, of the sample sizes of the experiment through the quantity $1/n_1 + 1/n_2$, and of $\mu_1 - \mu_2$. When $H_{0\mu}$ is false, the test statistic t has a so-called *noncentral* t-distribution with $f = n_. - 2$ degrees of freedom and *noncentrality parameter*

$$\lambda = \frac{\mu_1 - \mu_2}{\sigma\sqrt{1/n_1 + 1/n_2}}. \tag{5.5}$$

The probability of rejecting $H_{0\mu}$ as a function of λ is symmetric around 0, so in the following we shall consider only positive values of λ, i.e.,

$$\lambda = \frac{|\mu_1 - \mu_2|}{\sigma\sqrt{1/n_1 + 1/n_2}}. \tag{5.6}$$

The distribution function of a noncentral t-distribution is a fairly complicated function of λ and there is no point in giving its mathematical expression. It is available in tables, but more importantly it is available in statistical packages. Although the exact dependence of the power on λ and the degrees of freedom f is complicated, it is easy to describe qualitatively how the power depends on λ and f. For fixed f the power increases with λ and for fixed λ the power increases with f.

It appears from the expression for λ in (5.6) that λ increases, and hence the power, if σ^2 or $1/n_1 + 1/n_2$ is decreased or if $|\mu_1 - \mu_2|$ is increased. In addition, the power depends on the significance level. It is easier to reject $H_{0\mu}$ when the significance level is 5% than when the significance level is 1%.

From Figure 5.2 to Figure 5.5 it is possible to see in a bit more detail how the power depends on λ and the degrees of freedom f when using the significance levels $\alpha = 1\%$ or $\alpha = 5\%$. For selected values of the degrees of freedom, the power has been plotted as a function of λ. In Figure 5.2 and Figure 5.3 a linear vertical axis has been used. The two figures give a clear qualitative impression of how the power increases as a function of the noncentrality parameter λ and the degrees of freedom f, and also that for the same value of λ and f the power is higher for $\alpha = 5\%$ than for $\alpha = 1\%$. In Figure 5.4 and Figure 5.5 the same power functions are plotted with a logarithmic vertical axis, which gives an improved resolution in the power interval between 0.8 and 0.95, which is the interval for the power of greatest practical interest.

It is clear from Figures 5.2 to 5.5 that very small differences in means will be hard to detect because small differences in means will give small λ's, which in turn will give a small power of the test. Fortunately, it is often not important to detect small differences in means because a small difference will be of little practical importance, but on the other hand, it is essential to detect large differences in means. If one can find a borderline, a so-called *smallest relevant difference* between the unimportant and the important differences, one can design the experiment to have a reasonable large power at the smallest relevant difference. For larger differences than the smallest relevant difference the test will have even larger power.

In technology and medicine the concept of a smallest relevant difference is very natural. If, for example, a new drug is being developed to improve the quality of life of patients with a certain disease and is being tested against a known treatment, it will be possible to agree on a smallest relevant difference.

In science it may be argued that even the smallest difference is relevant, but it does not change the fact that the chance of obtaining a significant result depends on the size of the difference and if the difference is small, it may be necessary to design a completely different experiment to confirm the theory.

Example 5.1 (Continued)

The experimenter who wanted to establish the effect of mechanical stress on the foliage of soya beans might have gone through the following considerations in the planning of the experiment. From previous experiments he knew that the total area of the leaves of soya bean plants were between 200 cm^2 and 300 cm^2. It was considered essential to be able to confirm a difference in means at the two levels of stress if it was around 10% of the mean the area of the leaves. The smallest relevant difference was chosen to be 20 cm^2. The variance σ^2 of a single measurement was more difficult to guess. But then it occurred to the experimenter that in a normal distribution, close to 95% of the observations would fall in an interval of length 4σ. In this experiment the observations were expected to lie in the interval between 200 cm^2 to 300 cm^2, and this suggests that the standard deviation σ of a single measurement would be 25 cm^2.

Now λ could be calculated as

$$\lambda = \frac{20 \text{ cm}^2}{25 \text{ cm}^2\sqrt{1/13+1/13}} = 2.04.$$

If the significance level of 5% is selected, we can from Figure 5.2 see that the power of the t-test in that experiment is slightly smaller than 0.5, noting that the degrees of freedom for the t-test are 24.

This means that in a little more than half the experiments of this size, the t-test would not reject the hypothesis $H_{0\mu}$ of equal means even if the difference in means is as big as 20 cm^2. □

This is a fairly large risk of failing to get a significant result, and so the experimenter has hardly considered the power of his test before making the experiment. If we consider the power of the test of an important hypothesis, we should try to have a power of around 90%.

If the power of an experiment with a specified number of observations turns out to be too low, we must resist the temptation to argue that the variance is probably smaller than first thought and that the smallest relevant difference is higher. The only real way to improve the power of the experiment is to increase the number of observations or to actively do something to make the variance smaller.

A useful supplement to the plots of the power functions in Figure 5.2 to Figure 5.5 is the formula (5.7), which combines all the quantities that influence the power of the t-test:

$$(t_{1-\alpha/2}+t_{1-\beta})^2 = \frac{(\mu_1-\mu_2)^2}{\sigma^2(\frac{1}{n_1}+\frac{1}{n_2})}. \tag{5.7}$$

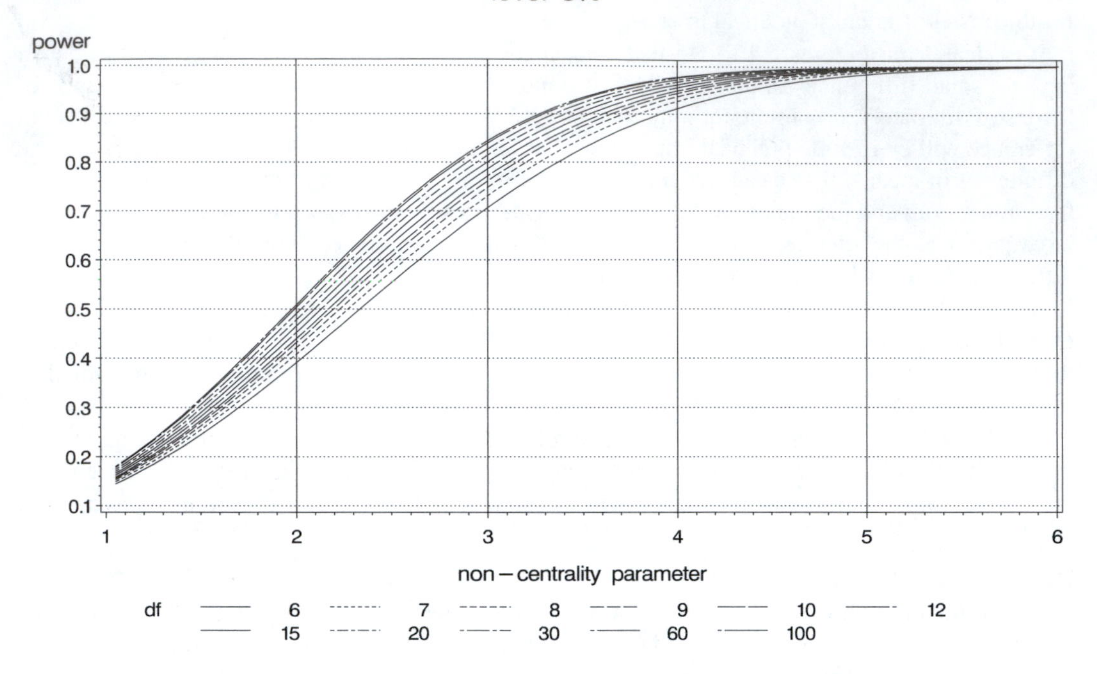

Figure 5.2 *The power function for the t-test with significance level $\alpha = 5\%$ for a range of degrees of freedom.*

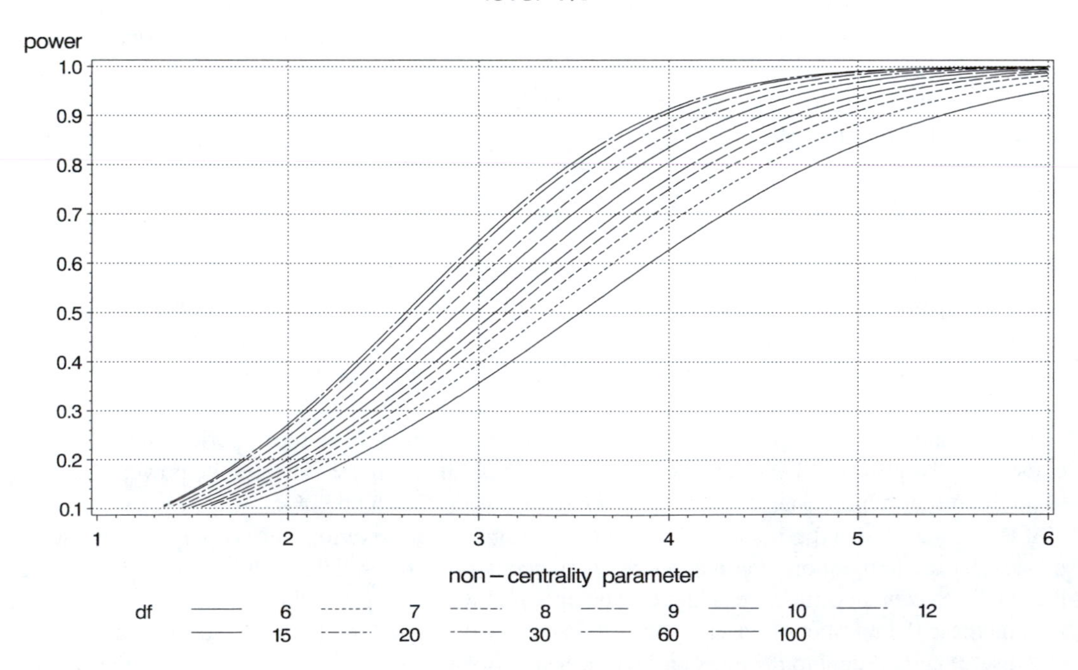

Figure 5.3 *The power function for the t-test with significance level $\alpha = 1\%$ for a range of degrees of freedom.*

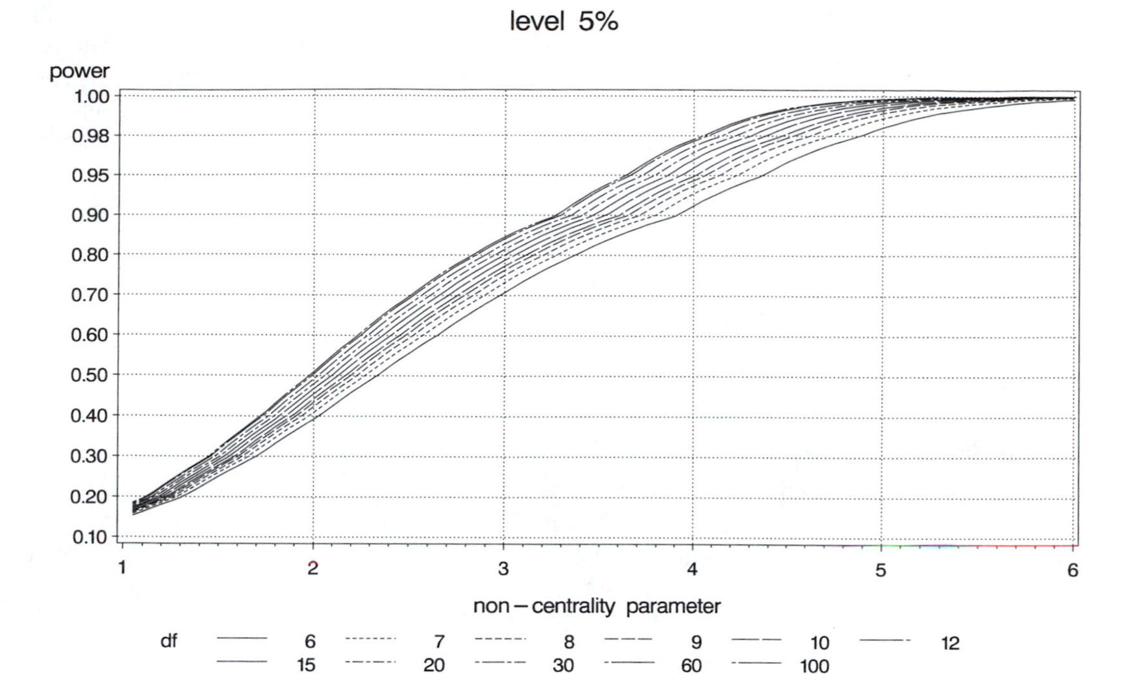

Figure 5.4 *The power function for the t-test with significance level* $\alpha = 5\%$ *for a range of degrees of freedom. Logarithmic vertical axis.*

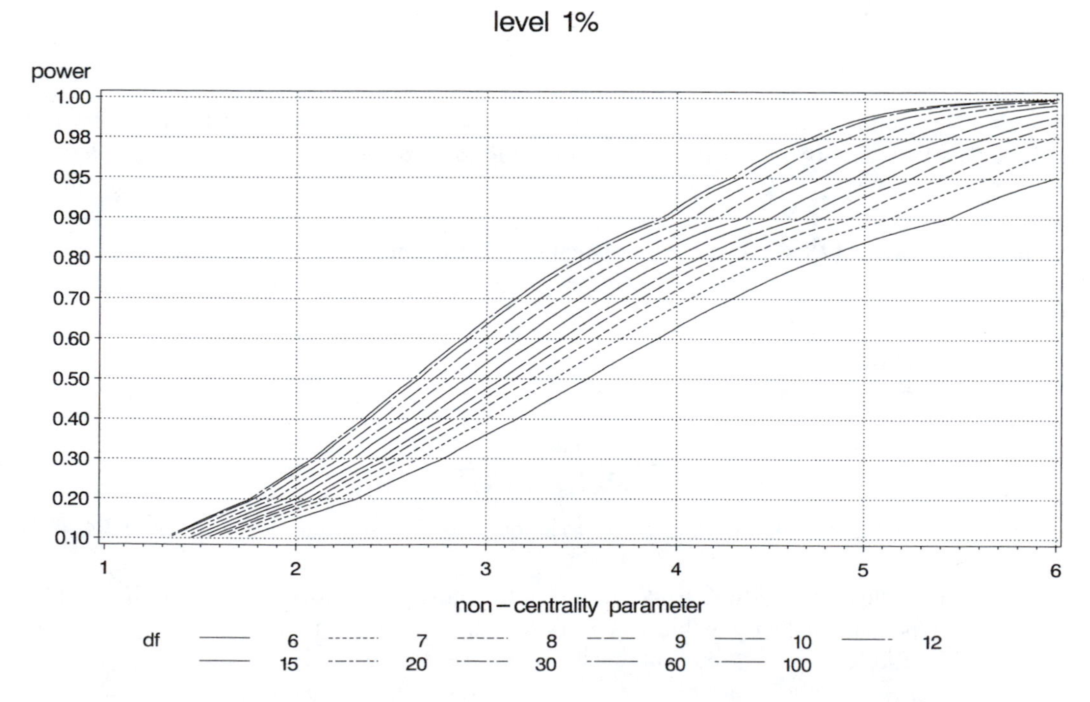

Figure 5.5 *The power function for the t-test with significance level* $\alpha = 1\%$ *for a range of degrees of freedom. Logarithmic vertical axis.*

In the formula the power appears as $1-\beta$, where β is the probability of *not* rejecting the null hypothesis when it is false. When the test fails to reject a null hypothesis that is wrong, an error has been made; this type of error is traditionally called an *error of type II* in statistics. Here $t_{1-\alpha/2}$ and $t_{1-\beta}$ denote the $1-\alpha/2$ and $1-\beta$ fractiles in the t-distribution with $n. - 2$ degrees of freedom.

If $n_1 = n_2 = n./2$, the formula can be rewritten to give the total number of observations as a function of the power, $1-\beta$, the significance level, α, the variance, σ^2, and the smallest relevant difference, $|\mu_1 - \mu_2|$:

$$n. = 4(t_{1-\alpha/2}+t_{1-\beta})^2 \frac{\sigma^2}{(\mu_1-\mu_2)^2}. \tag{5.8}$$

It is a slight complication that formulas (5.7) and (5.8) depend on $t_{1-\alpha/2}$ and $t_{1-\beta}$, which in turn depend on the degrees of freedom, which are $n. - 2$. Thus both sides of the equation depend on the total number of observations $n.$, and this means that we must use the formula a couple of times before the total number of observations $n.$ is obtained.

Example 5.1 (Continued)
How large should the experiment be to have a power of $1-\beta = 0.9$, when the level of the test is $\alpha = 0.05$, the smallest relative difference is 20 cm^2, and the standard deviations of a single measurement is 25 cm^2? As a first approximation one can use the fractiles $u_{1-\alpha/2}$ and $u_{1-\beta}$ in the normal distribution rather than $t_{1-\alpha/2}$ and $t_{1-\beta}$. Because $u_{1-\alpha/2} < t_{1-\alpha/2}$ and $u_{1-\beta} < t_{1-\beta}$, this leads to a suggestion for the number of observations which is too small. Now $u_{0.975} = 1.96$ and $u_{0.90} = 1.28$, and the suggested number of observations is that

$$n. = 4(1.96+1.28)^2 \frac{(25\text{ cm}^2)^2}{(20\text{ cm}^2)^2} = 65.61.$$

This number of observations is too small, but the formula can be applied once more with $66 - 2 = 64$ degrees of freedom. The fractiles are $t_{0.975}(64) = 1.998$ and $t_{0.90}(64) = 1.295$ and the new suggestion is

$$n. = 4(1.998+1.295)^2 \frac{(25\text{ cm}^2)^2}{(20\text{ cm}^2)^2} = 67.77.$$

A further application of (5.8) with $68-2 = 66$ degrees of freedom changes the fractiles very little and the answer is that 68 *observations are required to get a power of 0.9 under the specified circumstances.* □

Formulas (5.7) and (5.8) are only approximations, but the answer they provide can be checked using the power functions in Figure 5.4 or 5.5 depending on whether the significance level is 1% or 5%.

Example 5.1 (Continued)
With $n. = 68$ the noncentrality parameter becomes

$$\lambda = \frac{20\text{ cm}^2}{25\text{ cm}^2\sqrt{1/34+1/34}} = 3.30,$$

and according to Figure 5.4, this corresponds to a power of 0.90 at a significance level of 5%. □

The main consideration in this chapter has been to adjust the number of observations to obtain a satisfactory power. Another possibility is to try to reduce σ^2. In some cases this can be achieved using blocking, which plays a prominent role in design of experiments. This is the subject of Section 5.2.

The *error of type II* concept, i.e., to fail to reject a null hypothesis that is wrong, was mentioned above. For completeness we mention that the *error of type I* is to reject a null hypothesis that is true. The probability of an error of type I is equal to the significance level α.

5.2 Reduction of σ^2 – An Example of Blocking

In all the models for normal data that have been considered so far in this book, σ^2 is a measure of the unexplained variation. Reducing the unexplained variation for given data corresponds to exploiting knowledge of the experiment to make a more elaborate model. If we look at the explanation of Example 5.1 on page 241, we cannot find any information that makes this possible. We were only told that 26 plants had been randomly assigned to two groups that received different treatments. Part of the variation in the area of the leaves between plants that receive the same treatment may be interpreted as genetic variation between the plants. If there had been a need to reduce σ^2 in this situation, one could, in advance, have determined that the 26 seedlings were genetically identical in pairs and then apply different treatments to the two plant in each of 13 pairs. If that is the case, it is said that the experiment is performed in *blocks*. Here the blocks are the pairs of genetically identical plants.

If the experiment had been performed in blocks, the analysis of the data in Example 5.1 in Section 5.1 would have been wrong. In the following we shall see how the analysis proceeds if the experiment is performed in blocks of size two.

Example 5.2
Let the situation be as described in Example 5.1 on page 241 but with the extra information that the plants were genetically identical in pairs, and that this had been exploited in the experiment to give genetically identical plants different treatments. Data have been displayed in Table 5.2 with the information about the pairs.

Table 5.2 *The total area of the leaves ordered after the genetically identically pairs and the treatment*

Total area of leaves in cm^2		
Pair	*No stress*	*Stress*
1	205	193
2	220	168
3	230	187
4	234	210
5	235	207
6	237	200
7	246	210
8	258	219
9	261	217
10	269	240
11	273	235
12	293	277
13	294	261

Let x_{ij} denote the response for the plant in the ith pair that received the jth treatment. The treatment $j = 1$ is *no stress and* treatment $j = 2$ is *stress*. A possible model for the data in Table 5.2 in the light of the information about the experiment is that the observations x_{ij} are realizations of independent, normally distributed random variables with the same variance but with the means depending on the pair and the treatment, i.e., *the model,*

$$M_2\colon x_{i1} \sim\sim N(\alpha_i, \sigma^2), \qquad x_{i2} \sim\sim N(\alpha_i + \delta, \sigma^2), \quad i = 1, \ldots, n, \tag{5.9}$$

where as usual the independence of the observations is subsumed. Here α_i is the mean of the plant in the ith pair that is *not* exposed to stress and $\alpha_i + \delta$ is the mean of the plant in the ith pair that is exposed to stress. Thus δ is the difference between the means of the observations at the two levels of the treatment.

It is an important aspect of the model M_2 that the influence of the treatment is additive and the same for all pairs regardless of α_i. It is by no means obvious that this must be so, and it is essential to check that the assumption holds or, at least, is not contradicted by the data; for the further analysis relies heavily on the model. This is done by plotting the observations of each pair $(X_{i1}, X_{i2}), i = 1, \ldots, n$, in a coordinate system. In Figure 5.6 this has been done for the data in Table 5.2. If the variance σ^2 had been 0 and supposing that M_2 had been true, the points in the graph would all be on a straight line with a slope of 1, which intersects the vertical axis in δ. With a positive variance the points will not lie exactly on a straight line with slope 1, but they will lie around the line in an area of constant width if the variance of all observations is the same.

This is exactly the picture seen in Figure 5.6. First, the points clearly lie around a line with slope 1 and, secondly, the spread around the line seems constant. Obviously the points in Figure 5.6 do not lie around the identity line. All the points lie below the identity line and rather around a line that is translated about 30 units (cm^2) along the horizontal axis. This is an indication that the effect of stress is significant and also that the mean area of the leaves of plants that are not stressed is about 30 cm^2 higher than genetically identical plants that are stressed.

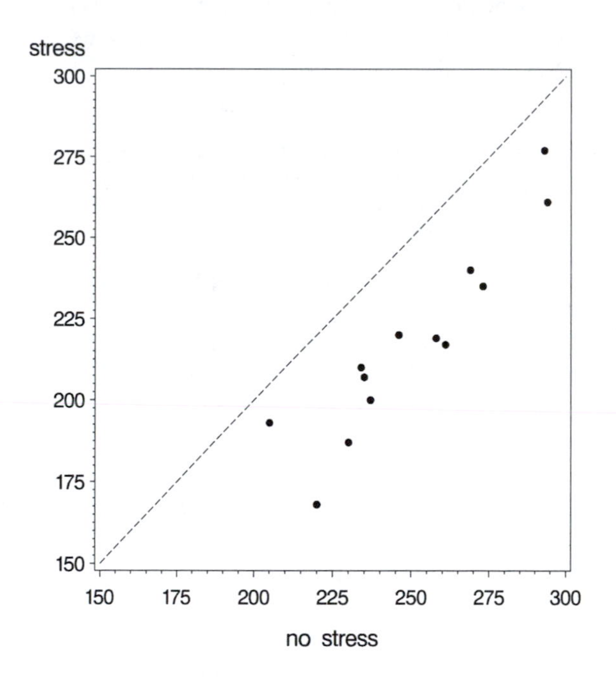

Figure 5.6 *Plot of stress against no stress.*

As a consequence of the model M_2 the differences are a normal sample, i.e.,

$$d_i = x_{i1} - x_{i2} \sim\sim N(-\delta, 2\sigma^2), \quad i = 1, \ldots, n, \tag{5.10}$$

and this provides an opportunity to check the assumption of a normal distribution by making a normal fractile plot based on the observed differences. This fractile plot is shown in Figure 5.7 and it does not contradict the assumption that the differences are a normal sample.

In M_2 the hypothesis of no effect of stress is

$$H_0\colon \delta = 0,$$

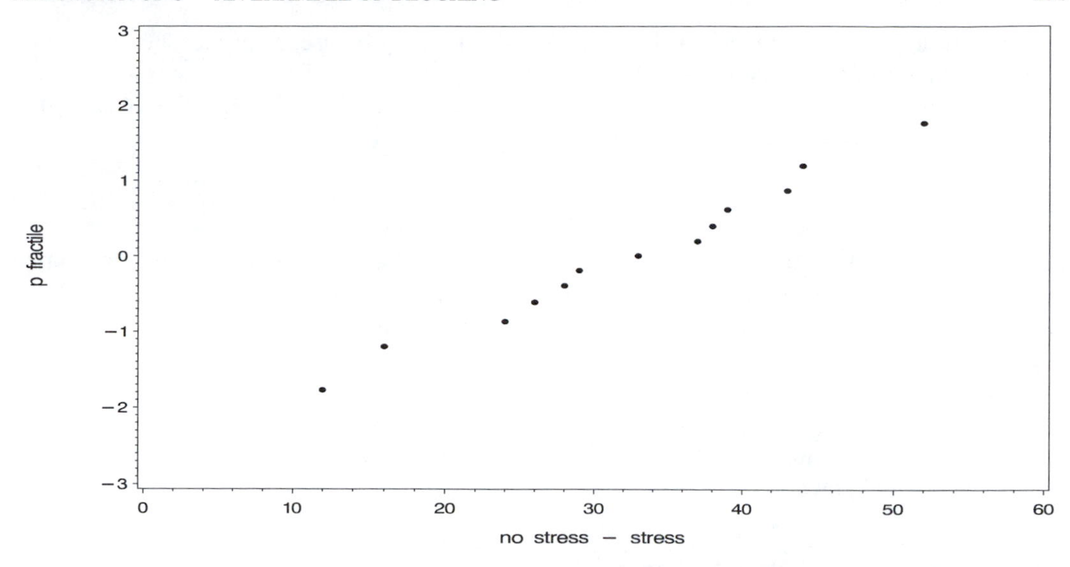

Figure 5.7 *Normal fractile plot of the difference stress − no stress.*

and it is tested in the sample of differences using the t-test for the mean being equal to zero in a normal sample:

$$t = \frac{\bar{d}.}{\sqrt{s_d^2/n}} = \frac{\bar{x}_{.1} - \bar{x}_{.2}}{\sqrt{s_d^2/n}} = \frac{250.38 - 218.00}{\sqrt{64.7949/13}} = \frac{32.38}{3.1573} = 10.26, \tag{5.11}$$

where s_d^2 is the estimate of the variance for the differences.

The test statistic is evaluated in a t-distribution with 12 degrees of freedom and corresponds to a significance probability below 0.0001. Thus the null hypothesis is rejected and the effect of stress that is visible in Figure 5.6 is significant. A 95% confidence interval for $-\delta$ is

$$\begin{aligned} \bar{d}. \pm t_{1-\alpha/2}(n-1)\sqrt{s_d^2/n} &= 32.38 \text{ cm}^2 \pm 2.179 \times 3.1573 \text{ cm}^2 \\ &= [25.50, 39.26] \text{ cm}^2. \end{aligned} \tag{5.12}$$

This confidence interval should be compared with the one in (5.3) based on the two-sample t-test. We are here able to give a far more precise statement on the effect of stress, the length of the confidence interval being close to 14 cm^2 as compared to 46 cm^2 when the two-sample t-test is used. □

The test in (5.11) is called *the paired t-test.* It is a test that is closely tied to the planning of the investigation or the design of the experiment. It is **not** a test that can be used just because the number of observations in the two samples happens to be the same.

Notice that the σ^2 appearing in the model (5.1) on page 242 is not the same as the σ^2 in the model (5.9) on page 249. The first one includes the biological variation in the total area of the leaves, and it is the purpose of the blocked or paired experiment to avoid the biological variation in the comparison of the two treatments.

The idea of blocking is so compelling that it may lead people to think that a blocked design and hence a paired t-test should always be used. This is not so. Note that the two-sample t-test in (5.2) has 24 degrees of freedom, while the paired t-test only has 12. Fewer degrees of freedom means lower power, so one should only choose the blocked design if the expected decrease in the variance will be sufficiently large to compensate for the fewer degrees of freedom and make the blocked design the better one.

If it is decided to use the blocked design, formula (5.13) may be useful to calculate the number of blocks or pairs of observations, n, to achieve a desired power $1-\beta$. The σ^2 that appears in (5.13) is the same as in model (5.9), in other words, $2\sigma^2$ is the variance of a difference. The quantity δ in (5.13) denotes the smallest relevant difference:

$$n = (t_{1-\alpha/2} + t_{1-\beta})^2 \frac{2\sigma^2}{\delta^2}. \tag{5.13}$$

As in the previous formulas, $t_{1-\alpha/2}$ and $t_{1-\beta}$ denote the $1-\alpha/2$ the $1-\beta$ fractiles in the t-distribution with $n-1$ degrees of freedom. When applying (5.13) there is the slight complication that both the right-hand side and the left-hand side depend on the number of observations n. Therefore, a first approximation to n is obtained using the fractiles of the normal distribution $u_{1-\alpha/2}$ and $u_{1-\beta}$ in place of $t_{1-\alpha/2}$ and $t_{1-\beta}$ in (5.13). This value of n is used to calculate the $t_{1-\alpha/2}$ and $t_{1-\beta}$ fractiles with $n-1$ degrees of freedom, which are used in a second application of (5.13), which will give a sufficiently accurate approximation to the number of pairs of observations that must be used to achieve the desired power.

5.3 Control Plot for the Paired t-Test

The paired t-test is only justified in the model M_2, so it is important to make sure that M_2 is consistent with the data. The only way to do this is through appropriate plots of the data as was done in connection with Example 5.2. In this Section we will comment on what to look for in the control plots.

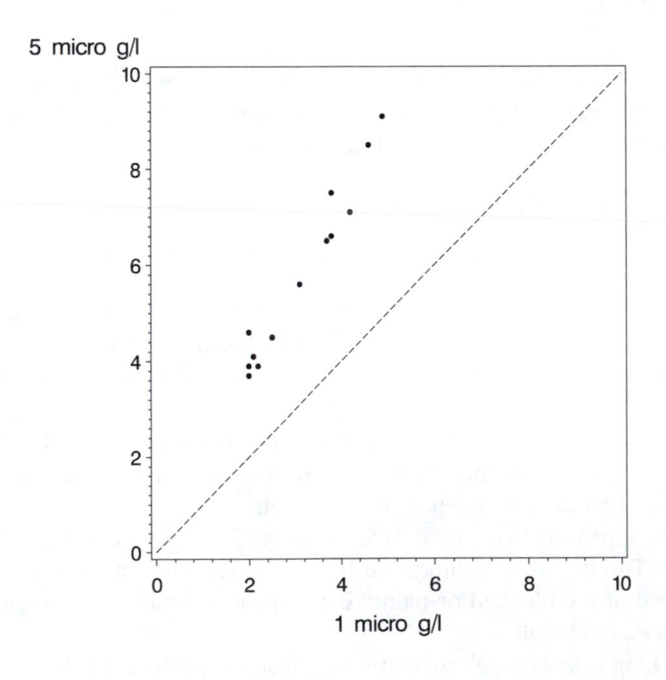

Figure 5.8 *Control plot of M_P. The observations are measurements of the concentration of Crome in eggs of trout. The eggs have been lying in incubators with a concentration of Crome of 1 or 5 μg Cr/L, respectively. A pair consists of observations of one egg from each incubator. The eggs in a pair have been lying in the incubators for the same number of days and the pairs differ with respect to the number of days they have been in the incubators.*

The data consist of n paired observations $X_{i1}, X_{i2}, i = 1, \ldots, n$. The application of the paired t-test is justified if the differences $X_{i2} - X_{i1}, i = 1, \ldots, n$, is a normal sample. This will be the case if the model is

$$M_P: X_{i1} \sim N(\alpha_i, \sigma_1^2) \quad \text{and} \quad X_{i2} \sim N(\alpha_i + \delta, \sigma_2^2),$$

where all random variables are independent. Under the model M_P the differences will be independent and identically normally distributed,

$$D_i = X_{i1} - X_{i2} \sim N(-\delta, \sigma_1^2 + \sigma_2^2).$$

Notice that the variances of X_{i1} and X_{i2} need not be the same. A typical application of this model is in the comparison of two measurement methods to investigate if they are identical, i.e., measure with the same mean ($\delta = 0$), and in that situation it would be unrealistic to assume that the variances of the two methods were identical.

In order to check the model M_P, the observations on the pairs $(X_{i1}, X_{i2}), i = 1, \ldots, n$ are plotted. This has been done in Figure 5.6 for the data in Table 5.2. If the variances in the model were 0, the points would be lying on a straight line that intersects the ordinate axis in δ and has a slope of 1. With positive variances the points will be lying in a strip with constant width around this line.

One deviation from the model M_P will be that the variances are not constant, but are a function of the means. It is quite common to see that the variances increase with the means and in that case the observations will not be lying in a strip of constant width, but rather in a fan-like structure. When this phenomenon is encountered, as it is in Figure 5.8, it would be wrong to use the paired t-test on the data.

When the fan-like structure is observed, it is often possible to fit a line through (0,0) with a slope that is different from 1 through the cloud of observations. This may indicate that the model M_P might be valid for the logarithm of the observations. This conjecture must be checked by a similar plot based on the logarithm of the observations. This is done in Figure 5.9, which does not contradict the assumption that the logarithm of the observations are consistent with M_P.

Once it has been confirmed or, rather, has not been rejected that the mean and the variance of the observations has the structure specified by the model M_P, the question of whether the observations are normally distributed can be addressed. The obvious choice is to make a normal fractile plot of the sample of differences.

5.4 The Paired t-Test and Two-Way Analysis of Variance

When the paired t-test is applicable, the data is in exactly the same form as for a two-way analysis of variance. The two factors are the treatment with two levels and the factor that divides the observations into pairs. In Table 5.2 rows has been used to represent the pairs and columns to represent the treatment factor. The model M_2 in (5.9) is the additivity model of two-way analysis of variance, but parameterized slightly differently, $-\delta$ in (5.9) being $\beta_1 - \beta_2$ in Section 4.3. We shall see now that using the paired t-test to test $H_{0\delta}$: $\delta = 0$ is equivalent to using the F-test of two-way analysis of variance to test the reduction to M_3^r, the model with row effects only.

We consider a two-way analysis of variance with n rows representing the pairs and look at the reduction from the additivity model,

$$M_2: x_{ij} \sim\sim N(\alpha_i + \beta_j, \sigma^2), \qquad i = 1, \ldots, n, \quad j = 1, 2,$$

to the model with row effects only,

$$M_3^r: x_{ij} \sim\sim N(\alpha_i, \sigma^2), \qquad i = 1, \ldots, n, \quad j = 1, 2,$$

and calculate the differences between the observations of the pairs,

$$d_i = x_{i1} - x_{i2}, \qquad i = 1, \ldots, n.$$

It can be shown that the SSD_{02} of two-way analysis of variance is half the SSD of the sample of

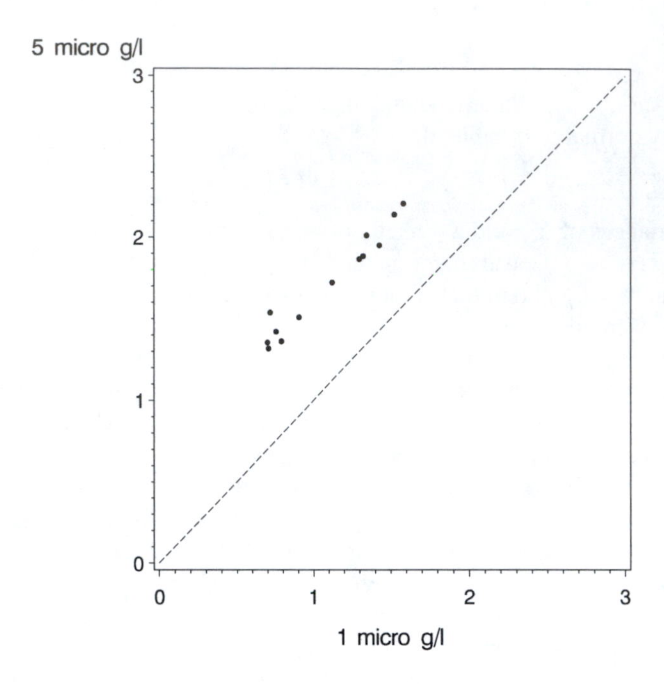

Figure 5.9 *Control plot of M_P for the logarithm of the Crome concentrations in Figure 5.8.*

differences that we shall denote by SSD_d, i.e.,

$$SSD_{02} = \frac{1}{2}\sum_{i=1}^{n}(d_i - \bar{d}_{\cdot})^2 = \frac{1}{2}SSD_d,$$

and, furthermore, that

$$SSD_3^r = \frac{1}{2}n(\bar{x}_{\cdot 1} - \bar{x}_{\cdot 2})^2 = \frac{1}{2}n\bar{d}_{\cdot}^2.$$

It follows that the F-test for the reduction from M_2 to M_3^r – or the test of no treatment effect – is

$$F = \frac{SSD_3^r/(2-1)}{SSD_{02}/(n-1)(2-1)} = \frac{n\bar{d}_{\cdot}^2}{s_d^2} = \left(\frac{\bar{d}_{\cdot}}{\sqrt{s_d^2/n}}\right)^2,$$

which is the squared paired t-test in (5.11).

Annex to Chapter 5

Calculations in SAS

The paired t-test is a t-test for mean 0 in a normal sample, and it is shown in Annex to Section 3.1 how *PROC MEANS* can be used for the calculations. Furthermore, the paired t-test is equivalent to the F-test for the reduction to the model with block effect only in a two-way analysis of variance, and this is treated in Annex to Section 4.3, where the calculation are performed with *PROC GLM*. For easy reference both situations are considered here. In SAS 8 *PROC TTEST* has been enhanced with a `PAIRED` statement to handle the paired t-test.

Whether *PROC MEANS*, *PROC TTEST* or *PROC GLM* is the easiest to use depends on how the data set is organized. It is also shown how one can create either form of the data set from the other.

Example 5.2 (Continued)
If the data in the data set are organized in the same way as in Table 5.2, i.e., as two variables each containing the observations from one of the two treatments and ordered with respect to the pairs as it is the case with the data set `stress` in the program below, it will be easiest to use *PROC MEANS* or *PROC TTEST*.

```
TITLE1 'Example 5.2';
DATA stress;
INPUT pair nostress stress;
dif=nostress-stress;
DATALINES;
   1       205         193
   2       220         168
   3       230         187
   4       234         210
   5       235         207
   6       237         200
   7       246         220
   8       258         219
   9       261         217
  10       269         240
  11       273         235
  12       293         277
  13       294         261
;
```

The commands

```
PROC MEANS DATA=stress ALPHA =0.05 VARDEF=DF MAXDEC=3
            N MEAN VAR STDERR T PRT CLM;
VAR dif;
LABEL dif='nostress-stress';
RUN;
```

gives in addition to the paired t-test also the 95% confidence interval for $-\delta$.

```
                             The MEANS Procedure

                   Analysis Variable : dif nostress-stress

                                                          Lower 95%     Upper 95%
 N      Mean     Variance    Std Error   t Value   Pr > |t|   CL for Mean   CL for Mean
-------------------------------------------------------------------------------------
13    32.385     129.590        3.157     10.26     <.0001        25.505        39.264
-------------------------------------------------------------------------------------
```

To use *PROC TTEST* to compute a paired t-test the data set must be organized in the same way as when using *PROC MEANS*. The SAS code is simple:

```
PROC TTEST DATA=stress;
PAIRED nostress*stress;
RUN;
```

and gives the following output listing:

```
                              The TTEST Procedure

                                  Statistics

                                 Lower CL                  Upper CL   Lower CL
Difference                  N        Mean        Mean         Mean    Std Dev    Std Dev

nostress - stress          13      25.505      32.385       39.264     8.1631     11.384

                                  Statistics

                                 Upper CL
        Difference                Std Dev     Std Err      Minimum      Maximum

        nostress - stress          18.792      3.1573           12           52


                                   T-Tests

              Difference                 DF     t Value     Pr > |t|

              nostress - stress          12       10.26       <.0001
```

There are a few advantages in using *PROC TTEST* with the `PAIRED` statement compared to using *PROC MEANS*. One is that several paired t-test can easily be requested with *PROC TTEST* without having to compute the differences first, see the syntax of *PROC TTEST* for the details. Another advantage is that *PROC TTEST* gives the confidence interval for the standard deviation of the difference.

In order to use a two-way analysis of variance, the observations must all be in one variable and there must be two further variables that contain the information about treatment and the pair of each observation. A data set like that can be constructed from the data set `stress`:

```
DATA long;
SET stress;
   area=nostress;
   block=par;
   treatment='nostress';
   OUTPUT;
   area=stress;
   block=pair;
   treatment='stress';
   OUTPUT;
KEEP area block treatment;
RUN;
```

Then *PROC GLM* can be used with the data set `long`:

```
PROC GLM DATA=long;
CLASS block treatment;
MODEL area=block treatment/SS1 SOLUTION;
ESTIMATE 'the no stress effect' treatment 1 -1;
CONTRAST 'the no stress effect' treatment 1 -1;
RUN;
```

with the result:

4

The GLM Procedure

Class Level Information

Class	Levels	Values
block	13	1 2 3 4 5 6 7 8 9 10 11 12 13
treatment	2	nostress stress

Number of observations 26

5

The GLM Procedure

Dependent Variable: area

Source	DF	Sum of Squares	Mean Square	F Value	Pr > F
Model	13	25592.50000	1968.65385	30.38	<.0001
Error	12	777.53846	64.79487		
Corrected Total	25	26370.03846			

R-Square	Coeff Var	Root MSE	area Mean
0.970514	3.437144	8.049526	234.1923

Source	DF	Type I SS	Mean Square	F Value	Pr > F
block	12	18775.53846	1564.62821	24.15	<.0001
treatment	1	6816.96154	6816.96154	105.21	<.0001

Contrast	DF	Contrast SS	Mean Square	F Value	Pr > F
the no stress effect	1	6816.961538	6816.961538	105.21	<.0001

Parameter	Estimate	Standard Error	t Value	Pr > |t|
the no stress effect	32.3846154	3.15728393	10.26	<.0001

Parameter	Estimate	Standard Error	t Value	Pr > |t|
Intercept	261.3076923 B	5.90673737	44.24	<.0001
block 1	-78.5000000 B	8.04952618	-9.75	<.0001
block 2	-83.5000000 B	8.04952618	-10.37	<.0001
block 3	-69.0000000 B	8.04952618	-8.57	<.0001
block 4	-55.5000000 B	8.04952618	-6.89	<.0001
block 5	-56.5000000 B	8.04952618	-7.02	<.0001
block 6	-59.0000000 B	8.04952618	-7.33	<.0001
block 7	-44.5000000 B	8.04952618	-5.53	0.0001
block 8	-39.0000000 B	8.04952618	-4.85	0.0004
block 9	-38.5000000 B	8.04952618	-4.78	0.0004
block 10	-23.0000000 B	8.04952618	-2.86	0.0144
block 11	-23.5000000 B	8.04952618	-2.92	0.0129
block 12	7.5000000 B	8.04952618	0.93	0.3698
block 13	0.0000000 B	.	.	.
treatment nostress	32.3846154 B	3.15728393	10.26	<.0001
treatment stress	0.0000000 B	.	.	.

NOTE: The X'X matrix has been found to be singular, and a generalized inverse was used to solve the normal equations. Terms whose estimates are followed by the letter 'B' are not uniquely estimable.

Here the estimate $-\bar{d}.$ of $-\delta = \beta_1 - \beta_2$, the paired t-test, the significance probability, and the estimate of the standard deviation of $\bar{d}.$ are found in the line `treatment nostress` at the bottom of the output listing. The confidence interval for $-\delta$ must be calculated by hand from the formula

$$\texttt{Estimate} \pm \texttt{Std Error of Estimate} \cdot t_{1-\alpha/2}(f).$$

If on the other hand the data are organized with the response (here area) in one variable, as is the case with the data set `long` (in the variable `area`), and if one wishes to split the response into two variables corresponding to the two treatments and in such a way that the order of the pairs is the same in the two variables, it may be done in the following way.

```
PROC SORT DATA=long;
BY block treatment;
RUN;

DATA split;
ARRAY split {2} c1-c2;
DO I=1 TO 2;
   SET long;
   split{i}=area;
   dif=c1-c2;
DROP i area treatment;
END;
RUN;
```

□

Supplement to Chapter 5

5.5 Noncentral t-, χ^2-, and F-Distributions

In previous sections we have seen that for normally distributed data, the t-, χ^2- and F-distributions appear as distributions of the test statistics for reductions in the models that we consider. This is only when the reductions of the models are true that the test statistics have those distributions. Occasionally we use the expression that they are the distributions *under the null hypothesis*.

When a reduction of a model is not justified or, in other words, the null hypothesis is not true, the test statistics have distributions that in addition to the degrees of freedom they have under the null hypothesis also depend on how far the null hypothesis is from being true. Those distributions are the noncentral t-, χ^2-, and F-distributions that are defined below. The formulas for the densities and the distribution functions of the noncentral distributions are quite complicated and there is an extensive literature that deal with approximations and tables.

Today many questions of the power of tests can be answered using distribution functions and density functions – both noncentral and "central" – that are available in statistical packages.

5.5.1 The Noncentral t-Distribution

Definition 5.1 If U and Z are two independent random variables such that $U \sim N(\lambda, 1)$ and $Z \sim \chi^2(f)/f$, the ratio,

$$t = \frac{U}{\sqrt{Z}}, \tag{5.14}$$

is *noncentrally* t-distributed *with* f *degrees of freedom and noncentrality parameter* λ, which is written as $t \sim t(f, \lambda)$. ▲

The definition of the noncentral t-distribution can be briefly written as

$$t(f, \lambda) = \frac{N(\lambda, 1)}{\sqrt{\chi^2(f)/f}}$$

where the numerator and the denominator represent *independent* random variables.

The central and the noncentral t-distribution appears in situations where a ξ is estimated by a random variable X, which is distributed as $N(\xi, k\sigma^2)$, where k is a known constant, whereas σ^2 is unknown, but an estimate s^2 of σ^2 is available that is independent of X and has a $\sigma^2\chi^2(f)/f$-distribution. The hypothesis $H_{0\xi}$: $\xi = \xi_0$ about the unknown parameter ξ is tested with the statistic,

$$t = \frac{X - \xi_0}{\sqrt{s^2 k}} = \frac{(X - \xi_0)/\sigma}{\sqrt{s^2 k/\sigma^2}} \sim \frac{N\left(\frac{\xi - \xi_0}{\sigma\sqrt{k}}, 1\right)}{\sqrt{\chi^2(f)/f}},$$

which is distributed as a $t(f)$-distribution if $H_{0\xi}$ is true, but is distributed as a noncentral $t(f, \lambda)$-distribution with noncentrality parameter $\lambda = (\xi - \xi_0)/(\sigma\sqrt{k})$ if $H_{0\xi}$ is not true.

5.5.2 The Noncentral χ^2-Distribution

Definition 5.2 If $U_1, \ldots, U_f$ are independent, normally distributed random variables, $U_i \sim N(\mu_i, 1)$ for $i = 1, \ldots, f$, the distribution of

$$X = \sum_{i=1}^{f} U_i^2 \tag{5.15}$$

depends on the means μ_i through $\sum_{i=1}^{f}\mu_i^2$ only, and the distribution of (5.15) is a *noncentral* χ^2-*distribution with* f *degrees of freedom and noncentrality parameter* $\lambda = \sum_{i=1}^{f}\mu_i^2$, which we write as $X \sim \chi^2(f,\lambda)$. ▲

5.5.3 The Noncentral F-Distribution

Definition 5.3 If Z_1 and Z_2 are two independent random variables with $Z_1 \sim \chi^2(f_1,\lambda)/f_1$ and $Z_2 \sim \chi^2(f_2)/f_2$, the ratio,

$$F = \frac{Z_1}{Z_2}, \tag{5.16}$$

is *noncentrally* F*-distributed with* f_1 *degrees of freedom in the numerator and* f_2 *degrees of freedom in the denominator and with noncentrality parameter* λ, which we write as $F \sim F(f_1,f_2,\lambda)$. ▲

Using the symbols of the central and the noncentral χ^2-distribution the definition of the noncentral F-distribution can be briefly stated as

$$F(f_1,f_2,\lambda) = \frac{\chi^2(f_1,\lambda)/f_1}{\chi^2(f_2)/f_2}$$

if we remember that numerator and denominator are *independent* random variables.

The noncentral F-distribution is important in power calculations in connection with linear normal models. In a linear normal we consider the reduction from

$$M_1\colon \boldsymbol{\mu} \in L_1 \text{ with } \dim L_1 = d_1$$

to

$$M_2\colon \boldsymbol{\mu} \in L_2 \text{ with} \dim L_2 = d_2,$$

where $L_2 \subset L_1$. If M_2 is not true, i.e. if $\boldsymbol{\mu} \in L_1 \setminus L_2$, the F-test,

$$F = \frac{\|P_1\mathbf{x} - P_2\mathbf{x}\|^2/(d_1-d_2)}{\|\mathbf{x} - P_1\mathbf{x}\|^2/(n-d_1)},$$

will have a noncentral $F(d_1-d_2, n-d_2, \lambda)$-distribution with a noncentrality parameter,

$$\lambda = \frac{1}{\sigma^2}\|\boldsymbol{\mu} - P_2\boldsymbol{\mu}\|^2.$$

Exercises for Chapter 5

SAS programs for the examples can be found at the address
`http://www.imf.au.dk/biogeostatistics/`

Exercise 5.1 In the planning of a clinical trial to investigate a drug that is supposed to reduce blood pressure, it was argued that the smallest relevant difference is 5 mm Hg and that the standard deviation of a single measurement is 10 mm Hg. The significance level is chosen to be 5%.

(1) How many patients must be enrolled into the study if a power of 0.9 is desired?

(2) How many patients will do if the power only has to be 0.8?

The new drug is only available for 50 patients.

(3) How many patients must be given the old treatment if the power must be 0.8?

Exercise 5.2 Use (5.8) to make a small table of the total number of observations $n.$ in the two-sample t as a function of relevant values of $|\mu_1 - \mu_2| / \sigma$ when the power must be 0.8 or 0.9.

Exercise 5.3 A manufacturer of plastics wanted to investigate how suitable two types, A and B, would be for soles on sports shoes.

For that purpose 14 pairs of shoes were made with one sole of type A and the other sole of type B. The shoes were given to 14 boys and after two months, the wear of each sole was measured as the weight loss and expressed as a percentage of the initial weight.

The data are found in Table 5.3.

Table 5.3 *Data for Exercise 5.3*

Boy no.	A	B	$B-A$
1	13.9	14.6	0.7
2	10.5	9.1	−1.4
3	10.1	8.9	−1.2
4	9.8	9.2	−0.6
5	13.7	13.1	−0.6
6	12.6	12.5	−0.1
7	11.3	10.0	−1.3
8	8.8	7.0	−1.8
9	14.1	12.9	−1.2
10	9.5	9.5	0.0
11	9.4	9.0	−0.4
12	6.9	7.4	0.5
13	16.9	16.4	−0.5
14	9.8	10.7	0.9

(1) Is there a difference in the wear of soles of the two types?

(2) Compute a 95% confidence interval for the difference in mean wear for the two types of soles.

CHAPTER 6

Correlation

6.1 Introduction

In probability theory the term correlation is used for a precise aspect (6.2) of the joint distribution of two random variables. For a bivariate sample of data the term correlation is used in connection with several empirical measures that are intended to quantify the association between the two variables. The emphasis in this chapter will be on the empirical counterpart to the correlation defined in probability theory. We concentrate on continuous variables, but it is also meaningful to talk about the correlation between pairs of ordinal random variables, see the definition of the Spearman correlation, page 509, Chapter 12, on nonparametric statistics.

Section 6.2 contains the definitions of correlation and empirical correlation together with some of their basic properties as well as the warning that calculating correlations without first plotting the data is meaningless. In Section 6.3 we present the example that is used to illustrate the theory throughout this chapter. The statistical test of zero correlation and the confidence interval for a correlation coefficient is developed within the framework of the statistical model of the bivariate normal distribution, which is introduced in Section 6.4. Based on the properties of the bivariate normal distribution, the assumption that a sample comes from that distribution can be checked; this is explained in Section 6.5 and is considered again on page 288 in connection with calculations in SAS. Inference on the correlation coefficient includes test of the hypothesis that the correlation coefficient is equal to 0 and the calculation of a $(1-\alpha)$ confidence interval and both tools are given in Section 6.6. Section 6.7 presents the technique for testing the identity of the correlations from k bivariate normal distributions based on independent samples from those distributions. Finally, Section 6.8 and Section 6.9 deal with aspects of the interpretation of correlation.

6.2 Definitions

For two random variables X and Y, the *covariance* of X and Y is

$$Cov(X,Y) = E((X-EX)(Y-EY)), \tag{6.1}$$

and the *correlation* or the *correlation coefficient* between X and Y is the covariance divided by the standard deviations of X and Y, i.e.,

$$Cor(X,Y) = \frac{E((X-EX)(Y-EY))}{\sqrt{VarX}\sqrt{VarY}}. \tag{6.2}$$

The correlation coefficient is a dimensionless quantity that is bounded by -1 and 1, i.e.,

$$-1 \leq Cor(X,Y) \leq 1. \tag{6.3}$$

If the correlation coefficient is -1 or 1, there is an exact straight-line relationship between X and Y in the sense that constants α and β exist such that with probability one,

$$Y = \alpha + \beta X. \tag{6.4}$$

If $Cor(X,Y) = -1$, then β is negative and Y decreases as X increases, and if $Cor(X,Y) = 1$, then β is positive and X and Y increase together.

For a sample $(x_1,y_1),\ldots,(x_n,y_n)$ from a bivariate distribution the empirical counterpart of (6.1) is

$$Cov(\mathbf{x},\mathbf{y}) = \frac{1}{n}\sum_{i=1}^{n}(x_i - \bar{x}.)(y_i - \bar{y}.) \tag{6.5}$$
$$= \frac{1}{n}SPD_{\mathbf{xy}},$$

and the counterpart of (6.2) is the *empirical correlation coefficient* or the *sample correlation coefficient,*

$$r = r_{\mathbf{xy}} = \frac{\sum_{i=1}^{n}(x_i - \bar{x}.)(y_i - \bar{y}.)}{\sqrt{\sum_{i=1}^{n}(x_i - \bar{x}.)^2}\sqrt{\sum_{i=1}^{n}(y_i - \bar{y}.)^2}} \tag{6.6}$$

$$= \frac{SPD_{\mathbf{xy}}}{\sqrt{SSD_{\mathbf{x}}}\sqrt{SSD_{\mathbf{y}}}}. \tag{6.7}$$

The statistic in (6.6) is often referred to as the *Pearson correlation coefficient* after Karl Pearson (1857–1937). Often the word "coefficient" is omitted, and we speak simply of the *empirical correlation*, the *sample correlation*, the *estimated correlation,* or the *Pearson correlation*.

The connections between (6.1) and (6.5) and between (6.2) and (6.6) are that the sample defines a bivariate distribution, the empirical distribution, which assigns the probability $\frac{1}{n}$ to each of the sample values and (6.5) is the covariance (6.1) of X and Y of the empirical distribution and (6.6) is the correlation (6.2) of X and Y of the empirical distribution.

This means that the bounds in (6.3) apply to the empirical correlation as well, thus

$$-1 \leq r_{\mathbf{xy}} \leq 1 \tag{6.8}$$

and if $r_{\mathbf{xy}} = -1$ or $r_{\mathbf{xy}} = 1$, there is an exact linear relationship between the $x_1,\ldots,x_n$ and the $y_1,\ldots,y_n$. Thus

$$y_i = \alpha + \beta x_i, \qquad i = 1,\ldots,n, \tag{6.9}$$

where β is negative if $r_{\mathbf{xy}} = -1$ and where β is positive if $r_{\mathbf{xy}} = 1$.

Another special value of the correlation coefficient is 0. If X and Y are independent, then the covariance is 0 and, consequently, the correlation is 0. But a correlation equal to 0 does not in general imply that X and Y are independent. For the bivariate normal distribution, which will be treated in some detail later, zero correlation implies independence.

The perfect linear relationship (6.9) between the variables of the sample when $r_{\mathbf{xy}} = \pm 1$ should not lead to the belief that a numerical value of $|r_{\mathbf{xy}}|$ between 0 and 1 expresses a degree of linear relationship between the variables. Several examples have been constructed to illustrate this. An often quoted example was presented by Anscombe (1973) and is reproduced in Figure 6.1. The four artificial sets of data all have the value of $r_{\mathbf{xy}}$ equal to 0.816, but the plots tell very different stories and only for the data set A can the correlation be considered to be a reasonable summary of the association between the two random variables.

Another point is illustrated by the two data sets shown in Figure 6.2. For both data sets the empirical correlation is close to 0. The conclusion often drawn from a small observed correlation is that there is *no* relationship between the two variables. Obviously, this conclusion only makes sense for the data set in plot B of Figure 6.2. In plot A of Figure 6.2, there is a clear nonlinear relationship between the two variables.

Both examples emphasize the point that calculation and interpretation of empirical correlations are meaningless unless they are supported by plots of the data.

There is one model where the empirical correlation coefficient is particularly meaningful: the bivariate normal distribution. In this model the empirical correlation coefficient is the maximum likelihood estimate of the correlation coefficient (6.2) of the bivariate normal distribution, and the numerical value of the correlation coefficient of the bivariate normal distribution can be interpreted as the strength of the association between the two variables.

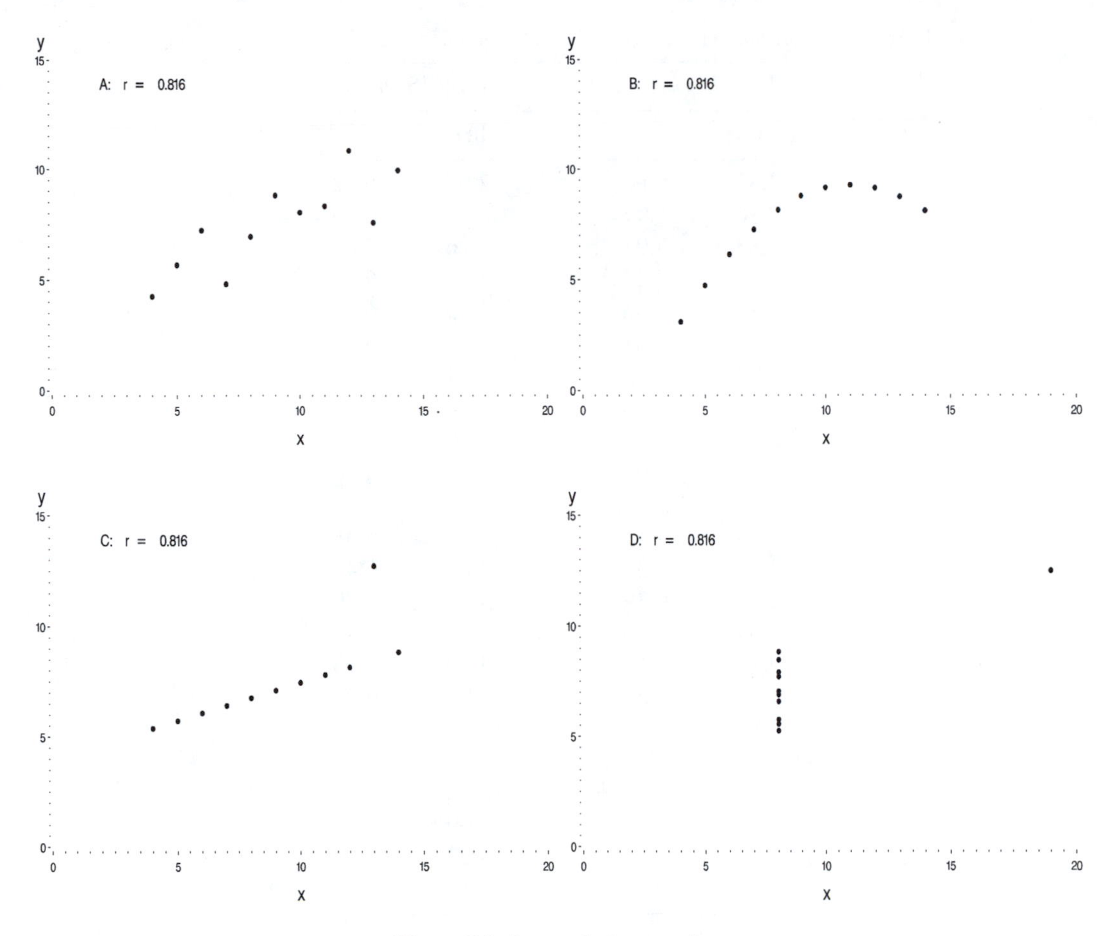

Figure 6.1 *Anscombe's examples.*

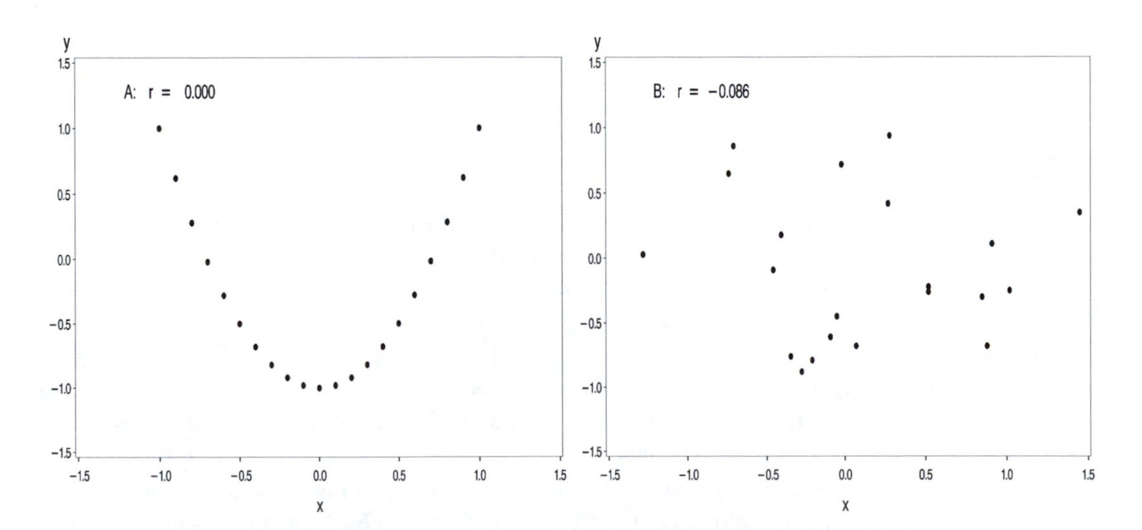

Figure 6.2 *The dataset in plot A has $r = 0$ and the dataset in plot B has $r = -0.086$.*

Table 6.1 *Natural logarithm of the length in mm* (L) *and the weight in mg* (W) *for young toads*

24/7-75 ($n_1 = 18$)		16/8-75 ($n_2 = 23$)	
$\ln L$	$\ln W$	$\ln L$	$\ln W$
2.83	5.86	2.97	6.22
2.86	5.97	3.00	6.23
2.83	5.97	3.26	7.03
2.71	5.47	3.07	6.50
2.89	5.96	3.09	6.68
2.67	5.31	3.14	6.65
2.74	5.72	3.14	6.84
2.71	5.50	3.14	6.55
2.80	5.79	3.00	6.27
2.83	5.93	3.09	6.60
2.77	5.65	3.11	6.59
2.74	5.59	3.00	6.38
2.86	5.96	3.04	6.44
2.67	5.39	3.20	6.79
2.89	6.06	3.04	6.48
2.92	6.13	3.20	6.93
2.71	5.50	3.00	6.23
3.00	6.27	3.16	6.82
		3.07	6.58
		3.00	6.36
		3.09	6.64
		3.07	6.64
		3.11	6.68

6.3 Examples

Example 6.1
The data in this example are from Keiding (1976).

On 24 July and 16 August, 18 and 23 young specimens of the European common toad, *Bufo bufo,* were caught in the same area. The data were obtained in order to describe the size distribution of young toads at the times when the two samples were taken and in order to describe the relationship between the size distributions of the toads at the two dates.

Here size refers to both length and weight, and we shall describe the association between ln length and ln weight in both samples as well as the change in means of the two variables between the two samples. Later, we shall return to this example and explain why we consider the logarithms of length and weight rather than just length and weight.

Table 6.1 contains the natural logarithm of the length L (in mm) and the natural logarithm of the weight W (in mg). The data are plotted in Figure 6.3 together with 50% and 90% contour ellipses of bivariate normal distributions fitted to the samples. The contour ellipses will be explained in the following section. □

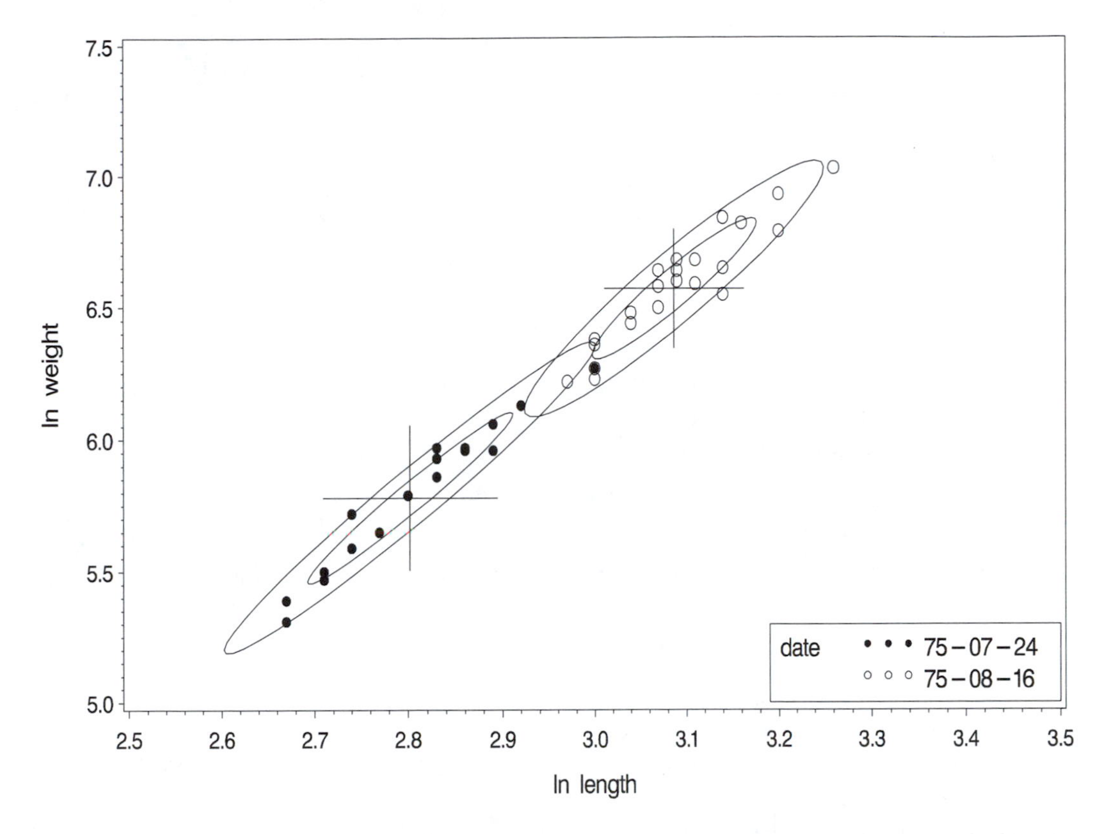

Figure 6.3 *Scatter plot of the data and 50% and 90% contour curves for the density functions of a bivariate normal distribution with parameters equal to the estimated values from the two samples in Table 6.1. The crosses are centered at the estimated means and they are formed by lines with lengths equal to two estimated standard deviations.*

6.4 The Bivariate Normal Distribution

In this section we will list a number of the basic properties of the bivariate normal distribution that are absolutely essential for understanding correlation and for using the bivariate normal distribution.

6.4.1 Definition and basic properties

Suppose X is normally distributed $N(\mu_x, \sigma_x^2)$ and Y is normally distributed $N(\mu_y, \sigma_y^2)$. If X and Y are independent, the joint density function, f, of (X,Y) is the product of the marginal density functions,

$$f(x,y) = \frac{1}{2\pi\sigma_x\sigma_y} e^{-\frac{1}{2}\left\{\left(\frac{x-\mu_x}{\sigma_x}\right)^2 + \left(\frac{y-\mu_y}{\sigma_y}\right)^2\right\}}, \quad (x,y) \in \mathbb{R}^2. \tag{6.10}$$

The joint distribution of X and Y is a bivariate normal distribution with parameters,

$$\begin{pmatrix} \mu_x \\ \mu_y \end{pmatrix}, \left\{ \begin{array}{cc} \sigma_x^2 & 0 \\ 0 & \sigma_y^2 \end{array} \right\},$$

where we use the convention to quote first the vector of means of the distribution and then the matrix of variances and covariances. The matrix of variances and covariances is referred to as the variance-covariance matrix.

If X and Y are correlated with correlation coefficient ρ, (X,Y) is bivariate normally distributed if (X,Y) has joint density function,

$$f(x,y) = \frac{1}{2\pi\sigma_x\sigma_y\sqrt{1-\rho^2}} e^{-\frac{1}{2(1-\rho^2)}\left\{\left(\frac{x-\mu_x}{\sigma_x}\right)^2 - 2\rho\frac{(x-\mu_x)}{\sigma_x}\frac{(y-\mu_y)}{\sigma_y} + \left(\frac{y-\mu_y}{\sigma_y}\right)^2\right\}}, \tag{6.11}$$

which is written as

$$\begin{pmatrix} X \\ Y \end{pmatrix} \sim N_2\left(\begin{pmatrix} \mu_x \\ \mu_y \end{pmatrix}, \begin{Bmatrix} \sigma_x^2 & \rho\sigma_x\sigma_y \\ \rho\sigma_x\sigma_y & \sigma_y^2 \end{Bmatrix}\right). \tag{6.12}$$

Notice that x and y enter symmetrically in the expression for the density function (6.11).

If $\rho = 0$, the expression in (6.11) simplifies to (6.10) and X and Y are independent.

The contour curves of the density function of the bivariate normal looks like ellipses in Figure 6.3 and in Figure 6.4 and indeed they are. Therefore the contour curves of the bivariate normal distribution are referred to as *contour ellipses*. A contour curve of a bivariate density function f is the set of points that satisfy the equation $f(x,y) = k$. For the bivariate normal this equation is equivalent to an equation,

$$\frac{1}{1-\rho^2}\left\{(\frac{x-\mu_x}{\sigma_x})^2 - 2\rho\frac{x-\mu_x}{\sigma_x}\frac{y-\mu_y}{\sigma_y} + (\frac{y-\mu_y}{\sigma_y})^2\right\} = c^2, \tag{6.13}$$

which is the equation of an ellipse. For $\rho = 0$ this equation simplifies to the equation,

$$(\frac{x-\mu_x}{\sigma_x})^2 + (\frac{y-\mu_y}{\sigma_y})^2 = c^2, \tag{6.14}$$

and the set of point satisfying this equation is the ellipse with center in (μ_x,μ_y) and semiaxes of length $(c\sigma_x, c\sigma_y)$ that are parallel to the coordinate axes.

For general $\rho \in]-1,1[$ the contour curves, i.e., the set of points (x,y) that satisfy (6.13) is still an ellipse with center in (μ_x,μ_y) but for $\rho \neq 0$ the axes are not parallel to the coordinate axes. Moreover, the lengths of the semiaxes depend on the standard deviations as well as the value of ρ.

A useful rewriting of (6.11) is

$$f(x,y) = \frac{1}{\sqrt{2\pi}\sigma_x} e^{-\frac{1}{2\sigma_x^2}(x-\mu_x)^2} \frac{1}{\sqrt{2\pi}\sigma_y\sqrt{1-\rho^2}} e^{-\frac{1}{2\sigma_y^2(1-\rho^2)}\left(y-\mu_y-\frac{\rho\sigma_y}{\sigma_x}(x-\mu_x)\right)^2}, \tag{6.15}$$

which shows the joint density function as the product of the density function of the $N(\mu_x, \sigma_x^2)$ distribution and the density function of the $N(\mu_y + \rho\frac{\sigma_y}{\sigma_x}(x-\mu_x), \sigma_y^2(1-\rho^2))$ distribution.

The factorization (6.15) has several applications. It can be used to show that when X and Y are bivariate normally distributed as in (6.12), then

$$E((X-EX)(Y-EY)) = E((X-\mu_x)(Y-\mu_y)) = \rho\sigma_x\sigma_y,$$

and consequently that

$$Cor(X,Y) = \rho.$$

Thus the parameter ρ of the bivariate normal distribution in (6.12) is the correlation.

Moreover, (6.15) can be used to show that if X and Y has the bivariate normal distribution in (6.12), then the marginal distribution of X is the $N(\mu_x, \sigma_x^2)$ distribution and the conditional distribution of Y given $X = x$ is the $N(\mu_y + \rho\frac{\sigma_y}{\sigma_x}(x-\mu_x), \sigma_y^2(1-\rho^2))$ distribution. Thus the factorization (6.15) is the factorization $f(x,y) = f(x)f(y|x)$ of the joint density function of (X,Y) as the product of the marginal density function of X and the conditional density function of Y given $X = x$.

Due to the symmetry of X and Y in the bivariate normal distribution, we have the analogous results that the marginal distribution of Y is the $N(\mu_y, \sigma_y^2)$ distribution and the conditional distribution of X given $Y = y$ is the $N(\mu_x + \rho\frac{\sigma_x}{\sigma_y}(y-\mu_y), \sigma_x^2(1-\rho^2))$ distribution.

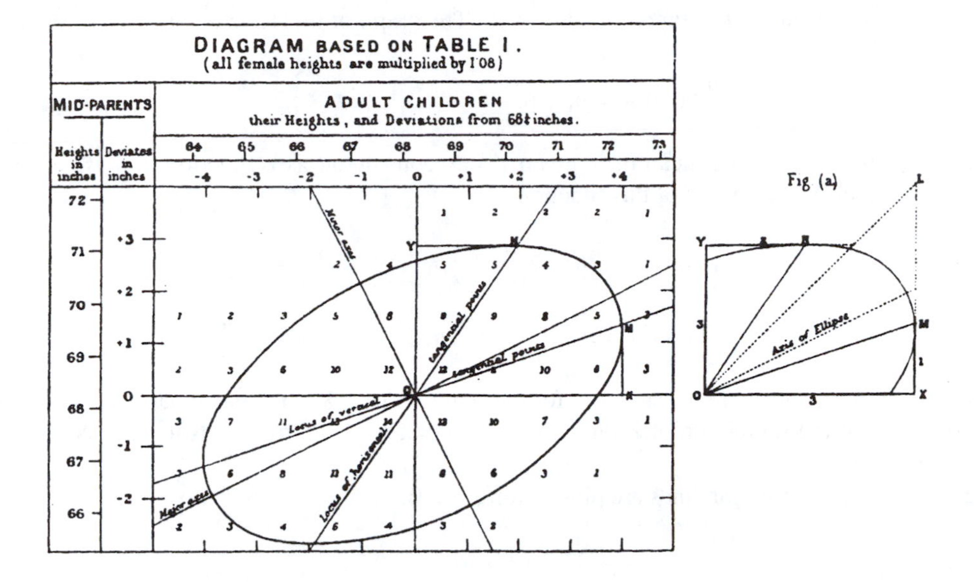

Figure 6.4 *Contour ellipse with main axes and regression lines. The regression of Y on X is the straight line through the two points where the contour ellipse has a vertical tangent and the regression of X on Y is the straight line through the two points where the contour ellipse has a horizontal tangent.*

Because these results are so important in understanding the bivariate normal distribution, they will be displayed on separate lines for easy reference. If

$$\begin{pmatrix} X \\ Y \end{pmatrix} \sim N_2\left(\begin{pmatrix} \mu_x \\ \mu_y \end{pmatrix}, \begin{Bmatrix} \sigma_x^2 & \rho\sigma_x\sigma_y \\ \rho\sigma_x\sigma_y & \sigma_y^2 \end{Bmatrix}\right),$$

then

$$X \sim N(\mu_x, \sigma_x^2), \tag{6.16}$$

$$Y \mid X = x \sim N(\mu_y + \rho\frac{\sigma_y}{\sigma_x}(x - \mu_x), \sigma_y^2(1 - \rho^2)), \tag{6.17}$$

$$Y \sim N(\mu_y, \sigma_y^2), \tag{6.18}$$

$$X \mid Y = y \sim N(\mu_x + \rho\frac{\sigma_x}{\sigma_y}(y - \mu_y), \sigma_x^2(1 - \rho^2)), \tag{6.19}$$

and, moreover,

$$\frac{1}{1-\rho^2}\left\{(\frac{X-\mu_x}{\sigma_x})^2 - 2\rho\frac{X-\mu_x}{\sigma_x}\frac{Y-\mu_y}{\sigma_y} + (\frac{Y-\mu_y}{\sigma_y})^2\right\} \sim \chi^2(2). \tag{6.20}$$

The last result shows that the distribution of the probability mass inside the contour ellipses is determined by the $\chi^2(2)$ distribution.

Figure 6.4 is borrowed from a classical paper from 1885 by Francis Galton with the title, *Regression toward Mediocrity in Hereditary Stature*.

Even if Francis Galton used the word "regression" in that paper in the original sense of returning to a previous state it is nevertheless due to its appearance in this paper that the word regression is now used for a host of techniques where the variation of one variable is being explained by the variation of one or more variables.

The conditional mean of Y given $X = x$ as a function of x, i.e., the function that maps x into

$\mu_y + \rho \frac{\sigma_y}{\sigma_x}(x - \mu_x)$ is called the regression of Y on X. The graph of this function is

$$\left\{ (x,y) \,|\, y = \mu_y + \rho \frac{\sigma_y}{\sigma_x}(x - \mu_x), \quad x \in \mathbb{R} \right\}.$$

Analogously, the conditional mean of X given $Y = y$ considered as a function of y is called the regression of X on Y. The graph of this function is

$$\left\{ (x,y) \,|\, x = \mu_x + \rho \frac{\sigma_x}{\sigma_y}(y - \mu_y), \quad y \in \mathbb{R} \right\} =$$
$$\left\{ (x,y) \,|\, y = \mu_y + \frac{\sigma_y}{\rho \sigma_x}(x - \mu_x), \quad x \in \mathbb{R} \right\}.$$

For $\rho > 0$ we note that the regression line for the regression of X on Y has greater slope than the regression line of the regression of Y on X. Note also that $\rho = 0$ corresponds to zero slope of the regression of Y on X.

In Figure 6.4 the regression lines are plotted for a $\rho > 0$.

6.4.2 Estimation

Let $(x_1, y_1), \ldots, (x_n, y_n)$, with $n \geq 3$, be a sample from the bivariate normal distribution. Thus (x_i, y_i), $i = 1, \ldots, n$, are realizations of random variables (X_i, Y_i), $i = 1, \ldots n$, which are independent and have the distribution,

$$N_2\left(\left(\begin{array}{c} \mu_x \\ \mu_y \end{array} \right), \left\{ \begin{array}{cc} \sigma_x^2 & \rho\sigma_x\sigma_y \\ \rho\sigma_x\sigma_y & \sigma_y^2 \end{array} \right\} \right). \tag{6.21}$$

The estimates of the five parameters of the bivariate normal distribution that are used are

$$\mu_x \leftarrow \bar{x}. = \frac{1}{n}\sum_{i=1}^{n} x_i,$$

$$\sigma_x^2 \leftarrow s_{\mathbf{x}}^2 = \frac{1}{n-1}\sum_{i=1}^{n}(x_i - \bar{x}.)^2 = \frac{1}{n-1} SSD_{\mathbf{x}},$$

$$\mu_y \leftarrow \bar{y}. = \frac{1}{n}\sum_{i=1}^{n} y_i,$$

$$\sigma_y^2 \leftarrow s_{\mathbf{y}}^2 = \frac{1}{n-1}\sum_{i=1}^{n}(y_i - \bar{y}.)^2 = \frac{1}{n-1} SSD_{\mathbf{y}},$$

$$\rho \leftarrow r = r_{\mathbf{xy}} = \frac{SPD_{\mathbf{xy}}}{\sqrt{SSD_{\mathbf{x}} SSD_{\mathbf{y}}}},$$

where *SPD* just as in linear regression denotes the *Sum of Products of Deviations* and is given by the formula,

$$SPD_{\mathbf{xy}} = \sum_{i=1}^{n}(x_i - \bar{x}.)(y_i - \bar{y}.).$$

It is beyond the scope of this book to give the details of the derivation of the estimates, but we remark that the estimates are maximum likelihood estimates except for the estimates of the variances, which have been rescaled to have the right means. It should also be noted that when $(x_1, y_1), \ldots, (x_n, y_n)$ is a sample from the bivariate normal distribution in (6.11), then $x_1, \ldots, x_n$ is a sample from the $N(\mu_x, \sigma_x^2)$ distribution and the estimates of μ_x and σ_x^2 are the usual estimates of the mean and variance based on a single normal sample. Similarly, $y_1, \ldots, y_n$ is a sample from the $N(\mu_y, \sigma_y^2)$ distribution and the same remark applies to the estimates of μ_y and σ_y^2. In this light the estimates of means and variances are the natural ones. We also note that the correlation coefficient ρ is estimated by the empirical correlation coefficient r in (6.7).

We use bold-face suffixes $\mathbf{x}$ and $\mathbf{y}$ on estimated variances $s_{\mathbf{x}}^2$, $s_{\mathbf{y}}^2$, on sums of squares of deviations $SSD_{\mathbf{x}}$, $SSD_{\mathbf{y}}$, on $SPD_{\mathbf{xy}}$, and on $r_{\mathbf{xy}}$ to emphasize that the quantities are functions of the sample values $\mathbf{x} = (x_1, \ldots, x_n)$ or $\mathbf{y} = (y_1, \ldots, y_n)$ or both. When we wish to emphasize that we consider their properties as random variables, i.e., as functions of the random variables of the sample, we will use bold-face capital letters $\mathbf{X}$ and $\mathbf{Y}$ as suffixes.

6.4.3 *The distribution of the estimates*

The estimates of means and variances are the usual estimates based either on the normal sample, $x_1, \ldots, x_n$, or the normal sample, $y_1, \ldots, y_n$ and, therefore the marginal distributions of the estimates are the well known:

$$\mu_x \leftarrow \bar{x}_{\bullet} \sim\sim N(\mu_x, \frac{\sigma_x^2}{n}),$$

$$\sigma_x^2 \leftarrow s_{\mathbf{x}}^2 \sim\sim \frac{\sigma_x^2}{n-1}\chi^2(n-1),$$

$$\mu_y \leftarrow \bar{y}_{\bullet} \sim\sim N(\mu_y, \frac{\sigma_y^2}{n}),$$

$$\sigma_y^2 \leftarrow s_{\mathbf{y}}^2 \sim\sim \frac{\sigma_y^2}{n-1}\chi^2(n-1).$$

The distribution of the estimate r of ρ is more complicated and so is the joint distribution of the five estimates. We will only give the exact distribution of r when $\rho = 0$.

Here we consider the empirical correlation coefficient as a function of the random variables of the sample so as mentioned before we use bold-face capital letters $\mathbf{X}$ and $\mathbf{Y}$ as suffixes and write the estimated correlation coefficient as

$$r_{\mathbf{XY}} = \frac{SPD_{\mathbf{XY}}}{\sqrt{SSD_{\mathbf{X}} SSD_{\mathbf{Y}}}}.$$

As a function of random variables, it is itself a random variable whose distribution we are going to describe.

Distribution of $r_{\mathbf{XY}}$ for $\rho = 0$

We shall not give the distribution of $r_{\mathbf{XY}}$ itself, but of a monotonically increasing function of $r_{\mathbf{XY}}$, which has a $t(n-2)$ distribution. The result can be stated as

$$T = T(r_{\mathbf{XY}}) = \sqrt{n-2}\frac{r_{\mathbf{XY}}}{\sqrt{1 - r_{\mathbf{XY}}^2}} \sim\sim t(n-2). \tag{6.22}$$

The symmetry of the t-distribution around 0 and the antisymmetry $(T(-r_{\mathbf{XY}}) = -T(r_{\mathbf{XY}}))$ of the T transformation shows that the distribution of $r_{\mathbf{XY}}$ for $\rho = 0$ is symmetric around 0. It is beyond the scope of this book to prove a result like (6.22), but it is useful to realize that

$$T = \sqrt{n-2}\frac{r_{\mathbf{XY}}}{\sqrt{1-r_{\mathbf{XY}}^2}} \tag{6.23}$$
$$= \sqrt{n-2}\frac{SPD_{\mathbf{XY}}}{\sqrt{(SSD_{\mathbf{X}}SSD_{\mathbf{Y}} - SPD_{\mathbf{XY}}^2)}}$$
$$= \frac{SPD_{\mathbf{XY}}/\sqrt{SSD_{\mathbf{X}}}}{\sqrt{(SSD_{\mathbf{Y}} - SPD_{\mathbf{XY}}^2/SSD_{\mathbf{X}})/(n-2)}}.$$

The last expression is the same function of **Y** and **X** as the t-test statistic for the hypothesis that the slope is 0 in the regression of **Y** on **X**, and in the regression model the distribution is also $t(n-2)$. The explanation of this result can be found in the conditional distribution of **Y** given $\mathbf{X} = \mathbf{x}$. It follows from (6.17) that conditional on the observed value **x** of **X**, the model for **Y** is a linear regression model, so we know that

$$T(r_{\mathbf{xY}}) = \frac{SPD_{\mathbf{xY}}/\sqrt{SSD_{\mathbf{x}}}}{\sqrt{(SSD_{\mathbf{Y}} - SPD_{\mathbf{xY}}^2/SSD_{\mathbf{x}})/(n-2)}}$$

has a $t(n-2)$ distribution. It is a result in probability theory that because the conditional distribution of $T(r_{\mathbf{xY}})$ does not depend on **x**, then the unconditional distribution of $T(r_{\mathbf{XY}})$ has the same $t(n-2)$ distribution.

The primary application of the result in (6.22) is to test the hypothesis of zero correlation, i.e., $\rho = 0$. This test is considered in Section 6.6.

Distribution of $r_{\mathbf{XY}}$ for $\rho \neq 0$

The distribution of $r_{\mathbf{XY}}$ from a sample of n independent observations from a bivariate normal distribution depends in a complicated way on ρ. The distribution was derived by R.A. Fisher (1915). For $\rho = 0$ the distribution is symmetric around 0 as was noted below (6.22) but for $\rho \neq 0$ the distribution becomes increasingly asymmetric as the numerical value $|\rho|$ of the correlation coefficient ρ increases. The density of $r_{\mathbf{XY}}$ for $n = 8$ is shown in Figure 6.5 for $\rho = 0$ and for $\rho = 0.8$.

For most practical purposes we do not need the distribution of $r_{\mathbf{XY}}$ because there is a monotone transformation of $r_{\mathbf{XY}}$, which is approximately normally distributed with a mean that to a good approximation already for small n depends on ρ only and a variance that depends on the sample size only. This transformation was given by R.A. Fisher (1915) and is called Fisher's Z, which is

$$Z = \frac{1}{2}\ln\frac{1+r_{\mathbf{XY}}}{1-r_{\mathbf{XY}}} = \tanh^{-1}(r_{\mathbf{XY}}). \tag{6.24}$$

The density of Z is shown in Figure 6.5 for $n = 8$ and for $\rho = 0$ and for $\rho = 0.8$. Letting ζ denote the similar transformation of ρ,

$$\zeta = \frac{1}{2}\ln\frac{1+\rho}{1-\rho}, \tag{6.25}$$

an accurate and much used approximation is

$$(Z-\zeta)\sqrt{n-3} \approx N(0,1). \tag{6.26}$$

Another way of expressing the approximation is to say that Z is approximately normally distributed with mean ζ and a variance of $1/(n-3)$,

$$Z \approx N(\zeta, \frac{1}{n-3}). \tag{6.27}$$

This approximation can be applied for sample sizes as small as 10.

A slightly more accurate approximation than (6.26) is

$$(Z-\zeta-\frac{\rho}{2(n-1)})\sqrt{n-3}\approx N(0,1). \tag{6.28}$$

For moderate values of n the term $\rho/(2(n-1))$ is small in comparison to the standard deviation $1/\sqrt{n-3}$ and therefore that term is often ignored and one simply uses the approximation (6.26).

The results in (6.26) and (6.27) are used to test hypotheses H: $\rho=\rho_0$ or rather to construct confidence intervals of ρ. This is explained in Section 6.6.

6.5 Model Checking

The most important features of the bivariate normal distribution are listed in equations (6.16) to (6.20), and they are the basis for checking whether a sample from a bivariate distribution can be assumed to be a sample from a bivariate normal distribution.

First we check if the marginal distributions can be assumed to be normal using fractile plots. If this is not the case, we try to find transformations of the variables such that the transformed variables are normal.

Once it is decided that the marginal distributions can be assumed to be normal, the joint distribution of the variables is investigated. The starting point is always a scatter plot of the data to see if the scatter of points has the shape of an ellipse. A more accurate investigation is based on the result in (6.20). If

$$\binom{X}{Y}\sim N_2\left(\binom{\mu_x}{\mu_y},\left\{\begin{array}{cc}\sigma_x^2 & \rho\sigma_x\sigma_y\\ \rho\sigma_x\sigma_y & \sigma_y^2\end{array}\right\}\right),$$

the quantity,

$$W=\frac{1}{1-\rho^2}\left[(\frac{X-\mu_x}{\sigma_x})^2-2\rho\frac{X-\mu_x}{\sigma_x}\frac{Y-\mu_y}{\sigma_y}+(\frac{Y-\mu_y}{\sigma_y})^2\right], \tag{6.29}$$

is χ^2 distributed with 2 degrees of freedom. If the parameters were known, (6.29) could be calculated for each pair (x_i,y_i) of sample values, thus obtaining n independent observations from the $\chi^2(2)$ distribution. In applications of this result the parameters are unknown and will have to be replaced by their estimates, i.e., for $i=1,\ldots,n$ one calculates

$$w_i=\frac{1}{1-r^2}\left[\left(\frac{x_i-\bar{x}.}{s_x}\right)^2-2r\frac{x_i-\bar{x}.}{s_x}\frac{y_i-\bar{y}.}{s_y}+\left(\frac{y_i-\bar{y}.}{s_y}\right)^2\right]. \tag{6.30}$$

The empirical distribution of $w_1,\ldots,w_n$ may be compared to the distribution function of the $\chi^2(2)$ with a fractile plot. This is particularly easy because the $\chi^2(2)$ distribution has distribution function $F(x)=1-e^{-\frac{1}{2}x}$ for $x>0$, so the fractiles can be found simply solving the equation,

$$p=F(x_p)=1-e^{-\frac{1}{2}x_p},$$

for x_p giving

$$x_p=-2\ln(1-p).$$

An example of a fractile plot is shown for ln weight and ln length in the July sample from Example 6.1 in Figure 6.8 on page 292. The bottom row of Figure 6.6 contains the fractile plots for both samples of Example 6.1 and for both length and weight as well as for the logarithms of those variables.

If the number n of observations is large, this investigation may be supplemented by an investigation of the conditional distributions given X (6.17) and given Y (6.19). The conditional distributions must be normal. Moreover, it may be checked that the conditional distributions given X all have the same variance and similarly for the conditional distributions given Y. Finally, the linearity of the regression lines may be checked using a regression analysis.

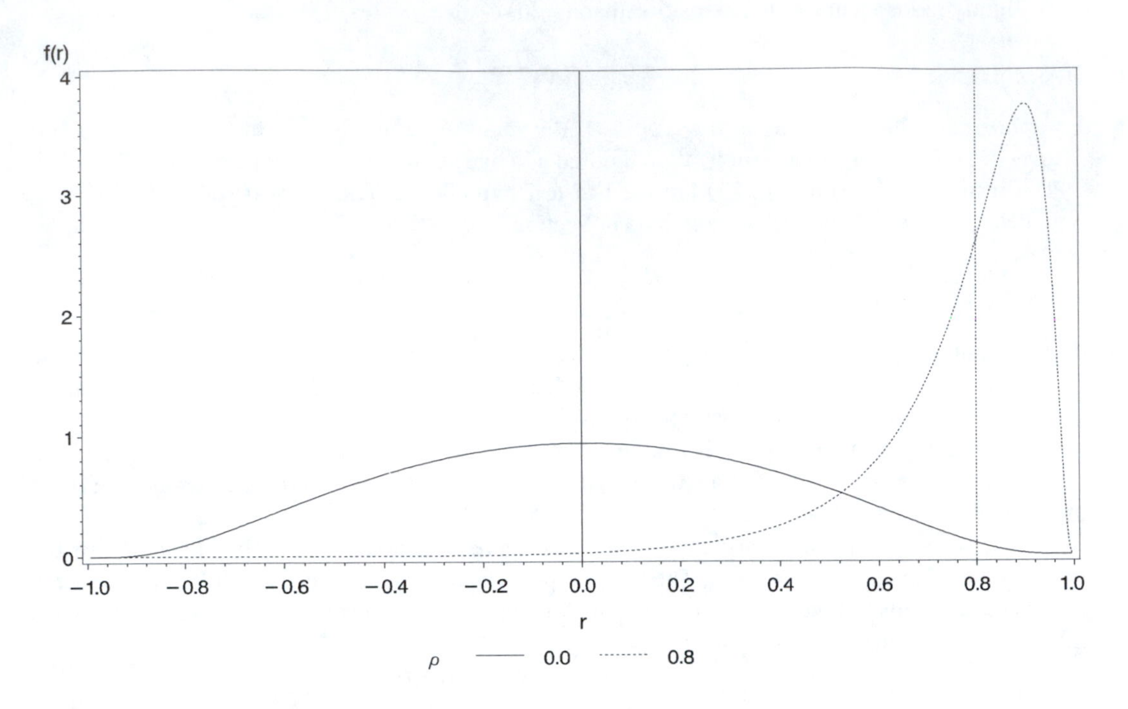

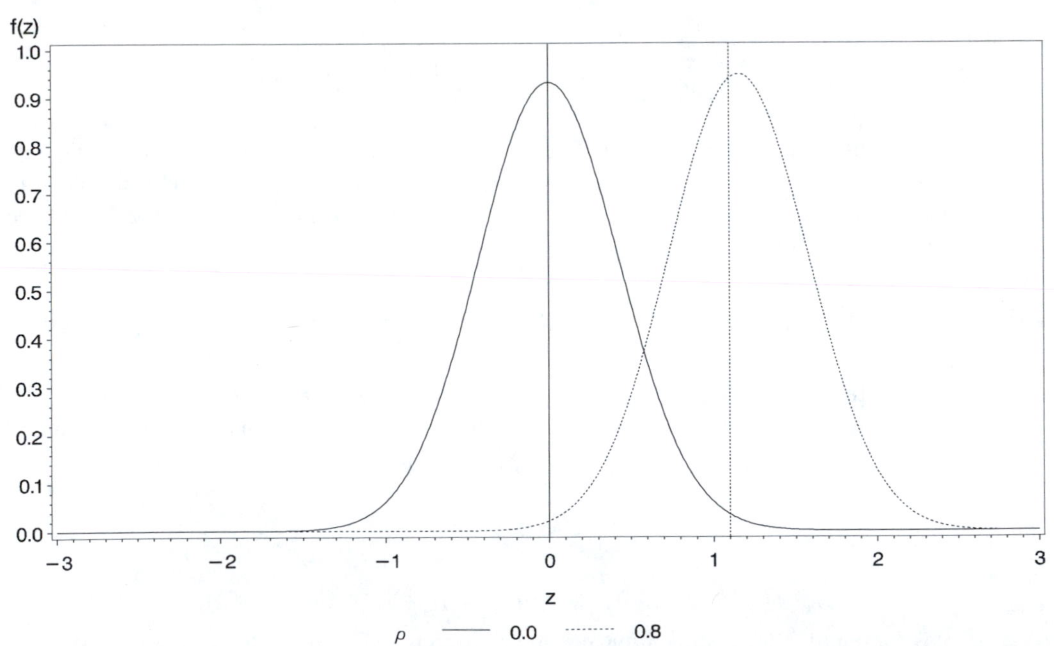

Figure 6.5 *Density functions for r (top) and Z (bottom) shown for* $n = 8$, $\rho = 0$, *and* $\rho = 0.8$

Checking for normality of marginal and conditional distributions using fractile plots has been discussed in Chapter 2, testing for homogeneity of variances using Bartlett's test has been explained in Section 3.2.2 on page 101, and testing linearity of regression lines has been treated in Section 3.3.8. The only new aspect is checking for bivariate normality using (6.29). How to do this using SAS is explained in Section 6.10.2.

Example 6.1 (Continued)
Figure 6.6 shows the plots that are considered when checking for bivariate normality of the distribution of length and weight in each of the two samples and also for bivariate normality of the distribution of the natural logarithms of length and weight. The top plots are the scatterplots with the regression lines of weight (ln weight) on length (ln length). Both plots are consistent with the assumption of two bivariate normal distributions, but for the original measurements the regression lines for the two samples have different slopes.

The four plots in the middle show the fractile plots for the marginal distributions. None of the plots contradict the assumption of normal distributions of the original as well as the transformed variables. Noting that the largest observation in each sample is less than 3 times the smallest observation in the sample, this is to be expected. Observe, however, that for the weight, the slope of the normal fractile plot for the August sample is lower than for the July sample, indicating a larger variance of weight for the August sample.

Finally, the two plots at the bottom show the fractile plots for bivariate normality based on (6.29). None of the plots show a marked devation from the identity line, so these plots confirm the first-hand impression from the scatter plots, that both the original and the transformed data are consistent with the assumption of bivariate normal distributions for the two samples.

We choose to work with the logarithms of the measurements for a number of reasons. First, for larger data sets it is often seen that a logarithmic transformation is needed to obtain a normal distribution. This applies to measurements of length and weight for a wide range of species. Secondly, for the original data the variance of weight of the August sample is significantly larger than the variance of weight in the July sample. Finally, the description of the growth of the toad is much simpler in terms of the logarithm of length and the logarithm of weight than in terms of the untransformed variables. This will be illustrated when we analyze the means of the samples in Example 6.1 starting on page 282.

Using x to denote the logarithm of length and y to denote the logarithm of weight we can formulate the model for the two samples as follows. For the July sample, $(x_{11},y_{11}),\ldots,(x_{1n_1},y_{1n_1})$, are observations of independent, identically distributed random variables with a bivariate normal distribution,

$$N_2\left(\begin{pmatrix}\mu_{1x}\\ \mu_{1y}\end{pmatrix},\left\{\begin{array}{cc}\sigma_{1x}^2 & \rho_1\sigma_{1x}\sigma_{1y}\\ \rho_1\sigma_{1x}\sigma_{1y} & \sigma_{1y}^2\end{array}\right\}\right), \tag{6.31}$$

and for the August sample, $(x_{21},y_{21}),\ldots,(x_{2n_2},y_{2n_2})$, are observations of independent identically distributed random variables with a bivariate normal distribution,

$$N_2\left(\begin{pmatrix}\mu_{2x}\\ \mu_{2y}\end{pmatrix}\left\{\begin{array}{cc}\sigma_{2x}^2 & \rho_2\sigma_{2x}\sigma_{2y}\\ \rho_2\sigma_{2x}\sigma_{2y} & \sigma_{2y}^2\end{array}\right\}\right). \tag{6.32}$$

Furthermore the two samples are independent.

The estimates of means, variances, and correlation coefficients of the two samples are are given below.
July sample:

$$\begin{array}{llll}\bar{x}_{1\bullet} & = 2.80 & \bar{y}_{1\bullet} & = 5.78\\ s_{(1)x}^2 & = 0.00863 & s_{(1)y}^2 & = 0.0758\end{array}$$
$$r_1 = 0.978$$

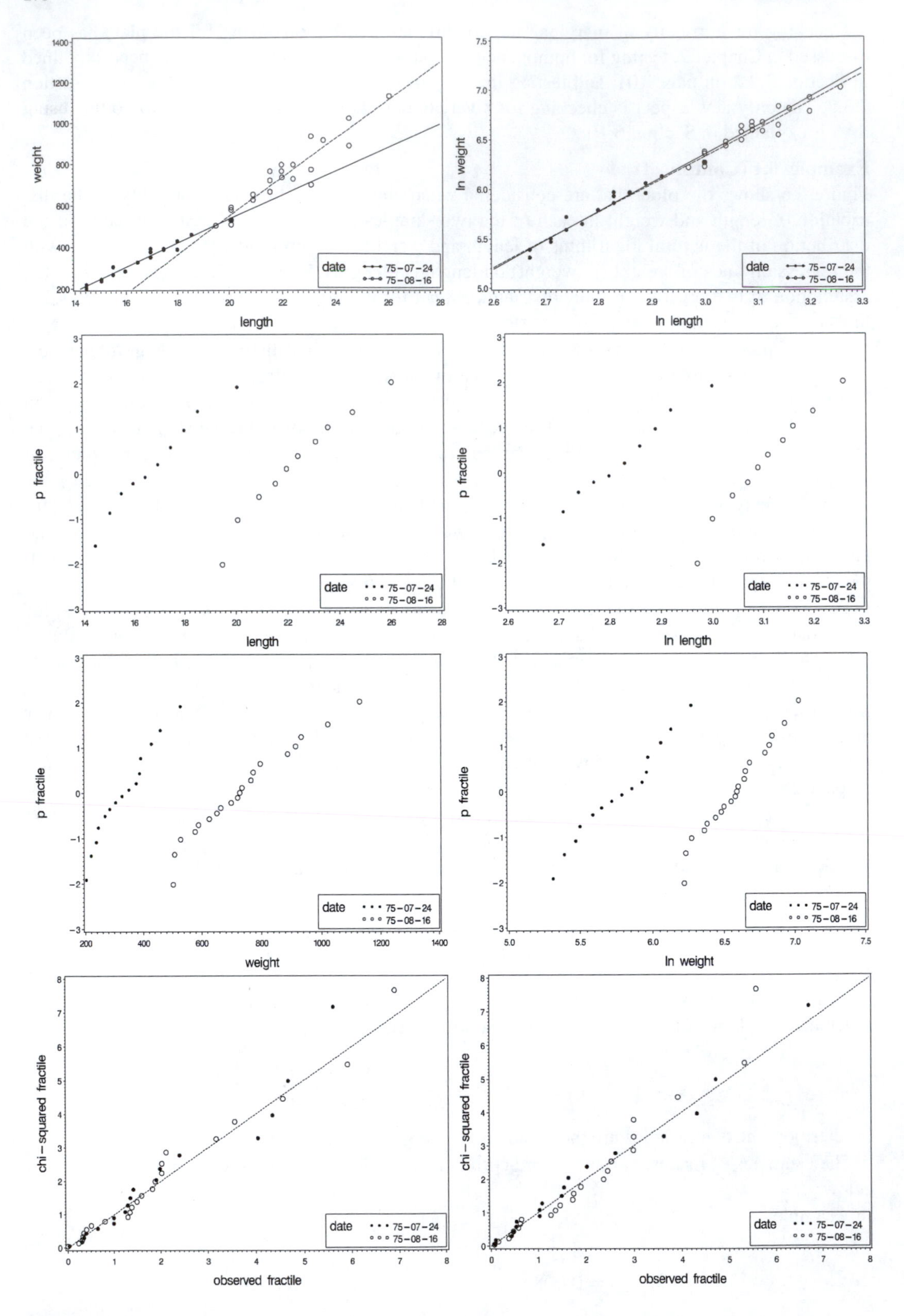

Figure 6.6 *Various checks of the assumption of a bivariate normal distribution of length and weight (to the left) and of ln length and ln weight (to the right).*

August sample:

$$\bar{x}_{2\bullet} = 3.09 \qquad \bar{y}_{2\bullet} = 6.57$$
$$s^2_{(2)x} = 0.00578 \qquad s^2_{(2)y} = 0.0509$$
$$r_2 = 0.942$$

Here we adopt the notation that was introduced on page 80 and use parentheses around the sample indices on the estimates of the variances. □

6.6 Inference on ρ Based on a Single Bivariate Normal Sample

6.6.1 *Test of the hypothesis* $\rho = 0$

An intuitively reasonable test of the hypothesis H: $\rho = 0$ is based on the estimate $r_{\mathbf{xy}}$ of ρ. The distribution of $r_{\mathbf{XY}}$ is symmetic around 0, so the values of r that are more critical for the hypothesis than the observed value $r_{\mathbf{xy}}$ are those that are numerically greater than or equal to $r_{\mathbf{xy}}$. Therefore the significance probability is

$$\begin{aligned} p_{obs}(\mathbf{x},\mathbf{y}) &= P(|r_{\mathbf{XY}}| \geq |r_{\mathbf{xy}}|) \\ &= P(|T(r_{\mathbf{XY}})| \geq |T(r_{\mathbf{xy}})|) \\ &= 2\left(1 - F_{t(n-2)}(|T(r_{\mathbf{xy}})|)\right), \end{aligned} \tag{6.33}$$

where we have used the T-transformation in (6.22) and the fact that $T(r_{\mathbf{XY}})$ has a $t(n-2)$-distribution.

The statistics $r_{\mathbf{xy}}$ and $T(r_{\mathbf{xy}})$ are equivalent as test statistics for the hypothesis H: $\rho = 0$, but if calculations are made by hand, $T(r_{\mathbf{xy}})$ is the natural choice because the significance probability is calculated using its $t(n-2)$-distribution.

In the subsection where the T-transformation was introduced on page 271, we noted that $T(r_{\mathbf{xy}})$ was equal to the t-test statistic for the hypothesis that the slope of the regression line was equal to 0 in the regression of Y on X. The latter test statistic also has a $t(n-2)$ distribution so the significance probabilities of the two tests are identical. By symmetry, the test for zero correlation is also identical to the test for zero slope in the linear regression of X on Y.

6.6.2 *Confidence interval of* ρ

The approximating normal distribution in (6.26) can be used to test if an observered correlation coefficient r deviates significantly from a theoretical value ρ_0, or, in other words, to test the hypothesis H_0: $\rho = \rho_0$. Transforming to ζ this corresponds to the hypothesis,

$$H_0\colon \zeta = \zeta_0 = \frac{1}{2}\ln\frac{1+\rho_0}{1-\rho_0},$$

which is tested using the test statistic,

$$u = (z - \zeta_0)\sqrt{n-3}, \tag{6.34}$$

and calculating the significance probability,

$$p_{obs} = 2(1 - \Phi(u)).$$

We have used u to denote the test statistic in (6.34) in recognition of the fact that it is an example of the u statistic introduced in (3.5); in this case with only one observation z from a normal distribution with variance $1/(n-3)$.

The most important application of the test of a fixed value of the correlation coefficient is in finding confidence intervals for the correlation coeffcient. Using (6.26) a $(1-\alpha)$ confidence interval for ζ can immediately be obtained as

$$\left[z - \frac{1}{\sqrt{n-3}}u_{1-\alpha/2}, z + \frac{1}{\sqrt{n-3}}u_{1-\alpha/2}\right], \tag{6.35}$$

which is an application of (3.20) with one observation from a normal distribution with a known variance of $1/(n-3)$. Using the inverse transformation of (6.25) to the limits of this confidence interval, it will be transformed into a confidence interval for ρ. The inverse transformation of (6.25) is

$$\rho = \frac{e^{2\zeta}-1}{e^{2\zeta}+1}. \tag{6.36}$$

Example 6.1 (Continued)

We are now able to evaluate if there is a significant correlation between ln length and ln weight in the two samples and to calculate the confidence intervals for the correlation coefficients to get an idea of the precision of the estimates.

July sample:

The estimate of the correlation coefficient is $r_1 = 0.978$ and the test statistic of zero correlation is

$$\sqrt{n_1-2}\frac{r_1}{\sqrt{1-r_1^2}} = \sqrt{16}\frac{0.978}{\sqrt{1-0.978^2}} = 18.753,$$

and since $P(|T| > 18.753) < 0.001$, where T is t-distributed with 16 degrees of freedom, the hypothesis of zero correlation in the July population is rejected. The estimate of the correlation coefficient is positive, so the conclusion is that there is a positive association between ln length and ln weight in the July population.

A confidence interval for ρ_1 based on the sample can be calculated using Fisher's Z in (6.24) and the formula (6.35), which gives a confidence interval for $\zeta_1 = \frac{1}{2}\ln\frac{1+\rho_1}{1-\rho_1}$, which in turn is transformed into a confidence interval for ρ_1. Fisher's Z is

$$z_1 = \frac{1}{2}\ln\frac{1+r_1}{1-r_1} = 2.249;$$

the 95% confidence interval for ζ_1 is

$$\left[z_1 - \frac{1}{\sqrt{n_1-3}}u_{0.975}, z_1 + \frac{1}{\sqrt{n_1-3}}u_{0.975}\right] =$$

$$\left[2.249 - \frac{1}{\sqrt{15}}1.96, 2.249 + \frac{1}{\sqrt{15}}1.96\right] =$$

$$[1.743, 2.755],$$

and the 95% confidence interval for ρ_1 is obtained applying the transformation in (6.36) to the limits of this interval,

$$\left[\frac{e^{2\times1.743}-1}{e^{2\times1.743}+1}, \frac{e^{2\times2.755}-1}{e^{2\times2.755}+1}\right] = [0.941, 0.992].$$

All the calculations have been based on the estimated correlation coefficient $r_1 = 0.978$, but as mentioned on page 281 when it comes to interpreting the the size of the correlation coefficient the squared correlation coefficient is more relevant. In this case $r_1^2 = 0.956$ and we can state that 95.6% of the variation in ln length is explained by the variation in ln weight and conversely.

August sample:

The results for the August sample are very similar to those found for the July sample. The T-transformation of $r_2 = 0.942$ is 12.862 and evaluated in a t-distribution with 21 degrees of freedom, this gives a significance probability smaller than 0.001. The estimate of the correlation coefficient is positive, so the conclusion is that there is a postive association between ln length and ln weight in the August population. A 95% confidence interval for ρ_2 is $[0.866, 0.975]$. As mentioned later on page 281, the size of the correlation coefficient the squared correlation coefficient is more relevant when it comes to interpreting the size of the correlation. In this case $r_2^2 = 0.887$ and we can state that 88.7% of the variation in ln length is explained by the variation in ln weight and conversely. □

6.7 Inference on ρ Based on Several Bivariate Normal Samples

If we observed k correlation coefficients r_i, $i = 1, \dots, k$, based on independent samples of size n_i from bivariate normal distributions with correlation coefficients ρ_i, we may use (6.26) to test the hypothesis,

$$H\colon \rho_1 = \cdots = \rho_k = \rho,$$

that the correlation coefficients are identical. The test statistic is (see Exercise 6.3 for the derivation)

$$X^2 = \sum_{i=1}^{k} (z_i - \bar{z})^2 (n_i - 3), \tag{6.37}$$

where

$$\bar{z} = \frac{\sum_{i=1}^{k} z_i (n_i - 3)}{\sum_{i=1}^{k} (n_i - 3)}. \tag{6.38}$$

For accurate calculation by hand a more useful expression of X^2 is

$$X^2 = \sum_{i=1}^{k} z_i^2 (n_i - 3) - \frac{\left(\sum_{i=1}^{k} z_i (n_i - 3) \right)^2}{\sum_{i=1}^{k} (n_i - 3)}. \tag{6.39}$$

Under the hypothesis of a common correlation coefficient, X^2 is approximately χ^2 distributed with $k-1$ degrees of freedom, and large values of X^2 are critical for the hypothesis H, so the significance probability is

$$p_{obs} = 1 - F_{\chi^2(k-1)}(X^2).$$

If H is not rejected, $\bar{z}$ given by (6.38) is used as an estimate of ζ, and an estimate of ρ is obtained using the transformation (6.36) with ζ replaced by $\bar{z}$. The approximate distribution of $\bar{z}$ is

$$\bar{z} \sim\approx N(\zeta, 1 / \sum_{i=1}^{k} (n_i - 3)),$$

and therefore a $(1 - \alpha)$ confidence interval for ζ can be obtained as

$$\left[\bar{z} - \frac{1}{\sqrt{\sum_{i=1}^{k} (n_i - 3)}} u_{1-\alpha/2},\ \bar{z} + \frac{1}{\sqrt{\sum_{i=1}^{k} (n_i - 3)}} u_{1-\alpha/2} \right], \tag{6.40}$$

and it can in turn be transformed to a $(1 - \alpha)$ confidence interval for ρ using the transformation (6.36).

If the number of samples is large, it may be important to use the more accurate approximation in (6.28) and include the term $\rho_i / (2(n_i - 1))$ in EZ_i. If we decide to do this, we must consider the model where $Z_1, \dots, Z_k$ are independent and

$$Z_i \sim N\left(\frac{1}{2} \ln \frac{1 + \rho_i}{1 - \rho_i} + \frac{\rho_i}{2(n_i - 1)}, \frac{1}{n_i - 3}\right),$$

and in this model estimate ρ and test the hypothesis H. The likelihood equation for ρ is

$$\sum_{i=1}^{k}(n_i-3)(\frac{1}{1-\rho^2}+\frac{1}{2(n_i-1)})(z_i-\frac{1}{2}\ln\frac{1+\rho}{1-\rho}-\frac{\rho}{2(n_i-1)})=0,$$

which must be solved iteratively. The likelihood ratio statistic for the hypothesis H is

$$-2\ln Q=\sum_{i=1}^{k}(n_i-3)(z_i-\frac{1}{2}\ln\frac{1+\hat{\rho}}{1-\hat{\rho}}-\frac{\hat{\rho}}{2(n_i-1)})^2,$$

where $\hat{\rho}$ denotes the maximum likelihood estimate of ρ under the hypothesis H. $-2\ln Q$ is approximately χ^2 distributed with $k-1$ degrees of freedom and, as always with likelihood ratio tests, with large values being critical for the hypothesis.

Example 6.1 (Continued)
The hypothesis of a common correlation in the July and August populations, H: $\rho_1=\rho_2=\rho$, can be tested using the test statistic X^2 in (6.37). Using the alternative expression of X^2 in (6.39),

$$X^2=2.091,$$

the significance probability is

$$p_{obs}=1-F_{\chi^2(1)}(2.091)=0.15.$$

The hypothesis H of a common correlation is not rejected. The estimate of the common correlation coefficient is obtained from the estimate $\bar{z}=1.96796$ of the common value of ζ via the transformation (6.36) as

$$\frac{e^{2\times 1.96796}-1}{e^{2\times 1.96796}+1}=0.96169\rightarrow\rho.$$

A 95% confidence interval of the common value of ζ is obtained from the formula (6.40) as

$$\left[\bar{z}-\frac{1}{\sqrt{\sum_{i=1}^{k}(n_i-3)}}u_{1-\alpha/2},\ \bar{z}+\frac{1}{\sqrt{\sum_{i=1}^{k}(n_i-3)}}u_{1-\alpha/2}\right]=$$

$$\left[1.96796-\frac{1}{\sqrt{15+20}}1.960,\ 1.96796+\frac{1}{\sqrt{15+20}}1.960\right]=$$

$$[1.6367,\ 2.2993],$$

which in turn may be transformed into a 95% confidence interval for the common value of ρ:

$$\left[\frac{e^{2\times 1.6367}-1}{e^{2\times 1.6367}+1},\frac{e^{2\times 2.2993}-1}{e^{2\times 2.2993}+1}\right]=[0.92701,\ 0.98007].$$

The estimate of ρ^2 is $0.96169^2=0.925$ with the interpretation that 92.5% of the variation in ln length is explained by the variation in ln weight and conversely. □

6.8 Correlation and Regression

When there is data consisting of a number of pairs of observations $(x_1,y_1),\ldots,(x_n,y_n)$, the correlation coefficient will give an indication of the association between the two variables *provided that the values are a sample from a bivariate normal distribution*. If we wish to predict the y from the corresponding x, then we may apply a regression analysis in the conditional distribution of Y for given X.

Very often the assumption of a bivariate normal distribution is not tenable, and a regression analysis is the only reasonable way to proceed. This is where it is very convenient that the test of zero correlation is equivalent to the test of zero slope in either of the two regession analyses (X on Y or Y on X) one may consider instead of the correlation analysis. Thus if someone reports a correlation analysis where the assumption of a bivariate normal distribution is unwarranted, perhaps because one of the variables cannot be considered to be random or because its variation has been restricted, but a regression model is reasonable, then the test of zero correlation is still informative, because it tells us whether the slope of the regression line can be considered to be different from zero.

Concerning the interpretation of the size of the correlation coefficient we notice that

$$V(Y) = \sigma_y^2 = \rho^2\sigma_y^2 + (1-\rho^2)\sigma_y^2,$$

which is a trivial decomposition of the variance of Y. But the last term is the variance of the conditional distribution of Y given $X = x$, see (6.17), and this gives rise to the interpretation that ρ^2 is the fraction of the variation in Y, which is explained by the variation in X. Thus r^2 rather than r is the relevant statistic to quote at the end of a correlation analysis.

6.9 Interpretation of Correlation

If the hypothesis $\rho = 0$ is rejected for two variables, it means that the two variables are not stochastically independent or, in other words, that they are stochastically dependent. This does not imply that the two variables are causally related in the sense that a change in one of the variables will cause a change in the other. Height and weight, for example, will be positively correlated in most samples from any human population, and this means that knowing either height or weight of an individual from the population will make it possible to predict the value of the other variable more precisely than without this information. But it does not mean that the variables are causally related such that an increase in weight of an individual will cause an increase in height of the same individual.

Another phenomenon has been called *spurious correlation*. This expression has been coined for the situation where two variables display a significant correlation, but it is unwarranted to conclude that the two variables are related. Two kinds of spurious correlations can be distinguished. The first covers the situation where it has been ignored that other variables have varied and this variation is the explanation of the significant correlation. The second kind occurs when the variables that are being correlated are manipulated variables and the correlation is due to the manipulation of the variables. Examples will follow below.

Consider first Example 6.1 to see that it may lead to a higher correlation if other variables are ignored. In this example there is in addition to ln length and ln weight, a third variable giving the date of the measurement. It is obvious from Figure 6.3 that if the information of the third variable was ignored and all observations were to be considered as one sample the we would estimate a higher correlation between ln length and ln weight than the one we have found in the two samples.

Spurious correlation is easily avoided if we know that other variables are varying and if the values of those values are available. Then the data in divided into subgroups where the other variables are constant or varies only a little. The correlation is calculated in each subgroup, and it can be tested by the method of Section 6.7 whether the correlations are significantly different in the subgroups. If not, a common estimate and a confidence interval may be given as explained in Section 6.7. This was exactly the way the analysis proceeded in Example 6.1 on page 280.

The second type of spurious correlation is the following. We observe three random variables U,V and W, where U and V are independent for fixed value of W. The technical term is that U and V are conditionally indpendent given W. Instead of analyzing all three variables we decide to manipulate the data in such a way that there are only two variables. For example, we may choose to analyze $X = U/W$ and $Y = V/W$ and they may often show a significant correlation due to the common divisor W. In this way a correlation may be introduced simply by manipulations of the data. Variables as X and Y are often called rates or index numbers or densities, so beware of spurious correlation when correlations between variables with those names are being reported. In Exercise 6.4 an artificial data set due to Neyman (1952) is presented, where U is the number of storks, V is the number of babies, and W is the number of women for a number of counties. In this data set spurious correlations are found both between the birth rate, V/W, and the density of storks, U/W, and between the number of storks U, and the birth rate, V/W.

6.10 Further Topics in Bivariate Normal Distribution

So far we have only considered the association between two variables as quantified by the correlation coefficient. Another important aspect of bivariate normal data is to make inferences about the mean of a bivariate normal distribution and to compare the means of two bivariate normal distributions based on samples from the distributions.

Before going on to describe those techniques, we will conclude the analysis of Example 6.1 using only univariate techniques, which in this example are the most appropriate ones.

Example 6.1 (Continued)
A full treatment of this example includes a description of the changes, if any, in the size distribution of toads between the two dates.

On page 275 we formulated the model for the two samples in (6.31) and in (6.32) as two independent samples from bivariate normal distributions with no restrictions on the parameters.

In the continuation of Example 6.1 on page 280, we found that we could assume that the correlation coefficients were identical and we denoted the common correlation coefficient by ρ. This is the only reduction in the model considered formally so far. It appears from the fractile plot of the transformed variables in Figure 6.6 that neither the assumption of a common variance of ln length in the two samples nor the assumption of a common variance of ln weight in the two samples will be contradicted by the data. Indeed, the hypotheses $\sigma_{1x}^2 = \sigma_{2x}^2 = \sigma_x^2$ and $\sigma_{1y}^2 = \sigma_{2y}^2 = \sigma_y^2$ are not rejected by the formal F-tests of identical variances of Section (3.2.1), see page 94. The calculations are not reproduced here.

Thus the model has been reduced to two independent samples from two bivariate normal distributions with common variance-covariance matrix. For the July sample, the bivariate normal distribution is

$$N_2\left(\begin{pmatrix} \mu_{1x} \\ \mu_{1y} \end{pmatrix}, \begin{Bmatrix} \sigma_x^2 & \rho\sigma_x\sigma_y \\ \rho\sigma_x\sigma_y & \sigma_y^2 \end{Bmatrix}\right), \tag{6.41}$$

and for the August sample the bivariate normal distribution is

$$N_2\left(\begin{pmatrix} \mu_{2x} \\ \mu_{2y} \end{pmatrix}, \begin{Bmatrix} \sigma_x^2 & \rho\sigma_x\sigma_y \\ \rho\sigma_x\sigma_y & \sigma_y^2 \end{Bmatrix}\right). \tag{6.42}$$

The estimates of the variances are the usual estimates based on two independent univariate normal samples,

$$\sigma_x^2 \leftarrow s_x^2 = 0.00702 \sim\sim \sigma_x^2\chi^2(39)/(39),$$
$$\sigma_y^2 \leftarrow s_y^2 = 0.0618 \sim\sim \sigma_y^2\chi^2(39)/(39),$$

where $39 = n_1 + n_2 - 2$.

The estimate of ρ was given on page 280 as 0.962, but we will make no use of the estimate here.

The estimates of the means are the averages that were given at the bottom of page 275 and on the top of page 277.

We can start by testing whether the distribution of ln length have the same mean in the two samples, H_{0x}: $\mu_{1x} = \mu_{2x}$. The t-test statistic is

$$t = \frac{\bar{x}_{1\cdot} - \bar{x}_{2\cdot}}{\sqrt{s_x^2\left(\frac{1}{n_1} + \frac{1}{n_2}\right)}} = -10.8,$$

and since t has a $t(39)$-distribution the significance probability is smaller than 0.001 and H_{0x} is rejected. The estimate of the mean is greater for the August sample so the toads are significantly longer in August than in July.

We could go on and test the hypothesis H_{0y}: $\mu_{1y} = \mu_{2y}$ for ln weight and this would show that the toads are significantly heavier in August than in July, but this is not really satisfactory. Due to the strong correlation between ln length and ln weight, we would still wonder whether the significant difference in the means of ln weight can be explained simply by the increase in ln length and the correlation.

The solution is to use the factorization (6.15) of the bivariate normal distribution. For the two distributions in (6.41) and in (6.42) the factoriztions give

$$x_{1i} \sim\sim N(\mu_{1x}, \sigma_x^2), \tag{6.43}$$

$$y_{1i} | X_{1i} = x_{1i} \sim\sim N(\mu_{1y} + \rho\frac{\sigma_y}{\sigma_x}(x_{1i} - \mu_{1x}), \sigma_y^2(1-\rho^2)), \tag{6.44}$$

and

$$x_{2i} \sim\sim N(\mu_{2x}, \sigma_x^2), \tag{6.45}$$

$$y_{2i} | X_{2i} = x_{2i} \sim\sim N(\mu_{2y} + \rho\frac{\sigma_y}{\sigma_x}(x_{2i} - \mu_{2x}), \sigma_y^2(1-\rho^2)). \tag{6.46}$$

Rearranging the terms in the means of the conditional distributions in (6.44) and (6.46) they can be written as

$$y_{1i} | X_{1i} = x_{1i} \sim\sim N(\alpha_1 + \beta x_{1i}, \sigma_y^2(1-\rho^2))$$

and

$$y_{2i} | X_{2i} = x_{2i} \sim\sim N(\alpha_2 + \beta x_{2i}, \sigma_y^2(1-\rho^2))$$

where

$$\beta = \rho\frac{\sigma_y}{\sigma_x},$$

$$\alpha_1 = \mu_{1y} - \rho\frac{\sigma_y}{\sigma_x}\mu_{1x} = \mu_{1y} - \beta\mu_{1x},$$

$$\alpha_2 = \mu_{2y} - \rho\frac{\sigma_y}{\sigma_x}\mu_{2x} = \mu_{2y} - \beta\mu_{2x}.$$

Note that the regression models for ln weight given ln length have the same slope because the variances and the correlation are assumed to be the same in the two samples. The top right-hand plot in Figure 6.6 shows the data with the regression lines plotted and the slopes are very close. The only parameters that may differ between the two samples are the intercept and it is very natural to test the hypothesis that they are identical, i.e., H_0: $\alpha_1 = \alpha_2$, because when formulated in terms of the means this hypothesis is

$$\mu_{1y} - \beta\mu_{1x} = \mu_{2y} - \beta\mu_{2x}$$

or

$$\mu_{1y} - \mu_{2y} = \beta(\mu_{1x} - \mu_{2x}).$$

Thus, the hypothesis states that the change in the means of ln weight between the two samples $(\mu_{1y} - \mu_{2y})$ is fully explained by the change in the means of ln length $(\mu_{1x} - \mu_{2x})$ between the two samples.

The hypothesis is tested using the techniques for comparison of regression lines in Section 4.2. The following SAS program provides the necessary calculations.

```
PROC GLM DATA=bufobufo;
CLASS date;
MODEL ln_weight=ln_length date/SS1 SOLUTION;
RUN;
```

The variable `date` index the two samples and since `date` appears in the `CLASS` statement, the `MODEL` statement specifies two parallel regression lines with possible different intercepts. If `date` is deleted from the `MODEL` statement, a single regression line is specified and therefore the F-test of H_0: $\alpha_1 = \alpha_2$ can be found in the bottom line of the `SS1` table in the output listing, see the explanation of type I sums of squares on page 202. The equivalent t-test can be found in the line `date 75-07-24` in the `Parameter Estimate` table of the output listing.

Source	DF	Type I SS	Mean Square	F Value	Pr > F
ln_length	1	8.55086377	8.55086377	1774.28	<.0001
date	1	0.00112318	0.00112318	0.23	0.6320

Parameter		Estimate	Standard Error	t Value	Pr > |t|
Intercept		-2.232896441 B	0.40976991	-5.45	<.0001
ln_length		2.852325935	0.13267820	21.50	<.0001
date	75-07-24	0.021074390 B	0.04365397	0.48	0.6320
date	75-08-16	0.000000000 B	.	.	.

The F-test statistic is 0.23 and this corresponds to a significance probability of 0.632, so the hypothesis is not rejected. This is not very surprising in view of the top right-hand plot in Figure 6.6, which shows that the regression lines fitted to the two samples almost coincide. Thus the data do not contradict the assumption that the change in means of ln weight is fully explained by the change in the means of ln length.

This is the simple description of the growth of the toad we had in mind on page 275 when we decided to analyze the logarithms rather than the original variables. □

We conclude this chapter with a few results that are useful for the analysis of a single sample from a bivariate normal distribution and for two independent samples from bivariate normal distributions. Here we need to use some notation from linear algebra, and we will consider vectors as column vectors and use the * to denote the transpose of a vector. The results will be given without proof, but the results are similar to the results for one and two univariate normal samples given in Sections 3.1 and 3.2 in Chapter 3.

6.10.1 One bivariate normal sample

Let $(x_1,y_1)^*,\ldots,(x_n,y_n)^*$, with $n \geq 3$, be a sample from the bivariate normal distribution. Thus $(x_i,y_i)^*$, $i = 1,\ldots,n$, are realizations of random variables $(X_i,Y_i)^*$, $i = 1,\ldots n$, which are independent and have the distribution,

$$N_2\left(\begin{pmatrix} \mu_x \\ \mu_y \end{pmatrix}, \left\{\begin{array}{cc} \sigma_x^2 & \rho\sigma_x\sigma_y \\ \rho\sigma_x\sigma_y & \sigma_y^2 \end{array}\right\}\right). \tag{6.47}$$

Estimation in this model has been treated in Subsections 6.4.2 and 6.4.3 but the results are repeated here in the shorthand notation, thus

$$\begin{pmatrix} \mu_x \\ \mu_y \end{pmatrix} \leftarrow \begin{pmatrix} \bar{x}. \\ \bar{y}. \end{pmatrix} \sim\sim N_2\left(\begin{pmatrix} \mu_x \\ \mu_y \end{pmatrix}, \frac{1}{n}\left\{\begin{array}{cc} \sigma_x^2 & \rho\sigma_x\sigma_y \\ \rho\sigma_x\sigma_y & \sigma_y^2 \end{array}\right\}\right). \tag{6.48}$$

Note that although the variance of the estimates of the mean decrease with the sample size, the correlation between the estimates remains constant and is equal to the correlation ρ between any pair of observations in the sample.

The estimate of the variance-covariance matrix is

$$\left\{ \begin{array}{cc} \sigma_x^2 & \rho\sigma_x\sigma_y \\ \rho\sigma_x\sigma_y & \sigma_y^2 \end{array} \right\} \leftarrow \frac{1}{n-1}\sum_{i=1}^{n}\left(\begin{array}{c} x_i - \bar{x}. \\ y_i - \bar{y}. \end{array} \right)(x_i - \bar{x}., y_i - \bar{y}.) = \frac{1}{n-1}\left\{ \begin{array}{cc} SSD_{\mathbf{x}} & SPD_{\mathbf{xy}} \\ SPD_{\mathbf{xy}} & SSD_{\mathbf{y}} \end{array} \right\}.$$

It will be convenient to have symbols for the two by two matrices and we will use the notation

$$\Sigma = \left\{ \begin{array}{cc} \sigma_x^2 & \rho\sigma_x\sigma_y \\ \rho\sigma_x\sigma_y & \sigma_y^2 \end{array} \right\}, \quad SSD = \left\{ \begin{array}{cc} SSD_{\mathbf{x}} & SPD_{\mathbf{xy}} \\ SPD_{\mathbf{xy}} & SSD_{\mathbf{y}} \end{array} \right\} \quad \text{and} \quad \mathbf{S} = \frac{1}{n-1}SSD. \tag{6.49}$$

Here $\mathbf{S}$ denotes the estimate of Σ and we use the bold-face symbol to distinguish it from the symbol we have used to denote a sum of observations.

The null hypothesis,

$$H_{0\mu}\colon (\mu_x, \mu_y) = (\mu_{0x}, \mu_{0y}),$$

which specifies that the vector of means is completely known can be tested with the statistic,

$$T^2 = n(\bar{x}. - \mu_{0x}, \bar{y}. - \mu_{0y})\mathbf{S}^{-1}\left(\begin{array}{c} \bar{x}. - \mu_{0x} \\ \bar{y}. - \mu_{0y} \end{array} \right).$$

The test statistic T^2 generalizes the square of the one sample t-test statistic, $\sqrt{n}(\bar{x}. - \mu_{0x})/s_{\mathbf{x}}$, and the distribution of T^2 can be scaled to an $F(2, n-2)$ distribution, i.e.,

$$\frac{n-2}{2(n-1)}T^2 \sim F(2, n-2). \tag{6.50}$$

Large values of T^2 are critical for the null hypothesis, so the significance probability is

$$p_{obs} = 1 - F_{F(2,n-2)}\left(\frac{n-2}{2(n-1)}T^2\right).$$

The test statistic T^2 is called *Hotelling's* T^2 after Harold Hotelling who derived the distribution of T^2 in Hotelling (1931). Hotelling's T^2 is equivalent to the likelihood ratio test statistic.

An important application of Hotelling's T^2 is to provide confidence regions for the vector of means of a bivariate normal distribution based on a sample. The $(1-\alpha)$ confidence regions consists of those values that would not be rejected if they were used to specify the null hypothesis. Thus the $(1-\alpha)$ confidence region for (μ_x, μ_y) is

$$\left\{ (\mu_x, \mu_y) \,\middle|\, \frac{n(n-2)}{2(n-1)}(\bar{x}. - \mu_x, \bar{y}. - \mu_y)\mathbf{S}^{-1}\left(\begin{array}{c} \bar{x}. - \mu_x \\ \bar{y}. - \mu_y \end{array} \right) \le F_{1-\alpha}(2, n-2) \right\}.$$

The $(1-\alpha)$ confidence region for (μ_x, μ_y) is an ellipse with the center at $(\bar{x}., \bar{y}.)$ with orientation and semiaxes determined by the confidence coefficient $(1-\alpha)$, the sample size n, and the estimated variance-covariance matrix. The shape of the confidence ellipse will be identical to the contours of the fitted bivariate normal distribution. The 95% confidence regions for the means based on the the two samples of Example 6.1 are given in Figure 6.7.

6.10.2 Two independent bivariate normal samples

The model can be specified as follows.

The first sample $(x_{11}, y_{11})^*, \ldots, (x_{1n_1}, y_{1n_1})^*$ consists of observations of independent, identically distributed random variables with a bivariate normal distribution,

$$N_2\left(\left(\begin{array}{c} \mu_{1x} \\ \mu_{1y} \end{array} \right), \left\{ \begin{array}{cc} \sigma_{1x}^2 & \rho_1\sigma_{1x}\sigma_{1y} \\ \rho_1\sigma_{1x}\sigma_{1y} & \sigma_{1y}^2 \end{array} \right\}\right), \tag{6.51}$$

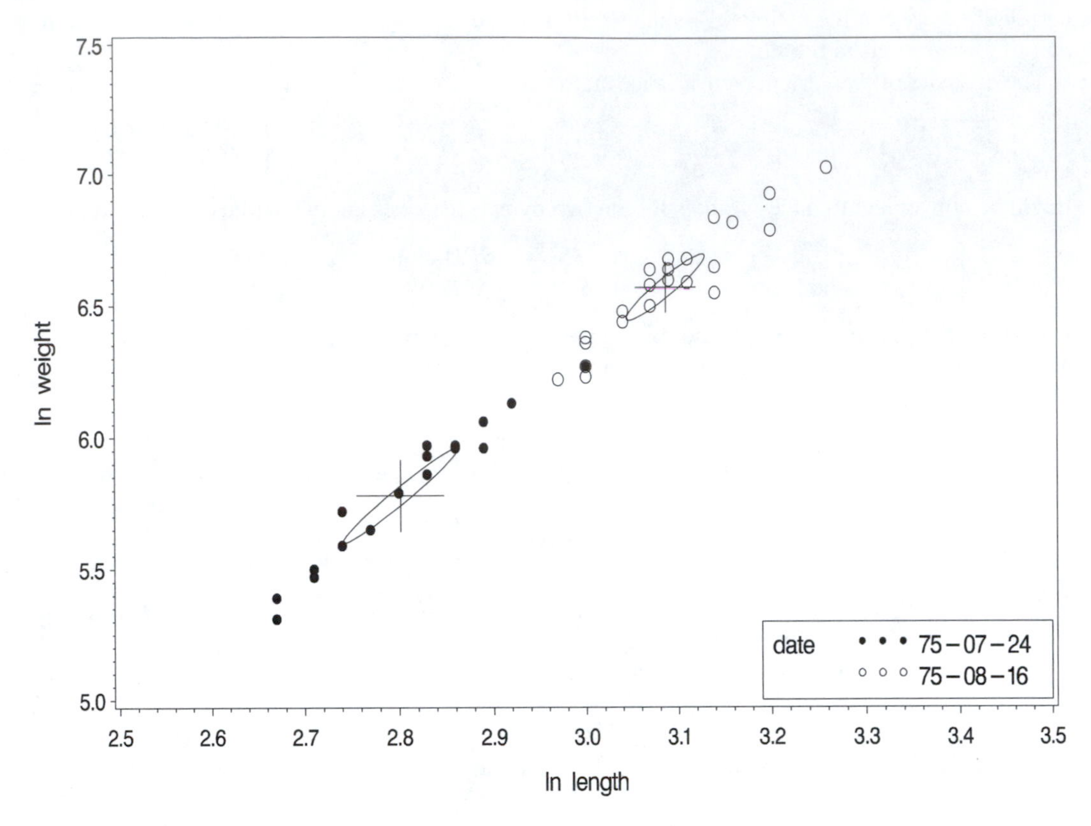

Figure 6.7 *Scatter plot of the data and 95% confidence regions for the means of the two samples in Example 6.1. For comparison, 95% confidence intervals for the means based on the one sample t-test statistics for the marginal samples are plotted.*

and the second sample $(x_{21}, y_{21})^*, \ldots, (x_{2n_2}, y_{2n_2})^*$ consists of observations of independent, identically distributed random variables with a bivariate normal distribution,

$$N_2\left(\left(\begin{array}{c}\mu_{2x}\\ \mu_{2y}\end{array}\right), \left\{\begin{array}{cc}\sigma_{2x}^2 & \rho_2\sigma_{2x}\sigma_{2y}\\ \rho_2\sigma_{2x}\sigma_{2y} & \sigma_{2y}^2\end{array}\right\}\right). \tag{6.52}$$

Furthermore, the two samples are independent.

This is the model we formulated for the two samples of ln length and ln weight in Example 6.1 on page 275. Estimation of the parameters of the two distribution have been treated in the previous section, and we will use the notation that was introduced there, simply adding an index to identify the sample. The variance-covariance matrices, for example, will be denoted by Σ_1 and Σ_2 and their estimates by $\mathbf{S}_{(1)}$ and $\mathbf{S}_{(2)}$, respectively.

We will focus on the test of the hypothesis that the mean vectors are the same in the two distribution, i.e.,

$$H_{0\mu}\colon (\mu_{1x}, \mu_{1y}) = (\mu_{2x}, \mu_{2y}).$$

For the univariate two-sample problem there is an exact t-test of the hypothesis of identical means when the variances are the same in the two distributions, but only an approximate one when the variances are different, see the main points in Subsection 3.2.1 on page 94, for example. The situation is very similar for bivariate normal samples, but we will only give the test of $H_{0\mu}$ when a common variance-covariance matrix can be assumed.

The assumption of a common variance-covariance matrix can be checked with a test of the hypothesis

$$H_{0\Sigma} : \Sigma_1 = \Sigma_2.$$

We will not derive the test here, but only mention that the test is a generalization of Bartlett's test for homogeneity of variances in (3.44) to the multivariate normal distribution. It is explained on page 292 how the test is obtained using SAS.

If it is decided to adopt the model with a common variance-covariance matrix, the estimate of the common variance-covariance matrix Σ is a weighted average of the estimates based on the two samples,

$$\Sigma \leftarrow \frac{1}{n_1+n_2-2}\big((n_1-1)\mathbf{S}_{(1)} + (n_2-1)\mathbf{S}_{(2)}\big) = \mathbf{S}_1.$$

Returning to the hypothesis $H_{0\mu}$: $(\mu_{1x}, \mu_{1y}) = (\mu_{2x}, \mu_{2y})$ in the model with a common variance, it is natural to look at the vector of differences $(\bar{x}_{1\cdot} - \bar{x}_{2\cdot}, \bar{y}_{1\cdot} - \bar{y}_{2\cdot})$ between the estimates of the means in the two samples. Under $H_{0\mu}$ this vector of differences is a realization of a random variable with a distribution,

$$\begin{pmatrix} \bar{x}_{1\cdot} - \bar{x}_{2\cdot} \\ \bar{y}_{1\cdot} - \bar{y}_{2\cdot} \end{pmatrix} \sim\sim N_2\left(\begin{pmatrix} 0 \\ 0 \end{pmatrix}, \left(\frac{1}{n_1} + \frac{1}{n_2}\right)\Sigma\right),$$

and normalizing this vector by the estimated variance-covariance matrix of its distribution leads to the test statistic,

$$T^2 = \frac{n_1 n_2}{n_1+n_2}(\bar{x}_{1\cdot} - \bar{x}_{2\cdot}, \bar{y}_{1\cdot} - \bar{y}_{2\cdot})\mathbf{S}_1^{-1}\begin{pmatrix} \bar{x}_{1\cdot} - \bar{x}_{2\cdot} \\ \bar{y}_{1\cdot} - \bar{y}_{2\cdot} \end{pmatrix}.$$

This is also a Hotelling's T^2 test statistic and it can be shown that it can be scaled to have an F-distribution. In this model,

$$\frac{(n_1+n_2-3)}{2(n_1+n_2-2)}T^2 \sim F(2, n_1+n_2-3).$$

Large values of T^2 are critical for the null hypothesis, so the significance probability is

$$p_{obs} = 1 - F_{F(2,n_1+n_2-3)}\left(\frac{(n_1+n_2-3)}{2(n_1+n_2-2)}T^2\right).$$

Just as for one sample, Hotelling's T^2 test statistic can be used to construct a confidence region for the vector of differences between means $(\mu_{1x} - \mu_{2x}, \mu_{1y} - \mu_{2y})$. The confidence region will be an ellipse centered at $(\bar{x}_{1\cdot} - \bar{x}_{2\cdot}, \bar{y}_{1\cdot} - \bar{y}_{2\cdot})^*$.

Annex to Chapter 6

Calculations in SAS

The bivariate normal distribution is just a special case of the multivariate normal distribution so no procedures in SAS are designed especially to deal with data from the bivariate normal distribution.

Based on Example 6.1 we show how to

1. compute correlation coefficients and estimate the parameters of a bivariate distribution
2. check the assumptions of a bivariate normal distribution as described in Section 6.5
3. obtain Bartlett's test of homogeneity of variance-covariance matrices
4. obtain Hotelling's test of a simple null hypothesis
5. obtain Hotelling's test of the identity of the means based on two bivariate normal samples

The data for Example 6.1 is in the dataset `bufobufo`, which is created in the following way.

```
PROC FORMAT;
   VALUE $datefmt    '1'='75-07-24' '2'='75-08-16';
RUN;

DATA bufobufo;
INPUT date $ ln_length ln_weight@@;
FORMAT date $datefmt.;
length=exp(ln_length);
weight=exp(ln_weight);
DATALINES;
1 2.83 5.86 1 2.86 5.97 1 2.83 5.97 1 2.71 5.47 1 2.89 5.96
1 2.67 5.31 1 2.74 5.72 1 2.71 5.50 1 2.80 5.79 1 2.83 5.93
1 2..77 5.65 1 2.74 5.59 1 2.86 5.96 1 2.67 5.39 1 2.89 6.06
1 2.92 6.13 1 2.71 5.50 1 3.00 6.27
2 2.97 6.22 2 3.00 6.23 2 3.26 7.03 2 3.07 6.50 2 3.09 6.68
2 3.14 6.65 2 3.14 6.84 2 3.14 6.55 2 3.00 6.27 2 3.09 6.60
2 3.11 6.59 2 3.00 6.38 2 3.04 6.44 2 3.20 6.79 2 3.04 6.48
2 3.20 6.93 2 3.00 6.23 2 3.16 6.82 2 3.07 6.58 2 3.00 6.36
2 3.09 6.64 2 3.07 6.64 2 3.11 6.68
;
```

The data set contains the lengths, weights, ln lengths, ln weights, and the dates of the two samples in the variables `length`, `weight`, `ln_length`, `ln_weight`, and `date`.

1. *Computing the correlation coefficient*

The easiest way to compute the correlation coeffcient between two or more variables in a dataset is to use *PROC CORR*.

```
PROC CORR DATA=bufobufo;
VAR ln_weight;
WITH ln_length;
BY date;
RUN;
```

If the `VAR` and the `WITH` statements are omitted, *PROC CORR* computes the correlation coefficient between all variables in the data set, but the way to control which correlation coefficients are computed is to use the `VAR` and the `WITH` statements. Both statements may contain a list of variables in which case the correlation coefficients between all the variables in the `VAR` statement and all the variables in the `WITH` statement are computed.

The program produces the following output listing.

```
------------------------ date=75-07-24 ------------------------

                    The CORR Procedure

              1 With Variables:    ln_length
              1      Variables:    ln_weight


                    Simple Statistics

Variable            N          Mean       Std Dev           Sum

ln_length          18       2.80167       0.09288      50.43000
ln_weight          18       5.77944       0.27541     104.03000

                    Simple Statistics

          Variable          Minimum         Maximum

          ln_length         2.67000         3.00000
          ln_weight         5.31000         6.27000


        Pearson Correlation Coefficients, N = 18
               Prob > |r| under H0: Rho=0

                              ln_weight

              ln_length         0.97782
                                 <.0001
------------------------ date=75-08-16 ------------------------

                    The CORR Procedure

              1 With Variables:    ln_length
              1      Variables:    ln_weight


                    Simple Statistics

Variable            N          Mean       Std Dev           Sum

ln_length          23       3.08652       0.07601      70.99000
ln_weight          23       6.57087       0.22573     151.13000

                    Simple Statistics

          Variable          Minimum         Maximum

          ln_length         2.97000         3.26000
          ln_weight         6.22000         7.03000


        Pearson Correlation Coefficients, N = 23
               Prob > |r| under H0: Rho=0

                              ln_weight

              ln_length         0.94219
                                 <.0001
```

Below the Pearson correlation coefficient the listing gives the significance probability for the test that the population correlation coefficient is zero. This is the significance probability in (6.33), which is based on the assumption that the two variables in question are bivariate normally distributed. Obviously, the significance probability is only meaningful if the assumption of a bivariate normal distribution is justified.

Note that the means and the standard deviations are also given above so the estimates of the parameters of a bivariate normal distribution can be found from the output listing from *PROC CORR*.

2. *Model checking for the bivariate normal distribution*

As mentioned in Section 6.5, the only new technique is the check for the bivariate aspect of the bivariate normal distribution, which is to calculate the statistic,

$$w_i = \frac{1}{1-r^2}\left[\left(\frac{x_i - \bar{x}.}{s_x}\right)^2 - 2r\frac{x_i - \bar{x}.}{s_x}\frac{y_i - \bar{y}.}{s_y} + \left(\frac{y_i - \bar{y}.}{s_y}\right)^2\right], \tag{6.53}$$

for each pair of observations and make a fractile plot of the w_i, $i = 1,\ldots,n$ based on a $\chi^2(2)$ distribution. The notation is tied to the bivariate normal distribution, but the method applies in general to the multivariate normal distribution so in equation (6.54) the notation is changed to a notation that applies for the general p-dimensional normal distribution.

An alternative way of writing (6.53) is

$$w_i = (x_i - \bar{x}., y_i - \bar{y}.)\begin{Bmatrix} s_x^2 & rs_xs_y \\ rs_xs_y & s_y^2 \end{Bmatrix}^{-1}\begin{pmatrix} x_i - \bar{x}. \\ y_i - \bar{y}. \end{pmatrix}$$

or

$$w_i = (\mathbf{x}_i - \bar{\mathbf{x}}.)^*\mathbf{S}^{-1}(\mathbf{x}_i - \bar{\mathbf{x}}.), \tag{6.54}$$

where $\mathbf{S}$ denotes the estimated variance-covariance matrix, and $(\mathbf{x}_i - \bar{\mathbf{x}}.)^* = (x_i - \bar{x}., y_i - \bar{y}.)$. In the calculation of w_i, $i = 1,...,n$, we write

$$w_i = (\mathbf{x}_i - \bar{\mathbf{x}}.)^* UL^{-1}U^*(\mathbf{x}_i - \bar{\mathbf{x}}.) \tag{6.55}$$

$$= \sum_{j=1}^{2} y_{ij}^2, \tag{6.56}$$

where

$$y_{ij} = \frac{1}{\sqrt{l_j}}(\mathbf{x}_i - \bar{\mathbf{x}}.)^* u_{(j)}. \tag{6.57}$$

Here $L = \text{diag}(l_1, l_2)$ is the diagonal matrix, where $l_1 \geqq l_2 > 0$ are the eigenvalues of $\mathbf{S}$ and U is the 2×2 matrix,

$$U = \{\, u_{(1)} \quad u_{(2)} \,\},$$

whose jth column $u_{(j)}$ is the standardized eigenvector corresponding to the eigenvalue l_j, i.e., the length of the vector $u_{(j)}$ is 1.

In the $n \times 2$ matrix,

$$\left\{\sqrt{l_j}y_{ij}\right\}_{i=1\,j=1}^{n\quad 2}$$

the jth column is called the jth *principal component*, and $\sqrt{l_j}y_{ij}$ is called the value of the jth principal component on the ith observation.

We illustrate the calculations in SAS with the July sample. The advantage of the representation (6.55) or (6.56) is that *PROC PRINCOMP* can be used to compute the principal components with the statements:

```
DATA a;
SET bufobufo(WHERE=(date='1'));
   xcol1=ln_length;
   xcol2=ln_weight;
RUN;

PROC PRINCOMP DATA=a COV OUT=b NOPRINT;
VAR xcol1 xcol2;
RUN;

DATA qq;
SET b;
```

```
w=USS(OF prin1 prin2);
RUN;
```

The central parts in the computations are the application of `PROC PRINCOMP` to compute the principal components, which are stored in `prin1` and `prin2`, and the command `w=USS(OF prin1 prin2);` in the following data step, which corresponds to (6.56).

It is important to notice that the two options `COV` and `STD` are essential in order to obtain what we want; namely that the estimated covariance matrix is used in the calculation (`COV`), and that the principal components are standardized (`STD`), which means that they are divided by the square root of the corresponding eigenvalue as it is done in (6.57).

In order to make the fractile plot, the empirical fractiles and the corresponding fractiles in the $\chi^2(2)$ distribution must be found. The observations are sorted in ascending order and the ith smallest observation is the empirical p-fractile for $p = (i - \frac{1}{2})/n$ where n is the sample size. The corresponding p-fractiles of the $\chi^2(2)$ distribution are then calculated using the SAS function `CINV(p,f)`, which gives $F^{-1}_{\chi 2(f)}(p)$, the inverse of the distribution function of the χ^2-distribution with f degrees of freedom. The SAS statements are as follows.

```
PROC SORT DATA=qq;
BY w;

DATA qq;
SET qq NOBS=totn;
chisq=CINV(((_n_-.5)/totn),2);
RUN;
```

We use two tricks to perform the calculations. First, the total number of observations in the sample is allocated to a variable `ntot` with the option `NOBS=ntot` to the `SET` statement. Secondly, the SAS variable `_n_` contains the number of the observations in the data set, so for a variable sorted in ascending order `_n_` contains the ranks of the observations in that variable.

Finally, the fractiles are plotted with *PROC GPLOT* with the following statements.

```
DATA annoline;
  XSYS='2';
  YSYS='2';
  LINE=3;
INPUT function $ x y;
DATALINES;
MOVE 0 0
DRAW 8 8
;

AXIS5 ORDER=(0 TO 8 BY 1)
   COLOR=black  WIDTH=2  VALUE=(H=2.7)
   LABEL=(H=3 'observed fractile' )
   ;

AXIS6 ORDER=(0 TO 8 BY 1)
   COLOR=black  WIDTH=2  VALUE=(h=2.7)
   LABEL=(H=3 A=90 'chi-squared fractile' )
   ;

SYMBOL V=dot I=none H=2;

PROC GPLOT DATA=qq1;
PLOT chisq*w/
   haxis=axis5 vaxis=axis6
   annotate=annoline;
RUN; QUIT;
TITLE1;
```

The identity line from $(0,0)$ to $(8,8)$ in the plot is plotted with the annotate option using the data set `annotate`.

The fractile plot is shown in Figure 6.8.

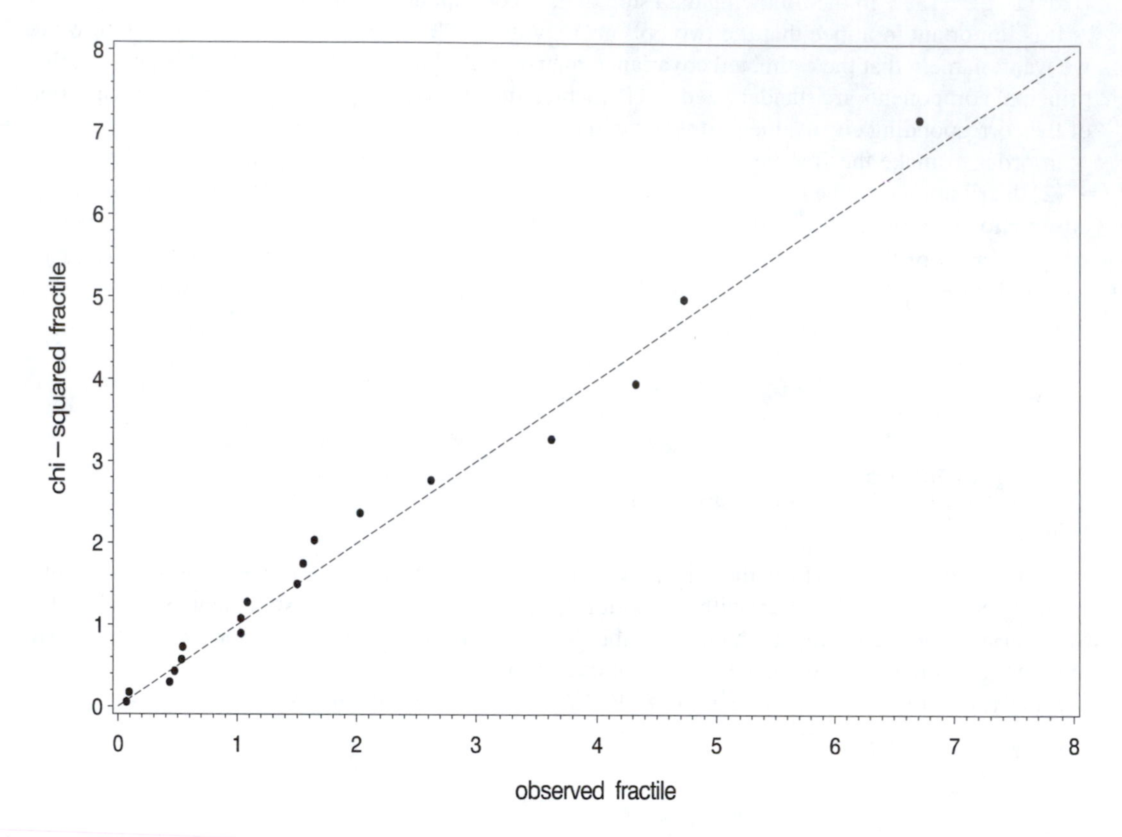

Figure 6.8 $\chi^2(2)$ *fractile plot to check for bivariate normal distribution of ln length and ln weight of the July sample.*

3. *Bartlett's test of homogeneity of variance-covariance matrices*

Bartlett's test of homogeneity of variance-covariance matrices is found in the output from *PROC DISCRIM*, which performs a discriminant analysis. Discriminant analysis is a host of methods for finding a rule for discriminating between populations based on multivariate data and possibly a model for the data. One of the methods is based on multivatiate normal samples with a common covariance, and Bartlett's test is included in the output to give the opportunity to check that aspect of the model.

```
PROC DISCRIM DATA=bufobufo WCOV WSSCP METHOD=NORMAL
                                POOL=TEST MANOVA;
CLASS date;
VAR ln_length ln_weight;
RUN;
```

The program that produces Bartlett's test for the two samples in Example 6.1 is given above and Bartlett's test is only a small part of the output listing. The output listing even gives the formula for the test statistic and it shows great similarity with Bartlett's test for homogeneity of variances in (3.44). We note that the degrees of freedom of the approximating χ^2-distribution in this example is 3, which is consistent with the fact that the variance-covariance matrix of a bivariate distribution has 3 parameters, and we test the reduction from a model with two unknown variance-covariance matrices to one with a common variance-covariance matrix.

```
                   The DISCRIM Procedure
       Test of Homogeneity of Within Covariance Matrices

Notation: K     = Number of Groups

          P     = Number of Variables

          N     = Total Number of Observations - Number of Groups

          N(i)  = Number of Observations in the i'th Group - 1

                                                N(i)/2
                     ∏   |Within SS Matrix(i)|
          V     = -----------------------------------
                                              N/2
                         |Pooled SS Matrix|

                         _                      _       2
                        |           1       1    |   2P  + 3P - 1
          RHO   = 1.0 - |  SUM -----   -  ---    |   -------------
                        |_          N(i)     N  _|   6(P+1)(K-1)

          DF    = .5(K-1)P(P+1)

                                               _                      _
                                              |        PN/2            |
                                              |      N          V      |
Under the null hypothesis:      -2 RHO ln     |   ------------------   |
                                              |               PN(i)/2  |
                                              |_      ∏   N(i)        _|

is distributed approximately as Chi-Square(DF).

                Chi-Square        DF     Pr > ChiSq

                  2.159134         3         0.5400
```

4. *Hotelling's test of a simple null hypothesis*

We will illustrate the calculation of Hotelling's test of a simple null hypothesis with the July sample of Example 6.1. The calculation uses *PROC GLM* and in addition we obtain the estimates of the parameters of the bivariate normal distribution and the test of zero correlation between ln length and ln height. Note that both variables appear on the left-hand side of the MODEL equation. The variable date is constant for all the observations in the data set because of the restriction imposed by WHERE=(date='1'), so the model being specified is that of a single sample from a bivariate normal distribution. The model statement implies the same model for each variable on the left-hand side of the model equation and the NOUNI option to the MODEL statement is used to ensure that the *PROC GLM* output from each of the univariate analysis is not included in the output listing. The MEANS date statement gives the estimates of the means of the two variables and their estimated standard deviations. Finally, the MANOVA statement requests the test of the hypothesis that the intercept can be removed from the model, i.e., the test of the hypothesis that the mean is zero. This is not a very meaningful null hypothesis for these data, but it will do as an example. The PRINTE option adds information about the estimate of the variance-covariance matrix to the output listing.

```
PROC GLM DATA=bufobufo(WHERE=(date='1'));
CLASS date;
MODEL ln_length ln_weight=date/NOUNI;
MEANS date;
MANOVA H=INTERCEPT/PRINTE;
RUN;
```

The MEANS date statement gives the following output.

```
                                                                          2
                              The GLM Procedure

Level of               ----------ln_length---------   ----------ln_weight---------
date          N              Mean         Std Dev             Mean         Std Dev

75-07-24     18        2.80166667      0.09287880       5.77944444      0.27541353
```

The PRINTE option to the MANOVA statement adds the following to the output listing. The first part gives the SSCP (Sum of Squares and Cross Products) matrix which is the *SSD* matrix of (6.49) so its entries have to be divided by $n-1$ to give the estimate of the variance-covariance matrix. The second part gives the estimate of the correlation between ln length and ln weight and the significance probability of the test of zero correlation.

```
                                                                          3
                              The GLM Procedure
                      Multivariate Analysis of Variance

                           E = Error SSCP Matrix

                              ln_length           ln_weight

           ln_length            0.14665        0.4252166667
           ln_weight       0.4252166667        1.2894944444


   Partial Correlation Coefficients from the Error SSCP Matrix / Prob > |r|

               DF = 17        ln_length        ln_weight

               ln_length       1.000000         0.977821
                                                 <.0001

               ln_weight       0.977821         1.000000
                                 <.0001
```

The output from the MANOVA H=INTERCEPT command is given below. We will not comment on the output in detail but will concentrate on explaining where the information about Hotelling's T^2 is hidden. At the bottom of the output four test statistics for the hypothesis are listed. In this simple case they are all equivalent as can be seen from the fact the F-values are the same for all four test statistics. In spite of this equivalence, Hotelling's T^2 is most closely related to the test statistic which is called Wilks' Lambda; for Wilk's Lambda, W, is the likelihood ratio test statistic raised to the power of $2/n$, and the relationship between Wilk's Lambda and Hotelling's T^2 is

$$W = Q^{\frac{2}{n}} = \frac{1}{1 + \frac{T^2}{n-1}}. \tag{6.58}$$

This relationship is analogous to the one found between the one sample t-test statistic and the likelihood ratio test statistic in (3.27) on page 74.

```
                                                                          4
                              The GLM Procedure
                      Multivariate Analysis of Variance

          Characteristic Roots and Vectors of: E Inverse * H, where
                   H = Type III SSCP Matrix for Intercept
                           E = Error SSCP Matrix

          Characteristic                 Characteristic Vector  V'EV=1
                    Root      Percent         ln_length         ln_weight

              2711.98135       100.00        11.3457339        -3.3761673
                 0.00000         0.00        -5.1697193         2.5060939


              MANOVA Test Criteria and Exact F Statistics for
               the Hypothesis of No Overall Intercept Effect
```

```
                H = Type III SSCP Matrix for Intercept
                        E = Error SSCP Matrix

                          S=1     M=0     N=7

Statistic                       Value    F Value   Num DF   Den DF   Pr > F

Wilks' Lambda               0.0003686    21695.9        2       16   <.0001
Pillai's Trace              0.9996314    21695.9        2       16   <.0001
Hotelling-Lawley Trace   2711.9813509    21695.9        2       16   <.0001
Roy's Greatest Root      2711.9813509    21695.9        2       16   <.0001
```

5. *Hotelling's test of the identity of the means based on two bivariate normal samples*
The program is very similar to the one giving Hotelling's T^2 for a simple hypothesis:

```
PROC GLM DATA=bufobufo;
CLASS date;
MODEL ln_length ln_weight=date/NOUNI;
MEANS date;
MANOVA H=date/PRINTE;
RUN;
```

The output listing is given below.

```
                             The SAS System                            1

                            The GLM Procedure

                         Class Level Information

                  Class         Levels    Values

                  date               2    75-07-24 75-08-16


                    Number of observations    41
                             The SAS System                            2

                            The GLM Procedure

 Level of          ---------ln_length--------   ---------ln_weight---------
 date         N         Mean          Std Dev          Mean          Std Dev

 75-07-24    18     2.80166667     0.09287880    5.77944444     0.27541353
 75-08-16    23     3.08652174     0.07601487    6.57086957     0.22572938
                             The SAS System                            3

                            The GLM Procedure
                   Multivariate Analysis of Variance

                          E = Error SSCP Matrix

                             ln_length          ln_weight

                ln_length    0.2737717391       0.7808862319
                ln_weight    0.7808862319       2.4104770531


  Partial Correlation Coefficients from the Error SSCP Matrix / Prob > |r|

                  DF = 39        ln_length       ln_weight

                  ln_length       1.000000        0.961262
                                                    <.0001

                  ln_weight       0.961262        1.000000
                                    <.0001
                             The SAS System                            4

                            The GLM Procedure
                   Multivariate Analysis of Variance

          Characteristic Roots and Vectors of: E Inverse * H, where
```

```
                H = Type III SSCP Matrix for date
                      E = Error SSCP Matrix

Characteristic                  Characteristic Vector  V'EV=1
          Root     Percent        ln_length         ln_weight

    3.01727494      100.00       2.50388598       -0.21051569
    0.00000000        0.00      -6.46592474        2.32725928

            MANOVA Test Criteria and Exact F Statistics
            for the Hypothesis of No Overall date Effect
                H = Type III SSCP Matrix for date
                      E = Error SSCP Matrix

                      S=1     M=0      N=18

Statistic                        Value    F Value    Num DF    Den DF    Pr > F

Wilks' Lambda               0.24892496      57.33         2        38    <.0001
Pillai's Trace              0.75107504      57.33         2        38    <.0001
Hotelling-Lawley Trace      3.01727494      57.33         2        38    <.0001
Roy's Greatest Root         3.01727494      57.33         2        38    <.0001
```

Again Hotelling's T^2 is related to Wilks' statistic and the relationship is given by

$$W = Q^{\frac{2}{n}} = \frac{1}{1 + \frac{T^2}{n_1+n_2-2}},$$

where $n = n_1 + n_2$, W denotes Wilks' statistic and Q denotes the likelihood ratio test statistic.

Main Points in Section 6.10

Data consist of pairs of observations, (x_i, y_i), $i = 1, \ldots, n$.

The empirical correlation is an estimate of the correlation between any two random variables with a bivariate distribution. However, interpretation of the correlation as the degree of association between the random variables, and application of statistical techniques, such as the test of zero correlation, confidence intervals and test of identity of correlations, are based on the bivariate normal distribution.

Checking for bivariate normality:
At least a scatter plot of the pairs of observations (x_i, y_i), $i = 1, \ldots, n$. If $n > 10$ the scatter plot is supplemented by the $\chi^2(2)$ fractile plot considered in Section 6.5 to check for bivariate normality and normal fractile plots to check for normality of the marginal distributions.

I. A single sample.
Estimation:

$$r = r_{\mathbf{xy}} = \frac{SPD_{\mathbf{xy}}}{\sqrt{SSD_{\mathbf{x}}}\sqrt{SSD_{\mathbf{y}}}}.$$

Test of H_0: $\rho = 0$:
Test statistic:

$$T(r_{\mathbf{xy}}) = \sqrt{n-2}\frac{r_{\mathbf{xy}}}{\sqrt{1 - r_{\mathbf{xy}}^2}}.$$

Significance probability:

$$p_{obs}(\mathbf{x}, \mathbf{y}) = 2\left(1 - F_{t(n-2)}(|T(r_{\mathbf{xy}})|)\right).$$

$(1-\alpha)$ confidence interval for ρ:

First calculate the $(1-\alpha)$ confidence interval for $\zeta = \frac{1}{2}\ln\frac{1+\rho}{1-\rho}$ based on the value of Fisher's Z,

$$z = \frac{1}{2}\ln\frac{1 + r_{\mathbf{xy}}}{1 - r_{\mathbf{xy}}},$$

as

$$\left[z - \frac{1}{\sqrt{n-3}}u_{1-\alpha/2}, z + \frac{1}{\sqrt{n-3}}u_{1-\alpha/2}\right] = [z_l, z_u],$$

where $u_{1-\alpha/2}$ is the $1-\alpha/2$ fractile in the $N(0,1)$-distribution and then transform the limits of this interval to obtain the $(1-\alpha)$ confidence interval for ρ as

$$\left[\frac{e^{2z_l} - 1}{e^{2z_l} + 1}, \frac{e^{2z_u} - 1}{e^{2z_u} + 1}\right].$$

II. k samples.
Test of H_0: $\rho_1 = \cdots = \rho_k = \rho$:

The test is based on the values $z_1, \ldots, z_k$ of Fisher's Z-transformation in the k samples and the sizes $n_1, \ldots, n_k$ of the k samples. The test statistic is

$$X^2 = \sum_{i=1}^{k} z_i^2(n_i - 3) - \frac{\left(\sum_{i=1}^{k} z_i(n_i - 3)\right)^2}{\sum_{i=1}^{k}(n_i - 3)}$$

and the significance probability is

$$p_{obs} = 1 - F_{\chi^2(k-1)}(X^2).$$

Exercises for Chapter 6

Exercise 6.1 Consider the bivariate normal distribution:

$$\binom{X}{Y} \sim N_2\left(\left(\begin{array}{c}\mu_x\\ \mu_y\end{array}\right), \left\{\begin{array}{cc}\sigma_x^2 & \rho\sigma_x\sigma_y\\ \rho\sigma_x\sigma_y & \sigma_y^2\end{array}\right\}\right).$$

Show that the density function can be written as

$$f(x,y) = \frac{1}{2\pi|\Sigma|^{\frac{1}{2}}} e^{\frac{1}{2}(\mathbf{z}-\boldsymbol{\mu})^*\Sigma^{-1}(\mathbf{z}-\boldsymbol{\mu})},$$

where $\mathbf{z} = (x,y)^*$, $\boldsymbol{\mu} = (\mu_x, \mu_y)^*$ and $\Sigma = \left\{\begin{array}{cc}\sigma_x^2 & \rho\sigma_x\sigma_y\\ \rho\sigma_x\sigma_y & \sigma_y^2\end{array}\right\}$.

Exercise 6.2 Let (X,Y) have the bivariate normal distribution,

$$\binom{X}{Y} \sim N_2\left(\left(\begin{array}{c}\mu_x\\ \mu_y\end{array}\right), \left\{\begin{array}{cc}\sigma_x^2 & \rho\sigma_x\sigma_y\\ \rho\sigma_x\sigma_y & \sigma_y^2\end{array}\right\}\right).$$

(1) Prove the factorization (6.15) of the density function of the bivariate normal distribution and use it to find the marginal distributions of X and Y and the conditional distribution of Y given $X = x$.

(2) Use the factorization to show that

$$E((X-EX)(Y-EY)) = E((X-\mu_x)(Y-\mu_y)) = \rho\sigma_x\sigma_y$$

and consequently that

$$Cor(X,Y) = \rho.$$

(3) Use the factorization to show that

$$\frac{1}{1-\rho^2}\left\{\left(\frac{X-\mu_x}{\sigma_x}\right)^2 - 2\rho\frac{X-\mu_x}{\sigma_x}\frac{X_2-\mu_y}{\sigma_y} + \left(\frac{X_2-\mu_y}{\sigma_y}\right)^2\right\} \sim \chi^2(2).$$

Exercise 6.3 (One-way analysis of variance with known variance)

This exercise gives the theory behind the comparison of several correlation coefficients based on Fisher's Z.

Let X_{ij}, $j = 1,\ldots,n_i$, $i = 1,\ldots,k$, denote independent independent and normally distributed random variables,

$$X_{ij} \sim N(\mu_i, \sigma_{0i}^2), \quad \mu_i \in \mathbb{R}, \quad \sigma_{0i}^2 > 0,$$

where the variances σ_{0i}^2 are known constants.

(1) Show that the maximum likelihood estimator $(\hat{\mu}_{1k},\ldots,\hat{\mu}_k)$ of $(\mu_1,\ldots,\mu_k)$ is

$$(\hat{\mu}_1,\ldots,\hat{\mu}_k) = (\bar{X}_{1\cdot},\ldots,\bar{X}_{k\cdot}),$$

where $\bar{X}_{i.} = \frac{1}{n_i}\sum_{j=1}^{n_i} X_{ij}$. Specify the distribution of the maximum likelihood estimator.

Consider the hypotheses of a common mean of all observations, i. e.,

$$H_1\colon \mu_i = \mu, \quad i = 1,\ldots,k.$$

(2) Show that the maximum likelihood estimator $\hat{\mu}$ of μ under the hypothesis H_1 is

$$\hat{\mu} = \frac{\sum_{i=1}^{k} \frac{n_i \bar{X}_{i\cdot}}{\sigma_{0i}^2}}{\sum_{i=1}^{k} \frac{n_i}{\sigma_{0i}^2}}.$$

Thus $\hat{\mu}$ is the weighted mean of the estimators $\bar{X}_{i\cdot}$ from the individual samples with the reciprocal variances as weights. Show that the distribution of $\hat{\mu}$ under H_1 is

$$\hat{\mu} \sim N(\mu, 1/\sum_{i=1}^{k} \frac{n_i}{\sigma_{0i}^2}).$$

(3) Show that the the likelihood ratio test statistic of H_1 under the hypothesis H_1 is

$$Q_1(X) = e^{-\frac{1}{2}\sum_{i=1}^{k} \frac{n_i}{\sigma_{0i}^2}(\bar{X}_{i\cdot} - \hat{\mu})^2}.$$

Show furthermore that $-2\ln Q_1(X)$ is distributed as χ^2 with $k-1$ degrees of freedom. (This question requires linear normal models techniques.)

Exercise 6.4 The data in this exercise were first published in a paper with the ironical title, *On a most powerful method of discovering statistical regularities* by Neyman (1952). The data were used by Kronmal (1993).

```
DATA storks;
INPUT county women storks babies @@;
DATALINES;
 1 1   2 10     2 1   2 15     3 1   2 20
 4 1   3 10     5 1   3 15     6 1   3 20
 7 1   4 10     8 1   4 15     9 1   4 20

10 2   4 15    11 2   4 20    12 2   4 25
13 2   5 15    14 2   5 20    15 2   5 25
16 2   6 15    17 2   6 20    18 2   6 25

19 3   5 20    20 3   5 25    21 3   5 30
22 3   6 20    23 3   6 25    24 3   6 30
25 3   7 20    26 3   7 25    27 3   7 30

28 4   6 25    29 4   6 30    30 4   6 35
31 4   7 25    32 4   7 30    33 4   7 35
34 4   8 25    35 4   8 30    36 4   8 35

37 5   7 30    38 5   7 35    39 5   7 40
40 5   8 30    41 5   8 35    42 5   8 40
43 5   9 30    44 5   9 35    45 5   9 40

46 6   8 35    47 6   8 40    48 6   8 45
49 6   9 35    50 6   9 40    51 6   9 45
52 6  10 35    53 6  10 40    54 6  10 45
RUN;
```

The table gives the number of women ($\times 10000$), the number of storks, and the number of babies for 54 counties.

(1) Is there any association between the number of storks and the number of babies born? Is there any association between the number of storks and the number of babies in counties with the same number of women.

(2) Describe the association if any between the birth rate (number of babies per 10000 women) and the density of storks, i.e., the number of storks per 10000 women.

(3) Describe the association if any between the birth rate and the number of storks.

Exercise 6.5 Let X_1 and X_2 be independent and $N(0,1)$ distributed. Define (Z_1, Z_2) as follows:

$$(Z_1, Z_2) = \begin{cases} (X_1, X_2) \text{ if } X_1 \text{ and } X_2 \text{ have the same sign} \\ (-X_1, X_2) \text{ if } X_1 \text{ and } X_2 \text{ have opposite signs} \end{cases}$$

Show that the marginal distributions of Z_1 and Z_2 are normal and that (Z_1, Z_2) is not bivariate normally distributed.

CHAPTER 7

The Multinomial Distribution

The multinomial distribution may be introduced in the following way. Consider an experiment for which the following four conditions are satisfied:

(a) The experiment consists of n identical trials,

(b) Each trial may result in precisely one of k events, $B_1,\ldots,B_j,\ldots,B_k$,

(c) The probability of each of the k events is the same in all n trials, $P(B_1)=\pi_1,\ldots,P(B_j)=\pi_j,\ldots,P(B_k)=\pi_k$,

(d) The outcomes of the n trials are stochastically independent.

If $\mathbf{X}$ denotes the discrete k-dimensional random vector, $(X_1,\ldots,X_j,\ldots,X_k)^*$, where the jth component X_j gives the number of times the event B_j has occurred, then $\mathbf{X}$ has a multinomial distribution with *number of trials n* and *probability vector* $\boldsymbol{\pi}=(\pi_1,\ldots,\pi_j,\ldots,\pi_k)^*$ i.e., the probability function of $\mathbf{X}$ is

$$P(\mathbf{X}=\mathbf{x})=\binom{n}{x_1\ldots x_j\ldots x_k}\pi_1^{x_1}\cdots\pi_j^{x_j}\cdots\pi_k^{x_k}. \tag{7.1}$$

Here $\mathbf{x}=(x_1,\ldots,x_j,\ldots,x_k)^*$ is a vector such that

$$x_j\in\{0,1,\ldots,n\},\quad j=1,\ldots,k \qquad \text{and} \qquad \sum_{j=1}^{k}x_j=n.$$

In (7.1) $\pi_1^{x_1}\cdots\pi_j^{x_j}\cdots\pi_k^{x_k}$ is the probability of a particular outcome of the experiment in which B_j has occurred x_j times, $j=1,\ldots,k$ and

$$\binom{n}{x_1\ldots x_j\ldots x_k}=\frac{n!}{x_1!\cdots x_j!\cdots x_k!}$$

is the number of different outcomes of the experiment in which B_j has occurred x_j times, $j=1,\ldots,k$. This number is often referred to as the *multinomial coefficient*. The events, $B_1,\ldots, B_j,\ldots, B_k$, are sometimes called the *categories* of the multinomial distribution.

In biology and geology there are numerous examples of data from experiments that satisfy the conditions (a) – (d) above and the statistical analysis of such data can therefore without further model checking be carried out by means of a model based on the multinomial distribution. In Section 7.1 we introduce four data sets from biology or geology that will be used as illustrations later in this chapter. Section 7.2 is concerned with statistical inference based on one multinomial distribution only and among other things, it illustrates a test of a simple hypothesis, a test of the hypothesis of Hardy-Weinberg proportions, and a test of the hypothesis of independence in two-way tables. In Section 7.3 the theory of a model based on several independent multinomial distributions is illustrated with the test of identity of the probability vectors in independent multinomial distributions. All the above-mentioned tests are based on likelihood theory and the χ^2-approximation to the significance probability discussed in Section 11.3. Section 7.4 gives an example of the application of Fisher's exact test in a situation where conditions for applying the χ^2-approximation to the significance probability are not satisfied. In Section 7.6 we give and illustrate the results from the theory for testing a number of successive hypotheses in the model for a single multinomial distribution.

In Chapter 2 we considered various graphical methods for checking the assumptions concerning the distributions in a statistical model. Sometimes these methods can be supplemented with significance tests often referred to as test for goodness of fit. The topic in Section 7.5 is tests of this type.

All the tests in this chapter may be performed with a statistical computer package. In an annex to this chapter we give examples of calculations performed in SAS.

7.1 Examples

In this section we introduce the four examples that will be used to illustrate statistical inference in models based on the multinomial distribution.

Example 7.1

Mendel (1866) crossed two strains of peas, one with yellow and round seeds, the other with green and wrinkled seeds. All the resulting peas were yellow and round. All these peas were grown and self-fertilized. The distribution of 556 of the resulting peas according to color and form was:

Yellow&Round	*Yellow&Wrinkled*	*Green&Round*	*Green&Wrinkled*
315	101	108	32

In another experiment Mendel showed that the form of the pea is determined by a gene with two allelic forms A and a. Each pea possesses a pair of the gene, one received through the pollen and one through the ovule, and three genotypes AA, Aa, and aa are therefore possible. The strain with round seeds has genotype AA and the wrinkled strain has genotype aa. The cross between these gives hybrid peas of genotype Aa, which has the phenotype round. The allele A is thus dominant and allele a is recessive, and the peas of the genotypes AA and Aa are round and those of genotype aa have the phenotype wrinkled. In the above experiment the two phenotypes are expected to segregate in the Mendelian proportions 3:1. The genotypes AA and aa are called homozygotes and Aa heterozygote. Similarly, two alleles B and b of another gene determine the color. The dominant allele B gives yellow seeds and the recessive allele b gives green seeds. If the two genes are carried on different chromosomes the inheritance of the two characters is independent, and segregation into the four phenotypes will be in the proportions 9:3:3:1.

In this situation we may consider the observations as the result of 556 trials corresponding to the 556 peas. For each pea it is recorded which of the four possible combinations of color and form has occurred. Assuming that the probability of the four combinations is the same for all peas and, in addition, that the color and form of a pea have no influence of the color and form of the rest of the peas, the conditions (a) – (d) above are satisfied. Thus we may consider the multinomial model with $k = 4$ and in this model to test the simple hypothesis:

$$H_0: \quad (\pi_1, \pi_2, \pi_3, \pi_4) = (\frac{9}{16}, \frac{3}{16}, \frac{3}{16}, \frac{1}{16}). \tag{7.2}$$

□

Example 7.2

In 1974 200 eelpouts (or viviparous blennies *Zoarces viviparus*) were caught in Mariager Fjord and their esterase types were determined by electrophoresis. The esterase types are determined by 2 allelic genes that will be denoted A and a, i.e., the genotypes are AA, Aa, and aa. According to Christiansen, Frydenberg, and Simonsen (1984) the result was:

AA	*Aa*	*aa*
56	107	37

In this situation we consider the multinomial model with $k = 3$. If p denotes the frequency* of the allele A, the hypothesis concerning *Hardy-Weinberg proportions* may in this model be specified as the composite hypothesis,

$$H_0\colon (\pi_1, \pi_2, \pi_3) = (p^2, 2p(1-p), (1-p)^2), \qquad p \in\ ]0,1[\,. \tag{7.3}$$

Hardy-Weinberg proportions describe the phenomenon that the probability of sampling an individual of a given genotype is the same as independently drawing the two genes from the population of genes. These proportions occur when the organism breeds by random mating with respect to the variation being studied. □

Example 7.3

As part of an investigation of stones, a student at the Department of Earth Sciences, University of Aarhus, was interested in determining if there is a connection between the shape of a stone and its weight. He therefore measured weight (in g), breadth (in cm), and length (in cm) of 179 stones. As a measure of the shape of a stone we consider the ratio between its breadth and its length. The three variables considered are all continuous, but in order to get a first crude impression of if and how the shape depends on the weight, the stones were classified as shown in the table below:

		Breadth/Length		
		< 0.6	0.6–0.8	> 0.8
	< 25	6	31	11
Weight	25–50	8	29	19
	> 50	10	39	26

We could of course write the observed frequencies as a vector of length 9, but it turns out to be convenient to choose a notation in accordance with the way the observations are given in the table above. Consequently, we let x_{ij} denote the number of stones in the ith category of the variable *weight* and in the jth category of the variable *breadth/length.* With this notation it seems reasonable to assume that the matrix $\{x_{ij}\}$ is a realization of a random matrix $\{X_{ij}\}$, which has a multinomial distribution with number of trials $n = 179$ and probability matrix $\{\pi_{ij}\}$.

We want to examine whether the shape (breadth/length) is independent of the weight. Let ρ_i denote the probability that a stone belongs to the ith weight category and, similarly, let σ_j denote the probability that the stone belongs to the jth shape category. Since π_{ij} denotes the probability that a stone belongs to the ith weight category and the jth shape category, the question of independence between weight and shape may be formulated as a hypothesis in the multinomial model in the following way:

$$H_0\colon \pi_{ij} = \rho_i \sigma_j, \qquad i = 1,2,3, \quad j = 1,2,3. \tag{7.4}$$

□

Example 7.4

The data in this example are from the county of Aarhus and they are concerned with the concentration of the pesticide *2.6-dichlorbenzamid* in ground water and in drinking water. The limit value for the concentration of this pesticide is 0.10 μg/l and the most important factors influencing the concentration are the content of clay in the soil and the depth from which the water is taken. The data are from major waterworks in 2 areas in the county, one in the central part and one in the northeastern part. In both areas samples of water from 2 municipalities have been investigated. The

* In genetics the "frequency of an allele A" is the proportion of genes in a population that are of allele type A and therefore a number between 0 and 1. This is in contrast to the meaning in statistics where a "frequency" is an integer. Although it would be more appropriate to use the term "relative frequency" in this context, we adopt the terminology from genetics in examples from this scientific area.

results are seen in the table below:

Area	*Municipality*	*Concentration* ≤ 0.01	0.01–0.10	>0.10	*Total*
1	*Hadsten*	23	12	6	41
	Hammel	20	5	9	34
2	*Nørre Djurs*	37	6	4	47
	Randers	47	4	3	54

It seems reasonable to consider a model consisting of 4 independent multinomial distributions, each with 3 categories. For each of the areas it is of interest to examine if the two samples are homogeneous. In the proposed model this may be done by investigating for each area if the two probability vectors within the area are identical. Furthermore, it is of interest to investigate if the distributions vary from area to area. Thus, the scientific problem indicates a statistical analysis involving three hypotheses of identity of the probability vectors in independent multinomial distributions. □

7.2 Inference in One Multinomial Distribution

By way of introduction, we review those properties of the multinomial distribution that will be used. A discrete random vector, $\mathbf{X} = (X_1, \ldots, X_j, \ldots, X_k)^*$, has a multinomial distribution with number of trials n and probability vector $\boldsymbol{\pi} = (\pi_1, \ldots, \pi_j, \ldots \pi_k)^*$, $\mathbf{X} \sim m(n, \boldsymbol{\pi})$ for short, if the probability density function of $\mathbf{X}$ is

$$P(\mathbf{X} = \mathbf{x}) = \frac{n!}{x_1! \cdots x_j! \cdots x_k!} \pi_1^{x_1} \cdots \pi_j^{x_j} \cdots \pi_k^{x_k}, \tag{7.5}$$

where $\mathbf{x} = (x_1, \ldots, x_j, \ldots, x_k)^*$ is a vector such that

$$x_j \in \{0, 1, \ldots, n\}, \quad j = 1, \ldots, k \qquad \text{and} \qquad \sum_{j=1}^{k} x_j = n.$$

Sometimes we also use the notation $(X_1, \ldots, X_j, \ldots, X_k) \sim m(n, (\pi_1, \ldots, \pi_j, \ldots \pi_k))$ to indicate that $\mathbf{X} \sim m(n, \boldsymbol{\pi})$.

The probability vector $\boldsymbol{\pi}$ belongs to the set,

$$\boldsymbol{\Pi}^{(k)} = \{\boldsymbol{\pi} \in \mathbb{R}^k \mid \pi_j > 0,\ j = 1, \ldots, k,\ \sum_{j=1}^{k} \pi_j = 1\}. \tag{7.6}$$

Note that even though $\boldsymbol{\pi}$ is a k-dimensional vector its components do not vary freely. If, for instance, we know $\pi_1, \ldots, \pi_{k-1}$, then π_k may be calculated as $1 - \pi_1 - \cdots - \pi_{k-1}$. In the terminology, which will be introduced on page 306, the multinomial model has $k-1$ *free parameters*.

From probability theory it is well known that the vector of means of $\mathbf{X}$ is

$$E\mathbf{X} = n\boldsymbol{\pi} = (n\pi_1, \ldots, n\pi_j, \ldots, n\pi_k)^* \tag{7.7}$$

and that the covariance matrix of $\mathbf{X}$ has elements,

$$\begin{aligned} (Cov\,\mathbf{X})_{jj} &= VarX_j = n\pi_j(1-\pi_j), && j = 1, \ldots, k \\ (Cov\,\mathbf{X})_{ij} &= Cov(X_i, X_j) = -n\pi_i\pi_j, && i \neq j,\ i, j = 1, \ldots, k. \end{aligned} \tag{7.8}$$

Finally, the jth component X_j of $\mathbf{X}$ has a binomial distribution with number of trials n and probability parameter π_j, i.e.,

$$X_j \sim b(n, \pi_j), \qquad j = 1, \ldots, k. \tag{7.9}$$

When maximizing likelihood functions in the following, we use the mathematical result given in Proposition 7.1 several times. We omit the proof of the proposition.

Proposition 7.1 Suppose that $x_j > 0$ for $j = 1, \ldots, k$ and that $x_1 + \cdots + x_k = n$. Then the function,

$$g : \boldsymbol{\Pi}^{(k)} \to \mathbb{R}$$
$$\boldsymbol{\pi} \to \pi_1^{x_1} \cdots \pi_j^{x_j} \cdots \pi_k^{x_k},$$

attains its maximal value at the point,

$$\hat{\boldsymbol{\pi}} = (\frac{x_1}{n}, \cdots, \frac{x_j}{n}, \cdots, \frac{x_k}{n})^*.$$

♦

After these preliminary remarks we are now ready to consider statistical inference in the multinomial model,

$$M_0\colon \mathbf{X} = (X_1, \ldots, X_j, \ldots, X_k)^* \sim m(n, \boldsymbol{\pi}), \quad \boldsymbol{\pi} \in \boldsymbol{\Pi}^{(k)}.$$

Estimation

The probability of the vector of observations is given in (7.5) and considered as a function of the probability vector $\boldsymbol{\pi}$, it is the likelihood function, i.e.,

$$L(\boldsymbol{\pi}) = \frac{n!}{x_1! \cdots x_j! \cdots x_k!} \pi_1^{x_1} \cdots \pi_j^{x_j} \cdots \pi_k^{x_k}, \tag{7.10}$$

and it follows from Proposition 7.1 that the maximum likelihood estimate of $\boldsymbol{\pi}$ is

$$\boldsymbol{\pi} \leftarrow \hat{\boldsymbol{\pi}}(\mathbf{x}) = (\frac{x_1}{n}, \cdots, \frac{x_j}{n}, \cdots, \frac{x_k}{n})^*; \tag{7.11}$$

in other words, the maximum likelihood estimate $\hat{\pi}_j$ of π_j is the *relative frequency* with which the event B_j occurs in the n trials. Most often the distribution of $\hat{\boldsymbol{\pi}}$ is given as follows:

$$n\hat{\boldsymbol{\pi}} = \mathbf{X} \sim m(n, \boldsymbol{\pi}).$$

Hypotheses

We shall consider simplifications of the basic model M_0. The model M_0 is sometimes referred to as the *full* model or the *saturated* model to emphasize that no restrictions are imposed on the parameters of M_0. The simplifications of M_0 will be formulated as models and they will be submodels of M_0 in the sense that they have fewer parameters than M_0. In Section 7.6 we will consider a sequence of models,

$$M_0 \to M_1 \to \cdots \to M_{i-1} \to M_i \to \cdots,$$

where an arrow points from a model to its submodel. We shall consider reductions from a model to the next model in the sequence, e.g.,

$$M_{i-1} \to M_i.$$

A reduction from a model to a submodel is called a hypothesis and the reduction $M_{i-1} \to M_i$ is referred to as H_{0i}. With this notation H_{0i} is the hypothesis that the model M_i may replace the model M_{i-1} as explanation of the data. The models all assume a multinomial distribution of the frequencies $(X_1, \ldots, X_k)$ and therefore a model will be completely specified by the domain of variation of the probability vector of the multinomial distribution, and we will use the notation,

$$M_i\colon \boldsymbol{\pi} \in \boldsymbol{\Pi}_i.$$

For M_0 the parameter set $\boldsymbol{\Pi}_0$ is identical to $\boldsymbol{\Pi}^{(k)}$ defined in (7.6). So far the notation is analogous to the notation used for linear normal models, see Subsection 3.3.9 on page 134 and Section 4.1 on page 177, the vector of probabilities $\boldsymbol{\pi}$ plays the role of the vector of means $\boldsymbol{\mu}$ and the subset $\boldsymbol{\Pi}_i$ plays the role of the linear subspace L_i. An essential feature of the linear normal models is the number of unknown parameters of the vector of means, which plays an important role in determining the degrees of freedom of the F-test of the reduction. In linear normal models it is the dimension of the linear subspace which makes the number of parameters of the mean precise.

In the models for the multinomial distribution, the number of parameters in the model also play a crucial part in determining the distribution of the test statistics, but for the multinomial distribution we must be particularly careful when we define the number of parameters; for it is not just the number of parameters that is important, but the number of *free* parameters, or the smallest number of parameters needed to specify the vector of probabilities. A precise definition of the number of free parameters of a model M is as follows.

Let $\boldsymbol{\pi}$ be a one-to-one mapping of a domain $\boldsymbol{\Theta}$ in $\mathbb{R}^d$ onto a subset $\boldsymbol{\Pi}$ of the parameter space $\boldsymbol{\Pi}_0$,

$$\begin{aligned} \boldsymbol{\pi} : \boldsymbol{\Theta} \subseteq \mathbb{R}^d &\to \boldsymbol{\Pi} \subset \boldsymbol{\Pi}_0 \, (= \boldsymbol{\Pi}^{(k)}) \\ \boldsymbol{\theta} = (\theta_1, \ldots, \theta_d)^* &\to \boldsymbol{\pi}(\boldsymbol{\theta}) = (\pi_1(\boldsymbol{\theta}), \ldots, \pi_j(\boldsymbol{\theta}), \ldots, \pi_k(\boldsymbol{\theta}))^*. \end{aligned} \tag{7.12}$$

The model,

$$M\colon \mathbf{X} = (X_1, \ldots, X_j, \ldots, X_k)^* \sim m(n, \boldsymbol{\pi}), \quad \boldsymbol{\pi} \in \boldsymbol{\Pi} = \pi(\boldsymbol{\Theta}) \ (\subset \boldsymbol{\Pi}_0), \tag{7.13}$$

is then said to have *d free parameters.*

Note that this definition has already been applied to $\boldsymbol{\Pi}_0$ after (7.6) when it was pointed out that M_0 has $k-1$ free parameters.

When we consider a sequence of models, the mapping $\boldsymbol{\pi}$, the number of free parameters d and the parameter sets $\boldsymbol{\Theta}$ and $\boldsymbol{\Pi}$ will have the same index as the model, see page 332.

The general definition is illustrated in Figure 7.1 and Figure 7.2 on page 314 shows the picture in connection with the model corresponding to the hypothesis of Hardy-Weinberg proportions in Example 7.2. Since the one-to-one mapping $\boldsymbol{\pi}$ is defined on $\boldsymbol{\Theta}$ and has range set $\boldsymbol{\Pi}$, it simply means that for every element $\boldsymbol{\theta}$ in $\boldsymbol{\Theta}$ there exists one and only one element $\boldsymbol{\pi}(\boldsymbol{\theta})$ in $\boldsymbol{\Pi}$ and vice versa; in other words, the set $\boldsymbol{\Theta}$ is used for naming of the elements in $\boldsymbol{\Pi}$.

In the following we will consider the reduction from the full model to the model,

$$M\colon \mathbf{X} = (X_1, \ldots, X_j, \ldots, X_k)^* \sim m(n, \boldsymbol{\pi}), \quad \boldsymbol{\pi} \in \boldsymbol{\Pi},$$

or, in other words, the hypothesis,

$$H_0\colon \boldsymbol{\pi} \in \boldsymbol{\Pi}.$$

We have defined a hypothesis as the reduction of one model to a simpler one. Often the submodel is defined by some constraints on the parameters of a model and then we will denote those constraints as the hypothesis. When the submodel is fully specified, for example, this is formulated as the simple hypothesis,

$$H_0\colon \boldsymbol{\pi} = \boldsymbol{\pi}_0.$$

Incidentally, the model corresponding to a simple hypothesis has 0 free parameters and this is what we have in mind when we use the term "fully specified."

When we estimate in a submodel or specify a distribution in a submodel, we may equivalently say that we estimate under the hypothesis or specify the distribution under the hypothesis. Moreover, because a hypothesis defines the submodel, we will say that a hypothesis has d free parameters if the corresponding submodel has d free parameters.

Estimation under a Hypothesis

We now consider maximum likelihood estimation in the model M. In M the probability vector is of the form $\boldsymbol{\pi}(\boldsymbol{\theta}) = (\pi_1(\boldsymbol{\theta}), \ldots, \pi_j(\boldsymbol{\theta}), \ldots, \pi_k(\boldsymbol{\theta}))^*$, where $\boldsymbol{\theta} = (\theta_1, \ldots, \theta_i, \ldots, \theta_d)^*$. From (7.10) we get

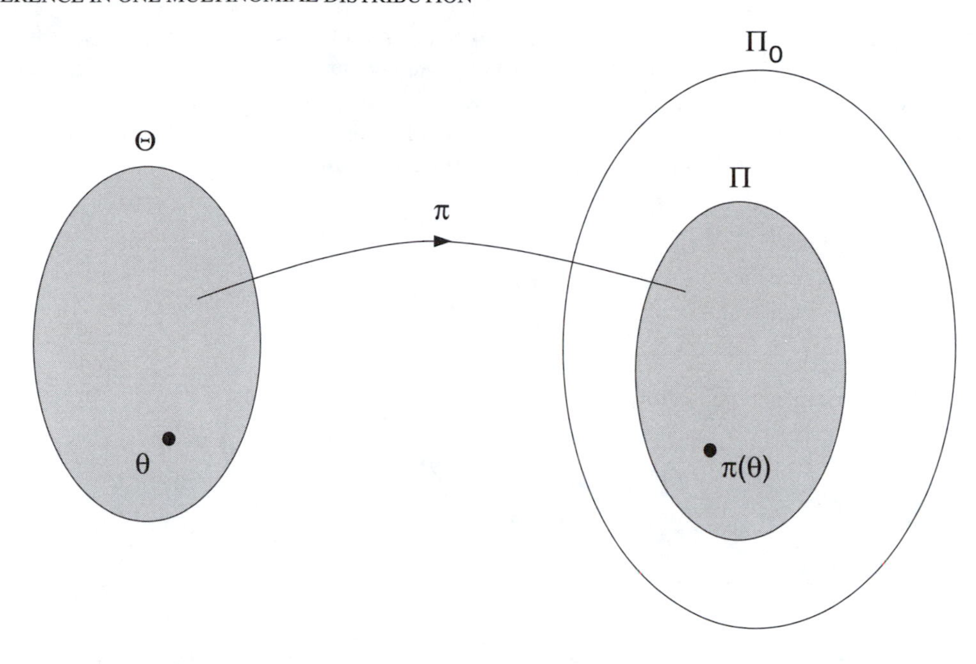

Figure 7.1 *Schematic illustration of the definition of a model M with d free parameters. The set* $\boldsymbol{\Theta}$ *is assumed to be a domain in* $\mathbb{R}^d$. *The set* $\boldsymbol{\Pi}_0$ *symbolizes the parameter space in the basic model* M_0 *and* $\boldsymbol{\Pi}$ *symbolizes the parameter space corresponding to M. In general, it is not possible to illustrate that the dimension d of the sets* $\boldsymbol{\Theta}$ *and* $\boldsymbol{\Pi}$ *is less than the dimension* $k-1$ *of* $\boldsymbol{\Pi}_0$, *which is a subset of* $\mathbb{R}^k$. *This is only possible if* $d=1$ *and* $k=3$ *as in Figure 7.2 on page 314.*

that the likelihood function for $\boldsymbol{\theta}$ is

$$L(\boldsymbol{\theta}) = \frac{n!}{x_1!\cdots x_j!\cdots x_k!}\,\pi_1(\boldsymbol{\theta})^{x_1}\cdots\pi_j(\boldsymbol{\theta})^{x_j}\cdots\pi_k(\boldsymbol{\theta})^{x_k}. \tag{7.14}$$

Therefore, the log likelihood function and the likelihood equations become, respectively,

$$l(\boldsymbol{\theta}) = \ln\left(\frac{n!}{x_1!\cdots x_j!\cdots x_k!}\right) + \sum_{j=1}^{k} x_j \ln(\pi_j(\boldsymbol{\theta}))$$

and

$$\frac{\partial l}{\partial\theta_i}(\boldsymbol{\theta}) = \sum_{j=1}^{k} x_j \frac{1}{\pi_j(\boldsymbol{\theta})}\frac{\partial\pi_j}{\partial\theta_i}(\boldsymbol{\theta}), \qquad i=1,\ldots,d.$$

How the likelihood equations are solved depends of course on the hypothesis H_0 and can therefore not be discussed in general. However, it is often possible, as illustrated in the following, to maximize the likelihood function $L(\boldsymbol{\theta})$ using Proposition 7.1 and in such cases the likelihood equations are without interest.

Let $\hat{\boldsymbol{\theta}}$ denote the maximum likelihood estimate of $\boldsymbol{\theta}$. The vector of means of $\mathbf{X}$ calculated in the distribution with probability vector $\boldsymbol{\pi}(\hat{\boldsymbol{\theta}})$ is called the *vector of expected frequencies under* H_0. According to (7.7) it is

$$\mathbf{e} = (e_1,\ldots,e_j,\ldots,e_k)^* = (n\pi_1(\hat{\boldsymbol{\theta}}),\ldots,n\pi_j(\hat{\boldsymbol{\theta}}),\ldots,n\pi_k(\hat{\boldsymbol{\theta}}))^*, \tag{7.15}$$

Test of a Hypothesis

Recall that $\hat{\boldsymbol{\theta}}$ is the value of the parameter $\boldsymbol{\theta}$, which assigns the largest probability to the observation $\mathbf{x}$, and that the vector of expected frequencies $\mathbf{e} = n\boldsymbol{\pi}(\hat{\boldsymbol{\theta}})$ is the maximum likelihood estimate, under

H_0, of the vector of means of $\mathbf{X}$. The question of whether the hypothesis is rejected may therefore be decided by investigating whether the vector of expected frequencies $\mathbf{e}$ "deviates too much from" the observation $\mathbf{x}$. Only one question remains: "How should the comparison between $\mathbf{e}$ and $\mathbf{x}$ be performed?" Let us consider which answer the likelihood method gives to this question.

From (7.10), (7.11), and (7.14), the likelihood ratio test statistic of H_0 is

$$\begin{aligned} Q(\mathbf{x}) &= \frac{L(\hat{\boldsymbol{\theta}})}{L(\hat{\boldsymbol{\pi}})} \\ &= \frac{\pi_1(\hat{\boldsymbol{\theta}})^{x_1}\cdots\pi_j(\hat{\boldsymbol{\theta}})^{x_j}\cdots\pi_k(\hat{\boldsymbol{\theta}})^{x_k}}{\left(\frac{x_1}{n}\right)^{x_1}\cdots\left(\frac{x_j}{n}\right)^{x_j}\cdots\left(\frac{x_k}{n}\right)^{x_k}} \\ &= \left(\frac{n\pi_1(\hat{\boldsymbol{\theta}})}{x_1}\right)^{x_1}\cdots\left(\frac{n\pi_j(\hat{\boldsymbol{\theta}})}{x_j}\right)^{x_j}\cdots\left(\frac{n\pi_k(\hat{\boldsymbol{\theta}})}{x_k}\right)^{x_k} \\ &= \left(\frac{e_1}{x_1}\right)^{x_1}\cdots\left(\frac{e_j}{x_j}\right)^{x_j}\cdots\left(\frac{e_k}{x_k}\right)^{x_k}, \end{aligned}$$

and thus we have

$$-2\ln Q(\mathbf{x}) = 2\sum_{j=1}^{k} x_j \ln\left(\frac{x_j}{e_j}\right). \tag{7.16}$$

If *the expected frequencies are all greater than or equal to* 5, we may use the approximation in (11.33), i.e., we have the following approximation of the p-value (the significance probability) $p_{obs}(\mathbf{x})$,

$$p_{obs}(\mathbf{x}) \doteq 1 - F_{\chi^2(k-1-d)}(-2\ln Q(\mathbf{x})). \tag{7.17}$$

The degrees of freedom are $k-1-d$ because M_0 has $k-1$ free parameters and M has d free parameters. Note that it is possible to test the hypothesis only when $d < k-1$, since the degrees of freedom in the approximating χ^2-distribution have to be positive.

A couple of remarks concerning the calculation of the $-2\ln Q$-test statistic on a pocket calculator are called for. The most frequent mistake is that the **factor 2** on the right-hand side in formula (7.16) is forgotten. Furthermore, it is important to emphasize that since $Q(\mathbf{x})$ is the likelihood ratio test statistic, $0 < Q(\mathbf{x}) \leq 1$ and, consequently, $-2\ln Q(\mathbf{x}) > 0$. If a calculation results in a negative value of $-2\ln Q(\mathbf{x})$, there is only one explanation: **miscalculation**.

Sometimes the comparison of the observed frequencies $\mathbf{x}$ and the expected frequencies $\mathbf{e}$ is performed using the X^2-test statistic (read: chi-square test statistic),

$$X^2(\mathbf{x}) = \sum_{j=1}^{k} \frac{(x_j - e_j)^2}{e_j}. \tag{7.18}$$

If *the expected frequencies are all greater than or equal to* 5, the p-value (the significance probability) $p^*_{obs}(\mathbf{x})$ of the X^2-test of a hypothesis H_0 with d free parameters may be approximated using the χ^2-distribution:

$$p^*_{obs}(\mathbf{x}) \doteq 1 - F_{\chi^2(k-1-d)}(X^2(\mathbf{x})). \tag{7.19}$$

This is exactly the same approximation as the approximation of the p-value of the $-2\ln Q$-test statistic. It can be shown, but it is beyond the scope of this book to do so, that the X^2-test statistic is an approximation of the $-2\ln Q$-test statistic. This is the explanation behind the fact that the X^2 statistic and the $-2\ln Q$ statistic are very similar. In Remark 7.5 on page 312 it is shown in a particular case that the X^2-test statistic is an approximation of the $-2\ln Q$-test statistic.

Of the two tests of H_0 we have a slight preference for the $-2\ln Q$-test since the X^2-test statistic is only an approximation of the $-2\ln Q$-test statistic. In the literature, especially older literature, one often finds applications of the X^2-test. A possible explanation may be that the ln-key was not available on pocket calculators at the time, so it was more difficult to calculate the $-2\ln Q$-test, and also that historically, the X^2 statistic was developed before the likelihood ratio test. For the sake of illustration in some of the examples we calculate both X^2-test and the $-2\ln Q$-test, but the conclusions are drawn by means of the $-2\ln Q$-test.

Remark 7.1 Neither the $-2\ln Q$-test in (7.16) nor the X^2-test in (7.18) takes the numbering of the k categories into account, and sometimes these tests are supplemented with a study of the *residuals* $\mathbf{r}$ in the model M, which are the differences between the observed frequencies $\mathbf{x}$ and the expected frequencies $\mathbf{e}$ in M, i.e., $r_j = x_j - e_j$, $j = 1, \ldots, k$. Note that the sum of the residuals is equal to 0, and that the contribution to X^2-test statistic from the jth observation is r_j^2/e_j.

Both the signs of the residuals and their magnitudes are of interest.

If there is a natural ordering of the categories, for instance, if the categories are defined in terms of a numerical quantity, the signs of the residual are studied. If a systematic pattern of the signs is found, for instance, a sequence of positive or negative signs, the hypothesis H_0 is occasionally rejected even though the test in (7.16) or in (7.18) does not reject the hypothesis. In order to detect significant deviations in terms of the signs, it is required that the number of categories k is not too small and it should be taken into account that the sums of the residuals is equal to 0.

When the hypothesis H_0 is rejected, the residuals are of interest in interpreting what aspects of H_0 or the model reduction that were not supported by the data. Both the signs of the residuals and their magnitudes are important, and often the contributions to the X^2-test statistic, r_j^2/e_j, $j = 1, \ldots, k$, are used to find those observations x_j, which deviate significantly from the corresponding expected frequencies e_j. As a rule of thumb this is the case if $r_j^2/e_j > \chi^2_{0.95}(1) = 3.84$. The rule is based on the approximation $r_j^2/e_j \sim\approx \chi^2(1)$, which is correct for a simple hypothesis H_0, but the rule is applied in connection with composite hypotheses also. ▼

Remark 7.2 The condition that the expected frequencies should all be larger than or equal to 5 in order to apply the approximation based on the χ^2-distribution to the significance probability in (7.17) or in (7.19) is sometimes too restrictive. Numerical simulations of the distribution of the $-2\ln Q$-test statistic and of the distribution of the X^2-test statistic show that the condition may be weakened in some situations and that the approximations in (7.17) and (7.19) are reliable even though some of the expected frequencies are somewhat less than 5. A general discussion of this point is difficult (impossible), because the condition imposed on the size of the expected frequencies to ensure that the approximations in (7.17) and (7.19) are valid depends on the hypothesis H_0.

On page 323 we consider an example where 3 out of 6 expected frequencies are less than 5.

In Section 7.4 we discuss Fisher's exact test, which may be used instead of the $-2\ln Q$-test, and the X^2-test in the test of independence in a two-way table, see Subsection 7.2.3, and in the test of homogeneity, see Subsection 7.3.1, when some of the expected frequencies are less than 5. ▼

Confidence Intervals

We will not discuss confidence regions for the probability vector $\boldsymbol{\pi}$ in the model M_0, but restrict ourselves to giving the confidence interval for the jth component π_j of $\boldsymbol{\pi}$. Since the jth component X_j of $\mathbf{X}$ has a binomial distribution with number of trials n and probability parameter π_j, according to formula (7.9), the problem is reduced to finding the confidence interval for the probability parameter π in a binomial model,

$$M_b\colon X \sim b(n, \pi).$$

Since a $(1-\alpha)$ confidence interval are those values of the parameter π that will not be rejected as null hypothesis by a test with significance level α, we have to start by discussing a test of the hypothesis $H_0\colon \pi = \pi_0$ on the basis of the observation x of X. This test will be called the *binomial test.*

First, suppose that $x < n\pi_0$. Observations less than x will be more critical or at least as critical for H_0 as the observation x. This also applies to observations on the other side of the mean $n\pi_0$.

However, since the binomial distribution is symmetrical around the mean only if $\pi_0 = \frac{1}{2}$, we simply define the p-value $p^b_{obs}(\mathbf{x})$ as twice the probability that an observation is less than or equal to x, i.e.,

$$p^b_{obs}(x) = 2P_{b(n,\pi_0)}(X \leq x) = 2\sum_{i=0}^{x} b(i;n,\pi_0). \tag{7.20}$$

According to the normal approximation of the binomial distribution, also known as de Moivre-Laplace's theorem, the tail probabilities in the distribution $b(n,\pi_0)$ may be approximated by the tail probabilities in the normal distribution with the same mean and the same variance, i.e., in the distribution $N(n\pi_0, n\pi_0(1-\pi_0))$. Thus we may approximate $p^b_{obs}(x)$ by

$$p^n_{obs}(x) = 2\Phi\left(\frac{x - n\pi_0}{\sqrt{n\pi_0(1-\pi_0)}}\right).$$

If $x > n\pi_0$, it is the values larger than or equal to x that are more critical or at least as critical for H_0 as the observation x, so in that case the significance probability is defined as

$$p^b_{obs}(x) = 2P_{b(n,\pi_0)}(X \geq x) = 2\sum_{i=x}^{n} b(i;n,\pi_0), \tag{7.21}$$

and the normal approximation of the significance probability is

$$\begin{aligned} p^n_{obs}(x) &= 2\left[1 - \Phi\left(\frac{x - n\pi_0}{\sqrt{n\pi_0(1-\pi_0)}}\right)\right] \\ &= 2\Phi\left(-\frac{x - n\pi_0}{\sqrt{n\pi_0(1-\pi_0)}}\right). \end{aligned}$$

To sum up, we may define the normal approximation of the significance probability as

$$p^n_{obs}(x) = 2\Phi\left(-\frac{|x - n\pi_0|}{\sqrt{n\pi_0(1-\pi_0)}}\right), \tag{7.22}$$

In order to calculate the confidence interval, we note that H_0 is not rejected if $p^n_{obs}(x) > \alpha$, i.e., if

$$-\frac{|x - n\pi_0|}{\sqrt{n\pi_0(1-\pi_0)}} > \Phi^{-1}(\alpha/2) = u_{a/2} = -u_{1-\alpha/2}$$

or if

$$-u_{1-\alpha/2} < \frac{x - n\pi_0}{\sqrt{n\pi_0(1-\pi_0)}} < u_{1-\alpha/2}. \tag{7.23}$$

Since the $(1-\alpha)$ confidence interval for the probability parameter π of a binomial distribution are those values that are not rejected by a test with significance level α, the confidence interval consists of precisely the values π_0, which satisfy the equation (7.23). That is

$$\begin{aligned} C_{1-\alpha} = [\pi_-, \pi_+] &= \{\pi_0 \,|\, H_0\colon \pi = \pi_0 \text{ is not rejected by a test with significance level } \alpha\} \\ &= \left\{\pi_0 \,\middle|\, -u_{1-\alpha/2} < \frac{x - n\pi_0}{\sqrt{n\pi_0(1-\pi_0)}} < u_{1-\alpha/2}\right\}. \end{aligned} \tag{7.24}$$

Here the lower limit of the confidence interval π_- is the solution to the equation,

$$\frac{x - n\pi_0}{\sqrt{n\pi_0(1-\pi_0)}} = u_{1-\alpha/2}, \tag{7.25}$$

and the upper limit π_+ is the solution to the equation,

$$\frac{x - n\pi_0}{\sqrt{n\pi_0(1-\pi_0)}} = -u_{1-\alpha/2}. \tag{7.26}$$

(7.25) and (7.26) may be rewritten as the same equation of second degree in π_0, where π_- is the smallest and π_+ the largest of the two roots. The result is

$$\pi_- = \frac{1}{n+u_{1-\alpha/2}^2}\left[x+\frac{u_{1-\alpha/2}^2}{2} - u_{1-\alpha/2}\sqrt{\frac{x(n-x)}{n}+\frac{u_{1-\alpha/2}^2}{4}}\right] \tag{7.27}$$

and

$$\pi_+ = \frac{1}{n+u_{1-\alpha/2}^2}\left[x+\frac{u_{1-\alpha/2}^2}{2} + u_{1-\alpha/2}\sqrt{\frac{x(n-x)}{n}+\frac{u_{1-\alpha/2}^2}{4}}\right]. \tag{7.28}$$

Since the confidence interval is based on an approximation, one has to demonstrate either by exact calculations or by simulations that the confidence interval has the property that justifies its application in practice. The defining property of a $(1-\alpha)$ confidence interval is that the probability that it contains the parameter is $1-\alpha$. For a $(1-\alpha)$ confidence interval the number $1-\alpha$ was originally called the confidence coefficient but the name *coverage probability* has gradually become common. For approximate confidence intervals we sometimes use the term *nominal coverage probability*, in order to emphasize that we are aiming to have a $(1-\alpha)$ confidence interval but realize that in reality this may not be achieved due to the approximations involved. In this particular case exact calculations suffice and they show that the confidence interval with the limits (7.27) and (7.28) have real coverage probabilities that depend on the parameter π, but that vary around the nominal coverage probability of $1-\alpha$.

Remark 7.3 The literature contains another version of the binomial test based on another approximation of the tail probabilities in the binomial distribution, which for $x < n\pi_0$ approximates $p_{obs}^b(x)$ by

$$p_{obs}^{nc}(x) = 2\Phi\left(\frac{x-n\pi_0+\frac{1}{2}}{\sqrt{n\pi_0(1-\pi_0)}}\right),$$

and which for every x is

$$p_{obs}^{nc}(x) = 2\Phi\left(\frac{-|x-n\pi_0|+\frac{1}{2}}{\sqrt{n\pi_0(1-\pi_0)}}\right). \tag{7.29}$$

This approximation is said to use the *continuity correction*. If this approximation is used to construct the confidence interval, then

$$\pi_{c-} = \frac{1}{n+u_{1-\alpha/2}^2}\left[(x-\frac{1}{2})+\frac{u_{1-\alpha/2}^2}{2} - u_{1-\alpha/2}\sqrt{\frac{(x-\frac{1}{2})(n-x+\frac{1}{2})}{n}+\frac{u_{1-\alpha/2}^2}{4}}\right]$$

and

$$\pi_{c+} = \frac{1}{n+u_{1-\alpha/2}^2}\left[(x+\frac{1}{2})+\frac{u_{1-\alpha/2}^2}{2} + u_{1-\alpha/2}\sqrt{\frac{(x+\frac{1}{2})(n-x-\frac{1}{2})}{n}+\frac{u_{1-\alpha/2}^2}{4}}\right],$$

for the limits of the confidence interval for π. This version of the confidence interval we call the confidence interval with continuity correction. We cannot recommend using the confidence intervals with continuity correction because a closer study shows that for the confidence intervals with continuity correction the real coverage probability is systematically larger than the nominal coverage probability. ▼

Remark 7.4 In the literature one may also find a third version of the confidence interval for the probability parameter of the binomial distribution. It has the limits,

$$\frac{1}{n}\left[x \pm u_{1-\alpha/2}\sqrt{\frac{x(n-x)}{n}}\right].$$

Use of this confidence interval cannot be recommended. It has much worse coverage probabilities than the confidence interval with the limits in (7.27) and (7.28). ▼

Remark 7.5 In connection with the confidence interval for the probability parameter of a binomial distribution we introduced the binomial test. Before that we have considered the likelihood ratio test statistic in (7.16) and the X^2-test statistic in (7.18). It follows from the remarks preceding formula (7.23) that the binomial test rejects H_0: $\pi = \pi_0$ if

$$\left(\frac{x - n\pi_0}{\sqrt{n\pi_0(1-\pi_0)}}\right)^2 > u_{1-\alpha/2}^2 = \chi_{1-\alpha}^2(1).$$

It may be seen that

$$\left(\frac{x - n\pi_0}{\sqrt{n\pi_0(1-\pi_0)}}\right)^2 = \frac{(x-n\pi_0)^2}{n\pi_0} + \frac{(n-x-n(1-\pi_0))^2}{n(1-\pi_0)},$$

which is the X^2-test statistic for H_0. Thus we have shown how de Moivre-Laplace's theorem in this situation ensures the approximative $\chi^2(1)$-distribution of the X^2-test statistic for H_0: $\pi = \pi_0$ in the binomial model.

For this model it is also feasible to show that the approximating second-order Taylor polynomial around $x = n\pi_0$ of

$$-2\ln Q((x, n-x)) = 2\left(x\ln\frac{x}{n\pi_0} + (n-x)\ln\frac{n-x}{n-n\pi_0}\right)$$

is the X^2-test statistic. In general, the X^2-test statistic is a quadratic approximation of the likelihood ratio test statistic $-2\ln Q$ and this is the reason why these two quantities not only have the same approximating distribution, but also that numerically they are close to each other for every data set. ▼

There exist many applications of the theory for one multinomial distribution as described in this section. Here we restrict ourselves to illustrating the theory by discussing a test of a simple hypothesis, test of the hypothesis of Hardy-Weinberg proportions, and test of independence in a two-way table. This is done in the three subsections of Section 7.2 by means of Example 7.1 to Example 7.3.

7.2.1 Test of a Simple Hypothesis

In Example 7.1 the hypothesis is that the probability vector is totally specified and, consequently, it is a so-called simple hypothesis.

Example 7.1 (Continued)
In the full multinomial model with $k = 4$,

$$M_0\colon (X_1, X_2, X_3, X_4) \sim m(556, (\pi_1, \pi_2, \pi_3, \pi_4)), \quad \boldsymbol{\pi} \in \boldsymbol{\Pi}^{(4)},$$

the maximum likelihood estimate of the probability vector $\boldsymbol{\pi}$ is

$$\hat{\boldsymbol{\pi}}^* = (\frac{315}{556}, \frac{101}{556}, \frac{108}{556}, \frac{32}{556}) = (0.5665,\ 0.1817,\ 0.1942,\ 0.0576).$$

The hypothesis that the segregation is in the proportions 9:3:3:1 corresponds in terms of the probability vector to the hypothesis,

$$H_0\colon \boldsymbol{\pi} = \boldsymbol{\pi}_0 = (\frac{9}{16}, \frac{3}{16}, \frac{3}{16}, \frac{1}{16})^* = (0.5625, 0.1875, 0.1875, 0.0625)^*$$

and it is seen that $\hat{\boldsymbol{\pi}}$ is close to this value. The hypothesis is simple, so no estimation under the hypothesis is necessary, and the expected frequencies under H_0 can be calculated immediately as $\mathbf{e} = 556\boldsymbol{\pi}_0$. We find:

	Yellow&Round	*Yellow&Wrinkled*	*Green&Round*	*Green&Wrinkled*	*Total*
Observed $\mathbf{x}$	315	101	108	32	556
Expected $\mathbf{e}$	312.75	104.25	104.25	34.75	556

Using formula (7.16),

$$-2\ln Q(\mathbf{x}) = 0.4754.$$

The expected frequencies are all greater than 5 and $d = 0$ because the hypothesis is simple, so according to (7.17) the significance probability becomes

$$p_{obs}(\mathbf{x}) = 1 - F_{\chi^2(3)}(0.4754) = 0.924,$$

and the hypothesis is not rejected.

Consequently, the experiment does not contradict Mendel's theories on dominance and independent heredity of form and color. We later return to this example for a further evaluation of theories of Mendel in light of this experiment. □

7.2.2 *Hardy-Weinberg Proportions*

In the full multinomial model with $k = 3$,

$$M_0\colon (X_1, X_2, X_3) \sim m(n, (\pi_1, \pi_2, \pi_3)), \quad \boldsymbol{\pi} \in \boldsymbol{\Pi}^{(3)},$$

the hypothesis of Hardy-Weinberg proportions may be formulated as the following hypothesis concerning the probability vector $\boldsymbol{\pi}$:

$$H_{01}\colon (\pi_1, \pi_2, \pi_3) = (p^2, 2p(1-p), (1-p)^2), \qquad p \in\,]0,1[\,.$$

Obviously, this hypothesis has one free parameter p, so $d_1 = 1$. Formally, this may be seen by defining the mapping $\boldsymbol{\pi}_1$ in (7.12) as

$$\begin{aligned} \boldsymbol{\pi}_1 :\ ]0,1[\ &\rightarrow \boldsymbol{\Pi}_0\ (= \boldsymbol{\Pi}^{(3)}) \\ p &\rightarrow (p^2, 2p(1-p), (1-p)^2)^*. \end{aligned}$$

The mapping $\boldsymbol{\pi}_1$ is illustrated in Figure 7.2.

The likelihood function under H_{01} is according to (7.14):

$$\begin{aligned} L(p) &= \frac{n!}{x_1!x_2!x_3!}(p^2)^{x_1}(2p(1-p))^{x_2}((1-p)^2)^{x_3} \\ &= \frac{n!}{x_1!x_2!x_3!}2^{x_2}p^{2x_1+x_2}(1-p)^{x_2+2x_3}. \end{aligned}$$

Using Proposition 7.1 we find that

$$\begin{aligned} \hat{p} &= \frac{2x_1 + x_2}{(2x_1 + x_2) + (x_2 + 2x_3)} \\ &= \frac{2x_1 + x_2}{2n}; \end{aligned}$$

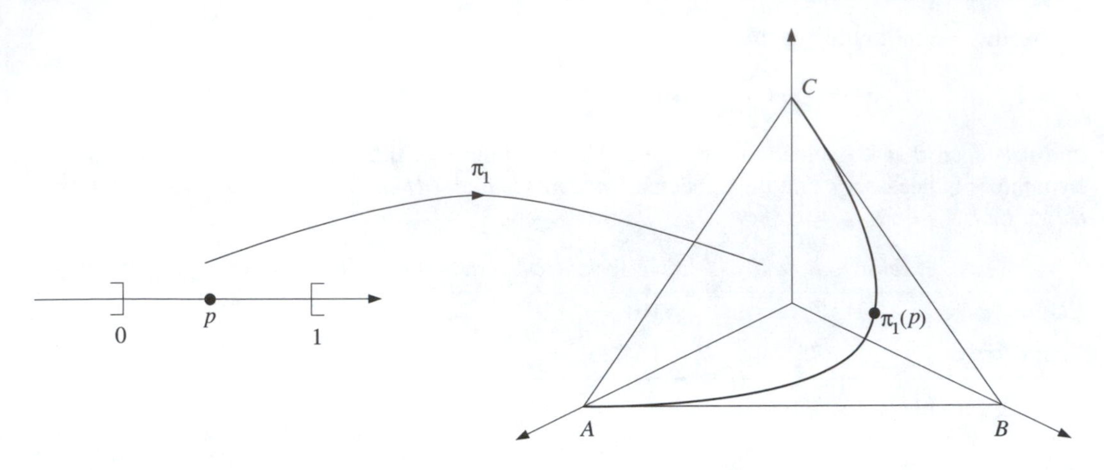

Figure 7.2 *Illustration of the mapping $\boldsymbol{\pi}_1$, which defines the hypothesis H_{01} of Hardy-Weinberg proportions. The parameter space for the basic model, the three-dimensional multinomial distribution, is the triangle determined by the points $A = (1,0,0)$, $B = (0,1,0)$, and $C = (0,0,1)$. The parameter space for H_{01} is the curve from A to C.*

in other words, the maximum likelihood estimate $\hat{p}$ of the frequency p of the allele A is the *relative frequency* of the A gene among the total number of genes, $2n$. The expected frequencies $\mathbf{e}$ under H_{01} may now be calculated as

$$\mathbf{e} = (n\hat{p}^2, 2n\hat{p}(1-\hat{p}), n(1-\hat{p})^2)^*. \tag{7.30}$$

The $-2\ln Q$-test statistic of H_{01} has $k-1-d_1 = 3-1-1 = 1$ degree of freedom. If the hypothesis of Hardy-Weinberg proportions is not rejected, the model M_0 is reduced to

$$M_1\colon (X_1, X_2, X_3) \sim m(n, (p^2, 2p(1-p), (1-p)^2)).$$

In this model it may be shown that the number of A genes has a binomial distribution with number of trials $2n$ and probability parameter p, i.e.,

$$2X_1 + X_2 \sim b(2n, p), \tag{7.31}$$

a result which may be used to test that the frequency p of the allele A has a particular value p_0 or to find the confidence interval for p, by means of the formulas (7.27) and (7.28).

Example 7.2 (Continued)
For these data the maximum likelihood estimate of the frequency p of the allele A is

$$\hat{p} = \frac{2x_1 + x_2}{2n} = \frac{2 \times 56 + 107}{2 \times 200} = \frac{219}{400} = 0.5475,$$

and the expected frequencies $\mathbf{e}$ may now be calculated using formula (7.30). We find:

	AA	*Aa*	*aa*	*Total*
Observed $\mathbf{x}$	56	107	37	200
Expected $\mathbf{e}$	59.951	99.098	40.951	200

Since the expected frequencies are greater than 5 and

$$-2\ln Q(\mathbf{x}) = 1.274,$$

the significance probability is

$$p_{obs}(\mathbf{x}) = 1 - F_{\chi^2(1)}(1.274) = 0.259,$$

so the hypothesis of Hardy-Weinberg proportions is not rejected.

To find the confidence interval for the frequency p of the allele A, we note that according to (7.31) the number of A genes, which is observed to be 219, has a binomial distribution with number of trials $2n$, here 400, and probability parameter p. Using the formulas (7.27) and (7.28) with $n = 400$ and $x = 219$ we find that the 95% confidence interval for p is

$$[0.4985, 0.5956].$$

□

7.2.3 *Independence in a Two-Way Table*

The data set in Example 7.3 is a particular case of the following general situation. Suppose that n objects are classified according to two criteria having r and s categories, respectively. The data $\mathbf{x}$ may then be considered as an $r \times s$ matrix $\{x_{ij}\}$, where x_{ij} denotes the number of objects in the (i, j)th class corresponding to the ith category of the first criterion and to the jth category of the second criterion, $i = 1, \ldots, r$ and $j = 1, \ldots, s$.

	1	$\cdots$	j	$\cdots$	s	Σ
1	x_{11}	$\cdots$	x_{1j}	$\cdots$	x_{1s}	$x_{1\bullet}$
$\vdots$	$\vdots$		$\vdots$		$\vdots$	$\vdots$
i	x_{i1}	$\cdots$	x_{ij}	$\cdots$	x_{is}	$x_{i\bullet}$
$\vdots$	$\vdots$		$\vdots$		$\vdots$	$\vdots$
r	x_{r1}	$\cdots$	x_{rj}	$\cdots$	x_{rs}	$x_{r\bullet}$
Σ	$x_{\bullet 1}$	$\cdots$	$x_{\bullet j}$	$\cdots$	$x_{\bullet s}$	n

In the table $x_{i\bullet}$ and $x_{\bullet j}$ denote the sum of the observations in the ith row and in the jth column, respectively, i.e.,

$$x_{i\bullet} = \sum_{j=1}^{s} x_{ij} \quad \text{and} \quad x_{\bullet j} = \sum_{i=1}^{r} x_{ij}.$$

The full multinomial model for the rs-dimensional discrete random matrix $\mathbf{X} = \{X_{ij}\}$,

$$M_0\colon \mathbf{X} = \{X_{ij}\} \sim m(n, \{\pi_{ij}\}), \quad \boldsymbol{\pi} \in \boldsymbol{\Pi}^{(rs)}, \tag{7.32}$$

has $rs - 1$ free parameters.

Let ρ_i denote the probability that an object belongs to the ith category of the first criterion, $i = 1, \ldots, r$ and, similarly, let σ_j denote the probability of the jth category of the second criterion, $j = 1, \ldots, s$. The hypothesis,

$$H_{01}\colon \pi_{ij} = \rho_i \sigma_j, \qquad i = 1, \ldots, r, \quad j = 1, \ldots, s, \tag{7.33}$$

is called the hypothesis of *independence in a two-way table* and it has $d_1 = (r-1) + (s-1) = r + s - 2$ free parameters, since

$$\sum_{i=1}^{r} \rho_i = 1 \quad \text{and} \quad \sum_{j=1}^{s} \sigma_j = 1.$$

In the model M_0 the likelihood function for $\boldsymbol{\pi} = \{\pi_{ij}\}$ is

$$L(\boldsymbol{\pi}) = \frac{n!}{x_{11}! \cdots x_{rs}!} \prod_{i=1}^{r} \prod_{j=1}^{s} \pi_{ij}^{x_{ij}} \tag{7.34}$$

and according to Proposition 7.1 on page 305 the maximum likelihood estimate of $\boldsymbol{\pi}$ is given by

$$\hat{\pi}_{ij} = \frac{x_{ij}}{n}, \qquad i = 1, \ldots, r, \quad j = 1, \ldots, s.$$

The likelihood function for $\boldsymbol{\rho} = (\rho_1, \ldots \rho_i, \ldots, \rho_r)^*$ and $\boldsymbol{\sigma} = (\sigma_1, \ldots, \sigma_j, \ldots \sigma_s)^*$ is found from (7.34) by inserting $\pi_{ij} = \rho_i \sigma_j$ and we find:

$$\begin{aligned} L(\boldsymbol{\rho}, \boldsymbol{\sigma}) &= \frac{n!}{x_{11}! \cdots x_{rs}!} \prod_{i=1}^{r} \prod_{j=1}^{s} (\rho_i \sigma_j)^{x_{ij}} \\ &= \frac{n!}{x_{11}! \cdots x_{rs}!} \prod_{i=1}^{r} \rho_i^{x_{i\bullet}} \prod_{j=1}^{s} \sigma_j^{x_{\bullet j}}. \end{aligned}$$

It is seen that $L(\boldsymbol{\rho}, \boldsymbol{\sigma})$ has a factor depending on $\boldsymbol{\rho}$ only and a factor depending on $\boldsymbol{\sigma}$ only. Furthermore, since $\boldsymbol{\rho}$ and $\boldsymbol{\sigma}$ are variation independent we may apply Proposition 7.1 on each of the factors in order to find the maximum likelihood estimates of $\boldsymbol{\rho}$ and $\boldsymbol{\sigma}$. We find:

$$\hat{\rho}_i = \frac{x_{i\bullet}}{n}, \quad i = 1, \ldots, r \qquad \text{and} \qquad \hat{\sigma}_j = \frac{x_{\bullet j}}{n}, \quad j = 1, \ldots, s, \tag{7.35}$$

that is, the *relative frequencies* of the ith category of the first criterion and the jth category of the second criterion, respectively.

The matrix $\mathbf{e} = \{e_{ij}\}$ of expected frequencies under H_{01} has the elements,

$$\begin{aligned} e_{ij} &= n \hat{\rho}_i \hat{\sigma}_j \\ &= \frac{x_{i\bullet} x_{\bullet j}}{n}. \end{aligned} \tag{7.36}$$

Thus, the expected frequency in the (i, j)th class is found as the product of the ith row sum $x_{i\bullet}$ and the jth column sum $x_{\bullet j}$ divided by the total n.

Using (7.36) we obtain the following formula for the calculation of the $-2\ln Q$-test statistic of H_{01}:

$$-2\ln Q(\mathbf{x}) = 2 \sum_{i=1}^{r} \sum_{j=1}^{s} x_{ij} \ln\left(\frac{x_{ij}}{e_{ij}}\right) \tag{7.37}$$

$$\begin{aligned} &= 2 \sum_{i=1}^{r} \sum_{j=1}^{s} x_{ij} \ln\left(\frac{x_{ij}}{x_{i\bullet} x_{\bullet j}/n}\right) \\ &= 2 \sum_{i=1}^{r} \sum_{j=1}^{s} x_{ij} \left[\ln(x_{ij}) - \ln(x_{i\bullet}) - \ln(x_{\bullet j}) + \ln(n)\right] \end{aligned}$$

$$= 2\left[\sum_{i=1}^{r} \sum_{j=1}^{s} x_{ij} \ln(x_{ij}) - \sum_{i=1}^{r} x_{i\bullet} \ln(x_{i\bullet}) - \sum_{j=1}^{s} x_{\bullet j} \ln(x_{\bullet j}) + n \ln(n)\right]. \tag{7.38}$$

Thus $-2\ln Q$ may be calculated using two different formulas. Formula (7.38) is recommended for calculation of $-2\ln Q$ using a pocket calculator. The advantage of (7.38) over (7.37) is that the expected frequencies need not be calculated. However, this does not mean that calculation of the expected frequencies should be omitted. Comparing these with the observed frequencies we get a first impression of whether the hypothesis is rejected. If the expected frequencies resemble the observed ones, it indicates that the hypothesis is not rejected by the $-2\ln Q$-test. Furthermore, the expected frequencies should be sufficiently large in order to apply the χ^2-approximation of the significance probability in (7.39). For both purposes it is only required that the expected frequencies are recorded with one or two decimal places, whereas exact calculation of $-2\ln Q$ using (7.37) usually requires more decimal places. Finally, note that if the primary interest in the expected frequencies is to check that they are sufficiently large to apply the χ^2-approximation, it may be achieved by computing the smallest expected frequency, which is e_{ij}, where i is the row with the smallest row sum and j is the column with the smallest column sum.

By calculation of $-2\ln Q$ on a pocket calculator using formula (7.38), it is very useful to have a small program that calculates the quantity $\Sigma\, x\ln(x)$, since the number in the square bracket is obtained from the table with the observed frequencies by calculating the quantity for the insides of the table minus the quantity calculated for the row sums minus the quantity calculated for the column sums plus the quantity calculated for the total sum. Again it is important to remember the **factor 2** on the right-hand side in this formula.

As mentioned above, the number of free parameters in the full model M_0 is $rs-1$ and since the number of free parameters in H_{01} is $r+s-2$, the number of degrees of freedom for the $-2\ln Q$-test statistic of H_{01} is $f=(rs-1)-(r+s-2)=(r-1)(s-1)$. Consequently, if the expected frequencies under the hypothesis of independence are all greater than or equal to 5, the significance probability can be calculated as

$$p_{obs}(\mathbf{x}) \doteq 1-F_{\chi^2(r-1)(s-1)}(-2\ln Q(\mathbf{x})). \tag{7.39}$$

If H_{01} is not rejected, the model,

$$M_1\colon \mathbf{X}=\{X_{ij}\}\sim m(n,(\{\rho_i\sigma_j\})), \quad \boldsymbol{\rho}\in\boldsymbol{\Pi}^{(r)},\boldsymbol{\sigma}\in\boldsymbol{\Pi}^{(s)}, \tag{7.40}$$

replaces M_0 as the model for the data. Thus further inference is based on the model M_1.

In M_1 it may be shown that the vector of row sums $\mathbf{X}_{*\bullet}=(X_{1\bullet},\ldots,X_{i\bullet},\ldots,X_{r\bullet})^*$ and the vector of column sums $\mathbf{X}_{\bullet *}=(X_{\bullet 1},\ldots,X_{\bullet j},\ldots,X_{\bullet s})^*$ are stochastically independent and distributed according to the multinomial distribution. More precisely,

$$\begin{array}{l}\mathbf{X}_{*\bullet}^{*}=(X_{1\bullet},\ldots,X_{i\bullet},\ldots,X_{r\bullet})\sim m(n,(\rho_1,\ldots,\rho_i,\ldots,\rho_r))\\ \mathbf{X}_{\bullet *}^{*}=(X_{\bullet 1},\ldots,X_{\bullet j},\ldots,X_{\bullet s})\sim m(n,(\sigma_1,\ldots,\sigma_j,\ldots,\sigma_s))\\ \mathbf{X}_{*\bullet}\text{ and }\mathbf{X}_{\bullet *}\text{ are stochastically independent.}\end{array} \tag{7.41}$$

Inference in M_1 concerning the vector of row probabilities $\boldsymbol{\rho}$ and the vector of column probabilities $\boldsymbol{\sigma}$ may therefore be performed by considering the distribution of row sums $\mathbf{X}_{*\bullet}$ and the distribution of column sums $\mathbf{X}_{\bullet *}$, respectively.

If there is a natural ordering of both the r categories of the first criterion and the s categories of the second criterion in the $r\times s$ table, the $-2\ln Q$-test of the hypothesis of independence can be supplemented with a nonparametric test based on the Spearman rank correlation coefficient, see Section 12.2.4, page 508.

Example 7.3 (Continued)
If the table with the observed frequencies is supplemented with the row sums, the column sums, and the total sum, we get the following table:

	$\mathbf{x}$	*Breadth /Length* <0.6	0.6–0.8	>0.8	*Total*
	<25	6	31	11	48
Weight	25–50	8	29	19	56
	>50	10	39	26	75
	Total	24	99	56	179

From this table the expected frequencies **e** may be calculated by means of (7.36) and recording results with an accuracy of one decimal place we find:

	e	*Breadth /Length* < 0.6	$0.6–0.8$	> 0.8	*Total*
	< 25	6.4	26.5	15.0	47.9
Weight	25–50	7.5	31.0	17.5	56.0
	> 50	10.1	41.5	23.5	75.1
	Total	24.0	99.0	56.0	179.0

According to (7.38),

$$-2\ln Q(\mathbf{x}) = 2[564.428 - 735.049 - 756.610 + 928.542] = 2.622,$$

and since the expected frequencies are all greater than 5, one finds, using (7.39), that the significance probability of the test of independence is

$$p_{obs}(\mathbf{x}) = 1 - F_{\chi^2(4)}(2.622) = 0.623,$$

and the hypothesis is not rejected. On the basis of this investigation we are not able to prove any connection between the shape and the weight of a stone.

In this example there are natural orderings of the categories of both criteria. The nonparametric test of hypothesis of independence based on the Spearman rank correlation coefficient may be found on page 510. □

Example 7.1 (Continued)
As mentioned earlier the 9:3:3:1 segregation is a consequence of the following three assumptions:

(1) form and color are transmitted independently; the two genes are located on different chromosomes

(2) 3:1 segregation in plants with round and wrinkled seeds; round (A) is dominant

(3) 3:1 segregation in plants with yellow and green seeds; yellow (B) is dominant

We now consider Mendel's data again in order to investigate each of these assumptions separately. The data are arranged in a 2×2 table:

Observed **x**		*Color* *Yellow*	*Green*	*Total*
Form	*Round*	315	108	423
	Wrinkled	101	32	133
	Total	416	140	556

In the full model,

$$M_0\colon \mathbf{X} = \{X_{ij}\} \sim m(556, (\{\pi_{ij}\})),$$

assumption (1) may be formulated as a hypothesis, here denoted by H_{0I}, of independence of the two criteria form and color, i.e., as

$$H_{0I}\colon \pi_{ij} = \rho_i \sigma_j, \qquad i, j = 1, 2.$$

According to (7.36) the expected frequencies under H_{0I} are:

Expected **e**		*Color* *Yellow*	*Green*	*Total*
Form	*Round*	316.5	106.5	423.0
	Wrinkled	99.5	33.5	133.0
	Total	416.0	140.0	556.0

Since the expected frequencies are greater than 5, we get from (7.38) and (7.39) that

$$-2\ln Q_I(\mathbf{x}) = 0.1171,$$

and

$$p_{obs}(\mathbf{x}) = 1 - F_{\chi^2(1)}(0.1171) = 0.732.$$

The data do not reject the hypothesis of independence and we decide to adopt the model,

$$M_I : \mathbf{X} = \{X_{ij}\} \sim m(556, (\{\rho_i \sigma_j\})),$$

as the model for the data. This means that we will work under the assumption that the two genes are located on different chromosomes.

To investigate the assumptions (2) and (3), we use the result in (7.41) and consider the hypotheses of 3:1 segregation in the marginal model for the row sums and the column sums, respectively:

$$M_0^A: (X_{1\cdot}, X_{2\cdot}) \sim m(556, (\rho_1, \rho_2)) \qquad M_0^B: (X_{\cdot 1}, X_{\cdot 2}) \sim m(556, (\sigma_1, \sigma_2))$$
$$H_{01}^A: (\rho_1, \rho_2) = (\frac{3}{4}, \frac{1}{4}) \qquad H_{01}^B: (\sigma_1, \sigma_2) = (\frac{3}{4}, \frac{1}{4})$$

The expected frequencies are found as

$$n(\frac{3}{4}, \frac{1}{4}) = 556(\frac{3}{4}, \frac{1}{4}) = (417, 139)$$

for both marginal models. (The expected frequencies are integers because n is a multiple of 4.) The observed and expected frequencies are given in the tables below for easy comparison.

Form	*Round*	*Wrinkled*	*Total*
Observed **x**	423	133	556
Expected **e**	417	139	556

Color	*Yellow*	*Green*	*Total*
Observed **x**	416	140	556
Expected **e**	417	139	556

In both cases the expected frequencies are greater than 5, and therefore the significance probabilities may be calculated by means of the formulas (7.16) and (7.17). We find:

$$-2\ln Q_A = 0.3487 \qquad p_{obs}^A = 1 - F_{\chi^2(1)}(0.3487) = 0.555$$
$$-2\ln Q_B = 0.0096 \qquad p_{obs}^B = 1 - F_{\chi^2(1)}(0.0096) = 0.922.$$

In both cases the hypothesis of a 3:1 segregation is not rejected.

In this example we have calculated $-2\ln Q$ with four decimal places, which is not necessary; usually two or three decimal places suffice. It has been done here in order to illustrate the following partition of the $-2\ln Q$-test statistic of the hypothesis of 9:3:3:1 segregation:

$$\begin{array}{ccccccc} -2\ln Q & = & -2\ln Q_I & + & -2\ln Q_A & + & -2\ln Q_B \\ 0.4754 & = & 0.1171 & + & 0.3487 & + & 0.0096 \end{array}$$

and the corresponding partition of the degrees of freedom:

$$\begin{array}{ccccccc} f & = & f_I & + & f_A & + & f_B \\ 3 & = & 1 & + & 1 & + & 1. \end{array}$$

Thus we have obtained a partition of the $-2\ln Q$-test statistic of the hypothesis of 9:3:3:1 segregation, which has 3 degrees of freedom into a sum of three $-2\ln Q$-test statistics, each with 1 degree of freedom, of the three hypotheses (assumptions), which together constitute the hypothesis of 9:3:3:1 segregation. This partition is exact for the $-2\ln Q$-test statistic but not for the X^2-statistic. □

7.3 Inference in Several Multinomial Distributions

The theory of statistical inference in one multinomial distribution, considered in Section 7.2, may quite easily be generalized to several multinomial distributions. We will only touch upon this theory briefly and illustrate it by means of a single example.

7.3.1 Homogeneity of Several Multinomial Distributions

The problem outlined in Example 7.4 is a particular case of the following general situation. Suppose that the data may be described as outcomes of r independent, s-dimensional discrete random vectors $\mathbf{X}_i = (X_{i1},\ldots,X_{ij},\ldots,X_{is})^*$, which has a multinomial distribution with number of trials n_i and probability vector $\boldsymbol{\pi}_i = (\pi_{i1},\ldots,\pi_{ij},\ldots,\pi_{is})^*$, $i = 1,\ldots,r$, and that we wish to investigate whether the probability vectors of the r distributions are identical. The observations may then be arranged in an $r \times s$ table,

	1	$\cdots$	j	$\cdots$	s	Σ
1	x_{11}	$\cdots$	x_{1j}	$\cdots$	x_{1s}	n_1
$\vdots$	$\vdots$		$\vdots$		$\vdots$	$\vdots$
i	x_{i1}	$\cdots$	x_{ij}	$\cdots$	x_{is}	n_i
$\vdots$	$\vdots$		$\vdots$		$\vdots$	$\vdots$
r	x_{r1}	$\cdots$	x_{rj}	$\cdots$	x_{rs}	n_r
Σ	$x_{\bullet 1}$	$\cdots$	$x_{\bullet j}$	$\cdots$	$x_{\bullet s}$	$n_\bullet$

where

$$x_{\bullet j} = \sum_{i=1}^{r} x_{ij} \quad \text{and} \quad n_\bullet = \sum_{i=1}^{r} n_i.$$

In the model,

$$M_0\colon\quad \begin{array}{l} \mathbf{X}_i = (X_{i1},\ldots,X_{ij},\ldots,X_{is})^* \sim m(n_i,\boldsymbol{\pi}_i) = m(n_i,(\pi_{i1},\ldots,\pi_{ij},\ldots,\pi_{is})^*) \\ \mathbf{X}_1,\ldots,\mathbf{X}_i,\ldots,\mathbf{X}_r \text{ are stochastically independent,} \end{array} \tag{7.42}$$

the hypothesis,

$$H_{01}\colon \boldsymbol{\pi}_1 = \cdots = \boldsymbol{\pi}_i = \cdots = \boldsymbol{\pi}_r = \boldsymbol{\pi} = (\pi_1,\ldots,\pi_j,\ldots,\pi_s)^*, \tag{7.43}$$

is referred to as the hypothesis of *homogeneity*.

Since an s-dimensional multinomial distribution has $s-1$ free parameters and since the model M_0 consists of r independent distributions of this kind, the total number of free parameters in M_0 is $r(s-1)$. The likelihood function in M_0 is

$$L(\boldsymbol{\pi}_1,\ldots,\boldsymbol{\pi}_i,\ldots,\boldsymbol{\pi}_r) = \prod_{i=1}^{r} \frac{n_i!}{x_{i1}!\cdots x_{ij}!\cdots x_{is}!} \pi_{i1}^{x_{i1}} \cdots \pi_{ij}^{x_{ij}} \cdots \pi_{is}^{x_{is}}, \tag{7.44}$$

and the maximum likelihood estimate of π_{ij} in M_0 is given by

$$\hat{\pi}_{ij} = \frac{x_{ij}}{n_i}, \qquad i=1,\ldots,r, \quad j=1,\ldots,s.$$

The model corresponding to the hypothesis of homogeneity is

$$\begin{aligned} M_1\colon\quad & \mathbf{X}_i = (X_{i1},\ldots,X_{ij},\ldots,X_{is})^* \sim m(n_i,\boldsymbol{\pi}) = m(n_i,(\pi_1,\ldots,\pi_j,\ldots,\pi_s)^*) \\ & \mathbf{X}_1,\ldots,\mathbf{X}_i,\ldots,\mathbf{X}_r \text{ are stochastically independent} \end{aligned} \tag{7.45}$$

This model has $s-1$ free parameters and the likelihood function for $\boldsymbol{\pi}$ is obtained by setting $\pi_{ij} = \pi_j$ in (7.44), i.e.,

$$L(\boldsymbol{\pi}) = \left\{ \prod_{i=1}^{r} \frac{n_i!}{x_{i1}!\cdots x_{ij}!\cdots x_{is}!} \right\} \pi_1^{x_{\bullet 1}} \cdots \pi_j^{x_{\bullet j}} \cdots \pi_s^{x_{\bullet s}}$$

Using Proposition 7.1 we find that the maximum likelihood estimate of the jth component of the common probability vector $\boldsymbol{\pi}$ is given by

$$\hat{\pi}_j = \frac{x_{\bullet j}}{n_\bullet}, \qquad j=1,\ldots,s. \tag{7.46}$$

Consequently, the expected frequencies under M_1 become

$$\begin{aligned} e_{ij} &= n_i \hat{\pi}_j \\ &= \frac{n_i x_{\bullet j}}{n_\bullet}, \qquad i=1,\ldots,r, \quad j=1,\ldots,s; \end{aligned} \tag{7.47}$$

i.e., the expected frequency in the jth category of the ith distribution is the product of the ith row sum and the jth column sum divided by the total sum $n_\bullet$.

The formula for calculating the $-2\ln Q$-test statistic is

$$\begin{aligned} -2\ln Q(\mathbf{x}) &= 2\sum_{i=1}^{r}\sum_{j=1}^{s} x_{ij} \ln\left(\frac{x_{ij}}{e_{ij}}\right) \qquad (7.48) \\ &= 2\sum_{i=1}^{r}\sum_{j=1}^{s} x_{ij} \ln\left(\frac{x_{ij}}{n_i x_{\bullet j}/n_\bullet}\right) \\ &= 2\sum_{i=1}^{r}\sum_{j=1}^{s} x_{ij} \left[\ln(x_{ij}) - \ln(n_i) - \ln(x_{\bullet j}) + \ln(n_\bullet)\right] \\ &= 2\left[\sum_{i=1}^{r}\sum_{j=1}^{s} x_{ij}\ln(x_{ij}) - \sum_{i=1}^{r} n_i \ln(n_i) - \sum_{j=1}^{s} x_{\bullet j}\ln(x_{\bullet j}) + n_\bullet \ln(n_\bullet)\right]. \qquad (7.49) \end{aligned}$$

Remarks similar to those on page 316 concerning calculation of $-2\ln Q$ also apply here.

The number of degrees of freedom of the $-2\ln Q$-test statistic is $f = r(s-1)-(s-1) = (r-1)(s-1)$, so, if the expected frequencies are all greater than or equal to 5, the significance probability of the hypothesis of homogeneity may be calculated as

$$p_{obs}(\mathbf{x}) \doteq 1 - F_{\chi^2((r-1)(s-1))}(-2\ln Q(\mathbf{x})). \tag{7.50}$$

In the model M_1 it may be shown that the vector sum, $\mathbf{X}_\bullet = \mathbf{X}_1 + \cdots + \mathbf{X}_r$, has a multinomial distribution with number of trials $n_\bullet$ and probability vector $\boldsymbol{\pi}$,

$$\mathbf{X}_\bullet = (X_{\bullet 1},\ldots,X_{\bullet j},\ldots,X_{\bullet s})^* \sim m(n_\bullet,\boldsymbol{\pi}) = m(n_\bullet,(\pi_1,\ldots,\pi_j,\ldots,\pi_s)^*). \tag{7.51}$$

This result may be used if we wish to make inference about the common probability vector $\boldsymbol{\pi}$.

If there is a natural ordering of the s categories of the multinomial distribution, the $-2\ln Q$-test of the hypothesis of homogeneity can be supplemented with the nonparametric Kruskal-Wallis test in Section 12.2.3 on page 505.

Similarities and Differences Between the Tests of Independence and Homogeneity

A comparison of the formulas (7.36) to (7.39) and (7.47) to (7.50) shows that for the two tests the *calculations are identical.*

The tests concern *different hypotheses* and are performed in *different models*. The test of independence is carried out in a model that involves one multinomial distribution only, the only nonrandom quantity being the total number of observations n. The test of homogeneity is executed in a model that involves r multinomial distributions and in this model the nonrandom quantities are the number of observations, $n_1,\ldots,n_i,\ldots,n_r$, in the r distributions. In other words, different strategies have been applied when collecting the data in the two situations.

Mathematically, the two models are related by conditioning. If we condition on the row sums, $X_{i\cdot}$, in the full model in (7.32) for an $r\times s$ table based on a single multinomial distribution, i.e.,

$$M_0\colon \{X_{ij}\}\sim m(n,\{\pi_{ij}\}),$$

it can be shown that we obtain the conditional model,

$$M_0^*\colon (X_{i1},\ldots,X_{ij},\ldots,X_{is})\,|\,X_{i\cdot}=x_{i\cdot}\sim m(x_{i\cdot},(\frac{\pi_{i1}}{\pi_{i\cdot}},\ldots,\frac{\pi_{ij}}{\pi_{i\cdot}},\ldots,\frac{\pi_{is}}{\pi_{i\cdot}})),\quad i=1,\ldots,r,$$

where $\pi_{i\cdot}=\pi_{i1}+\cdots+\pi_{ij}+\cdots+\pi_{is}$, and, furthermore, that the conditional distributions are independent. Thus, the model M_0^* corresponds to the model in (7.42), which is the basic model for the testing the hypothesis of homogeneity. The hypothesis of independence is in M_0 formulated as

$$H_{01}\colon \pi_{ij}=\rho_i\sigma_j,\quad i=1,\ldots,r,\ j=1,\ldots,s.$$

Since $\sum_{i=1}^{s}\sigma_j=1$, we have under H_{01} that $\pi_{i\cdot}=\rho_i$ and that

$$(\frac{\pi_{i1}}{\pi_{i\cdot}},\ldots,\frac{\pi_{ij}}{\pi_{i\cdot}},\ldots,\frac{\pi_{is}}{\pi_{i\cdot}})=(\sigma_1,\ldots,\sigma_j,\ldots,\sigma_s),\quad i=1,\ldots,r,$$

and it is seen that the hypothesis H_{01} corresponds to the hypothesis of homogeneity in the conditional model M_0^*.

Example 7.4 (Continued)

In order to determine if the distributions in area 1 are homogeneous we consider the model,

$$M_0^1\colon\quad \mathbf{X}_{1i}\sim m(n_{1i},\boldsymbol{\pi}_{1i}),\quad i=1,2,$$
$$\mathbf{X}_{11}\text{ and }\mathbf{X}_{12}\text{ are stochastically independent,}$$

where the superscript 1 indicates that data are from area 1, and in this model we test the hypothesis,

$$H_{01}^1\colon \boldsymbol{\pi}_{11}=\boldsymbol{\pi}_{12}=\boldsymbol{\pi}_1.$$

From the table with the observed frequencies $\mathbf{x}$ in area 1

x	*Concentration*			
Municipality	≤0.01	0.01–0.10	>0.10	*Total*
Hadsten	23	12	6	41
Hammel	20	5	9	34
Total	43	17	15	75

we find using (7.47) that the expected frequencies **e** under H_{01}^1 are:

e	*Concentration*			
Municipality	$\leq$0.01	0.01–0.10	>0.10	*Total*
Hadsten	23.5	9.3	8.2	41.0
Hammel	19.5	7.7	6.8	34.0
Total	43.0	17.0	15.0	75.0

The observed and expected frequencies are very similar, so it is to be expected that the hypothesis will not be rejected. Since the expected frequencies are greater than 5 and

$$-2\ln Q_1 = 3.129,$$

we find that the p-value is

$$p_{obs}^1(\mathbf{x}) = 1 - F_{\chi^2(2)}(3.129) = 0.209.$$

Thus, there is no significant difference between the concentrations in the two municipalities in area 1.

In area 2 the observed frequencies are

x	*Concentration*			
Municipality	$\leq$0.01	0.01–0.10	>0.10	*Total*
Nørre Djurs	37	6	4	47
Randers	47	4	3	54
Total	84	10	7	101

from which we find that the expected frequencies under the hypothesis of homogeneity are

e	*Concentration*			
Municipality	$\leq$0.01	0.01–0.10	>0.10	*Total*
Nørre Djurs	39.1	4.7	3.3	47.1
Randers	44.9	5.3	3.7	53.9
Total	84.0	10.0	7.0	101.0

The observed and expected frequencies are very similar and this results in a small value of the test statistic:

$$-2\ln Q_2 = 1.254.$$

However, note that the expected frequencies are not all greater than 5; in fact, half of them are less than 5, so the p-value calculated using the χ^2-approximation in (7.50) may not be reliable. The p-value is

$$p_{obs}^2(\mathbf{x}) = 1 - F_{\chi^2(2)}(1.254) = 0.534.$$

This value is *so far from the significance level* (5%) that we have no reservations about claiming that there is no significant difference between the distributions of the concentration in the two municipalities in area 2.

An exact test of this hypothesis given in the continuation of this example on page 328 confirms this conclusion.

Since there is no difference between the two municipalities within an area, we consider the totals when the areas are to be compared. This is precisely the meaning of formula (7.51) in this situation. Consequently, the comparison of the areas is performed in the model,

$$M_0: \quad \mathbf{X}_{i\cdot} \sim m(n_{i\cdot}, \boldsymbol{\pi}_i), \quad i = 1,2.$$
$$\mathbf{X}_{1\cdot} \text{ and } \mathbf{X}_{2\cdot} \text{ are stochastically independent,}$$

by testing the hypothesis of identity of the probability vector in the two areas, i.e., the hypothesis

$$H_{01}: \boldsymbol{\pi}_1 = \boldsymbol{\pi}_2 = \boldsymbol{\pi}.$$

The observed frequencies in the model M_0 are

x	*Concentration*			
Area	≤0.01	0.01–0.10	>0.10	*Total*
1	43	17	15	75
2	84	10	7	101
Total	127	27	22	176

and the expected frequencies under H_{01} become

e	*Concentration*			
Area	≤0.01	0.01–0.10	>0.10	*Total*
1	54.1	11.5	9.4	75.0
2	72.9	15.5	12.6	101.0
Total	127.0	27.0	22.0	176.0

There is a relatively large difference between the observed and the expected frequencies. This impression is confirmed by calculating

$$-2\ln Q(\mathbf{x}) = 14.434$$

and the corresponding p-value,

$$p_{obs}(\mathbf{x}) = 1 - F_{\chi^2(2)}(14.434) = 0.0007.$$

Thus, the hypothesis H_{01} is rejected. The maximum likelihood estimates in M_0 with an accuracy of three decimal places are:

$$\hat{\boldsymbol{\pi}}_1 = (0.573,\ 0.227,\ 0.200)^*$$

$$\hat{\boldsymbol{\pi}}_2 = (0.832,\ 0.099,\ 0.069)^*.$$

Here the third component of the probability vector is of particular interest since it is the estimate of the probability that the limit value of the concentration of *2.6-dichlorbenzamid* is exceeded. The 95% confidence interval for this probability is $[0.1251, 0.3041]$ and $[0.0340, 0.1362]$ in area 1 and area 2, respectively. The confidence intervals are not disjoint and therefore it may be relevant to consider a test for the hypothesis that the probability of exceeding the limit value is the same in the two areas. It is seen from the output listing on page 342 that the hypothesis is rejected since the p-value is 0.0097.

Clearly, in this example there is a natural ordering of the three categories of the multinomial distribution. The Kruskal-Wallis tests of the three hypotheses of homogeneity may be found in the continuation of the example on page 511. □

Example 7.2 (Continued)

In addition to the investigation of eelpouts in Mariager Fjord, which is mentioned earlier in the example, similar investigations were conducted in 1974 on 595 eelpouts from Roskilde Fjord and on 537 eelpouts from Slien. After these investigations it was concluded that in each of these two locations the population was in Hardy-Weinberg proportions. In order to investigate whether the frequency of the allele A varies from location to location, we may, because of (7.31), base the inference on the observed counts of A and a genes at the three locations. These counts were:

y	A	a	*Total*
Mariager Fjord	219	181	400
Roskilde Fjord	520	670	1190
Slien	176	898	1074
Total	915	1749	2664

A comparison of the three locations may be made by testing whether the probability parameters in three independent binomial distribution are identical, or, equivalently, whether the probability vectors in three independent, two-dimensional multinomial distributions are identical, i.e., we consider the model,

$$M_0: \quad Y_i \sim b(n_i, p_i) \quad i = M, R, S,$$
$$Y_M, Y_R \text{ og } Y_S \text{ are stochastically independent,}$$

where the y's denote the observed counts of A genes and in this model we test the hypothesis,

$$H_{01}: p_M = p_R = p_S = p.$$

Using (7.47) the expected frequencies under the hypothesis calculated with an accuracy of one decimal place become:

e	A	a	*Total*
Mariager Fjord	137.4	262.6	400.0
Roskilde Fjord	408.7	781.3	1190.0
Slien	368.9	705.1	1074.0
Total	915.0	1749.0	2664.0

The expected frequencies show appreciable deviations from the observed frequencies so rejection of H_{01} is expected. By means of (7.49) and (7.50) we find:

$$-2\ln Q(\mathbf{y}) = 287.829,$$

and

$$\varepsilon(\mathbf{y}) = 1 - F_{\chi^2(2)}(287.829) = 3.15 \times 10^{-63}.$$

The hypothesis H_{01} is rejected and we can conclude that the frequency of the A allele is different at the three locations. The maximum likelihood estimates and the 95% confidence intervals for these frequencies are:

$$\hat{p}_M = 0.5475 \quad [0.4950, 0.5956]$$
$$\hat{p}_R = 0.4370 \quad [0.4090, 0.4652]$$
$$\hat{p}_S = 0.1639 \quad [0.1429, 0.1872].$$

□

7.4 Fisher's Exact Test

All the tests we have applied in Section 7.2 and Section 7.3 have been approximate tests based on the χ^2-approximation to the significance probability mentioned in Section 11.3. In order to apply these tests we have used the criterion that the expected frequencies should all be greater than or equal to 5. This criterion is based on numerical simulations, and for some models the criterion may be eased, since the calculated significance probability is reliable even though some of the expected frequencies are somewhat less that 5. However, the question of what to do when the expected frequencies are too small often appears in applications of the theory. We now mention Fisher's exact test, which is often used in connection with two-way tables where some of the expected frequencies are too small. The method may be used in connection with a test of independence and a test of homogeneity. We give a detailed description of the method for 2×2 tables for the principle used in the general case. The calculation of Fisher's exact test is often too extensive to be done manually, but most statistical computer packages are able to execute the test. Let us consider an example.

Example 7.5

A survey carried out by the Cancer Registry of the Danish Cancer Society considers the question whether children living close to a high-voltage transmission line have an increased risk of cancer. The survey is conducted as a so-called case-control study. In the Cancer Registry 1707 cases of the diseases leukemia, brain tumor, and lymphoma are registered in the period 1968–1986 among children who at the time of diagnosis were under 15 years of age. These children constitute the case group. For each child in this group a number of children of the same sex and age have been chosen at random. These children constitute the control group. For all the children it has been assessed whether they have been living so close to a high-voltage line that they have been exposed to an average magnetic field of 0.10 μT (microTesla) or more on an annual basis. Children who have been exposed to a magnetic filed of 0.10 μT or more will be said to be *exposed*. Here we consider the case group for the disease lymphoma and the corresponding control group, which was chosen to be five times as large. For this group the following observed frequencies may be found in Olsen, Nielsen, and Schulgen (1993):

Exposure	*Exposed*	*Not exposed*	*Total*
Case	3	247	250
Control	3	1247	1250
Total	6	1494	1500

Because of the sampling strategy of a case-control study, an essential ingredient in the statistical analysis is a test of homogeneity of two binomial distributions. The totals of the case group and of the control group are fixed, so the model is two independent binomial distributions. We focus on the probability of being exposed in the two groups and let X_i denote the frequency of exposed in the ith group, $i =$ case, control.

$$\begin{aligned} M_0\colon \quad & X_i \sim b(n_i, p_i) \quad i = \text{case, control}, \\ & X_{\text{case}} \text{ and } X_{\text{control}} \text{ are stochastically independent} \end{aligned} \tag{7.52}$$

Here p_i is the probability of being exposed in group i, $i =$ case, control. If there is no connection between exposure and disease, the probability of being exposed is the same in the two groups. On the other hand, if $p_{\text{case}} > p_{\text{control}}$, the probability of being exposed in the case group is larger than in the control group and it will be interpreted as an increased risk of getting the disease among children who are exposed. It is therefore important to test the hypothesis,

$$H_{01}\colon p_{\text{case}} = p_{\text{control}}. \tag{7.53}$$

From the table with the observed frequencies, it is seen that $\hat{p}_{\text{case}} = 0.0120$ and $\hat{p}_{\text{control}} = 0.0024$, i.e., the relative frequency of the exposed children in the case group is five times as large as the relative frequency in the control group. The question is now if this difference is significant.

The expected frequencies under H_{01} are, using (7.47), found to be

Exposure	*Exposed*	*Not exposed*	*Total*
Case	1	249	250
Control	5	1245	1250
Total	6	1494	1500

One of the expected frequencies is 1, so we cannot rely on the χ^2-approximation to the significance probability. □

Fisher's Exact Test in 2×2 *Tables*

In order to facilitate the discussion of Fisher's exact test for a general $r \times s$ table, we formulate the model in (7.52) and the hypothesis in (7.53) by means of the multinomial distribution rather than the binomial distribution. Thus we consider the model,

$$\begin{aligned} M_0\colon\quad & \mathbf{X}_i = (X_{i1}, X_{i2}) \sim m(n_i, (\pi_{i1}, \pi_{i2})), \quad i = 1,2, \\ & \mathbf{X}_1 \text{ and } \mathbf{X}_2 \text{ are stochastically independent,} \end{aligned}$$

and in this model the hypothesis of identity of the probability vectors, i.e.,

$$H_{01}\colon \boldsymbol{\pi}_1 = \boldsymbol{\pi}_2 = \boldsymbol{\pi} = (\pi_1, \pi_2)^*.$$

Fisher proposed testing the hypothesis H_{01} by considering the conditional distribution under H_{01} of $(\mathbf{X}_1, \mathbf{X}_2)$ given $\mathbf{X}_\bullet = \mathbf{X}_1 + \mathbf{X}_2$. Using (7.51) this conditional distribution is found by the following calculations:

$$\begin{aligned} P((\mathbf{X}_1, \mathbf{X}_2) = (\mathbf{x}_1, \mathbf{x}_2) \mid \mathbf{X}_\bullet = \mathbf{x}_\bullet) &= \frac{P((\mathbf{X}_1, \mathbf{X}_2) = (\mathbf{x}_1, \mathbf{x}_2))}{P(\mathbf{X}_\bullet = \mathbf{x}_\bullet)} \\ &= \frac{\dfrac{n_1!}{x_{11}!(n_1 - x_{11})!} \pi_1^{x_{11}} \pi_2^{n_1 - x_{11}} \dfrac{n_2!}{x_{21}!(n_2 - x_{21})!} \pi_1^{x_{21}} \pi_2^{n_2 - x_{21}}}{\dfrac{n_\bullet!}{x_{\bullet 1}!(n_\bullet - x_{\bullet 1})!} \pi_1^{x_{\bullet 1}} \pi_2^{n_\bullet - x_{\bullet 1}}}. \end{aligned}$$

After suitable reductions this calculation may be written in terms of binomial coefficients in the following way:

$$P((\mathbf{X}_1, \mathbf{X}_2) = (\mathbf{x}_1, \mathbf{x}_2) \mid \mathbf{X}_\bullet = \mathbf{x}_\bullet) = \frac{\binom{x_{\bullet 1}}{x_{11}} \binom{n_\bullet - x_{\bullet 1}}{n_1 - x_{11}}}{\binom{n_\bullet}{n_1}}.$$

Note that since we have conditioned on $\mathbf{X}_\bullet = \mathbf{x}_\bullet$, the quantities $x_{\bullet 1}, n_1, n_2$ and, consequently, $n_\bullet$, are fixed in this expression and therefore x_{11} is the only quantity which varies.

The discrete distribution with probability function,

$$h(x; M, N, n) = \frac{\binom{M}{x} \binom{N - M}{n - x}}{\binom{N}{n}}, \qquad x = \max\{0, n + M - N\}, \ldots, \min\{n, M\},$$

is called the *hypergeometric distribution with parameters* M, N, and n. The hypergeometric distribution appears in connection with drawing a sample of size n from a population with N elements of which a subpopulation with M elements is of particular interest. Then $h(x; M, N, n)$ is the probability that the sample contains precisely x elements from the subpopulation of particular interest.

It follows that the conditional distribution of X_{11} given the column sums and the row sum is the hypergeometric distribution with parameters $x_{\cdot 1}$, $n_{\cdot}$ and $n_{1\cdot}$.

The significance probability of Fisher's exact test is

$$p^F_{obs}(\mathbf{x}) = \sum_y^* h(y; x_{\cdot 1}, n_{\cdot}, n_1), \tag{7.54}$$

where * over the summation sign indicates that the summation is performed over all y for which $h(y; x_{\cdot 1}, n_{\cdot}, n_1) \le h(x_{11}, x_{\cdot 1}, n_{\cdot}, n_1)$. This definition of the significance probability may be explained in the following way. If the probabilities in the conditional distribution are used as a measure of how critical the various observations are, then we have the usual interpretation of a significance probability as the probability of all outcomes y, which are just as critical as or more critical than the observed outcome x_{11}.

As noted earlier, the calculations in the test of independence in a two-way table are identical to those in the test of homogeneity. It is therefore not surprising that the calculations above give the same result if we wish to test independence in a 2×2 table. The only difference is that the parameters of the conditional hypergeometric distribution become $x_{\cdot 1}$, n and $x_{1\cdot}$ instead of $x_{\cdot 1}$, $n_{\cdot}$ and $n_{1\cdot}$.

The calculation of the significance probability in (7.54) may be too laborious to do manually, but the most common statistical computer packages have a procedure for calculating this probability.

Example 7.5 (Continued)
The relevant hypergeometric distribution for this data set has the parameters 6, 1500, and 250. The distribution is shown in Figure 7.3. In this case the significance probability is

$$p^F_{obs}(\mathbf{x}) = \sum_{y=3}^{6} h(y; 6, 1500, 250) = 0.062,$$

which gives no reason to reject the hypothesis (7.53). In other words, the group with the disease lymphoma is not significantly different from the control group with respect to exposure from magnetic fields. Finally, let us consider which conclusions we would have obtained if we had incorrectly used the $-2\ln Q$-test or the X^2-test. From the quantities,

$$\begin{aligned} -2\ln Q(\mathbf{x}) &= 3.546 \qquad & p_{obs}(\mathbf{x}) &= 1 - F_{\chi^2(1)}(3.546) = 0.060 \\ X^2(\mathbf{x}) &= 4.819 \qquad & p^*_{obs}(\mathbf{x}) &= 1 - F_{\chi^2(1)}(4.819) = 0.028, \end{aligned}$$

it is seen that the $-2\ln Q$-test gives the same conclusion as Fisher's exact test. The use of the X^2-test implies, however, incorrect rejection of the hypothesis H_{01}: $p_{\text{case}} = p_{\text{control}}$, i.e., it incorrectly concludes that there is a significant difference between the case group and the control group. □

Fisher's Exact Test in $r \times s$ Tables

The principle behind this test is the same as that for a 2×2 table. The significance probability is calculated in the conditional distribution of the $r \times s$ table given its row sums and its column sums by summing up the conditional probabilities over all tables that have a conditional probability less than or equal to the conditional probability of the observed table. Here the calculations are nearly always so laborious that it is necessary to use a statistical computer package in order to execute the test.

Example 7.4 (Continued)
On page 323 it is seen that under the hypothesis of homogeneity, 3 of 6 expected frequencies in area 2 are less than 5. From the output from SAS on page 341, it follows that the significance probability of Fisher's exact test in this situation is $p^F_{obs}(\mathbf{x}) = 0.5312$, so the test does not reject the hypothesis.

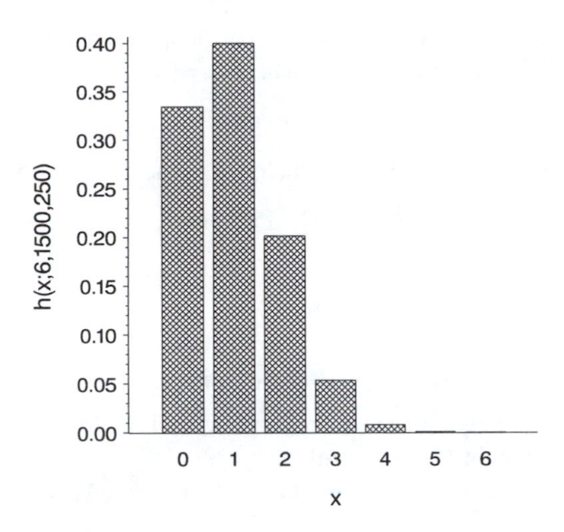

Figure 7.3 *The probability function of the hypergeometric distribution with parameters* $(M,N,n) = (6,1500,250)$.

This conclusion agrees with the one reached on page 323. □

7.5 Test for Goodness of Fit

In Section 2.3 we saw that by means of fractile diagrams, it is possible to investigate graphically if the data $\mathbf{x} = (x_1, \ldots, x_n)$ may be considered as a sample from a class of distributions characterized by a position parameter and/or a scale parameter. The evaluation of fractile diagrams and other graphical methods for control of a statistical model is of course to a certain extent subjective, even though it is possible by means of simulations, as in Appendix A1, to obtain insight into how the relevant figures should be evaluated. If the size n of the sample is sufficiently large, the graphical control may be supplemented by a numerical test, a so-called test for *goodness of fit*.

This kind of model checking is general, i.e., it may also be applied in situations where the class of distributions considered is not a location-scale family. More precisely, we want to examine whether the data may be considered as a sample from a distribution $F_{\boldsymbol{\theta}}$ belonging to the class of distributions

$$\mathcal{F} = \{F_{\boldsymbol{\theta}} \,|\, \boldsymbol{\theta} \in \boldsymbol{\Theta}\},$$

which is parameterized by a d-dimensional parameter $\boldsymbol{\theta}$, i.e., $\boldsymbol{\theta} \in \boldsymbol{\Theta} \subseteq \mathbb{R}^d$. If, for instance, we wish to investigate if $\mathbf{x}$ is a sample from the normal distribution, then $d = 2$, since the normal distribution is parametrized by (μ, σ^2), the mean, and the variance. If the class of distributions $\mathcal{F}$ is the set of Poisson distributions, then $d = 1$ since these distributions are parameterized by the mean λ, etc. Despite the general formulation above, the goodness-of-fit test is most frequently considered in connection with classes of discrete distributions, for instance, the Poisson distributions in Chapter 8.

Suppose that $-\infty \leq y_0 < y_1 < \ldots < y_j < \ldots < y_k \leq \infty$ determine a partition of $\mathbb{R}$ into k intervals $I_1, \ldots, I_j, \ldots, I_k$, where

$$I_j = \,]y_{j-1}, y_j], \qquad j = 1, \ldots, k.$$

Let a_j denote the frequency of observations in the jth interval, i.e.,

$$a_j = \#\{i \mid x_i \in I_j\}, \qquad j = 1,\ldots,k.$$

Since the observations, $x_1,\ldots,x_n$, are assumed to be independent and identically distributed, we have the following model for the *observed frequencies* $\mathbf{a} = (a_1,\ldots,a_j,\ldots,a_k)^*$:

$$M_0\colon (a_1,\ldots,a_j,\ldots,a_k) \sim\sim m(n,(\pi_1,\ldots,\pi_j,\ldots,\pi_k)) \tag{7.55}$$

where π_j is the probability that the observation belongs to the jth interval, I_j i.e.,

$$\pi_j = P(X_1 \in I_j), \qquad j = 1,\ldots,k.$$

Under the hypothesis,

$$H_0\colon X_i \sim F_{\boldsymbol{\theta}}, \qquad i = 1,\ldots,n,$$

we have

$$\pi_j = \pi_j(\boldsymbol{\theta}) = F_{\boldsymbol{\theta}}(y_j) - F_{\boldsymbol{\theta}}(y_{j-1}), \qquad j = 1,\ldots,k, \tag{7.56}$$

so H_0 may be considered as a hypothesis in the multinomial model M_0. Since $\boldsymbol{\theta}$ is d-dimensional, the number of free parameters in H_0 is precisely d.

The likelihood function under H_0 corresponding to the observed frequencies $\mathbf{a}$ is

$$L(\boldsymbol{\theta}) = \frac{n!}{a_1!\cdots a_j!\cdots a_k!}\pi_1(\boldsymbol{\theta})^{a_1}\cdots\pi_j(\boldsymbol{\theta})^{a_j}\cdots\pi_k(\boldsymbol{\theta})^{a_k}.$$

The expressions in (7.56) are often complicated and therefore we choose to estimate $\boldsymbol{\theta}$ on the basis of the original observations, $x_1,\ldots x_n$, i.e., by means of the likelihood function,

$$L(\boldsymbol{\theta}) = \prod_{i=1}^{n} f(x_i;\boldsymbol{\theta}), \tag{7.57}$$

where $f(\cdot;\boldsymbol{\theta})$ denotes the probability density function corresponding to $F_{\boldsymbol{\theta}}$. Let $\hat{\boldsymbol{\theta}}$ denote the maximum likelihood estimate of $\boldsymbol{\theta}$, which maximizes (7.57). The estimates of the parameters in the multinomial distribution then become

$$\pi_j(\hat{\boldsymbol{\theta}}) = F_{\hat{\boldsymbol{\theta}}}(y_j) - F_{\hat{\boldsymbol{\theta}}}(y_{j-1}),$$

and, consequently, the *expected frequencies* $\mathbf{e}$ under H_0 are

$$\begin{aligned} e_j &= n\pi_j(\hat{\boldsymbol{\theta}}) \qquad (7.58)\\ &= n\{F_{\hat{\boldsymbol{\theta}}}(y_j) - F_{\hat{\boldsymbol{\theta}}}(y_{j-1})\}, \qquad j = 1,\ldots,k. \end{aligned}$$

The significance probability of the $-2\ln Q$-test statistic of H_0,

$$-2\ln Q(\mathbf{a}) = 2\sum_{j=1}^{k} a_j \ln\left(\frac{a_j}{e_j}\right), \tag{7.59}$$

is approximated by

$$p_{obs}(\mathbf{a}) \doteq 1 - F_{\chi^2(k-1-d)}(-2\ln Q(\mathbf{a})) \tag{7.60}$$

if the expected frequencies $\mathbf{e}$ are all greater than or equal to 5.

As mentioned before the X^2-test was developed before the $-2\ln Q$-test and therefore the X^2-test is often considered in the literature in connection with a test for goodness of fit. With the notation here the X^2-test statistic and the corresponding significance probability are:

$$X^2(\mathbf{a}) = \sum_{j=1}^{k} \frac{(a_j - e_j)^2}{e_j} \tag{7.61}$$

$$p^*_{obs}(\mathbf{a}) \doteq 1 - F_{\chi^2(k-1-d)}(X^2(\mathbf{a})), \tag{7.62}$$

where the approximation may be applied if the expected frequencies are greater than or equal to 5.

Example 7.6
In order to illustrate the calculations in the goodness-of-fit test, suppose that we want to investigate whether the observations, $x_1, \ldots, x_n$, may be considered as a sample from the normal distribution. This distribution is parameterized by its mean μ and its variance σ^2, i.e., $d = 2$. First we estimate these two parameters:

$$\mu \leftarrow \bar{x}. = \frac{1}{n} \sum_{i=1}^{n} x_i$$

$$\sigma^2 \leftarrow s^2 = \frac{1}{n-1} \sum_{i=1}^{n} (x_i - \bar{x}.)^2.$$

Since the distribution function of a normal distribution with mean $\bar{x}.$ and variance s^2 may be expressed in terms of the distribution function of the standard normal distribution in the following way,

$$F_{(\bar{x}.,s^2)}(y) = \Phi(\frac{y - \bar{x}.}{s}),$$

the expected frequencies are, according to (7.58),

$$e_j = n\{\Phi(\frac{y_j - \bar{x}.}{s}) - \Phi(\frac{y_{j-1} - \bar{x}.}{s})\}, \qquad j = 1, \ldots, k. \tag{7.63}$$

As an illustration of the test for goodness of fit for the normal distribution we again consider the measurements in Example 2.1 of the height by 247 asthmatic girls in the age of 10 to 12 years. We consider the grouped version of these data in Table 2.2, but replace the intervals $]112, 116]$ and $]164, 168]$ with $]-\infty, 116]$ and $]164, \infty[$, respectively. From the discussion of Example 2.1 in Section 3.1 we know that

$$\mu \leftarrow \bar{x}. = 140.13$$

$$\sigma^2 \leftarrow s^2 = 85.8317.$$

The observed and expected frequencies calculated using (7.63) are given in Table 7.1.

As a start we consider 14 intervals, but in order to meet the requirement that the expected frequencies should be greater than or equal to 5 we have to combine some of the intervals as indicated. After this we have $k = 10$ intervals and since $d = 2$ the number of degrees of freedom of the test for goodness of fit is equal to $k - 1 - d = 10 - 1 - 2 = 7$.

For the $-2\ln Q$-test we have from (7.59) and (7.60) that

$$-2\ln Q(\mathbf{a}) = 7.2653$$

and

$$p_{obs}(\mathbf{a}) = 1 - F_{\chi^2(7)}(7.2653) = 0.4018.$$

The formulas (7.61) and (7.62) imply that for the X^2-test we find

$$X^2(\mathbf{a}) = 7.5472$$

and, consequently,

$$p^*_{obs}(\mathbf{a}) = 1 - F_{\chi^2(7)}(7.5472) = 0.3742.$$

Thus neither of these tests gives reasons to doubt that the observations may be considered as a sample from the normal distribution in accordance with the fractile diagram in Figure 2.14. □

Table 7.1 *Calculation of the test for goodness of fit of the normal distribution for the data in Table 2.2*

Interval	**a**		**e**	
$]-\infty,116]$	1		1.136	
$]116,120]$	0	9	2.544	10.087
$]120,124]$	8		6.407	
$]124,128]$	20		13.432	
$]128,132]$	24		23.435	
$]132,136]$	32		34.031	
$]136,140]$	49		41.132	
$]140,144]$	41		41.378	
$]144,148]$	26		34.646	
$]148,152]$	21		24.145	
$]152,156]$	14		14.005	
$]156,160]$	6		6.761	
$]160,164]$	4	11	2.716	10.710
$]164,\infty[$	1		1.233	

7.6 Sequence of Models

In this section we give the results from the theory for testing a number of successive hypotheses in the model for a single multinomial distribution,

$$M_0\colon \mathbf{X} \sim m(n,\boldsymbol{\pi}), \quad \boldsymbol{\pi} \in \boldsymbol{\Pi}^{(k)} \ (=\boldsymbol{\Pi}_0),$$

where $\boldsymbol{\Pi}^{(k)}$ is the set in (7.6) on page 304.

The procedure may also be applied to models for several multinomial distributions after obvious modifications.

Suppose that for $i = 1,2,\ldots$, the one-to-one mapping,

$$\begin{array}{llcl} \boldsymbol{\pi}_i\colon & \boldsymbol{\Theta}_i \subseteq \mathbb{R}^{d_i} & \to & \boldsymbol{\Pi}_i\,(\subseteq \boldsymbol{\Pi}_0) \\ & \boldsymbol{\theta}_i & \to & \boldsymbol{\pi}_i(\boldsymbol{\theta}_i) \end{array}$$

specifies a hypothesis,

$$H_{0i}\colon \boldsymbol{\pi} \in \boldsymbol{\Pi}_i$$

with d_i free parameters and let M_i denote the corresponding model, i.e.,

$$M_i\colon \mathbf{X} \sim m(n,\boldsymbol{\pi}), \quad \boldsymbol{\pi} \in \boldsymbol{\Pi}_i.$$

Assume, in addition, that

$$k-1 = d_0 > d_1 > \cdots > d_{i-1} > d_i > \cdots$$

and that

$$\boldsymbol{\Pi}_0 \supset \boldsymbol{\Pi}_1 \supset \cdots \supset \boldsymbol{\Pi}_{i-1} \supset \boldsymbol{\Pi}_i \supset \cdots,$$

corresponding to the model reductions,

$$M_0 \to M_1 \to \cdots \to M_{i-1} \to M_i \to \cdots.$$

If Q_{0i} denotes the likelihood ratio test statistic for testing H_{0i} in the basic model corresponding to the reduction $M_0 \to M_i$, we have from Section 7.2 the well-known approximation,

$$-2\ln Q_{0i}(\mathbf{x}) = 2\sum_{j=1}^{k} x_j \ln(\frac{x_j}{n\pi_{ij}(\hat{\boldsymbol{\theta}}_i)}) = 2\sum_{j=1}^{k} x_j \ln(\frac{x_j}{e_{ij}}) \sim\sim \chi^2(f_{0i}). \tag{7.64}$$

Here $\hat{\boldsymbol{\theta}}_i$ is the maximum likelihood estimate of $\boldsymbol{\theta}_i$,

$$\boldsymbol{\pi}_i(\hat{\boldsymbol{\theta}}_i) = (\pi_{i1}(\hat{\boldsymbol{\theta}}_i), \dots, \pi_{ik}(\hat{\boldsymbol{\theta}}_i))^*$$

is the maximum likelihood estimate in the model M_i of the probability vector $\boldsymbol{\pi}_i$,

$$\mathbf{e}_i = n\boldsymbol{\pi}_i(\hat{\boldsymbol{\theta}}_i) = (e_{i1}, \dots, e_{ik})^* = (n\pi_{i1}(\hat{\boldsymbol{\theta}}_i), \dots, n\pi_{ik}(\hat{\boldsymbol{\theta}}_i))^*$$

is the vector of expected frequencies in M_i and

$$f_{0i} = d_0 - d_i = k - 1 - d_i$$

is the number of degrees of freedom of the $-2\ln Q_{0i}$ test statistic.

The result in (7.64) may be used if $e_{ij} = n\pi_{ij}(\hat{\boldsymbol{\theta}}_i) \geq 5,\ j = 1, \dots, k$.

If Q_i denotes the likelihood ratio test statistic for testing H_{0i} in the model M_{i-1} corresponding to the reduction $M_{i-1} \to M_i$, we have the following approximation:

$$-2\ln Q_i(\mathbf{x}) = 2\sum_{j=1}^{k} x_j \ln(\frac{n\pi_{i-1,j}(\hat{\boldsymbol{\theta}}_{i-1})}{n\pi_{ij}(\hat{\boldsymbol{\theta}}_i)}) = 2\sum_{j=1}^{k} x_j \ln(\frac{e_{i-1,j}}{e_{ij}}) \sim\sim \chi^2(f_i), \tag{7.65}$$

where

$$f_i = f_{0i} - f_{0,i-1}\ (= d_{i-1} - d_i). \tag{7.66}$$

The result in (7.65) may be applied if $e_{ij} = n\pi_{ij}(\hat{\boldsymbol{\theta}}_i) \geq 5,\ j = 1, \dots, k$.

Notice that

$$-2\ln Q_i(\mathbf{x}) = -2\ln Q_{0i}(\mathbf{x}) - (-2\ln Q_{0,i-1}(\mathbf{x})), \tag{7.67}$$

$$-2\ln Q_{0i}(\mathbf{x}) = \sum_{m=1}^{i} -2\ln Q_m(\mathbf{x}) \tag{7.68}$$

and

$$f_{0i} = \sum_{m=1}^{i} f_m(\mathbf{x}). \tag{7.69}$$

The formulas (7.66) and (7.67) show how the degrees of freedom and the value of the $-2\ln Q$-test statistic for the reduction $M_{i-1} \to M_i$ can be calculated if we know the corresponding quantities for the reductions $M_0 \to M_i$ and $M_0 \to M_{i-1}$. This observation will be used repeatedly below and in Chapter 9.

Example 7.7

The data below are from a genetic survey of eelpouts conducted at the University of Aarhus. 782 pregnant eelpouts were caught at Kaloe Cove in 1969. By electrophoresis three esterase types of eelpouts – denoted 1, 2, and 3 – are found. For each mother and one of her offspring the esterase type was determined. The result is seen in Table 7.2 (from Christiansen, Frydenberg, and Simonsen, 1973).

Let x_{ij} be the number of times it was observed that the type of mother and offspring was of types i and j, respectively. As the basic model we assume that the matrix of the corresponding random variables has a multinomial distribution with number of trials $n = 782$ and probability matrix $\{\pi_{ij}\}_{i,j=1,2,3}$ where $\sum_{i=1}^{3}\sum_{j=1}^{3}\pi_{ij} = 1$ with $\pi_{13} = \pi_{13} = 0$ and $\pi_{ij} > 0$, otherwise, i.e.,

$$M_0\colon \{X_{ij}\} \sim m(782, \{\pi_{ij}\}).$$

Here we exclude the two apparently impossible cases where the mother-offspring combination is either (3,1) or (1,3). The remaining seven cases are considered as the outcome of a multinomial

Table 7.2 *The observed distribution of number of mother-offspring combinations of esterase types*

		Type of offspring			
		1	2	3	Total
	1	41	70	0	111
Type of mother	2	65	173	119	357
	3	0	127	187	314
	Total	106	370	306	782

experiment and the basic model has 6 free parameters. Below and in Exercise 7.9 several hypotheses are tested successively with the purpose of investigating if the population of eelpouts may be considered to be in Hardy-Weinberg equilibrium.

As a first step we investigate whether the esterase types may be described by two allelic genes A and a such that the types 1-3 correspond to the genotypes AA, Aa, aa. That is, if $r_i > 0$ is the probability that the mother is of type i and $s_i > 0$ is the conditional probability that the father gives sperm with the gene A while mating, given that the mother is of type i, the probabilities of the possible genotypic combinations of mothers and offspring are given in Table 7.3. These probabilities correspond to a hypothesis H_{01} with $d_{01} = 5$ free parameters, since $r_1 + r_2 + r_3 = 1$, whereas there are no restrictions on s_1, s_2 and s_3.

Table 7.3 *The probabilities of the possible mother-offspring genotype combinations under the hypothesis* H_{01}

		Type of offspring			
		AA	Aa	aa	Total
	AA	$\pi_{11} = r_1 s_1$	$\pi_{12} = r_1(1-s_1)$		r_1
Type of mother	Aa	$\pi_{21} = r_2 s_2/2$	$\pi_{22} = r_2/2$	$\pi_{23} = r_2(1-s_2)/2$	r_2
	aa		$\pi_{32} = r_3 s_3$	$\pi_{33} = r_3(1-s_3)$	r_3

In Exercise 7.9 it is shown that the likelihood function under H_{01} corresponds to the product of four likelihood functions: one for a multinomial experiment with three categories and probability vector (r_1, r_2, r_3) and three for binomial experiments where the probability parameters are s_1, s_2, and s_3, respectively. The parameters of the four likelihood functions, (r_1, r_2, r_3), s_1, s_2, and s_3, are variation independent and in Exercise 7.9 it is seen that the maximum likelihood estimates under H_{01} are

$$\hat{r}_1 = \frac{x_{1\cdot}}{n} = \frac{111}{782} = 0.142, \quad \hat{r}_2 = \frac{x_{2\cdot}}{n} = \frac{357}{782} = 0.457\,, \quad \hat{r}_3 = \frac{x_{3\cdot}}{n} = \frac{314}{782} = 0.402, \tag{7.70}$$

and

$$\hat{s}_1 = \frac{x_{11}}{x_{1\cdot}} = \frac{41}{111} = 0.369, \; \hat{s}_2 = \frac{x_{21}}{x_{21} + x_{23}} = \frac{65}{184} = 0.353, \; \hat{s}_3 = \frac{x_{32}}{x_{3\cdot}} = \frac{127}{314} = 0.404. \tag{7.71}$$

Here and in the rest of the example the results are given with three decimal places. The expected frequencies under H_{01} calculated using (7.70) and (7.71) are given in Table 7.4.

From Table 7.2 and Table 7.4 it is seen that the observed and expected frequencies are the same for mothers of type AA and type aa. Using these tables the $-2\ln Q$-test statistic for testing H_{01} in the model M_0 becomes

$$-2\ln Q_{01}(\mathbf{x}) = 0.339 \sim\sim \chi^2(1), \tag{7.72}$$

Table 7.4 *The expected value of the possible mother-offspring genotype combinations under the hypothesis* H_{01}

		Type of offspring			
		AA	*Aa*	*aa*	*Total*
	AA	41.000	70.000	0.000	111.000
Type of mother	*Aa*	63.057	178.500	115.443	357.000
	aa	0.000	127.000	187.000	314.000

and since the expected frequencies in Table 7.4 are greater than 5, the significance probability may be calculated as

$$p_{obs}(\mathbf{x}) = 1 - F_{\chi^2(1)}(0.339) = 0.560.$$

The hypothesis H_{01} is not rejected and we adopt the model M_1 corresponding to the hypothesis H_{01}.

Next, we investigate if there is a random and independent choice of partners in the parent generation with respect to the locus considered, i.e., the hypothesis,

$$H_{02}: s_1 = s_2 = s_3 = s, \quad 0 < s < 1,$$

in the model M_1.

The hypothesis H_{02} has $d_{02} = 3$ free parameters. In Exercise 7.9 it is shown that the maximum likelihood estimates of the r_i's are the same as under H_{01} and that

$$\hat{s} = \frac{x_{11} + x_{21} + x_{32}}{n - x_{22}} = \frac{233}{609} = 0.383. \tag{7.73}$$

From (7.70) and (7.73) it follows that the expected frequencies under H_{02} are as shown in Table 7.5.

Table 7.5 *The expected value of the possible mother-offspring genotype combinations under the hypothesis* H_{02}

		Type of offspring			
		AA	*Aa*	*aa*	*Total*
	AA	42.468	68.532	0.000	111.000
Type of mother	*Aa*	68.293	178.500	110.207	357.000
	aa	0.000	120.135	193.865	314.000

From Table 7.2 and Table 7.5 we find that the $-2\ln Q$-test statistic for testing H_{02} in the model M_0 is

$$-2\ln Q_{02}(\mathbf{x}) = 1.730,$$

and using (7.67) and (7.72) the $-2\ln Q$-test statistic for testing H_{02} under the hypothesis H_{01} is seen to be

$$-2\ln Q_2(\mathbf{x}) = -2\ln Q_{02}(\mathbf{x}) - (-2\ln Q_{01}(\mathbf{x})) = 1.730 - 0.339 = 1.391 \sim\sim \chi^2(2),$$

because $d_{01} - d_{02} = 5 - 3 = 2$. Since the expected frequencies in Table 7.5 are greater than 5, the significance probability of the test of H_{02} in the model M_1 is

$$p_{obs}(\mathbf{x}) = 1 - F_{\chi^2(2)}(1.391) = 0.499.$$

The hypothesis H_{02} is not rejected and we adopt the corresponding model M_2.

In Exercise 7.9 two further hypotheses are tested. First the hypotheses of Hardy-Weinberg proportions among the mothers,

$$H_{03}\colon (r_1, r_2, r_3) = (p^2, 2p(1-p), (1-p)^2), \quad 0 < p < 1,$$

is tested in the model M_2 and, finally, the hypothesis that the frequency of the A gene is the same for mothers and fathers,

$$H_{04}\colon s = p, \quad 0 < p < 1,$$

is tested in the model M_3 corresponding to H_{03}. □

Annex to Chapter 7

Calculations in SAS

The calculation of the $-2\ln Q$-test and/or the X^2-test on the basis of the observed and expected frequencies may be performed using the macro `fittest`, which we have made for this purpose.

The SAS programs made for this book as well as SAS macros can be found at the address
`http://www.imf.au.dk/biogeostatistics`

Input to the macro `fittest` is a data set with the two variables `observed` and `expected` containing the observations and the expected values under the hypothesis and, in addition, a non-negative integer giving the number of free parameters d in the hypothesis on the basis of which the expected frequencies are calculated. The output from the macro needs no explanation.

Example 7.1 (Continued)
The following program segment makes the relevant calculations:

```
DATA ex7_1;
INPUT observed expected;
DATALINES;
315 312.75
101 104.25
108 104.25
 32  34.75
;
RUN;

%fittest(DATA=ex7_1,d=0);
```

The output listing from this program is:

```
                              fittest 1

                                                  observed-
               group     observed     expected     expected

                 1          315        312.75        2.25
                 2          101        104.25       -3.25
                 3          108        104.25        3.75
                 4           32         34.75       -2.75


                              fittest 2

                          number
  total       total         of                            Prob >      chi-         Prob >
observed    expected      groups     d      -2lnQ       -2lnQ      square      chi-square

   556         556           4       0     0.4754      0.9243      0.4700        0.9254
```

Alternatively, the X^2-test – but not the $-2\ln Q$-test – may be calculated by means of the statement `TABLES` in *PROC FREQ* using the option `TESTF` or `TESTP`, where the expected frequencies or the probabilities under the hypothesis are stated. The program segment

```
TITLE1 'PROC FREQ; TESTF option';
PROC FREQ DATA=ex7_1 ORDER=DATA;
WEIGHT observed;
TABLES observed/TESTF=(312.75 104.25 104.25 34.75) CHISQ;
RUN;
```

produces the output listing:

```
                    PROC FREQ; TESTF option

                     The FREQ Procedure

                              Test                               Cumulative     Cumulative
observed     Frequency     Frequency      Percent       Frequency        Percent
-------------------------------------------------------------------------------
     315           315        312.75        56.65             315          56.65
     101           101        104.25        18.17             416          74.82
     108           108        104.25        19.42             524          94.24
      32            32         34.75         5.76             556         100.00

                     Chi-Square Test
                  for Specified Frequencies
                  -------------------------
                  Chi-Square          0.4700
                  DF                       3
                  Pr > ChiSq          0.9254

                     Sample Size = 556
```

and

```
TITLE1 'PROC FREQ; TESTP option';
PROC FREQ DATA=ex7_1 ORDER=DATA;
WEIGHT observed;
TABLES observed/TESTP=(56.25 18.75 18.75 6.25) CHISQ;
RUN;
```

produces the output listing:

```
                    PROC FREQ; TESTP option

                     The FREQ Procedure

                                            Test      Cumulative     Cumulative
observed     Frequency      Percent      Percent       Frequency        Percent
-------------------------------------------------------------------------------
     315           315        56.65        56.25             315          56.65
     101           101        18.17        18.75             416          74.82
     108           108        19.42        18.75             524          94.24
      32            32         5.76         6.25             556         100.00

                     Chi-Square Test
                  for Specified Proportions
                  -------------------------
                  Chi-Square          0.4700
                  DF                       3
                  Pr > ChiSq          0.9254

                     Sample Size = 556
```

□

The calculations in the test of the hypothesis of Hardy-Weinberg proportions may be performed using the macro hw. In addition to the observed and expected frequencies the $-2\ln Q$-test and the X^2-test of the hypothesis are calculated. Furthermore, the output from hw contains the necessary information for further calculations concerning the frequency of the dominant allele. Input to hw is the observed distribution of the genotypes.

Example 7.2 (Continued)

The following small program segment makes the relevant calculations:

```
%hw(56,107,37);
```

The output listing is:

```
                    Hardy-Weinberg 1

                                              observed-
        genotype    observed    expected      expected

           AA           56       59.9513      -3.95125
           Aa          107       99.0975       7.90250
           aa           37       40.9513      -3.95125

                    Hardy-Weinberg 2

                    number                Prob >     chi-       Prob >
  n          p     A-genes      -2lnQ     -2lnQ     square    chi-square

 200     0.5475      219       1.2745    0.2589     1.2718      0.2594
```

□

The calculation of the $-2\ln Q$-test and the X^2-test in $r \times s$ tables, i.e., the test of independence in a two-way table and the test of homogeneity, may be obtained using the `TABLES` statement in *PROC FREQ*. The output has a table with the observed and expected frequencies and, furthermore, information on the $-2\ln Q$-test and the X^2-test is given under the heading `Statistic` in the lines `Likelihood Ratio Chi-Square` and `Chi-Square`, respectively. The additional lines in this summary are not interesting in this context. If some of the expected frequencies are less than or equal to 5, the output contains a notice concerning this and if $r = s = 2$, information concerning Fisher's exact test appears in a table with the heading `Fisher's Exact Test`. In the general case Fisher's exact test may be calculated by adding `EXACT` in the sequence of options to `TABLES`, see the continuation of Example 7.4 below.

Example 7.3 (Continued)

First the data set is created by means of the following program segment:

```
DATA ex7_3;
INPUT weight$1-5 shape$7-14 count;
DATALINES;
  <25     <0.6  6
  <25 0.6-0.8  31
  <25     >0.8 11
25-50     <0.6  8
25-50 0.6-0.8  29
25-50     >0.8 19
  >50     <0.6 10
  >50 0.6-0.8  39
  >50     >0.8 26
;
RUN;
```

The calculation of the test of independence may be performed by the following lines:

```
PROC FREQ DATA=ex7_3 ORDER=DATA;
WEIGHT count;
TABLES weight*shape/EXPECTED
                    CHISQ
                    NOROW
                    NOCOL
                    NOPERCENT;
RUN;
```

In the `INPUT` line `shape$7-14` indicates that the value of this variable (breadth/length) are the characters in the positions 7-14 in the lines after `DATALINES;` The option `ORDER=DATA` in the call of *PROC FREQ* ensures that the entries of the frequency table that is requested in the `TABLES` statement are given in the same order as they are read into the data set. The output listing is:

```
                    The FREQ Procedure

             Table of weight by width_length

         weight     shape

         Frequency|
         Expected |<0.6    |0.6-0.8 |>0.8    |  Total
         ---------+--------+--------+--------+
         <25      |      6 |     31 |     11 |     48
                  | 6.4358 | 26.547 | 15.017 |
         ---------+--------+--------+--------+
         25-50    |      8 |     29 |     19 |     56
                  | 7.5084 | 30.972 |  17.52 |
         ---------+--------+--------+--------+
         >50      |     10 |     39 |     26 |     75
                  | 10.056 |  41.48 | 23.464 |
         ---------+--------+--------+--------+
         Total          24       99       56     179

        Statistics for Table of weight by width_length

     Statistic                     DF        Value       Prob
     ------------------------------------------------------
   ->Chi-Square                     4       2.5564     0.6346
   ->Likelihood Ratio Chi-Square    4       2.6220     0.6229
     Mantel-Haenszel Chi-Square     1       0.7679     0.3809
     Phi Coefficient                        0.1195
     Contingency Coefficient                0.1187
     Cramer's V                             0.0845

                    Sample Size = 179
```

On the output we have indicated the X^2 and $-2\ln Q$ statistics by the arrows ->. □

Example 7.4 (Continued)

The calculation of Fisher's exact test is performed in SAS by using the option EXACT in the TABLES statement in *PROC FREQ*. The following program segment calculates the test of homogeneity in the two areas:

```
DATA dichlor;
LENGHT concentration$8;
INPUT area municipality$3-10 concentration$12-20 count;
DATALINES;
1 Hadsten    <=0.01 23
1 Hadsten  0.01-0.1 12
1 Hadsten      >0.1  6
1 Hammel     <=0.01 20
1 Hammel   0.01-0.1  5
1 Hammel       >0.1  9
2 Nr Djurs   <=0.01 37
2 Nr Djurs 0.01-0.1  6
2 Nr Djurs     >0.1  4
2 Randers    <=0.01 47
2 Randers  0.01-0.1  4
2 Randers      >0.1  3
;
RUN;

PROC FREQ DATA=dichlor ORDER=data;
TABLES municipality*concentration/EXPECTED CHISQ
                                  NOPERCENT NOCOL NOROW
                                  EXACT;
WEIGHT count;
BY area;
RUN;
```

In the output for area 2 below it is seen in the line `Pr <= P` that the p-value of Fisher's exact test is 0.5312. Note that the p-value of the $-2\ln Q$-test is 0.5342 indicating that this test is reliable in this situation despite the fact that half of the expected frequencies are less than 5.

```
------------------------------------ area=2 -----------------------------------

                        The FREQ Procedure

              Table of municipality by concentration

           municipality
                     concentration
           Frequency|
           Expected |<=0.01  |0.01-0.1|>0.1    |  Total
           ---------+--------+--------+--------+
           Nr Djurs |     37 |      6 |      4 |     47
                    | 39.089 | 4.6535 | 3.2574 |
           ---------+--------+--------+--------+
           Randers  |     47 |      4 |      3 |     54
                    | 44.911 | 5.3465 | 3.7426 |
           ---------+--------+--------+--------+
           Total          84       10        7      101


      Statistics for Table of municipality by concentration

      Statistic                     DF       Value      Prob
      ------------------------------------------------------
      Chi-Square                     2      1.2542    0.5341
      Likelihood Ratio Chi-Square    2      1.2538    0.5342
      Mantel-Haenszel Chi-Square     1      0.9880    0.3202
      Phi Coefficient                       0.1114
      Contingency Coefficient               0.1108
      Cramer's V                            0.1114

       WARNING: 50% of the cells have expected counts less
                than 5. Chi-Square may not be a valid test.


                      Fisher's Exact Test
                ----------------------------------
                Table Probability (P)       0.0433
                Pr <= P                     0.5312
```

Test of the hypothesis that the probability of exceeding the limit value is the same for the two areas is obtained from the program segment:

```
DATA dichlor1;
SET dichlor;
IF concentration NE '>0.1' THEN concentration='<=0.1';
RUN;

PROC FREQ DATA=dichlor1 ORDER=data;
TABLES area*concentration/EXPECTED
                          CHISQ
                          NOPERCENT NOCOL NOROW;
WEIGHT count;
RUN;
```

Part of the output listing is:

```
                    The FREQ Procedure

              Table of area by concentration

         area      concentration

         Frequency|
         Expected |<=0.1    |>0.1     |  Total
         ---------+--------+--------+
                1 |     60 |     15 |     75
                  | 65.625 |  9.375 |
         ---------+--------+--------+
                2 |     94 |      7 |    101
                  | 88.375 | 12.625 |
         ---------+--------+--------+
         Total         154       22      176


                    The FREQ Procedure

        Statistics for Table of area by concentration

Statistic                        DF       Value      Prob
---------------------------------------------------------
Chi-Square                        1      6.7214    0.0095
Likelihood Ratio Chi-Square       1      6.6905    0.0097


                   Fisher's Exact Test
          ----------------------------------
          Table Probability (P)       0.0069
          Two-sided Pr <= P           0.0116

                    Sample Size = 176
```

□

Main Points in Chapter 7

General Models and Hypotheses

Model:
The full model is based on the k-dimensional multinomial distribution with number of trials n and probability vector $\boldsymbol{\pi} = (\pi_1, \dots, \pi_j, \dots, \pi_k)^*$

$$M_0: \mathbf{X} = (X_1, \dots, X_j, \dots, X_k)^* \sim m(n, \boldsymbol{\pi}), \qquad \boldsymbol{\pi} \in \boldsymbol{\Pi}^{(k)}.$$

Model Checking:
Check that the conditions (a) – (d) on page 301 are fulfilled.
Estimate:
On the basis of the observed frequencies $\mathbf{x} = (x_1, \dots, x_j, \dots, x_k)^*$, the probability vector $\boldsymbol{\pi}$ is estimated as the vector of relative frequencies:

$$\boldsymbol{\pi} \leftarrow \hat{\boldsymbol{\pi}}(\mathbf{x}) = (\frac{x_1}{n}, \dots, \frac{x_j}{n}, \dots, \frac{x_k}{n})^*.$$

The distribution of the estimate is given by

$$n\hat{\boldsymbol{\pi}} = \mathbf{X} \sim m(n, \boldsymbol{\pi}).$$

Confidence Intervals:
Under M_0 we have $X_j \sim b(n, \pi_j)$ and the confidence interval for π_j may therefore be calculated in the same way as the confidence interval for the probability parameter π in the binomial model $X \sim b(n, \pi)$ is calculated on the basis of the observation x. This interval, which is based on an approximation, is

$$C_{1-\alpha}(x) = [\pi_-, \pi_+],$$

where

$$\pi_- = \frac{1}{n + u_{1-\alpha/2}^2} \left[x + \frac{u_{1-\alpha/2}^2}{2} - u_{1-\alpha/2} \sqrt{\frac{x(n-x)}{n} + \frac{u_{1-\alpha/2}^2}{4}} \right],$$

and

$$\pi_+ = \frac{1}{n + u_{1-\alpha/2}^2} \left[x + \frac{u_{1-\alpha/2}^2}{2} + u_{1-\alpha/2} \sqrt{\frac{x(n-x)}{n} + \frac{u_{1-\alpha/2}^2}{4}} \right].$$

Submodels and Hypotheses:
A submodel,

$$M: \mathbf{X} = (X_1, \dots, X_j, \dots, X_k)^* \sim m(n, \boldsymbol{\pi}), \qquad \boldsymbol{\pi} \in \boldsymbol{\Pi},$$

of the model M_0 has d free parameters if $\boldsymbol{\Pi}$ is the range set of one-to-one mapping $\boldsymbol{\pi}$ from a domain $\boldsymbol{\Theta}$ of $\mathbb{R}^d$ into $\boldsymbol{\Pi}_0 = \boldsymbol{\Pi}^{(k)}$, i.e., if $\boldsymbol{\Pi} = \boldsymbol{\pi}(\boldsymbol{\Theta}) (\subset \boldsymbol{\Pi}_0)$. The sets $\boldsymbol{\Theta}$ and $\boldsymbol{\Pi}$ are both referred to as the parameter space of the model M.

The hypothesis corresponding to the reduction from the full model M_0 to the model M,

$$H_0: \boldsymbol{\pi} \in \boldsymbol{\Pi} = \boldsymbol{\pi}(\boldsymbol{\Theta}) (\subset \boldsymbol{\Pi}_0),$$

is also said to have d free parameters.

When a sequence of submodels is considered, indices are used to distinguish between the models. The mapping $\boldsymbol{\pi}$, the number of free parameters d, and the hypothesis H_0 and parameter sets $\boldsymbol{\Theta}$ and $\boldsymbol{\Pi}$ have the same index as the model.
Test of the Hypothesis H_0 (Reduction from M_0 to M):

If $\hat{\boldsymbol{\theta}}$ is the maximum likelihood estimate of the parameter $\boldsymbol{\theta}$ under H_0, the vector $\mathbf{e}$ of expected frequencies under H_0 is given as

$$\mathbf{e}^* = (e_1, \ldots, e_j, \ldots, e_k) = (n\pi_1(\hat{\boldsymbol{\theta}}), \ldots, n\pi_j(\hat{\boldsymbol{\theta}}), \ldots, n\pi_k(\hat{\boldsymbol{\theta}})).$$

Of the two approximative tests, the $-2\ln Q$-test and the X^2-test, of H_0 we prefer the $-2\ln Q$-test. Both tests are based on a comparison between the observed frequencies $\mathbf{x}$ and the expected frequencies $\mathbf{e}$. If **the expected frequencies are all greater than or equal to 5,** the following test statistics and the corresponding approximative significance probabilities may be used.
$-2\ln Q$-test:

$$-2\ln Q(\mathbf{x}) = 2\sum_{j=1}^{k} x_j \ln(\frac{x_j}{e_j})$$

$$p_{obs}(\mathbf{x}) \doteq 1 - F_{\chi^2(k-1-d)}(-2\ln Q(\mathbf{x})).$$

X^2-test:

$$X^2(\mathbf{x}) = \sum_{j=1}^{k} \frac{(x_j - e_j)^2}{e_j}$$

$$p^*_{obs}(\mathbf{x}) \doteq 1 - F_{\chi^2(k-1-d)}(X^2(\mathbf{x})).$$

Special Models and Hypotheses

The Hypothesis of Hardy-Weinberg Proportions:
The model for the description of the observed counts of the genotypes AA, Aa and aa is

$$M_0\colon (X_1, X_2, X_3) \sim m(n, (\pi_1, \pi_2, \pi_3)).$$

The hypothesis of Hardy-Weinberg proportions is

$$H_{01}\colon (\pi_1, \pi_2, \pi_3) = (p^2, 2p(1-p), (1-p)^2), \qquad p \in \,]0,1[\,,$$

where p denotes the frequency of the allele A. The maximum likelihood estimate of p is

$$\hat{p} = \frac{2x_1 + x_2}{2n},$$

i.e., the relative frequency of the gene of allele type A among the total number of genes, $2n$. The expected frequencies under H_{01} are

$$\mathbf{e} = (n\hat{p}^2, 2n\hat{p}(1-\hat{p}), n(1-\hat{p})^2)^*.$$

The $-2\ln Q$-test statistic is

$$-2\ln Q(\mathbf{x}) = 2\sum_{j=1}^{3} x_j \ln(\frac{x_j}{e_j}),$$

and the significance probability may be calculated as

$$p_{obs}(\mathbf{x}) \doteq 1 - F_{\chi^2(1)}(-2\ln Q(\mathbf{x}))$$

if the expected frequencies **e are all greater than or equal to 5.**

If H_{01} is not rejected, the model M_0 may be replaced by

$$M_1\colon (X_1, X_2, X_3) \sim m(n, (p^2, 2p(1-p), (1-p)^2)),$$

and in this model the number of A genes has a binomial distribution,

$$2X_1 + X_2 \sim b(2n, p).$$

Independence in a Two-Way Table:
The data $\mathbf{x}$ are given as an $r \times s$ table $\{x_{ij}\}$ and the full model is

$$M_0\colon \mathbf{X} = \{X_{ij}\} \sim m(n, (\{\pi_{ij}\})).$$

The hypothesis,

$$H_{01}\colon \pi_{ij} = \rho_i \sigma_j, \qquad i = 1,\dots,r, \quad j = 1,\dots,s,$$

is called the hypothesis of independence (in a two-way table) and it has $d_1 = r + s - 2$ free parameters. The maximum likelihood estimates of the vectors $\boldsymbol{\rho}$ and $\boldsymbol{\sigma}$ of row probabilities and column probabilities, respectively, are given by

$$\hat{\rho}_i = \frac{x_{i\bullet}}{n}, \quad i = 1,\dots,r \qquad \text{and} \qquad \hat{\sigma}_j = \frac{x_{\bullet j}}{n}, \quad j = 1,\dots,s,$$

and the expected frequencies $\mathbf{e} = \{e_{ij}\}$ are calculated as

$$e_{ij} = \frac{x_{i\bullet} x_{\bullet j}}{n}.$$

If these **are all greater than or equal to 5**, the test statistic (remember the factor 2) and the corresponding significance probability are

$$-2\ln Q(\mathbf{x}) = 2[\sum_{i=1}^{r}\sum_{j=1}^{s} x_{ij}\ln(x_{ij}) - \sum_{i=1}^{r} x_{i\bullet}\ln(x_{i\bullet}) - \sum_{j=1}^{s} x_{\bullet j}\ln(x_{\bullet j}) + n\ln(n)],$$

$$p_{obs}(\mathbf{x}) \doteq 1 - F_{\chi^2((r-1)(s-1))}(-2\ln Q(\mathbf{x})).$$

If H_{01} is not rejected, the model M_0 may be replaced by

$$M_1\colon \mathbf{X} = \{X_{ij}\} \sim m(n, (\{\rho_i \sigma_j\}),$$

in which case we have

$$\begin{aligned}
&\mathbf{X}^*_{*\bullet} = (X_{1\bullet},\dots,X_{i\bullet},\dots,X_{r\bullet}) \sim m(n,(\rho_1,\dots,\rho_i,\dots,\rho_r))\\
&\mathbf{X}^*_{\bullet *} = (X_{\bullet 1},\dots,X_{\bullet j},\dots,X_{\bullet s}) \sim m(n,(\sigma_1,\dots,\sigma_j,\dots,\sigma_s))\\
&\mathbf{X}_{*\bullet} \text{ and } \mathbf{X}_{\bullet *} \text{ are stochastically independent.}
\end{aligned}$$

Homogeneity of Several Multinomial Distributions:
In the model,

$$\begin{aligned}
M_0\colon \quad &\mathbf{X}_i = (X_{i1},\dots,X_{ij},\dots,X_{is})^* \sim m(n_i,\boldsymbol{\pi}_i) = m(n_i,(\pi_{i1},\dots,\pi_{ij},\dots,\pi_{is})^*)\\
&\mathbf{X}_1,\dots,\mathbf{X}_i,\dots,\mathbf{X}_r \text{ are stochastically independent,}
\end{aligned}$$

the hypothesis to be tested deals with homogeneity, or identity, of the parameters of r multinomial distributions:

$$H_{01}\colon \boldsymbol{\pi}_1 = \cdots = \boldsymbol{\pi}_i = \cdots = \boldsymbol{\pi}_r = \boldsymbol{\pi} = (\pi_1,\dots,\pi_j,\dots,\pi_s)^*.$$

The maximum likelihood estimate of the components of the common probability vector is

$$\hat{\pi}_j = \frac{x_{\bullet j}}{n_\bullet}, \quad j = 1,\dots,s,$$

and the expected frequencies are calculated as

$$e_{ij} = \frac{n_i x_{\bullet j}}{n_\bullet}.$$

The $-2\ln Q$-test statistic is (remember the factor 2)

$$-2\ln Q(\mathbf{x}) = 2[\sum_{i=1}^{r}\sum_{j=1}^{s} x_{ij}\ln(x_{ij}) - \sum_{i=1}^{r} n_i\ln(n_i) - \sum_{j=1}^{s} x_{\bullet j}\ln(x_{\bullet j}) + n_\bullet \ln(n_\bullet)]$$

and the significance probability may be calculated as

$$p_{obs}(\mathbf{x}) \doteq 1 - F_{\chi^2((r-1)(s-1))}(-2\ln Q(\mathbf{x}))$$

provided the expected frequencies **are all greater than or equal to 5.**

If H_{01} is not rejected, the model is reduced to

$$\begin{aligned}
M_1\colon \quad &\mathbf{X}_i = (X_{i1},\dots,X_{ij},\dots,X_{is})^* \sim m(n_i,\boldsymbol{\pi}) = m(n_i,(\pi_1,\dots,\pi_j,\dots,\pi_s)^*)\\
&\mathbf{X}_1,\dots,\mathbf{X}_i,\dots,\mathbf{X}_r \text{ are stochastically independent,}
\end{aligned}$$

and in this model, we have

$$\mathbf{X}_{\bullet} = (X_{\bullet 1}, \ldots, X_{\bullet j}, \ldots, X_{\bullet s})^* \sim m(n_{\bullet}, \boldsymbol{\pi}) = m(n_{\bullet}, (\pi_1, \ldots, \pi_j, \ldots, \pi_s)^*).$$

Exercises for Chapter 7

Exercise 7.1 The shape of a radish may be long, oval, or round and its color red, purple, or white. A cultivated long and white variant was crossed with a cultivated round and red variant. All the resulting plants from this crossing were oval and purple. A possible explanation for this is that two allelic genes A and a determine the shape such that the genotypes AA, Aa and aa correspond to long, oval, and round radishes, respectively, and, similarly, that two allelic genes B and b determine the color such that the genotypes BB, Bb, and bb correspond to red, purple, and white radishes, respectively. The oval and purple radishes, which under this assumption have the genotypes Aa and Bb, were grown and self-fertilized. The distribution of the resulting 132 radishes according to shape and color was:

		Color			
		Red	*Purple*	*White*	*Total*
	Long	9	15	7	31
Shape	*Oval*	18	32	17	67
	Round	10	16	8	34
	Total	37	63	32	132

(1) Show that it may be assumed that shape and color are inherited independently, i.e., the two genes are located on different chromosomes.

(2) Investigate if there is a 1:2:1 segregation in long, oval, and round radishes.

(3) Investigate if there is a 1:2:1 segregation in red, purple, and white radishes.

Exercise 7.2 All the cats on a small island were caught in traps and the following distribution of the fur color was observed:

	Fur color			
Sex	*Dark*	*Medium*	*Light*	*Total*
Female	92	70	18	180
Male	102	0	53	155

A genetic explanation of the observed phenotype numbers is that a gene with alleles A and a on the cats X chromosome determines the fur color. The females have two X chromosomes such that the genotypes AA, Aa and aa correspond to *dark, medium, and light*, respectively, whereas the males have one X chromosome and A and a correspond to *dark* and *light*, respectively.

(1) Show that it may be assumed that the genotypic frequencies among female cats are in Hardy-Weinberg proportions.

(2) Show that it may be assumed that the frequency of the A allele is the same for males and females.

(3) Give an estimate of and a 95% confidence interval for the frequency of the A allele in the population.

Exercise 7.3 A particular gene in the beetle *Tetraopes tetraophthalmus* has three codominant alleles A, B, and C. Consequently, there exist six different genotypes AA, AB, AC, BB, BC, and CC. These six genotypes may be distinguished by electrophoresis also giving six phenotypes.

From a population on Long Island, 287 beetles were collected and classified according to genotype (Eanes et al., 1977). The result was:

AA	*AB*	*AC*	*BB*	*BC*	*CC*	*total*
9	85	16	99	66	12	287

An obvious model for the observed numbers, $x_1,\ldots,x_6$, is the multinomial model with number of trials $n = 287$ and probability vector $(\pi_1,\ldots,\pi_6)$, where $\pi_1,\ldots,\pi_6$ are the frequencies of the genotypes in the population of *Tetraopes tetraophthalmus* on Long Island, i.e., $(x_1,\ldots,x_6)$~~ $m(287,(\pi_1,\ldots,\pi_6))$.

If there is random mating and no selection in the population with respect to this system of genes, it is to be expected that probabilities of the genotype frequencies are

$$(\pi_1,\ldots,\pi_6) = (p^2, 2pq, 2pr, q^2, 2qr, r^2),$$

where $p+q+r = 1$ and where p, q, and r denotes the frequency of the alleles A, B, and C, respectively. If this hypothesis is satisfied, the population is said to be in Hardy-Weinberg equilibrium.

(1) Show that under the hypothesis of Hardy-Weinberg proportions the likelihood function for (p,q,r) is

$$L(p,q,r) = \binom{n}{x_1 \ldots x_6} p^{2x_1}(2pq)^{x_2}(2pr)^{x_3} q^{2x_4}(2qr)^{x_5} r^{2x_6}$$

and that the maximum likelihood estimate of (p,q,r) is given by

$$\hat{p} = \frac{2x_1+x_2+x_3}{2n}, \quad \hat{q} = \frac{x_2+2x_4+x_5}{2n}, \quad \hat{r} = \frac{x_3+x_5+2x_6}{2n}.$$

(Hint: Rewrite the likelihood function and use Proposition 7.1.)

(2) Test the hypothesis of Hardy-Weinberg proportions.

Under the hypothesis of Hardy-Weinberg proportions the observed numbers of A, B, and C genes has a multinomial distribution with number of trials $2n$ and probability vector (p,q,r).

On another locality on Long Island 274 beetles were collected and the hypothesis that the population there is in Hardy-Weinberg equilibrium was not rejected. The observed numbers of genes were:

A	*B*	*C*	*Total*
74	297	177	548

(3) Examine if it can be assumed that the gene frequencies are the same in the two populations.

Exercise 7.4 In 1990–1991 the concentration of Barium was measured in samples of ground water in the county of South Jutland, 21 samples from the northern part, and 31 samples from the southern part. For each of the samples it was investigated if the recommended limit value for the concentration 100 μg/l was exceeded. The result was:

	$\leq 100\ \mu$g/l	$> 100\ \mu$g/l	*Total*
Northern part	11	10	21
Southern part	14	17	31

(1) Show that there is no evidence of a difference between the distributions on the categories ≤ 100 μg/l and > 100 μg/l in the two parts of the county.

(2) Find an estimate of and a 95% confidence interval for the probability that the concentration of Barium in a sample of ground water does not exceed 100 μg/l .

In 1995–1996 72 samples of ground water were taken in the County of South Jutland and their concentrations of Barium were determined. The result was:

$\leq 100\ \mu$g/l	$> 100\ \mu$g/l	*Total*
32	40	72

(3) Examine whether the distribution on the two categories has changed from 1990–1991 to 1995–1996.

Exercise 7.5 The data below is from the report "Drikkevandskvaliteten i Danmark" (the quality of drinking water in Denmark) published by the Danish Environmental Protection Agency (Report number 156, 1990). In one of the tables we find for each of the counties the number of waterworks with a yearly production larger than 10000 m^3 together with the number of waterworks for which at least one of the limit values, as laid down by the Danish Ministry of Environment and Energy, for the parameters NO_3, Cl, Na, K, and F has been exceeded. Here we consider the data for a geographical grouping of some of the counties. In the table below, + and – indicate whether or not the limit values have been exceeded.

Group	*County*	+	–	*Total*
1	*Sønderjylland*	10	238	248
	Ribe	7	110	117
	Vejle	10	231	241
	Ringkøbing	5	140	145
2	*Århus*	62	364	426
	Nordjylland	51	315	366
3	*Roskilde*	15	127	142
	Vestsjælland	25	223	248
4	*København*	3	50	53
	Frederiksborg	5	128	133

(1) Show for each of the 4 groups of counties that it may be assumed that there is no difference between the distributions of waterworks on the categories + and – for the counties within the group considered.

(2) Investigate whether the distributions of the waterworks on the categories + and – are the same for the 4 groups of counties.

Exercise 7.6 When a botanist wants to determine how frequently a particular plant occurs on a locality, he often uses a method known as *Raunkiær circling*. Within a sample site n circular plots of the same area are placed at random positions and the number of plots where the plant occurs is counted. The plots are referred to as Raunkiær circles, hence the name of the method.

The data here are from a larger investigation undertaken by Esbern Warncke, the Department of Plant Ecology, University of Aarhus, who investigated the occurrence of marsh thistle, *Cirsium palustra*, at various locations with springs in different geographical areas of Jutland. All the sample sites were quadratic with a side length of 50 m and within each locality $n = 100$ Raunkiær circles were laid out, each circle with an area of 0.1 m^2. Here we consider 3 areas, the first in the southern part of Jutland with 4 localities, Høllund Bro, Sillerup, Tinnerup, and Tvilho, the second in the central part of Jutland with 3 localities, Bredsgårde, Hald, and Vinkel and the third in the northern part of Jutland with 2 localities, Krogensmølle and Øster Vrå. The counts are seen in the table below where, for instance, it is shown that the plant marsh thistle occurred (+) in 24 of 100 plots at Høllund

Bro.

Area	*Locality*	+	−	*Total*
1	*Høllund Bro*	24	76	100
	Sillerup	22	78	100
	Tinnet	18	82	100
	Tvilho	23	77	100
2	*Bredsgårde*	41	59	100
	Hald	33	67	100
	Vinkel	44	56	100
3	*Krogensmølle*	57	43	100
	Øster Vrå	45	55	100

(1) Show that it may be assumed that, for each of the 3 areas in Jutland, marsh thistle occurs with the same probability in each locality within the area.

(2) Investigate whether it can be assumed that the occurrence of marsh thistle is the same in the 3 areas.

Exercise 7.7 A student at the Institute of Biological Sciences, University of Aarhus, has performed a large experiment with recapture of 2- to 3-year-old sea trouts (*Salmo trutta*). At 5 stations along the brook Bygholm Å placed 13210 m, 11460 m, 9660 m, 8160 m, and 3360 m, respectively, above the point where the brook enters the lake Bygholm Sø, a number of trout were released after being marked with a marker that identified the station. Some of the fish were caught again in a net placed at the opposite end of Bygholm Sø. At the 5 stations the numbers of trout released were 838, 940, 487, 486, and 987, respectively, and the column + in the table below indicates how many of these fish were caught again.

Station	+	−	*Total*
1	183	655	838
2	231	709	940
3	107	380	487
4	96	390	486
5	218	769	987

(1) Show that it may be assumed that the probability of recapture is the same for the 5 stations.

(2) Give an estimate of the common probability of recapture together with a 95% confidence interval for this probability.

Exercise 7.8 A student from the Institute of Biological Sciences, University of Aarhus, has studied the migration behavior of the green toad *Bufo viridis* on the island Samsø. As part of the large investigation the student watched 4 toads closely for 17 days and 19 nights and recorded the number of migrations undertaken by the toads. (The position of a toad was observed at the beginning and at the end of a time period. If the positions were different, the animal had migrated.) Unfortunately, the recordings for each single toad are not available so the registrations of the 4 toads have to be considered together.

In 10 of the 17 days and in 7 of the 19 nights there was rainfall. The number of trials in a binomial model for the number of migrations for the 4 toads is therefore 40 in days with rain and 28 in nights with rain, respectively. Similarly, the number of trials is 28 during the 7 dry days and 48 during the 12 dry nights, respectively. We will examine whether and how the precipitation (rain/dry) and time (night/day) influence the probability of migration. The result of the investigation is seen in the table below, where + indicates that a migration has taken place.

Time	*Precipitation*	*Migrations* +	−	*Total*
Night	*Rain*	17	11	28
	Dry	20	28	48
Day	*Rain*	2	38	40
	Dry	4	24	28

(1) Show for each of the times (night/day) that it may be assumed that the probability of a migration is the same in rain and in dry weather.

(2) Investigate whether it can be assumed that the probability of a migration is the same during the day as at night.

(3) Investigate whether it may be assumed that the probability of a migration at nighttime is 0.5.

Exercise 7.9 Consider the data and the hypotheses in Example 7.7 on page 333.

(1) Show that the likelihood function under H_{01} is

$$\begin{aligned} L(r_1,r_2,r_3,s_1,s_2,s_3) \propto & (r_1 s_1)^{x_{11}} (r_1(1-s_1))^{x_{12}} \\ & (r_2 s_2/2)^{x_{21}} (r_2/2)^{x_{22}} (r_2(1-s_2)/2)^{x_{23}} \\ & (r_3 s_3)^{x_{32}} (r_3(1-s_3))^{x_{33}} \\ \propto & \left[r_1^{x_{1\cdot}} r_2^{x_{2\cdot}} r_3^{x_{3\cdot}}\right] \left[s_1^{x_{11}}(1-s_1)^{x_{12}}\right] \left[s_2^{x_{21}}(1-s_2)^{x_{23}}\right] \left[s_3^{x_{32}}(1-s_3)^{x_{33}}\right], \end{aligned}$$

where $x_{i\cdot} = x_{i1} + x_{i2} + x_{i3}$ and where $\propto$ denotes proportionality.

The likelihood function corresponds to the product of 4 likelihood functions for a multinomial experiment and 3 binomial experiments where the parameters of the 4 experiments (r_1, r_2, r_3), s_1, s_2, and s_3 are variation independent.

(2) Show that the maximum likelihood estimates under H_{01} are

$$\hat{r}_1 = \frac{x_{1\cdot}}{n}, \quad \hat{r}_2 = \frac{x_{2\cdot}}{n}, \quad \hat{r}_3 = \frac{x_{3\cdot}}{n},$$

and

$$\hat{s}_1 = \frac{x_{11}}{x_{1\cdot}}, \quad \hat{s}_2 = \frac{x_{21}}{x_{21}+x_{23}}, \quad \hat{s}_3 = \frac{x_{32}}{x_{3\cdot}}.$$

(3) Perform a test of H_{01} in M_0.

Next, we investigate whether females choose partners independently of their genotypes, i.e., the hypothesis,

$$H_{02}\colon s_1 = s_2 = s_3 = s, \quad 0 < s < 1,$$

in the model M_1 corresponding to the hypothesis H_{01}.

(4) Show that the likelihood function under H_{02} is

$$L(r_1,r_2,r_3,s) \propto \left[r_1^{x_{1\cdot}} r_2^{x_{2\cdot}} r_3^{x_{3\cdot}}\right] \left[s^{x_{11}+x_{21}+x_{32}}(1-s)^{x_{12}+x_{23}+x_{33}}\right]$$

and that the maximum likelihood estimates of the r_i's are as under H_{01}, while

$$\hat{s} = \frac{x_{11}+x_{21}+x_{32}}{n-x_{22}}.$$

Table 7.6 *The probabilities for the possible genotypic combinations of mothers and offspring in model M_4 corresponding to the hypothesis H_{04}*

		Type of offspring			
		AA	*Aa*	*aa*	*Total*
	AA	$\pi_{11} = p^3$	$\pi_{12} = p^2(1-p)$		p^2
Type of mother	*Aa*	$\pi_{21} = p^2(1-p)$	$\pi_{22} = p(1-p)$	$\pi_{23} = p(1-p)^2$	$2p(1-p)$
	aa		$\pi_{32} = (1-p)^2 p$	$\pi_{33} = (1-p)^3$	$(1-p)^2$
	Total	p^2	$2p(1-p)$	$(1-p)^2$	1

(5) Perform a test of H_{02} in M_1.

As the third point it is investigated whether the genotype frequencies among the mothers are in Hardy-Weinberg proportions, i.e., the hypothesis,

$$H_{03}\colon (r_1, r_2, r_3) = (p^2, 2p(1-p), (1-p)^2), \quad 0 < p < 1,$$

in the model M_2 corresponding to the hypothesis H_{02}.

(6) Show that the maximum likelihood estimate of s is the same as under H_{02}, whereas

$$\hat{p} = (2x_{1\cdot} + x_{2\cdot})/2n.$$

(7) Perform a test of H_{03} in M_2.

Finally, it is investigated if the relative frequency of A alleles is the same in mothers and fathers, i.e.,

$$H_{04}\colon s = p, \quad 0 < p < 1,$$

in the model M_3 corresponding to the hypothesis H_{03}.

(8) Show that the hypothesis H_{04} has one free parameter and that the probabilities of the possible genotype combinations of mothers and offspring are as shown in Table 7.6. Furthermore, show that the likelihood function under H_{04} is

$$L(p) \propto p^{2x_{1\cdot}+x_{2\cdot}+x_{11}+x_{21}+x_{32}}(1-p)^{x_{2\cdot}+2x_{3\cdot}+x_{12}+x_{23}+x_{33}}$$

and that the maximum likelihood estimate of p is

$$\hat{p} = \frac{2x_{1\cdot} + x_{2\cdot} + x_{11} + x_{21} + x_{32}}{3n - x_{22}}.$$

(9) Perform a test of H_{04} in M_3.

Exercise 7.10 Based on morphological characters, the cod has been divided into a number of geographic races or subspecies. In Danish waters two subspecies are distinguished: the Baltic cod (*Gadus morhua calarias*) living in the eastern Baltic and the Atlantic cod (*Gadus morhua morhua*) inhabiting the western Baltic and the other Danish waters. The difference between the subspecies may reflect divergence of reproductively isolated populations, but alternatively it may just be a direct consequence of the difference in environment in the two areas. The aim of the following genetic investigation was to resolve this question.

At three locations, Fehmarn Belt, Bornholm, and the Åland Islands, marked by the numbers 1, 2, and 3 on the map in Figure 7.4, a number of cods were caught and their hemoglobin types were determined by electrophoresis. In cods the hemoglobin types are determined by two allelic genes A

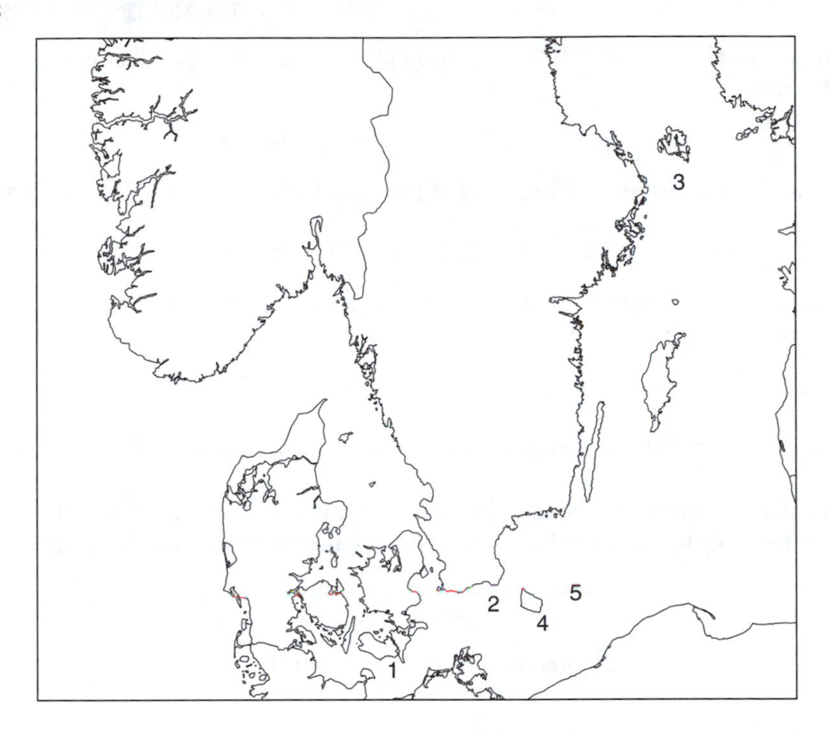

Figure 7.4 *The fishing grounds in Fehmarn Belt and near Bornholm and the Åland Islands.*

and a resulting in the genotypes AA, Aa, and aa. The results from Sick (1965) are reviewed in the table below:

	AA	Aa	aa
Fehmarn Belt	27	30	12
Bornholm	14	20	52
Åland Islands	0	5	75

The data were considered in Example 1.1 on page 1. Here we consider tests of some of the hypotheses mentioned in the example.

If there is random mating and no selection in a population, then the population genotypes are expected to occur in Hardy-Weinberg proportions, i.e., the genotypes are in the ratio $p^2 : 2p(1-p) : (1-p)^2$, where p is the frequency of the allele A in the population.

(1) Show that populations in Fehmarn Belt and near the Åland Islands may be assumed to safisfy the hypothesis of Hardy-Weinberg proportions in contrast to the population near Bornholm. The estimated frequencies of the allele A and the Hardy-Weinberg frequencies, i.e., the expected frequencies under the hypothesis of Hardy-Weinberg proportions, are:

		AA	Aa	aa
Fehmarn Belt	$\hat{p} = 0.609$	25.6	32.9	10.6
Bornholm	$\hat{p} = 0.279$	6.7	34.5	44.7
Åland Islands	$\hat{p} = 0.0313$	0.1	4.8	75.1

A possible explanation of the fact that the population near Bornholm does not satisfy the Hardy-Weinberg law is that the population is a mixture of the two cod populations such that α percent of the population are Baltic cods and $100-\alpha$ percent of the population are Atlantic cods. To examine this hypothesis, we let

$$\boldsymbol{\pi}_W = (p_W^2, 2p_W(1-p_W), (1-p_W)^2),$$

where $p_W = 0.609$ is the estimated frequency of the gene A in Fehmarn Belt and, similarly, we let

$$\boldsymbol{\pi}_E = (p_E^2, 2p_E(1-p_E), (1-p_E)^2),$$

where $p_E = 0.0313$ is the estimated frequency of the gene A near the Åland Islands. The mixture hypothesis may then be formulated as

$$H_0\colon \boldsymbol{\pi}_{\text{Bornholm}} = (1-\alpha)\boldsymbol{\pi}_W + \alpha\boldsymbol{\pi}_E,$$

where $\boldsymbol{\pi}_{\text{Bornholm}}$ is the probability vector of the genotypes *AA*, *Aa* and *aa* near Bornholm.

(2) Show that the maximum likelihood estimate $\hat{\alpha}$ of α and the expected frequencies under H_0 are as given in the table below (in order to determine $\hat{\alpha}$ numerical methods are needed):

	AA	*Aa*	*aa*	$\hat{\alpha}$
Bornholm	13.6	20.4	51.9	57.4

Furthermore, show that the hypothesis H_0 is not rejected.

To further investigate the hypothesis of a mixed population near Bornholm, cods were caught at two additional places near Bornholm indicated by the numbers 4 and 5 in Figure 7.4.

The results at the three localities near Bornholm are:

Locality	*AA*	*Aa*	*aa*	*Total*
2	14	20	52	86
4	12	12	54	78
5	2	4	66	72

(3) Show that the estimates of α at the three localities near Bornholm are:

$$\hat{\alpha}_2 = 57.4 \quad \hat{\alpha}_4 = 67.5 \quad \hat{\alpha}_5 = 94.9.$$

For each of the localities calculate the expected frequencies under H_0 and show that the data do not reject the hypothesis.

Notice that the percentage of Baltic cods increases as the position of the fishing ground becomes more easterly.

The conclusion of this investigation is that with respect to reproduction, the Atlantic cods and the Baltic cods form two isolated breeding populations, and the cods caught near Bornholm is a mixture of individuals from these two populations.

Exercise 7.11 The following data in Andersen and Keiding (1983) are from an investigation of the criminality of twins. The first table concerns identical twins and fraternal twins of the same sex. The twins have been divided into three groups according to whether both twins are noncriminal (*nn*), one twin is criminal (*nc*), or both twins are criminal (*cc*). In the table, MZ denotes identical (monozygote) twins, DZ fraternal (dizygote) twins, M male, and F female.

Criminality of twins of the same sex

	Type			
Group	*MZMM*	*DZMM*	*MZFF*	*DZFF*
nn	200	420	306	552
nc	81	154	18	38
cc	50	37	4	3
Total	331	611	328	593

(1) Investigate for each sex if the distributions for the identical twins and for the fraternal twins can be assumed to be identical.

(2) Investigate for DZMM and DZFF if the data may be described by the model below and interpret the model.

$$\begin{aligned} P\{\text{both noncriminal}\} &= p^2 \\ P\{\text{one criminal}\} &= 2p(1-p) \\ P\{\text{both criminal}\} &= (1-p)^2. \end{aligned}$$

(3) In the table below fraternal twins of the type MF have been divided into four groups corresponding to the criminal status of both twins. Propose a model similar to that in (2) and investigate if the data may be described by this model.

	Female		
Male	*Noncriminal*	*Criminal*	*Total*
Noncriminal	1219	24	1243
Criminal	286	16	302
Total	1505	40	1545

CHAPTER 8

The Poisson Distribution

One of the reasons the Poisson distribution often appears in practice is the Poisson process, which is a probabilistic model describing completely random occurrence of events in, for instance, time, plane, or space. According to the model the number of events occurring in a subset of the set considered, for instance, in a time interval or a region in the plane or in the space, is Poisson distributed.

Section 8.2 describes the Poisson process briefly and, furthermore, lists a few well-known properties of the Poisson distribution that are used when analyzing data using a statistical model based on this distribution. Section 8.1 introduces the examples used to illustrate the theory in this chapter. Statistical analysis of a sample by means of the Poisson distribution is discussed in Section 8.3, and in Section 8.4 we discuss models for several Poisson distributions with possibly different parameters but nevertheless with some structure on the parameters. Important examples are the Poisson model with proportional parameters and the multiplicative Poisson model, which are treated in some detail. Finally, Section 8.5 illustrates the use of the square root transformation in connection with the Poisson distribution.

We conclude this introduction with a general remark concerning samples from *discrete distributions*. If the size n of a sample, $x_1,\ldots,x_n$, is very large, the observations are often presented in *table form*, which for each observed value j records the number a_j of x's in the sample with the value j, i.e.,

$$a_j = \#\{i \,|\, x_i = j\}.$$

It should be noted that the table form preserves the information on *which values* have been observed. But the sequence of the original observations in the sample, $x_1,\ldots,x_n$, cannot be reconstructed from the table form. As long as one believes in the model that the observations are independent and identically distributed, this is of no consequence, because the table form has all the information that is necessary for estimating and testing in the model. The original sequence of observations is needed to check the assumptions of independence and identical distribution, for example, with a plot of the observation against serial order, i.e., a plot of x_i against i, $i = 1,\ldots,n$, to look for dependencies and trends in the observations. The advice is always to keep a record of the original data.

Recording discrete observations in table form may be considered as a kind of grouping, which in contrast to the usual grouping of continuous data, does not imply loss of information on the values of the single observations.

8.1 Examples

In this section we introduce the data sets that will be used to illustrate statistical analysis in models based on the Poisson distribution.

Example 8.1

The data in this example originates from an exercise in a course of radioisotopic methods for biologists, where the students measure β-radiation from a radioactive source. In a series of consecutive time intervals of the same length, the number of particles emitted by the radioactive source are recorded with a Geiger-counter. In this experiment the radioactive source is the isotope ^{90}Sr. We consider the measurements made by two students under similar experimental conditions. From the results, given in table form in Table 8.1, it is seen that student No. 1 performed measurements in 141 time intervals and that in 7 of these intervals, no particles were recorded while in 18 of the intervals 1 particle was recorded, and so on.

Table 8.1 *Counts, in table form, of the β-particles from the isotope ^{90}Sr made by two students. In addition, the expected frequencies* **e** *in the fitted Poisson distribution are given for each student.*

Student No. 1					Student No. 2				
j	a_j		e_j		j	a_j		e_j	
0	7		5.756		0	6		7.424	
1	18		18.410		1	30		22.559	
2	31		29.443		2	29		34.276	
3	24		31.392		3	34		34.718	
4	31		25.102		4	19		26.374	
5	14		16.058		5	24		16.029	
6	10		8.561		6	8		8.118	
7	3	} 6	3.912	} 6.278	7	3	} 5	3.524	} 5.502
8	3		1.564		8	2		1.339	
≥9	0		0.802		≥9	0		0.639	
Σ	141		141.000		Σ	155		155.000	

One of the classical illustrations of the Poisson process to be introduced in Section 8.3 is precisely concerned with recordings of the number of β- or α- particles from a radioactive source. Since the counting of the particles is performed in disjoint intervals of the same length, it should according to this model be possible to consider the data as a sample from a Poisson distribution with a parameter which, among other things, depends on the radioactive source and the common length of the time intervals. Since the two students performed the measurements under the same conditions, it is also of interest to investigate whether the parameters in the two Poisson distributions, one for each student, are identical. □

Example 8.2

The data in this example are counts of nematodes obtained by using a technique similar to that used when counting bacteria. The nematode is a roundworm that may be found in soil. The counts, here made by Lars Stubsgaard at the Institute of Biological Sciences, University of Aarhus, are concerned with the nematode *feltiae*, which is 0.5 to 1.0 mm long. From a stock solution of nematodes a volume of 20 μL was taken with a pipette and diluted with a certain volume of water. Three dilutions were prepared in exactly the same way. Next, the number of nematodes in small volumes that are sampled from the dilutions are counted. The sample volume is taken with a pipette and is placed in a Petri dish and the number of nematodes is counted through a magnifying glass. From each of the three dilutions 15 sample volumes were taken and counted once. From one dilution all the volumes were 40 μL and from the other two dilutions all volumes were 20 μL. The counts are shown in Table 8.2.

There is some variation between the counts from the same dilution. The variation in bacteria counts can often be described by means of the Poisson distribution. We will return to this example later and check that the three series of counts may be considered as samples from three Poisson distributions. Next we will investigate whether the parameters of these three distributions reflect the

Table 8.2 *Counts of the nematode feltiae in three dilutions. Each row is a sample from one dilution.*

Volume	*Number of feltiae*														
40 μL	31	28	33	38	28	32	39	27	28	39	21	39	45	37	41
20 μL	14	16	18	9	21	21	14	12	13	13	14	20	24	15	24
20 μL	18	13	19	14	15	16	14	19	25	16	16	18	9	10	9

fact that the volumes of the dilution used were 40 μL, 20 μL, and 20 μL, respectively, in the three samples. □

Example 8.3

In April 1979 an investigation of the occurrence of the water flea of the species *Calanus hyperboreus* in different depths of the sea was undertaken by the drift ice station "Fram 1" in Greenland Sea north of the straits between northeast Greenland and Svalbard.

In each of the depths 35 to 125 m, 125 to 250 m, and 250 to 300 m, a number of hauls were made with a net with a mechanism that ensured that the net could be opened or closed at the desired depth. The data collection lasted for 24 hours. At each depth 6 hauls were made during the day and 6 hauls were made during the night. The same net was used in all 12 hauls, which lasted for the same amount of time. The data have have been placed at our disposal by Lars Haumann, Zoological Museum, University of Copenhagen. Table 8.3 shows the catch numbers for water fleas classified according to depth and time. It is of interest to examine how the catch depends on these two factors. □

Table 8.3 *The number of water fleas of the species Calanus hyperboreus caught at different depths and time periods*

		Time	
		Day	*Night*
	035–125 m	59	72
Depth	125–250 m	371	392
	250–300 m	91	100

8.2 Probabilistic Results for the Poisson Distribution

In this section we review those probabilistic results for the Poisson distribution that will be used in the statistical analysis of models based on this distribution. Furthermore, we give a brief introduction to the Poisson process, which is a mathematical model for the description of occurrence of random events in time, plane, or space. Finally, in Proposition 8.1 we mention a mathematical result that will be used several times in this chapter in connection with maximum likelihood estimation.

A discrete random variable X is distributed according to the Poisson distribution with parameter $\lambda > 0$, $X \sim po(\lambda)$, if the probability function of X is

$$po(x;\lambda) = e^{-\lambda}\frac{\lambda^x}{x!}, \qquad x = 0,1,2,\ldots. \tag{8.1}$$

If $X \sim po(\lambda)$, the mean and the variance of X are

$$EX = \lambda \tag{8.2}$$

and

$$Var\, X = \lambda, \tag{8.3}$$

respectively. Consequently, the index of dispersion (or the coefficient of dispersion) is

$$\frac{Var\, X}{EX} = 1. \tag{8.4}$$

With respect to the dispersion index, the Poisson distribution differs from other discrete distributions. For instance the index of dispersion of the binomial distribution $b(n, \pi)$ equals $1 - \pi$ and is therefore less than 1, whereas for the negative binomial distribution* $b^-(\kappa, \pi)$ it is $(1-\pi)^{-1}$ and it is therefore greater than 1.

The following result relates the probability function of the binomial distribution $b(n, \pi)$ to the probability function of the Poisson distribution:

$$b(x; n, \pi) = \binom{n}{x} \pi^x (1-\pi)^{n-x} \rightarrow e^{-\lambda} \frac{\lambda^x}{x!} = po(x; \lambda), \text{ for } n \rightarrow \infty \text{ and } \pi \rightarrow 0 \text{ such that } n\pi \rightarrow \lambda. \tag{8.5}$$

The result is sometimes used in model considerations to change from a model based on the binomial distribution to a model based on the Poisson distribution.

Loosely speaking, the result in (8.5) may be formulated: if $X \sim b(n, \pi)$, one has the approximation,

$$X \approx po(n\pi),$$

if n is large and π is small.

Tail probabilities of the Poisson distribution may be approximated by tail probabilities of the normal distribution with the same mean and variance. If $X \sim po(\lambda)$ and Φ denotes the distribution function of the standard normal distribution $N(0,1)$, then

$$\lim_{\lambda \rightarrow \infty} P\left(a \leq \frac{X - \lambda}{\sqrt{\lambda}} \leq b\right) = \Phi(b) - \Phi(a), \tag{8.6}$$

which is a result of one version of the central limit theorem in probability theory. In the short form, we write

$$X \sim po(\lambda) \quad \text{and } \lambda \text{ large} \quad \Rightarrow \quad X \approx N(\lambda, \lambda). \tag{8.7}$$

Many of the approximate results in this chapter may be understood by thinking of calculations in the normal distribution, which approximates the Poisson distribution. In practice the approximating normal distribution may be applied if $\lambda > 5$.

Suppose that $X_1, \ldots, X_i, \ldots, X_n$ are independent random variables such that $X_i \sim po(\lambda_i)$, $i = 1, \ldots, n$. Let $X.$ denote the sum of the variables, i.e., $X. = X_1 + \cdots + X_i + \cdots + X_n$, and, similarly, let

* The *negative binomial distribution* with *index parameter* $\kappa > 0$ and *probability parameter* $\pi \in (0,1)$, denoted $b^-(\kappa, \pi)$, has the probability function,

$$b^-(x; \kappa, \pi) = \binom{\kappa + x - 1}{x} (1-\pi)\pi^x, \quad x = 0, 1, 2, \ldots,$$

where

$$\binom{\kappa + x - 1}{x} = \frac{\kappa(\kappa+1)\cdots(\kappa+x-1)}{x!}$$

is the *generalized binomial coefficient*. Properties, applications, and tables of the negative binomial distribution may be found in Williamson and Bretherton (1963).

If $X \sim b^-(\kappa, \pi)$, it may be shown that

$$EX = \frac{\kappa\pi}{1-\pi} \quad \text{and} \quad Var X = \frac{\kappa\pi}{(1-\pi)^2}.$$

$\lambda_.$ denote the sum of the parameters, $\lambda_. = \lambda_1 + \cdots + \lambda_i + \cdots + \lambda_n$. This yields the following result for the distribution of the sum and for the conditional distribution of the variables given the sum:

$$X_. \sim po(\lambda_.) \tag{8.8}$$

and

$$(X_1, \ldots, X_i, \ldots, X_n) \mid X_. = x_. \sim m(x_., \frac{\lambda_1}{\lambda_.}, \cdots, \frac{\lambda_i}{\lambda_.}, \cdots, \frac{\lambda_n}{\lambda_.})). \tag{8.9}$$

The conditioning result in (8.9) is the key to appreciating many similarities between tests in the multinomial distribution and tests in the Poisson distribution.

We now give a short description of the *Poisson process*, which is one of the reasons the Poisson distribution often is considered in practice. The Poisson process is a probabilistic model for occurrence of random events. Suppose we consider events in a subset S of the real axis, the plane, or the space, for instance, time points for registrations on a Geigercounter, points of impact of meteorites, positions of bacteria colonies on an agar plate, positions at which fish are caught, positions at which stones are collected, etc. Let $N(A)$ denote the number of events in the set $A \subseteq S$. We will use $|A|$ to denote the length, the area or the volume of A depending on whether A is a subset of the real line, the plane, or the space. Suppose that the following three assumptions are fulfilled:

(a) The probability that precisely n events occur in A depends on $|A|$ and n only, i.e., $P(N(A) = n)$ depends on $|A|$ and n only.

(b) The number of events in disjoint sets are independent, i.e., $N(A)$ and $N(B)$ are independent random variables if the sets A and B are disjoint, i.e.,
$$P(N(A) = n, N(B) = m) = P(N(A) = n)P(N(B) = m) \quad \text{if } A \cap B = \emptyset.$$

(c) The probability that more than one event occurs in A is small if $|A|$ is small, or more precisely,
$$\frac{P(N(A) \geq 2)}{|A|} \to 0, \qquad \text{for } |A| \to 0.$$

It may then be shown that there exists a number $\lambda > 0$ such that for all subsets A of S, the number of events in A has a Poisson distribution with parameter $\lambda|A|$, i.e.,

$$N(A) \sim po(\lambda |A|). \tag{8.10}$$

The parameter λ is referred to as the *intensity* of the Poisson process on S.

Using the formulas (8.8) and (8.9) together with condition (b), it may be shown that if

$$A = \bigcup_{i=1}^{k} A_i, \qquad \text{where } A_i \cap A_j = \emptyset \text{ if } i \neq j,$$

then

$$(N(A_1), \ldots, N(A_k)) \mid N(A) = n \sim m(n, (\frac{|A_1|}{|A|}, \cdots, \frac{|A_k|}{|A|}));$$

in other words, given that n events occur in A, the number of events in the disjoint subsets, $A_1, \ldots,$ A_k, (which together constitute A) has a multinomial distribution with number of trials n and a probability vector that specifies how large a part the single subsets, $A_1, \ldots, A_k$, constitute of A.

The mathematical result in Proposition 8.1 will be used several times in the following in connection with maximum likelihood estimation, and even though the proposition is easy to prove we omit the proof.

Proposition 8.1 Suppose that $x > 0$ and $c > 0$. Then the function,

$$g :]0,\infty[\rightarrow \mathbb{R}$$
$$\lambda \rightarrow e^{-c\lambda}\lambda^x,$$

attains its maximum value at the point,

$$\hat{\lambda} = \frac{x}{c}.$$

♦

8.3 One Sample

In this section we consider a single sample from the Poisson distribution. That is, we assume that the observations, $x_1, \ldots, x_n$, may be considered as realizations of independent random variables, $X_1, \ldots,$ X_n, which all have a Poisson distribution with parameter λ. We write this model as

$$M_0\colon X_i \sim po(\lambda), \qquad i = 1, \ldots, n, \tag{8.11}$$

where the independence of the observations is subsumed.

Estimation

The likelihood function for λ is

$$\begin{aligned} L(\lambda) &= \prod_{i=1}^{n} e^{-\lambda} \frac{\lambda^{x_i}}{x_i!} \\ &= e^{-n\lambda}\lambda^{x_\cdot} \prod_{i=1}^{n} \frac{1}{x_i!}, \end{aligned}$$

where $x_\cdot = x_1 + \cdots + x_n$. Using Proposition 8.1 we find that the maximum likelihood estimate $\hat{\lambda}$ of λ is

$$\hat{\lambda} = \bar{x}_\cdot = \frac{1}{n} x_\cdot = \frac{1}{n} \sum_{i=1}^{n} x_i. \tag{8.12}$$

Thus, the parameter λ, which according to (8.2) is the mean of the $po(\lambda)$-distribution, is estimated by the empirical mean $\bar{x}_\cdot$. From (8.8) we see that one has the following result concerning the distribution of the maximum likelihood estimator:

$$n\hat{\lambda} = X_\cdot \sim po(n\lambda). \tag{8.13}$$

Inference concerning the value of λ, for instance, the test of the hypothesis $\lambda = \lambda_0$, may be performed by considering the distribution of $X_\cdot$; either the exact $po(n\lambda)$ distribution or the approximating $N(n\lambda, n\lambda)$ distribution. The formulas (8.22) and (8.23) give the limits of the $(1 - \alpha)$ confidence interval for λ based on the $N(n\lambda, n\lambda)$ distribution. Exercise 8.1 on page 393 is concerned with the test of the hypothesis $\lambda = \lambda_0$.

Model Checking

The model M_0 may be checked using a χ^2-test for goodness of fit as described in Section 7.5 if the sample size n is sufficiently large.

An alternative control of the model M_0 is based on the result that the index of dispersion of the Poisson distribution is 1, see formula (8.4). Thus it has to be expected that the ratio,

$$id = \frac{s^2}{\bar{x}_\cdot}, \tag{8.14}$$

between the empirical variance s^2 and the empirical mean $\bar{x}_\cdot$ is close to 1. For large values of λ or for large values of n, we have the following approximation of the distribution of the random variable corresponding to id,

$$id \sim\approx \chi^2(n-1)/(n-1).$$

This statement is read "id is a realization of a random variable whose distribution may be approximated by the $\chi^2(n-1)/(n-1)$-distribution." The approximation may be used if $n \geq 15$ or if $\bar{x}. \geq 5$. The result is often used to check the model M_0. The model is rejected with significance level α if

$$id < \chi^2_{\alpha/2}(n-1)/(n-1) \tag{8.15}$$

or if

$$id > \chi^2_{1-\alpha/2}(n-1)/(n-1). \tag{8.16}$$

The test statistic in (8.14) is referred to as *Fisher's index of dispersion* for the Poisson distribution as it was considered in Fisher et al. (1922).

If the Poisson model M_0 is rejected because the observed value of id is too large, because of the remark after formula (8.4), we may try to describe the sample by means of a model based on the negative binomial distribution. If M_0 is rejected because of a too small value of id, the remark points at a binomial model. However, in order to apply a binomial model the way the data have been collected should be reconsidered. Each of the observations, x_i, $i = 1, \ldots, n$, should be the result of an experiment satisfying the conditions (a) to (d) on page 301 for $k = 2$. In addition, the number of trials in the experiment should be the same for all the observations.

The calculation of the empirical mean and variance in terms of which the index of dispersion is defined depends on whether all the individual observations are available or the observations are given in table form. In obvious notation we have

$$S = \sum_{i=1}^{n} x_i = \sum_j j a_j,$$

$$USS = \sum_{i=1}^{n} x_i^2 = \sum_j j^2 a_j$$

and

$$\bar{x}. = \frac{1}{n} S \qquad \text{og} \qquad s^2 = \frac{1}{n-1}\left(USS - \frac{S^2}{n}\right).$$

In the next section two further approximate tests of the model M_0 are mentioned, namely the $-2\ln Q$-test in formula (8.39) on page 368 and the equivalent X^2-test, which is shown to be related to the test based on the index of dispersion, cf. formula (8.40) and the remarks following (8.40).

Example 8.1 (Continued)

For the counts of β-particles made by the two students, we have the following calculations with an accuracy of four decimal places:

Table 8.4 *Calculation of Fisher's index of dispersion for the two samples in Table 8.1*

Student	n	S	USS	$\bar{x}.(\hat{\lambda})$	s^2	id
1	141	451	1903	3.1986	3.2889	1.0282
2	155	471	1919	3.0387	3.1673	1.0423

In both cases the sample size n is sufficiently large to ensure that Fisher's index of dispersion id may be applied. For student No. 1 id is to be evaluated in a $\chi^2(f)/f$-distribution with $f = 140$. The 2.5% fractile of this distribution is 0.7795 and the 97.5% fractile is 1.2475. The observed value of id falls between these fractiles and according to (8.15) and (8.16), the model M_0 is not rejected. For student No. 2 the value of id is assessed in the $\chi^2(f)/f$-distribution with $f = 154$ degrees of freedom. The

2.5% and the 97.5% fractile of this distribution are 0.7892 and 1.2354, respectively. Thus, also for student No. 2, Fisher's index of dispersion gives no reason for doubting the model M_0.

Since the sample size in both cases is large, we may in this situation also use tests for goodness of fit as a model check. The expected frequencies under M_0, which are calculated as

$$e_j = ne^{-\hat{\lambda}}\hat{\lambda}^j/j!, \quad j = 0, 1, \ldots, 8,$$

and

$$e_{\geq 9} = n(1 - \sum_{j=0}^{8} e^{-\hat{\lambda}}\hat{\lambda}^j/j!) = n - \sum_{j=0}^{8} e_j,$$

are seen in Table 8.1. For both students the groups 7, 8 and ≥ 9 are combined in order to meet the requirement that the expected frequencies should be greater than or equal to 5. In both cases the number of groups is reduced to $k = 8$ and since the model M_0 has one free parameter λ, the degrees of freedom in the test for goodness of fit become $f = k - 1 - 1 = 6$. Using Table 8.1 we find

$$-2\ln Q_1 = 4.044 \qquad p_{obs\,1} = 1 - F_{\chi^2(6)}(4.044) = 0.671$$
$$-2\ln Q_2 = 9.155 \qquad p_{obs\,2} = 1 - F_{\chi^2(6)}(9.155) = 0.165.$$

The test for goodness of fit gives the same conclusion as Fisher's index of dispersion for both samples. There is no evidence against the model M_0. □

Example 8.2 (Continued)
For the three samples in Table 8.2, we have the calculations with an accuracy of four decimal places in Table 8.5.

Table 8.5 *Calculation of Fisher's index of dispersion for the three samples in Table 8.2*

Volume	*n*	*S*	*USS*	$\bar{x}.(\hat{\lambda})$	s^2	*id*
40 μL	15	506	17678	33.7333	43.4952	1.2894
20 μL	15	248	4390	16.5333	20.6952	1.2517
20 μL	15	231	3811	15.4000	18.1143	1.1763

The sample size is 15 for all three samples, so we may apply Fisher's index of dispersion. The observed values of *id* are to be assessed in a $\chi^2(f)/f$-distribution with $f = 14$ degrees of freedom. All three values of *id* fall between $\chi^2_{0.025}(14)/14 = 0.4021$ and $\chi^2_{0.975}(14)/14 = 1.8656$, so according to (8.15) and (8.16) there is no evidence against the model M_0 in any of the three samples.

Since the sample size is only 15 in each of the three samples, it is not possible to make the test for goodness of fit in this situation due to the requirement that the expected frequencies must be larger than or equal to 5. □

Confidence Interval

We start by giving the formula for an approximative $(1-\alpha)$ confidence interval for the mean λ, based on a single observation x of a random variable X, which is $po(\lambda)$- distributed. The confidence interval is approximative since it is based on the approximating $N(\lambda, \lambda)$ distribution, cf. (8.7). In this distribution the inequality,

$$-u_{1-\alpha/2} < \frac{X - \lambda}{\sqrt{\lambda}} < u_{1-\alpha/2}, \tag{8.17}$$

holds with a probability of $1-\alpha$. Solving the inequality (8.17) with respect to λ we obtain the equivalent inequality,

$$X+\frac{1}{2}u_{1-\alpha/2}^2-u_{1-\alpha/2}\sqrt{X+\frac{1}{4}u_{1-\alpha/2}^2}<\lambda<X+\frac{1}{2}u_{1-\alpha/2}^2+u_{1-\alpha/2}\sqrt{X+\frac{1}{4}u_{1-\alpha/2}^2}, \quad (8.18)$$

which also holds with a probability of $1-\alpha$. Insertion of the actual observation x in (8.18) gives a $(1-\alpha)$ confidence interval for the mean λ of a Poisson distribution of the form,

$$C_{1-\alpha}(x)=[\lambda_-,\lambda_+] \quad (8.19)$$

where

$$\lambda_-=x+\frac{1}{2}u_{1-\alpha/2}^2-u_{1-\alpha/2}\sqrt{x+\frac{1}{4}u_{1-\alpha/2}^2} \quad (8.20)$$

and

$$\lambda_+=x+\frac{1}{2}u_{1-\alpha/2}^2+u_{1-\alpha/2}\sqrt{x+\frac{1}{4}u_{1-\alpha/2}^2}. \quad (8.21)$$

Note that the formula (8.18) emphasizes that it is the limits of the confidence interval that are random and that an interpretation of the confidence interval based on the observation x is that either λ is in the confidence interval or an event with a probability less than α has occurred.

Furthermore, since $(X-\lambda)/\sqrt{\lambda}$ is a test statistic of the hypothesis that the parameter of the Poisson distribution has the particular value λ, the confidence interval also has the interpretation as those values of the parameter that are not rejected as a null hypothesis on the basis of the observation x according to (8.17).

Remark 8.1 Sometimes we are interested in calculating the confidence interval for a parameter λ in situations where the random variable X has a Poisson distribution with parameter $c\lambda$, where c denotes a known constant. In such cases the confidence interval for the mean $c\lambda$ is calculated using the formulas (8.20) and (8.21) and this interval is then transformed to a confidence interval for λ. ▼

The first example of the situation in Remark 8.1 is one sample from the Poisson distribution, where $x. \sim\sim po(n\lambda)$ and (8.20) and (8.21) are the limits of the $(1-\alpha)$ confidence interval for $n\lambda$, which is transformed to a confidence interval for λ with the limits,

$$\lambda_-=\frac{1}{n}(n\lambda)_-=\frac{1}{n}\left[x.+\frac{1}{2}u_{1-\alpha/2}^2-u_{1-\alpha/2}\sqrt{x.+\frac{1}{4}u_{1-\alpha/2}^2}\right] \quad (8.22)$$

and

$$\lambda_+=\frac{1}{n}(n\lambda)_+=\frac{1}{n}\left[x.+\frac{1}{2}u_{1-\alpha/2}^2+u_{1-\alpha/2}\sqrt{x.+\frac{1}{4}u_{1-\alpha/2}^2}\right]. \quad (8.23)$$

8.4 Several Samples

In this section we offer examples of statistical analysis of models involving several Poisson distributions. Furthermore, it is shown that because of the result in formula (8.9), there is an intimate connection between the analysis of such models and the analysis of models based on the multinomial distribution.

8.4.1 The Poisson Model with Proportional Parameters

The premise for the following discussion is that the data set $\mathbf{x}$ consists of the observations, $x_1,\ldots,x_k$, which may be considered as realizations of independent random variables, $X_1,\ldots,X_k$, which are all Poisson distributed but with different parameters, i.e., the basic model is

$$M_0\colon X_i\sim po(\lambda_i), \qquad i=1,\ldots,k. \quad (8.24)$$

Suppose that associated with each observation x_i we have a known constant m_i and that we are interested in the hypothesis,

$$H_{01}: \lambda_i = m_i\lambda, \qquad i = 1,\dots,k, \tag{8.25}$$

where λ is an unknown parameter. This means that the vector of means is proportional to the vector of known constants, i.e.,

$$(\mu_1,\dots,\mu_k) = \lambda(m_1,\dots,m_k),$$

and it explains that the name of the corresponding model,

$$M_1: X_i \sim po(m_i\lambda), \qquad i = 1,\dots,k, \tag{8.26}$$

is the *Poisson model with proportional parameters*. The parameter λ is referred to as the *intensity* and the known constants m_i are called *proportionality constants*. A more general model in which the hypothesis H_{01} can be tested is given in Remark 8.2 on page 368.

M_1 has a single free parameter λ.

The model M_1 is very useful when comparing counts where an adjustment is necessary before the comparison is justified. A typical example is a comparison of the number of accidents on a road one year and the first six months of the following year. Here the m_i's might be 12 months and 6 months, or 365 days and 181 days, if we are very meticulous and neither of the two years are leap years.

A special case of the hypothesis H_{01} is $m_1 = \cdots = m_k = 1$. This corresponds to investigating if the observations can be considered to be identically distributed assuming that they are independent and Poisson distributed.

The likelihood function under M_0 is

$$\begin{aligned} L(\lambda_1,\dots,\lambda_k) &= \prod_{i=1}^{k} e^{-\lambda_i}\frac{\lambda_i^{x_i}}{x_i!} \\ &= e^{-\lambda.}\prod_{i=1}^{k}\lambda_i^{x_i}\prod_{i=1}^{k}\frac{1}{x_i!}, \end{aligned}$$

where $\lambda. = \lambda_1 + \cdots + \lambda_k$. The log likelihood function under M_0 therefore becomes

$$l(\lambda_1,\dots,\lambda_k) = -\lambda. + \sum_{i=1}^{k} x_i \ln(\lambda_i) - \sum_{i=1}^{k}\ln(x_i!). \tag{8.27}$$

Since the parameters, $\lambda_1,\dots,\lambda_k$, are variation independent, using Proposition 8.1 on the first expression for the likelihood function, the maximum likelihood estimate of λ_i under M_0 is

$$\hat{\lambda}_i = x_i, \qquad i = 1,\dots,k.$$

The log likelihood function for λ under H_{01} is obtained by replacing λ_i with $m_i\lambda$ in (8.27). We find

$$\begin{aligned} l(\lambda) &= -\sum_{i=1}^{k}\lambda m_i + \sum_{i=1}^{k} x_i\ln(m_i\lambda) - \sum_{i=1}^{k}\ln(x_i!) \qquad (8.28) \\ &= -\lambda m. + x.\ln(\lambda) + \sum_{i=1}^{k} x_i \ln(m_i) - \sum_{i=1}^{k}\ln(x_i!), \end{aligned}$$

where $m. = m_1 + \cdots + m_k$. Thus, the likelihood equation for λ becomes

$$\frac{dl}{d\lambda} = -m. + \frac{x.}{\lambda} = 0,$$

which has the solution $\lambda = \frac{x.}{m.}$. Since

$$\frac{d^2 l}{(d\lambda)^2} = -\frac{x.}{\lambda^2} < 0,$$

it is seen that if $x. > 0$, the maximum likelihood estimate of λ under M_1 is

$$\hat{\lambda} = \frac{x.}{m.}. \tag{8.29}$$

The expected frequency corresponding to the observation x_i – i.e., the mean of X_i calculated under the hypothesis H_0 and using $\hat{\lambda}$ for the unknown intensity – is

$$e_i = m_i \hat{\lambda} = x. \frac{m_i}{m.}. \tag{8.30}$$

From the formulas (8.27) to (8.30), it is seen that the $-2\ln Q$-test statistic of H_{01} is

$$\begin{aligned} -2\ln Q(\mathbf{x}) &= 2[l(\hat{\lambda}_1,\ldots,\hat{\lambda}_k) - l(\hat{\lambda})] \\ &= 2[\sum_{i=1}^{k} x_i \ln(x_i) - \sum_{i=1}^{k} x_i \ln(m_i\hat{\lambda})] \\ &= 2\sum_{i=1}^{k} x_i \ln(\frac{x_i}{e_i}). \end{aligned} \tag{8.31}$$

The degrees of freedom of the χ^2-distribution, which approximates the distribution of $-2\ln Q$ under H_{01}, are $k-1$ since there are k free parameters in M_0 and a single free parameter in M_1. If *the expected frequencies are all greater than or equal to* 5, we have the following approximation of the significance probability of H_{01}:

$$p_{obs}(\mathbf{x}) \doteq 1 - F_{\chi^2(k-1)}(-2\ln Q(\mathbf{x})). \tag{8.32}$$

The hypothesis H_{01} may also be tested by means of the X^2-test statistic, which is

$$X^2(\mathbf{x}) = \sum_{i=1}^{k} \frac{(x_i - e_i)^2}{e_i}. \tag{8.33}$$

The corresponding approximation of the significance probability is

$$p^*_{obs}(\mathbf{x}) \doteq 1 - F_{\chi^2(k-1)}(X^2(\mathbf{x})). \tag{8.34}$$

The distribution of the maximum likelihood estimator of λ under H_{01} is usually given in the following way:

$$m.\hat{\lambda} = X. \sim po(m.\lambda). \tag{8.35}$$

Replacing x with $x.$ in (8.20) and (8.21) we obtain the limits of the $(1-\alpha)$ confidence interval for $m.\lambda$. This interval is then transformed to a confidence interval for λ with the limits,

$$\lambda_- = \frac{1}{m.}(m.\lambda)_- = \frac{1}{m.}\left[x. + \frac{1}{2}u^2_{1-\alpha/2} - u_{1-\alpha/2}\sqrt{x. + \frac{1}{4}u^2_{1-\alpha/2}}\right] \tag{8.36}$$

and

$$\lambda_+ = \frac{1}{m.}(m.\lambda)_+ = \frac{1}{m.}\left[x. + \frac{1}{2}u^2_{1-\alpha/2} + u_{1-\alpha/2}\sqrt{x. + \frac{1}{4}u^2_{1-\alpha/2}}\right]. \tag{8.37}$$

If $m_1 = \cdots = m_k = 1$, the model M_1 corresponds to a sample from the Poisson distribution. This model was denoted by M_0 in Section 8.3. In this case,

$$\hat{\lambda} = \frac{x.}{k} = \bar{x}.,$$

and the expected frequencies are identical since

$$e_i = \bar{x}., \qquad i = 1,\ldots,k, \tag{8.38}$$

and formula (8.31) may therefore be reduced to

$$-2\ln Q(\mathbf{x}) = 2[\sum_{i=1}^{k} x_i \ln(x_i) - x.\ln(\bar{x}.)]. \tag{8.39}$$

Furthermore, in this situation,

$$X^2(\mathbf{x}) = \sum_{i=1}^{k} \frac{(x_i - \bar{x}.)^2}{\bar{x}.} = \frac{(k-1)s^2}{\bar{x}.} = (k-1)id, \tag{8.40}$$

where *id* is Fisher's index of dispersion defined in formula (8.14). Thus, there is a connection between Fisher's index of dispersion *id* and $X^2(\mathbf{x})$; however, it is noteworthy that *id* rejects for both small and large values, whereas the X^2-test rejects for large values of $X^2(\mathbf{x})$ only. The explanation is that Fisher's index of dispersion and the X^2-test are derived in different models and test different hypotheses. For the X^2-test the model is that the observations are independent and Poisson distributed but not necessarily identically distributed and in this model the hypothesis to be tested is precisely that the observations are identically distributed. In contrast, Fisher's index of dispersion is derived in a model where the observations are independent and identically distributed according to an unspecified distribution. Here the hypothesis to be tested is that this common distribution is the Poisson distribution.

An illustration of the use of the $-2\ln Q$-test for checking the model of one sample from the Poisson distribution will be given below in the continuation of Example 8.2.

Relation to the Multinomial Model

There is a close connection between tests in the Poisson model and tests in the multinomial model. For the two tests in (8.31) and (8.33), this may be explained by means of formula (8.9). If we condition on the sum of the observations $x.$ in the model M_0 and use the result in (8.9), we get the conditional model,

$$M_0^*\colon (X_1,\ldots,X_k) \mid X. = x. \sim m(x.,(\pi_1,\ldots,\pi_k)), \tag{8.41}$$

where

$$(\pi_1,\ldots,\pi_k) = (\frac{\lambda_1}{\lambda.},\ldots,\frac{\lambda_k}{\lambda.}).$$

Since the λ's vary freely, there are no restrictions on the variation of the probability vector $(\pi_1,\ldots,\pi_k)$ in (8.41) either; in other words, the conditional model in (8.41) is the basic multinomial model with k categories and number of trials $x.$. In this model the hypothesis H_{01} corresponds to the simple hypothesis,

$$H_{01}^*\colon (\pi_1,\ldots,\pi_k) = (\frac{m_1}{m.},\ldots,\frac{m_k}{m.}).$$

From (8.30) it is seen that the expected frequencies under the hypothesis H_{01} in the model M_0 are precisely the same as the expected frequencies under the hypothesis H_{01}^* in the model M_{01}^* and, consequently, the $-2\ln Q$-test statistics (or the X^2-test statistics) are identical.

Even though the calculations in the two models are identical the models are different. The difference between the models is due to the way the data have been collected. For the multinomial model, we consider, in advance, observations with a given sum, which is specified by the number of trials, usually denoted by n, but in the model M_{01}^* denoted by $x.$. In contrast, in the Poisson model no restrictions have been made in advance on the sum of the observations.

Remark 8.2 The model H_0 in (8.24) on page 365 seems to be a special case of k samples, where each sample consists of just one observation. However, the situation with a varying number of observations in the k samples is covered by the formulation on page 365. If we have k samples with n_i observations in the ith sample, i.e.,

$$M_0\colon X_{ij} \sim po(\lambda_i), \quad j = 1,\ldots,n_i,\ i = 1,\ldots,k,$$

then it follows from (8.8) that the sums $X_{i\cdot}$ of the samples will be distributed according to the model,

$$\tilde{M}_0\colon X_{i\cdot} \sim po(n_i\lambda_i), \quad i = 1,\ldots,k.$$

If we wish to investigate the relationship,

$$H_{01}\colon \lambda_i = m_i\lambda, \quad i = 1,\ldots,k,$$

for the intensities, it corresponds to the relationship,

$$n_i\lambda_i = n_i m_i \lambda,$$

for the means of $X_{i\cdot}$.

Thus, we simply replace the individual samples by their sums $X_{i\cdot}$ and replace the known constants m_i by the sample size n_i multiplied by m_i, i.e., by $\tilde{m}_i = n_i m_i$. ▼

Example 8.2 (Continued)
Applying the $-2\ln Q$-test in formula (8.39) to each of the three samples we obtain the results in the table below. As an illustration of the agreement between the $-2\ln Q$-test and the X^2-test the results for the X^2-test are also given. (In a concrete situation it is of course not necessary to calculate both tests.)

Volume	k	$x_{\cdot}$	$\bar{x}_{\cdot}$	$-2\ln Q$	p_{obs}	X^2	p^*_{obs}
40 μL	15	506	33.7333	18.481	0.186	18.051	0.205
20 μL	15	248	16.5333	17.458	0.233	17.524	0.229
20 μL	15	231	15.4000	16.588	0.279	16.468	0.286

It is seen from the table that in this situation the tests may be applied since the expected frequencies $\bar{x}_{\cdot}$ are greater than 5 and, furthermore, that the tests do not challenge the assumption of identical distributions of the 15 counts in each of the three samples. Note that the question of applying the Poisson distribution is not discussed. We test the hypothesis of identical distributions in a model based on the Poisson distribution.

According to (8.13), if we want to draw inference on the parameters of the three Poisson distributions, we may consider the following model for the sums:

$$M_0\colon X_{i\cdot} \sim po(\tilde{\lambda}_i), \qquad i = 1,2,3,$$

$$X_{1\cdot},\ X_{2\cdot} \text{ and } X_{3\cdot} \text{ are stochastically independent.}$$

Here $\tilde{\lambda}_i = 15\lambda_i$, where λ_i denotes the parameter of the Poisson distribution for the ith sample. Due to the construction of the experiment, it is of interest to examine whether the counts are homogeneous in the sense that λ_i is proportional to the applied volume of the dilution, i.e., to examine whether $\lambda_1 = 40\lambda$ and $\lambda_2 = \lambda_3 = 20\lambda$, where λ denotes the mean of the number of nematodes corresponding to 1 μL of the dilution. If this hypothesis is correct, the model M_0 is reduced to:

$$M_1\colon X_{i\cdot} \sim po(m_i\lambda), \qquad i = 1,2,3$$

$$X_{1\cdot},\ X_{2\cdot} \text{ and } X_{3\cdot} \text{ are stochastically independent,}$$

where $m_1 = 15 \times 40 = 600$ and $m_2 = m_3 = 15 \times 20 = 300$. Note that m_1, m_2 and m_3 are the total volumes counted in the three samples. Using formula (8.29) the maximum likelihood estimate of λ in the model M_1 is seen to be

$$\hat{\lambda} = \frac{985}{1200} = 0.8208.$$

The expected frequencies may then be calculated by means of (8.30). We find

Volume	m_i	$x_i.$	e_i
40 μL	600	506	492.50
20 μL	300	248	246.25
20 μL	300	231	246.25
total	1200	985	985.00

and from (8.31) and (8.32) we get

$$-2\ln Q = 1.344$$

and

$$p_{obs} = 1 - F_{\chi^2(2)}(1.344) = 0.511$$

and so we conclude that there is no evidence against M_1 in the data.

In M_1 the sum of all the observations has a Poisson distribution. More precisely,

$$X_{\cdot\cdot} \sim po(m_{\cdot}\lambda),$$

a result that may be used to find a 95% confidence interval for λ. Since $x_{\cdot\cdot} = 985$ and $m_{\cdot} = 1200$, by means of (8.36) and (8.37), we find that the confidence interval is

$$[\lambda_-, \lambda_+] = [0.7711, 0.8737].$$

The practical implication of this experiment and the analysis is that it is the total volume that determines the precision of the estimated mean number per unit volume (λ). If the experimenter decides that counting 400 μL will give a satisfactory precision, then it is immaterial whether 20 volumes of 20 μL are counted or 10 volumes of 40μL are counted. □

Example 8.1 (Continued)

To investigate whether the parameters of the two Poisson distributions – one for each student – are identical, we use (8.13) and consider the model,

$$M_0: X_{i\cdot} \sim po(\tilde{\lambda}_i), \qquad i = 1,2,$$

$$X_{1\cdot} \text{ and } X_{2\cdot} \text{ are stochastically independent.}$$

Here $\tilde{\lambda}_i = n_i\lambda_i$, where n_i denotes the number of observations performed by ith student and λ_i denotes the parameter in the corresponding Poisson distribution. The hypothesis,

$$H_0: \lambda_1 = \lambda_2,$$

is therefore of the general form considered in (8.25) with $m_i = n_i$, $i = 1,2$. From the calculations in the table

Student	m_i	$x_i.$	e_i
1	141	451	439.196
2	155	471	482.804
Total	296	922	922.000

and the formulas (8.29) to (8.32) we find that

$$\hat{\lambda} = \frac{922}{296} = 3.1149,$$

$$-2\ln Q = 0.6054$$

and

$$p_{obs} = 1 - F_{\chi^2(1)}(0.6054) = 0.437.$$

Consequently, we accept the hypothesis that the parameters of the two Poisson distributions are identical, which was to be expected since the experiments, as mentioned, were carried out under the same circumstances.

Under H_0 the sum of all the observations has a Poisson distribution with parameter $m_{\bullet}\lambda$, i.e.,

$$X_{\bullet\bullet} \sim po(m_{\bullet}\lambda).$$

A 95% confidence interval for λ may be obtained using the formulas (8.36) and (8.37) for the confidence limits, the sum of the counts $x_{\bullet\bullet} = 922$ and the total number of intervals $m_{\bullet} = 296$:

$$[\lambda_-, \lambda_+] = [2.9202,\ 3.3225].$$

The half-life period of the isotope ^{90}Sr is 28 years. The mean of the number of particles emitted during a time interval depends on the volume of the radioactive source, its half-life period, the distance between the source and the Geiger-counter, and the length of the time interval. The experiment considered here was planned such that the mean of the number of counts in a time interval was approximately 3. This is in accordance with the 95% confidence interval for λ above. □

8.4.2 The Multiplicative Poisson Model

This model is used in situations where the observations, as shown below, may be presented in an $r \times s$ table corresponding to two factors with r and s levels, respectively. The observation corresponding to the ith level of the first factor and the jth level of the second factor is denoted by x_{ij}, $i = 1, \ldots, r$, $j = 1, \ldots, s$.

The data were also presented in an $r \times s$ table on page 315 in connection with a test of independence in a two-way table in a multinomial model. In that case a fixed number n of items were classified according to two criteria with r and s categories, respectively. In contrast, for the Poisson model the sum of all the observations is not known in advance but is random. However, it turns out that the analysis of the multinomial model corresponding to independence is very similar to the analysis of the multiplicative Poisson model.

Furthermore, the structure of the data is similar to that considered in a two-way analysis of variance without repetitions and as will be shown, there are certain similarities between this model and the multiplicative Poisson model. However, the models are very different. The first mentioned is a model for continuous data where a hypothesis of *additive* structure of the means is considered, whereas the Poisson model is a model for discrete data where, as seen below, we consider a hypothesis of a *multiplicative* structure of the means.

We illustrate the theory by means of Example 8.3 where the data are presented as a 3×2 table in Table 8.3.

For a general $r \times s$ table the data are of the form:

	1	$\cdots$	j	$\cdots$	s	Σ
1	x_{11}	$\cdots$	x_{1j}	$\cdots$	x_{1s}	$x_{1\cdot}$
$\vdots$	$\vdots$		$\vdots$		$\vdots$	$\vdots$
i	x_{i1}	$\cdots$	x_{ij}	$\cdots$	x_{is}	$x_{i\cdot}$
$\vdots$	$\vdots$		$\vdots$		$\vdots$	$\vdots$
r	x_{r1}	$\cdots$	x_{rj}	$\cdots$	x_{rs}	$x_{r\cdot}$
Σ	$x_{\cdot 1}$	$\cdots$	$x_{\cdot j}$	$\cdots$	$x_{\cdot s}$	$x_{\cdot\cdot}$

In the table $x_{i\cdot}$ and $x_{\cdot j}$ denotes the sum of the observations in the ith row and the jth column, respectively, and $x_{\cdot\cdot}$ is the sum of all the observations,

$$x_{i\cdot} = \sum_{j=1}^{s} x_{ij}, \qquad x_{\cdot j} = \sum_{i=1}^{r} x_{ij}, \qquad x_{\cdot\cdot} = \sum_{i=1}^{r}\sum_{j=1}^{s} x_{ij}.$$

If we assume that all the observations are outcomes of independent random variables, the models we will consider may be written in the following way.
The basic model:

$$M_0\colon x_{ij} \sim\sim po(\lambda_{ij}), \quad i = 1,\ldots,r, \quad j = 1,\ldots,s.$$

The *multiplicative* model or the model with *no interaction*:

$$M_1\colon x_{ij} \sim\sim po(\alpha_i\beta_j), \quad i = 1,\ldots,r, \quad j = 1,\ldots,s.$$

The model with *row effect only*:

$$M_2^r\colon x_{ij} \sim\sim po(\alpha_i), \quad i = 1,\ldots,r, \quad j = 1,\ldots,s.$$

The model with *column effect only*:

$$M_2^c\colon x_{ij} \sim\sim po(\beta_j), \quad i = 1,\ldots,r, \quad j = 1,\ldots,s.$$

The model for *homogeneity*:

$$M_3\colon x_{ij} \sim\sim po(\lambda), \quad i = 1,\ldots,r, \quad j = 1,\ldots,s.$$

The last four of these models all correspond to hypotheses concerning the influence of two factors on the distributions in the basic model.

The model M_1 corresponds to the hypothesis $H_{01}\colon \lambda_{ij} = \alpha_i\beta_j$. The name of the model M_1 is due to the fact that in M_1 the mean of the random variable X_{ij} in the ith row and the jth column is obtained multiplying a parameter α_i associated with the ith row by a parameter β_j associated with the jth column. Note that in M_1 the ratio of means in two different rows g and h does not depend on the column j in which the ratio is determined, since

$$\frac{EX_{gj}}{EX_{hj}} = \frac{\alpha_g\beta_j}{\alpha_h\beta_j} = \frac{\alpha_g}{\alpha_h}$$

is independent of j. Similarly, in M_1,

$$\frac{EX_{il}}{EX_{im}} = \frac{\alpha_i\beta_l}{\alpha_i\beta_m} = \frac{\beta_l}{\beta_m},$$

so the ratio of means in two different columns l and m is independent of the row i in which the ratio is calculated.

The interpretation of the models M_2^r, M_2^c, and M_3 in relation to the two criteria is obvious. They correspond to r independent samples (the rows), s independent samples (the columns), and 1 sample (all observations) from the Poisson distribution, respectively.

Parametrization of the Models

The model M_1 appears to have $r+s$ free parameters, namely the r α's and the s β's. However, this is not true. Multiplying the α's with a positive constant and dividing the β's with the same constant will give a different set of parameters that corresponds to the same model. Therefore M_1 has a most $r+s-1$ free parameters. In fact, the model M_1 has $r+s-1$ free parameters. This will be clear from a slightly different parameterization introduced below. This parameterization will be very convenient for maximum likelihood estimation and also for interpreting the results.

Let $\alpha.$, $\beta.$, and $\lambda..$ denote the sum of the α's, the β's, and the λ's, respectively, and, furthermore, let

$$\rho_i = \frac{\alpha_i}{\alpha.}, \quad i = 1, \ldots, r \qquad \text{and} \qquad \sigma_j = \frac{\beta_j}{\beta.}, \quad j = 1, \ldots, s.$$

Since

$$\lambda.. = \sum_{i=1}^{r} \sum_{j=1}^{s} \lambda_{ij} = \sum_{i=1}^{r} \sum_{j=1}^{s} \alpha_i \beta_j = \sum_{j=1}^{r} \alpha_i \sum_{j=1}^{s} \beta_j = \alpha.\beta.,$$

the mean of X_{ij} in the model M_1 may be written as

$$\alpha_i \beta_j = \alpha.\beta. \frac{\alpha_i}{\alpha.} \frac{\beta_j}{\beta.} = \lambda..\rho_i \sigma_j,$$

and $\lambda..$, $\boldsymbol{\rho} = (\rho_1, \ldots, \rho_r)^*$ and $\boldsymbol{\sigma} = (\sigma_1, \ldots, \sigma_s)^*$ will be the set of parameters that we will use. Note that $\lambda..$, $\boldsymbol{\rho}$, and $\boldsymbol{\sigma}$ vary independently and, in addition, that $\boldsymbol{\rho}$ varies freely in the set of r-dimensional probability vectors and, similarly, that $\boldsymbol{\sigma}$ varies freely in the set of s-dimensional probability vectors. (Each of the two vectors $\boldsymbol{\rho}$ and $\boldsymbol{\sigma}$ has positive components whose sum is 1.) The number of free parameters of the model M_1 is therefore $d_1 = 1 + (r-1) + (s-1) = r+s-1$.

The relationship between the models M_0, M_1, M_2^r, M_2^c, and M_3 is shown in the following diagram where the parameters $\lambda..$, $\boldsymbol{\rho}$, and $\boldsymbol{\sigma}$ are used:

$$\begin{array}{ccccc}
 & & M_2^r\colon X_{ij} \sim po(\lambda..\rho_i \tfrac{1}{s}) & & \\
 & \nearrow & & \searrow & \\
M_0\colon X_{ij} \sim po(\lambda_{ij}) \longrightarrow M_1\colon X_{ij} \sim po(\lambda..\rho_i\sigma_j) & & & & M_3\colon X_{ij} \sim po(\lambda..\tfrac{1}{rs}) \\
 & \searrow & & \nearrow & \\
 & & M_2^c\colon X_{ij} \sim po(\lambda..\tfrac{1}{r}\sigma_j) & &
\end{array}$$

It is always the question of reducing the basic model M_0 to the multiplicative Poisson model M_1 that is of primary interest. If this can be done, it may, depending of the scientific problem, be of interest to test for a possible effect of the two criteria. That is to test the hypothesis of *no row effect*,

$$H_{0\rho}\colon \boldsymbol{\rho} = (\frac{1}{r}, \ldots, \frac{1}{r})^*,$$

and/or the hypothesis of *no column effect*,

$$H_{0\sigma}\colon \boldsymbol{\sigma} = (\frac{1}{s}, \ldots, \frac{1}{s})^*.$$

As it appears from the overview of the models, both hypotheses may be tested in two models. For instance, both the reduction $M_1 \to M_2^r$ and the reduction $M_2^c \to M_3$ correspond to the hypothesis $H_{0\sigma}$ of no column effect. In other words, the hypothesis $H_{0\sigma}$ can be tested both in M_1 and in M_2^c, and if the hypothesis is not rejected, it corresponds to the reduction to M_2^r and M_3, respectively. We shall see below that irrespective of whether the test of the hypothesis $H_{0\sigma}$ of no column effect is performed in M_1 or in M_2^c, the test is the same.

Similar remarks apply to the hypothesis $H_{0\rho}$ of no row effect.

Estimation

Tests of the various reductions of the models are carried out using the $-2\ln Q$-test described in Section 11.3. The basic information that we need for each model is the following:

(1) maximum likelihood estimates of the parameters,

(2) $\hat{l}$, the maximum value of the log likelihood function l,

(3) d, the number of free parameters of the model,

(4) $\mathbf{e}$, the expected frequencies in the model.

Before we go through all the models we indicate where the information is needed in the test of the reduction $M_a \rightarrow M_b$, where M_a is any of the models M_0, M_1, M_2^r, and M_2^c and M_b is any model that can be reached from M_a following the arrows from M_a.

The test statistic is

$$-2\ln Q = 2(\hat{l}_a - \hat{l}_b),$$

but in order to find the significance probability via the χ^2-distribution, we need to know its degrees of freedom,

$$f = d_a - d_b,$$

and we need to check that the expected frequencies $\mathbf{e}_b$ in the model M_b are sufficiently large, so that we can use the χ^2-approximation to the distribution of $-2\ln Q$. In that case, the significance probability of the test of the reduction, $M_a \rightarrow M_b$ can be calculated as

$$p_{obs} = 1 - F_{\chi^2(f)}(-2\ln Q).$$

The maximum likelihood estimates are used to calculate the maximum value of the log likelihood function and the expected frequencies.

In the following the quantities in (1) to (4) are calculated for the five models M_0, M_1, M_2^r, M_2^c, and M_3.

M_0:

The likelihood function is

$$L(\{\lambda_{ij}\}) = \prod_{i=1}^{r}\prod_{j=1}^{s} e^{-\lambda_{ij}}\lambda_{ij}^{x_{ij}}\frac{1}{x_{ij}!} \tag{8.42}$$

$$= e^{-\lambda_{\cdot\cdot}}\prod_{i=1}^{r}\prod_{j=1}^{s}\lambda_{ij}^{x_{ij}}\prod_{i=1}^{r}\prod_{j=1}^{s}\frac{1}{x_{ij}!}. \tag{8.43}$$

Since the λ_{ij}'s are variation independent, we find applying Proposition 8.1 to (8.42) that the maximum likelihood estimate of λ_{ij} is

$$\hat{\lambda}_{ij} = x_{ij},$$

and from (8.43), it is seen that the corresponding value of the log likelihood function is

$$\hat{l}_0 = -x_{\cdot\cdot} + \sum_{i=1}^{r}\sum_{j=1}^{s} x_{ij}\ln(x_{ij}) - \sum_{i=1}^{r}\sum_{j=1}^{s}\ln(x_{ij}!). \tag{8.44}$$

Finally, the number of free parameters is $d_0 = rs$ and the expected frequencies are $(\mathbf{e}_0)_{ij} = x_{ij}$.

M_1:

The likelihood function corresponding to the model is

$$L(\lambda_{\cdot\cdot},\boldsymbol{\rho},\boldsymbol{\sigma}) = \prod_{i=1}^{r}\prod_{j=1}^{s} e^{-\lambda_{\cdot\cdot}\rho_i\sigma_j}(\lambda_{\cdot\cdot}\rho_i\sigma_j)^{x_{ij}}\frac{1}{x_{ij}!}$$
$$= \left(e^{-\lambda_{\cdot\cdot}}\lambda_{\cdot\cdot}^{x_{\cdot\cdot}}\right)\prod_{i=1}^{r}\rho_i^{x_{i\cdot}}\prod_{j=1}^{s}\sigma_j^{x_{\cdot j}}\prod_{i=1}^{r}\prod_{j=1}^{s}\frac{1}{x_{ij}!}. \tag{8.45}$$

Since $\lambda_{\cdot\cdot}$, $\boldsymbol{\rho}$ and $\boldsymbol{\sigma}$ vary independently of each other, we find by using Proposition 8.1 on the first factor and Proposition 7.1 on the next two factors in (8.45) that the maximum likelihood estimate is given by

$$\hat{\lambda}_{\cdot\cdot} = x_{\cdot\cdot}, \qquad \hat{\rho}_i = \frac{x_{i\cdot}}{x_{\cdot\cdot}}, \qquad \hat{\sigma}_j = \frac{x_{\cdot j}}{x_{\cdot\cdot}}.$$

The value of the log likelihood function at the maximum likelihood estimate is

$$\hat{l}_1 = -x_{\cdot\cdot} + x_{\cdot\cdot}\ln(x_{\cdot\cdot}) + \sum_{i=1}^{r} x_{i\cdot}\ln\left(\frac{x_{i\cdot}}{x_{\cdot\cdot}}\right) + \sum_{j=1}^{s} x_{\cdot j}\ln\left(\frac{x_{\cdot j}}{x_{\cdot\cdot}}\right) - \sum_{i=1}^{r}\sum_{j=1}^{s}\ln(x_{ij}!)$$
$$= -x_{\cdot\cdot} + \sum_{i=1}^{r} x_{i\cdot}\ln(x_{i\cdot}) + \sum_{j=1}^{s} x_{\cdot j}\ln(x_{\cdot j}) - x_{\cdot\cdot}\ln(x_{\cdot\cdot}) - \sum_{i=1}^{r}\sum_{j=1}^{s}\ln(x_{ij}!). \tag{8.46}$$

The number of free parameters is $d_1 = r+s-1$ and the expected frequencies are

$$(\mathbf{e}_1)_{ij} = \frac{x_{i\cdot}x_{\cdot j}}{x_{\cdot\cdot}}; \tag{8.47}$$

i.e., the expected frequency in the (i,j)th cell is the product of the ith row sum and the jth column sum divided by the total sum.

M_2^r:

If Proposition 8.1 is applied to the first factor and Proposition 7.1 to the second factor of the likelihood function in (8.48), then

$$L(\lambda_{\cdot\cdot},\boldsymbol{\rho}) = \prod_{i=1}^{r}\prod_{j=1}^{s} e^{-\lambda_{\cdot\cdot}\rho_i/s}(\lambda_{\cdot\cdot}\rho_i/s)^{x_{ij}}\frac{1}{x_{ij}!}$$
$$= \left(e^{-\lambda_{\cdot\cdot}}\lambda_{\cdot\cdot}^{x_{\cdot\cdot}}\right)\prod_{i=1}^{r}\rho_i^{x_{i\cdot}}\left(\frac{1}{s}\right)^{x_{\cdot\cdot}}\prod_{i=1}^{r}\prod_{j=1}^{s}\frac{1}{x_{ij}!}, \tag{8.48}$$

the maximum likelihood estimate in M_2^r is found to be

$$\hat{\lambda}_{\cdot\cdot} = x_{\cdot\cdot}, \qquad \hat{\rho}_i = \frac{x_{i\cdot}}{x_{\cdot\cdot}}.$$

The maximum value of the log likelihood function in M_2^r is

$$\hat{l}_2^r = -x_{\cdot\cdot} + x_{\cdot\cdot}\ln(x_{\cdot\cdot}) + \sum_{i=1}^{r} x_{i\cdot}\ln\left(\frac{x_{i\cdot}}{x_{\cdot\cdot}}\right) + x_{\cdot\cdot}\ln\left(\frac{1}{s}\right) - \sum_{i=1}^{r}\sum_{j=1}^{s}\ln(x_{ij}!)$$
$$= -x_{\cdot\cdot} + \sum_{i=1}^{r} x_{i\cdot}\ln(x_{i\cdot}) - x_{\cdot\cdot}\ln(s) - \sum_{i=1}^{r}\sum_{j=1}^{s}\ln(x_{ij}!). \tag{8.49}$$

The number of free parameters is $d_2^r = r$ and the expected frequencies become

$$(\mathbf{e}_2^r)_{ij} = \frac{x_{i\cdot}}{s}; \tag{8.50}$$

thus the expected frequencies in the ith row are all equal to the average number of observations in the ith row.

M_2^c:

For this model we find in analogy with M_2^r that

$$\hat{\lambda}_{\cdot\cdot} = x_{\cdot\cdot}, \qquad \hat{\sigma}_j = \frac{x_{\cdot j}}{x_{\cdot\cdot}},$$

$$\hat{l}_2^c = -x_{\cdot\cdot} + \sum_{j=1}^{s} x_{\cdot j}\ln(x_{\cdot j}) - x_{\cdot\cdot}\ln(r) - \sum_{i=1}^{r}\sum_{j=1}^{s}\ln(x_{ij}!), \tag{8.51}$$

$d_2^c = s$ and the expected frequencies in the jth column are all equal to the average number of observations in the jth column, i.e.,

$$(\mathbf{e}_2^c)_{ij} = \frac{x_{\cdot j}}{r}. \tag{8.52}$$

M_3:

Applying Proposition 8.1 to the first factor of the likelihood function in (8.53),

$$\begin{aligned} L(\lambda_{\cdot\cdot}) &= \prod_{i=1}^{r}\prod_{j=1}^{s} e^{-\lambda_{\cdot\cdot}/(rs)}(\lambda_{\cdot\cdot}/(rs))^{x_{ij}}\frac{1}{x_{ij}!} \\ &= \left(e^{-\lambda_{\cdot\cdot}}\lambda_{\cdot\cdot}^{x_{\cdot\cdot}}\right)\left(\frac{1}{r}\right)^{x_{\cdot\cdot}}\left(\frac{1}{s}\right)^{x_{\cdot\cdot}}\prod_{i=1}^{r}\prod_{j=1}^{s}\frac{1}{x_{ij}!}, \end{aligned} \tag{8.53}$$

it is seen that

$$\hat{\lambda}_{\cdot\cdot} = x_{\cdot\cdot},$$

and, furthermore, that

$$\hat{l}_3 = -x_{\cdot\cdot} + x_{\cdot\cdot}\ln(x_{\cdot\cdot}) - x_{\cdot\cdot}\ln(r) - x_{\cdot\cdot}\ln(s) - \sum_{i=1}^{r}\sum_{j=1}^{s}\ln(x_{ij}!). \tag{8.54}$$

Finally, $d_3 = 1$ and

$$(\mathbf{e}_3)_{ij} = \frac{x_{\cdot\cdot}}{rs}; \tag{8.55}$$

in other words, the expected frequencies in all the cells are equal to the average of all observations.

Test of Hypotheses

From the formulas (11.32), (8.44), and (8.46) we obtain that the hypothesis of *multiplicative* effect (or *no interaction*) of the two criteria,

$$H_{01}\colon \lambda_{ij} = \lambda_{\cdot\cdot}\rho_i\sigma_j, \qquad i = 1,\ldots,r, \quad j = 1,\ldots,s, \tag{8.56}$$

is tested by means of the test statistic,

$$\begin{aligned} -2\ln Q(\mathbf{x}) &= 2[\hat{l}_0 - \hat{l}_1] \\ &= 2\Big[\sum_{i=1}^{r}\sum_{j=1}^{s} x_{ij}\ln(x_{ij}) - \sum_{i=1}^{r} x_{i\cdot}\ln(x_{i\cdot}) - \sum_{j=1}^{s} x_{\cdot j}\ln(x_{\cdot j}) + x_{\cdot\cdot}\ln(x_{\cdot\cdot})\Big], \end{aligned} \tag{8.57}$$

which has to be evaluated in a χ^2-distribution with $d_0 - d_1 = (r-1)(s-1)$ degrees of freedom. If the expected frequencies $\mathbf{e}_1$ in (8.47) are greater than or equal to 5, the significance probability may be calculated as

$$p_{obs}(\mathbf{x}) = 1 - F_{\chi^2((r-1)(s-1))}(-2\ln Q(\mathbf{x})). \tag{8.58}$$

The hypothesis of *no row effect* may be specified by

$$H_{0\sigma}\colon \boldsymbol{\sigma}^* = (\sigma_1,\ldots,\sigma_j,\ldots,\sigma_s) = \left(\frac{1}{s},\cdots,\frac{1}{s},\cdots,\frac{1}{s}\right).$$

In the model M_1 the hypothesis $H_{0\sigma}$ corresponds to the reduction to M_2^r and is tested by considering the quantity,

$$-2\ln Q(\mathbf{x}) = 2[\hat{l}_1 - \hat{l}_2^r]$$
$$= 2[\sum_{j=1}^{s} x_{\cdot j}\ln(x_{\cdot j}) - x_{\cdot\cdot}\ln(\frac{x_{\cdot\cdot}}{s})]. \tag{8.59}$$

A comparison with (8.39) shows that $-2\ln Q(\mathbf{x})$ is identical with the test statistic of the hypothesis of identity of the parameters for the s column sums, $X_{\cdot 1},\ldots,X_{\cdot s}$. Consequently, the significance probability of the hypothesis of no column effect corresponding to the reduction $M_1 \to M_2^r$ may be calculated as

$$p_{obs}(\mathbf{x}) = 1 - F_{\chi^2(s-1)}(-2\ln Q(\mathbf{x})), \tag{8.60}$$

provided that the common expected frequency $x_{\cdot\cdot}/s$ of the column sums is greater than or equal to 5.

The hypothesis of *no row effect* is

$$H_{0\rho}\colon \boldsymbol{\rho}^* = (\rho_1,\ldots,\rho_i,\ldots,\rho_r) = (\frac{1}{r},\cdots,\frac{1}{r},\cdots,\frac{1}{r}).$$

In the model M_2^r the hypothesis $H_{0\rho}$ corresponds to the reduction to M_3 and is tested in this model by considering the quantity,

$$-2\ln Q(\mathbf{x}) = 2[\hat{l}_2^r - \hat{l}_3]$$
$$= 2[\sum_{i=1}^{r} x_{i\cdot}\ln(x_{i\cdot}) - x_{\cdot\cdot}\ln(\frac{x_{\cdot\cdot}}{r})]. \tag{8.61}$$

From (8.39) it is seen that this test statistic is identical with the test statistic for identity of the parameters of the r row sums. If the common expected frequency $x_{\cdot\cdot}/r$ of the row sums is greater than or equal to 5, the significance probability is calculated as

$$p_{obs}(\mathbf{x}) = 1 - F_{\chi^2(r-1)}(-2\ln Q(\mathbf{x})). \tag{8.62}$$

We have shown how we can make reductions in the model M_0 via the "route" $M_0 \to M_1 \to M_2^r \to M_3$. From the formulas (8.44), (8.46), (8.49), (8.51), and (8.54) above it is seen that we have the following identities:

$$\begin{aligned} 2[\hat{l}_1 - \hat{l}_2^r] &= 2[\hat{l}_2^c - \hat{l}_3] & d_1 - d_2^r &= d_2^c - d_3 \\ &= 2[\sum_{j=1}^{s} x_{\cdot j}\ln(x_{\cdot j}) - x_{\cdot\cdot}\ln(\frac{x_{\cdot\cdot}}{s})], & &= s-1, \\ 2[\hat{l}_1 - \hat{l}_2^c] &= 2[\hat{l}_2^r - \hat{l}_3] & d_1 - d_2^c &= d_2^r - d_3 \\ &= 2[\sum_{i=1}^{r} x_{i\cdot}\ln(x_{i\cdot}) - x_{\cdot\cdot}\ln(\frac{x_{\cdot\cdot}}{r})], & &= r-1. \end{aligned}$$

Hence it is seen that the test of the hypothesis for no column effect is given by the formulas (8.59) and (8.60) irrespective of which of the routes $M_0 \to M_1 \to M_2^r \to M_3$ or $M_0 \to M_1 \to M_2^c \to M_3$ we consider. In other words, if we have accepted the model of multiplicative effects M_1, then a possible row effect has no influence on the test of no column effect; the test is the same despite the fact that it is performed in the two different models M_1 and M_2^c.

A similar remark holds, of course, for the test of the hypothesis of no row effect.

By means of formula (8.8), it may be shown that in the model M_1 the *components* $X_{i\cdot}$ of the random vector consisting of the row sums $\mathbf{X}_{*\cdot} = (X_{1\cdot},\ldots,X_{i\cdot},\ldots,X_{r\cdot})^*$ are stochastically independent and, in addition, that

$$X_{i\cdot} \sim po(\lambda_{\cdot\cdot}\rho_i), \qquad i = 1,\ldots,r. \tag{8.63}$$

Similarly, in M_1 the *components* $X_{\cdot j}$ of the vector of column sums $\mathbf{X}_{\cdot *} = (X_{\cdot 1},\ldots,X_{\cdot j},\ldots,X_{\cdot s})^*$

are stochastically independent and

$$X_{\cdot j} \sim po(\lambda_{\cdot\cdot}\sigma_j), \qquad j = 1,\ldots,s. \tag{8.64}$$

In the model for the column sums $\mathbf{X}_{\cdot *}$ corresponding to (8.64) the hypothesis of no column effect $H_{0\sigma}$: $\boldsymbol{\sigma} = (\frac{1}{s},\ldots,\frac{1}{s})^*$ may be tested using the formulas (8.32) and (8.39). In this model the expected frequencies are $e_j = x_{\cdot\cdot}/s$ according to (8.38). Since the formulas (8.39) and (8.32) for the column sum are identical to (8.59) and (8.60), the hypothesis of no column effect corresponding to the reduction $M_1 \to M_2^r$ or to the reduction $M_2^c \to M_3$ may be tested by considering the column sums $\mathbf{X}_{\cdot *}$ only. The expected frequencies $(\mathbf{e}_2^r)_{ij} = x_{i\cdot}/s$ in M_2^r and $(\mathbf{e}_3)_{ij} = x_{\cdot\cdot}/(rs)$ in M_3 are less that $e_j = x_{\cdot\cdot}/s$. The condition that the expected frequencies should all be larger than or equal to 5 in order to apply the χ^2-approximation to the significance probability is therefore most easily satisfied in the model based on the column sums $\mathbf{X}_{\cdot *}$. Consequently, in the following the hypothesis $H_{0\sigma}$ of no column effect is tested in the marginal model (8.64) for the column sums $\mathbf{X}_{\cdot *}$.

Similarly, we perform the test of the hypothesis $H_{0\rho}$ of no row effect in the marginal model (8.63) for the row sums $\mathbf{X}_{*\cdot}$.

Distributional Results and Relation to the Multinomial Model

If we condition on the sum $x_{\cdot\cdot}$ of all the observations in the model M_0, we obtain according to (8.9) the conditional model,

$$M_0^*\colon \{X_{ij}\} \mid X_{\cdot\cdot} = x_{\cdot\cdot} \sim m(x_{\cdot\cdot},(\{\frac{\lambda_{ij}}{\lambda_{\cdot\cdot}}\})). \tag{8.65}$$

Since the λ's vary freely, there are no restrictions on the probability matrix in the multinomial distribution and the model M_0^* corresponds to the basic model based on the multinomial distribution for an $r \times s$ table with number of trials $x_{\cdot\cdot}$. In this model the hypothesis H_{01}: $\lambda_{ij} = \lambda_{\cdot\cdot}\rho_i\sigma_j$ turns into the hypothesis,

$$H_{01}^*\colon \frac{\lambda_{ij}}{\lambda_{\cdot\cdot}} = \rho_i\sigma_j, \qquad i = 1,\ldots,r, \quad j = 1,\ldots,s.$$

The hypothesis H_{01} of multiplicative effect of the rows and columns in the Poisson model M_0 therefore corresponds to the hypothesis H_{01}^* of independence in the conditional multinomial model M_0^*, and is seen from the formulas (7.38), (7.39), (8.59), and (8.60) that the test of H_{01}^* and the test of H_{01} are identical. Thus, we have yet another example showing how conditioning on the sum of all the observations in a model based on the Poisson distribution results in a "return" to a well-known multinomial model.

In a concrete situation, it should be decided whether to use the Poisson model or the multinomial model. We should use information on how the observations in the $r \times s$ table have been collected. As mentioned earlier, the multinomial model should be used if it has been decided in advance to consider how a *given* number of objects are classified according to the two criteria. In contrast, if the sum of the observations is not known in advance, the Poisson model should be used.

Since the analysis of the data is carried out in the same way in the two models, it is not important to recognize which of the two models is considered regarding the structure of the data. The only difference between the two models is that in the Poisson model the total sum of observations $x_{\cdot\cdot}$ contains information on the intensity of the phenomenon studied, whereas the total sum of the observations n in the multinomial model does not contain information.

Remark 8.3 If the hypothesis H_{01}^* is not rejected, the model M_0^* may be reduced to another conditional model,

$$M_1^*\colon \{X_{ij}\} \mid X_{\cdot\cdot} = x_{\cdot\cdot} \sim m(x_{\cdot\cdot},(\{\rho_i\sigma_j\})).$$

It follows from formula (7.41) that in M_1^* the vector of row sums $\mathbf{X}_{*\cdot}$ and the vector of column sums

$\mathbf{X}_{\cdot *}$ are *conditionally independent* given $X_{\cdot\cdot} = x_{\cdot\cdot}$ and, furthermore, that

$$\mathbf{X}_{*\cdot} \mid X_{\cdot\cdot} = x_{\cdot\cdot} \sim m(x_{\cdot\cdot}, \boldsymbol{\rho}) \tag{8.66}$$

and

$$\mathbf{X}_{\cdot *} \mid X_{\cdot\cdot} = x_{\cdot\cdot} \sim m(x_{\cdot\cdot}, \boldsymbol{\sigma}). \tag{8.67}$$

In the model based on (8.67), it may be shown that the $-2\ln Q$ statistic of the hypothesis $H_{0\sigma}$: $\boldsymbol{\sigma} = (\frac{1}{s}, \ldots, \frac{1}{s})^*$ and the corresponding significance probability are given by (8.59) and (8.60), respectively, and furthermore, that the expected frequencies under the hypothesis are all equal to $x_{\cdot\cdot}/s$. Thus, the test of $H_{0\sigma}$ in the conditional model for $\mathbf{X}_{\cdot *}$ given $X_{\cdot\cdot} = x_{\cdot\cdot}$ is precisely the same as the test of $H_{0\sigma}$ in the (unconditional) model for $\mathbf{X}_{\cdot *}$.

Similar remarks hold for the test of the hypothesis $H_{0\rho}$ of no row effect. ▼

Example 8.3 (Continued)

If we supplement Table 8.3 with row sums, column sums, and total sum we have:

Observed	*Day*	*Night*	*Total*
035–125 m	59	72	131
125–250 m	371	392	763
250–300 m	91	100	191
Total	521	564	1085

From the description of how the data has been collected, in this situation it seems reasonable to consider a model for the 3×2 table based on the Poisson distribution. The reason the multinomial model does not appear to be the right model for these data is that the total number of water fleas definitely was not fixed in advance.

Thus, as the basic model we consider the model M_0 with $r = 3$ and $s = 2$. From the table above and formula (8.47) we find that the expected frequencies in the multiplicative model M_1 are:

Expected	*Day*	*Night*	*Total*
035–125 m	62.9	68.1	131.0
125–250 m	366.4	396.6	763.0
250–300 m	91.7	99.3	191.0
Total	521.0	564.0	1085.0

Since the expected frequencies are all greater than 5, the reduction to the multiplicative model M_1 may be tested by means of the formulas (8.57) and (8.58). We find

$$-2\ln Q(\mathbf{x}) = 0.590$$

and

$$p_{obs}(\mathbf{x}) = 1 - F_{\chi^2(2)}(0.590) = 0.745.$$

This is not critical for the multiplicative model M_1, which is adopted as the model for the data.

As mentioned above the hypothesis of no column effect may be examined by testing the identity of the parameter for the column sums. In this situation no column effect means that the mean number of fleas caught is the same by day and by night. Since the expected frequency in the jth column under the hypothesis of no column effect is $x_{\cdot\cdot}/s$ – here $x_{\cdot\cdot}/2$ since $s = 2$ – the observed and expected frequencies are as follows:

Column sums	*Day*	*Night*	*Total*
Observed	521	564	1085
Expected	542.5	542.5	1085.0

From (8.59) and (8.60) we obtain that

$$-2\ln Q(\mathbf{x}) = 1.705$$

and

$$p_{obs}(\mathbf{x}) = 1 - F_{\chi^2(1)}(1.705) = 0.192;$$

this does not contradict the hypothesis that the mean number of fleas caught is the same by day and by night.

To evaluate if the catch depends on the depth, the row sums are considered and it is investigated whether the parameters of the distributions of the row sums are identical. We find with an accuracy of one decimal place:

Row sums	35–125 m	125–250 m	250–300 m	*Total*
Observed	131	763	191	1085
Expected	361.7	361.7	361.7	1085.1

By means of (8.61) and (8.62), we get that

$$-2\ln Q(\mathbf{x}) = 629.261$$

and

$$p_{obs}(\mathbf{x}) = 1 - F_{\chi^2(2)}(629.261) \approx 0,$$

and the hypothesis of no row effect is rejected. Thus, the mean number of water fleas caught depends on the depth.

The final model for these data is therefore

$$M_2^r\colon x_{ij} \sim\sim po(\alpha_i) = po(\lambda_{\cdot\cdot}\rho_i/s), \quad i = 1,2,3, \quad j = 1,2,$$

with $s = 2$. In this model the row sums are independent and

$$x_{i\cdot} \sim\sim po(\lambda_{\cdot\cdot}\rho_i), \quad i = 1,2,3.$$

The estimates are $\hat{\lambda}_{\cdot\cdot} = x_{\cdot\cdot}$ and $\hat{\rho}_i = x_{i\cdot}/x_{\cdot\cdot}$, $i = 1,2,3$, i.e.,

$$\hat{\lambda}_{\cdot\cdot} = 1085, \quad \hat{\rho}_1 = \frac{131}{1085} = 0.121, \quad \hat{\rho}_2 = \frac{763}{1085} = 0.703, \quad \hat{\rho}_3 = \frac{191}{1085} = 0.176.$$

Here $\hat{\lambda}_{\cdot\cdot}$ contains information on the intensity or the occurrence of water fleas in the investigated waters. The estimate $\hat{\lambda}_{\cdot\cdot}$ may be used for comparison of the occurrence of water fleas in other waters where similar investigations have been made.

The confidence interval for $\lambda_{\cdot\cdot}$ can be obtained from (8.20) and (8.21) assuming that $X_{\cdot\cdot} \sim po(\lambda_{\cdot\cdot})$. For $i = 1,2,3$, the confidence interval for ρ_i can be calculated from the formulas (7.27) and (7.28) since $X_{i\cdot} \mid X_{\cdot\cdot} = x_{\cdot\cdot} \sim b(x_{\cdot\cdot}, \rho_i)$. □

8.5 Transformation

In connection with the Poisson distribution, the square root transformation is often considered. It may be shown that if

$$X \sim po(\lambda),$$

for large values of λ we have

$$Y = \sqrt{X} \approx N(\sqrt{\lambda}, \frac{1}{4}). \tag{8.68}$$

The result in (8.68) is often used in situations where the statistical analysis of the original data by means of a model based on the Poisson distribution is troublesome and where an analysis of the square root of the counts in a model based on the normal distribution may be more convenient. An example is given in Exercise 3.6.

According to Bartlett (1936) the approximation in (8.68) may be used if λ is greater than or equal to 10. Anscombe (1948) suggests consideration of the transformation,

$$Z = \sqrt{X + 3/8} \tag{8.69}$$

for values of λ between 3 and 10.

The need for transforming Poisson-distributed data and analyzing them using the normal distribution does not exist anymore because of the development of the theory for generalized linear models that include the Poisson distribution. Generalized linear models is the subject of Chapter 9.

We therefore discuss the square root transformations (8.68) or (8.69) mainly for historical reasons and we restrict ourselves to illustrate them in connection with Example 8.1 and Example 8.2.

Example 8.2 (Continued)
According to the result of the analysis of these data, the two samples corresponding to a volume of 20 μL may be described by the same Poisson distribution and may therefore be considered as a single sample from the Poisson distribution. From Table 8.5 it is seen that the sum of the 30 observations in the two rows is $248 + 231 = 479$. Consequently, the maximum likelihood estimate of the parameter in the Poisson distribution is $479/30 = 15.967$. Since the estimate is greater than 10, there are reasons to believe that the transformation (8.68) will be valid for these data.

Later in this chapter we consider the data in this example and plot on page 385 the fractile diagram of the square root of the number of nematodes corresponding to a volume of 20 μL as an illustration of calculations in SAS. The fractile diagram gives no justification for rejecting a normal model for the transformed data. Furthermore, from the output from SAS on page 386 it is seen that the empirical variance of the transformed data is 0.3008352. Using the test of the hypothesis, $\sigma^2 = \sigma_0^2$, in *Main Points in Section 3.1* on page 78, it may be established that this value does not differ significantly from a theoretical variance of 0.25. □

Example 8.1 (Continued)
Since the hypothesis that the parameters of the two Poisson-distributed samples are identical was not rejected, the data will be considered as a single sample of size 296 from the Poisson distribution. As mentioned earlier the estimate of the parameter in the common distribution is 3.1149, so the approximation (8.68) may be questionable in this case. It appears from the fractile diagram of the square root transformed data in Figure 8.1 that there are problems with the small values.

The figure also shows the fractile diagram of the data obtained by the transformation (8.69). This diagram gives no reason to doubt that these data may be considered as a single sample from the normal distribution. □

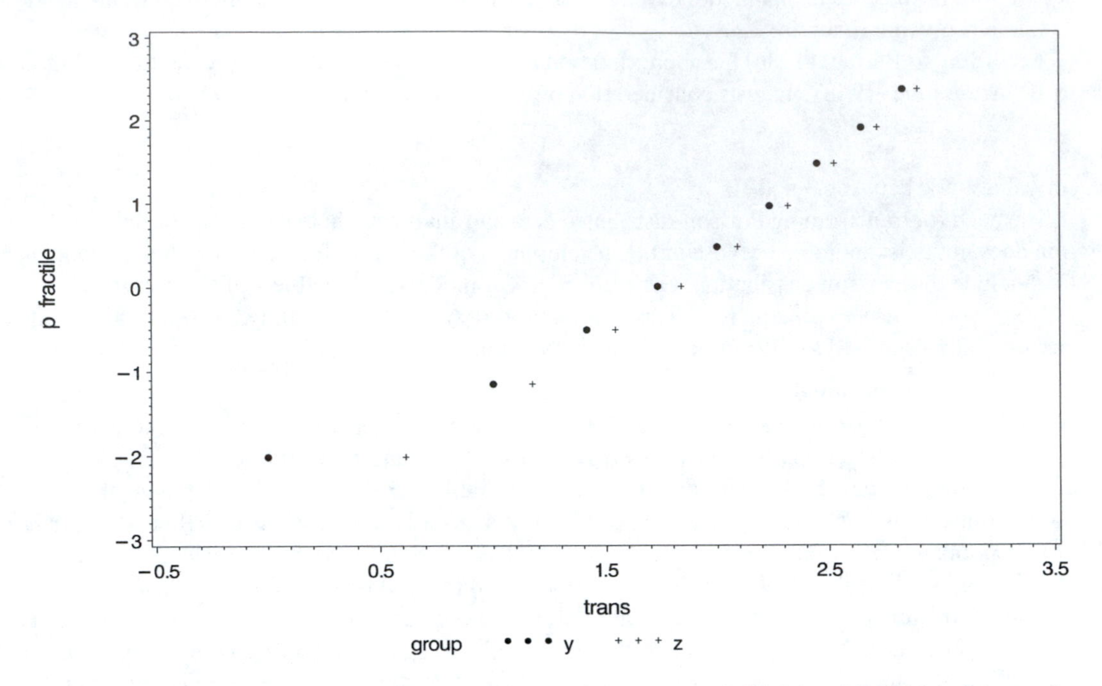

Figure 8.1 *Fractile diagram of data obtained by the transformations in (8.68) (y) and (8.69) (z) of the number of particles in the 296 time intervals in Table 8.1.*

Annex to Chapter 8

Calculations in SAS

The calculation of the $-2\ln Q$-test and/or the X^2-test of the hypothesis of proportional parameters may be performed by means of the macro `poisson`. The data set and the variable to be considered are specified by means of `data=` and `obs=`, while the vector that will be tested for proportionality with the vector of means is given by `m=`. If no variable is assigned to `m`, the macro assumes that all elements of `m` are equal to 1, and, consequently, the test performed is the test for identical parameters. The output listing from the macro needs no explanation.

Example 8.2 (Continued)

The program segment below makes the calculations for each of the three samples in Table 8.2 that are relevant in connection with the Poisson model. Note that when reading the data we use the symbol @@, which implies that a single line may contain information on several observations. This facility is very useful if the data set contains many observations.

```
DATA felti;
INPUT sample x @@;
DATALINES;
1 31 1 28 1 33 1 38 1 28 1 32 1 39 1 27
1 28 1 39 1 21 1 39 1 45 1 37 1 41
2 14 2 16 2 18 2  9 2 21 2 21 2 14 2 12
2 13 2 13 2 14 2 20 2 24 2 15 2 24
3 18 3 13 3 19 3 14 3 15 3 16 3 14 3 19
3 25 3 16 3 16 3 18 3  9 3 10 3  9
;
RUN;

%poisson(data=felti(WHERE=(sample=1)),obs=x);
```

The output listing from the statement
`%poisson(data=felti(WHERE=(sample=1)),obs=x)`
is seen below. This statement is now repeated for the observations in the second and third sample corresponding to the values 2 and 3 of the variable `sample`. In order to save space these lines are not included.

```
                    Macro "POISSON"

Dataset:              FELTI(WHERE=(SAMPLE=1))
Observations:         X
Weights:              All weights are set to 1
Number of groups: 15


group              X     expected
-------------------------------
    1             31        33.73
    2             28        33.73
    3             33        33.73
    4             38        33.73
    5             28        33.73
    6             32        33.73
    7             39        33.73
    8             27        33.73
    9             28        33.73
   10             39        33.73
   11             21        33.73
   12             39        33.73
   13             45        33.73
   14             37        33.73
   15             41        33.73
```

```
-------------------------------
   sum          506       506.00


Test of the hypothesis of proportional parameters:

statistic        value       Prob > value
-------------------------------------------
-2lnQ          18.4814            0.1857
chi-square     18.0514            0.2045


Estimate and 95% confidence interval for lambda
under the hypothesis of proportional parameters:

estimate   lower limit        upper limit
-----------------------------------------
33.7333        30.9193           36.8034
```

As seen from formula (8.40) the value of Fisher's index of dispersion *id* can be found by dividing the value of the X^2-statistic by the number of degrees of freedom, here as $18.0514/14 = 1.2894$ in accordance with the first row in Table 8.5 on page 364.

The estimate and the confidence interval are not particularly interesting because they concern the mean number of nematodes in 40 μL dilution.

The test mentioned on page 370 of the hypothesis that the mean number of nematodes per 1 μL is the same for the three dilutions may be obtained from the `poisson` macro, but first the appropriate data set is constructed.

```
DATA sums;
INPUT x mm;
DATALINES;
506 600
248 300
231 300
;
RUN;

%poisson(data=sums,obs=x,m=mm);
```

The output listing from the macro `poisson` in this case is:

```
                    Macro "POISSON"

Dataset:             SUMS
Observations:        X
Weights:             MM
Number of groups: 3


group              MM           X     expected
------------------------------------------
     1            600         506       492.50
     2            300         248       246.25
     3            300         231       246.25
------------------------------------------
   sum           1200         985       985.00


Test of the hypothesis of proportional parameters:

statistic        value       Prob > value
-------------------------------------------
-2lnQ           1.3437            0.5108
chi-square      1.3269            0.5151


Estimate and 95% confidence interval for lambda
under the hypothesis of proportional parameters:

estimate   lower limit        upper limit
-----------------------------------------
```

```
   0.8208          0.7711            0.8737
```

Here the estimate and the confidence interval are of interest since they concern the mean number of nematodes in a volume of 1 μL of the three dilutions.

We will now illustrate the square root transformation of a Poisson sample. The analysis showed that we could assume that the intensity was the same for all three samples and since the sample volumes are the same for sample 2 and for sample 3, we can merge these two samples into one sample.

```
DATA trans23;
SET felti(WHERE=(sample NE 1));
y=sqrt(x);
RUN;

%normgraph(data=trans23,var=y,table=no,graphics=no);

PROC MEANS DATA=trans23 VARDEF=DF N MEAN VAR;
   var y;
RUN;
```

The resulting fractile diagram becomes:

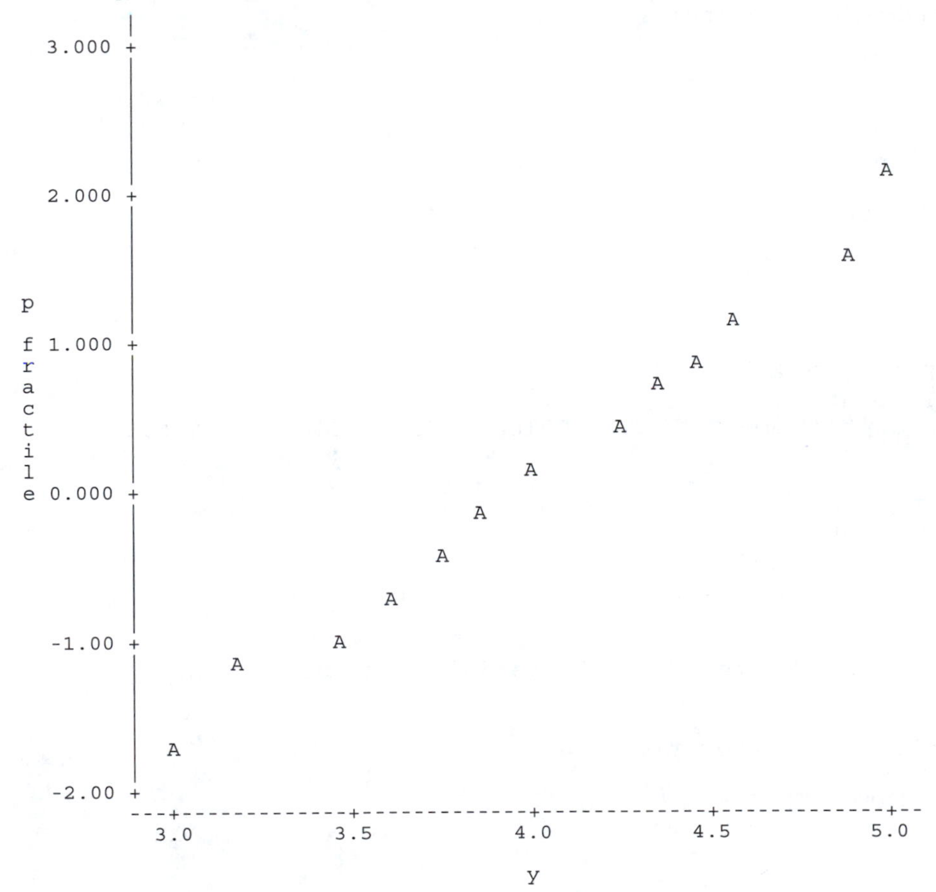

From the output listing from *PROC MEANS*,

```
                The MEANS Procedure

               Analysis Variable : y

          N            Mean          Variance
          ----------------------------------
         30       3.9592751         0.3008352
          ----------------------------------
```

it is seen that the estimated variance of the square root transformed data is 0.3008352, which is not a significant deviation from the theoretical variance of 0.25. □

Since for an $r \times s$ table the calculations of the test for the multiplicative Poisson model are identical with the calculations of the test of independence in the multinomial model, the calculations may be performed by means of the `TABLES` statement in the SAS-procedure *PROC FREQ*. Test of no row effect or no column effect may also be obtained from *PROC FREQ* with the limitation, however, that the results concern the X^2-test only. Alternatively the macro `poisson` may be used and in addition to the $-2\ln Q$-test of the hypothesis, it gives a 95% confidence interval for the parameter.

Example 8.3 (Continued)

The SAS calculations in the multiplicative Poisson model for data in Table 8.3 may be done by means of the following program:

```
DATA water_flea;
INPUT depth$1-7 time$9-13 count;
DATALINES;
035-125 day    59
035-125 night  72
125-250 day   371
125-250 night 392
250-300 day    91
250-300 night 100
;
RUN;

PROC FREQ DATA=water_flea;
WEIGHT count;
TABLES depth*time /EXPECTED CHISQ NOROW NOCOL NOPERCENT;
RUN;

PROC FREQ DATA=water_flea;
WEIGHT count;
TABLES depth /TESTP=(0.3333 0.3333 0.3333)
             OUT=row;
RUN;

/* row sums are stored in the variable 'count' in the data set
   'row'*/

PROC FREQ DATA=water_flea;
WEIGHT count;
TABLES time /TESTP=(0.5 0.5)
            OUT=column;
RUN;

%poisson(data=column,obs=count);

%poisson(data=row,obs=count);
```

The output listing from the three calls of *PROC FREQ* is:

```
                    The SAS System                                   1

                  The FREQ Procedure

                Table of depth by time

           depth     time

           Frequency|
           Expected |day     |night   |  Total
           ---------+--------+--------+
           035-125  |     59 |     72 |    131
                    | 62.904 | 68.096 |
           ---------+--------+--------+
           125-250  |    371 |    392 |    763
                    | 366.38 | 396.62 |
           ---------+--------+--------+
           250-300  |     91 |    100 |    191
                    | 91.715 | 99.285 |
           ---------+--------+--------+
           Total         521      564     1085


          Statistics for Table of depth by time

  Statistic                     DF       Value      Prob
  ------------------------------------------------------
  Chi-Square                     2      0.5889    0.7449
  Likelihood Ratio Chi-Square    2      0.5899    0.7446
  Mantel-Haenszel Chi-Square     1      0.1277    0.7208
  Phi Coefficient                       0.0233
  Contingency Coefficient               0.0233
  Cramer's V                            0.0233

                  Sample Size = 1085
                    The SAS System                                   2

                  The FREQ Procedure

                                          Test    Cumulative    Cumulative
depth       Frequency     Percent      Percent     Frequency       Percent
--------------------------------------------------------------------------
035-125          131       12.07        33.33           131         12.07
125-250          763       70.32        33.33           894         82.40
250-300          191       17.60        33.33          1085        100.00


                    Chi-Square Test
               for Specified Proportions
               -------------------------
               Chi-Square       673.0701
               DF                      2
               Pr > ChiSq         <.0001

                  Sample Size = 1085
                    The SAS System                                   3

                  The FREQ Procedure

                                        Test    Cumulative    Cumulative
time        Frequency     Percent    Percent     Frequency       Percent
------------------------------------------------------------------------
day              521       48.02       50.00           521         48.02
night            564       51.98       50.00          1085        100.00
```

```
        Chi-Square Test
   for Specified Proportions
   -------------------------
   Chi-Square         1.7041
   DF                      1
   Pr > ChiSq         0.1917

       Sample Size = 1085
```

The output listing from the two calls of the macro `poisson` follows below:

```
               Macro "POISSON"

Dataset:           ROW
Observations:      COUNT
Weights:           All weights are set to 1
Number of groups: 3


group          COUNT    expected
-------------------------------
     1          131      361.67
     2          763      361.67
     3          191      361.67
-------------------------------
   sum         1085     1085.00


Test of the hypothesis of proportional parameters:

statistic       value       Prob > value
-------------------------------------------
-2lnQ       629.2570            0.0000
chi-square  673.0028            0.0000


Estimate and 95% confidence interval for lambda
under the hypothesis of proportional parameters:

estimate   lower limit      upper limit
-----------------------------------------
361.667      340.7770         383.8368

               Macro "POISSON"

Dataset:           COLUMN
Observations:      COUNT
Weights:           All weights are set to 1
Number of groups: 2


group          COUNT    expected
-------------------------------
     1          521      542.50
     2          564      542.50
-------------------------------
   sum         1085     1085.00


Test of the hypothesis of proportional parameters:

statistic       value       Prob > value
-------------------------------------------
-2lnQ          1.7046           0.1917
chi-square     1.7041           0.1917


Estimate and 95% confidence interval for lambda
under the hypothesis of proportional parameters:

estimate   lower limit      upper limit
-----------------------------------------
542.500      511.1656         575.7552
```

□

Main Points in Chapter 8

One Sample

Model:
The observations, $x_1, \ldots, x_n$, are considered as realizations of independent random variables, $X_1, \ldots,$ X_n, which are all Poisson distributed with parameter λ, i.e.,

$$M_0\colon X_i \sim po(\lambda), \qquad i = 1, \ldots, n.$$

Estimate:
The maximum likelihood estimate $\hat{\lambda}$ of λ is

$$\hat{\lambda} = \bar{x}_{\bullet} = \frac{1}{n} x_{\bullet} = \frac{1}{n} \sum_{i=1}^{n} x_i.$$

The distribution of the maximum likelihood estimator is given as

$$n\hat{\lambda} = X_{\bullet} \sim po(n\lambda).$$

Model Checking:
If the sample size n is sufficiently large, M_0 may be checked using a χ^2-test for goodness of fit as described in Section 7.5.

An alternative control of the model M_0 is based on *Fisher's index of dispersion*, which is the ratio,

$$id = \frac{s^2}{\bar{x}_{\bullet}},$$

between the empirical variance s^2 and the empirical mean $\bar{x}_{\bullet}$. The calculation of these quantities depends on whether the single observations are available or the observations are given in table form. With obvious notation we have

$$S = \sum_{i=1}^{n} x_i = \sum_j j a_j,$$

$$USS = \sum_{i=1}^{n} x_i^2 = \sum_j j^2 a_j$$

and

$$\bar{x}_{\bullet} = \frac{1}{n} S \qquad \text{and} \qquad s^2 = \frac{1}{n-1}\left(USS - \frac{S^2}{n}\right).$$

The model M_0 is *rejected* by a test of significance level α if

$$id < \chi^2_{\alpha/2}(n-1)/(n-1)$$

or if

$$id > \chi^2_{1-\alpha/2}(n-1)/(n-1).$$

The test is based on an approximation that may be used if $\mathbf{n \geq 15}$ **or** $\bar{\mathbf{x}}_{\bullet} \geq \mathbf{5}$.

Confidence Intervals:
The mean of a Poisson-distributed random variable:
An approximate $(1-\alpha)$ confidence interval for the parameter λ based on a single observation x from the $po(\lambda)$ distribution is of the form,

$$C_{1-\alpha}(x) = [\lambda_-, \lambda_+],$$

where

$$\lambda_- = x + \frac{1}{2}u_{1-\alpha/2}^2 - u_{1-\alpha/2}\sqrt{x + \frac{1}{4}u_{1-\alpha/2}^2}$$

and

$$\lambda_+ = x + \frac{1}{2}u_{1-\alpha/2}^2 + u_{1-\alpha/2}\sqrt{x + \frac{1}{4}u_{1-\alpha/2}^2}.$$

In the formulas $u_{1-\alpha/2}$ denotes the $(1-\alpha/2)$-fractile of the u-distribution. If $\alpha = 0.05$, the fractile is $u_{0.975} = 1.960$.

The mean in a single sample from the Poisson distribution:

Here the sum is $x. \sim\sim po(n\lambda)$ and the $(1-\alpha)$ confidence interval for λ has the limits,

$$\lambda_- = \frac{1}{n}\left[x. + \frac{1}{2}u_{1-\alpha/2}^2 - u_{1-\alpha/2}\sqrt{x. + \frac{1}{4}u_{1-\alpha/2}^2}\right]$$

and

$$\lambda_+ = \frac{1}{n}\left[x. + \frac{1}{2}u_{1-\alpha/2}^2 + u_{1-\alpha/2}\sqrt{x. + \frac{1}{4}u_{1-\alpha/2}^2}\right].$$

Several Samples

Poisson Model with Proportional Parameters:

The data set consists of the observations, $x_1, \ldots, x_k$, which are considered as realizations of independent random variables, $X_1, \ldots, X_k$, all Poisson-distributed but with different parameters, i.e., the basic model is

$$M_0\colon X_i \sim po(\lambda_i), \qquad i = 1, \ldots, k.$$

To each observation x_i is associated a known constant m_i and we are interested in testing the hypothesis that the parameters in the model M_0 are proportional to the known constants $m_1, \ldots, m_k$, i.e., the hypothesis,

$$H_{01}\colon \lambda_i = m_i\lambda, \qquad i = 1, \ldots, k.$$

If the hypothesis is not rejected, M_0 may be reduced to the model,

$$M_1\colon X_i \sim po(m_i\lambda), \qquad i = 1, \ldots, k.$$

Note that in the model M_0 it may be examined if $x_1, \ldots, x_k$ can be considered as a sample from the $po(\lambda)$-distribution by testing the hypothesis corresponding to $m_1 = \cdots = m_k = 1$.

The maximum likelihood estimate of λ in M_1 is

$$\hat{\lambda} = \frac{x.}{m.},$$

where $x. = x_1 + \cdots + x_k$ and $m. = m_1 + \ldots + m_k$. The expected frequency in M_1 corresponding to the observation x_i is

$$e_i = m_i\hat{\lambda} = x.\frac{m_i}{m.}$$

and the $-2\ln Q$-test statistic of H_{01} is

$$-2\ln Q(\mathbf{x}) = 2\sum_{i=1}^{k} x_i \ln(\frac{x_i}{e_i}).$$

If **the expected frequencies are all greater than or equal to 5**, the following approximation of the significance probability of H_{01} is valid:

$$p_{obs}(\mathbf{x}) \doteq 1 - F_{\chi^2(k-1)}(-2\ln Q(\mathbf{x})).$$

The distribution of the maximum likelihood estimator of λ under H_{01} is usually given in the following way:

$$m.\hat{\lambda} = X. \sim po(m.\lambda).$$

Confidence intervals for the parameter in the Poisson model with proportional parameters:

The $(1-\alpha)$ confidence interval for λ has the limits,

$$\lambda_{-} = \frac{1}{m.}\left[x. + \frac{1}{2}u^2_{1-\alpha/2} - u_{1-\alpha/2}\sqrt{x. + \frac{1}{4}u^2_{1-\alpha/2}}\right]$$

and

$$\lambda_{+} = \frac{1}{m.}\left[x. + \frac{1}{2}u^2_{1-\alpha/2} + u_{1-\alpha/2}\sqrt{x. + \frac{1}{4}u^2_{1-\alpha/2}}\right].$$

The Multiplicative Poisson Model:

As shown on page 372, the observations may be displayed in an $r \times s$ table corresponding to two factors with r and s levels, respectively. The observation corresponding to the ith level of the first factor and the jth level of the second factor is denoted by x_{ij}, $i = 1,\ldots,r$, $j = 1,\ldots,s$. Furthermore, $x_{i.}$ and $x_{.j}$ denotes the sum of the observations in the ith row and the jth column, respectively, and $x_{..}$ is the sum of all observations, i.e.,

$$x_{i.} = \sum_{j=1}^{s} x_{ij}, \qquad x_{.j} = \sum_{i=1}^{r} x_{ij}, \qquad x_{..} = \sum_{i=1}^{r}\sum_{j=1}^{s} x_{ij}.$$

The observations are assumed to be outcomes of independent random variables and the following models are considered:

The basic model:

$$M_0\colon x_{ij} \sim\sim po(\lambda_{ij}), \quad i = 1,\ldots,r, \quad j = 1,\ldots,s.$$

The *multiplicative* model or the model for *no interaction*:

$$M_1\colon x_{ij} \sim\sim po(\alpha_i\beta_j), \quad i = 1,\ldots,r, \quad j = 1,\ldots,s.$$

The model for *row effect only*:

$$M_2^r\colon x_{ij} \sim\sim po(\alpha_i), \quad i = 1,\ldots,r, \quad j = 1,\ldots,s.$$

The model for *column effect only*:

$$M_2^c\colon x_{ij} \sim\sim po(\beta_j), \quad i = 1,\ldots,r, \quad j = 1,\ldots,s.$$

The model for *homogeneity*:

$$M_3\colon x_{ij} \sim\sim po(\lambda), \quad i = 1,\ldots,r, \quad j = 1,\ldots,s.$$

Rewriting the parameter in M_1 as

$$\alpha_i\beta_j = \alpha.\beta.\frac{\alpha_i}{\alpha.}\frac{\beta_j}{\beta.} = \lambda_{..}\rho_i\sigma_j,$$

the models M_0, M_1, M_2^r, M_2^c, and M_3 and their mutual relations are as follows :

$$M_2^r\colon X_{ij} \sim po(\lambda_{..}\rho_i\tfrac{1}{s})$$

$$M_0\colon X_{ij} \sim po(\lambda_{ij}) \longrightarrow M_1\colon X_{ij} \sim po(\lambda_{..}\rho_i\sigma_j) \quad \nearrow \searrow \quad M_3\colon X_{ij} \sim po(\lambda_{..}\tfrac{1}{rs})$$

$$M_2^c\colon X_{ij} \sim po(\lambda_{..}\tfrac{1}{r}\sigma_j)$$

The hypothesis of *multiplicative* effect (or no interaction) of the two factors,

$$H_{01}\colon \lambda_{ij} = \lambda_{\bullet\bullet}\rho_i\sigma_j, \qquad i=1,\dots,r, \quad j=1,\dots,s,$$

corresponds to the reduction from M_0 to M_1 and it is tested by means of the test statistic,

$$-2\ln Q(\mathbf{x}) = 2[\sum_{i=1}^{r}\sum_{j=1}^{s} x_{ij}\ln(x_{ij}) - \sum_{i=1}^{r} x_{i\bullet}\ln(x_{i\bullet}) - \sum_{j=1}^{s} x_{\bullet j}\ln(x_{\bullet j}) + x_{\bullet\bullet}\ln(x_{\bullet\bullet})].$$

If the expected frequencies under M_1,

$$(\mathbf{e}_1)_{ij} = \frac{x_{i\bullet}x_{\bullet j}}{x_{\bullet\bullet}},$$

are all greater than or equal to 5, the significance probability may be calculated as

$$p_{obs}(\mathbf{x}) = 1 - F_{\chi^2((r-1)(s-1))}(-2\ln Q(\mathbf{x})).$$

The hypothesis of *no column effect,*

$$H_{0\sigma}\colon \boldsymbol{\sigma}^* = (\sigma_1,\dots,\sigma_j,\dots,\sigma_s) = (\frac{1}{s},\cdots,\frac{1}{s},\cdots,\frac{1}{s}),$$

corresponds in the model M_1 to the reduction to M_2^r and here it is tested by means of the quantity,

$$-2\ln Q(\mathbf{x}) = 2[\sum_{j=1}^{s} x_{\bullet j}\ln(x_{\bullet j}) - x_{\bullet\bullet}\ln(\frac{x_{\bullet\bullet}}{s})].$$

The significance probability may be calculated as

$$p_{obs}(\mathbf{x}) = 1 - F_{\chi^2(s-1)}(-2\ln Q(\mathbf{x}))$$

provided that $\mathbf{x_{\bullet\bullet}/s \geq 5}$.

The hypothesis of *no row effect,*

$$H_{0\rho}\colon \boldsymbol{\rho}^* = (\rho_1,\dots,\rho_i,\dots,\rho_r) = (\frac{1}{r},\cdots,\frac{1}{r},\cdots,\frac{1}{r}),$$

corresponds in the model M_1 to the reduction to M_2^c and here it is tested by considering

$$-2\ln Q(\mathbf{x}) = 2[\sum_{i=1}^{r} x_{i\bullet}\ln(x_{i\bullet}) - x_{\bullet\bullet}\ln(\frac{x_{\bullet\bullet}}{r})].$$

If $\mathbf{x_{\bullet\bullet}/r \geq 5}$, the significance probability may be calculated as

$$p_{obs}(\mathbf{x}) = 1 - F_{\chi^2(r-1)}(-2\ln Q(\mathbf{x})).$$

In the model M_2^r the hypothesis of *no row effect* corresponds to the reduction to M_3 and it is tested in this model by considering the quantity,

$$-2\ln Q(\mathbf{x}) = 2[\sum_{i=1}^{r} x_{i\bullet}\ln(x_{i\bullet}) - x_{\bullet\bullet}\ln(\frac{x_{\bullet\bullet}}{r})].$$

If $\mathbf{x_{\bullet\bullet}/r \geq 5}$, the significance probability is calculated as

$$p_{obs}(\mathbf{x}) = 1 - F_{\chi^2(r-1)}(-2\ln Q(\mathbf{x})).$$

In the model M_2^c the hypothesis of *no column effect* corresponds to the reduction to M_3. If $\mathbf{x_{\bullet\bullet}/s \geq 5}$, the test statistic,

$$-2\ln Q(\mathbf{x}) = 2[\sum_{j=1}^{s} x_{\bullet j}\ln(x_{\bullet j}) - x_{\bullet\bullet}\ln(\frac{x_{\bullet\bullet}}{s})],$$

is used and the corresponding significance probability is

$$p_{obs}(\mathbf{x}) = 1 - F_{\chi^2(s-1)}(-2\ln Q(\mathbf{x})).$$

Exercises for Chapter 8

Exercise 8.1 Assume that $X_1, \ldots, X_n$ are independent and identically Poisson distributed with parameter λ, $X_j \sim po(\lambda)$, $j = 1, \ldots, n$.

(1) Show that the $-2\ln Q$-test statistic of the point hypothesis,

$$H_0\colon \lambda = \lambda_0$$

is

$$-2\ln Q(\mathbf{x}) = 2\left[x.\ln(\frac{\bar{x}.}{\lambda_0}) + n\lambda_0 - n\bar{x}.\right] \sim\sim \chi^2(1).$$

(2) Show that this quantity is also the $-2\ln Q$-test statistic of the hypothesis $H_0\colon \lambda = \lambda_0$ in the model $X. \sim po(n\lambda)$.

The message in this exercise is that as long as we believe that the sample is Poisson distributed, all the information about the mean λ is contained in the sum of the observations and the sample size.

Exercise 8.2 The data are from E. Rutherford, and M. Geiger (1910). Rutherford and Geiger counted the number of scintillations caused by radioactive decay of a quantity of the element *polonium* in a total of 2608 time intervals all of length 72 seconds. Data are seen in Table 8.6 below.

Table 8.6 *The number of scintillations in 2608 time intervals all of length 72 seconds is given in table form. Furthermore, the number of observations n, the sum S, and the uncorrected sum of squares USS are given.*

j	a_j
0	57
1	203
2	383
3	525
4	532
5	408
6	273
7	139
8	45
9	27
10	10
11	4
12	0
13	1
14	1
n	2608
S	10097
USS	48727

(1) Show by considering Fisher's index of dispersion or the goodness-of-fit test that it may be assumed that the 2608 observations may be considered as a sample from the Poisson distribution.

(2) Give an estimate of the number of scintillations in a time interval of 72 seconds together with a 95% confidence interval for this number.

Assume that in a similar experiment with the quantity of polonium, 3151 scintillations were observed in 1000 time intervals all of length 60 seconds.

(3) Examine if it may be assumed that the number of scintillations per time *unit* is the same in the two experiments.

Exercise 8.3 In order to investigate how physical and geological conditions influence the transport of sand by the wind, Keld Rømer Rasmussen, Department of Earth Sciences, University of Aarhus, and Michael Sørensen, Department of Mathematical Sciences, University of Aarhus, conducted a series of experiments in the wind tunnel at the Department of Earth Sciences. By means of a laser Doppler anemometer (a speed indicator) connected to a computer, the time and the velocity with which grains of sand pass through a control volume of 1 mm^3 are recorded. The registration is carried out under different velocities v of the wind and in different heights h (in mm) over the surface of the sand.

Here we will only consider a few results from the very extensive experiment. We consider measurements corresponding to two velocities A and B of the wind of which A is the smallest. For the velocity A we have measurements in three heights ($h = 12$, 15, and 30), whereas for B we have measurements from the height $h = 12$ only. In all four cases the registrations are made in a period of 60 seconds.

Consider a particular combination of wind speed and height. Divide the period of 60 seconds into 60 successive intervals of 1 second and let X_i denote the number of sand grains passing the control volume in the ith interval, $i = 1,2,\dots,60$. If it is assumed that the time of the passages may be described by a Poisson process, then X_i, $i = 1,2,\dots,60$, will be independent and identically Poisson distributed. The observed frequencies x_i, $i = 1,2,\dots,60$, are given in table form for the four combinations of wind speed and height in Table 8.7 below.

(1) Show that for each combination of wind speed and height that the 60 observations may be considered as a sample from the Poisson distribution.

The number of grains passing the control volume is expected to depend on the wind speed and the height over the surface of the sand since the number is expected to increase with velocity and decrease with height.

(2) Examine by considering the two sample in the height 12 mm if the number of grains passing the control volume depends on the velocity of the wind.

(3) Examine by considering the three samples corresponding to the velocity A if it may be assumed that the number of grains passing the control volume is inversely proportional with the height.

Exercise 8.4 A biology student interested in the occurrence of great titmouse (*Parus major*) at three different localities counts the number of great titmice in time periods of the same length for all three localities. In an area of 15 hectares on Vestre Kirkegård 13 great titmice are observed. The next day 16 great titmice are observed in an area of 16 hectares in Universitetsparken. A week later the student observes 22 great titmice in an area of 36 hectares at Gellerup.

When answering the three questions below, it may be assumed that the observed numbers of great titmice are outcomes of three independent random variables that are Poisson distributed but not necessarily identically distributed.

(1) Show that it may be assumed that the mean number of great titmice per hectare is the same for Vestre Kirkegård and for Universitetsparken.

(2) Show that it may be assumed that the mean number of great titmice per hectare is same for the area at Gellerup as for the combined areas Vestre Kirkegård and Universitetsparken.

Table 8.7 *For each of the four combinations considered of velocity v and height h the number grains passing the control volume in 60 successive time intervals of length 1 second are given in table form. Furthermore, the number of observations n, the sum S, and the uncorrected sum of squares USS are given for the four samples.*

v	A	A	A	B
h	12	15	30	12
j	a_j	a_j	a_j	a_j
0	1	5	10	0
1	7	6	17	1
2	10	22	18	5
3	13	10	7	5
4	10	8	6	6
5	6	2	2	10
6	6	3	0	9
7	4	2	0	6
8	1	2	0	4
9	2	0	0	7
10	0	0	0	3
11	0	0	0	2
12	0	0	0	2
n	60	60	60	60
S	226	170	108	367
USS	1112	696	298	2683

(3) Give an estimate and the 95% confidence interval for the mean number of great titmice per hectare at the three localities and examine if it may be assumed that this number is 1.

Exercise 8.5 The data in this exercise are derived from a data set from The Open University (1981) and are concerned with "serious" earthquakes in a period of 75 years, from 1903 through 1977. An earthquake is considered as "serious" if its magnitude is at least 7.5 on the Richter scale, or if more than 1000 people are killed. Table 8.8 below shows the annual number of earthquakes for the 75-year period.

(1) Show by considering Fisher's index of dispersion or the goodness-of-fit test that it may be assumed that the 75 observations may be considered as a sample from the Poisson distribution.

(2) Give an estimate of the mean of the annual number of "serious" earthquakes in the period considered and a 95% confidence interval for this number.

Assume that in the next 25 years 23 "serious" earthquakes will occur.

(3) Examine if it may be assumed that the annual number of "serious" earthquakes in the next 25 years is the same as in the period from 1903 through 1977.

Exercise 8.6 Table 8.9 below shows for 5 Petri dishes the number of bacteria colonies in 16 small quadratic regions of the same area in the central part of the dish.

(1) Show by considering Fisher's index of dispersion that for each of the 5 Petri dishes is may be assumed that the 16 observations constitute a sample from the Poisson distribution.

(2) Show that it may be assumed that the mean of the number of bacteria colonies in the small quadratic regions is the same for all 5 Petri dishes.

Table 8.8 *The annual number of "serious" earthquakes in the 75 years from 1903 through 1977, both years included, is shown in table form. Furthermore, the number of observations n, the sum S, and the uncorrected sum of squares USS are given.*

j	a_j
0	31
1	28
2	14
3	1
4	1
n	75
S	63
USS	109

Table 8.9 *For 5 Petri dishes the number of bacteria colonies in 16 small quadratic regions of the same area in the central part of the dish are given. Furthermore, the number of observations n, the sum S, and the uncorrected sum of squares USS are given for each of the 5 Petri dishes.*

Dish	*Bacteria colonies*				n	S	USS
	2	3	4	2			
1	5	0	1	3	16	44	168
	1	2	4	2			
	4	3	1	7			
	6	1	5	1			
2	2	3	4	2	16	42	144
	4	1	2	3			
	2	3	2	1			
	6	2	3	2			
3	1	4	2	4	16	42	144
	2	3	3	1			
	2	1	1	5			
	2	2	1	4			
4	4	3	2	3	16	38	116
	2	3	5	3			
	1	2	1	0			
	2	4	0	1			
5	5	1	3	2	16	35	111
	4	2	0	2			
	4	3	1	1			

(3) Give an estimate of and a 95% confidence interval for the common mean of the number of bacteria colonies in the small quadratic regions.

Exercise 8.7 In connection with an investigation of the common tern (*S. fluviatilis*), the type of surroundings of the nest was recorded together with an indication of the quality of the nest.

For the surroundings of the nest, 5 levels based on color were considered: *S1*: green plants; *S2*: grey-blue, mostly pebbles and chert; *S3*: gray, mostly sand; *S4*: mottled, water-rolled gravel; *S5*: brown, mostly sand. By the quality of the nest, 3 levels were considered: *Q1*: no nest-building; *Q2*: attempts at nest-building; *Q3*: finished nest. Table 8.10 is a result of the classification:

Table 8.10 *Classification of the common terns (S. fluviatilis) according to the surroundings and quality of the nest*

		Nest-Building			
		Q1	*Q2*	*Q3*	*Total*
	S1	13	14	29	56
	S2	20	31	21	72
Surroundings	*S3*	17	24	15	56
	S4	52	45	49	146
	S5	30	13	32	75
	Total	132	127	146	405

(1) Examine whether there is independence between the quality of the nest-building and the surroundings of the nest.

Exercise 8.8 In connection with an investigation of allergic children, the occurrence of spores of different fungi was observed in their homes.

The experiment was carried out by placing three Petri dishes in the home of each child for 20 minutes. This was done on a particular day every month during a year. The spores deposited in the dishes during the 20 minutes were grown, and for each child and each month, it was investigated which fungi occurred and the number of colonies for each fungus was recorded.

The data are from Andersen and Keiding (1983), and in Table 8.11 the total number of colonies of *Penicillium sp.* in the three dishes are given for three children living in Brønshøj, Lyngby, and Slangerup, respectively. (The fungus *Penicillium sp.* flourishes in house dust, insulating materials, upholstery, etc., and has spores that are very small and light.)

It is expected that the number of colonies vary between locations and, furthermore, that the number of colonies vary with the time of year.

(1) Show that it may be assumed that the data can be described by the multiplicative Poisson model.

(2) Examine if the number of colonies vary according to the season.

(3) Examine if the number of colonies vary with the location.

Exercise 8.9 The data in Table 8.12 have been placed at our disposal by Christian Kronborg, Department of Earth Sciences, University of Aarhus, and they come from a collection of fine-gravel particles in a profile at the Melbjerg Hoved location. In three different layers of the profile, two collections have been made. The layer from 18 to 20 m, here referred to as the upper layer, and the layer from 11 to 14 m, the middle layer, both consist of moraine sand, whereas the lower layer, from 4 to 11 m consists of moraine clay. The fine-gravel particles in each of the six samples have

Table 8.11 *The total number of colonies of Penicillium sp. in three Petri dishes for three children living in three different locations*

Month	Location *Brønshøj*	*Lyngby*	*Slangerup*	*Total*
January	14	8	20	42
February	26	10	19	55
March	16	12	12	40
April	12	11	30	53
May	10	6	18	34
June	17	6	26	49
July	10	3	16	29
August	20	8	19	47
September	15	15	19	49
October	17	9	22	48
November	14	7	14	35
December	8	9	23	40
Total	179	104	238	521

Table 8.12 *The composition of two samples of fine-gravel particles in the upper layer, the middle layer, and the lower layer, respectively, of a profile*

Sample	*Cryst.*	*Sedim.*	*Chert*	*Quarts*	*Total*
*U*1	159	19	94	41	313
*U*2	142	12	100	51	305
*M*1	87	23	46	133	289
*M*2	104	14	64	127	309
*L*1	207	48	15	30	300
*L*2	205	61	12	28	306

been classified according to the categories fragments of crystalline rocks, fragments of sedimentary rocks, chert and quarts. The observed numbers are shown in Table 8.12:

(1) Show for each of the three layers that the data may de described by the multiplicative Poisson model.

(2) Examine whether the composition of the samples varies from layer to layer.

Exercise 8.10 Creutzfeldt-Jakob disease is the human variant of *bovine spongifom encephalopathy* (BSE) in cattle. Creutzfeldt-Jakob disease may be hereditary and most frequently occurs among elderly people. It is not known with certainty if, and if so how, BSE may be transmitted from cattle to humans.

In England the systematic surveillance of Creutzfeldt-Jakob disease was reinstituted in 1990 in connection with the epidemic occurrence of BSE at that time. Special attention was paid to cases among younger people and to the so-called sporadic cases, i.e., cases not ascribable to heredity or treatment with growth hormone.

Table 8.13 below is from Will et al. (1996) and it shows all sporadic cases of Creutzfeldt-Jakob disease among persons under 45 years in UK in the period 01.01.1970 - 31.01.1996.

Table 8.13 *32 cases of Creutzfeldt-Jakob disease classified according to age and year of diagnosis*

	Age at diagnosis		
Year of diagnosis	≤34	35–44	*Total*
1970–1989	4	15	19
1990–1996*	9	4	13
Total	13	19	32

**Until and including January* 1996

(1) Examine if there is a connection between age at diagnosis and period of diagnosis for the 32 cases of Creutzfeldt-Jakob disease.

2) Examine for the age group 35 to 44 years if there is difference between the occurrence of the disease in the two periods of diagnosis.

(3) Let X_{ij}, $i = 1,2$, $j = 1,2$, be independent and Poisson distributed random variables with mean $EX_{ij} = \lambda_{ij} m_i$ where m_1 and m_2 are known positive numbers, while all the λ_{ij}'s are unknown parameters. This defines the basic model M_0.

Consider the hypotheses below:

H_{1a}: $\quad EX_{i1} = \gamma_{i1} m_i, EX_{i2} = \gamma_2 m_i, \gamma_{i1} > 0, \gamma_2 > 0, \quad i = 1,2.$

H_{1b}: $\quad EX_{ij} = \lambda \alpha_i \beta_j m_i, \lambda > 0, \alpha_i m_i > 0$ with $\alpha_1 m_1 + \alpha_2 m_2 = 1$ and $\beta_j > 0$ with $\beta_1 + \beta_2 = 1, \; i = 1,2, \; j = 1,2.$

H_2: $\quad EX_{ij} = \gamma_j m_i, \gamma_j > 0, \; i = 1,2 \; j = 1,2,$

which together with the assumption of independent Poisson distributions define the models M_{1a}, M_{1b}, and M_2. The following relation exists between the models:

$$M_0 \to M_{1a} \to M_2$$

and

$$M_0 \to M_{1b} \to M_2.$$

Find the maximum likelihood estimators and the likelihood ratio test statistics for the two sequences of models and discuss how these tests are related to the tests in (1) and (2).

CHAPTER 9

Generalized Linear Models

In Chapter 4 we got an impression of the versatility of the linear normal model. From a theoretical point of view, this class of models is relatively simple and this may also be said about the calculations required by the users of those models provided that they can use a statistical computer package, for example, the procedure *PROC GLM* in SAS. However, as the name of this class of models indicates, they can be applied only in connection with normally distributed data. It is therefore natural to investigate whether the relatively simple structure in linear normal models may be generalized to data that are not normally distributed, such as, for instance, discrete data. In this chapter we give a brief introduction to a generalization of the linear normal model, which in literature is known as *the generalized linear model*. Some statistical computer packages contain procedures that can perform the calculations in this class of models. This also applies to SAS where the calculations may be carried out by means of the procedure *PROC GENMOD*.

In Section 9.1 we introduce the classes of distributions that may be used to build a generalized linear model, and it is shown that the normal distributions, the Poisson distributions, the binomial distributions, the gamma distributions, and the inverse Gaussian distributions are among these classes of distributions. There are several others than the five mentioned, but these will suffice here. The generalized linear model is defined and discussed in Section 9.2. It is not only with respect to the distributional assumption that the generalized linear model is more extensive than the linear normal model. The latter is specified by assuming that the vector of means belongs to a linear subspace. A generalized linear model is specified by assuming that a vector obtained by a transformation of the vector of means by the so-called *link function* belongs to an affine subspace. Since the link function may be chosen in several ways, the generalized linear model is a very extensive model type. Estimation, tests and calculations by means of *PROC GENMOD* are discussed in the subsections 9.2.1 to 9.2.3, respectively, and in 9.2.4 *PROC GENMOD* is compared with *PROC GLM*.

In Section 9.3 we give a few examples of application of the generalized linear model. Among the examples are the multiplicative Poisson model, the Poisson model with proportional parameters, and regression analysis in the Poisson model. Furthermore, it is shown how calculations in the multinomial distribution may be performed by means of *PROC GENMOD*. Finally, the logistic dose-response model is mentioned.

9.1 Classes of Distributions

In this section we define and give examples of the classes of distributions that are considered in connection with generalized linear models, and that include both classes of continuous distributions and classes of discrete distributions.

If X is a discrete random variable with probability function f, or if X is a continuous random variable with probability density function f, we consider a class of distributions determined by the assumption that f may be written in the form,

$$f(x;\theta,\phi) = \exp\{(x\theta - b(\theta))/a(\phi) + c(x,\phi)\}, \tag{9.1}$$

where θ and ϕ are parameters and where a, b and c are functions that are characteristic for the class of distributions in question. The parameter θ is called the *canonical* parameter, while the parameter

ϕ is referred to as *the dispersion* in cases where the function a is of the form,

$$a(\phi) = \frac{\phi}{w}, \tag{9.2}$$

where w is a constant.

Several well-known classes of distributions that are often considered in applications are of the form (9.1) for certain choices of the functions a, b, and c. Below we give five examples in Example 9.1 to Example 9.5.

We use the notation $X \sim F(\theta, \phi)$ if X has probability function or probability density function f of the form (9.1), and we let $\mathcal{F}$ denote the class of distributions of this type, i.e.,

$$\mathcal{F} = \{F(\theta, \phi) \,|\, \theta \in \Theta, \phi \in \Phi\},$$

where $\Theta \subseteq \mathbb{R}$ and $\Phi \subseteq]0, \infty[$ and we implicitly assume that the functions a, b, and c are given.

If $X \sim F(\theta, \phi)$, it may be shown that

$$\mu = EX = b'(\theta) \tag{9.3}$$

and

$$Var X = a(\phi) b''(\theta). \tag{9.4}$$

Furthermore, it may be shown that there is a one-to-one correspondence between the *mean* parameter μ and the canonical parameter θ, i.e., the function b' is one-to-one. The inverse function of b' is denoted by

$$\theta = \theta(\mu). \tag{9.5}$$

The function,

$$V(\mu) = b''(\theta(\mu)),$$

is called the *variance function*.

The *linear* parameter in a generalized linear model is defined as

$$\eta = g(\mu) \tag{9.6}$$

where g is a one-to-one function referred to as the *link function*. If the linear parameter η is equal to the canonical parameter θ, the corresponding link function, i.e., the function in (9.5), is called the *canonical link function*.

The five distributions considered in Example 9.1 to Example 9.5 below are optional in *PROC GENMOD*. Furthermore, the procedure allows both user-defined distributions and link functions.

Example 9.1

The class of all normal distributions may be applied in connection with generalized linear models. If $X \sim N(\mu, \sigma^2)$, it is seen from

$$\begin{aligned} f(x; \mu, \sigma^2) &= (2\pi\sigma^2)^{-1/2} \exp\{-(x-\mu)^2/(2\sigma^2)\} \\ &= \exp\{(x\mu - \mu^2/2)/\sigma^2 - 1/2(x^2/\sigma^2 + \ln(2\pi\sigma^2))\}, \quad x \in (-\infty, \infty), \end{aligned} \tag{9.7}$$

that the probability density function of X is of the form (9.1) with

$$\theta = \mu, \tag{9.8}$$

$\phi = \sigma^2$, $a(\phi) = \phi$, $b(\theta) = \theta^2/2$ and $c(x, \phi) = -1/2(x^2/\phi + \ln(2\pi\phi))$.

Since $EX = \mu$, it follows from (9.8) that the canonical link function for the normal distribution is the identity, i.e., $\eta = \theta = \mu$. Note that the domain of variation of η is $\mathbb{R}$.

If $X_1, \ldots, X_n$ are independent and identically $N(\mu, \sigma^2)$-distributed, then the average $\bar{X}.$ is $N(\mu, \sigma^2/n)$-distributed. Thus, the probability density function of $\bar{X}.$ is of the form (9.1) with $w = n$. □

Example 9.2
The probability function of the Poisson distribution with parameter λ may be rewritten in the following way:

$$\begin{aligned} f(x;\lambda) &= \exp\{-\lambda\}\frac{\lambda^x}{x!} \\ &= \exp\{x\ln(\lambda) - \lambda - \ln(x!)\}, \quad x \in \{0,1,2,\ldots\}. \end{aligned} \tag{9.9}$$

Consequently, the probability function is of the form (9.1) with

$$\theta = \ln(\lambda), \tag{9.10}$$

$\phi = 1$, $a(\phi) = 1$, $b(\theta) = \lambda = e^{\theta}$ and $c(x,\phi) = -\ln(x!)$.

Since $\mu = EX = \lambda$, it follows from formula (9.10) that the canonical link function for the Poisson distribution is the natural logarithmic function ln, i.e., $\theta(\mu) = \ln(\mu)$. (It should be noted that the natural logarithm often is denoted by log, whereas the symbol $\log_{10}$ is used for logarithms to the base 10. SAS uses `log` and `log10` to denote these functions.) Since λ varies in $]0,\infty[$, we have that $\eta = \ln(\lambda)$ varies in $\mathbb{R}$. □

Example 9.3
The binomial distribution is also considered in connection with generalized linear models, however, first after a simple transformation of the distribution. If $Y \sim b(n,\pi)$, i.e., if Y denotes the number of successes in the n trials in a binomial experiment, one considers $X = Y/n$, the *relative* number of successes in the n trials. The probability function of X is

$$\begin{aligned} f(x;n,\pi) &= P(X = x) \\ &= P(Y = nx) \\ &= \binom{n}{nx}\pi^{nx}(1-\pi)^{n-nx} \\ &= \exp\{xn\ln(\frac{\pi}{1-\pi}) + n\ln(1-\pi) + \ln(\binom{n}{nx})\}, \quad x \in \{0,\frac{1}{n},\frac{2}{n},\ldots,1\}. \end{aligned} \tag{9.11}$$

Setting

$$\theta = \ln(\frac{\pi}{1-\pi}), \tag{9.12}$$

we have that $\pi = e^{\theta}/(1+e^{\theta})$ and $1-\pi = 1/(1+e^{\theta})$ and it follows that

$$f(x;n,\theta) = \exp\{(\theta x - \ln(1+e^{\theta}))/(1/n) + \ln(\binom{n}{nx})\};$$

i.e., the probability function of X is of the form (9.1) with $\phi = 1/n$, $a(\phi) = \phi$, $b(\theta) = \ln(1+e^{\theta})$ and $c(x,\phi) = \ln(\binom{\phi^{-1}}{\phi^{-1}x})$.

Since $\mu = EX = \pi$, it is seen from (9.12) that the canonical link function is

$$\theta(\mu) = \ln(\frac{\mu}{1-\mu}). \tag{9.13}$$

This function is referred to as the *logit* function. Since $\mu = \pi \in]0,1[$, formula (9.13) implies that the domain of variation of $\eta = \ln(\mu/(1-\mu))$ is $\mathbb{R}$. □

Example 9.4
The *gamma* distribution $\Gamma(v,\lambda)$ with parameters $v > 0$ and $\lambda > 0$ has the probability density func-

tion,

$$f(x;\nu,\lambda) = \frac{\lambda^\nu}{\Gamma(\nu)} x^{\nu-1} \exp\{-\lambda x\}, \quad x \in (0,\infty). \tag{9.14}$$

If $X \sim \Gamma(\nu,\lambda)$, then

$$\mu = EX = \frac{\nu}{\lambda} \quad \text{and} \quad Var X = \frac{\nu}{\lambda^2}.$$

Rewriting the probability density function,

$$\begin{aligned} f(x;\nu,\lambda) &= \frac{\lambda^\nu}{\Gamma(\nu)} x^{\nu-1} \exp(-\lambda x) \\ &= \frac{1}{\Gamma(\nu)x} (\frac{\nu x}{\mu})^\nu \exp(-\frac{\nu}{\mu} x) \\ &= \exp\{\nu(-\frac{1}{\mu}x + \ln(\frac{1}{\mu})) + \nu \ln(\nu x) - \ln(x) - \ln(\Gamma(\nu))\}, \end{aligned}$$

it is seen that the probability density function is of the form (9.1) with the canonical link function,

$$\theta(\mu) = -\frac{1}{\mu}, \tag{9.15}$$

and with $a(\phi) = \phi = \nu^{-1}$, $b(\theta) = -\ln(-\theta)$ and $c(x,\phi) = \phi^{-1}\ln(\phi^{-1}x) - \ln(x) - \ln(\Gamma(\phi^{-1}))$, respectively.

Since $b'(\theta) = -\theta^{-1}$ and $b''(\theta) = \theta^{-2}$, it follows that the variance function of the gamma distribution is $V(\mu) = \mu^2$.

If $X_1, \ldots, X_n$ are independent and identically $\Gamma(\nu,\lambda)$-distributed, it may be shown that the average $\bar{X}.$ is $\Gamma(n\nu, n\lambda)$-distributed. Thus, the probability density function of $\bar{X}.$ is of the form (9.1) with $w = n$.

Important special cases of the gamma distribution are the *exponential distribution* $e(\lambda)$, which is the distribution $\Gamma(1,\lambda)$, the $\chi^2(f)$-distribution, which is the distribution $\Gamma(f/2, 1/2)$, and the $\sigma^2\chi^2(f)/f$-distribution, which is the distribution $\Gamma(f/2, f/(2\sigma^2))$. □

Example 9.5

A positive continuous random variable X is said to be *inverse Gaussian*-distributed with parameters $\chi > 0$ and $\psi > 0$, $X \sim N^-(\chi,\psi)$, if the probability density function of X is

$$f(x;\chi,\psi) = \frac{\sqrt{\chi}}{\sqrt{2\pi}} x^{-3/2} \exp\{\sqrt{\chi\psi} - \frac{1}{2}(\chi x^{-1} + \psi x)\}, \quad x \in (0,\infty). \tag{9.16}$$

If $X \sim N^-(\chi,\psi)$, we have

$$\mu = EX = \frac{\sqrt{\chi}}{\sqrt{\psi}} \quad \text{and} \quad Var X = \frac{\sqrt{\chi}}{\sqrt{\psi^3}}.$$

Rewriting the probability density function as

$$f(x;\chi,\psi) = (2\pi\chi^{-1}x^3)^{-1/2} \exp\{-\frac{1}{2}\chi x^{-1}\} \exp\{\chi(-\frac{1}{2}\frac{\psi}{\chi}x + \frac{\sqrt{\psi}}{\sqrt{\chi}})\},$$

it follows that the probability density function of X is of the form (9.1) with the canonical link function,

$$\theta(\mu) = -\frac{1}{2}\frac{\psi}{\chi} = -\frac{1}{2}\mu^{-2}, \tag{9.17}$$

$a(\phi) = \phi = \chi^{-1}$, $b(\theta) = -(-2\theta)^{1/2}$ and $c(x,\phi) = -1/2(\ln(2\pi\phi x^3) + \phi^{-1}x^{-1})$. The class of inverse

Some characteristic quantities of the 5 distributions most commonly used in connection with generalized linear models

Distribution	*Normal*	*Poisson*	*Binomial*	*Gamma*	*Inverse Gaussian*
Notation	$N(\mu,\sigma^2)$	$po(\lambda)$	$b(n,\pi)/n$	$\Gamma(v,\lambda)$	$N^-(\chi,\psi)$
Density (formula)	(9.7)	(9.9)	(9.11)	(9.14)	(9.16)
Support	$(-\infty,\infty)$	$0(1)\infty$	$0(1)n/n$	$(0,\infty)$	$(0,\infty)$
Mean: μ	μ	λ	π	v/λ	$(\chi/\psi)^{1/2}$
Dispersion: ϕ	σ^2	1	$1/n$	v^{-1}	χ^{-1}
Canonical link: $\theta(\mu)$	μ	$\ln(\mu)$	$\ln(\mu/(1-\mu))$	$-\mu^{-1}$	$-1/2\mu^{-2}$
$a(\phi)$	ϕ	ϕ	ϕ	ϕ	ϕ
$b(\theta)$	$\theta^2/2$	$\exp(\theta)$	$\ln(1+\exp(\theta))$	$-\ln(-\theta)$	$-(-2\theta)^{1/2}$
$c(x,\phi)$	$-1/2(x^2/\phi+\ln(2\pi\phi))$	$-\ln(x!)$	$\ln(\binom{\phi^{-1}}{\phi^{-1}x})$	$\phi^{-1}\ln(\phi^{-1})-\ln(x)$ $-\ln(\Gamma(\phi^{-1}))$	$-1/2(\ln(2\pi\phi x^3)+\phi^{-1}x^{-1})$
$\mu(\theta)$	θ	$\exp(\theta)$	$\exp(\theta)/(1+\exp(\theta))$	$-\theta^{-1}$	$(-2\theta)^{-1/2}$
Variance function: $V(\mu)$	1	μ	$\mu(1-\mu)$	μ^2	μ^3
"Scale parameter"	$\phi^{1/2}$	1	1	ϕ^{-1}	$\phi^{1/2}$

Gaussian distributions may therefore be applied in connection with generalized linear models. Expressed in terms of the mean μ and the dispersion ϕ the probability density function of X is

$$f(x;\mu,\phi) = (2\pi\phi x^3)^{-1/2}\exp\{-\frac{1}{2}\phi^{-1}x^{-1}\}\exp\{\phi^{-1}(-\frac{1}{2}\mu^{-2}x+\mu^{-1})\}.$$

Furthermore, using the expression for the function b, it may be shown that the variance function of the inverse Gaussian distribution is $V(\mu) = \mu^3$.

If $\bar{X}.$ denotes the average of n independent and identically $N^-(\chi,\psi)$-distributed random variables, it may be shown that $\bar{X}. \sim N^-(n\chi, n\psi)$, and it follows that the probability density function of $\bar{X}.$ is of the form (9.1) with $w = n$. □

In the table on page 405, characteristic quantities of the five distributions in Example 9.1–Example 9.5 are given. The parameters in the row "scale parameter" are taken from Chapter 29, "The GENMOD Procedure", in the SAS manual. The name scale parameter seems to be misleading since this parameter is not always the scale parameter in the usual sense. For instance, in the gamma distribution the "scale parameter" is $\phi^{-1} = \nu$. Usually, this parameter is referred to as the index parameter of the gamma distribution, whereas the parameter λ is the scale parameter.

9.2 The Generalized Linear Model

Let $x_1,\ldots,x_n$ be realizations of independent random variables, $X_1,\ldots,X_n$, such that

$$X_i \sim F(\theta_i,\phi), \quad i = 1,\ldots,n,$$

where F belongs to one and the same of the distribution classes $\mathcal{F}$ introduced in Section 9.1. We assume that the condition in (9.2) is satisfied and, furthermore, that the domain of variation of the linear parameter η is $\mathbb{R}$.

Note that it is assumed that all the observations have the same dispersion ϕ. This assumption has to be carefully checked if possible. Sometimes this control is superfluous since the class of distributions considered has a dispersion that is constant. This applies, for instance, to the class of Poisson distributions and to the class of binomial distributions.

The observations, $x_1,\ldots,x_n$, are considered as a vector in $\mathbb{R}^n$,

$$\mathbf{x} = (x_1,\ldots,x_n)^*,$$

and, similarly, the means and the linear parameters are considered as vectors,

$$\boldsymbol{\mu} = (\mu_1,\ldots,\mu_n)^*$$

and

$$\boldsymbol{\eta} = (\eta_1,\ldots,\eta_n)^*,$$

which are called *the vector of means* and *the vector of linear parameters*, respectively.

The models are specified by assuming that the vector of linear parameters $\boldsymbol{\eta}$ belongs to an *affine subspace* A of $\mathbb{R}^n$ that is a subset of $\mathbb{R}^n$ of the form,

$$A = \boldsymbol{\omega} + L = \{\mathbf{y} \in \mathbb{R}^n \,|\, \mathbf{y} = \boldsymbol{\omega} + \mathbf{l},\ \mathbf{l} \in L\},$$

where $\boldsymbol{\omega}$ is a vector in $\mathbb{R}^n$ and L a linear subspace of $\mathbb{R}^n$. The vector $\boldsymbol{\omega}$ is often (for instance in SAS) referred to as the vector of *offset* values. The dimension of the affine subspace A, i.e., the dimension of the linear subspace L is denoted by d.

By the generalized linear model M specified by the affine subspace A, we understand that $x_1,\ldots,$ x_n are realizations of independent random variables, $X_1,\ldots,X_n$, whose distributions are $X_i \sim F(\theta_i,\phi)$ and, furthermore, that

$$\boldsymbol{\eta} = (\eta_1,\ldots,\eta_n)^* \in A.$$

In brief notation we write

$$M\colon \boldsymbol{\eta} \in A,$$

since we implicitly assume that the observations are independent and distributed according to (9.1).

9.2.1 Estimation

In the generalized linear model $M\colon \boldsymbol{\eta} \in A$, the vector $\boldsymbol{\theta}$ of canonical parameters may be considered as a function of $\boldsymbol{\eta}$, the vector of linear parameters, i.e.,

$$\boldsymbol{\theta}(\boldsymbol{\eta}) = (\theta_1(\boldsymbol{\eta}),\ldots,\theta_n(\boldsymbol{\eta}))^*.$$

With this notation the likelihood function is

$$\begin{aligned} L(\boldsymbol{\eta},\phi) &= \prod_{i=1}^{n} \exp\{w_i[x_i\theta_i(\boldsymbol{\eta}) - b(\theta_i(\boldsymbol{\eta}))]/\phi + c(x_i,\phi)\} \\ &= \exp(\sum_{i=1}^{n} \{w_i[x_i\theta_i(\boldsymbol{\eta}) - b(\theta_i(\eta))]/\phi + c(x_i,\phi)\}), \end{aligned}$$

and therefore the log likelihood function becomes

$$l(\boldsymbol{\eta},\boldsymbol{\phi}) = \sum_{i=1}^{n} \{w_i[x_i\theta_i(\boldsymbol{\eta}) - b(\theta_i(\eta))]/\phi + c(x_i,\phi)\}.$$

In order to find the maximum likelihood estimates $\hat{\boldsymbol{\eta}}$ of $\boldsymbol{\eta}$ and $\hat{\phi}$ of ϕ, the log likelihood function has to be maximized for $\boldsymbol{\eta}$ varying in A and for ϕ varying in $]0,\infty[$. In most cases it is not possible to find explicit solutions to the likelihood equations that therefore have to be solved numerically. The reason for treating many different distributions and models under the framework of generalized linear models is that a common estimation algorithm for distributions of the form (9.1) was discovered. The task of solving the likelihood equations is handed over to statistical computer packages, in our case to SAS. The maximal value of the log likelihood function in the model M we will denote by $\hat{l}$, i.e.,

$$\hat{l} = l(\hat{\boldsymbol{\eta}},\hat{\phi}).$$

9.2.2 Test

A sequence of models is considered. For the ith model M_i we use the symbols A_i for its affine subspace, d_i for the dimension of A_i, and $\hat{l}_i$ for the maximal value of the log likelihood function. This is similar to the notation used for sequences of linear normal models, see page 134, and to the notation used for sequences of models in the multinomial distribution, see page 305.

In the basic model M_0 there are no restrictions on the domain of variation of $\boldsymbol{\eta}$, i.e.,

$$M_0\colon \boldsymbol{\eta} \in \mathbb{R}^n;$$

in others words $A_0 = \mathbb{R}^n$ and $d_0 = n$. M_0 is referred to as the full model or the saturated model, because there is one unknown parameter for each observation.

The sequence of models is

$$M_0 \to M_1 \to M_2 \to \cdots \to M_{i-1} \to M_i \to \cdots,$$

and it gives a gradually simpler description of the vector $\boldsymbol{\eta}$ of linear parameters or, equivalently, the corresponding affine subspaces constitute a decreasing sequence,

$$\mathbb{R}^n \supset A_1 \supset A_2 \supset \cdots \supset A_{i-1} \supset A_i \supset \cdots.$$

The hypothesis corresponding to the reduction,

$$M_{i-1} \to M_i,$$

is as usual denoted by H_{0i}. This hypothesis is tested by means of the $-2\ln Q$-test statistic as described in Section 11.3. In the notation introduced above the test statistic of H_{0i} is

$$-2\ln Q_i(\mathbf{x}) = 2(\hat{l}_{i-1} - \hat{l}_i), \tag{9.18}$$

and the corresponding significance probability may be approximated by

$$p_{obs}(\mathbf{x}) \doteq 1 - F_{\chi^2(d_{i-1}-d_i)}(-2\ln Q_i(\mathbf{x})). \tag{9.19}$$

Since the degrees of freedom in a χ^2-distribution is positive, it is seen from (9.19) that the following condition concerning the dimensions of the affine subspaces is necessary in order to be able to test all reductions:

$$n > d_1 > d_2 > \cdots > d_{i-1} > d_i > \cdots.$$

9.2.3 Calculations

Let $-2\ln Q_{0i}$ and f_{0i} denote the $-2\ln Q$-test statistic and the corresponding degrees of freedom for the reduction $M_0 \rightarrow M_i$, respectively, i.e.,

$$-2\ln Q_{0i}(\mathbf{x}) = 2(\hat{l}_0 - \hat{l}_i) \tag{9.20}$$

and

$$f_{0i} = n - d_i. \tag{9.21}$$

From (9.18) and (9.19) it is seen that the approximation of the significance probability of the $-2\ln Q$-test statistic for the reduction $M_{i-1} \rightarrow M_i$ may be calculated if we know the quantities in (9.20) and (9.21) for the two models M_{i-1} and M_i, since

$$\begin{aligned} -2\ln Q_i(\mathbf{x}) &= 2(\hat{l}_{i-1} - \hat{l}_i) \\ &= 2(\hat{l}_0 - \hat{l}_i) - 2(\hat{l}_0 - \hat{l}_{i-1}) \\ &= -2\ln Q_{0i}(\mathbf{x}) - (-2\ln Q_{0i-1}(\mathbf{x})) \end{aligned} \tag{9.22}$$

and

$$\begin{aligned} f_i &= d_{i-1} - d_i \\ &= (n - d_i) - (n - d_{i-1}) \\ &= f_{0i} - f_{0i-1}. \end{aligned} \tag{9.23}$$

Note that d, the dimension of the affine subspace A, may be interpreted as the number of (free) parameters in the model. Consequently, the degrees of freedom f_i in (9.23) of the $-2\ln Q$-test statistic in (9.22) is the number of free parameters in M_{i-1} (the model in which we test) minus the number of free parameters in M_i (the model under test), in agreement with formula (11.32).

The two quantities in (9.20) and (9.21) can be found in the output from the estimation of the model M_i in a statistical computer package that is able to analyze generalized linear models. In Table 9.1 and Table 9.2 the relevant part of the output from the SAS procedure *PROC GENMOD* is given for M_1 and M_2, respectively. This part of the output contains further information indicated in the tables by $\cdots$ since we are not making use of it here. The quantities (9.20) and (9.21) are found in the line `Deviance` under `Value` and `DF`, respectively. Furthermore, the maximal value of the log likelihood function under the model considered is found in the line `Log Likelihood` under `Value`.

9.2.4 PROC GLM or PROC GENMOD

The formulas in Section 9.2.3 show that there are formal points of resemblance between the analysis of linear normal models by means of the SAS procedure *PROC GLM* described in Chapter 4 and the

Table 9.1 *Part of the output from the SAS procedure PROC GENMOD for model M_1*

Criteria For Assessing Goodness Of Fit

	Criterion	DF	Value	Value/DF
$\rightarrow$	Deviance	$n-d_1$	$2(\hat{l}_0-\hat{l}_1)$	$2(\hat{l}_0-\hat{l}_1)/(n-d_1)$
	Scaled Deviance	...	...	...
	Pearson Chi-Square	...	...	...
	Scaled Pearson X2	...	...	...
$\rightarrow$	Log Likelihood	.	$\hat{l}_1$	.

Table 9.2 *Part of the output from the SAS procedure PROC GENMOD for model M_2*

Criteria For Assessing Goodness Of Fit

	Criterion	DF	Value	Value/DF
$\rightarrow$	Deviance	$n-d_2$	$2(\hat{l}_0-\hat{l}_2)$	$2(\hat{l}_0-\hat{l}_2)/(n-d_2)$
	Scaled Deviance	...	...	...
	Pearson Chi-Square	...	...	...
	Scaled Pearson X2	...	...	...
$\rightarrow$	Log Likelihood	.	$\hat{l}_2$	.

analysis of generalized linear models by means of the procedure *PROC GENMOD*. In the treatment of the examples in Section 9.3, it will be clear that the model for the vector of linear parameters is specified in *PROC GENMOD* in the same way as the model for the vector of means is specified in *PROC GLM*. In addition, there are further similarities in the interpretation of the estimates of the parameters in the output from the two procedures. It should be emphasized that there is a great difference between linear normal models and generalized linear models. As the name indicates, the first type of models can be applied only in connection with normally distributed data, whereas the second type of models may be used in the more general situation where the data may be described by a distribution of the form (9.1). Another important difference is that in linear normal models, the vector of means is assumed to belong to a linear subspace, whereas in generalized linear models the vector of linear parameters – not necessarily the vector of means – is assumed to belong to an affine subspace. Finally, in linear normal models the significance probabilities of the likelihood ratio tests can be calculated exactly using an F-distribution, whereas the significance probabilities of the likelihood ratio tests in generalized linear models rely on the approximation with the χ^2-distribution as shown in (9.19).

It is seen from Example 9.1 that a linear normal model may be considered as a generalized linear model. However, one should **not** do this since by doing so one misses the opportunity to use exact significance probabilities.

9.3 Examples

In 5 subsections we give examples of applications of the generalized linear model in simple situations. First we consider the multiplicative Poisson model in Section 9.3.1 and the Poisson model with proportional parameters in Section 9.3.2. Then a regression model for Poisson-distributed data is discussed in Section 9.3.3. In Section 9.3.4 it is shown how multinomial distributed data may be analyzed by means of *PROC GENMOD*. Finally, the logistic dose-response model is discussed in Section 9.3.5.

9.3.1 The Multiplicative Poisson Model

The multiplicative Poisson model was introduced in Section 8.4.2. Recall that the data was given as an $r \times s$ table for which we considered the basic model,

$$M_0 \colon X_{ij} \sim po(\lambda_{ij}), \quad i = 1,\ldots,r,\ j = 1,\ldots,s.$$

Furthermore, we considered the submodels M_1, M_2^r, M_2^c, and M_3. Here M_1 is the multiplicative Poisson model, M_2^r is the model with row effects only, M_2^c is the model with column effect only, and M_3 is the model for a single sample from the Poisson distribution. The relation between the models is shown in the following diagram, where we use the parametrization that was used to introduce the models on page 372 in Section 8.4.2:

$$\begin{array}{ccccc} & & & M_2^r\colon X_{ij} \sim po(\alpha_i) & \\ & & \nearrow & & \searrow \\ M_0\colon X_{ij} \sim po(\lambda_{ij}) & \longrightarrow & M_1\colon X_{ij} \sim po(\alpha_i\beta_j) & & M_3\colon X_{ij} \sim po(\lambda) \\ & & \searrow & & \nearrow \\ & & & M_2^c\colon X_{ij} \sim po(\beta_j) & \end{array}$$

We want to show that all five models are generalized linear models. The Poisson distribution is one of the distributions that can be used in generalized linear models as pointed out in Example 9.2. It remains for us to show that the restrictions on the means imposed by the models correspond to affine subspaces for the vector of linear parameters, which is obtained from the vector of means applying a suitable link function to each of its coordinates. The first choice is the natural logarithm, which is the canonical link function for the Poisson distribution, and with this choice the models are generalized linear models.

In M_0 the affine subspace A_0 is $\mathbb{R}^{rs}$, because

$$\boldsymbol{\eta} = (\ln\lambda_{11},\ldots,\ln\lambda_{rs})$$

maps $\boldsymbol{\mu} = (\lambda_{11},\ldots,\lambda_{rs}) \in]0,\infty[^{rs}$ onto $\mathbb{R}^{rs}$.

In M_1 the affine subspace A_1 is a vector space in $\mathbb{R}^{rs}$:

$$A_1 = \{\boldsymbol{\eta} \in \mathbb{R}^{rs} \,|\, \eta_{ij} = \ln\alpha_i + \ln\beta_j = \tilde{\alpha}_i + \tilde{\beta}_j,\ \tilde{\alpha}_i \in \mathbb{R},\ \tilde{\beta}_j \in \mathbb{R},\ i = 1,\ldots,r,\ j = 1\ldots,s\}.$$

The linear subspace A_1 is the subspace associated with the additivity model in a two-way analysis of variance, see page 208. The dimension of A_1 is $d_1 = r+s-1$.

Similarly, the affine subspaces A_2^r and A_2^c are the subspaces,

$$A_2^r = \{\boldsymbol{\eta} \in \mathbb{R}^{rs} \,|\, \eta_{ij} = \ln\alpha_i = \tilde{\alpha}_i,\ \tilde{\alpha}_i \in \mathbb{R},\ i = 1,\ldots,r\}$$

and

$$A_2^c = \{\boldsymbol{\eta} \in \mathbb{R}^{rs} \,|\, \eta_{ij} = \ln\beta_j = \tilde{\beta}_j,\ \tilde{\beta}_j \in \mathbb{R},\ j = 1\ldots,s\},$$

which are the linear subspaces associated with the models of row effect only and column effect only in the two-way analysis of variance on page 208. The dimension of A_2^r is $d_2^r = r$ and the dimension of A_2^c is $d_2^c = s$.

Finally, the affine subspace A_3 of M_3 is the one-dimensional linear subspace,

$$A_3 = \{\boldsymbol{\eta} \in \mathbb{R}^{rs} \,|\, \eta_{ij} = \ln\lambda = \tilde{\lambda},\ \tilde{\lambda} \in \mathbb{R}\}.$$

Example 9.6

We consider again the data in Example 8.3. The starting point for the analysis is the model,

$$M_0 \colon X_{ij} \sim po(\lambda_{ij}), \quad i = 1,2,3, \quad j = 1,2,$$

where i refers to the three depths (35 to 125 m, 125 to 250 m, 250 to 300 m) and j to the two times (day, night).

SAS program and output may be found in the Annex to Subsection 9.3.1 starting on page 413. Table 9.1 for these data is on page 414. It is seen from the table `Criteria For Assessing Goodness Of Fit` that the $-2\ln Q$-test statistic of the hypothesis $H_{01}: \eta_{ij} = \tilde{\alpha}_i + \tilde{\beta}_j$ corresponding to the hypothesis of no interaction is 0.5899 with 2 degrees of freedom. This is exactly what was found in the previous treatment of the example on page 379. The significance probability is not given in the SAS output. As before we find that it is 0.745 and we decide to adopt M_1 as the model. (If the calculations are made in the saturated model M_0, SAS gives this significance probability, see page 416.)

The estimates of the parameters in M_1 do not appear directly from the SAS output on page 414, but may be found in the table `Analysis Of Parameter Estimates` from the lines marked by `->` in the same somewhat troublesome way as when analyzing linear normal models with *PROC GLM*. In the line `Intercept` we find the estimate of $\tilde{\alpha}_3 + \tilde{\beta}_2$ (4.5980). From the lines `depth 035-125` and `depth 125-150` we obtain the estimates of $\tilde{\alpha}_1 - \tilde{\alpha}_3$ (-0.3771) and $\tilde{\alpha}_2 - \tilde{\alpha}_3$ (1.3850), respectively. Finally, in the line `time day` we find the estimate of $\tilde{\beta}_1 - \tilde{\beta}_2$ (-0.0793). From these quantities the estimate of $\boldsymbol{\eta}$ under M_1 may be calculated. For instance, since

$$\eta_{11} = \tilde{\alpha}_1 + \tilde{\beta}_1 = (\tilde{\alpha}_1 - \tilde{\alpha}_3) + (\tilde{\beta}_1 - \tilde{\beta}_2) + (\tilde{\alpha}_3 + \tilde{\beta}_2),$$

we get that

$$\eta_{11} \leftarrow -0.3771 + (-0.0793) + 4.5980 = 4.1416;$$

i.e., the expected frequency under M_1 in the cell (1,1) is

$$\lambda_{11} = e^{\eta_{11}} \leftarrow e^{4.1416} = 62.903,$$

which we also found on page 379.

As by *PROC GLM* the way of giving the estimates is actually clever since the values given in addition to the estimate under `Chi-Square` and `Pr > Chi` are Wald's X_W^2-test statistic, see formula (11.38) on page 492, and the significance probability of the hypothesis that the parameter in question is 0. Thus in the line `time day` we see that the significance probability of Wald's X_W^2-test statistic for the reduction from the model M_1 to M_2^r – corresponding to the hypothesis H_{02}^r: $\tilde{\beta}_1 = \tilde{\beta}_2$ – is 0.1919, so the hypothesis is not rejected.

The reason we can test the hypothesis H_{02}^r of no column effect using only *PROC GENMOD*'s table with the estimates in M_1 is of course that in this example we have two columns only. We therefore now show how the hypothesis may be tested using the $-2\ln Q$-test by means of the formulas (9.22) and (9.23). Using these and the table `Criteria For Assessing Goodness Of Fit` for M_1 and M_2^2 on pages 414 and 415, respectively, we find that the $-2\ln Q$-test statistic of H_{02}^r is

$$-2\ln Q(\mathbf{x}) = 2.2945 - 0.5899 = 1.7046.$$

The degree of freedom is $3-2=1$ so we find, just as on page 380, that the significance probability is 0.192.

The estimates in the model M_2^r may be found from the output listing from M_2^r on page 416. Note that the estimate of $\tilde{\alpha}_3$ (4.5591) is found in the line `Intercept`, $\tilde{\alpha}_1 - \tilde{\alpha}_3 \leftarrow -0.3771$ `(depth 035-125)` and $\tilde{\alpha}_2 - \tilde{\alpha}_3 \leftarrow 1.3850$ `(depth 125-250)`, so for example $\tilde{\alpha}_1 \leftarrow -0.3771 + 4.5591 = 4.1820$ and $\eta_{11} = \eta_{12} = \exp(4.1820) = 65.5$.

All relevant information concerning the test of the hypothesis H_{02}^r may actually be obtained directly from the output from M_1 starting on page 414 by using `TYPE1` or `TYPE3` as an option to the `MODEL` statement in *PROC GENMOD*. If the option `TYPE3` is used, we get the table `LR Statistics For Type 3 Analysis` on page 415. In this table there is a row for each variable entering the model specification. The numbers in a row under `DF`, `Chi-Square` and `Pr > Chi`, are the degrees of freedom, the $-2\ln Q$-test statistic, and the significance probability of the

$-2\ln Q$-test for the reduction from the specified model to the model, where the variable in question has be removed from the model. In our case the variable `time` corresponds to the columns and we see that the results of our calculations concerning the hypothesis of no column effect are summarized in this row. Furthermore, in the line `depth` in the table, it is seen that this variable cannot be omitted from the model M_1.

If the option `TYPE1` is used in the `MODEL` statement for M_1 in *PROC GENMOD*, we get a table with the heading `LR Statistics For Type 1 Analysis`, here reproduced on page 415. The quantities in the table depend on the order of the terms in the model formula. The number in the column `Deviance` in the ith row is the $-2\ln Q$-test statistic for the reduction from the specified model to the model, which contains all terms in the rows above and including the ith row. This is precisely similar to the type I sums of squares in *PROC GLM* as shown on page 203. Consequently, the difference between the numbers in the $(i-1)$th and the ith row in the column `Deviance` is the $-2\ln Q$-test statistic of the hypothesis that the term in the ith row can be removed from a model that contains all terms in the rows above and including the ith row. This test statistic, the corresponding degrees of freedom, and the corresponding significance probability is found in the ith row under `Chi-Square`, `DF` and `Pr > Chi`. In the example here the model M_1 includes all terms in the table and the variable `time` corresponds to the columns. In the line `time` it is seen that the $-2\ln Q$-test statistic of the hypothesis of no column effect is 1.70 with 1 degree of freedom corresponding to a significance probability of 0.1917. The model M_2^r comprises all terms above and including the row `depth`, and it is seen that the reduction $M_2^r \to M_3$ is not possible.

The latter may also be seen by means of the `TYPE1`-table for the model M_2^r on page 417.

A closing remark is that `TYPE1`-tables are generally more informative than `TYPE3`-tables. Using a convenient/clever way of choosing the order of the terms in the model formula, there is a chance that all lines in the `TYPE1`-table are relevant, whereas it is typically only a single line that is of interest in a `TYPE3`-table. Furthermore with our approach to testing in a sequence of models, some of the lines in a `TYPE3`-table may sometimes be difficult to interpret. This is illustrated on page 416 where the calculations are made in the saturated model M_0 using the model statement,

```
MODEL count= depth time depth*time;
```

which corresponds to writing the linear parameter of M_0 as

$$\eta_{ij} = \alpha_i + \beta_j + \gamma_{ij}.$$

Here the parameter γ_{ij} is referred to as the interaction between the ith value of `depth` and the jth value of `time` and α_i and β_j are the (main) effects of ith value of `depth` and the jth value of `time`, respectively. In SAS the interactions are specified as `depth*time`, and the reduction $M_0 \to M_1$ may be expressed as the hypothesis $\gamma_{ij} = 0$ for $i = 1,2,3$ and $j = 1,2$. The `TYPE1`-table on page 416 summarizes our calculations above for the reductions $M_0 \to M_1 \to M_2^r \to M_3$. The value of the $-2\ln Q$-test statistic and the corresponding p-value for the reductions $M_0 \to M_1$, $M_1 \to M_2^r$ and $M_2^r \to M_3$ are found in the lines `depth*time`, `time` and `depth`, respectively. (In contrast to page 414, SAS here gives the p-value 0.7446 for the reduction $M_0 \to M_1$.) The line `time` in the `TYPE3`-table gives the $-2\ln Q$-test statistic for the hypothesis that the term `time` may be removed from the model formula, i.e., for the reduction from M_0 to a model with linear parameter,

$$\eta_{ij} = \alpha_i + \gamma_{ij}.$$

However, models containing interactions without the corresponding main effects are difficult to interpret and therefore only the last line of the `TYPE3`-table, which contains information concerning the reduction $M_0 \to M_1$, is relevant. □

Annex to Subsection 9.3.1

Calculations in SAS

We show how *PROC GENMOD* may be used in connection with the multiplicative Poisson model by presenting the program that performs the calculations in Example 9.6.

Example 9.6 (Continued)
In the following program the data set is constructed and then the models M_1 and M_2^r are specified in *PROC GENMOD*.

```
OPTIONS NODATE PAGENO=1 LS=80 PS=100;

TITLE1 'Example 9.6 - multiplicative Poisson model';
/* same data as in Example 8.3 */

DATA water_flea;
INPUT depth$1-7 time$9-13 count;
DATALINES;
035-125 day    59
035-125 night  72
125-250 day   371
125-250 night 392
250-300 day    91
250-300 night 100
;
RUN;

TITLE2 'multiplicative model';
TITLE3 'M1';
PROC GENMOD DATA=water_flea;
     CLASS depth time;
     MODEL count= depth time /DIST=poisson
                              LINK=log
                              TYPE1
                              TYPE3;
RUN;

TITLE2 'row effect only';
TITLE3 'M2r';
PROC GENMOD DATA=water_flea;
     CLASS depth time;
     MODEL count= depth /DIST=poisson
                         LINK=log
                         TYPE1;
RUN;
```

The structure of the statements in *PROC GENMOD* is in many ways similar to the structure of the statements in *PROC GLM*. There is of course the important distinction that the class of distributions and the link function must be specified in *PROC GENMOD*. The is done as options to the `MODEL` statement. The class of Poisson distributions is selected with the keyword `poisson` in the `DIST` option and the natural logarithmic function is selected as the link function with the `log` keyword in the `LINK` option.

The model specification is completed when the affine subspace for the vector of linear parameters

η has been specified. Recall that the affine subspace is defined as

$$A = \omega + L,$$

where ω is an arbitrary vector in A and L is a linear subspace. The linear subspace L of the model is specified exactly as in *PROC GLM*. In this case the linear subspace is the additivity subspace of the two-way analysis of variance, and the subspace is specified using the column variable and the row variable as `CLASS` variables. The vector ω is specified naming the variable that contains the coordinates of ω in the `OFFSET` option. In this case A is a linear subspace and therefore no `OFFSET` option is necessary.

The options `TYPE1` and `TYPE3` are explained on page 411.

The output listing from the calculations in M_1 is:

```
                Example 9.6 - multiplicative Poisson model              1
                           multiplicative model
                                    M1

                           The GENMOD Procedure

                             Model Information

                  Data Set                      WORK.WATER_FLEA
                  Distribution                          Poisson
                  Link Function                             Log
                  Dependent Variable                      count
                  Observations Used                           6

                          Class Level Information

              Class       Levels     Values

              depth            3     035-125 125-250 250-300
              time             2     day night

                  Criteria For Assessing Goodness Of Fit

       Criterion                    DF           Value         Value/DF

       Deviance                      2          0.5899           0.2950
       Scaled Deviance               2          0.5899           0.2950
       Pearson Chi-Square            2          0.5889           0.2945
       Scaled Pearson X2             2          0.5889           0.2945
       Log Likelihood                        4869.8506

Algorithm converged.

                    Analysis Of Parameter Estimates

                                     Standard   Wald 95% Confidence       Chi-
Parameter              DF   Estimate      Error         Limits          Square

->Intercept             1     4.5980     0.0780     4.4451     4.7509  3473.23
->depth       035-125   1    -0.3771     0.1134    -0.5994    -0.1547    11.05
->depth       125-250   1     1.3850     0.0809     1.2264     1.5436   293.02
  depth       250-300   0     0.0000     0.0000     0.0000     0.0000      .
->time        day       1    -0.0793     0.0608    -0.1984     0.0398     1.70
  time        night     0     0.0000     0.0000     0.0000     0.0000      .
  Scale                 0     1.0000     0.0000     1.0000     1.0000

                    Analysis Of Parameter Estimates

                    Parameter                  Pr > ChiSq

                  ->Intercept                      <.0001
                  ->depth         035-125          0.0009
                  ->depth         125-250          <.0001
                    depth         250-300           .
                  ->time          day              0.1919
                    time          night             .
                    Scale

NOTE: The scale parameter was held fixed.
```

```
                    LR Statistics For Type 1 Analysis

                                                    Chi-
        Source            Deviance          DF     Square     Pr > ChiSq

        Intercept         631.5515
        depth               2.2945           2     629.26         <.0001
        time                0.5899           1       1.70         0.1917


                    LR Statistics For Type 3 Analysis

                                             Chi-
               Source              DF      Square     Pr > ChiSq

               depth                2      629.26         <.0001
               time                 1        1.70         0.1917
```

The calculations in the model M_2^r, corresponding to row effect only, are carried out by means of the program segment:

```
TITLE2 'row effect only';
TITLE3 'M2r';
PROC GENMOD DATA=water_flea;
     CLASS depth time;
     MODEL count= depth /DIST=poisson
                         LINK=log
                         TYPE1;
RUN;
```

The output listing from this program segment is:

```
                Example 9.6 - multiplicative Poisson model                      2
                             row effect only
                                  M2r

                          The GENMOD Procedure

                           Model Information

                Data Set                 WORK.WATER_FLEA
                Distribution                     Poisson
                Link Function                        Log
                Dependent Variable                 count
                Observations Used                      6


                        Class Level Information

             Class       Levels     Values

             depth            3     035-125 125-250 250-300
             time             2     day night


                  Criteria For Assessing Goodness Of Fit

        Criterion                   DF           Value           Value/DF

        Deviance                     3          2.2945             0.7648
        Scaled Deviance              3          2.2945             0.7648
        Pearson Chi-Square           3          2.2921             0.7640
        Scaled Pearson X2            3          2.2921             0.7640
        Log Likelihood                       4868.9983

  Algorithm converged.
```

Analysis Of Parameter Estimates

Parameter		DF	Estimate	Standard Error	Wald 95% Confidence Limits		Chi-Square
->Intercept		1	4.5591	0.0724	4.4173	4.7009	3970.06
->depth	035-125	1	-0.3771	0.1134	-0.5994	-0.1547	11.05
->depth	125-250	1	1.3850	0.0809	1.2264	1.5436	293.02
depth	250-300	0	0.0000	0.0000	0.0000	0.0000	.
Scale		0	1.0000	0.0000	1.0000	1.0000	

Analysis Of Parameter Estimates

Parameter		Pr > ChiSq
->Intercept		<.0001
->depth	035-125	0.0009
->depth	125-250	<.0001
depth	250-300	.
Scale		

NOTE: The scale parameter was held fixed.

LR Statistics For Type 1 Analysis

Source	Deviance	DF	Chi-Square	Pr > ChiSq
Intercept	631.5515			
depth	2.2945	2	629.26	<.0001

We close this example by showing TYPE1 and TYPE3 tables in the saturated model M_0. The program segment,

```
PROC GENMOD DATA=water_flea;
     CLASS depth time;
     MODEL count= depth time depth*time /DIST=poisson
                                         LINK=log
                                         TYPE1
                                         TYPE3;
RUN;
```

gives the output listing:

LR Statistics For Type 1 Analysis

Source	Deviance	DF	Chi-Square	Pr > ChiSq
Intercept	631.5515			
depth	2.2945	2	629.26	<.0001
time	0.5899	1	1.70	0.1917
depth*time	0.0000	2	0.59	0.7446

LR Statistics For Type 3 Analysis

Source	DF	Chi-Square	Pr > ChiSq
depth	2	629.64	<.0001
time	1	2.13	0.1440
depth*time	2	0.59	0.7446

□

9.3.2 The Poisson Model with Proportional Parameters

Recall that the Poisson model with proportional parameters was introduced in Section 8.4.1 on page 365 as a model for k pairs (x_i, m_i) where x_i, $i = 1 \ldots, k$, are observations of independent and Poisson-distributed random variables, whereas the m_i's are known numbers that are related to the means by $\lambda_i = m_i\lambda$. Here we will consider this model again, but we wish to allow for the situation where the observations corresponding to m_i are replicated n_i times. Thus we will consider k independent samples of observations from the Poisson distribution, x_{ij}, $j = 1, \ldots, n_i$, $i = 1, \ldots, k$, where j index the n_i observations in the ith sample. The model can be written as

$$M_2\colon x_{ij} \sim\sim po(m_i\lambda), \quad j = 1, \ldots, n_i, \quad i = 1, \ldots, k.$$

The model is labeled M_2 because we can identify two models that are more general than M_2, namely,

$$M_1\colon x_{ij} \sim\sim po(m_i\lambda_i), \quad j = 1, \ldots, n_i, \quad i = 1, \ldots, k,$$

and

$$M_0\colon x_{ij} \sim\sim po(\lambda_{ij}), \quad j = 1, \ldots, n_i, \quad i = 1, \ldots, k.$$

We consider the models in the sequence,

$$M_0 \rightarrow M_1 \rightarrow M_2,$$

as a check of the model M_2. Note that M_0 is the saturated model, and that M_1 specifies that the k groups of observations indexed with i are k samples, so M_1 might have been specified as

$$M_1\colon x_{ij} \sim\sim po(\tilde{\lambda}_i), \quad j = 1, \ldots, n_i, \quad i = 1, \ldots, k,$$

but it is advantageous to introduce the m_i's in M_1 because the m_i's usually reflect differences in the sampling units (for instance, area, time or volume) and then the λ_i's in the parametrization,

$$\tilde{\lambda}_i = m_i\lambda_i, \quad i = 1, \ldots, k,$$

will be directly comparable.

We now show that M_1 and M_2 are generalized linear models with the canonical link function of the Poisson distribution. We already know that the saturated model M_0 is a generalized linear model, and then the model reductions can be tested using generalized linear models.

Let $\mathbf{e}_1, \ldots, \mathbf{e}_k$ denote k vectors corresponding to the k rows, i.e.,

$$(\mathbf{e}_i)_{lj} = \begin{cases} 1 & \text{if } i = l \\ 0 & \text{otherwise.} \end{cases} \tag{9.24}$$

In M_1 the components of the vector $\boldsymbol{\eta}$ of linear parameters have the form,

$$\eta_{ij} = \ln(\lambda_{ij}) = \ln(m_i) + \ln(\lambda_i) = \ln(m_i) + \eta_i,$$

i.e., $\boldsymbol{\eta}$ may be written as

$$\boldsymbol{\eta} = \sum_{i=1}^{k} \ln(m_i)\, \mathbf{e}_i + \sum_{i=1}^{k} \eta_i \mathbf{e}_i, \tag{9.25}$$

and we see that

$$\boldsymbol{\eta} \in A_1 = \boldsymbol{\omega} + L_1,$$

where the offset vector $\boldsymbol{\omega}$ has components $\omega_{ij} = \ln(m_i)$ and where $L_1 = \text{span}\{\mathbf{e}_1, \ldots, \mathbf{e}_k\}$. Note that L_1 is the familiar subspace from k normal samples as a linear normal model. It follows from (9.25) that $\boldsymbol{\omega} \in L_1$ so A_1 is the linear subspace L_1, but as argued above, it is an advantage to consider the parametrization $\tilde{\lambda}_i = m_i\lambda_i$.

Similarly, it is seen that the affine subspace corresponding to M_2 is

$$A_2 = \boldsymbol{\omega} + L_2,$$

where $L_2 = \text{span}\{\mathbf{e}_1 + \cdots + \mathbf{e}_k\}$.

Example 9.7
We consider again the data in Example 8.2 consisting of three groups of counts of nematodes. In each group 15 counts have been made and the volumes that have been counted in the three groups are 40μL, 20μL, and 20μL, respectively. According to the model M_1 above, the observations in the ith group are identically distributed with mean $m_i\lambda_i$, where m_i denotes the volume of the dilution in the ith group and where λ_i is the concentration per μL in the ith group. The model M_2 specifies that the concentration per μL is the same in the three groups.

SAS program and output listing may be found in the Annex to Subsection 9.3.2 starting on page 420. From the table `Criteria For Assessing Goodness Of Fit` for these data on page 421, it is seen that the $-2\ln Q$-test statistic for M_1 is 52.5276 with 42 degrees of freedom. By means of (9.19) we find that the significance probability is 0.128 so M_1 is not rejected. Note that the test of M_1 is a simultaneous test of the hypothesis of having three samples from the Poisson distribution. Thus the $-2\ln Q$-test statistic and the degrees of freedom is the sum of the corresponding quantities in the three individual tests that were reported in the table on page 369. These tests could have been obtained by adding the statement `BY sample` to the program segment for M_1.

The estimates of the parameters in M_1 may be found from the three lines on 421 marked with `->`. In the line `Intercept` the estimate of $\eta_3 = \ln(\lambda_3)$ `(-0.2614)` is found, whereas the estimates of $\eta_1 - \eta_3 = \ln(\lambda_1/\lambda_3)$ `(0.0910)` and $\eta_2 - \eta_3 = \ln(\lambda_2/\lambda_3)$ `(0.0710)` are found in the lines `sample 1` and `sample 2`, respectively. From this we find that

$$\lambda_1 \leftarrow \exp\{(\eta_1 - \eta_3) + \eta_3\} = \exp\{0.0910 - 0.2614\} = 0.8433,$$

$$\lambda_2 \leftarrow \exp\{(\eta_2 - \eta_3) + \eta_3\} = \exp\{0.0710 - 0.2614\} = 0.8266$$

and

$$\lambda_3 \leftarrow \exp\{\eta_3\} = \exp\{-0.2614\} = 0.7700.$$

If the λ_i's are of primary interest, the estimates of these quantities may obtained directly by means of the option `NOINT` to the `MODEL` statement.

Furthermore, in the line `INTERCEPT` it is seen that Wald's 95% confidence interval for η_3 is `[-0.3903, -0.1324]` and applying the exponential function to the limits of this interval, we get that the 95% confidence interval for λ_3 is $[0.6769, 0.8760]$. Similarly, in the line `sample 1` it is seen that the 95% confidence interval for $\ln(\lambda_1/\lambda_3) = \eta_1 - \eta_3$ is `[-0.0647, 0.2466]` which implies that the 95% confidence interval for λ_1/λ_3 is $[0.9373, 1.2797]$.

Finally, note that the estimate of the mean of the observations in, for instance the first group, is

$$m_1\lambda_1 \leftarrow \exp\{\ln(40) + 0.0910 - 0.2614\} = \exp\{3.5185\} = 33.7334,$$

which we also found on page 369.

From the table `Criteria For Assessing Goodness Of Fit` for M_1 and M_2 on page 421 and 422, respectively, we find that

$$-2\ln Q(\mathbf{x}) = 53.8712 - 52.5276 = 1.3436$$

with $44 - 42 = 2$ degrees of freedom. Since M_1 has `TYPE1` among the options to `PROC GENMOD`, these quantities may also be found in the table `LR Statistics For Type 1 Analysis` on page 421. Here, in addition, it is seen that the corresponding significance probability is 0.5108, i.e., the reduction is not rejected. Again there is agreement with the results on page 370.

Finally, it is seen on page 422 in the line `Intercept` that in the model M_2 the estimate of the common value of the linear parameter η is equal to `-0.1974`, i.e., the common value of λ is estimated by

$$\hat{\lambda} = e^{-0.1974} = 0.82086,$$

in agreement with what was found on page 369. In the same line it is seen that Wald's 95% confidence interval for η is `[-0.2599, -0.1350]`, so the corresponding confidence interval for λ

is

$$[\lambda_-, \lambda_+] = [0.7711, 0.8737],$$

which is precisely the same as the confidence interval for λ we found on page 369. □

Annex to Subsection 9.3.2

Calculations in SAS

The application of *PROC GENMOD* in connection with the Poisson model with proportional parameters is illustrated by means of the data in Example 8.2. The calculations have been commented on in Example 9.7.

Example 9.7 (Continued)
In connection with the recording of the data two variables are calculated, `quantity`, which contains the values of volume (in μL) of the quantity of dilution in which the nematodes are counted, and `ln_quantity`, the corresponding (natural) logarithm.

```
OPTIONS NODATE PAGENO=1 LS=80 PS=65;

TITLE1 'Example 9.7 proportional parameters in Poisson model';
/* same data as in Example 8.2 */

DATA felti;
INPUT sample count @@;
quantity=(2-int(sample/2))*20;
ln_quantity=log(quantity);
DATALINES;
1 31 1 28 1 33 1 38 1 28 1 32 1 39 1 27
1 28 1 39 1 21 1 39 1 45 1 37 1 41
2 14 2 16 2 18 2  9 2 21 2 21 2 14 2 12
2 13 2 13 2 14 2 20 2 24 2 15 2 24
3 18 3 13 3 19 3 14 3 15 3 16 3 14 3 19
3 25 3 16 3 16 3 18 3  9 3 10 3  9
;
RUN;

TITLE2 '3 samples';
TITLE3 'M1';
PROC GENMOD DATA=felti;
CLASS sample;
MODEL count=sample/DIST=poisson
                   LINK=log
                   OFFSET=ln_quantity
                   TYPE1;

RUN;
```

In this example the linear subspace of the model corresponds to the three samples defined by values of the `CLASS` variable `sample`, whereas the vector of offset values - `ln_quantity` - is specified by means of an option to `MODEL` statement namely `OFFSET=ln_quantity`. The output listing from this program is given below.

```
                    Example 9.7 proportional parameters in Poisson model                    1
                                          3 samples
                                             M1

                                    The GENMOD Procedure

                                     Model Information

                         Data Set                       WORK.FELTI
                         Distribution                      Poisson
                         Link Function                         Log
                         Dependent Variable                  count
                         Offset Variable               ln_quantity
                         Observations Used                      45

                                  Class Level Information

                            Class         Levels     Values

                            sample             3     1 2 3


                         Criteria For Assessing Goodness Of Fit
Criterion                           DF            Value          Value/DF

          Deviance                          42           52.5276            1.2507
          Scaled Deviance                   42           52.5276            1.2507
          Pearson Chi-Square                42           52.0431            1.2391
          Scaled Pearson X2                 42           52.0431            1.2391
          Log Likelihood                               2122.7269


  Algorithm converged.


                             Analysis Of Parameter Estimates

                                    Standard       Wald 95%              Chi-
  Parameter       DF   Estimate        Error    Confidence Limits     Square   Pr > ChiSq

->Intercept        1    -0.2614       0.0658    -0.3903    -0.1324     15.78       <.0001
->sample      1    1     0.0910       0.0794    -0.0647     0.2466      1.31       0.2519
->sample      2    1     0.0710       0.0914    -0.1082     0.2502      0.60       0.4374
  sample      3    0     0.0000       0.0000     0.0000     0.0000       .            .
  Scale            0     1.0000       0.0000     1.0000     1.0000

NOTE: The scale parameter was held fixed.


                              LR Statistics For Type 1 Analysis

                                                              Chi-
           Source             Deviance          DF          Square       Pr > ChiSq

           Intercept           53.8712
           sample              52.5276           2            1.34           0.5108
```

The program and the corresponding output listing for the calculations in M_2 are:

```
TITLE2 'proportional parameters';
TITLE3 'M2';
PROC GENMOD DATA=felti;
CLASS sample;
MODEL count= /DIST=poisson
              LINK=log
              OFFSET=ln_quantity;
RUN;
```

```
            Example 9.7 proportional parameters in Poisson model             2
                          proportional parameters
                                    M2

                           The GENMOD Procedure

                            Model Information

                    Data Set                      WORK.FELTI
                    Distribution                     Poisson
                    Link Function                        Log
                    Dependent Variable                 count
                    Offset Variable              ln_quantity
                    Observations Used                     45


                         Class Level Information

                       Class         Levels     Values

                       sample             3     1 2 3


                  Criteria For Assessing Goodness Of Fit

          Criterion                 DF           Value        Value/DF

          Deviance                  44         53.8712          1.2243
          Scaled Deviance           44         53.8712          1.2243
          Pearson Chi-Square        44         52.9695          1.2039
          Scaled Pearson X2         44         52.9695          1.2039
          Log Likelihood                     2122.0551


 Algorithm converged.


                     Analysis Of Parameter Estimates

                              Standard       Wald 95%          Chi-
   Parameter  DF  Estimate       Error   Confidence Limits   Square  Pr > ChiSq

 ->Intercept   1   -0.1974      0.0319   -0.2599   -0.1350    38.40      <.0001
   Scale       0    1.0000      0.0000    1.0000    1.0000

NOTE: The scale parameter was held fixed.
```

□

9.3.3 Poisson Regression

We will briefly discuss regression models for Poisson-distributed data. The structure of the data is the same as in normal linear regression in Section 3.3.8 on page 133. The observations are divided into k groups and, moreover, they come in pairs consisting of a response variable and an explanatory variable that has the same value within each group. In the usual notation we have data,

$$(t_{ij}, x_{ij}), \quad j = 1, \ldots, n_i, \quad i = 1, \ldots, k,$$

where the group is indexed by i and x_{ij} denotes the response and t_{ij} denotes the explanatory variable, where

$$t_{ij} = t_i, \quad j = 1, \ldots, n_i, \quad i = 1, \ldots, k.$$

The *Poisson regression model* is

$$M_2\colon x_{ij} \sim\sim po(\exp\{\alpha + \beta t_i\}), \quad j = 1, \ldots n_i, \quad i = 1, \ldots, k,$$

where the independence of the observations is tacitly subsumed. In addition to M_2 we consider the saturated model,

$$M_0\colon x_{ij} \sim\sim po(\lambda_{ij}), \quad j=1,\ldots n_i, \quad i=1,\ldots,k,$$

the model for k independent samples from the Poisson distribution,

$$M_1\colon x_{ij} \sim\sim po(\lambda_i), \quad j=1,\ldots n_i, \quad i=1,\ldots,k,$$

and the sequence of model reductions,

$$M_0 \to M_1 \to M_2,$$

in order to check the regression model M_2.

Note that in a Poisson regression the mean is the exponential function applied to a linear function of the explanatory variable. This may be seen as a convenient way to have the simplicity of a linear function of the explanatory variable and at the same time overcome the problem that a linear function will attain negative values so the positive mean of a Poisson distribution cannot be expressed simply as a linear function. This was indeed the consideration behind the first applications of the Poisson regression. Here we use Poisson regression as an example of the generalized linear model; for the exponential function is the inverse of the canonical link function of the Poisson distribution giving the linear expression for the linear parameter,

$$\eta_{ij} = \ln\lambda_{ij} = \alpha + \beta t_i, \quad j=1,\ldots,n_i, \quad i=1,\ldots,k.$$

This corresponds to a generalized linear model with the affine subspace identical to the linear subspace used when the normal regression model is considered as a linear normal model.

As in the case of normally distributed data the line with the equation, $\eta(t) = \alpha + \beta t$, is referred to as the *regression line*, the parameter α as the *intercept* or the *position*, and the parameter β as the *slope* or the *regression coefficient.*

In order to interpret the slope parameter β of the Poison regression model, observe that the ratio of means corresponding to two explanatory variables is

$$\frac{\lambda_{ij}}{\lambda_{i'j'}} = \exp\{\beta(t_i - t_{i'})\}, \tag{9.26}$$

i.e., the ratio of means depends on β and difference between the explanatory variables only.

In M_2 the hypothesis of known regression coefficient is

$$H_{03}\colon \beta = \beta_0,$$

and the hypothesis of known intercept is

$$H_{03}^*\colon \alpha = \alpha_0,$$

The corresponding models are

$$M_3\colon x_{ij} \sim\sim po(\exp\{\alpha + \beta_0 t_i\}), \quad j=1,\ldots,n_i, \quad i=1,\ldots,k,$$

and

$$M_3^*\colon x_{ij} \sim\sim po(\exp\{\alpha_0 + \beta t_i\}), \quad j=1,\ldots,n_i, \quad i=1,\ldots,k.$$

If H_{03} is not rejected, the hypothesis $\alpha = \alpha_0$ may be tested in M_3. The hypothesis is then denoted by H_{04}, since it corresponds to the reduction from M_3 to

$$M_4\colon x_{ij} \sim\sim po(\exp\{\alpha_0 + \beta_0 t_i\}), \quad j=1,\ldots,n_i, \quad i=1,\ldots,k.$$

Similarly, the hypothesis $H_{04}^*\colon \beta = \beta_0$ may be tested in M_3^* if the reduction to this model has not been rejected in M_2.

Let $\mathbf{e}$ and $\mathbf{t}$ denote the vectors defined from the vectors in (9.24) in the following way:

$$\mathbf{e} = \mathbf{e}_1 + \cdots + \mathbf{e}_k, \quad \mathbf{t} = t_1\mathbf{e}_1 + \cdots + t_k\mathbf{e}_k.$$

With this notation the affine subspaces (for $\boldsymbol{\eta}$) corresponding to the models M_0, M_1, M_2, M_3, M_3^*, and M_4 are, respectively,

$$\begin{aligned} A_0 &= \mathbb{R}^n, \\ A_1 &= \text{span}\{\mathbf{e}_1,\ldots,\mathbf{e}_k\}, \\ A_2 &= \text{span}\{\mathbf{e},\mathbf{t}\}, \\ A_3 &= \beta_0\mathbf{t} + \text{span}\{\mathbf{e}\}, \\ A_3^* &= \alpha_0\mathbf{e} + \text{span}\{\mathbf{t}\}, \end{aligned}$$

and

$$A_4 = \alpha_0\mathbf{e}+\beta_0\mathbf{t}.$$

Actually, the affine subspaces A_0, A_1, and A_2 are linear subspaces and may be specified without an offset variable. If $\beta_0 \neq 0$, then A_3 is a genuine affine subspace and to specify the model the variable, $\beta_0\mathbf{t}$ must be computed and specified as an offset variable. A similar remark applies to A_3^* and α_0.

The relations between the models and the hypotheses connecting them appear from the graphical representation:

$$\begin{array}{ccc} & M_3\colon \eta_{ij} = \alpha + \beta_0 t_i & \\ H_{03}\colon \beta = \beta_0 \nearrow & & \searrow H_{04}\colon \alpha = \alpha_0 \\ M_2\colon \eta_{ij} = \alpha + \beta t_i & & M_4\colon \eta_{ij} = \alpha_0 + \beta_0 t_i \\ H_{03}^*\colon \alpha = \alpha_0 \searrow & & \nearrow H_{04}^*\colon \beta = \beta_0 \\ & M_3^*\colon \eta_{ij} = \alpha_0 + \beta t_i & \end{array}$$

Comparing with page 139 it is seen that this structure is precisely similar to one we considered in connection with regression analysis for normally distributed data.

Before we consider an example of an application of the regression model for Poisson-distributed data, we note that the model corresponding to the hypothesis of proportional parameters in the Poisson model – $\lambda_{ij} = m_i\lambda$ – or

$$\eta_{ij} = \ln(m_i) + \ln(\lambda) = \ln(m_i) + \eta,$$

may be considered as the special case of the regression model M_3 corresponding to $t_i = \ln(m_i)$, $\beta_0 = 1$ and $\alpha = \eta = \ln(\lambda)$.

Example 9.8

We consider the data in Example 8.2, which was also used in Example 9.7. Here we use the model for Poisson regression with the volumes as the explanatory variables, i.e., $t_i = \ln(m_i)$. We consider the sequence of models,

$$M_0 \to M_1 \to M_2 \to M_3,$$

where β_0 in M_3 is equal to 1. As explained above, the model M_3 with $\beta_0 = 1$ is the Poisson model with proportional parameters and in Example 9.7 we considered the sequence of models,

$$M_0 \to M_1 \to M_3,$$

where the model M_3 was denoted by M_2 in Example 9.7. The distinction between the treatment in Example 9.7 and here is the introduction of the regression model as an intermediate step between M_1 and M_3.

Incidentally, note that the interpretation of β in the Poisson regression is particularly simple, because (9.26) can be simplified to

$$\frac{\lambda_{ij}}{\lambda_{i'j'}} = \exp\{\beta(t_i - t_{i'})\} = \exp\{\beta(\ln m_i - \ln m_{i'})\} = \left(\frac{m_i}{m_{i'}}\right)^{\beta}.$$

Part of the SAS program and the output listing from it is given in Annex to Subsection 9.3.3 starting on page 426.

The output from M_1 is almost identical to the corresponding output in Example 9.7.

From the table `Criteria For Assessing Goodness Of Fit` for M_1 and M_2 on page 427 and page 428, respectively, using the formulas (9.22) and (9.23) we find the following information concerning the test of the hypothesis of linearity H_{02}:

$$-2\ln Q(\mathbf{x}) = 53.1310 - 52.5276 = 0.6034$$

and

$$f = 43 - 42 = 1.$$

According to (9.19) the significance probability of H_{02} is

$$p_{obs}(\mathbf{x}) = 1 - F_{\chi^2(1)}(0.6034) = 0.437,$$

and the hypothesis is not rejected.

The estimates of the parameters in M_2 may be found on page 428. In the line `Intercept` the estimate of α is found to be `-0.4622` and in the line `ln_quantity` it is seen that $\beta \leftarrow$ `1.0791`. In the same line it is seen that the 95% confidence interval for β based on Wald's test is

$$[\beta_-, \beta_+] = [0.8989,\ 1.2594].$$

Since the confidence interval contains the value 1, the hypothesis H_{03}: $\beta = 1$ is not rejected by Wald's test. We now consider the $-2\ln Q$ test of H_{03}. By means of `Criteria For Assessing Goodness Of Fit` on page 428 and page 429 together with the formulas (9.22) and (9.23), we find the test statistic,

$$-2\ln Q(\mathbf{x}) = 53.8712 - 53.1310 = 0.7402,$$

with $44 - 43 = 1$ degree of freedom. Thus the significance probability of H_{03}: $\beta = 1$ becomes

$$p_{obs}(\mathbf{x}) = 1 - F_{\chi^2(1)}(0.7402) = 0.390,$$

and the hypothesis is accepted. Recalling the remark immediately before the example we see that we have once again adopted the model of proportional parameters for this data set.

Furthermore, from the line `Intercept` in `Analysis Of Parameter Estimates` for M_3 page 429 it is seen that $\alpha \leftarrow$ `-0.1974`; i.e.,

$$\lambda = e^{\alpha} \leftarrow e^{-0.1974} = 0.82086,$$

which we have seen a couple of times before where we have considered the Poisson model with proportional parameters.

Since there are no scientific reasons for reducing the model M_3, the analysis is complete. However, for the sake of illustration we end the example by testing the hypothesis H_{04}: $\alpha = \ln(0.75) = (-0.2877)$ in the model M_3. As noted immediately before the example the intercept α in M_3 is equal to $\ln(\lambda)$, where λ is the number of nematodes per unit volume of the dilution. Thus H_{04} is equivalent to the hypothesis $\lambda = 0.75$. From page 429 and page 430 by means of (9.22) and (9.23) we find that

$$-2\ln Q(\mathbf{x}) = 61.6576 - 53.8712 = 7.7863$$

with $44 - 43 = 1$ degree of freedom. Consequently, the significance probability is

$$p_{obs}(\mathbf{x}) = 1 - F_{\chi^2(1)}(7.7863) = 0.0053,$$

and the hypothesis H_{04} is rejected. (This is in accordance with the output from M_3, where it is seen in the line `Intercept` that the Wald's 95% confidence interval for α is `[-0.2599, -0.1350]`.)

□

Annex to Subsection 9.3.3

Calculations in SAS

We now show how SAS by means of *PROC GENMOD* can carry out the calculations in a regression analysis for Poisson-distributed data. The calculations, performed on the data in Example 8.2, are discussed in Example 9.8.

Example 9.8 (Continued)
The data are recorded and the two variables `quantity` and `ln_quantity` are calculated in the same way as in Example 9.7 on page 420. In addition, an offset variable `offset_4`, which is used in the model M_4, is calculated as the sum of `ln_quantity` and a variable `alpha_0` with elements $\alpha_0 = \ln(0.75)$.

All calculations in the models M_1, M_2, M_3, and M_4 may be carried out by means of the following program:

```
TITLE1 'Example 9.8 - Poisson regression';
/* same data as in Example 8.2 */

TITLE2 '3 samples';
TITLE3 'M1';
PROC GENMOD DATA=felti;
CLASS sample;
MODEL count=sample/DIST=poisson
                   LINK=log;
RUN;

TITLE2 'regression';
TITLE3 'M2';
PROC GENMOD DATA=felti;
MODEL count=ln_quantity/DIST=poisson
                        LINK=log;
RUN;

TITLE2 'slope=1';
TITLE3 'M3';
PROC GENMOD DATA=felti;
MODEL count=/DIST=poisson
            LINK=log
            OFFSET=ln_quantity;
RUN;

TITLE2 'intercept=ln(0.75) and slope=1';
TITLE3 'M4';
PROC GENMOD DATA=felti;
MODEL count=/DIST=poisson
            LINK=log
            OFFSET=offset_4
            NOINT;
RUN;
```

Part of the output listing from this program is given on the following pages.

Program and output listing from M_1:

```
TITLE2 '3 samples';
TITLE3 'M1';
PROC GENMOD DATA=felti;
CLASS sample;
MODEL count=sample/DIST=poisson
                   LINK=log;
RUN;
```

```
                    Example 9.8 - Poisson regression                  1
                              3 samples
                                 M1

                          The GENMOD Procedure

                           Model Information

                   Data Set                    WORK.FELTI
                   Distribution                   Poisson
                   Link Function                      Log
                   Dependent Variable               count
                   Offset Variable            ln_quantity
                   Observations Used                   45


                        Class Level Information

                     Class         Levels     Values

                     sample             3     1 2 3


                 Criteria For Assessing Goodness Of Fit

        Criterion                  DF           Value         Value/DF

        Deviance                   42         52.5276           1.2507
        Scaled Deviance            42         52.5276           1.2507
        Pearson Chi-Square         42         52.0431           1.2391
        Scaled Pearson X2          42         52.0431           1.2391
        Log Likelihood                      2122.7269


  Algorithm converged.


                    Analysis Of Parameter Estimates

                              Standard      Wald 95%          Chi-
 Parameter     DF  Estimate     Error   Confidence Limits   Square  Pr > ChiSq

 Intercept      1   -0.2614    0.0658   -0.3903   -0.1324    15.78      <.0001
 sample     1   1    0.0910    0.0794   -0.0647    0.2466     1.31      0.2519
 sample     2   1    0.0710    0.0914   -0.1082    0.2502     0.60      0.4374
 sample     3   0    0.0000    0.0000    0.0000    0.0000      .           .
 Scale          0    1.0000    0.0000    1.0000    1.0000

NOTE: The scale parameter was held fixed.
```

Program and output listing for M_2:

```
TITLE2 'regression';
TITLE3 'M2';
PROC GENMOD DATA=felti;
MODEL count=ln_quantity/DIST=poisson
                        LINK=log;
RUN;
```

```
                    Example 9.8 - Poisson regression                    2
                              regression
                                  M2

                          The GENMOD Procedure

                           Model Information

                    Data Set                    WORK.FELTI
                    Distribution                   Poisson
                    Link Function                      Log
                    Dependent Variable               count
                    Observations Used                   45


                 Criteria For Assessing Goodness Of Fit

     Criterion                    DF           Value        Value/DF

     Deviance                     43         53.1310          1.2356
     Scaled Deviance              43         53.1310          1.2356
     Pearson Chi-Square           43         52.6839          1.2252
     Scaled Pearson X2            43         52.6839          1.2252
     Log Likelihood                        2122.4252


  Algorithm converged.


                    Analysis Of Parameter Estimates

                             Standard       Wald 95%          Chi-
  Parameter     DF  Estimate     Error  Confidence Limits   Square  Pr > ChiSq

->Intercept      1   -0.4622    0.3099  -1.0696    0.1452     2.22      0.1358
->ln_quantity    1    1.0791    0.0920   0.8989    1.2594   137.67      <.0001
  Scale          0    1.0000    0.0000   1.0000    1.0000

NOTE: The scale parameter was held fixed.
```

Program and output listing for M_3:

```
TITLE2 'slope=1';
TITLE3 'M3';
PROC GENMOD DATA=felti;
MODEL count=/DIST=poisson
            LINK=log
            OFFSET=ln_quantity;
RUN;
```

```
                    Example 9.8 - Poisson regression                    3
                                slope=1
                                  M3

                          The GENMOD Procedure

                           Model Information

                    Data Set                  WORK.FELTI
                    Distribution                 Poisson
                    Link Function                    Log
                    Dependent Variable             count
                    Offset Variable          ln_quantity
                    Observations Used                 45

                 Criteria For Assessing Goodness Of Fit

          Criterion                 DF           Value        Value/DF

          Deviance                  44         53.8712          1.2243
          Scaled Deviance           44         53.8712          1.2243
          Pearson Chi-Square        44         52.9695          1.2039
          Scaled Pearson X2         44         52.9695          1.2039
          Log Likelihood                     2122.0551

  Algorithm converged.

                    Analysis Of Parameter Estimates

                             Standard     Wald 95%            Chi-
   Parameter  DF  Estimate     Error   Confidence Limits     Square  Pr > ChiSq

 ->Intercept   1   -0.1974    0.0319   -0.2599   -0.1350      38.40      <.0001
   Scale       0    1.0000    0.0000    1.0000    1.0000

NOTE: The scale parameter was held fixed.
```

Program and output listing for M_4:

```
TITLE2 'intercept=ln(0.75) and slope=1';
TITLE3 'M4';
PROC GENMOD DATA=felti;
MODEL count=/DIST=poisson
            LINK=log
            OFFSET=offset_4
            NOINT;
RUN;
```

```
                    Example 9.8 - Poisson regression                     4
                     intercept=ln(0.75) and slope=1
                                   M4

                         The GENMOD Procedure

                           Model Information

                    Data Set                   WORK.FELTI
                    Distribution                  Poisson
                    Link Function                     Log
                    Dependent Variable              count
                    Offset Variable              offset_4
                    Observations Used                  45


                 Criteria For Assessing Goodness Of Fit

      Criterion                 DF           Value        Value/DF

      Deviance                  45         61.6576          1.3702
      Scaled Deviance           45         61.6576          1.3702
      Pearson Chi-Square        45         66.0000          1.4667
      Scaled Pearson X2         45         66.0000          1.4667
      Log Likelihood                     2118.1619

  Algorithm converged.


                   Analysis Of Parameter Estimates

                          Standard      Wald 95%           Chi-
 Parameter  DF  Estimate     Error  Confidence Limits    Square  Pr > ChiSq

 Intercept   0    0.0000    0.0000    0.0000    0.0000        .           .
 Scale       0    1.0000    0.0000    1.0000    1.0000

NOTE: The scale parameter was held fixed.


                    Lagrange Multiplier Statistics

                 Parameter      Chi-Square    Pr > ChiSq

                 Intercept          8.0278        0.0046
```

□

9.3.4 *PROC GENMOD and the Multinomial Model*

As illustrated in Chapter 8 the multinomial model may be considered as the result of appropriate conditioning in a Poisson model. We also saw that the results of calculating the $-2\ln Q$-test statistic, the corresponding degrees of freedom and the significance probability are identical for the two models. This fact implies that *PROC GENMOD* may be used to carry out calculations in the multinomial distribution, even though this class of distributions is not of the form (9.1).

Once again it has to be emphasized that even though the calculations in the two model classes are identical, the models are different. In the multinomial model a *given* number of objects, which are classified according to various criteria, are decided upon in advance. In contrast, in the Poisson model the sum of the observations is stochastic and, therefore, not known in advance.

In this section we illustrate how calculations in the multinomial distribution may be performed in the Poisson distribution by means of *PROC GENMOD*. When doing this, one important rule has to be taken into account. A variable that indexes the multinomial distribution(s) has to enter the `MODEL` model statement of *PROC GENMOD* as a `CLASS` variable.

Example 9.9

In Example 7.4 we considered the data given below from the county of Aarhus concerning the concentration of the pesticide, *2.6-dichlorbenzamid*, in ground and drinking water. The data were from major waterworks in 2 areas in the county. In both areas samples of water from 2 municipalities have been investigated with the following result:

		Concentration			
Area	*Municipality*	≤ 0.01	0.01–0.10	>0.10	*Total*
1	*Hadsten*	23	12	6	41
	Hammel	20	5	9	34
2	*Nørre Djurs*	37	6	4	47
	Randers	47	4	3	54

In Chapter 7 the data were analyzed with a model consisting of 4 independent multinomial distributions, each with 3 categories. For each of the areas we saw that the two samples within an area may be considered to be homogeneous, and, furthermore, that the two common probability vectors within the areas vary between areas. Thus the statistical analysis consisted of testing three hypotheses of identity of the probability vectors in independent multinomial distributions. We now show how the calculations may be carried out using *PROC GENMOD*.

First we instigate for each of the two areas if the two multinomial distributions within the area are homogeneous, i.e., if it may be assumed that they have the same probability vector. From Chapter 7 and Chapter 8, we know that the calculations of the test of homogeneity in the multinomial model are identical to the calculations of the test of the multiplicative model in the Poisson distribution, so the hypothesis for each of the two areas are tested in the Poisson distribution.

The SAS program performing the calculations for each of the areas may be seen on page 433 together with its output listing. In order to save space only part of the output is shown. For area 1 it is seen in the table `Criteria For Assessing Goodness Of Fit` that the $-2\ln Q$-test statistic is 3.1291 with 2 degrees of freedom, which we also found on page 323 where the significance probability was shown to be 0.209. For area 2 the $-2\ln Q$-test statistic is 1.2538 with 2 degrees of freedom. The values may also be found on page 341 where the hypothesis is tested using Fisher's exact test, since some of the expected frequencies under hypothesis are less or equal to 5 as seen on page 323. For area 2 the significance probability of the test of homogeneity is 0.53. Thus the hypothesis of homogeneity is not rejected in any of the two areas.

Considering both areas together, this means that we adopt a Poisson model M_1 in which the linear parameter is of the form,

$$M_1\colon \eta_{ij} = \ln(\lambda_{ij}) = \alpha_i + \beta_{h(i)j}, \quad i = 1,2,3,4,\ j = 1,2,3,$$

where $h(i)$ indicates the area, i.e.,

$$h(i) = \begin{cases} 1 & \text{if } i = 1,2 \\ 2 & \text{if } i = 3,4. \end{cases}$$

Note that the row sums, i.e., the number of waterworks considered in the four municipalities, are fixed. This explains why a term depending on the rows only, α_i, has to be included in the formula for the linear parameter. The term $\beta_{h(i)j}$ expresses that the column parameters, which corresponds to the probability vectors in the multinomial distributions, depend on the area such that the column parameters within area 1 are identical and, similarly, that the column parameters within area 2 are identical. Expressed in terms of the multinomial distribution, this means that the probability vectors within each of the two areas are identical, but that they may vary between areas.

On page 324 we investigated whether the probability vectors in the four multinomial distributions do not depend on the area. In the Poisson model this is formulated as the hypothesis,

$$H_{02}\colon \beta_{h(i)j} = \beta_j, \quad i = 1,2,3,4,\ j = 1,2,3,$$

corresponding to the reduction from M_1 to the model,

$$M_2\colon \eta_{ij} = \ln(\lambda_{ij}) = \alpha_i + \beta_j, \quad i = 1,2,3,4,\ j = 1,2,3.$$

From the `Type 1` analysis on page 435, it is seen in the line `area*concentration` that the $-2\ln Q$-test statistic of H_{02} is 14.43 with 2 degrees of freedom corresponding to a significance probability of 0.0007, which we also found on page 324. Thus H_{02} is rejected, and we can conclude that there is a significant difference between the distributions of the concentration in the two areas. □

Annex to Subsection 9.3.4

Calculations in SAS

We now illustrate how *PROC GENMOD* may be used in connection with the multinomial distribution. As an example the data in Example 7.4 are considered. The output listing from the program has been commented on in Example 9.9 above.

Example 9.9 (Continued)
The SAS program for the analysis of data in Example 7.4 by means of *PROC GENMOD* is:

```
OPTION PS=65 LS=80 PAGENO=1 NODATE;

TITLE1 'Example 9.9';
/* same data as in Example 7.4*/

DATA dichlor;
LENGHT concentration$8;
INPUT area municipality$3-10 concentration$12-20 count;
DATALINES;
1 Hadsten    <=0.01 23
1 Hadsten  0.01-0.1 12
1 Hadsten      >0.1  6
1 Hammel     <=0.01 20
1 Hammel   0.01-0.1  5
1 Hammel       >0.1  9
2 Nr Djurs   <=0.01 37
2 Nr Djurs 0.01-0.1  6
2 Nr Djurs     >0.1  4
2 Randers    <=0.01 47
2 Randers  0.01-0.1  4
2 Randers      >0.1  3
;
RUN;

PROC GENMOD DATA=dichlor;
CLASS municipality concentration;
MODEL count=municipality concentration/DIST=poisson
                                       LINK=log;
BY area;
RUN;

PROC GENMOD DATA=dichlor;
CLASS area municipality concentration;
MODEL count=municipality concentration
             area*concentration/DIST=poisson
                                  LINK=log
                                  TYPE1;
RUN;
```

In this example we consider four multinomial distributions, one for each municipality. Therefore the variable `minicipality` enters the `MODEL` statement as a `CLASS` variable.

Part of the output listing from the program is given below.

The lines

```
PROC GENMOD DATA=dichlor;
CLASS municipality concentration;
MODEL count=municipality concentration/DIST=poisson
                                        LINK=log;
BY area;
RUN;
```

gives the output listing

```
                              Example 9.9                               1

---------------------------------- area=1 ----------------------------------

                          The GENMOD Procedure

                           Model Information

                 Data Set                   WORK.DICHLOR
                 Distribution                    Poisson
                 Link Function                       Log
                 Dependent Variable                count
                 Observations Used                     6


                        Class Level Information

        Class                 Levels    Values

        municipality              2     Hadsten Hammel
        concentration             3     0.01-0.1 <=0.01 >0.1


                 Criteria For Assessing Goodness Of Fit

   Criterion                  DF           Value         Value/DF

   Deviance                    2          3.1291           1.5646
   Scaled Deviance             2          3.1291           1.5646
   Pearson Chi-Square          2          3.0650           1.5325
   Scaled Pearson X2           2          3.0650           1.5325
   Log Likelihood                       123.8581
---------------------------------- area=2 ----------------------------------

                          The GENMOD Procedure

                           Model Information

                 Data Set                   WORK.DICHLOR
                 Distribution                    Poisson
                 Link Function                       Log
                 Dependent Variable                count
                 Observations Used                     6


                        Class Level Information

        Class                 Levels    Values

        municipality              2     Nr Djurs Randers
        concentration             3     0.01-0.1 <=0.01 >0.1


                 Criteria For Assessing Goodness Of Fit

   Criterion                  DF           Value         Value/DF

   Deviance                    2          1.2538           0.6269
   Scaled Deviance             2          1.2538           0.6269
   Pearson Chi-Square          2          1.2542           0.6271
   Scaled Pearson X2           2          1.2542           0.6271
   Log Likelihood                       238.0707
```

and

```
PROC GENMOD DATA=dichlor;
CLASS area municipality concentration;
MODEL count=municipality concentration
            area*concentration/DIST=poisson
                               LINK=log
                               TYPE1;
RUN;
```

gives the output listing

```
                           Example 9.9                              3

                      The GENMOD Procedure

                        Model Information

              Data Set                 WORK.DICHLOR
              Distribution                  Poisson
              Link Function                     Log
              Dependent Variable              count
              Observations Used                  12


                    Class Level Information

     Class               Levels    Values

     area                     2    1 2
     municipality             4    Hadsten Hammel Nr Djurs Randers
     concentration            3    0.01-0.1 <=0.01 >0.1


             Criteria For Assessing Goodness Of Fit

      Criterion                 DF           Value        Value/DF

      Deviance                   4          4.3830          1.0957
      Scaled Deviance            4          4.3830          1.0957
      Pearson Chi-Square         4          4.3192          1.0798
      Scaled Pearson X2          4          4.3192          1.0798
      Log Likelihood                      361.9288


               LR Statistics For Type 1 Analysis

                                                     Chi-
  Source                   Deviance         DF     Square      Pr > ChiSq

  Intercept                134.9174
  municipality             129.9226          3       4.99          0.1722
  concentration             18.8168          2     111.11          <.0001
  area*concentration         4.3830          2      14.43          0.0007
```

□

9.3.5 *Dose-response Models*

We conclude this chapter with a brief discussion of the generalized linear model in connection with dose-response experiments. As seen below such data may sometimes by analyzed by means of a regression model for data distributed according to the binomial distribution.

In medicine and biology, including toxicology, one is often interested in knowing the biological effect of a preparation, for instance a vitamin, a vaccine or a poisonous substance. Typically, one wishes to investigate the effect of the substance on a population of individuals. A frequently used method consists of giving different doses of the substance in question to experimental subjects (for example, animals, plants, bacterial cultures). Assume the n_i subjects have been given the dose $d_i, i = 1, ..., m$. For each of the n_i subjects who received the dose d_i, a quantity that characterizes the effect of the substance on the subject is observed. This quantity is referred to as the *response*. If the response is measured on a continuous scale, we use the term *quantitative response*. An example of a dose-response experiment of this type is an experiment in which a fertilizer is applied to a plant and where the response is increase in height of the plant during a certain period. An important type of dose-response experiments has a *quantal* (or *binary*) *response*. In such experiments it is recorded for each individual whether or not an event that is characteristic for the effect of the substance occurs. If the event occurs, the individual is said to *respond*. In experiments with poisonous substances, as in Example 9.10 below, we are often only interested in whether the individual survives; if not, the individual is said to respond. Here we will consider quantal dose-response experiments only but for the sake of completeness, it should be mentioned that we sometimes also consider dose-response experiments in which the response is discrete and may take more than two values as in quantal experiments.

The purpose of a dose-response experiment is on the basis of the observed responses to determine *the dose-response curve*, which describes the relationship between dose and response.

Example 9.10
Jørgen Jespersen, Danish Pest Infestation Laboratory, performed a dose-response experiment with 709 females of the common housefly *Musca domestica*. In order to investigate the effect of the insecticide *dimethoat*, a test is carried out in which 1 μL of a fixed concentration of the substance is placed on the back of the fly and it is recorded whether the fly dies. Jørgen Jespersen considered 13 different concentrations of the insecticide and as it appears from Table 9.3, for each concentration there are two or three replications. The replications all include 20 flies, expect for one with 10 flies and one with 19 flies, and for each of the replications the number of dead flies is given. □

Table 9.3 *Dose-response experiment with 709 females of the common housefly Musca domestica. The number of flies considered in one replication is denoted by n and the number of flies that die due to the insecticide dimethoat is denoted by x.*

Concen-	*Replications*					
tration	x	n	x	n	x	n
0.0160	0	20	0	20		
0.0226	0	20	0	20		
0.0320	0	20	1	20	1	20
0.0453	0	20	2	20	2	20
0.0640	5	20	3	20	6	20
0.0905	4	20	5	20	7	20
0.1280	5	20	12	20	10	20
0.1810	15	20	12	20	16	20
0.2560	14	20	13	20	13	20
0.3620	18	20	19	20	17	20
0.5120	20	20	20	20	20	20
0.7240	20	20	20	20	9	10
1.0240	20	20	19	19		

As mentioned above, we will only consider quantal dose-response experiments in which data have the same structure as in Example 9.10. Suppose that corresponding to the ith dose d_i the number of replications is g_i, $i = 1,...,k$. Let x_{ij} and n_{ij} denote the number of individuals responding and the number of individuals considered, respectively, in the jth replication of the experiment corresponding to the ith dose, $d_i, j = 1,...,g_i, i = 1,...,k$. As the basic model we consider the model in which the observations are realizations of independent random variables that are distributed according to the binomial distribution, each observation with its parameter, i.e., the saturated model,

$$M_0\colon x_{ij} \sim\sim b(n_{ij}, \pi_{ij}), \quad j = 1, \ldots, g_i, \quad i = 1, \ldots, k.$$

As a starting point we investigate whether the data may be considered as k samples from the binomial distribution, one sample for each dose, i.e., we test the hypothesis, $H_{01}\colon \pi_{ij} = \pi_i$, corresponding to the reduction from M_0 to the model,

$$M_1\colon x_{ij} \sim\sim b(n_{ij}, \pi_i), \quad j = 1, \ldots, g_i, \quad i = 1, \ldots, k.$$

In other words, when testing the reduction from M_0 to M_1 we are checking that the replications are indeed replications.

In experiments with quantal response, the dose-response curve is the graph of the function $\pi(d)$, which gives the probability that an individual exposed to the dose d responds. As a measure of the strength of the substance or toxin in question one considers LD_{50} (**l**ethal **d**ose 50), which is the dose where half of the experimental individuals die,

$$\pi(LD_{50}) = 0.50. \tag{9.27}$$

The dose-response models considered here are specified by assuming that the linear parameter is a linear function of the (natural) *logarithm* of the dose, i.e.,

$$\eta(d) = g(\pi(d)) = \alpha + \beta \ln(d), \tag{9.28}$$

where g is a link function.

In most literature only two models for quantal dose-response experiments are considered corresponding to two different choices of the link function g, namely the logistic dose-response model and the probit model, which we now introduce.

The *logistic dose-response model* turns up by considering the canonical link function for the binomial distribution, which, according to Example 9.3, is the *logit* function. Since

$$\eta(d) = \operatorname{logit}(\pi(d)) = \ln(\frac{\pi(d)}{1 - \pi(d)}) = \alpha + \beta \ln(d), \tag{9.29}$$

the dose-response curve in this model is

$$\pi(d) = \frac{1}{1 + e^{-\alpha - \beta \ln(d)}}. \tag{9.30}$$

Equating $\pi(d)$ to 0.50 and solving for d gives that

$$LD_{50} = e^{-\alpha/\beta}. \tag{9.31}$$

The dose-response model, which somewhat misleadingly is referred to as the *probit model*, appears by using the inverse Φ^{-1} of the distribution function Φ of the standard normal distribution as link function, i.e.,

$$\eta(d) = \Phi^{-1}(\pi(d)) = \alpha + \beta \ln(d). \tag{9.32}$$

Thus the dose-response curve in this model is

$$\pi(d) = \Phi(\alpha + \beta \ln(d)) \tag{9.33}$$

so also in this model LD_{50} is given by (9.31).

From a mathematical point of view, the only difference between the dose-response models is that they are specified by means of two different link functions. Consequently, in principle the statistical analysis of the models by means of the theory of generalized linear models is the same. In the following we concentrate on the logistic dose-response model.

If η_i and η_{ij} denote logit corresponding to π_i and π_{ij}, respectively, it may be investigated whether or not the logistic model is reasonable by testing the hypothesis,

$$H_{02}\colon \eta_i = \eta(d_i) = \alpha + \beta \ln(d_i), \quad i = 1,\ldots,k,$$

in the model M_1 or the hypothesis,

$$\eta_{ij} = \eta(d_i) = \alpha + \beta \ln(d_i), \quad j = 1,\ldots,g_i, \quad i = 1,\ldots,k,$$

in the model M_0. For both hypotheses the corresponding model is

$$M_2\colon X_{ij} \sim\sim b(n_{ij}, (1 + e^{-\alpha - \beta \ln(d_i)})^{-1}), \quad j = 1,\ldots,g_i, \quad i = 1,\ldots,k.$$

The dose-response model M_2 is of primary interest and sometimes the data are sufficiently structured such that M_2 may be checked by testing the hypotheses corresponding to the reductions $M_0 \to M_1$ and $M_1 \to M_2$. Usually, however, the test of the reduction to M_2 is carried out in the model M_0. Often the dose-response experiment has not been made with replications, i.e., $g_i = 1$, $i = 1,\ldots,k$, and in that case the models M_0 are M_1 identical. More importantly, often the expected frequencies – usually corresponding to small doses – in M_1 as well as in M_2 are too small in order for the distribution of the $-2\ln Q$-test statistic to be approximated by a χ^2-distribution. In such cases a more extensive model checking is necessary and therefore we choose to concentrate on the model M_2. An example of such a more extensive check of M_2 will be mentioned in connection with the continuation of Example 9.10 below.

If the model M_2 is not rejected, we may test hypotheses concerning the position (intercept) α and/or the regression coefficient (slope) β in the same way as outlined for the Poisson regression on page 424.

Example 9.10 (Continued)
As the basic model for the data we consider the model,

$$M_0\colon x_{ij} \sim\sim b(n_{ij}, \pi_{ij}), \quad j = 1,\ldots,g_i, \quad i = 1,\ldots,13.$$

The relative responses r_{ij} are the maximum likelihood estimates of the probability parameters in M_0; i.e.,

$$\pi_{ij} \leftarrow r_{ij} = x_{ij}/n_{ij}, \quad j = 1,\ldots,g_i, \quad 1,\ldots,13.$$

In order to get a first impression of relation between the dose, in this case the concentration, and the response, the relative response is plotted against the logarithm of the dose. This is done in Figure 9.1.

The figure is slightly deceptive because of some overlapping points. As it appears from Table 9.3, there are, for instance, 2 identical observations corresponding to each of the two lowest values of the dose. Figure 9.1 indicates that the dose-response curve has the same S-shaped course as that of a distribution function of a continuous random variable. However, on the basis of the figure, it is not possible to decide whether the data favor the logistic distribution function in (9.30) or at the distribution function Φ in (9.33).

In order to look more closely at the logistic dose-response model M_2 we plot

$$\ln\left(\frac{r_{ij}}{1 - r_{ij}}\right)$$

against the logarithm of the dose. This is an empirical version of (9.29) with $\pi(d)$ replaced by the estimated probabilities and according to (9.29), the point should be lying around a straight line. The plot may be seen in Figure 9.2.

There are fewer points in Figure 9.2 than in Figure 9.1 because some of the relative responses r_{ij} are either 0 or 1, and so the corresponding logits cannot be calculated. The points in Figure 9.2 are lying around a straight line, and this motivates further analysis of data by means of the logistic dose-response model.

The SAS program that analyzes the data using a logistic dose-response model by means of *PROC GENMOD* and the corresponding output may be found in the Annex to Subsection 9.3.5 starting on

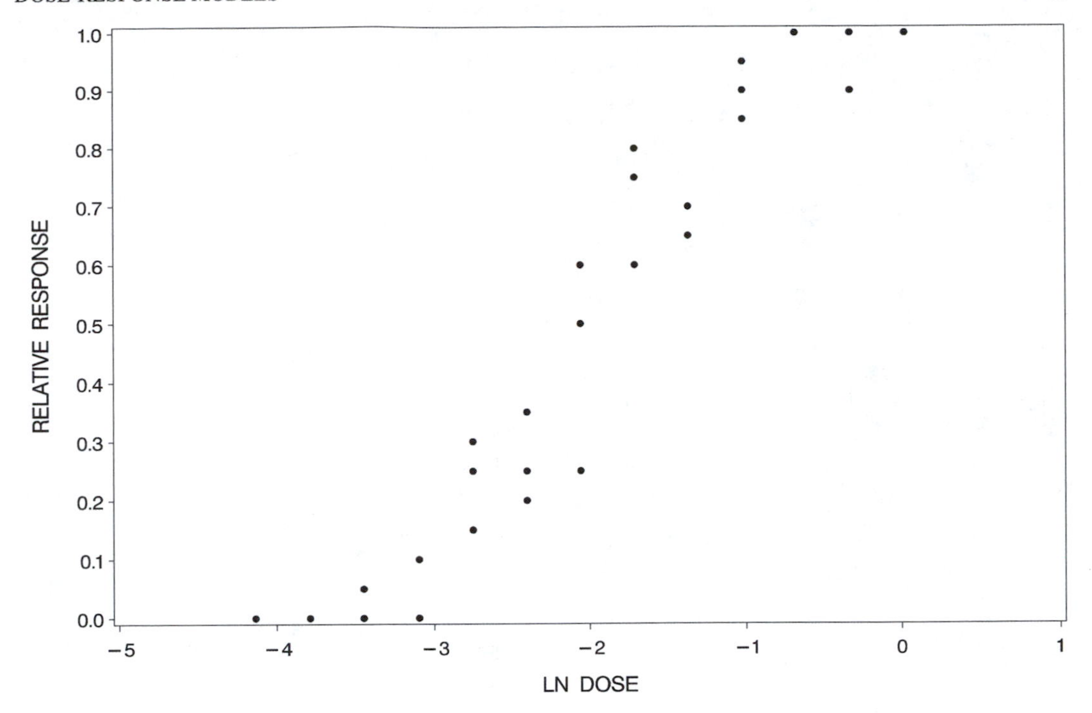

Figure 9.1 *The relative response r_{ij} plotted against the logarithm of dose for data in Table 9.3.*

page 445. In the table `Criteria For Assessing Goodness Of Fit` in the SAS output on page 446 it is seen that the $-2\ln Q$-test statistic (`Deviance`) for the reduction to M_2, the logistic dose-response model, is

$$-2\ln Q(\mathbf{x}) = 38.8137$$

with 34 degrees of freedom, which by means of (9.19), gives a significance probability of 0.262. However, the question is now if the approximation in (9.19) may be applied in this case, i.e., the question is whether the expected frequencies in M_2 are greater than or equal to 5.

This question may be examined because we added the line

```
MAKE 'obstats' OUT=tmp;
```

in the call of *PROC GENMOD* and, in addition, have `OBSTATS` among the options to the `MODEL` statement. This has the effect that some additional information on the model M_2 is transferred to a SAS data set `tmp`, which is automatically printed under the heading `Observation Statistics`. Due to the limited space we have abbreviated this output on page 446. In the column `Pred` the maximum likelihood estimates $\hat{\pi}_{ij}$ in the model M_2 are found. For instance, it is seen that

$$\hat{\pi}_{11} = 0.0077249,$$

and since the corresponding value of `total` - n_{11}- is 20, the expected frequency in M_2 is

$$e_{11} = n_{11}\hat{\pi}_{11} = 20 \times 0.0077249 = 0.154498.$$

Similarly, it may be seen that some of the other expected frequencies are less than or equal to 5. Consequently, we ought to be skeptical concerning the significance probability derived above and find alternative ways of checking the model M_2. Figure 9.3 shows the estimated response probabilities in M_2,

$$r_i = x_{i\cdot}/n_{i\cdot}, \quad i = 1, \ldots, 13,$$

plotted against the logarithm of dose and in the same figure the estimated dose-response curve is drawn. In Figure 9.4 the these quantities are plotted on the logit scale against the logarithm of dose

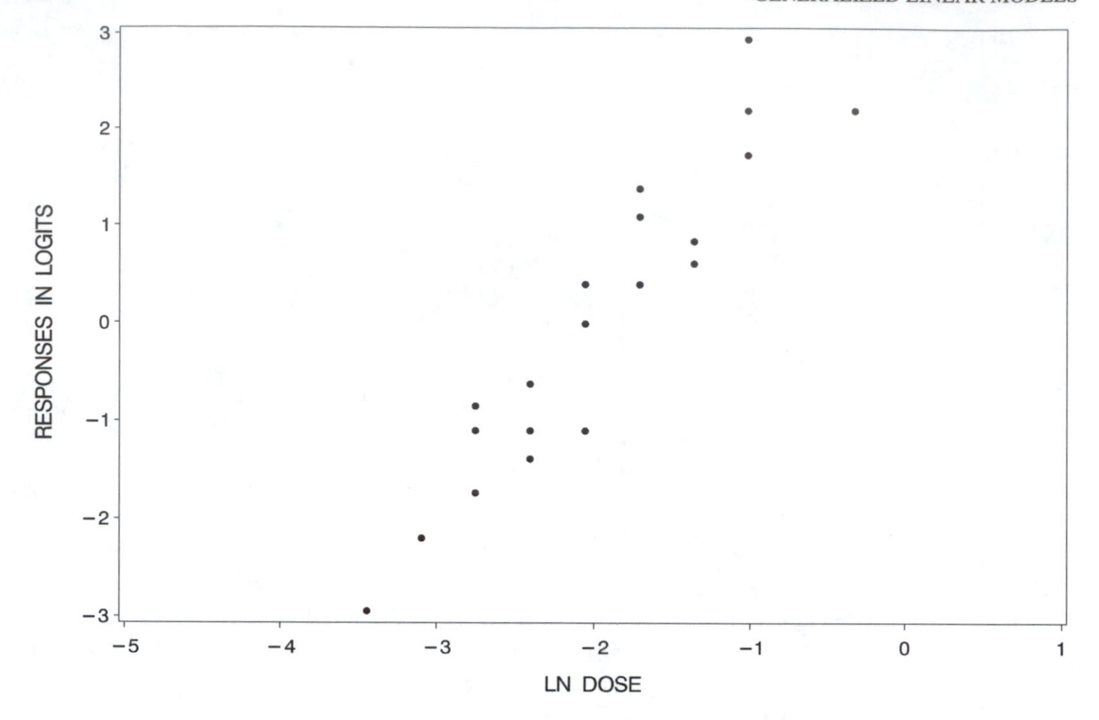

Figure 9.2 *Empirical responses in logits of the data in Table 9.3 plotted against the logarithm of dose.*

together with the estimated regression line. Both figures confirm the impression we got by means of the $-2\ln Q$-test, namely that M_2 gives good description of the data.

An additional check of M_2 is based on the so-called *deviance residuals*, which may be found on page 446 in the column `Resdev`. If e_{ij} denote the expected frequencies in M_2, these residuals are defined as

$$resdev_{ij} = \text{sign}(x_{ij} - e_{ij})\sqrt{2\{x_{ij}\ln(\frac{x_{ij}}{e_{ij}}) + (n_{ij} - x_{ij})\ln(\frac{n_{ij} - x_{ij}}{n_{ij} - e_{ij}})\}},$$

where $\text{sign}(y)$ denotes the sign of y. On closer inspection it appears that the name of the quantities is well-founded since except for the sign $resdev_{ij}$ is the square root of the contribution from the observation x_{ij} to the deviance, the $-2\ln Q$-test statistic. Under M_2 the deviance residuals are approximately independent and $N(0,1)$-distributed. Consequently, an additional check of the model M_2 consists of making a fractile diagram for the deviance residuals and in this diagram to evaluate whether the points exhibit a suitable linear pattern around the identity line. As usual the fractile diagram is made by means of the macro `normgraph` and it may be seen in Figure 9.5. Again we find no reason to question the logistic dose-response model.

On page 446 it is seen in the line `Intercept` that $\alpha \leftarrow 4.5119$ and in the line `lndose` that $\beta \leftarrow 2.2653$. Thus, by means of (9.31) we find that from this experiment the strength of the insecticide may be estimated as

$$LD_{50} = e^{-\alpha/\beta} \leftarrow e^{-4.5119/2.2653} = 0.1365$$

or, equivalently,

$$\ln(LD_{50}) \leftarrow -1.9917.$$

Note that a rough estimate of $\ln(LD_{50})$ may be obtained from Figure 9.4 as the value on the first axis (LN DOSE) corresponding to the value 0 on the second axis (LOGITS).

We conclude the discussion of the logistic dose-response model by noting that several interesting quantities may be found in the table `Observation Statistics` on page 446. For instance,

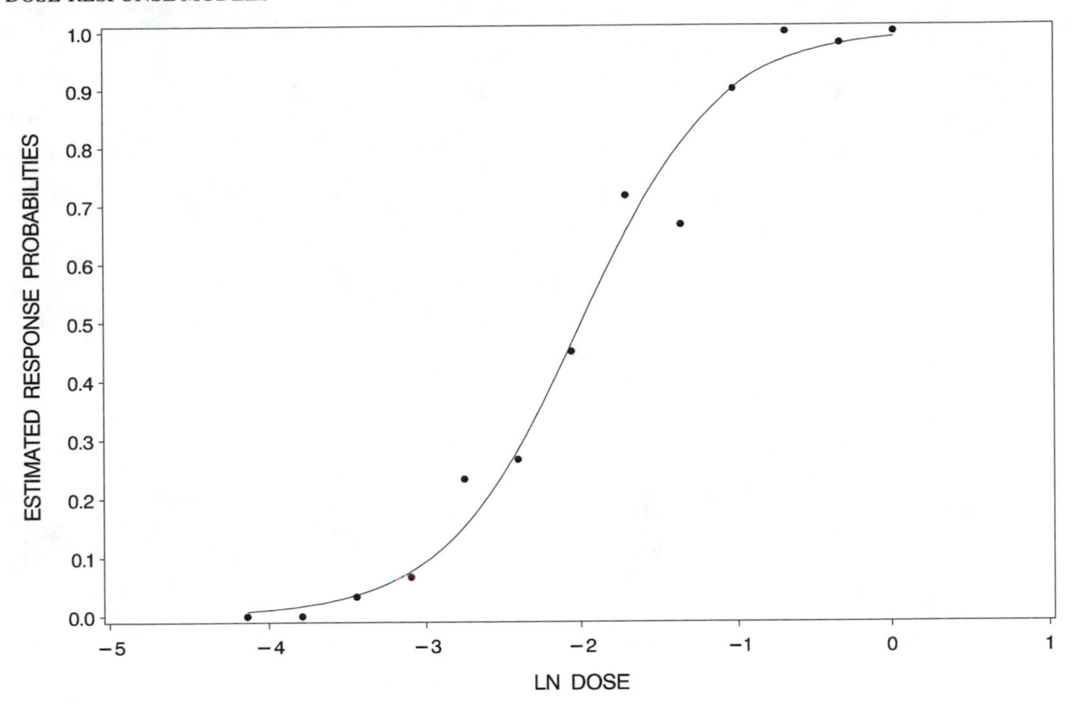

Figure 9.3 *The average relative response plotted against the logarithm of dose for the data in Table 9.3 and the estimated dose-response curve in the logistic dose-response model.*

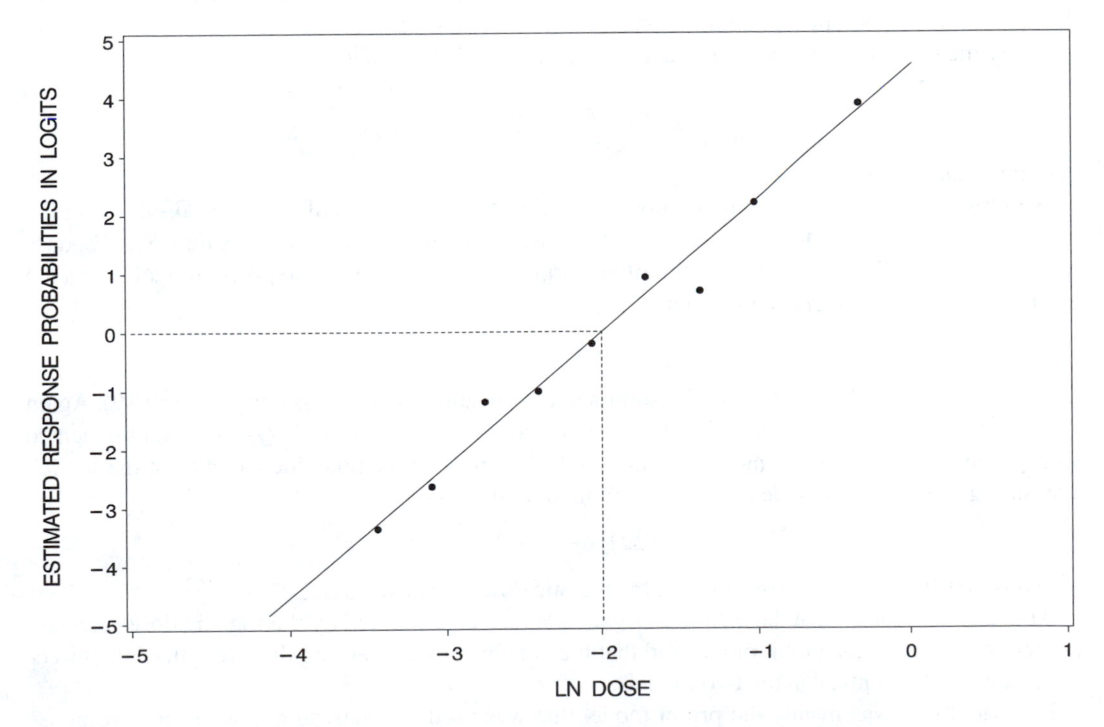

Figure 9.4 *Logits corresponding to the average relative response plotted against the logarithm of dose. Furthermore, the estimated regression line in the logistic dose-response model is drawn.*

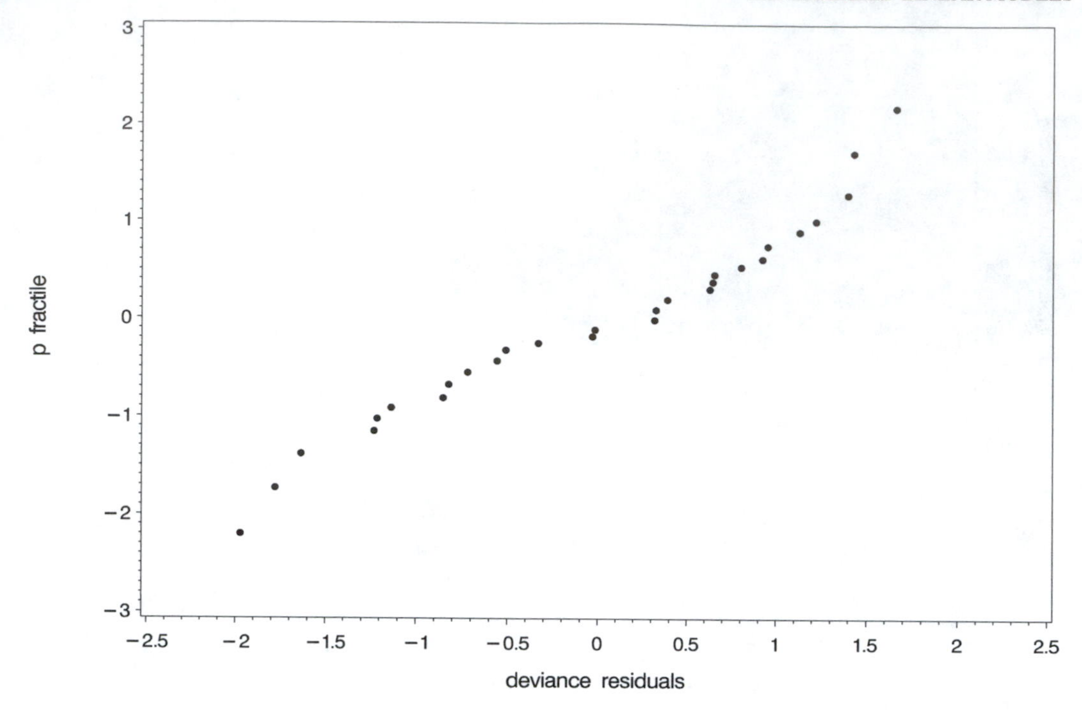

Figure 9.5 *The fractile diagram of the deviance residuals from the logistic dose-response model.*

the estimates of the response probabilities may be found under `Xbeta` and the estimated standard deviations of these quantities under `Std`. Furthermore, the 95% confidence intervals for π_{ij} are given by means of the columns `Lower` and `Upper`. Finally, the quantities,

$$r_{ij} - \hat{\pi} = \frac{x_{ij} - e_{ij}}{n_{ij}},$$

may be found under `Resraw`.

We close this example by showing how the analysis of the data could be carried out using the probit model. On page 446 it is indicated how the SAS program for the logistic dose-response model may easily be modified, such that it performs the calculations in the probit model. From the calculations in this model one finds that

$$-2\ln Q(\mathbf{x}) = 35.4882.$$

Since the degrees of freedom is 34, the significance probability is 0.398 according to (9.19). Again some of the expected frequencies are less than or equal to 5, so the $-2\ln Q$-test is supplemented with graphical control of the model. Figure 9.6 to Figure 9.8 give no evidence against the model. The strength of the insecticide is in the probit model estimated by

$$LD_{50} \leftarrow 0.1361,$$

which is nearly the same as we found in the logistic dose-response model.

Above we have seen that the data in this example may be described by the logistic dose-response model as well as by the probit model and that the conclusions concerning the strength of the insecticide are nearly identical in the two models.

Historically, it was mainly the probit model that was used for analyzing quantal dose-response experiments until the beginning of the 1940s where the logistic dose-response model emerged. The motivation behind the logistic model was partly that in physics as well as in biology we find curves of growth that may be described by the function in (9.30) and partly that the logistic model has nicer

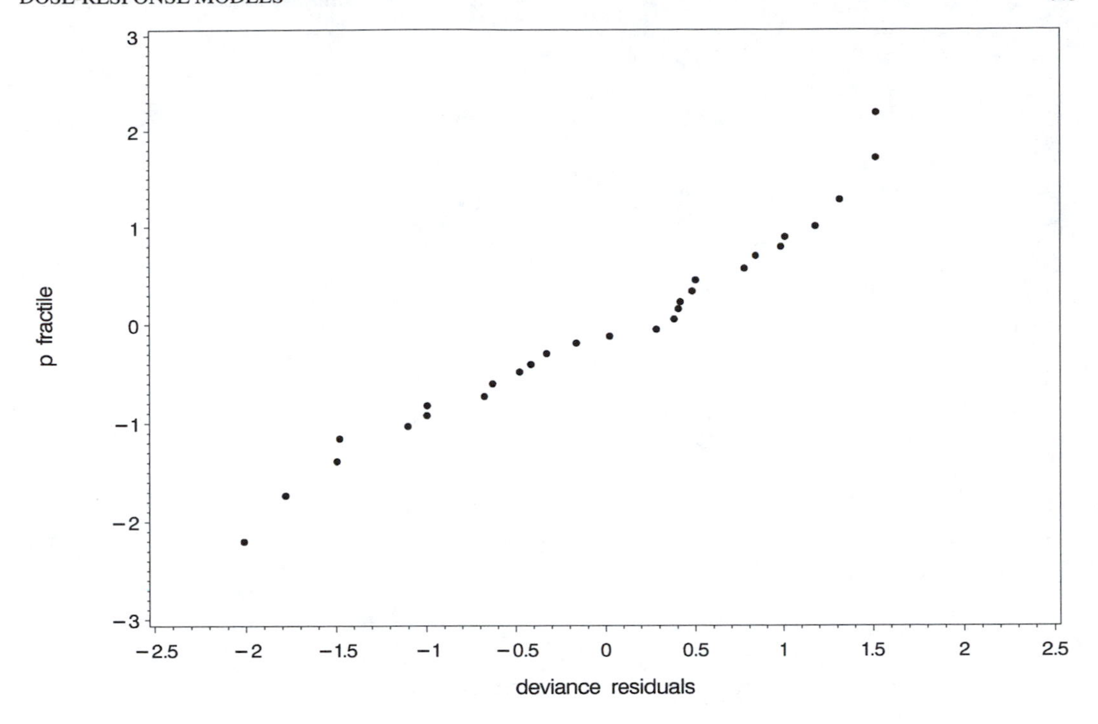

Figure 9.6 *The fractile diagram of the deviance residuals from the probit model.*

mathematical properties than the probit model, which primarily is due to the fact that the logistic model corresponds to the canonical link function of the binomial distribution. Calculations without a computer package are most easily carried out in the logistic model. However, with a computer package, as it appears from page 446, there is no difference between the models from a computational point of view. Consequently, a choice between the models may then be based on other criteria, such as, for instance, which kind of model one traditionally uses within the scientific field in question. □

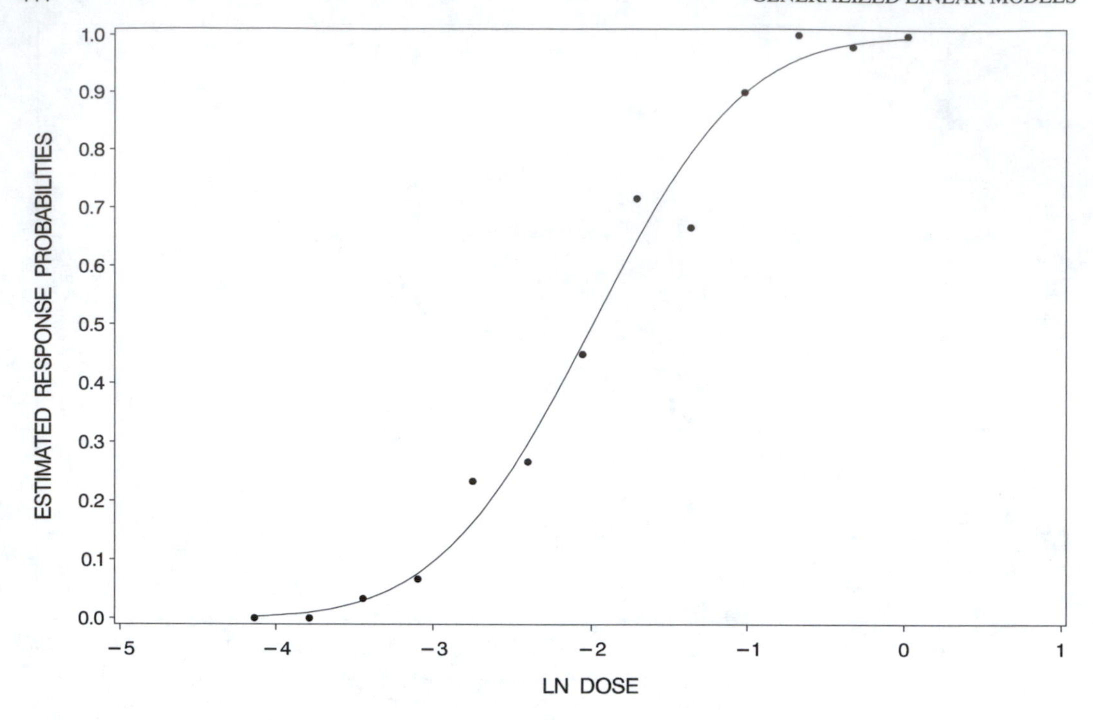

Figure 9.7 *The average relative response plotted against the logarithm of dose for data in Table 9.3 and the estimated dose-response curve in the probit model.*

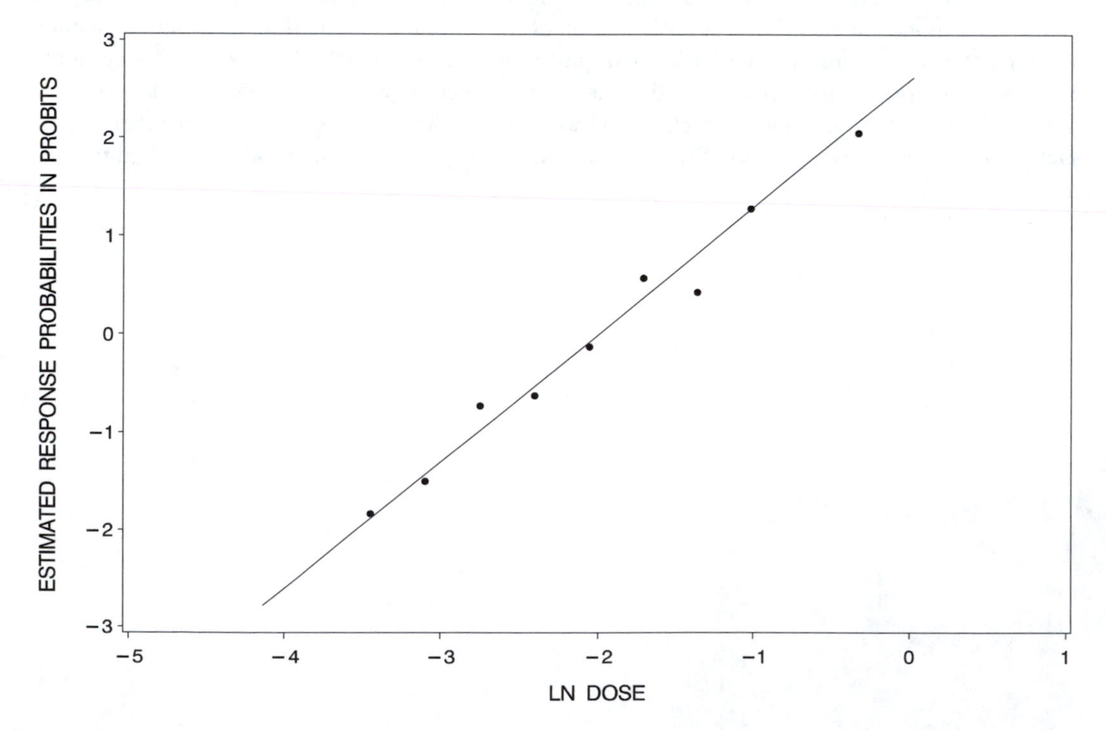

Figure 9.8 *Probits corresponding to the average relative response plotted against the logarithm of the dose. Furthermore, the estimated regression line in the probit model is drawn.*

Annex to Subsection 9.3.5

Calculations in SAS

We now show how *PROC GENMOD* performs the calculations in a regression analysis in the binomial distribution and in particular the calculations in the logistic model and in the probit model for analyzing dose-response experiments.

Example 9.10 (Continued)
The program below analyzes data by means of the logistic dose-response model. Furthermore, it is shown how the program may be modified in order to make the calculations in the probit model. Both program and corresponding output have been commented on earlier in the example.

The first part of the program records the data and calculates the logarithm of dose, the relative responses, and the corresponding values of logit. The last part contains the call of *PROC GENMOD* and the calculation of the fractile diagram of the deviance residuals.

```
TITLE1 'Example 9.10';
OPTION NODATE PS=100;

DATA ex9_10;
INPUT  dose response total @@;
relresp = response/total;
lndose=log(dose);
logit=log(relresp/(1-relresp));
CARDS;
0.0160  0  20  0.0160  0  20
0.0226  0  20  0.0226  0  20
0.0320  0  20  0.0320  1  20  0.0320  1  20
0.0453  0  20  0.0453  2  20  0.0453  2  20
0.0640  5  20  0.0640  3  20  0.0640  6  20
0.0905  4  20  0.0905  5  20  0.0905  7  20
0.1280  5  20  0.1280 12  20  0.1280 10  20
0.1810 15  20  0.1810 12  20  0.1810 16  20
0.2560 14  20  0.2560 13  20  0.2560 13  20
0.3620 18  20  0.3620 19  20  0.3620 17  20
0.5120 20  20  0.5120 20  20  0.5120 20  20
0.7240 20  20  0.7240 20  20  0.7240  9  10
1.0240 20  20  1.0240 19  19
;
RUN;

TITLE2 'the logistic dose-response model';
TITLE3 'M2';
PROC GENMOD DATA=ex9_10;
     MAKE 'obstats' OUT=tmp;
     MODEL response/TOTAL=lndose/
                                            DIST=bin
                                            LINK=logit
                                            OBSTATS;
RUN;

%normgraph(data=tmp,var=resdev);
```

The program analyzing the probit model may be obtained from the program above just by replacing `logit=log(relresp/(1-relresp))` by `probit=PROBIT(relresp)` in connection with the recording of data and `LINK=logit` by `LINK=probit` in the options to `MODEL`.

Part of the output listing from the program is listed below:

```
                          Example 9.10                                  1
                  the logistic dose-response model
                               M2

                       The GENMOD Procedure

                        Model Information

          Data Set                                WORK.EX8_8
          Distribution                              Binomial
          Link Function                                Logit
          Response Variable (Events)                response
          Response Variable (Trials)                   total
          Observations Used                               36
          Number Of Events                               348
          Number Of Trials                               709


               Criteria For Assessing Goodness Of Fit

     Criterion                 DF          Value        Value/DF

   ->Deviance                  34        38.8137          1.1416
     Scaled Deviance           34        38.8137          1.1416
     Pearson Chi-Square        34        34.1070          1.0031
     Scaled Pearson X2         34        34.1070          1.0031
     Log Likelihood                     -240.0496

Algorithm converged.


                Analysis Of Parameter Estimates

                             Standard      Wald 95%         Chi-
   Parameter  DF  Estimate    Error   Confidence Limits   Square  Pr > ChiSq

 ->Intercept   1    4.5119   0.3400   3.8454    5.1783    176.06      <.0001
 ->lndose      1    2.2653   0.1603   1.9511    2.5796    199.62      <.0001
   Scale       0    1.0000   0.0000   1.0000    1.0000

NOTE: The scale parameter was held fixed.


                     Observation Statistics

Observation   response      total      lndose        Pred       Xbeta         Std
                                      HessWgt       Lower       Upper      Resraw
                                       Reschi      Resdev    StResdev    StReschi
                                       Reslik

          1          0         20   -4.135167   0.0077249   -4.855552   0.3622948
                                    0.1533044   0.0038126   0.0155892   -0.154498
                                    -0.394589   -0.556952   -0.562641    -0.39862
                                    -0.559815
          2          0         20   -4.135167   0.0077249   -4.855552   0.3622948
                                    0.1533044   0.0038126   0.0155892   -0.154498
                                    -0.394589   -0.556952   -0.562641    -0.39862
                                    -0.559815
                                        .
                                        .
                                        .
         35         20         20   0.0237165   0.9897034   4.5655902   0.3436107
                                    0.2038119    0.980006   0.9947227   0.2059323
                                    0.4561521    0.643428   0.6513122   0.4617415
                                    0.6474026
         36         19         19   0.0237165   0.9897034   4.5655902   0.3436107
                                    0.1936213    0.980006   0.9947227   0.1956357
                                    0.4446021   0.6271361   0.6344297   0.4497728
                                    0.6308124
```

□

CHAPTER 10

Models for Directional Data

This chapter gives an introductory description of statistical analysis of data where the observations are directions in $\mathbb{R}^2$ or $\mathbb{R}^3$. In biology, geology, and related fields of natural science, there are numerous examples of such data and, historically, these data have been the most important source of inspiration for the development of statistical models for directions. As examples of two-dimensional directional data, one may think of recordings of wind directions or current directions and registrations of directions of the flying patterns of birds, whereas measurements of directions of remanent magnetism in specimens of lava is an example of a three-dimensional directional data set.

Mathematically, a direction may be represented as a unit vector. The necessary notation is introduced in Section 10.1 and the examples of directional data that will be considered in this chapter are introduced in Section 10.2. Here we shall only consider statistical analysis of directional data based on the *circular normal distribution*, which is defined in Section 10.3. Section 10.4 is concerned with the analysis of a single sample of directional data, whereas in Section 10.5 we briefly consider k samples. A primary reference concerning directional statistics is Mardia and Jupp (1999).

10.1 Notation

The models for directional data in $\mathbb{R}^2$ and $\mathbb{R}^3$ discussed here have so many points of resemblance that they may be considered together. We therefore consider directions in $\mathbb{R}^d$ where $d = 2$ or $d = 3$. Mathematically, a direction in $\mathbb{R}^d$ may be represented as a d-dimensional unit vector,

$$\mathbf{v} = (v_1, \ldots, v_d)^*,$$

belonging to the unit sphere,

$$E_d = \left\{ \mathbf{v} \in \mathbb{R}^d \mid \|\mathbf{v}\| = 1 \right\}.$$

Here $\|\mathbf{v}\|$ denotes the *length* of the vector $\mathbf{v}$, i.e.,

$$\|\mathbf{v}\| = \sqrt{v_1^2 + \cdots + v_d^2}.$$

The sum of n vectors, $\mathbf{v}_i = (v_{i1}, \ldots, v_{id})^*$, $i = 1, 2, \ldots, n$, is called the *resultant* and is denoted by $\mathbf{v}_{\bullet}$. The components of $\mathbf{v}_{\bullet}$ are obtained by adding the components of the n vectors,

$$\mathbf{v}_{\bullet} = (v_{\bullet 1}, \ldots, v_{\bullet d})^* = \left(\sum_{i=1}^{n} v_{i1}, \ldots, \sum_{i=1}^{n} v_{id} \right)^* . \tag{10.1}$$

The length of the vector $\mathbf{v}_{\bullet}$ is referred to as the *resultant length* and is denoted by R,

$$R = \|\mathbf{v}_{\bullet}\| = \sqrt{v_{\bullet 1}^2 + \cdots + v_{\bullet d}^2} \, . \tag{10.2}$$

Note that $0 \le R \le n$ and, furthermore, that $R = 0 \Leftrightarrow \mathbf{v}_{\bullet} = 0$ and $R = n \Leftrightarrow \mathbf{v}_1 = \cdots = \mathbf{v}_n$.
If $d = 2$, the *polar coordinates* (R, θ) of $\mathbf{v}_{\bullet}$ are given by

$$\begin{aligned} v_{\bullet 1} &= R\cos(\theta) \\ v_{\bullet 2} &= R\sin(\theta)\,, \end{aligned} \tag{10.3}$$

where $0^\circ \le \theta < 360^\circ$.

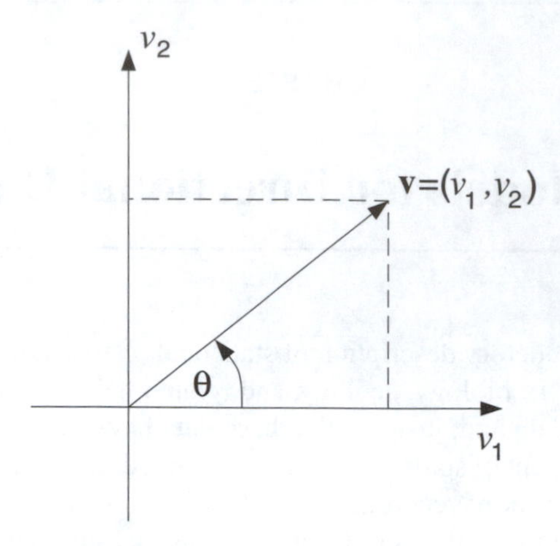

Figure 10.1 *The polar coordinates (R,θ) of the vector $\mathbf{v} = (v_1, v_2)$. R is the length of the vector $\mathbf{v}$ and the angle θ is shown on the figure.*

If $d = 3$, the *spherical coordinates* (R,θ,ϕ) of $\mathbf{v}_{\bullet}$ are given by

$$\begin{aligned} v_{\bullet 1} &= R\sin(\theta)\cos(\phi) \\ v_{\bullet 2} &= R\sin(\theta)\sin(\phi) \\ v_{\bullet 3} &= R\cos(\theta), \end{aligned} \tag{10.4}$$

where $0^\circ \leq \theta \leq 180^\circ$ and $0^\circ \leq \phi < 360^\circ$.

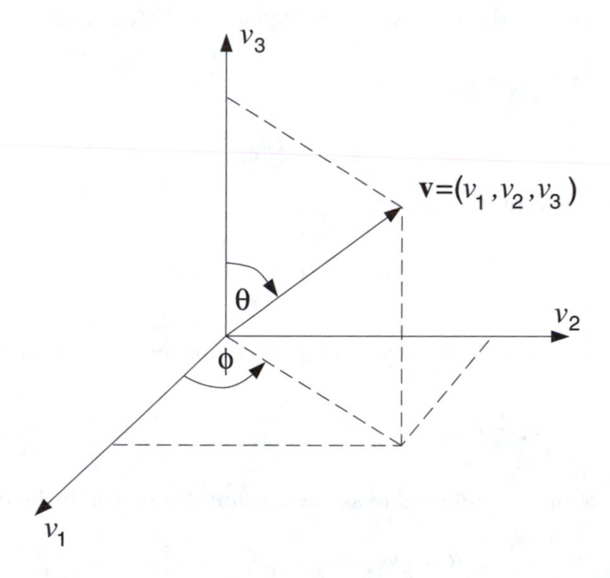

Figure 10.2 *The spherical coordinates (R,θ,ϕ) of the vector $\mathbf{v} = (v_1, v_2, v_3)$. R is the length of the vector $\mathbf{v}$ and the angles θ and ϕ are shown on the figure.*

Note that the usual geographical coordinates "degree of longitude" and "degree of latitude" (ϕ, θ') are obtained by setting $R = 1$ and $\theta' = 90^\circ - \theta$.

Some descriptive measures for directional data are discussed in Section 10.6 on page 474.

10.2 Examples

Example 10.1

$d = 3$. The three-dimensional circular normal distribution to be considered in Section 10.3 was introduced in Fisher (1953) in connection with an analysis of the directions of remanent magnetism in 9 specimens of Icelandic lava flows in 1947–1948. The measurements are given in Table 10.1.

Table 10.1 *9 measurements (in degrees) of the direction of remanent magnetism in lava flows on Iceland (Fisher, 1953). A and D denote azimuth and "dip"-angle, respectively.*

A	D	$v_1 = l$	$v_2 = m$	$v_3 = n$
5.7	73.0	0.2909	−0.0290	−0.9563
27.0	82.1	0.1225	−0.0624	−0.9905
36.9	70.1	0.2722	−0.2044	−0.9403
44.0	51.4	0.4488	−0.4334	−0.7815
50.4	69.3	0.2253	−0.2724	−0.9354
62.0	68.7	0.1705	−0.3207	−0.9317
343.2	66.1	0.3878	0.1171	−0.9143
357.6	58.8	0.5176	0.0217	−0.8554
359.0	79.5	0.1822	0.0032	−0.9833
	Σ	2.6178	−1.1803	−8.2887

Geologists do not use the spherical coordinates (θ, ϕ) in (10.4) when specifying the direction of magnetization but characterize a direction in $\mathbb{R}^3$ by means of two angles A called azimuth (declination) and D called the "dip"-angle (inclination). The relation to the spherical coordinates is given by

$$\begin{aligned} \theta &= D + 90^\circ, \quad -90^\circ < D < 90^\circ \\ \phi &= 360^\circ - A, \quad 0^\circ < A < 360^\circ. \end{aligned} \tag{10.5}$$

Traditionally, a unit vector is written as $\mathbf{v} = (l, m, n)^*$ and from (10.4) and (10.5) we have that

$$\begin{aligned} v_1 &= l = \cos(D)\cos(A) \\ v_2 &= m = -\cos(D)\sin(A) \\ v_3 &= n = -\sin(D). \end{aligned} \tag{10.6}$$

(In geology azimuth is traditionally measured clockwise from north such that east corresponds to 90°. As it appears from Figure 10.2, the angle ϕ is measured counterclockwise. Thus in this connection the v_2-axis of the spherical coordinates corresponds to west.)

On the basis of the 9 measurements that are shown in Figure 10.3, it is of interest to examine whether the directions of magnetization are uniformly distributed on the unit sphere in $\mathbb{R}^3$. If not, it seems obvious to search for a direction around which the measurements are distributed. □

Example 10.2

$d = 3$. The hypothesis that the orbital planes of the planets are random formed the basis of a prize subject arranged by l'Académie Royale des Sciences de Paris in 1732 and again in 1734, since no papers in 1732 were found worthy of receiving the prize.

Each orbital plane may be characterized by its directed unit normal (vector). The hypothesis that

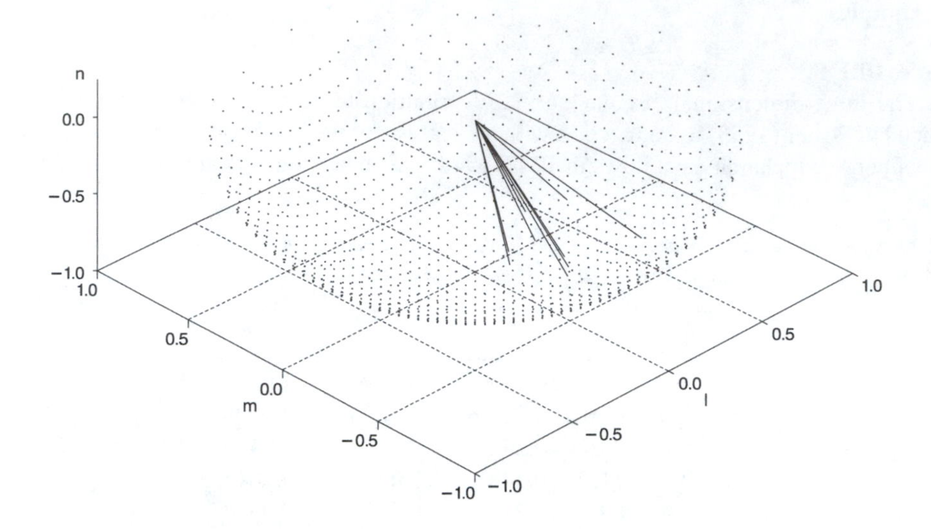

Figure 10.3 *The 9 unit vectors in Table 10.1 representing the direction of remanent magnetism in lava flows.*

the orbital planes are random is therefore equivalent to the hypothesis that the directed unit normals are uniformly distributed on the unit sphere E_3. In Table 10.2 the coordinates of the directed unit normals are given for the 9 planets in our solar system. Figure 10.4, with a graphical representation of the directed unit normals, strongly indicates that it is most unlikely that the hypothesis is true.

Table 10.2 *The directed unit normal vectors of the 9 planets (Watson, 1970)*

Planet	v_1	v_2	v_3
Mercury	0.0893	−0.0829	0.9925
Venus	0.0572	−0.0145	0.9983
Earth	0.0000	0.0000	1.0000
Mars	0.0243	−0.0213	0.9995
Jupiter	0.0227	0.0038	0.9997
Saturn	0.0402	0.0169	0.9990
Uranus	0.0128	−0.0038	0.9999
Neptune	0.0236	0.0203	0.9995
Pluto	0.2791	0.0961	0.9555
Σ	0.5492	0.0146	8.9439

From Watson, G.S., Orientation statistics in earth sciences, *Bull. Geol. Inst. Univ. Uppsala*, **2**, 1970. With permission.

As we shall see in Section 10.4, it is now possible to reject the hypothesis of random orbital planes by straightforward means.

The planets Uranus, Neptune, and Pluto were discovered in 1781, 1846, and 1930, respectively, and were therefore not known at the time of the prize subject. □

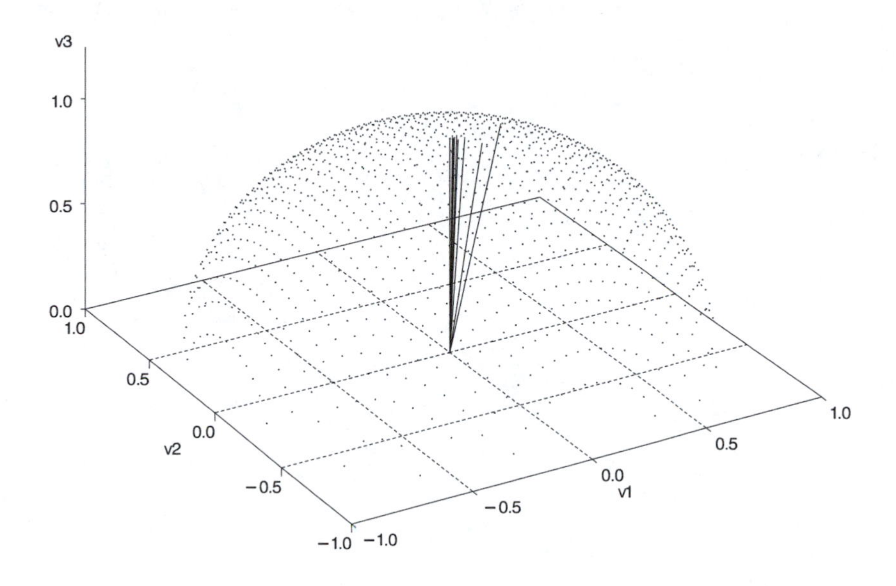

Figure 10.4 *The directed unit normal vectors in Table 10.2 of the 9 planets.*

Example 10.3
$d = 2$. The data in this example are from an experiment with 20 ants of the species *Cataglyhpis bicolor* that lives in the desert. The purpose of the experiment was to examine which influence the altitude of the sun has on the ability of the ants to find their way back to their nests. The ants are placed at a certain distance from their nests, and for each ant the angle between the direction to the nest and the direction it chooses is observed. Of the ants observed 10 were under an altitude of the sun that was artificially increased by means of mirrors, whereas the remaining 10 ants were observed under the natural altitude of the sun. The observations, ordered according to size, are reproduced in Table 10.3 and dot diagrams of the observations are shown in Figure 10.5.

Table 10.3 *The deviation between the direct way and the chosen way back to the nest for 20 ants of the species Cataglyhpis bicolor. The observations made under natural and artificial altitude of the sun, respectively, are ordered according to size (Duelli and Wehner, 1973).*

.

	Deviation (in degrees)									
Artificial altitude	−20	−10	−10	−10	−10	−10	−10	−10	0	0
Natural altitude	−10	−10	−10	0	0	0	10	10	10	20

□

Example 10.4
$d = 2$. Søren Toft, The Institute of Biological Sciences, University of Aarhus, and Yael Lubin performed a major investigation of the spider *Agelena lepida* in bushes on the mountain ridge Halukim at Sede Boker in the Negev desert. Among several other things it was observed how the spider places its web in relation to the center of the bush with the purpose of examining whether the placing of the

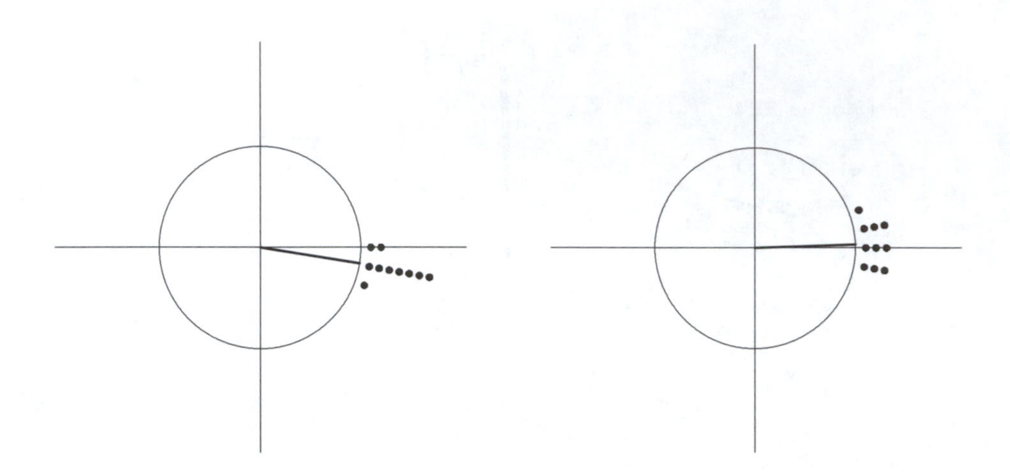

Figure 10.5 *Dot diagrams of the observations in Table 10.3.*

web depends on the direction of the slope on which the bush is located. A total of 448 spiders were observed. 103 observations were taken on slopes facing east, whereas the number of observations on slopes facing south, west, and north were 104, 63, and 96, respectively. Finally, 82 observations were taken on flat ground. The list of observations, too lengthy for this text, may be found on the file `example9_4.dat`, which can be downloaded from the site `http://www.imf.au.dk/biogeostatistics`. The grouping of the observations with respect to the direction of the slope is made, as in Table 10.4, by means of a variable with the values, 1, 2,..., 5, and the direction of the web in relation to the center of the bush is given by an angle measured clockwise from north such that, for instance, west corresponds to 270°. Dot diagrams of the five samples are shown in Figure 10.7 on page 468 and Table 10.4 gives the resultants of the observations in the different directions. □

Table 10.4 *The resultants of the measurements of the directions in which the spider Agelena lepida places its web in relation to the center of the bush. The observations are grouped according to the direction of the slope on which the bush is found*

Direction	i	n_i	$v_{i\cdot}$
East	1	103	(31.4758, 48.5000)
South	2	104	(−16.8829, 37.5614)
West	3	63	(14.6856, 2.2515)
North	4	96	(63.5078, 26.2288)
None	5	82	(−22.2580, 40.4016)

Example 10.5

$d = 3$. In Table 10.5 the resultants for 10, 11, and 15 measurements of the direction of magnetization at three different sites in "The Torridonian Sandstone Series" are given. For these data it is of interest to investigate whether there is a difference between the directions at the three sites. □

Table 10.5 *The resultants for 10, 11, and 15 measurements of the direction of magnetization at three sites in "The Torridonian Sandstone Series" (Watson, 1956)*

i	n_i	$v_{i\cdot}$
1	10	$(-2.693, -1.013, -6.370)$
2	11	$(-1.689, -3.701, -7.134)$
3	15	$(-2.369, -0.878, -11.929)$

From Watson, G.S., Analysis of dispersion on a sphere, *Monthly Notices Roy. Soc. Geophys. Suppl.*, **7**, 1956. With permission.

10.3 The Circular Normal Distribution

The class of distributions on E_d that we will consider is referred to as the circular normal distributions on E_d. A random vector $\mathbf{V}$ is said to have a *circular normal distribution* on E_d with *mean direction* $\boldsymbol{\mu} = (\mu_1, \ldots, \mu_d)^* \in E_d$ and *precision* $\kappa \geq 0$, $\mathbf{V} \sim C_d(\boldsymbol{\mu}, \kappa)$ for short, if the probability density function of $\mathbf{V}$ (with respect to the uniform measure on E_d) is

$$f(\mathbf{v}; \boldsymbol{\mu}, \kappa) = a_d(\kappa) \exp\{\kappa \boldsymbol{\mu} \cdot \mathbf{v}\}, \quad \mathbf{v} \in E_d. \tag{10.7}$$

Here $\boldsymbol{\mu} \cdot \mathbf{v}$ denotes the scalar product of the vectors, $\boldsymbol{\mu} = (\mu_1, \ldots, \mu_d)^*$ and $\mathbf{v} = (v_1, \ldots, v_d)^*$, i.e.,

$$\boldsymbol{\mu} \cdot \mathbf{v} = \mu_1 v_1 + \cdots + \mu_d v_d.$$

The scalar product $\boldsymbol{\mu} \cdot \mathbf{v}$ of two unit vectors $\boldsymbol{\mu}$ and $\mathbf{v}$ is the cosine of the angle $\angle(\boldsymbol{\mu}, \mathbf{v})$ between $\boldsymbol{\mu}$ and $\mathbf{v}$, i.e., $\boldsymbol{\mu} \cdot \mathbf{v} = \cos(\angle(\boldsymbol{\mu}, \mathbf{v}))$.

Furthermore, in (10.7) the norming constant $a_d(\kappa)$ is

$$a_d(\kappa) = \begin{cases} \dfrac{1}{2\pi I_0(\kappa)} & \text{if } d = 2 \\[2ex] \dfrac{\kappa}{4\pi \sinh(\kappa)} & \text{if } d = 3, \end{cases}$$

where I_0 is a Bessel function and where sinh denotes the hyperbolic sine function.

The circular normal distribution is often referred to as the *von Mises distribution* if $d = 2$ and as the *Fisher distribution* if $d = 3$. If $\kappa = 0$, it is seen from (10.7) that the probability density function is constant and the corresponding distribution is called the *uniform distribution* on E_d.

The probability density function (10.7) attains its maximal value $a_d(\kappa)e^{\kappa}$ at $\mathbf{v} = \boldsymbol{\mu}$ and, in addition, it is symmetric around this value in the sense that if $\mathbf{v}_1$ and $\mathbf{v}_2$ are two units vectors such that $\angle(\boldsymbol{\mu}, \mathbf{v}_1) = \pm\angle(\boldsymbol{\mu}, \mathbf{v}_2)$, then $f(\mathbf{v}_1; \boldsymbol{\mu}, \kappa) = f(\mathbf{v}_2; \boldsymbol{\mu}, \kappa)$. The parameter $\boldsymbol{\mu}$, the mean direction, plays the same role for the circular normal distribution as the mean does for the usual normal distribution. Furthermore, it may be shown that the maximal value $a_d(\kappa)e^{\kappa}$ of the probability density function increases as κ increases and therefore the probability density concentrates more and more around the value $\mathbf{v} = \boldsymbol{\mu}$ the larger κ is. This is the reason for referring to κ as the precision. In the usual normal distribution σ^{-2}, the reciprocal variance plays a similar role. Further analogies between the normal distribution and the circular normal distribution are mentioned in Section 10.7 on page 475. The von Mises distribution may be transformed to a distribution on the interval $(0°, 360°)$ or on the interval $(-180°, 180°)$. In Figure 10.6 four distributions on the interval $(-180°, 180°)$, with varying κ but all with mean direction $0°$, are shown. It is seen that the probability density function is symmetric around the mean direction, and the larger the precision κ is the more the distribution is concentrated around the mean direction.

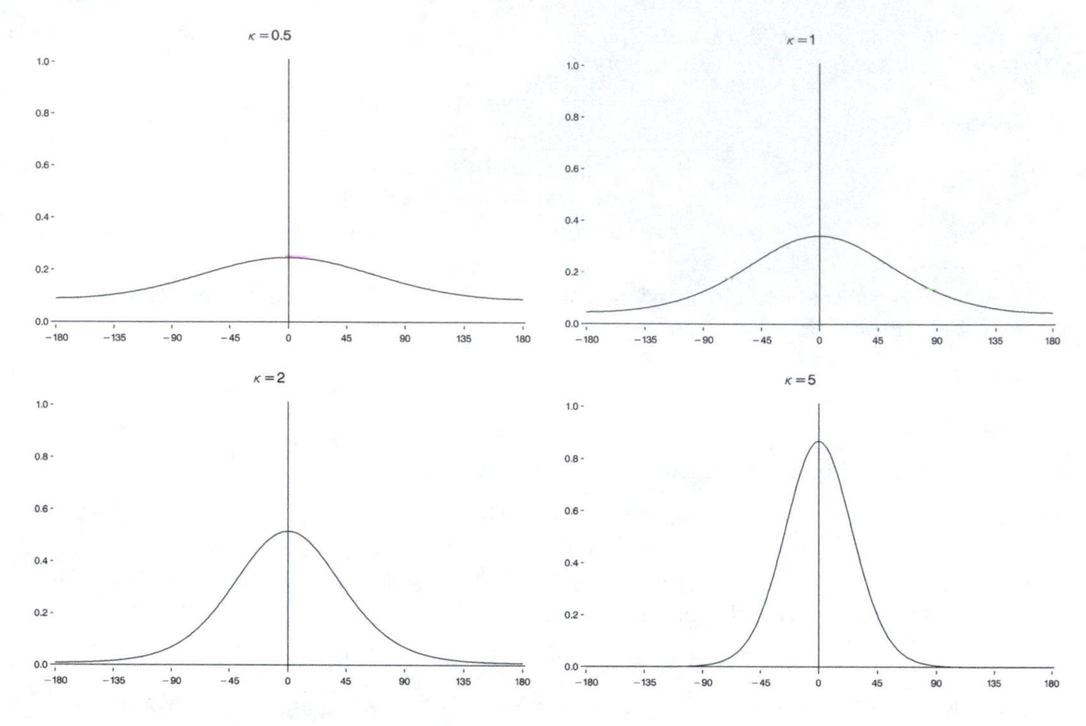

Figure 10.6 *The probability density functions of four von Mises distributions (transformed to the interval (-180°, 180°)) with mean direction 0° and with $\kappa = 0.5$, 1, 2, and 5, respectively.*

10.4 One Sample

Suppose that a set of directional data, $\mathbf{v}_1,...,\mathbf{v}_n$, may be considered as a single sample from the d-dimensional circular normal distribution, i.e., $\mathbf{v}_1,...,\mathbf{v}_n$ are realizations of independent random vectors, $\mathbf{V}_1,...,\mathbf{V}_n$, such that we have the following model:

$$M_0\colon \mathbf{V}_i \sim C_d(\boldsymbol{\mu},\kappa), \quad i = 1,2,\ldots,n.$$

Model Checking

If the sample size n is sufficiently large, the model M_0 can be checked using a goodness-of-fit test as described in Section 7.5. The test is illustrated in connection with the continuation of Example 10.4 on page 467.

Estimation

From (10.7) we get that the likelihood function for $(\boldsymbol{\mu},\kappa)$ is

$$\begin{aligned} L(\boldsymbol{\mu},\kappa) &= \prod_{i=1}^{n} a_d(\kappa)\exp\{\kappa\boldsymbol{\mu}\cdot\mathbf{v}_i\} \\ &= a_d(\kappa)^n \exp\{\kappa\boldsymbol{\mu}\cdot\mathbf{v}_{\bullet}\} \end{aligned} \tag{10.8}$$

and, consequently, that the log likelihood function is

$$l(\boldsymbol{\mu},\kappa) = n\ln(a_d(\kappa)) + \kappa\boldsymbol{\mu}\cdot\mathbf{v}_{\bullet}. \tag{10.9}$$

Since $\boldsymbol{\mu}.\mathbf{v}. \leq \frac{\mathbf{v}.}{\|\mathbf{v}.\|}.\mathbf{v}.$ for all $\boldsymbol{\mu} \in E_d$, it is seen for κ fixed that

$$\begin{aligned} l(\boldsymbol{\mu}, \kappa) &\leq l(\frac{\mathbf{v}.}{\|\mathbf{v}.\|}, \kappa) \\ &= n \ln(a_d(\kappa)) + \kappa \frac{\mathbf{v}.}{\|\mathbf{v}.\|}.\mathbf{v}. \\ &= n \ln(a_d(\kappa)) + \kappa R \,. \end{aligned}$$

Thus the maximum likelihood estimate $\hat{\boldsymbol{\mu}}$ of $\boldsymbol{\mu}$ is given by

$$\hat{\boldsymbol{\mu}} = \frac{\mathbf{v}.}{\|\mathbf{v}.\|}, \tag{10.10}$$

i.e., the estimate of the mean direction is the direction of the resultant $\mathbf{v}.$.

Furthermore, since

$$\frac{\partial l(\hat{\boldsymbol{\mu}}, \kappa)}{\partial \kappa} = n \frac{a_d'(\kappa)}{a_d(\kappa)} + R, \tag{10.11}$$

the maximum likelihood estimate $\hat{\kappa}$ of κ is found as a solution to the equation,

$$-\frac{a_d'(\kappa)}{a_d(\kappa)} = \frac{R}{n}, \tag{10.12}$$

which for $d = 2$ becomes

$$\frac{I_1(\kappa)}{I_0(\kappa)} = \frac{R}{n}, \tag{10.13}$$

where I_1 also denotes a Bessel function. For $d = 3$ the equation (10.12) becomes

$$\coth(\kappa) - \frac{1}{\kappa} = \frac{R}{n}. \tag{10.14}$$

Here coth denotes the hyperbolic cotangent function.

In order to solve equations (10.13) and (10.14), it is necessary to use numerical methods, see Appendix B. An approximative solution may be obtained by means of Table 10.6 and Table 10.7. Finally, we have the following approximations:

$d = 2$:

$$\hat{\kappa} \approx \begin{cases} \dfrac{1}{2(1 - R/n)} & \text{if } R/n \approx 1 \\ 2R/n & \text{if } R/n \approx 0. \end{cases} \tag{10.15}$$

$d = 3$:

$$\hat{\kappa} \approx \begin{cases} \dfrac{1}{1 - R/n} & \text{if } R/n \approx 1 \\ 3R/n & \text{if } R/n \approx 0. \end{cases} \tag{10.16}$$

Example 10.1 (Continued)

Suppose that the 9 measurements of the direction of the magnetization may be considered as a sample from the Fisher distribution, the three-dimensional circular normal distribution. From Table 10.1 we see that

$$\mathbf{v}. = (2.6178, -1.1803, -8.2887)^*,$$

so, using (10.2),

$$R = \|\mathbf{v}.\| = \sqrt{2.6178^2 + (-1.1803)^2 + (-8.2887)^2} = 8.7720$$

Table 10.6 *(continued on next page) $d = 2$ Maximum likelihood estimate $\hat{\kappa}$ corresponding to values of $\bar{R} = R/n$ in the interval [0.000, 0.495]*

$\bar{R}$	$\hat{\kappa}$	$\bar{R}$	$\hat{\kappa}$	$\bar{R}$	$\hat{\kappa}$	$\bar{R}$	$\hat{\kappa}$
0.000	0.00000	0.125	0.25198	0.250	0.51649	0.375	0.80988
0.005	0.01000	0.130	0.26223	0.255	0.52754	0.380	0.82254
0.010	0.02000	0.135	0.27250	0.260	0.53863	0.385	0.83528
0.015	0.03000	0.140	0.28279	0.265	0.54978	0.390	0.84812
0.020	0.04001	0.145	0.29310	0.270	0.56097	0.395	0.86105
0.025	0.05002	0.150	0.30344	0.275	0.57221	0.400	0.87408
0.030	0.06003	0.155	0.31380	0.280	0.58350	0.405	0.88721
0.035	0.07004	0.160	0.32419	0.285	0.59485	0.410	0.90043
0.040	0.08006	0.165	0.33460	0.290	0.60625	0.415	0.91376
0.045	0.09009	0.170	0.34503	0.295	0.61770	0.420	0.92720
0.050	0.10013	0.175	0.35550	0.300	0.62922	0.425	0.94075
0.055	0.11017	0.180	0.36599	0.305	0.64079	0.430	0.95440
0.060	0.12022	0.185	0.37652	0.310	0.65242	0.435	0.96818
0.065	0.13028	0.190	0.38707	0.315	0.66411	0.440	0.98207
0.070	0.14034	0.195	0.39766	0.320	0.67587	0.445	0.99608
0.075	0.15042	0.200	0.40828	0.325	0.68769	0.450	1.01022
0.080	0.16051	0.205	0.41893	0.330	0.69958	0.455	1.02449
0.085	0.17062	0.210	0.42962	0.335	0.71153	0.460	1.03889
0.090	0.18073	0.215	0.44034	0.340	0.72356	0.465	1.05342
0.095	0.19086	0.220	0.45110	0.345	0.73566	0.470	1.06810
0.100	0.20101	0.225	0.46190	0.350	0.74783	0.475	1.08292
0.105	0.21117	0.230	0.47273	0.355	0.76008	0.480	1.09788
0.110	0.22134	0.235	0.48361	0.360	0.77241	0.485	1.11300
0.115	0.23154	0.240	0.49453	0.365	0.78481	0.490	1.12828
0.120	0.24175	0.245	0.50549	0.370	0.79730	0.495	1.14372

and since $n = 9$ we find that

$$R/n = 0.9747.$$

Thus according to (10.10) we have that

$$\hat{\boldsymbol{\mu}} = (0.2984,\ -0.1346,\ -0.9449)^*$$

and from (10.6) we obtain

$$(\hat{A}, \hat{D}) = (24.2694,\ 70.8916).$$

In Example B.1 on page 535, it is shown that the solution $\hat{\kappa}$ to the equation (10.14), i.e., the maximum likelihood estimate of κ, is

$$\hat{\kappa} = 39.4791.$$

If we had used the approximation (10.16), the estimate of κ would have been 39.5257. □

Table 10.6 *(continued)* $d = 2$ *Maximum likelihood estimate* $\hat{\kappa}$ *corresponding to values of* $\bar{R} = R/n$ *in the interval [0.500, 0.995]*

$\bar{R}$	$\hat{\kappa}$	$\bar{R}$	$\hat{\kappa}$	$\bar{R}$	$\hat{\kappa}$	$\bar{R}$	$\hat{\kappa}$
0.500	1.15932	0.625	1.62255	0.750	2.36930	0.875	4.32651
0.505	1.17509	0.630	1.64506	0.755	2.41135	0.880	4.48876
0.510	1.19105	0.635	1.66798	0.760	2.45490	0.885	4.66548
0.515	1.20718	0.640	1.69134	0.765	2.50003	0.890	4.85871
0.520	1.22350	0.645	1.71516	0.770	2.54686	0.895	5.07082
0.525	1.24001	0.650	1.73945	0.775	2.59552	0.900	5.30469
0.530	1.25672	0.655	1.76423	0.780	2.64613	0.905	5.56376
0.535	1.27364	0.660	1.78953	0.785	2.69885	0.910	5.85223
0.540	1.29077	0.665	1.81536	0.790	2.75382	0.915	6.17529
0.545	1.30812	0.670	1.84177	0.795	2.81124	0.920	6.53939
0.550	1.32570	0.675	1.86876	0.800	2.87129	0.925	6.95270
0.555	1.34351	0.680	1.89637	0.805	2.93419	0.930	7.42572
0.560	1.36156	0.685	1.92463	0.810	3.00020	0.935	7.97217
0.565	1.37986	0.690	1.95357	0.815	3.06957	0.940	8.61035
0.570	1.39842	0.695	1.98322	0.820	3.14262	0.945	9.36520
0.575	1.41725	0.700	2.01363	0.825	3.21968	0.950	10.27169
0.580	1.43635	0.705	2.04482	0.830	3.30114	0.955	11.38029
0.585	1.45574	0.710	2.07685	0.835	3.38742	0.960	12.76676
0.590	1.47543	0.715	2.10976	0.840	3.47901	0.965	14.55016
0.595	1.49542	0.720	2.14359	0.845	3.57647	0.970	16.92891
0.600	1.51574	0.725	2.17840	0.850	3.68041	0.975	20.26016
0.605	1.53639	0.730	2.21425	0.855	3.79155	0.980	25.25810
0.610	1.55738	0.735	2.25120	0.860	3.91072	0.985	33.58892
0.615	1.57872	0.740	2.28930	0.865	4.03885	0.990	50.25065
0.620	1.60044	0.745	2.32865	0.870	4.17703	0.995	100.23161

Tests

Uniform Distribution: $\kappa = 0$

It is of course of interest to be able to decide whether a sample of directional data is concentrated around a certain direction or if it may be described by means of the uniform distribution corresponding to $\kappa = 0$.

If the directions are uniformly distributed, all directions are equally probable and therefore it has to be expected that the resultant length R is small. In contrast, if the data are concentrated around a certain direction, then R is large. Consequently, the hypothesis H_0: $\kappa = 0$ may be tested by considering the distribution of R or, equivalently, the distribution of R/n under H_0. Since large values of R/n are critical for H_0, the significance probability is

$$p_{obs}(r/n) = P(R/n \geq r/n) = 1 - F_{R/n}(r/n). \quad (10.17)$$

Here r denotes the observed resultant length and $F_{R/n}$ denotes the distribution function of R/n,

Table 10.7 *(continued on next page) $d = 3$ Maximum likelihood estimate $\hat{\kappa}$ corresponding to values of $\bar{R} = R/n$ in the interval [0.000, 0.495]*

$\bar{R}$	$\hat{\kappa}$	$\bar{R}$	$\hat{\kappa}$	$\bar{R}$	$\hat{\kappa}$	$\bar{R}$	$\hat{\kappa}$
0.000	0.00000	0.125	0.37857	0.250	0.77990	0.375	1.23466
0.005	0.01500	0.130	0.39402	0.255	0.79681	0.380	1.25459
0.010	0.03000	0.135	0.40951	0.260	0.81381	0.385	1.27469
0.015	0.04501	0.140	0.42503	0.265	0.83089	0.390	1.29497
0.020	0.06001	0.145	0.44060	0.270	0.84806	0.395	1.31542
0.025	0.07503	0.150	0.45621	0.275	0.86533	0.400	1.33605
0.030	0.09005	0.155	0.47186	0.280	0.88269	0.405	1.35688
0.035	0.10508	0.160	0.48756	0.285	0.90015	0.410	1.37789
0.040	0.12012	0.165	0.50330	0.290	0.91771	0.415	1.39911
0.045	0.13516	0.170	0.51909	0.295	0.93538	0.420	1.42053
0.050	0.15023	0.175	0.53493	0.300	0.95315	0.425	1.44216
0.055	0.16530	0.180	0.55083	0.305	0.97103	0.430	1.46401
0.060	0.18039	0.185	0.56678	0.310	0.98902	0.435	1.48609
0.065	0.19550	0.190	0.58278	0.315	1.00713	0.440	1.50839
0.070	0.21062	0.195	0.59884	0.320	1.02536	0.445	1.53093
0.075	0.22576	0.200	0.61497	0.325	1.04370	0.450	1.55372
0.080	0.24093	0.205	0.63115	0.330	1.06218	0.455	1.57676
0.085	0.25611	0.210	0.64740	0.335	1.08078	0.460	1.60005
0.090	0.27132	0.215	0.66371	0.340	1.09951	0.465	1.62362
0.095	0.28656	0.220	0.68009	0.345	1.11838	0.470	1.64745
0.100	0.30182	0.225	0.69654	0.350	1.13739	0.475	1.67157
0.105	0.31711	0.230	0.71306	0.355	1.15654	0.480	1.69599
0.110	0.33242	0.235	0.72965	0.360	1.17584	0.485	1.72070
0.115	0.34777	0.240	0.74632	0.365	1.19529	0.490	1.74573
0.120	0.36315	0.245	0.76307	0.370	1.21490	0.495	1.77108

where R is the resultant length of n independently and uniformly distributed unit vectors in $\mathbb{R}^d$. In Table 10.8 and Table 10.9 relevant fractiles of R/n for different values of n are found for $d = 2$ and $d = 3$, respectively. For large values of n the significance probability in (10.17) may be calculated using the approximation,

$$nd(R/n)^2 \approx \chi^2(d), \quad \text{for } n \to \infty. \tag{10.18}$$

The approximation in (10.18) may be applied if the sample size $n > 100$.

The test in (10.17), here derived by a heuristic argument, may be shown to be the likelihood ratio test of the hypothesis $\kappa = 0$ in M_0, the model for a single sample from the circular normal distribution.

Example 10.1 (Continued)
For the directions of magnetization we have from above that $n = 9$ and $r = 8.7720$ and, therefore, that $r/n = 0.975$. From Table 10.9 it is seen that the significance probability of the hypothesis H_0:

Table 10.7 *(continued)* $d = 3$ *Maximum likelihood estimate* $\hat{\kappa}$ *corresponding to values of* $\bar{R} = R/n$ *in the interval [0.500, 0.995]*

$\bar{R}$	$\hat{\kappa}$	$\bar{R}$	$\hat{\kappa}$	$\bar{R}$	$\hat{\kappa}$	$\bar{R}$	$\hat{\kappa}$
0.500	1.79676	0.625	2.58826	0.750	3.98905	0.875	7.99999
0.505	1.82278	0.630	2.62825	0.755	4.07197	0.880	8.33333
0.510	1.84915	0.635	2.66912	0.760	4.15819	0.885	8.69565
0.515	1.87589	0.640	2.71093	0.765	4.24793	0.890	9.09091
0.520	1.90300	0.645	2.75371	0.770	4.34143	0.895	9.52381
0.525	1.93051	0.650	2.79751	0.775	4.43894	0.900	10.00000
0.530	1.95842	0.655	2.84238	0.780	4.54076	0.905	10.52632
0.535	1.98674	0.660	2.88836	0.785	4.64719	0.910	11.11111
0.540	2.01550	0.665	2.93551	0.790	4.75857	0.915	11.76471
0.545	2.04470	0.670	2.98389	0.795	4.87528	0.920	12.50000
0.550	2.07437	0.675	3.03355	0.800	4.99772	0.925	13.33333
0.555	2.10451	0.680	3.08456	0.805	5.12635	0.930	14.28571
0.560	2.13515	0.685	3.13699	0.810	5.26167	0.935	15.38462
0.565	2.16630	0.690	3.19091	0.815	5.40422	0.940	16.66667
0.570	2.19799	0.695	3.24640	0.820	5.55463	0.945	18.18182
0.575	2.23023	0.700	3.30354	0.825	5.71357	0.950	20.00000
0.580	2.26304	0.705	3.36243	0.830	5.88181	0.955	22.22222
0.585	2.29646	0.710	3.42314	0.835	6.06021	0.960	25.00000
0.590	2.33049	0.715	3.48580	0.840	6.24971	0.965	28.57143
0.595	2.36516	0.720	3.55051	0.845	6.45141	0.970	33.33333
0.600	2.40050	0.725	3.61738	0.850	6.66652	0.975	40.00000
0.605	2.43654	0.730	3.68655	0.855	6.89645	0.980	50.00000
0.610	2.47331	0.735	3.75814	0.860	7.14279	0.985	66.66667
0.615	2.51083	0.740	3.83232	0.865	7.40737	0.990	100.00000
0.620	2.54914	0.745	3.90923	0.870	7.69228	0.995	200.00000

$\kappa = 0$ is less than 0,1%. Consequently, H_0 is rejected, which was to be expected since $\hat{\kappa} = 39.4791$. □

Example 10.2 (Continued)
Under the assumption that the directions of the directed unit normals in Table 10.2 may be considered as a sample from the three-dimensional circular normal distribution, $C_3(\boldsymbol{\mu}, \kappa)$, the hypothesis that the orbital planes are random may be formulated as H_0: $\kappa = 0$. From (10.2) and Table 10.2 we get that $n = 9$ and $r = 8.9608$ and so $r/n = 0.996$. Consequently, from Table 10.9 it is seen that the hypothesis of random orbital planes is rejected (even by a test with significance level 0.1%). With this we have given a solution to the prize subject from 1732 having, however, included the planets Uranus, Neptune, and Pluto. It is left to the reader to show that the conclusion is the same if these planets are excluded from the calculations. □

Table 10.8 $d = 2$ *Selected fractiles of the distribution of* $\bar{R} = R/n$, *where* R *is the resultant length of* n *independently and uniformly distributed unit vectors*

$n \setminus p$	0.900	0.950	0.975	0.990	0.999
2	0.9877	0.9969	0.9992	0.9999	1.0000
3	0.8828	0.9405	0.9699	0.9879	0.9997
4	0.7624	0.8456	0.9009	0.9454	0.9881
5	0.6726	0.7581	0.8239	0.8859	0.9629
6	0.6149	0.6901	0.7560	0.8251	0.9274
7	0.5710	0.6416	0.7016	0.7705	0.8873
8	0.5340	0.6022	0.6598	0.7246	0.8471
9	0.5037	0.5686	0.6244	0.6872	0.8087
10	0.4781	0.5402	0.5938	0.6550	0.7742
11	0.4560	0.5158	0.5674	0.6266	0.7436
12	0.4367	0.4943	0.5443	0.6017	0.7162
13	0.4196	0.4753	0.5237	0.5795	0.6916
14	0.4045	0.4584	0.5053	0.5595	0.6692
15	0.3908	0.4431	0.4887	0.5415	0.6488
16	0.3785	0.4293	0.4736	0.5251	0.6302
17	0.3672	0.4166	0.4599	0.5101	0.6130
18	0.3569	0.4051	0.4472	0.4963	0.5971
19	0.3474	0.3944	0.4356	0.4836	0.5824
20	0.3387	0.3846	0.4248	0.4718	0.5687
21	0.3305	0.3754	0.4148	0.4608	0.5560
22	0.3230	0.3669	0.4055	0.4505	0.5440
23	0.3159	0.3589	0.3967	0.4409	0.5327
24	0.3093	0.3514	0.3885	0.4319	0.5222
25	0.3030	0.3444	0.3808	0.4234	0.5122
30	0.2767	0.3147	0.3481	0.3874	0.4697
35	0.2562	0.2915	0.3227	0.3593	0.4362
40	0.2397	0.2728	0.3021	0.3365	0.4090
45	0.2260	0.2573	0.2850	0.3176	0.3863
50	0.2144	0.2442	0.2705	0.3015	0.3670
60	0.1958	0.2230	0.2471	0.2755	0.3358
70	0.1813	0.2065	0.2289	0.2553	0.3113
80	0.1696	0.1932	0.2142	0.2389	0.2916
90	0.1599	0.1822	0.2020	0.2254	0.2751
100	0.1517	0.1729	0.1917	0.2139	0.2612

Known Mean Direction: $\boldsymbol{\mu} = \boldsymbol{\mu}_0$

Just as it may be of interest for a normally distributed sample to test that the mean has a specific value, it is also of interest for a sample from the circular normal distribution to decide if it may be assumed that the mean direction has a specific value $\boldsymbol{\mu}_0$. Thus we now consider the hypothesis,

$$H_{0\mu}\colon \boldsymbol{\mu} = \boldsymbol{\mu}_0.$$

A general discussion of test of $H_{0\mu}$ is too extensive for this book, so we restrict our considerations to a test of $H_{0\mu}$ in the situation where κ is large.

For large values of κ it may be shown that we have the approximation,

$$2(n-R) \approx \kappa^{-1}\chi^2(f), \quad f = (n-1)(d-1) \quad \text{for } \kappa \to \infty. \tag{10.19}$$

Table 10.9 $d = 3$ *Selected fractiles of the distribution of* $\bar{R} = R/n$, *where* R *is the resultant length of* n *independently and uniformly distributed unit vectors*

$n \setminus p$	0.900	0.950	0.975	0.990	0.999
2	0.9487	0.9747	0.9874	0.9950	0.9995
3	0.8162	0.8728	0.9113	0.9445	0.9827
4	0.7119	0.7758	0.8246	0.8725	0.9419
5	0.6373	0.7002	0.7511	0.8046	0.8923
6	0.5828	0.6422	0.6921	0.7467	0.8432
7	0.5405	0.5969	0.6447	0.6983	0.7981
8	0.5062	0.5600	0.6059	0.6578	0.7579
9	0.4777	0.5291	0.5733	0.6236	0.7223
10	0.4535	0.5028	0.5454	0.5941	0.6909
11	0.4327	0.4801	0.5212	0.5684	0.6631
12	0.4145	0.4602	0.4999	0.5457	0.6382
13	0.3984	0.4426	0.4810	0.5255	0.6159
14	0.3840	0.4269	0.4642	0.5074	0.5957
15	0.3711	0.4127	0.4489	0.4910	0.5773
16	0.3594	0.3998	0.4351	0.4761	0.5605
17	0.3488	0.3881	0.4225	0.4625	0.5451
18	0.3390	0.3774	0.4109	0.4500	0.5308
19	0.3301	0.3675	0.4002	0.4384	0.5176
20	0.3218	0.3583	0.3903	0.4277	0.5054
21	0.3140	0.3498	0.3811	0.4177	0.4939
22	0.3069	0.3419	0.3725	0.4084	0.4832
23	0.3002	0.3344	0.3645	0.3997	0.4732
24	0.2939	0.3275	0.3570	0.3916	0.4638
25	0.2880	0.3209	0.3499	0.3839	0.4549
30	0.2630	0.2933	0.3199	0.3512	0.4169
35	0.2436	0.2717	0.2965	0.3257	0.3871
40	0.2279	0.2543	0.2776	0.3050	0.3629
45	0.2149	0.2398	0.2619	0.2878	0.3427
50	0.2039	0.2276	0.2485	0.2732	0.3255
60	0.1862	0.2079	0.2271	0.2497	0.2978
70	0.1724	0.1925	0.2103	0.2314	0.2760
80	0.1613	0.1801	0.1968	0.2165	0.2585
90	0.1520	0.1699	0.1856	0.2043	0.2439
100	0.1444	0.1601	0.1755	0.1939	0.2315

If we let $\tilde{\kappa}$ denote the maximum likelihood estimate of the precision κ calculated in the approximating distribution, it may be shown that $\tilde{\kappa}$ is given by

$$\tilde{\kappa}^{-1} = \frac{2(n-R)}{f} = \begin{cases} \dfrac{2(n-R)}{n-1} & \text{if } d = 2 \\[2ex] \dfrac{n-R}{n-1} & \text{if } d = 3, \end{cases} \tag{10.20}$$

compare with (10.15) and (10.16). From (10.19) and (10.20) we have that

$$\tilde{\kappa}^{-1} \approx \kappa^{-1}\chi^2(f)/f, \quad f = (n-1)(d-1) \quad \text{for } \kappa \to \infty. \tag{10.21}$$

The result in (10.21) is analogous to a well-known result in the usual normal distribution. Recall that the estimate s^2 of the variance σ^2 in a sample from a normal distribution has a $\sigma^2\chi^2(f)/f$ with $f=n-1$, and note also that the reciprocal variance, σ^{-2}, is the precision. The result for the normal distribution is exact, whereas the result in (10.21) is an approximation that is valid for large values of κ.

It seems natural to base a test of $H_{0\mu}$ on the difference, $\hat{\boldsymbol{\mu}}-\boldsymbol{\mu}_0$, between the estimate of the mean direction $\boldsymbol{\mu}$ and the value of the mean direction $\boldsymbol{\mu}_0$ to be tested. Since

$$(\hat{\boldsymbol{\mu}}-\boldsymbol{\mu}_0)\cdot\mathbf{v}. = (\frac{\mathbf{v}.}{\|\mathbf{v}.\|}-\boldsymbol{\mu}_0)\cdot\mathbf{v}.$$
$$= R-\boldsymbol{\mu}_0\cdot\mathbf{v}.$$

and, furthermore, since it may be shown that

$$2(R-\boldsymbol{\mu}_0\cdot\mathbf{v}.)/f_1 \approx \kappa^{-1}\chi^2(f_1)/f_1, \quad f_1=d-1, \quad \text{for } \kappa\to\infty, \tag{10.22}$$

and, in addition, that this quantity is independent of $\tilde{\kappa}^{-1}$ (for $\kappa\to\infty$), we obtain from (10.21) and (10.22) that

$$F \;=\; \frac{2(R-\boldsymbol{\mu}_0\cdot\mathbf{v}.)/f_1}{\tilde{\kappa}^{-1}} \approx F(f_1,f) \quad \text{for } \kappa\to\infty, \tag{10.23}$$

where $F(f_1,f)$ is the F distribution with (f_1,f) degrees of freedom, see page 166 for the definition of the F distribution.

If the quantity in (10.23) is used to test $H_{0\mu}$, large values of F are critical, i.e., the significance probability of $H_{0\mu}$ may be calculated as

$$p_{obs} = 1-F_{F(f_1,f)}(F), \tag{10.24}$$

where $f_1=d-1$ and $f=(n-1)(d-1)$.

For $d=2$ the significance probability (10.24) may be used if $n>3$ and $\boldsymbol{\mu}_0\cdot\mathbf{v}. > 5n/6$, and for $d=3$ if either $3\le n\le 8$ and $\boldsymbol{\mu}_0\cdot\mathbf{v}. > n/2$ or if $n>8$ and $\boldsymbol{\mu}_0\cdot\mathbf{v}. > 3n/5$.

Example 10.1 (Continued)

For the sake of illustration suppose that we want to investigate if azimuth A and the "dip"-angle D of the direction of magnetization have a particular value, for instance,

$$(A,D) = (0^\circ,60^\circ)$$

or, equivalently, because of (10.6) if

$$H_{0\mu}\colon \boldsymbol{\mu} = (0.5000, 0.0000, -0.8660)^*(=\boldsymbol{\mu}_0).$$

Since $\mathbf{v}. = (2.6178, -1.1803, -8.2887)^*$, we have that $\boldsymbol{\mu}_0\cdot\mathbf{v}. = 8.4869$, and since $R=8.7720$ we obtain from (10.20) that $\tilde{\kappa}^{-1}=0.0285$. Thus we find that

$$F = \frac{2(8.7720-8.4869)/2}{0.0285} = 10.004,$$

and since $n=9$ and $\boldsymbol{\mu}_0\cdot\mathbf{v}. > 3n/5 = 5.4$, the significance probability of $H_{0\mu}$ is

$$p_{obs} = 1-F_{F(2,16)}(10.004) = 0.0015$$

according to (10.24) and $H_{0\mu}$ is rejected. □

10.5 Several Samples

Suppose that we assume as a model for k samples of directional data in $\mathbb{R}^d$, $d = 2$ or $d = 3$,

$$\begin{array}{ccccc} \mathbf{v}_{11} & \cdots & \mathbf{v}_{1j} & \cdots & \mathbf{v}_{1n_1} \\ \vdots & \ddots & \vdots & \ddots & \vdots \\ \mathbf{v}_{i1} & \cdots & \mathbf{v}_{ij} & \cdots & \mathbf{v}_{in_i} \\ \vdots & \ddots & \vdots & \ddots & \vdots \\ \mathbf{v}_{k1} & \cdots & \mathbf{v}_{kj} & \cdots & \mathbf{v}_{kn_k}, \end{array}$$

that $\{\mathbf{V}_{ij} \mid j = 1,\ldots,n_i, i = 1,\ldots,k\}$ are independent random vectors such that

$$\mathbf{V}_{ij} \sim C_d(\boldsymbol{\mu}_i, \kappa_i), \ j = 1,\ldots,n_i, \ \ i = 1,\ldots,k,$$

i.e., we have k independent samples from the circular normal distribution. We denote this model by M_0. Here we will briefly discuss *"one-way analysis of variance"* for such data by which we mean test of the hypotheses of homogeneity of precisions, i.e.,

$$H_{0\kappa}\colon \kappa_1 = \cdots = \kappa_k$$

and test of the hypothesis of identical mean directions,

$$H_{0\boldsymbol{\mu}}\colon \boldsymbol{\mu}_1 = \cdots = \boldsymbol{\mu}_k.$$

We shall only test the hypothesis $H_{0\kappa}$ in the model M_0, but we will give the tests of $H_{0\boldsymbol{\mu}}$ both in M_0 and in the model for k samples with a common precision:

$$M_1\colon \mathbf{V}_{ij} \sim C_d(\boldsymbol{\mu}_i, \kappa), \quad j = 1,\ldots,n_i, i = 1,\ldots,k.$$

Notice that $H_{0\kappa}$ is the hypothesis corresponding to the reduction from M_0 to M_1.

Again we restrict ourselves to situations where the κ*'s are large.*

With $\mathbf{v}_{i\cdot}$ and R_i we denote the resultant and the resultant length in the ith sample, $i = 1,\ldots,k$, i.e.,

$$\mathbf{v}_{i\cdot} = \mathbf{v}_{i1} + \cdots + \mathbf{v}_{in_i}$$

and

$$R_i = \|\mathbf{v}_{i\cdot}\|.$$

The resultant and the resultant length of all the vectors are denoted by $\mathbf{v}_{\cdot\cdot}$ and R, i.e.,

$$\mathbf{v}_{\cdot\cdot} = \mathbf{v}_{1\cdot} + \cdots + \mathbf{v}_{k\cdot}\,.$$

and

$$R = \|\mathbf{v}_{\cdot\cdot}\|.$$

Furthermore, we let

$$n_{\cdot} = n_1 + \cdots + n_k$$

and

$$R_{\cdot} = R_1 + \cdots + R_k.$$

Calculation of the estimates in M_0 may be carried out by means of the template in Table 10.10, which is analogous to the template for calculations for normally distributed data on page 116.

From (10.20) and (10.21) we have that

$$\tilde{\kappa}_i^{-1} = \frac{2(n_i - R_i)}{f_{(i)}} \approx \kappa_i^{-1}\chi^2(f_{(i)})/f_{(i)}, \tag{10.25}$$

where

$$f_{(i)} = (n_i - 1)(d - 1).$$

Table 10.10 *Template for calculations in a one-way analysis of variance for directional data*

i	n_i	$\mathbf{v}^*_{i\cdot}$	R_i	$2(n_i - R_i)$	$f_{(i)}$	$\tilde{\kappa}_i^{-1}$
1	n_1	$\mathbf{v}^*_{1\cdot}$	$\lVert\mathbf{v}_{1\cdot}\rVert$	$2(n_1 - R_1)$	$(n_1-1)(d-1)$	$2(n_1 - R_1)/f_{(1)}$
...	...	...	...	...	...	...
i	n_i	$\mathbf{v}^*_{i\cdot}$	$\lVert\mathbf{v}_{i\cdot}\rVert$	$2(n_i - R_i)$	$(n_i-1)(d-1)$	$2(n_i - R_i)/f_{(i)}$
...	...	...	...	...	...	...
k	n_k	$\mathbf{v}^*_{k\cdot}$	$\lVert\mathbf{v}_{k\cdot}\rVert$	$2(n_k - R_k)$	$(n_k-1)(d-1)$	$2(n_k - R_k)/f_{(k)}$
Σ	$n.$	$\mathbf{v}^*_{\cdot\cdot}$	$R.$	$2(n. - R.)$	f_1	$\tilde{\kappa}^{-1}$

Test of Homogeneity of Precisions: $H_{0\kappa}$

If $k = 2$, the hypothesis $H_{0\kappa}$ is tested by means of an F-test,

$$F = \frac{\tilde{\kappa}_1^{-1}}{\tilde{\kappa}_2^{-1}} \sim F(f_{(1)}, f_{(2)}), \tag{10.26}$$

where both *small and large values are critical.*

If $k \geq 3$ and $f_{(i)} \geq 2$, $i = 1, \ldots, k$, the hypothesis $H_{0\kappa}$ is tested by means of Bartlett's test statistic,

$$Ba = \frac{1}{C}\left\{ f_1 \ln(\tilde{\kappa}^{-1}) - \sum_{i=1}^{k} f_{(i)} \ln(\tilde{\kappa}_i^{-1}) \right\}, \tag{10.27}$$

where

$$f_1 = \sum_{i=1}^{k} f_{(i)}, \tag{10.28}$$

$$\begin{aligned} \tilde{\kappa}^{-1} &= \frac{f_{(1)}\tilde{\kappa}_1^{-1} + \cdots + f_{(k)}\tilde{\kappa}_k^{-1}}{f_{(1)} + \cdots + f_{(k)}} \\ &= \frac{2(n_1 - R_1) + \cdots + 2(n_k - R_k)}{f_1} \\ &= \frac{2(n. - R.)}{f_1} \end{aligned} \tag{10.29}$$

and

$$C = 1 + \frac{1}{3(k-1)}\left\{ \left(\sum_{i=1}^{k} \frac{1}{f_{(i)}}\right) - \frac{1}{f_1} \right\}.$$

Here $\tilde{\kappa}$ denotes the estimate of the common precision κ in M_1. The significance probability of Bartlett's test is

$$p_{obs} = 1 - F_{\chi^2(k-1)}(Ba). \tag{10.30}$$

Test of Identical Mean Directions, $H_{0\mu}$*, in* M_1

In M_1 the hypothesis $H_{0\mu}$ is tested by means of an F-test using the test statistic,

$$F = \frac{2(R. - R)/f_2}{\tilde{\kappa}^{-1}} \approx F(f_2, f_1) \tag{10.31}$$

where $f_2 = (k-1)(d-1)$. Since large values of F are critical, the significance probability of $H_{0\mu}$ is

$$p_{obs} = 1 - F_{F(f_2, f_1)}(F). \tag{10.32}$$

The test (10.31) may be used if $R > 0.95n.$ when $d = 2$, and if $R > 2n./3$ when $d = 3$.

Test of Identical Mean Directions, $H_{0\mu}$, in M_0

A test of the hypothesis $H_{0\mu}$ of identical mean directions without assuming a common precision, i.e., in the model M_0, is based on the following quantities:

$$R^\star = \tilde{\kappa}_1 R_1 + \cdots + \tilde{\kappa}_k R_k, \tag{10.33}$$

$$\tilde{\mathbf{v}} = \tilde{\kappa}_1 \mathbf{v}_{1\cdot} + \cdots + \tilde{\kappa}_k \mathbf{v}_{k\cdot} \tag{10.34}$$

and

$$\tilde{R} = \|\tilde{\mathbf{v}}\|. \tag{10.35}$$

These quantities are easily calculated from Table 10.10. As a test statistic of the hypothesis $H_{0\mu}$ one considers

$$X^2 = 2(R^\star - \tilde{R}) \approx \chi^2(f_2), \tag{10.36}$$

where $f_2 = (k-1)(d-1)$. Since large values of X^2 are critical, the significance probability of $H_{0\mu}$ is:

$$p_{obs} = 1 - F_{\chi^2(f_2)}(X^2). \tag{10.37}$$

The test (10.36) may be used for large data sets only. There should be at least 25 observations in each sample, i.e., $n_i \geq 25$, and if $d = 2$, then, in addition, the conditions, $\tilde{\kappa}_i^{-1} \leq 0.5$, $i = 1, 2, \ldots, k$, should be fulfilled.

If the hypothesis $H_{0\mu}$ is not rejected in this situation, the common mean direction $\boldsymbol{\mu}$ is estimated by

$$\tilde{\boldsymbol{\mu}} = \frac{\tilde{\mathbf{v}}}{\tilde{R}}. \tag{10.38}$$

Example 10.3 (Continued)
$d = 2$. We start by calculating the unit vectors corresponding to the measurements of the angle between the direction to the nest and the directions the ants choose. If $i = 1$ and 2 denote artificial and natural altitude, respectively, and a_{ij} the angle measured for the jth ant, the corresponding unit vector is

$$\mathbf{v}_{ij} = (\cos(a_{ij}), \sin(a_{ij}))^*.$$

We consider the model M_0, i.e., we assume that the vectors $\mathbf{v}_{ij}$, $j = 1, \ldots, 10$, $i = 1, 2$, may be considered as two samples from the von Mises distribution, one sample for each of the two kinds of altitudes. For the data the calculations in Table 10.10 are:

i	n_i	$\mathbf{v}_{i\cdot}^*$	R_i	$2(n_i - R_i)$	$f_{(i)}$	$\tilde{\kappa}_i^{-1}$
1	10	(9.8333, −1.5576)	9.9559	0.0881	9	0.0098
2	10	(9.8485, 0.3420)	9.8545	0.2910	9	0.0323
Σ	20	(19.6818, −1.2156)	19.8104	0.3791	18	0.0211

Let θ_i denote the angle corresponding to the mean direction $\boldsymbol{\mu}_i$, i.e.,

$$\boldsymbol{\mu}_i = (\cos(\theta_i), \sin(\theta_i))^*, \quad i = 1, 2.$$

From the table with the calculations and formula (10.10), it is seen that the estimates and the mean directions and the corresponding angles are

$$\boldsymbol{\mu}_1 \leftarrow \frac{(9.8333, -1.5576)^*}{9.9559} = (0.9877, -0.1564)^*, \quad \theta_1 \leftarrow -9.00^\circ$$

$$\boldsymbol{\mu}_2 \leftarrow \frac{(9.8485, 0.3420)^*}{9.8545} = (0.9994, 0.0347)^*, \qquad \theta_2 \leftarrow 1.99^\circ.$$

The estimated mean directions are indicated in Figure 10.5 on page 10.5.

Since $k = 2$, the hypothesis $H_{0\kappa}$ of identical precisions may be tested using the F-test in (10.26). We find that

$$F = \frac{0.0323}{0.0098} = 3.296 \sim\sim F(9,9)$$

and since both small and large values are critical, the significance probability becomes 0.090 and $H_{0\kappa}$ is not contradicted by the data and we adopt the model M_1 as the basis for further analysis. By means of (10.29) the following estimate of the common precision is found:

$$\tilde{\kappa}^{-1} = \frac{0.3791}{18} = 0.0211.$$

In order to test the hypothesis $H_{0\mu}$ of identical mean directions, we first calculate the resultant length of the total sample,

$$R = \sqrt{19.6818^2 + (-1.2156)^2} = 19.7193.$$

Because $R > 19.00 = 0.95n.$, the hypothesis $H_{0\mu}$ may be tested by means of formula (10.31). Since

$$F = \frac{2(19.8104 - 19.7193)}{0.0211} = 8.635,$$

the significance probability is

$$p_{obs} = 1 - F_{F(1,18)}(8.635) = 0.0088$$

and $H_{0\mu}$ is rejected. Consequently, the experiment shows that the altitude of the sun somehow influences the ant's ability to find their way back to the nest. □

Example 10.5 (Continued)
$d = 3$. We assume that the measurements of the directions of magnetization may be considered as three samples from the Fisher distribution of sizes 10, 11, and 15, respectively. The calculations in the template in Table 10.10 are easily done using Table 10.5, as shown in the table below.

i	n_i	$\mathbf{v}_{i.}^*$	R_i	$2(n_i - R_i)$	$f_{(i)}$	$\tilde{\kappa}_i^{-1}$
1	10	$(-2.693, -1.013, -6.370)$	6.9897	6.0207	18	0.3345
2	11	$(-1.689, -3.701, -7.134)$	8.2124	5.5751	20	0.2788
3	15	$(-2.369, -0.878, -11.929)$	12.1936	5.6128	28	0.2005
Σ	36	$(-6.751, -5.592, -25.433)$	27.3957	17.2086	66	$\overline{0.2607}$

First, we examine the hypothesis $H_{0\kappa}$ of homogeneity of the precisions in the three samples. Since the estimate of the common precision according to (10.29) is

$$\tilde{\kappa}^{-1} = \frac{17.2086}{66} = 0.2607,$$

we find from (10.27) that Bartlett's test statistic for $H_{0\kappa}$ is

$$Ba = 1.5099.$$

Thus the significance probability of $H_{0\kappa}$ is

$$p_{obs} = 1 - F_{\chi^2(2)}(1.5099) = 0.470$$

and $H_{0\kappa}$ is not rejected and the model M_0 may be reduced to M_1.

Since $R = 26.9014 > 24 = 2 \times 36/3$, the test in (10.31) for the hypothesis $H_{0\mu}$ of identical mean directions may be used. We find that

$$F = \frac{2(27.3957 - 26.9014)/4}{0.2607} = 0.9480,$$

which has to be evaluated in the $F(4,66)$ distribution. The significance probability of $H_{0\boldsymbol{\mu}}$ is

$$p_{obs} = 1 - F_{(4,66)}(0.9480) = 0.442$$

and the hypothesis is not rejected. The common mean direction $\boldsymbol{\mu}$ and the common precision κ are estimated by

$$\hat{\boldsymbol{\mu}} = \frac{\mathbf{v}_{\cdot\cdot}}{R} = (-0.251, -0.208, -0.945)^*$$

and

$$\tilde{\kappa}_{02}^{-1} = \frac{2(n_{\cdot} - R)}{2(n_{\cdot} - 1)} = 0.2600,$$

respectively. □

Example 10.4 (Continued)

$d = 2$. Again we consider the model M_0 for the five samples that are shown in Figure 10.7 on page 468. Here the sample sizes are so large that the model may be checked by a test for goodness of fit as described in Section 7.5. This is done below. Table 10.11 below gives the calculations in Table 10.10 for this data set. The maximum likelihood estimates in M_0 are found using the resultants and the resultant lengths of the five samples. The estimates of the mean directions are found directly from Table 10.11 by means of formula (10.10), whereas the estimates of the precisions are determined from (10.13) by means of the numerical methods mentioned in Appendix B. The equation

$$\hat{\boldsymbol{\mu}} = (\cos(\hat{\theta}), \sin(\hat{\theta}))^* \tag{10.39}$$

gives the connection between the estimate of the mean direction $\boldsymbol{\mu}$ and the estimate of the corresponding angle θ. The estimated mean directions are shown in Figure 10.7 and the estimates of the angles and the precisions in the five samples may be seen in Table 10.12.

Table 10.11 *The calculations in Table 10.10 for the data in Example 10.4*

i	n_i	$\mathbf{v}_{i\cdot}^*$	R_i	$2(n_i - R_i)$	$f_{(i)}$	$\tilde{\kappa}_i^{-1}$
1	103	(31.4758, 48.5000)	57.8184	90.3631	102	0.8859
2	104	(−16.8829, 37.5614)	41.1812	125.6376	103	1.2198
3	63	(14.6856, 2.2515)	14.8572	96.2856	62	1.5530
4	96	(63.5078, 26.2288)	68.7109	54.5782	95	0.5745
5	82	(−22.2580, 40.4016)	46.1271	71.7459	81	0.8858
Σ	448	(70.5283, 154.9433)	228.6948	438.6104	443	$\overline{0.9901}$

As mentioned above, the sizes of the five samples in this example are so large that the model M_0 – five samples from the von Mises distribution – may be checked by a goodness-of-fit test as mentioned in Section 7.5. We show how this test may be calculated by considering the sample corresponding to the direction east. Let x denote the angle (measured in degrees [°]) corresponding to the vector $\mathbf{v}$, i.e.,

$$\mathbf{v} = (\cos(x), \sin(x))^*. \tag{10.40}$$

From (10.39) and (10.40) it is seen that

$$\hat{\boldsymbol{\mu}} \cdot \mathbf{v} = \cos(\hat{\theta})\cos(x) + \sin(\hat{\theta})\sin(x) = \cos(x - \hat{\theta}). \tag{10.41}$$

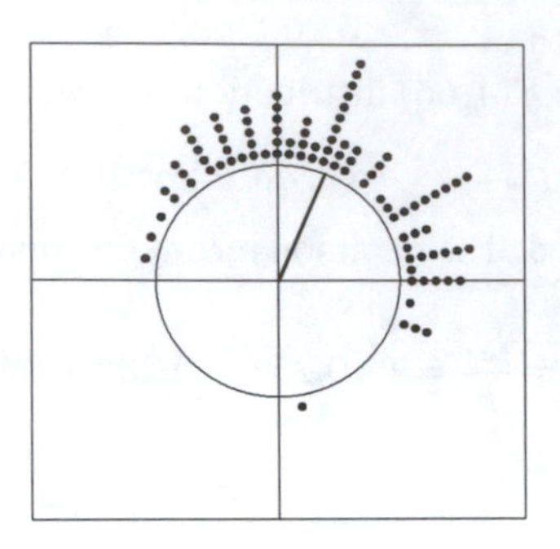

direction = west

direction = none

direction = east

direction = south

Figure 10.7 *5 dot diagrams of the observations in Example 10.4 corresponding to the different directions of the slope. The empirical mean directions are indicated in the diagrams.*

Table 10.12 *The maximum likelihood estimates of the angles corresponding to the mean directions and the precisions for the different directions according to which the data in Example 10.4 are grouped*

Direction	i	$\hat{\theta}_i$	$\hat{\kappa}_i$
East	1	57.02	1.3665
South	2	114.20	0.8636
West	3	8.72	0.4854
North	4	22.44	2.1147
None	5	118.85	1.3748

Consequently, if the numbers, $y_i, i = 0, 1, \ldots, m$, define a partition of the interval from 0° to 360° into m intervals, the expected frequency in the ith interval may be calculated according to (10.7) and (10.41) as

$$e_i = n \int_{y_{i-1}}^{y_i} \frac{1}{2\pi I_0(\hat{\kappa})} \exp\{\hat{\kappa}\cos(x - \hat{\theta})\} dx,$$

where n is the sample size. This integral may be calculated using numerical integration.

For the sample corresponding to the direction east we consider a partition of the interval from 0° to 360° into $m = 12$ intervals, i.e., $y_i = 30 \times i$, $i = 0, 1, \ldots, 12$. The observed frequencies **a** and the expected frequencies **e** are seen in Table 10.13. Before the $-2\ln Q$-test statistic in (7.59) is calculated, it is necessary to merge some of the intervals in order to meet the demand that the expected frequencies are larger than or equal to 5. If this is done as in Table 10.13, the number of groups is 8 and since we have estimated 2 parameters, κ and θ, the degrees of freedom is $f = 8 - 1 - 2 = 5$. By means of Table 10.13 and the formulas (7.59) and (7.60) we find that

$$-2\ln Q(\mathbf{a}) = 4.509$$

and

$$p_{obs}(\mathbf{a}) = 1 - F_{\chi^2(5)}(4.509) = 0.479.$$

Thus the goodness-of-fit test gives no reason for doubting that the sample corresponding to the direction east may be considered as von Mises distributed. Similar calculations for the four remaining samples give no occasion for rejecting the model M_0.

Figure 10.7 shows that we must use skepticism with respect to the hypothesis $H_{0\mu}$ of identity of the mean directions, in particular when the large sample sizes are taken into account. Bartlett's test statistic for the hypothesis $H_{0\kappa}$ of identical precisions calculated by SAS (see page 473) is

$$Ba = 22.5638 \sim\sim \chi^2(4),$$

which corresponds to a significance probability of 0.0002. Thus the hypothesis $H_{0\mu}$ cannot be tested by means of formula (10.31). The sample sizes are all larger than 25, but as it is seen from Table 10.11 the condition that $\tilde{\kappa}_i^{-1} \leq 0.5$, $i = 1, 2, \ldots, 5$, is not fulfilled, so the test in (10.36) may not be applied either. In other words, we are not able to test the hypothesis of identical mean directions using the approximative methods discussed here, which shows the limitations of the methods.

In order to illustrate the test in (10.36), we calculate the test, well aware that the condition for applying the test is not satisfied. Using (10.33) and Table 10.11 we obtain

$$R^\star = \frac{57.8184}{0.8859} + \frac{41.1812}{1.2198} + \frac{14.8572}{1.5530} + \frac{68.7109}{0.5745} + \frac{46.1271}{0.8858} = 280.2677.$$

Similarly, by means of (10.34) we find that

$$\tilde{\mathbf{v}} = (116.5622, 178.2547)^*$$

Table 10.13 *Calculation of the test for goodness of fit of the von Mises distribution for the sample corresponding to the east direction in Example 10.4*

Interval(°)	a		e	
0 – 30	17		15.508	
30 – 60	24		21.129	
60 – 90	13		20.397	
90 – 120	17		14.072	
120 – 150	10	13	7.612	11.394
150 – 180	3		3.782	
180 – 210	5		2.084	
210 – 240	0	5	1.504	5.149
240 – 270	0		1.561	
270 – 300	2	6	2.307	6.644
300 – 330	4		4.337	
330 – 360	8		8.708	

and therefore

$$\tilde{R} = \|\tilde{\mathbf{v}}\| = 212.9824.$$

Thus the test statistic in (10.36) becomes

$$X^2 = 2(280.2677 - 212.9824) = 134.5707.$$

If the conditions of the test were satisfied, this quantity should be evaluated in the $\chi^2(4)$-distribution which indicates a strongly significant deviation from $H_{0\mu}$.

Rounding off the example, we have demonstrated that the five samples may be described by means of the von Mises distribution and that the approximative theory in this chapter cannot be used to test the hypothesis of identical mean directions. However, on the basis of Figure 10.7 it seems reasonable to assume that the mean directions are not identical. Why the placing of the web in relation to the center of the bush depends on the direction of the slope on which the bush is placed is a question the biologist, not the statistician, must try to answer. □

Annex to Chapter 10

Calculations in SAS

Most of the calculations in this chapter may be carried out in SAS by means of the macro `direction`, which among others things makes the calculations in Table 10.10. Input to the macro is a data set, specified be means of `data=`, which for each sample contains the sample size and the resultant vector of the sample. These quantities are specified by means of `n=` and `v1= , v2= (,v3=)`, respectively. The dimension of the observations is specified by `d=`.

The output from the macro `direction` does not require many comments since it agrees with the notation in the rest of the chapter. Sometimes it is natural to give the directions in terms of the angles. Before the macro is used it is then necessary to transform the angles to unit vectors by means of the formula (10.3) if $d = 2$ or (10.4) if $d = 3$. How this may be done in SAS is illustrated below.

Example 10.5 (Continued)
$d = 3$. In this example we only have the size of the three samples and the corresponding resultants, so the data is read directly into the program by means of the command `DATALINES`. The following program makes the relevant calculations:

```
TITLE1 'Example 10.5';

DATA obs;
INPUT sample_size v1 v2 v3;
DATALINES;
10 -2.693 -1.013 -6.370
11 -1.689 -3.701 -7.134
15 -2.369 -0.878 -11.929
;
RUN;

%direction(data=obs,d=3,n=sample_size,v1=v1,v2=v2,v3=v3);
```

The output is:

```
                          Example 10.5                                    1

                        Macro "DIRECTION"

                        Dataset: OBS

                  Calculations in k samples:

 i    n     v1        v2        v3          R         2(n-R)    f  kappa inverse
----------------------------------------------------------------------------------
 1   10  -2.6930   -1.0130   -6.3700     6.9897     6.0207    18      0.3345
 2   11  -1.6890   -3.7010   -7.1340     8.2124     5.5751    20      0.2788
 3   15  -2.3690   -0.8780  -11.9290    12.1936     5.6128    28      0.2005
-----------------------------------------------------------------=================
     36  -6.7510   -5.5920  -25.4330    27.3957    17.2086    66      0.2607


              Bartlett's test for equality of precisions:

                 C           Bartlett statistic      p value
              1.0210              1.5099              0.4700

              F-test for equality of mean directions:

          R.              R              F-statistic  p value
       27.3957         26.9014             0.9479      0.4420
```

□

Example 10.3 (Continued)

$d = 2$. The two samples of measurements of angles are recorded by means of the following program segment in which the variable sun_height distinguishes between the two samples:

```
TITLE1 'Example 10.3';

DATA ex10_3;
INPUT sun_height angle @@;
DATALINES;
1 -20 1 -10 1 -10 1 -10 1 -10 1 -10 1 -10 1 -10 1   0 1   0
2 -10 2 -10 2 -10 2   0 2   0 2   0 2  10 2  10 2  10 2 20
;
RUN;
```

In the next program segment the unit vectors are calculated from the angles, which are given in degrees. Since SAS uses radians in connection with calculation of the trigonometric functions, it is necessary to convert degrees into radians before cos and sin are calculated.

```
DATA direct;
SET ex10_3;
pi=4*atan(1);
r1=cos(angle*pi/180);
r2=sin(angle*pi/180);
RUN;
```

The next program segment calculates the sample size and the resultant vector for each of the samples and write these to the data set obs, which is used as input to direction.

```
PROC SORT DATA=direct;
   BY sun_height;
RUN;

DATA directsum;
SET direct (KEEP= sun_height r1 r2);
RUN;

PROC MEANS DATA=directsum NOPRINT;
   VAR r1 r2;
   OUTPUT OUT=obs N=sample_size SUM=c s;
   BY sun_height;
RUN;
```

The last part of the program is the call of the macro direction:

```
%direction(data=obs,d=2,n=sample_size,v1=c,v2=s);
```

The output from the program is seen below.

```
                          Example 10.3                             1

                        Macro "DIRECTION"

                        Dataset: OBS


                    Calculations in k samples:

 i    n       v1        v2          R         2(n-R)      f   kappa inverse
---------------------------------------------------------------------------
 1   10     9.8333   -1.5576     9.9559      0.0881       9      0.0098
 2   10     9.8485    0.3420     9.8545      0.2910       9      0.0323
-------------------------------------------------------------=============
     20    19.6819   -1.2155    19.8104      0.3792      18      0.0211
```

```
                 F-test for equality of precisions:

        Degrees of freedom      F-statistic     p value
          9        9               3.3027        0.0898


                 F-test for equality of mean directions:

        R.              R                F-statistic  p value
      19.8104         19.7194              8.6426      0.0088
```

□

Example 10.4 (Continued)

$d = 2$. In this example the data set is so large that the recording of the data, i.e., a variable indicating the five groups and a variable with the measurements of the angles, is done from a permanent SAS data set `sas.example10_4`. Apart from this, the rest of the program including transformation of degrees to radians is identical to the program in Example 10.3. The recording is carried out by means of the lines:

```
TITLE1 'Example 10.4';

DATA data;
LIBNAME sas 'c:\biogeostatistics\chapter9\examples';
SET sas.example10_4;
RUN;
```

The output is:

```
                              Example 10.4                              1

                            Macro "DIRECTION"

                            Dataset: OBS


                      Calculations in k samples:

 i    n        v1         v2          R          2(n-R)       f   kappa inverse
----------------------------------------------------------------------------------
 1  103     31.4758    48.5000     57.8184     90.3631      102       0.8859
 2  104    -16.8829    37.5614     41.1812    125.6376      103       1.2198
 3   63     14.6856     2.2515     14.8572     96.2856       62       1.5530
 4   96     63.5078    26.2288     68.7109     54.5782       95       0.5745
 5   82    -22.2580    40.4016     46.1271     71.7459       81       0.8858
-------------------------------------------------------------------==============
    448     70.5283   154.9433    228.6948    438.6104      443       0.9901


                 Bartlett's test for equality of precisions:

                  C          Bartlett statistic      p value
               1.0047             22.5638             0.0002


                 F-test for equality of mean directions:

NOTE! The hypothesis that the precisions are equal is rejected.

            NOTE! The condition that R > 0.95*n is not satisfied.

        R.              R                F-statistic  p value
     228.6948        170.2401              29.5199     0.0000
```

Note that the output points out that the hypothesis of identical precisions is rejected and, furthermore, that the condition for using the test of identical mean directions is not fulfilled. □

Supplement to Chapter 10

10.6 Descriptive Measures for Directional Data

Let $\mathbf{v}_1,\ldots,\mathbf{v}_n$ be d-dimensional unit vectors and let

$$\mathbf{v}_{\bullet} = \mathbf{v}_1 + \cdots + \mathbf{v}_n,$$

$$R = \left\|\mathbf{v}_{\bullet}\right\|$$

and

$$\bar{R} = R/n$$

denote the *resultant*, the *resultant length*, and the *normed resultant length*, respectively. (The quantity $\left\|\mathbf{v}_{\bullet}\right\|$ and with it R is a realization of a random variable and, consequently, in accordance with the notation in the book, it should not be denoted by means of a capital letter. However, it is a tradition in the literature on directional data to use R as the designation of the resultant length. Occasionally, we write $R = r$, which means that the corresponding random has the value r).

The vector,

$$\bar{\mathbf{v}}_{\bullet} = \frac{\mathbf{v}_{\bullet}}{n},$$

is sometimes referred to as the *empirical mean vector* and the unit vector,

$$\tilde{\mathbf{v}}_{\bullet} = \frac{\mathbf{v}_{\bullet}}{\left\|\mathbf{v}_{\bullet}\right\|} = \frac{\mathbf{v}_{\bullet}}{R} = \frac{\bar{\mathbf{v}}_{\bullet}}{\bar{R}},$$

as the *empirical mean direction*.

For a set of data, $x_1,\ldots,x_n$, consisting of observations on the reel axis $\mathbb{R}$ the empirical mean,

$$\bar{x}_{\bullet} = \frac{1}{n}\sum_{i=1}^{n} x_i,$$

may be characterized as the value for which the function,

$$S(x) = \sum_{i=1}^{n} (x_i - x)^2, \quad x \in \mathbb{R},$$

attains its minimum value. We usually denote this value by *SSD*, i.e.,

$$SSD = S(\bar{x}_{\bullet}) = \sum_{i=1}^{n} (x_i - \bar{x}_{\bullet})^2,$$

and use *SSD* as a measure of the total variation of the data.

In order to obtain a measure of the total variation of a set of directional data, $\mathbf{v}_1,\ldots,\mathbf{v}_n$, let us consider the function,

$$S(\mathbf{v}) = \sum_{i=1}^{n} \|\mathbf{v}_i - \mathbf{v}\|^2, \quad \mathbf{v} \in E_d.$$

Since

$$\begin{aligned}
S(\mathbf{v}) &= \sum_{i=1}^{n} (\mathbf{v}_i - \mathbf{v}) \cdot (\mathbf{v}_i - \mathbf{v}) \\
&= \sum_{i=1}^{n} (\mathbf{v}_i \cdot \mathbf{v}_i + \mathbf{v} \cdot \mathbf{v} - 2\mathbf{v}_i \cdot \mathbf{v}) \\
&= \sum_{i=1}^{n} (1 + 1 - 2\mathbf{v}_i \cdot \mathbf{v}) \\
&= 2(n - \mathbf{v}_{\cdot} \cdot \mathbf{v}) \\
&\geq 2(n - \mathbf{v}_{\cdot} \cdot \frac{\mathbf{v}_{\cdot}}{\|\mathbf{v}_{\cdot}\|}) \\
&= 2(n - R),
\end{aligned}$$

the following analogies are obtained:

$$\begin{array}{ccc}
\textit{reel data} & & \textit{directional data} \\
\bar{x}_{\cdot} & \sim & \tilde{\mathbf{v}}_{\cdot} = \dfrac{\mathbf{v}_{\cdot}}{\|\mathbf{v}_{\cdot}\|} \\
SSD & \sim & 2(n-R).
\end{array}$$

10.7 Further Analogies

If we assume that $x_1, \ldots, x_n$ is a sample from normal distribution, i.e.,

$$X_i \sim N(\mu, \sigma^2), \quad i = 1, \ldots, n,$$

and that $\mathbf{v}_1, \ldots, \mathbf{v}_n$ is a sample from the circular normal distribution,

$$\mathbf{V}_i \sim C_d(\boldsymbol{\mu}, \kappa), \quad i = 1, \ldots, n,$$

we also have (as mentioned on page 453) the analogies,

$$\begin{array}{ccc}
N(\mu, \sigma^2) & & C_d(\boldsymbol{\mu}, \kappa) \\
\mu & \sim & \boldsymbol{\mu} \\
\sigma^2 & \sim & \kappa^{-1}.
\end{array}$$

Furthermore, in the normal distribution we have that

$$\bar{X}_{\cdot} \sim N(\mu, \frac{\sigma^2}{n})$$

and

$$SSD \sim \sigma^2 \chi^2(n-1).$$

The corresponding analogies for the circular normal distribution are

$$\tilde{\mathbf{V}}_{\cdot} \mid R = r \sim C_d(\boldsymbol{\mu}, r\kappa)$$

and

$$2(n-R) \approx \kappa^{-1} \chi^2((n-1)(d-1)) \text{ for } \kappa \to \infty.$$

Exercises for Chapter 10

Exercise 10.1 Consider the model for a sample from the $C_d(\boldsymbol{\mu}, \kappa)$ distribution,

$$M_0\colon \mathbf{V}_i \sim C_d(\boldsymbol{\mu}, \kappa), \quad i = 1, \ldots, n.$$

(1) Show that the $-2\ln Q$-test statistic of the hypothesis,

$$H_0\colon \kappa = 0,$$

is

$$-2\ln Q = 2[n \ln a_d(\hat{\kappa}) + \hat{\kappa} r - n \ln a_d(0)] \sim\sim \chi^2(d),$$

where r is the observed resultant length and $\hat{\kappa}$ is the maximum likelihood estimate of κ.

(2) Show that the test rejects for large values of $\hat{\kappa}$ or, equivalently, for large values of r.

Hint: Show that if

$$r = r(\hat{\kappa}) = -n \frac{a_d'(\hat{\kappa})}{a_d(\hat{\kappa})},$$

then one has that

$$\frac{dr(\hat{\kappa})}{d\hat{\kappa}} > 0,$$

and use this to prove that $-2\ln Q$ is an increasing function of $\hat{\kappa}$.

Exercise 10.2 (Mardia and Jupp, 1999) At Gorleston, England, the wind directions were recorded at 11 o'clock to 12 o'clock on 49 Sundays in 1968. (It was "calm" on the three remaining Sundays.) Representing the directions as unit vectors in $\mathbb{R}^2$, the resultant of the 49 vectors is $(2.1634, -5.4644)^*$.

Are the wind directions uniformly distributed?

Exercise 10.3 In a large investigation of sediment transport over a low sand dune at Ferring by the North Sea, Harald E. Mikkelsen, Department of Earth Sciences, University of Aarhus, measured among other things the direction of the wind by means of a wind hose placed on a mast. The wind directions are given in degrees and measured clockwise from north, such that east corresponds to 90°, etc. The wind directions measured every 10th minute for a period of 3 hours on September 26, 1985, by two wind hoses placed in the same mast but in the height of 10 m and 28 m, respectively, are given below.

In order to investigate whether there is a difference between the wind direction in the two heights, the table below also contains the difference θ between the direction in the height of 10 m and the direction in the height of 28 m and, in addition, the corresponding vector $\mathbf{v}^* = (v_1, v_2) = (\cos\theta, \sin$

θ).

hr.	10 m	28 m	θ	v_1	v_2
11.05	252.7	255.8	−3.1	0.9985	−0.0541
11.15	237.2	240.7	−3.5	0.9981	−0.0610
11.25	239.3	238.6	0.7	0.9999	0.0122
11.35	234.7	234.0	0.7	0.9999	0.0122
11.45	222.4	228.7	−6.3	0.9940	−0.1097
11.55	224.2	226.6	−2.4	0.9991	−0.0419
12.05	231.2	233.3	−2.1	0.9993	−0.0366
12.15	223.8	230.8	−7.0	0.9925	−0.1219
12.25	224.2	229.4	−5.2	0.9959	−0.0906
12.35	220.3	226.3	−6.0	0.9945	−0.1045
12.45	221.7	218.9	2.8	0.9988	0.0488
12.55	225.6	223.8	1.8	0.9995	0.0314
13.05	222.4	218.9	3.5	0.9981	0.0610
13.15	205.5	206.9	−1.4	0.9997	−0.0244
13.25	210.1	212.2	−2.1	0.9993	−0.0366
13.35	217.0	210.3	6.7	0.9932	0.1167
13.45	203.8	205.2	−1.4	0.9997	−0.0244
13.55	205.5	209.4	−3.9	0.9977	−0.0680

Assume that the 18 measurements of the differences, $\mathbf{v}_1, \ldots, \mathbf{v}_{18}$, may be considered as a sample from the von Mises distribution. Examine whether there is difference between the wind direction in the two heights.

By the calculations it may be used that $\mathbf{v}_{\cdot}^{*} = (17.9577, -0.4914)$.

Exercise 10.4 The data below are from Vistelius (1966) and consist of 10 measurements of the orientation of needle-shaped crystals in two exposures from the Yukspor Mountain, Kola Peninsula.

Assume that the data from an exposure may be considered as a sample from the Fisher distribution and examine whether the orientation is the same for the two exposures.

Exposure 1

A	*D*	v_1	v_2	v_3
18	26	0.8548	−0.2777	−0.4384
34	22	0.7687	−0.5185	−0.3746
64	10	0.4317	−0.8851	−0.1736
265	6	−0.0867	0.9907	−0.1045
314	8	0.6879	0.7123	−0.1392
334	6	0.8939	0.4360	−0.1045
340	24	0.8585	0.3125	−0.4067
342	20	0.8937	0.2904	−0.3420
345	14	0.9372	0.2511	−0.2419
355	8	0.9865	0.0863	−0.1392
Σ		7.2262	1.3980	−2.4646

Exposure 2

A	D	v_1	v_2	v_3
5	4	0.9938	−0.0869	−0.0698
31	10	0.8441	−0.5072	−0.1736
145	20	−0.7698	−0.5390	−0.3420
279	10	0.1541	0.9727	−0.1736
309	8	0.6232	0.7696	−0.1392
325	20	0.7698	0.5390	−0.3420
342	8	0.9418	0.3060	−0.1392
344	10	0.9467	0.2714	−0.1736
350	12	0.9633	0.1699	−0.2079
359	18	0.9509	0.0166	−0.3090
Σ		6.4179	1.9121	−2.0699

Exercise 10.5 (Poulsen, Rose-Hansen, and Springer, 1986) Data for strikes were measured in sandstone belonging to the Precambrian Eriksfjord formation in southern Greenland.

The Mâjût Sandstone Member in the lower part of the formation contains fanglomerats and fluvial sandstone where the strike varies in the range 83° to 102°, while the dip is south and varies in the range 10° to 15°. The overlaying Mussartût Member, which consists of sandstone, conglomerates, intrusive breccia, and tuff has similar values for the dip while the strike varies in the range 78° to 95°. The following measurements (in degrees or as a unit vector in $\mathbb{R}^2$) were obtained:

Mâjût Member			*Massartût Member*		
Strike	v_1	v_2	*Strike*	v_1	v_2
83	0.1219	0.9925	78	0.2079	0.9781
87	0.0523	0.9986	79	0.1908	0.9816
89	0.0175	0.9998	80	0.1736	0.9848
91	−0.0175	0.9998	80	0.1736	0.9848
93	−0.0523	0.9986	81	0.1564	0.9877
95	−0.0872	0.9962	83	0.1219	0.9925
96	−0.1045	0.9945	85	0.0872	0.9962
100	−0.1736	0.9848	85	0.0872	0.9962
102	−0.2079	0.9781	90	0.0000	1.0000
Σ	−0.4513	8.9429	92	−0.0349	0.9994
			95	−0.0872	0.9962
			Σ	1.0765	10.8975

Assume that the data may be considered as two samples from the von Mises distribution and examine whether the strike is the same in the two members.

Exercise 10.6 The data below are from a large investigation of sediment transport over a small dune at Ferring by the North Sea undertaken by Harald E. Mikkelsen, Department of Eath Sciences, University of Aarhus, in 1985.

As part of the investigation wind directions were recorded by means of wind hoses. The directions are given in degrees and measured clockwise from north, such that east corresponds to 90°, etc. The measurements here are from a wind hose placed at a height of 10 m on a mast. In four periods of one hour, each within two days, the wind direction was measured every 10th minute. In addition to the wind direction θ below the vector $\mathbf{v}^* = (v_1, v_2) = (\cos\,\theta,\, \sin\,\theta)$ is given and, furthermore, the resultant vector $\mathbf{v}^*_\cdot$ for each of the four hours.

September 30

Time	θ	v_1	v_2
15.05	200.6	−0.9361	−0.3518
15.15	194.6	−0.9677	−0.2521
15.25	193.9	−0.9707	−0.2402
15.35	199.5	−0.9426	−0.3338
15.45	197.1	−0.9558	−0.2940
15.55	191.4	−0.9803	−0.1977
	Σ	−5.7532	−1.6696

October 1

Time	θ	v_1	v_2
03.05	193.5	−0.9724	−0.2334
03.15	184.4	−0.9971	−0.0767
03.25	189.3	−0.9869	−0.1616
03.35	193.5	−0.9724	−0.2334
03.45	191.1	−0.9813	−0.1925
03.55	192.8	−0.9751	−0.2215
	Σ	−5.8852	−1.1191

October 1

Time	θ	v_1	v_2
15.05	218.9	−0.7782	−0.6280
15.15	211.1	−0.8563	−0.5165
15.25	209.7	−0.8686	−0.4955
15.35	209.7	−0.8686	−0.4955
15.45	210.1	−0.8652	−0.5015
15.55	215.4	−0.8151	−0.5793
	Σ	−5.0520	−3.2163

October 2

Time	θ	v_1	v_2
03.05	193.5	−0.9724	−0.2334
03.15	187.2	−0.9921	−0.1253
03.25	188.6	−0.9888	−0.1495
03.35	194.6	−0.9677	−0.2521
03.45	195.0	−0.9659	−0.2588
03.55	188.3	−0.9895	−0.1444
	Σ	−5.8764	−1.1635

Assume for each of the four hours that the six measurements of wind direction may be considered as a sample from the von Mises distribution and examine whether the wind direction is the same for the four hours.

(Using five decimal places in the calculations is recommended).

Exercise 10.7 At the Department of Mathematical Sciences, University of Aarhus, a small computer program for simulating samples from the von Mises distribution has been developed. The data in this exercise have been produced by this program.

First we consider a sample of size $n = 30$. For this sample the resultant is $\mathbf{v}. = (-2.9346, 2.6809)^*$.

(1) Examine whether it may be assumed that observations may be described by the uniform distribution.

In the rest of the exercise we consider two independent samples of size 15 and 25, respectively. The resultants for the two samples appear from the template for calculations below produced using the SAS macro `direction`. The template is to be used when answering the two questions below.

i	n	v1	v2	R	2(n-R)	f	kappa inverse
1	15	14.5955	0.2224	14.5972	0.8055	14	0.0575
2	25	24.1782	1.2599	24.2110	1.5779	24	0.0657
	40	38.7738	1.4824	38.8083	2.3835	38	0.0627

(2) Show that it may be assumed that the two samples have the same mean direction.

(3) Examine if the common mean direction for the two samples may be assumed to be the direction determined by the vector $(1,0)^*$.

CHAPTER 11

The Likelihood Method

Throughout this book, statistical inference is based on the likelihood method that was introduced in Section 3.1.4 on page 70 in connection with one sample from the normal distribution. Here we give a brief summary of the method without reference to particular examples. In Section 11.1 basic concepts, such as the likelihood function, the maximum likelihood estimator, the likelihood ratio test, and the significance probability are discussed. Sometimes it is necessary to use approximations in order to calculate the significance probability of the likelihood ratio test and Section 11.3 reviews the relevant results. Finally, Section 11.2 contains remarks concerning concepts from general test theory that are often mentioned in the literature.

11.1 Likelihood Inference

The ideas behind likelihood inference and the first basic developments of this concept are due to the English geneticist R. A. Fisher (Fisher, 1922, 1925). Likelihood inference is based on the *likelihood function*, which we now introduce and discuss.

In order to facilitate the discussion we assume, as in Section 1.3, that the data set from the observational study or the experiment only has one variable with values, $x_1, \ldots, x_n$, that are realizations of random variables and we refer to the column vector $\mathbf{x} = (x_1, \ldots, x_n)^*$ as the observation or the outcome. For data sets in which the randomness is represented by more than one variable, the discussion is very similar to that below with the exception of a more extensive sample space.

Recall from Section 1.3 that we consider parametric statistical models of the form,

$$(\mathcal{X}, \mathcal{A}; \mathcal{P}) = (\mathcal{X}, \mathcal{A}; \{P_{\boldsymbol{\omega}} \mid \boldsymbol{\omega} \in \Omega\}),$$

where $\mathcal{X}$ is the sample space, $\mathcal{A}$ is a system of events and $\mathcal{P} = \{P_{\boldsymbol{\omega}} \mid \boldsymbol{\omega} \in \Omega\}$ is a class of probability measures parameterized by a parameter $\boldsymbol{\omega} = (\omega_1, \ldots, \omega_k)^*$ varying in the parameter space Ω, which is a subset of $\mathbb{R}^k$. The class of probability measures is represented as a parameterized class of probability density functions, $\{f(\cdot;\boldsymbol{\omega}) \mid \boldsymbol{\omega} \in \Omega\}$. Finally, recall that the parameter $\boldsymbol{\omega}$ was chosen such that statements concerning the scientific problem considered can be formulated as statements about $\boldsymbol{\omega}$.

The *model function* is the probability density function considered as a function of both the outcome $\mathbf{x}$ and the parameter $\boldsymbol{\omega}$, i.e.,

$$\begin{array}{ccc} \mathcal{X} \times \Omega & \rightarrow & \mathbb{R} \\ (\mathbf{x}, \boldsymbol{\omega}) & \rightarrow & f(\mathbf{x}; \boldsymbol{\omega}). \end{array} \tag{11.1}$$

In order to make the mathematical considerations easier, we assume that the parameter set Ω is a *domain* in $\mathbb{R}^k$; i.e., Ω is an *open** and *connected*† subset of $\mathbb{R}^k$.

The Likelihood Function

It is seen from (11.1) that for fixed value of the parameter $\boldsymbol{\omega}$, the model function $f(\mathbf{x};\boldsymbol{\omega})$ is the probability density function of the random vector $\mathbf{X}$. Consequently, if $P_{\boldsymbol{\omega}}$ denotes the probability

* Ω is open if an arbitrary point $\boldsymbol{\omega} \in \Omega$ is the center of a sphere that is a subset of Ω.

† Ω is connected if two arbitrary points $\boldsymbol{\omega}$ og $\boldsymbol{\omega}$' in Ω can be connected by means of line segments that are all contained in Ω.

measure corresponding to the probability density function $f(\mathbf{x};\boldsymbol{\omega})$, we have

$$f(\mathbf{x};\boldsymbol{\omega}) = P_{\boldsymbol{\omega}}(\mathbf{X}=\mathbf{x}) \tag{11.2}$$

if $\mathbf{X}$ is discrete. If $\mathbf{X}$ is continuous, the relation between $f(\mathbf{x};\boldsymbol{\omega})$ and $P_{\boldsymbol{\omega}}$ is given by

$$f(\mathbf{x};\boldsymbol{\omega})\mathbf{dx} \approx P_{\boldsymbol{\omega}}(\mathbf{X}\in I_{\mathbf{x}}) \tag{11.3}$$

where $I_{\mathbf{x}}$ is a small set around $\mathbf{x}$, whose content is $\mathbf{dx}$.

Thus, for fixed value of $\boldsymbol{\omega}$, the model function describes the probabilities of all possible realizations of $\mathbf{X}$. The outcome $\mathbf{x}$, however, is a particular and fixed realization of $\mathbf{X}$, and since we want to make statements about the different values of $\boldsymbol{\omega}$ in light of the outcome $\mathbf{x}$, we could try to consider the model function as a function of $\boldsymbol{\omega}$ for fixed $\mathbf{x}$. We still have the interpretation that $f(\mathbf{x};\boldsymbol{\omega})$ is the probability of the observation $\mathbf{x}$ if the value of the parameter is $\boldsymbol{\omega}$. This follows directly from formula (11.2) if $\mathbf{X}$ is discrete, and via the interpretation in (11.3) if $\mathbf{X}$ is continuous. In this sense $f(\mathbf{x};\boldsymbol{\omega})$ expresses the ability of $\boldsymbol{\omega}$ to account for or to explain the observation $\mathbf{x}$, and we call this ability the *likelihood* of $\boldsymbol{\omega}$ in the light of $\mathbf{x}$ or, simply the likelihood of $\boldsymbol{\omega}$. By choosing the word "likelihood" rather than the word "probability," Fisher emphasized that we do not consider probabilities on the parameters.

As a function of $\boldsymbol{\omega}$ we refer to $f(\mathbf{x};\boldsymbol{\omega})$ as the *likelihood function* and denote it by

$$L(\boldsymbol{\omega}) = f(\mathbf{x};\boldsymbol{\omega}), \quad \boldsymbol{\omega}\in\Omega, \tag{11.4}$$

where the dependence of the observation is implicit. If we want to stress that we consider the likelihood function corresponding to the observation $\mathbf{x}$, we write $L(\boldsymbol{\omega};\mathbf{x})$ instead of $L(\boldsymbol{\omega})$.

An example of a likelihood function is seen in Figure 11.1.

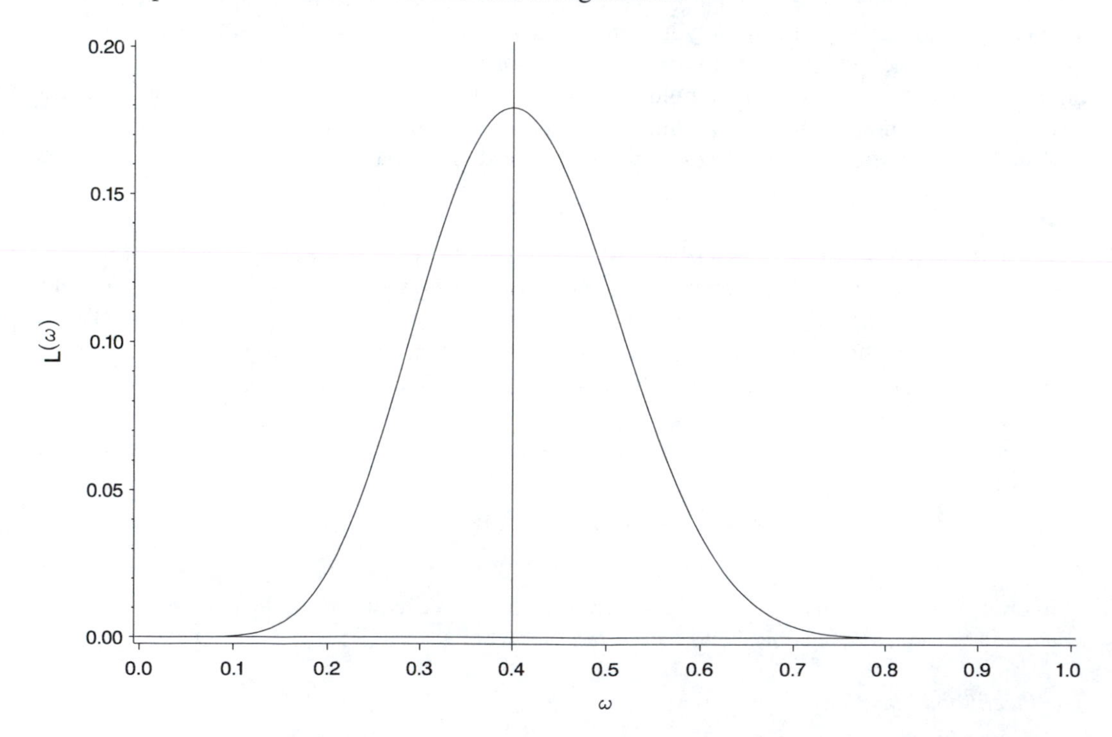

Figure 11.1 *The likelihood function $L(\omega)$ corresponding to the observation $x = 8$ in a statistical model based on the binomial distribution with probability parameter ω and number of trials $n = 20$.*

The likelihood function gives an ordering of the parameter set. If, for a moment, we only consider two parameter values $\boldsymbol{\omega}_1$ and $\boldsymbol{\omega}_2$ and on the basis of the observation $\mathbf{x}$, we were to choose which of those values gives the best explanation of the observation, it would be the value that assigns the largest probability to the observation, and that is exactly the value with the largest value of the likelihood function. We say that the value $\boldsymbol{\omega}_1$ is more *likely* than $\boldsymbol{\omega}_2$ in the light of the observation $\mathbf{x}$ if $L(\boldsymbol{\omega}_1) > L(\boldsymbol{\omega}_2)$.

Maximum Likehood Estimation

The ordering of the parameter set induced by the likelihood function immediately implies that if we wish to point out a single parameter value that is in best accordance with the observation $\mathbf{x}$, it has to be the value that makes the observed data most probable, i.e., the value for which the likelihood function attains its maximum. With these considerations we have introduced the concept of *maximum likelihood estimation.* If a uniquely determined value $\hat{\boldsymbol{\omega}}$ exists for which the likelihood function L attains its maximum, i.e.,

$$L(\hat{\boldsymbol{\omega}}) > L(\boldsymbol{\omega}) \quad \text{for all } \boldsymbol{\omega} \in \Omega \text{ such that } \boldsymbol{\omega} \neq \hat{\boldsymbol{\omega}},$$

this value $\hat{\boldsymbol{\omega}}$ of the parameter is called the *maximum likelihood estimate* of $\boldsymbol{\omega}$. In other words, the maximum likelihood estimate is the most likely value of the parameter $\boldsymbol{\omega}$ in light of the observation $\mathbf{x}$. The corresponding random vector $\hat{\boldsymbol{\omega}}(\mathbf{X})$ is referred to as the *maximum likelihood estimator.*

Sometimes it is easier to maximize the *log likelihood function*,

$$l(\boldsymbol{\omega}) = \ln L(\boldsymbol{\omega}), \quad \boldsymbol{\omega} \in \Omega, \tag{11.5}$$

rather than the likelihood function itself. In the models we consider, the log likelihood function is (at least) twice differentiable with continuous (partial) derivatives, and it facilitates the task of finding the value for which the likelihood function attains its maximum. Since the parameter set is assumed to be a domain, the maximum likelihood estimate $\hat{\boldsymbol{\omega}} = (\hat{\omega}_1, ..., \hat{\omega}_k)^*$ of $\boldsymbol{\omega}$ may be found as a solution to the equations,

$$\frac{\partial l}{\partial \omega_j}(\boldsymbol{\omega}) = 0, \quad j = 1, 2, \ldots, k. \tag{11.6}$$

These equations, which are called the *likelihood equations* or the *normal equations,* can sometimes be solved explicitly, but often numerical procedures must be used in order to find $\hat{\boldsymbol{\omega}}$. Furthermore, we should assess if a solution to the likelihood equations is a point for which the likelihood function attains its maximum. The equations may in principle have several solutions and among these, $\hat{\boldsymbol{\omega}}$ has to be found as the one that corresponds to the global maximum of the likelihood function or of the log likelihood function.

If the model is that for a sample, the values, $x_1, \ldots, x_n$, are realizations of independent random variables, $X_1, \ldots, X_n$, with a common distribution. The independence of the random variables implies that the probability density function of $\mathbf{X}$ is the product of the probability density functions of $X_1, \ldots, X_n$. In this situation the likelihood function L is

$$L(\boldsymbol{\omega}) = \prod_{i=1}^{n} f(x_i; \boldsymbol{\omega}) \tag{11.7}$$

and the log likelihood function l is

$$l(\boldsymbol{\omega}) = \sum_{i=1}^{n} \ln f(x_i; \boldsymbol{\omega}). \tag{11.8}$$

In (11.7) and (11.8) $f(x_i; \boldsymbol{\omega})$ denotes the probability density function of X_i evaluated at x_i.

Likelihood Ratio Test

Throughout this book relevant questions or hypotheses concerning the scientific context have been formulated as restrictions on the parameter of the model. Such a restriction, for instance, that two components of the parameter are equal or that a component of the parameter is equal to zero, was referred as a hypothesis, and mathematically a hypothesis was formulated as

$$H_0\colon \boldsymbol{\omega} \in \Omega_0,$$

where Ω_0 is a subset of Ω, i.e., $\Omega_0 \subset \Omega$.

The purpose of a test is to decide on the basis of the observation $\mathbf{x}$ whether the hypothesis is rejected. If not, the original model with parameter space Ω may be reduced to the simpler model with parameter space Ω_0.

We now discuss the test of a hypothesis based on the likelihood function. Initially, we notice two facts. First, the value of the likelihood function evaluated at the maximum likelihood estimate, $L(\hat{\boldsymbol{\omega}}(\mathbf{x}))$, is a function of the observation $\mathbf{x}$ and therefore it is a realization of the random variable $L(\hat{\boldsymbol{\omega}}(\mathbf{X}))$. Secondly, the interpretation of $L(\hat{\boldsymbol{\omega}}(\mathbf{x}))$ is that it is the maximum probability of the observation $\mathbf{x}$ in the statistical model.

Let $\hat{\boldsymbol{\omega}}(\mathbf{x})$ denote the maximum likelihood estimate of $\boldsymbol{\omega}$ in the original model, and let $\hat{\boldsymbol{\omega}}_0(\mathbf{x})$ denote the maximum likelihood estimate of $\boldsymbol{\omega}$ under H_0, i.e., in the statistical model with parameter space Ω_0. The *likelihood ratio test statistic* $Q(\mathbf{x})$, which was introduced in Neyman and Pearson (1928), is then defined as:

$$Q(\mathbf{x}) = \frac{\max\limits_{\boldsymbol{\omega}\in\Omega_0} L(\boldsymbol{\omega})}{\max\limits_{\boldsymbol{\omega}\in\Omega} L(\boldsymbol{\omega})} = \frac{L(\hat{\boldsymbol{\omega}}_0(\mathbf{x}))}{L(\hat{\boldsymbol{\omega}}(\mathbf{x}))}. \tag{11.9}$$

Note that $Q(\mathbf{x}) \leq 1$ because it is the same function that is maximized in the numerator and in the denominator, and because the maximization in the numerator is over a subset of the one in the denominator. Furthermore, $0 < Q(\mathbf{x})$ since $Q(\mathbf{x})$ is a ratio between two probabilities. Consequently, we have $0 < Q(\mathbf{x}) \leq 1$.

Next, we consider the interpretation of the likelihood ratio test statistic. $Q(\mathbf{x})$ is the ratio of likelihoods between the most likely value $\hat{\boldsymbol{\omega}}_0$ of $\boldsymbol{\omega}$ under H_0 and the most likely value $\hat{\boldsymbol{\omega}}$ of $\boldsymbol{\omega}$ at all. If $Q(\mathbf{x}) \approx 1$, then $L(\hat{\boldsymbol{\omega}}_0) \approx L(\hat{\boldsymbol{\omega}})$; thus there exists a value of the parameter under the hypothesis that is nearly as likely as the most likely value at all, and in this situation we do not consider $\mathbf{x}$ to be critical for the hypothesis and having observed $\mathbf{x}$ does not make us doubt H_0. In contrast, if $Q(\mathbf{x}) \approx 0$, then $L(\hat{\boldsymbol{\omega}}_0) << L(\hat{\boldsymbol{\omega}})$; so the most likely parameter under the hypothesis is much less likely than a value of the parameter under the model. This means that the observation $\mathbf{x}$ is *critical* for the hypothesis.

We have seen that small values of $Q(\mathbf{x})$ are critical for the hypothesis, and the smaller $Q(\mathbf{x})$ is, the more critical it is for the hypothesis. In this way $Q(\mathbf{x})$ induces an ordering in the sample space according to how critical the observations are in terms of $Q(\mathbf{x})$. We say that the observation $\mathbf{x}_1$ is *more critical or just as critical* for H_0 as the observation $\mathbf{x}_2$ if

$$Q(\mathbf{x}_1) \leq Q(\mathbf{x}_2).$$

Significance Probability of the Likelihood Ratio Test

In order to get an impression of how small $Q(\mathbf{x})$ should be before we reject H_0, we consider the set of all possible outcomes $\mathbf{y}$ of the observational study or the experiment which are at least as critical as the observed outcome $\mathbf{x}$, i.e., the set,

$$\{\mathbf{y} \in \mathcal{X} \mid Q(\mathbf{y}) \leq Q(\mathbf{x})\}. \tag{11.10}$$

To assess the size of the set in (11.10) relative to the size of the sample space $\mathcal{X}$, we use probability measures.

If the hypothesis H_0 is simple, i.e., if $\Omega_0 = \{\boldsymbol{\omega}_0\}$, the probability of the set in (11.10),

$$p_{obs}(\mathbf{x}) = P_{\boldsymbol{\omega}_0}(\{\mathbf{y} \in \mathcal{X} \mid Q(\mathbf{y}) \leq Q(\mathbf{x})\}), \tag{11.11}$$

is called *significance probability* of the likelihood ratio test. (The terms *observed significance level* and the *p-value* are synonymous.) From (11.11) it follows that the significance probability is *the* *probability (calculated under the hypothesis H_0) of all possible outcomes* $\mathbf{y}$, *which are at least as critical for H_0 as the observed outcome* $\mathbf{x}$. If $p_{obs}(\mathbf{x})$ is small, the probability of obtaining outcomes that are at least as critical for H_0 as the observed outcome $\mathbf{x}$ is small and, consequently, we reject H_0. That is, *if $p_{obs}(\mathbf{x})$ is small, then H_0 is rejected.* If $p_{obs}(\mathbf{x})$ is large, there is a high probability of obtaining outcomes that are at least as critical as the observed outcome $\mathbf{x}$ and, consequently, there is no reason for rejecting H_0. That is, *if $p_{obs}(\mathbf{x})$ is large, then H_0 is not rejected.* Notice that if we are not able to reject a hypothesis H_0, it in no way means that we have proved the correctness of H_0; it only means that we have found no significant deviations from H_0 in the present observation.

(In the literature it is often said that a hypothesis "is accepted" if it is not rejected. This terminology may arise from the application of statistics in quality control. Here the hypothesis to be tested is typically that some produced items satisfy some specifications and if the hypothesis is not rejected, the items are accepted.)

Let us emphasize the logic behind the rejection of the hypothesis H_0 when the p-value $p_{obs}(\mathbf{x})$ is small. The statistician considers two premisses: (1) "either the hypothesis H_0 is false or an event with small probability has occurred" and (2) "an event with small probability does not occur." Based on these premises the conclusion that "the hypothesis H_0 is false" must be drawn if an event with small probability under the hypothesis is observed.

One question is still unanswered. How small should the significance probability $p_{obs}(\mathbf{x})$ be before we reject H_0? In principle the answer depends on the nature of the hypothesis. In order to reject a well-reputed scientific hypothesis, such as, for instance, the second law of Newton, it is required that we find strongly significant deviations from the hypothesis, i.e., the significance probability should be very small, for instance, 0.1%. Less well-founded hypotheses, such as, for instance, "the content of a specific mineral in a piece of rock is 20%" or "the petal length of the two species *Iris versicolor* and *Iris virginica* are identical," are rejected for much higher probabilities (1% or 5%).

The likelihood ratio test with *significance level* α rejects the hypothesis H_0 if

$$p_{obs}(\mathbf{x}) \leq \alpha, \tag{11.12}$$

which implies that the corresponding region of rejection (or critical region) R_α is

$$R_\alpha = \{\mathbf{x} \in \mathcal{X} \mid p_{obs}(\mathbf{x}) \leq \alpha\} \tag{11.13}$$

and that the corresponding region of acceptance (or rather the region of no rejection) is

$$A_\alpha = \{\mathbf{x} \in \mathcal{X} \mid p_{obs}(\mathbf{x}) > \alpha\}. \tag{11.14}$$

In the statistical literature the preferred level of significance is traditionally 5% but the level 1% is also used in connection with more well-founded hypotheses. In all the examples and exercises in this book we have used the significance level 5%.

We conclude this section with remarks concerning the likelihood ratio test in situations where the hypothesis H_0 is composite, i.e., the subset Ω_0 specifying the hypothesis has more than one element. In this situation the significance probability of the likelihood ratio test statistic is defined as

$$p_{obs}(\mathbf{x}) = \sup_{\boldsymbol{\omega} \in \Omega_0} P_{\boldsymbol{\omega}}(\{\mathbf{y} \in \mathcal{X} \mid Q(\mathbf{y}) \leq Q(\mathbf{x})\}), \tag{11.15}$$

which means as the largest of the probabilities (under H_0) of the set in (11.10). The rejection region and the acceptance region are also in this case defined as in the formulas (11.13) and (11.14), respectively.

Very often it is difficult (or impossible) to calculate the exact values of the significance probabilities as defined in (11.11) or (11.15). In Section 11.3 we discuss how to calculate approximations of the significance probabilities in such situations. The approximations concern the $-2\ln Q$-test statistic rather than the likelihood ratio test statistic Q itself. In the case of a simple hypothesis concerning the k-dimensional parameter $\boldsymbol{\omega}$, the significance probability in (11.11) is approximated by

$$p_{obs}(\mathbf{x}) \doteq 1 - F_{\chi^2(k)}(-2\ln Q(\mathbf{x})), \tag{11.16}$$

where $F_{\chi^2(k)}$ denotes the distribution function of the χ^2-distribution with k degrees of freedom.

In Section 11.3 it is shown that if the parameter set Ω_0 can be specified by means of a d-dimensional parameter – or has d free parameters in the terminology of Section 11.3 – then the significance probability in (11.15) can be approximated by

$$p_{obs}(\mathbf{x}) \doteq 1 - F_{\chi^2(k-d)}(-2\ln Q(\mathbf{x})). \tag{11.17}$$

Note that the number of degrees of freedom $k-d$ of the approximating χ^2-distribution is the difference between the number of free parameters k in the basic model with parameter space Ω and the number of free parameters d in the model with parameter space Ω_0.

11.2 Concepts from General Test Theory

In this book we have primarily used the significance test, which only requires that one can identify the set of observations that are more critical for the hypothesis than the actual observation and that the probability of that set under the nul hypothesis can be computed. Obviously, we want the significance test to reject the hypothesis, when it is wrong. In this connection the concept of the *power of a test* is very important, and it is introduced here in connection with a few other concepts from the theory of statistical tests.

For given model and null hypothesis the *power function* is used to compare different tests so one can chose the most powerful test for the deviations from the hypothesis one is concerned about. The most important practical application of the concept of power is illustrated in Chapter 5 which focus on designing the experiment or the data collection such that adequate power is obtained for practically relevant deviations from the null hypothesis.

Significance Level of Test

The test of significance of the hypothesis H_0: $\boldsymbol{\omega} \in \Omega_0$ corresponding to the test statistic T is said to have *significance level* α if

$$\sup_{\boldsymbol{\omega}\in\Omega_0} P_{\boldsymbol{\omega}}(\mathbf{X} \in R) = \alpha, \tag{11.18}$$

i.e. if the largest probability of rejecting H_0 – calculated under H_0 – is α. In other words, the level of significance α is the probability of rejecting the null hypothesis when it is true. It is obvious that it would be desirable that α was 0, but such tests of significance do not exist.

It is a characteristic feature of statistical inference that it is not possible to state *with certainty* that the hypothesis H_0 is true or false. In this respect statistical inference differs from mathematics and logic. In these disciplines one draws conclusions based on fixed premisses. In statistical inference one infers from the outcome which is considered a realization of a random vector whose variation is described by means of a probabilistic model. Consequently, the conclusions in statistical inference are, of course, formulated by means of probabilities, for instance as significance probabilities or as confidence regions, see below.

Another important difference between the three disciplines is that mathematics and logic are *deductive*, that is, one concludes from the general to the specific. In contrast, statistical inference is *inductive* as one concludes from the specific (data) to the general (a scientific model).

Two Types of Errors and the Power Function

In test theory two kinds of errors may be committed. These are referred to as a type I error and a type II error, respectively. A type I error rejects a true hypothesis, and a type II error is committed if a false hypothesis is not rejected. The definition of the two kinds of errors are listed in Table 11.1. Notice that the probability of committing a type I error is equal to the significance level α.

The quality of a statistical test depends, among other things, on its power to reveal deviations from the hypothesis. This property may be expressed in terms of the *power function* of the test. With the

Table 11.1 *The different kinds of errors in test theory*

	H_0 is rejected	H_0 is not rejected
H_0 is true	type I	none
H_0 is false	none	type II

notation from Section 1.5 the power function of the test with test statistic T of the hypothesis H_0: $\boldsymbol{\omega} \in \Omega_0$ is defined as

$$pow(\boldsymbol{\omega}) = P_{\boldsymbol{\omega}}(\mathbf{X} \in R),$$

i.e., for each value of the parameter $\boldsymbol{\omega}$ the power $pow(\boldsymbol{\omega})$ is the probability – calculated by means of the probability measure corresponding to $\boldsymbol{\omega}$ – of rejecting the hypothesis H_0. Note that if the hypothesis is simple, $\Omega_0 = \{\boldsymbol{\omega}_0\}$, then $pow(\boldsymbol{\omega}_0)$ is exactly equal to the significance level α, and, furthermore, if we for $\boldsymbol{\omega} \neq \boldsymbol{\omega}_0$ let $\beta(\boldsymbol{\omega})$ denote the probability – corresponding to the parameter value $\boldsymbol{\omega}$ – of committing a type II error, then

$$\beta(\boldsymbol{\omega}) = 1 - pow(\boldsymbol{\omega}).$$

Consequently, from an ideal point of view the power function for a simple hypothesis H_0: $\boldsymbol{\omega} = \boldsymbol{\omega}_0$, should be constant and equal to 1 except for its value at $\boldsymbol{\omega}_0$, which should be 0. However, as mentioned above, no test statistics with such a power function exists. An example of a power function is shown in Figure 11.2.

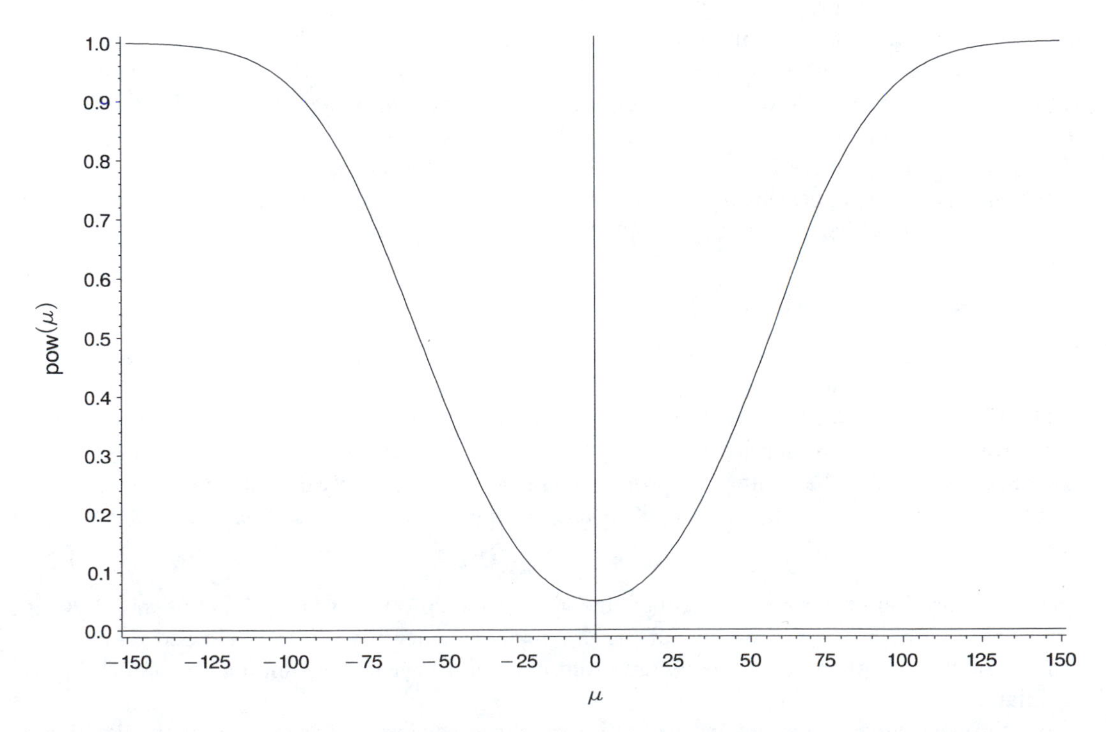

Figure 11.2 *The power function of the u-test of the hypothesis H_0: $\mu = 0$. The standard deviation σ is 115, corresponding to the problem considered in Example 2.2.*

Confidence Regions

We end this section by mentioning confidence regions, a concept whose definition is related to test theory and which is often used in applications. In light of the observation $\mathbf{x}$, the $(1-\alpha)$ *confidence region for the parameter* $\boldsymbol{\omega}$ is defined as

$$C_{1-\alpha}(\mathbf{x}) = \left\{ \boldsymbol{\omega}_0 \mid \begin{array}{l} \text{the hypothesis } H_0\text{: } \boldsymbol{\omega} = \boldsymbol{\omega}_0 \text{ is not rejected by} \\ \text{a test with significance level } \alpha \text{ on the basis of } \mathbf{x} \end{array} \right\}. \tag{11.19}$$

If the parameter is one-dimensional, the region is typically an interval, the $(1-\alpha)$ *confidence interval for the parameter* ω. It is obvious that the confidence region depends on the test statistic considered as well as on the chosen level of significance. Tests are usually conducted with a level of significance of 5% and in such cases the corresponding regions are 95% confidence regions.

An interpretation of 95% confidence regions is based on repeated random sampling, or more precisely, on the interpretation of probabilities as limit values of relative frequencies. Suppose that the observational study resulting in the observation $\mathbf{x}$ is repeated an infinite number of times and suppose that for the result $\mathbf{y}$ of each repetition of the study, one calculates the region $C_{1-\alpha}(\mathbf{y})$. The true value of the parameter $\boldsymbol{\omega}$ would then be included in the calculated region in 95% of the repetitions.

This interpretation is of course not very useful when one has the region $C_{1-\alpha}(\mathbf{x})$ calculated on the basis of the observation $\mathbf{x}$ at hand. However, it is the same interpretation as the one we use in connection with significance tests. Either the true value of the parameter is included in the region $C_{1-\alpha}(\mathbf{x})$ or an event with a probability less than α has occurred.

Sometimes the confidence interval for ω is referred to as an *interval estimate* of ω, because it gives a region of plausible values of the parameter. This is in contrast to an estimator, which only points out one value of the parameter as the estimate. An estimate is sometimes called a *point estimate* to stress this property. The confidence interval is useful in practice because the size of the confidence interval conveys an impression about how much information the observation $\mathbf{x}$ has about the parameter, and in this sense is an indication of the quality of the data. If the confidence interval is large, the information in the data about the parameter is rather vague, and if the confidence interval is small, the information in the data is precise. Here the descriptive terms "large" and "small" used about the confidence interval will depend on the practical situation.

For this reason it is good statistical practice to always accompany an estimate with a confidence interval, as it is has been done in the examples throughout this book.

11.3 Approximative Likelihood Theory

As noted toward the end of Section 11.1 on page 485, it is sometimes difficult or impossible to calculate the exact value of the significance probability $p_{obs}(\mathbf{x})$ of the likelihood ratio test in (11.11) or (11.15). In this section we discuss how the significance probability $p_{obs}(\mathbf{x})$ can be approximated. Furthermore, we consider approximations to the distribution of the maximum likelihood estimator, $\hat{\boldsymbol{\omega}} = \hat{\boldsymbol{\omega}}(\mathbf{X})$. Notice that the significance probability in (11.11) is precisely the value of the distribution function of the likelihood ratio testor $Q(\mathbf{X})$ evaluated at the observed value $Q(\mathbf{x})$, i.e.,

$$p_{obs}(\mathbf{x}) = F_{Q(\mathbf{X})}(Q(\mathbf{x})). \tag{11.20}$$

Thus, the question of how to approximate the significance probability in (11.11) or in (11.20) is equivalent to finding approximations of the distribution, under H_0, of the likelihood ratio testor. Similar remarks apply when H_0 is a composite hypothesis where the significance probability is calculated using formula (11.15).

We emphasize that there is no approximation or inaccuracy involved in specifying the likelihood function, in finding the maximum likelihood estimates or in computing the likelihood ratio statistic. The approximations concern the distribution of the maximum likelihood estimate and the distribution of the likelihood ratio test statistic.

The approximations we consider in the following are based on second-order Taylor expansions of the log likelihood function. The approximations are valid when the parameter space Ω is a domain in $\mathbb{R}^k$, and, furthermore, when the log likelihood function l is a least twice differentiable with continuous (partial) derivatives. All models considered in this book satisfy these conditions. More precisely, we have

$$l(\boldsymbol{\omega}) - l(\hat{\boldsymbol{\omega}}) \doteq \sum_{r=1}^{k} (\omega_r - \hat{\omega}_r) \frac{\partial l}{\partial \omega_r}(\hat{\boldsymbol{\omega}}) + \frac{1}{2} \sum_{r=1}^{k} \sum_{s=1}^{k} (\omega_r - \hat{\omega}_r)(\omega_s - \hat{\omega}_s) \frac{\partial^2 l}{\partial \omega_r \partial \omega_s}(\hat{\boldsymbol{\omega}}), \tag{11.21}$$

where $\doteq$ indicates approximation and where the expression on the right-hand side is the second-order Taylor polynomial of l around the maximum likelihood estimate $\hat{\boldsymbol{\omega}}$. Let $j(\boldsymbol{\omega};\mathbf{x})$ denote the symmetric $k \times k$ matrix with elements,

$$j_{rs}(\boldsymbol{\omega};\mathbf{x}) = -\frac{\partial^2 l}{\partial \omega_r \partial \omega_s}(\boldsymbol{\omega};\mathbf{x}), \quad r,s = 1,\ldots,k. \tag{11.22}$$

Since $\hat{\boldsymbol{\omega}}$ is a solution to the likelihood equations (11.6) it follows from (11.21) that

$$l(\boldsymbol{\omega}) - l(\hat{\boldsymbol{\omega}}) \doteq -\frac{1}{2}(\hat{\boldsymbol{\omega}} - \boldsymbol{\omega})^* j(\hat{\boldsymbol{\omega}};\mathbf{x})(\hat{\boldsymbol{\omega}} - \boldsymbol{\omega}), \tag{11.23}$$

in matrix notation where $*$ means transposition. The function $\bar{l}(\cdot) = l(\cdot) - l(\hat{\boldsymbol{\omega}})$ is called the *normed log likelihood function* and the matrix $j(\boldsymbol{\omega};\mathbf{x})$ is referred to as the *observed information matrix* corresponding to the observation $\mathbf{x}$. The mean of the corresponding random matrix $j(\boldsymbol{\omega};\mathbf{X})$,

$$i(\boldsymbol{\omega}) = E_{\boldsymbol{\omega}}\{j(\boldsymbol{\omega};\mathbf{X})\}, \tag{11.24}$$

is called the *expected information matrix* or *Fisher's information matrix.*

If the parameter ω is one-dimensional, i.e., $k = 1$, it follows from (11.23) that the normed log likelihood function $\bar{l}$ in a neighborhood of $\hat{\omega}$ can be approximated by the parabola,

$$p(\omega) = -\frac{1}{2} j(\hat{\omega};\mathbf{x})(\hat{\omega} - \omega)^2, \tag{11.25}$$

see Figure 11.3.

Distribution of Maximum Likelihood Estimator

We now want to discuss how the distribution of the maximum likelihood estimator $\hat{\boldsymbol{\omega}} = \hat{\boldsymbol{\omega}}(\mathbf{X})$ and the likelihood ratio testor $Q = Q(\mathbf{X})$, respectively, can be approximated in general when the parameter $\boldsymbol{\omega}$ is k-dimensional. It may be shown that the distribution of $\hat{\boldsymbol{\omega}}$ – calculated under the distribution corresponding to the parameter $\boldsymbol{\omega}$ – can be approximated by a k-dimensional normal distribution with vector of means $\boldsymbol{\omega}$ and covariance matrix $i(\boldsymbol{\omega})^{-1}$, which is the inverse matrix of the expected information matrix $i(\boldsymbol{\omega})$. We write this result in the following way:

$$\hat{\boldsymbol{\omega}} \approx N_k(\boldsymbol{\omega}, i(\boldsymbol{\omega})^{-1}). \tag{11.26}$$

It may be shown that the approximation is particularly good when $\mathbf{x}$ is a sample, $x_1,\ldots,x_n$, and the sample size n is large.

Approximation of Distribution and Significance Probability of the $-2\ln Q$-test Statistic: Simple Hypothesis

Using (3.114) it follows that (11.26) implies the following approximation:

$$(\hat{\boldsymbol{\omega}} - \boldsymbol{\omega})^* i(\boldsymbol{\omega})(\hat{\boldsymbol{\omega}} - \boldsymbol{\omega}) \approx \chi^2(k). \tag{11.27}$$

Furthermore, in this expression we may sometimes replace the expected information matrix $i(\boldsymbol{\omega})$ with the expected information matrix or with the observed information matrix evaluated at $\hat{\boldsymbol{\omega}}$, i.e.,

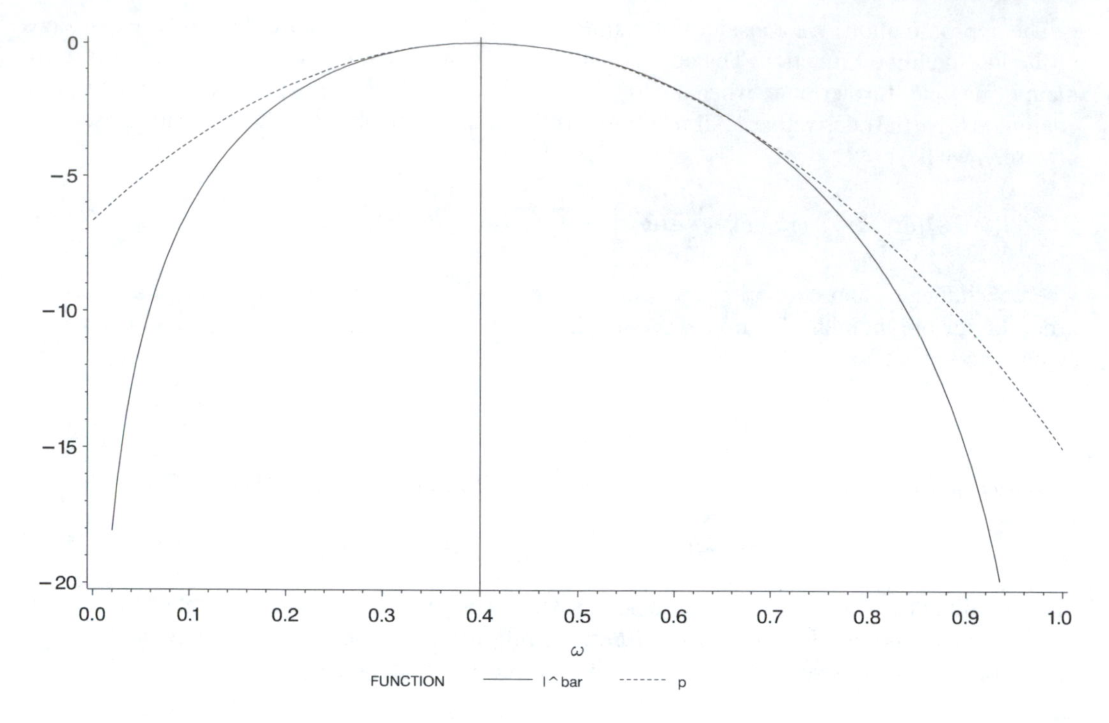

Figure 11.3 *The normed log likelihood function $\bar{l}$ and the approximating parabola p corresponding to the observation* $x = 8$ *in a binomial model with probability parameter* ω *and number of trials* $n = 20$ *(see also Figure 11.1).*

$i(\boldsymbol{\omega})$ is replaced by $i(\hat{\boldsymbol{\omega}})$ or by $j(\hat{\boldsymbol{\omega}}) = j(\hat{\boldsymbol{\omega}};\mathbf{x})$. Using the latter, we obtain the approximation:

$$(\hat{\boldsymbol{\omega}} - \boldsymbol{\omega})^* j(\hat{\boldsymbol{\omega}})(\hat{\boldsymbol{\omega}} - \boldsymbol{\omega}) \approx \chi^2(k). \tag{11.28}$$

Again the approximation is good if $\mathbf{x}$ is a sample, $x_1, ..., x_n$, and the sample size n is large.

Instead of approximating the distribution of the likelihood ratio testor $Q(\mathbf{X})$, we usually consider approximations of the quantity $-2\ln Q(\mathbf{X})$. The approximating distribution depends on whether the hypothesis is simple. More precisely, one has the following approximate result for the distribution of $-2\ln Q(\mathbf{X})$ in the case of is a simple hypothesis stating that the value of the k-dimensional parameter is $\boldsymbol{\omega}$:

$$-2\ln Q(\mathbf{X}) \approx \chi^2(k). \tag{11.29}$$

The approximation is a consequence of (11.23) and (11.28) since using these formulas one finds that

$$\begin{aligned} -2\ln Q(\mathbf{X}) &= -2\ln \frac{L(\boldsymbol{\omega})}{L(\hat{\boldsymbol{\omega}})} \\ &= -2(l(\boldsymbol{\omega}) - l(\hat{\boldsymbol{\omega}})) \\ &\doteq (\hat{\boldsymbol{\omega}} - \boldsymbol{\omega})^* j(\hat{\boldsymbol{\omega}})(\hat{\boldsymbol{\omega}} - \boldsymbol{\omega}) \\ &\approx \chi^2(k). \end{aligned}$$

Small values of the log likelihood ratio testor Q are critical for H_0, or, equivalently, large values of $-2\ln Q(\mathbf{X})$ are critical for H_0. From formula (11.29) we get the following important *approximation of the significance probability of the likelihood ratio test for the simple hypothesis* $\boldsymbol{\omega}$:

$$p_{obs}(\mathbf{x}) \doteq 1 - F_{\chi^2(k)}(-2\ln Q(\mathbf{x})), \tag{11.30}$$

since, using (11.11), we find that

$$\begin{aligned} p_{obs}(\mathbf{x}) &= P_{\boldsymbol{\omega}}(Q(\mathbf{X}) \leq Q(\mathbf{x})) \\ &= P_{\boldsymbol{\omega}}(-2\ln Q(\mathbf{X}) \geq -2\ln Q(\mathbf{x})) \\ &= 1 - P_{\boldsymbol{\omega}}(-2\ln Q(\mathbf{X}) < -2\ln Q(\mathbf{x})) \\ &\doteq 1 - F_{\chi^2(k)}(-2\ln Q(\mathbf{x})). \end{aligned}$$

Here we have used formula (11.29) and the fact that the distribution function of the $\chi^2(k)$-distribution is continuous.

Approximation of Distribution and Significance Probability of the $-2\ln Q$-test Statistic: Composite Hypothesis

For the likelihood ratio testor of a *composite hypothesis* H_0: $\boldsymbol{\omega} \in \Omega_0$, where $\Omega_0 \subseteq \Omega$, approximations analogous to those in (11.29) and (11.30) are valid. In order to formulate these results we need the following notation. A hypothesis H_0: $\boldsymbol{\omega} \in \Omega_0$ is said to have *d free parameters* if there exists a domain $\Theta \subset \mathbb{R}^d$ and a one-to-one mapping of Θ onto Ω_0, i.e.

$$\begin{array}{ccc} \Theta \subseteq \mathbb{R}^d & \rightarrow & \Omega_0 \subseteq \mathbb{R}^k \\ \boldsymbol{\theta} = (\theta_1, \ldots, \theta_d)^* & \rightarrow & \boldsymbol{\omega}(\boldsymbol{\theta}) = (\omega_1(\boldsymbol{\theta}), \ldots, \omega_k(\boldsymbol{\theta}))^*. \end{array} \tag{11.31}$$

Notice that since we have assumed that the parameter space Ω is a domain, the basic model may be considered as a hypothesis with k free parameters, and, furthermore, that for a simple hypothesis, $d = 0$.

Under certain regularity conditions, which are nearly always fulfilled in practice, we have the following *approximations of the likelihood ratio testor of a composite hypothesis with d free parameters:*

$$-2\ln Q(\mathbf{X}) \approx \chi^2(k-d), \tag{11.32}$$

and

$$p_{obs}(\mathbf{x}) \doteq 1 - F_{\chi^2(k-d)}(-2\ln Q(\mathbf{x})). \tag{11.33}$$

It is seen from (11.32) that the number of degrees of freedom for the approximating χ^2-distribution is equal to $k-d$, where k is the number of free parameters in the basic model (corresponding to Ω) and d is the number of free parameters in the hypothesis (corresponding to Ω_0).

The approximations in (11.32) and (11.33) have been used frequently throughout this book. The first application of the formulas was in connection with Bartlett's test for homogeneity in Section 3.2.2 on page 101 and nearly all the tests in Chapter 7 to Chapter 9 were based on these formulas.

Tests of Simple Hypotheses about Components of the Parameter Vector

Sometimes tests of a simple hypothesis concerning one of the components ω_i of the parameter vector $\boldsymbol{\omega} = (\omega_1, \ldots, \omega_k)^*$ are based on the approximation of the distribution of $\hat{\boldsymbol{\omega}}$ in (11.26). Among such tests is Wald's test, which we have considered in the analysis of generalized linear models in Chapter 9, in connection with the table `Analysis of Parameter Estimates` produced by the SAS procedure *PROC GENMOD*. We now briefly discuss Wald's test.

In combination with (3.113), formula (11.26) implies that for the distribution of the maximum likelihood estimator $\hat{\omega}_i$ of ω_i we have the approximation,

$$\hat{\omega}_i \approx N(\omega_i, \sigma_{ii}), \tag{11.34}$$

where σ_{ii} is the ith diagonal element of the matrix $i(\boldsymbol{\omega})^{-1}$. Replacing σ_{ii} with the ith diagonal element $\tilde{\sigma}_{ii}$ of the inverse of the observed information matrix $j(\hat{\boldsymbol{\omega}})$, it may be shown that

$$u = \frac{\hat{\omega}_i - \omega_i}{\sqrt{\tilde{\sigma}_{ii}}} \approx N(0,1), \tag{11.35}$$

or, equivalently, that

$$X^2 = \frac{(\hat{\omega}_i - \omega_i)^2}{\tilde{\sigma}_{ii}} \approx \chi^2(1). \tag{11.36}$$

From (11.35) and (11.36) it follows that the simple hypothesis about the ith component ω_i of the parameter vector $\boldsymbol{\omega}$,

$$H_0\colon \omega_i = \omega_{i0},$$

where ω_{i0} is a known number, may be tested by means of the statistic,

$$u_W = \frac{\hat{\omega}_i - \omega_{i0}}{\sqrt{\tilde{\sigma}_{ii}}}, \tag{11.37}$$

with significance probability,

$$p_{obs} \doteq 2(1 - \Phi(|u_W|))$$

or, equivalently, by means of the statistic,

$$X_W^2 = \frac{(\hat{\omega}_i - \omega_{i0})^2}{\tilde{\sigma}_{ii}}, \tag{11.38}$$

with significance probability,

$$p_{obs} \doteq 1 - F_{\chi^2(1)}(X_W^2).$$

The two tests are both referred to as Wald's test of the hypothesis H_0, which explains the index W on the test statistics.

The $(1-\alpha)$ confidence interval for ω_i based on Wald's u-test is

$$[\hat{\omega}_i - u_{1-\alpha/2}\sqrt{\tilde{\sigma}_{ii}},\, \hat{\omega}_i + u_{1-\alpha/2}\sqrt{\tilde{\sigma}_{ii}}], \tag{11.39}$$

where $u_{1-\alpha/2}$ is the $1-\alpha/2$ fractile of the $N(0,1)$-distribution, and it is called Wald's $(1-\alpha)$ confidence interval for ω_i.

Remark 11.1 Above we have referred to the hypothesis $H_0\colon \omega_i = \omega_{i0}$ as a simple hypothesis about the ith component ω_i of the parameter vector $\boldsymbol{\omega}$. Strictly speaking, considered as a hypothesis about the parameter vector $\boldsymbol{\omega}$, the hypothesis H_0 is a composite hypothesis with $k-1$ free parameters, namely the remaining $k-1$ components of $\boldsymbol{\omega}$. ▼

Remark 11.2 The approximation in (11.34) or, equivalently, the approximation,

$$\frac{\hat{\omega}_i - \omega_i}{\sqrt{\sigma_{ii}}} \approx N(0,1),$$

was first derived under the assumption that $\mathbf{x}$ is a sample, $x_1, \ldots, x_n$, and the approximation is good if the sample size $n \longrightarrow \infty$. Furthermore, for a sample it may be shown that $\tilde{\sigma}_{ii} \longrightarrow \sigma_{ii}$ when $n \longrightarrow \infty$. Combining these facts, the approximation in (11.35) appears. Thus, the result in (11.35) is valid for large samples. Nevertheless, Wald's tests are frequently used, for instance, in SAS, for small samples also without check of their validity.

Another version of Wald's tests is obtained by choosing $\tilde{\sigma}_{ii}$ as the ith diagonal element of the inverse of the estimate $i(\hat{\boldsymbol{\omega}})$ of Fisher's information matrix. The comments above also apply to these tests. ▼

CHAPTER 12

Some Nonparametric Tests

As it appears from Chapter 3 to Chapter 11 in this book, we consider statistical inference based on parameterized classes of distributions. The basis for the inference is that the data **x** are considered as the outcome of a random vector **X** whose distribution function is assumed to belong to a parameterized class of distribution functions, $\mathcal{F} = \{F_{\boldsymbol{\omega}} \mid \boldsymbol{\omega} \in \Omega\}$. Here the parameter $\boldsymbol{\omega}$ has been chosen such that it is of relevance for the scientific question that the data **x** should elucidate.

From time to time parametric statistical inference is criticized for being too sensitive with respect to deviations from the considered class of distributions. For instance, it is sometimes claimed that tests in statistical models based on the normal distribution are too sensitive with regard to deviations from the assumption of normality. Often the argument is that extreme values have a large influence on the empirical variance s^2. If, for instance, few observations have very extreme values, then the empirical variance s^2 increases. Since s or s^2 appears in the denominator in t- and F-tests, these test statistics become small, which means that significant deviations from the hypotheses in question are not revealed.

In contrast to a number of books on elementary statistics, in this book we have been dealing in great detail with the part of a statistical analysis, which in Chapter 1 was referred to as *model checking* and which is concerned exactly with the question of whether the data can be described by means of the parameterized class of distributions considered in the model. In all of the examples in this book the applicability of the distribution class has been evaluated by means of graphical techniques and/or statistical tests performed on the original observations or on the residuals from the model. If this check of the model does not turn out well, the model should of course not be used as a basis for the inference since conclusions based on a wrong model are hardly ever correct. If the model checking has given the impression that one or more observations are extreme compared to the others, it is natural to question the validity of these observations. If they are due to changes in the experimental or observational conditions, these extreme observations may be omitted from the calculations. However, if the validity of the extreme observations is confirmed, we may have to reject the model and search for new one.

If it turns out to be impossible to find a parameterized class of distributions describing the data, one possibility is to turn to nonparametric statistics, which is based on slightly less specific assumptions concerning the distribution of the observations. It is sometimes claimed that nonparametric statistics is free from assumptions, but this is not true. Most of the nonparametric tests are derived under the assumption that the observations are independent and identically distributed and sometimes even with the assumption that the distribution is symmetrical.

The purpose of this chapter is to give impression of ideas in nonparametric statistics. Section 12.1 is concerned with the *sign test*, which is the most simple of the nonparametric tests. In Section 12.2, which is based on Lehmann (1975), the most simple examples of *rank tests* are mentioned, that is, tests based on the ranks of the observations. Section 12.2.1 to Section 12.2.3 discuss rank tests for one, two, and several samples, respectively. In Section 12.2.4 the Spearman rank correlation test is introduced and in Section 12.2.5 some of the rank tests are considered in connection with two-way tables. Finally, in an annex to this chapter it is shown how some of the calculations can be made using SAS.

Table 12.1 *The quantity of calcium carbonate in the urine from 13 healthy elderly women measured during two diets (normal and increased intake of calcium) together with the differences of the two measurements, the absolute values of the differences, and the ranks of the absolute values of the differences*

Woman	*Normal*	*Calcium*	*Difference*	*|Difference|*	*Rank*
1	6.2	12.1	+5.9	5.9	13
2	3.3	4.6	+1.3	1.3	7
3	1.3	1.8	+0.5	0.5	3
4	5.6	8.2	+2.6	2.6	9
5	3.8	3.9	+0.1	0.1	1.5
6	5.1	7.9	+2.8	2.8	11.5
7	4.1	4.0	−0.1	0.1	1.5
8	1.9	4.6	+2.7	2.7	10
9	1.1	2.3	+1.2	1.2	5.5
10	7.2	10.0	+2.8	2.8	11.5
11	6.8	5.6	−1.2	1.2	5.5
12	2.7	3.5	+0.8	0.8	4
13	5.1	6.9	+1.8	1.8	8

12.1 Sign Test

The sign test will be introduced in connection with Example 12.1.

Example 12.1
The data are concerned with how different kinds of calcium are absorbed in the system of healthy elderly women. During the investigation 13 women were on a prescribed weekly diet and the quantity of calcium carbonate excreted in the urine was measured. The larger the quantity of calcium carbonate in the urine the more calcium carbonate was absorbed in the system. Here we consider two diets only. The women's normal diet (*normal*) supplemented with a placebo tablet without any calcium carbonate three times a day and their normal diet supplemented with a calcium tablet three times a day (*calcium*). For the women in the investigation the quantity of calcium carbonate in their urine corresponding to the two diets are shown in Table 12.1.

It is of interest to decide whether the diet has influence on the quantity of calcium carbonate in the system. □

The sign test for a sample, $x_1,\ldots,x_n$, is concerned with test of the hypothesis that the location of the underlying distribution F is a known value μ_0. Replacing the original observations with $x_1 - \mu_0,\ldots,x_n - \mu_0$, without loss of generality we may assume that $\mu_0 = 0$, which we do from now on. In the model for the sample,

$$M_0\colon X_i \sim F, \quad i = 1,\ldots,n,$$

one formulation of the hypothesis that the location is 0 is

$$H_0\colon P(X_i > 0 \mid X_i \neq 0) = P(X_i < 0 \mid X_i \neq 0) = 0.5.$$

Note that the hypothesis H_0 states that 0 belongs to the median of the distribution F, i.e., $0 \in x_{0.5} = \{x \in \mathbb{R} \mid F(x-) \leq 0.5 \leq F(x)\}$, and if F is strictly increasing, then 0 is the median of F, i.e., $x_{0.5} = \{0\}$.

Let n^+ and n^- denote the number of positive and negative values in the sample, respectively, i.e.,

$$n^+ = \#\{\, i \mid x_i > 0 \,\} \quad \text{and} \quad n^- = \#\{\, i \mid x_i < 0 \,\}.$$

Discarding observations in the sample, $x_1,\ldots,x_n$, with values equal to 0 (μ_0), it follows that under the hypothesis H_0 the number of positive values in the sample has a binomial distribution with number of trials $n^+ + n^-$ and probability parameter 0.5, i.e.,

$$N^+ \sim b(n^+ + n^-, 0.5).$$

Using the binomial test in on page 310 the p-value is calculated as, see (7.20) and (7.21),

$$p_{obs} = 2\sum_{i=0}^{\min(n^+,n^-)} b(i;n^+ + n^-, 0.5) = 0.5^{(n^+ + n^- - 1)} \sum_{i=0}^{\min(n^+,n^-)} \binom{n^+ + n^-}{i}. \tag{12.1}$$

In this situation the binomial test is referred to as the *sign test*.

Example 12.1 (Continued)
The hypothesis that the diets have no influence on the quantity of calcium carbonate in the system may be tested with a sign test on the differences. From Table 12.1 on page 494, it is seen that the observed numbers of positive and negative differences are $n^+ = 11$ and $n^- = 2$, respectively. From (12.1) it follows that the significance probability of H_0 is

$$p_{obs} = 2\sum_{i=0}^{2} b(i;13,0.5) = 0.0225$$

and the hypothesis of no influence of the diets is rejected. It is seen from Table 12.1 that the calcium diet gives the largest quantity of calcium carbonate in the urine. □

In Example 12.1 the hypothesis of a symmetric distribution around 0 was rejected by the sign test. The sign test, however, only uses very little of the information in the sample, namely the sign of the observations and not their size and therefore it is not a very powerful test. If the sign test fails to reject the hypothesis, we must always proceed with a more powerful test which uses the size of the observations as explained in the next section.

12.2 Rank Test

All the tests in this section are based on the ranks of the observations in a sample, $x_1,\ldots,x_n$. If all the observations are different, the *rank* of x_i is defined as the number of observations in the sample that are less than or equal to x_i, i.e.,

$$rank(x_i) = \#\{x_j \mid x_j \le x_i\}. \tag{12.2}$$

Thus, in this case the ranks are simply the numbers $1,2,\ldots,n$ and using formula (12.5) below, it is seen that the average of the ranks is $(n+1)/2$.

If k of the observations in the sample are equal to x_i, i.e., $k = \#\{x_j \mid x_j = x_i\}$, these observations are said to be *tied*. In that case the rank of x_i could in principle be defined as in (12.2), but a simple calculation shows that with this definition the average of the ranks is larger than $(n+1)/2$. We define the *rank* of x_i as

$$rank(x_i) = \#\{x_j \mid x_j < x_i\} + (k+1)/2 \tag{12.3}$$

and with this definition it follows from (12.9) the average of the ranks is $(n+1)/2$, i.e., the same as when all the n observations in the sample are different.

Clearly, the definition of the ranks of the observations in a sample is closely related to ordering the observations in the sample in ascending order. If

$$x_{(1)} \le \cdots \le x_{(i)} \le \cdots \le x_{(n)}$$

denotes the *ordered* sample corresponding to the sample, $x_1,\ldots,x_n$, the ranks of the observations can be calculated in the following way:

$$\begin{aligned} rank(x_{(i)}) &= i \quad \text{if} \quad x_{(i-1)} < x_{(i)} < x_{(i+1)} \\ rank(x_{(i)}) = \cdots = rank(x_{(i+k-1)}) &= i + (k-1)/2 \quad \text{if} \quad x_{(i)} = \cdots = x_{(i+k-1)}. \end{aligned} \tag{12.4}$$

Thus, the observation $x_{(i)}$ has rank i if $x_{(i)}$ is the only observation in the sample with this value, i.e., if $x_{(i-1)} < x_{(i)} < x_{(i+1)}$. If k observations, $x_{(i)}, x_{(i+1)}, \ldots, x_{(i+k-1)}$, have the same value, i.e., if $x_{(i)} = x_{(i+1)} = \cdots = x_{(i+k-1)}$, the observations are said to be tied and they are all assigned the rank $i + (k-1)/2$, which is the average of the k numbers $i, i+1, \ldots, i+k-1$. The ranks defined above are often referred to as "average ranks" or "midranks."

The theory of ranks test was developed for continuous distributions where the probability of a tie is 0. In practice, however, ties often occur.

In the calculation of the average above we have used the fact that the sum of the numbers 1, 2, ..., N is $N(N+1)/2$, i.e.,

$$\sum_{i=1}^{N} i = \frac{N(N+1)}{2}, \tag{12.5}$$

an identity that often appears in the discussion of rank tests. Two other useful identities are

$$\sum_{i=1}^{N} i^2 = \frac{N(N+1)(2N+1)}{6} \tag{12.6}$$

and

$$\sum_{i=1}^{N} \left(i - \frac{N+1}{2}\right)^2 = \frac{(N-1)N(N+1)}{12}. \tag{12.7}$$

In order to calculate the ranks of the observations in a sample we use the notation concerning samples introduced in Chapter 2. Suppose that there are m *different values* in the sample, $x_1, x_2, \ldots, x_n$, and let $y_1, y_2, \ldots, y_m$ denote the different values in ascending order, i.e.,

$$y_1 < y_2 < \ldots < y_m.$$

For $j = 1, 2, \ldots, m$ we let a_j denote the *number* of observations in the sample that are equal to y_j and, furthermore, we let k_j denote the *cumulative number*, i.e., $k_j = a_1 + \cdots + a_j$ so k_j is the number of observations in the sample that are less than or equal to y_j. Finally, we let $k_0 = 0$. With this notation the rank assigned to the a_j tied observations with value y_j is

$$rank(x_{(k_{j-1}+1)}) = \cdots = rank(x_{(k_{j-1}+a_j)}) = k_{j-1} + \frac{a_j+1}{2}. \tag{12.8}$$

The calculation of the ranks can be performed using Table 12.2.

If $r_1, \ldots, r_n$ denote the ranks of the observations in a sample, $x_1, \ldots, x_n$, it may be shown using (12.5) and (12.6) that

$$r. = \sum_{i=1}^{n} r_i = \frac{n(n+1)}{2}, \tag{12.9}$$

$$\sum_{i=1}^{n} r_i^2 = \frac{n(n+1)(2n+1)}{6} - \frac{1}{12}\sum_{j=1}^{m}(a_j^3 - a_j) \tag{12.10}$$

and

$$\sum_{i=1}^{n} (r_i - \bar{r}.)^2 = \frac{(n-1)n(n+1)}{12} - \frac{1}{12}\sum_{j=1}^{m}(a_j^3 - a_j), \tag{12.11}$$

where $\bar{r}. = r./n = (n+1)/2$.

In Section 12.2.1 to Section 12.2.3 we consider rank tests for one, two, and several samples. Section 12.2.4 is concerned with rank correlations and in Section 12.2.5 some of the ranks tests are applied in connection with two-way tables. In the discussion of the rank tests we assume by way of

Table 12.2 *Template for calculations of the ranks in a sample*

Observation y	*Number* a	*Cumulative number* k	*Rank* r
y_1	a_1	$k_1 = a_1$	$(a_1+1)/2$
y_2	a_2	$k_2 = a_1 + a_2$	$k_1 + (a_2+1)/2$
y_3	a_3	$k_3 = a_1 + a_2 + a_3$	$k_2 + (a_3+1)/2$
$\vdots$	$\vdots$	$\vdots$	$\vdots$
y_j	a_j	$k_j = a_1 + \dots + a_j$	$k_{j-1} + (a_j+1)/2$
$\vdots$	$\vdots$	$\vdots$	$\vdots$
y_m	a_m	$k_m = a_1 + \dots + a_m$	$k_{m-1} + (a_m+1)/2$

introduction that all observations are different and in remarks we comment on modifications of the tests in case of tied observations.

In order to find the significance probability of a rank test, it is important to describe the joint distribution of the ranks. Theorem 12.1 below gives this distribution and, in addition, the distribution of the ordered sample. In the theorem the set of all permutations of the set $\{1, 2, \dots, n\}$ is denoted by $\mathcal{S}(n)$.

Theorem 12.1 Let $X_1, \dots, X_n$ be independent and identically distributed continuous random variables with probability density function f and distribution function F. Furthermore, let $\mathbf{X}_{(*)} = (X_{(1)}, \dots, X_{(i)}, \dots, X_{(n)})$ be the ordered sample and $\mathbf{R} = (R_1, \dots, R_i, \dots, R_n)$ the vector of ranks. Then with probability one,

$$X_i = X_{(R_i)}. \tag{12.12}$$

Furthermore, the probability density function of $\mathbf{X}_{(*)}$ is

$$f_{\mathbf{X}_{(*)}}(y_1, \dots, y_n) = n! \prod_{i=1}^{n} f(y_i), \quad y_1 < y_2 < \cdots < y_n.$$

In addition, the ordered sample $\mathbf{X}_{(*)}$ is independent of the ranks $\mathbf{R}$, which has the probability function,

$$P(\mathbf{R} = \mathbf{r}) = \frac{1}{n!}, \quad \mathbf{r} \in \mathcal{S}(n), \tag{12.13}$$

implying that

$$P(R_i = s) = \frac{1}{n}, \quad s \in \{1, 2, \dots, n\}. \tag{12.14}$$

♦

The formula (12.13) shows that the ranks $\mathbf{R}$ has a uniform distribution on the set $\mathcal{S}(n)$ consisting of the $n!$ permutations of $\{1, 2, \dots, n\}$. As seen below, in most rank tests the test statistic is a sum of a subset of the ranks and intuitive arguments determine which possible values of the sum that are more critical for the hypothesis than the observed value. Thus, the exact significance probability is simply obtained by calculating the test statistic for all the $n!$ possible values of the ranks and count the number of times the value of the test statistic is more or just as extreme as the observed value of the test statistic. Obviously, if the sample size n is large, this is troublesome and the p-value is calculated using an approximation of the distribution of the test statistic.

Note that in Theorem 12.1 F is assumed to be the distribution function of a continuous random variable. This implies that the probability of ties is 0 and that the connection between the ordered sample and the ranks is as simple as stated in (12.12).

12.2.1 One Sample

The observations, $x_1, \ldots, x_n$, are assumed to be a sample from a continuous distribution F, i.e., we consider the model,

$$M_0\colon X_i \sim F, \quad i = 1, \ldots, n,$$

where we implicitly assume that the X's are independent. We want to test the hypothesis that the continuous distribution F is symmetric around a known value μ_0. Replacing the observations with $x_1 - \mu_0, \ldots, x_n - \mu_0$, without loss of generality we may assume that $\mu_0 = 0$. In that case the hypothesis of symmetry is

$$H_0\colon F(-x) = 1 - F(x), \quad \text{for all } x > 0.$$

Note that the hypothesis implies that the median of the distribution F is 0, and if the mean of F exists it is equal to the median so the hypothesis implies that the mean is 0.

A nonparametric test of H_0 is the *Wilcoxon signed rank test*. The test is based on the ranks of the absolute values $|x_1|, \ldots, |x_n|$ of the observations. The test only considers the observations that are different from 0. Let N denote the number of such observations, i.e., $N = \#\{i \mid x_i \neq 0\}$ and let r_i^+ denote the rank of $|x_i|$ in sample of absolute values. A test statistic is

$$w = \sum_{\{i:x_i>0\}} r_i^+, \tag{12.15}$$

i.e., the sum of the ranks calculated for the sample of absolute values but summed over those observations that are positive in the original sample. If the hypothesis H_0 is true, the positive and negative observations occur randomly between each other and they have approximately the same absolute values. The sum of the ranks r_i^+ of the positive observations in the original sample should therefore be approximately equal to the sum of the ranks r_i^+ of the negative observations in the original sample. If the sum of the ranks r_i^+ of the positive observations is considerably larger than the sum of the ranks r_i^+ of the negative observations, it is an indication that there are significantly more positive observations than negative observations or that the positive observations are significantly larger in absolute value than the negative observations. Thus, large values of w are critical for the hypothesis H_0. Similarly, an argument of symmetry shows that also small values of w are critical for H_0.

If all the observations different from 0 are negative, then $w = 0$, and if all observations different from 0 are positive, then $w = N(N+1)/2$ according to (12.5). The distribution of the corresponding random variable W under the hypothesis H_0 is a discrete distribution concentrated on the set $\{\, 0, 1, \ldots, N(N+1)/2 \,\}$, and it may be shown that the distribution of W does not depend on F, the common distribution of the X's, and, furthermore, that the mean of W is

$$EW = \frac{N(N+1)}{4}.$$

Often the mean of W is subtracted from W and we consider the statistic,

$$S = W - \frac{N(N+1)}{4} = \sum_{\{i:X_i>0\}} R_i^+ - \frac{N(N+1)}{4}. \tag{12.16}$$

Large absolute values of S are critical for H_0 so the significance probability of H_0 is calculated as

$$p_{obs} = P(|S| \geq |s|). \tag{12.17}$$

For small values of N the significance probability may be found using tables or a computer package. When $N > 20$, the significance probability is calculated using approximations. It may be shown that under the hypothesis H_0 the variance of S is

$$Var\,S = Var\,W = \frac{N(N+1)(2N+1)}{24}$$

and that the distribution of S may be approximated by a normal distribution for large values of N.

Since $ES = 0$, it follows that

$$U(\mathbf{X}) = \frac{W - \frac{N(N+1)}{4}}{\sqrt{\frac{N(N+1)(2N+1)}{24}}} = \frac{S}{\sqrt{Var S}} \approx N(0,1) \tag{12.18}$$

and the p-value of the Wilcoxon signed rank test is calculated as

$$p_{obs}(\mathbf{x}) = 2(1 - \Phi(|u(\mathbf{x})|)), \tag{12.19}$$

where $u(\mathbf{x})$ is the observed value of $U(\mathbf{X})$ and Φ is the distribution function of the $N(0,1)$-distribution.

Sometimes the significance probability in (12.17) is approximated using a t-distribution due to the fact that

$$T(\mathbf{X}) = \frac{S}{\sqrt{\frac{N}{N-1}\left(\frac{N(N+1)(2N+1)}{24} - \frac{S^2}{N}\right)}} \approx t(N-1). \tag{12.20}$$

In that case the p-value of the Wilcoxon signed rank test is calculated as

$$p_{obs}(\mathbf{x}) = 2(1 - F_{t(N-1)}(|t(\mathbf{x})|)), \tag{12.21}$$

where $t(\mathbf{x})$ is the observed value of $T(\mathbf{X})$ and $F_{t(N-1)}$ is the distribution function of the t-distribution with $N-1$ degrees of freedom.

The reason for using the words "signed rank" in the name of the test is that the statistics S, U, and T may be expressed in terms of the *signed ranks* q_i, which are defined in the following way:

$$q_i = \begin{cases} r_i^+ & \text{if } x_i > 0 \\ -r_i^+ & \text{if } x_i < 0. \end{cases} \tag{12.22}$$

Using (12.5) and (12.6) it follows after some calculations that the statistics S, U, and T may be written as

$$S = \frac{Q.}{2}, \tag{12.23}$$

$$U(\mathbf{X}) = \frac{Q.}{\sqrt{USS_Q}} = \frac{\bar{Q}.}{\sqrt{USS_Q/N^2}}, \tag{12.24}$$

where $Q. = \sum_{i=1}^{N} Q_i$, $\bar{Q}. = \frac{1}{N}Q.$ and $USS_Q = \sum_{i=1}^{N} Q_i^2$, and

$$T(\mathbf{X}) = \frac{\bar{Q}.}{\sqrt{s_Q^2/N}}, \tag{12.25}$$

where $s_Q^2 = \frac{1}{N-1}\sum_{i=1}^{N}(Q_i - \bar{Q}.)^2 = \frac{1}{N-1}(USS_Q - \frac{Q.^2}{N})$.

Consequently, a way to interpret the two approximative tests of H_0 is as follows: Consider the **signed ranks** $q_1, \ldots, q_N$ as a sample from a normal distribution and test the hypothesis that the mean of the distribution is 0 using a u-test if the variance of the signed ranks is assumed to be known (and equal to USS_Q/N), or using a t-test if the variance of the signed ranks is assumed to be unknown.

Remark 12.1 If there are tied observations, formula (12.24) is still valid and (12.6) implies that the statistic U in (12.18) takes the form,

$$U(\mathbf{X}) = \frac{W - \frac{N(N+1)}{4}}{\sqrt{\frac{N(N+1)(2N+1)}{24} - \frac{1}{48}\sum_j (a_j^3 - a_j)}}, \tag{12.26}$$

Table 12.3 *Calculation of the ranks of the absolute values of the differences in Table 12.1*

Observation y	*Number* a	*Cumulative number* k	*Rank* r
0.1	2	2	1.5
0.5	1	3	3
0.8	1	4	4
1.2	2	6	5.5
1.3	1	7	7
1.8	1	8	8
2.6	1	9	9
2.7	1	10	10
2.8	2	12	11.5
5.9	1	13	13

and the significance probability is calculated using (12.19).

Similarly, formula (12.25) may also be used when there are tied observations and in that case the statistic T is calculated as

$$T(\mathbf{X}) = \frac{S}{\sqrt{\frac{N}{N-1}\left(\frac{N(N+1)(2N+1)}{24} - \frac{1}{48}\sum_j(a_j^3 - a_j) - \frac{S^2}{N}\right)}} \qquad (12.27)$$

and the p-value is obtained using (12.21).

Note that the observed values y_j for which there are no tied observations, i.e., for which $a_j = 1$, do not contribute to the sum $\sum_j(a_j^3 - a_j)$ since for such values $a_j^3 - a_j = 1^3 - 1 = 0$. ▼

Example 12.1 (Continued)

We now consider the Wilcoxon signed rank test for the sample consisting of the differences in this example. From the fourth column in Table 12.1, it is seen that none of the differences are equal to 0, i.e., $N = 13$. The ranks of the absolute values of the differences are seen in the sixth column of the table. The ranks are found in Table 12.3 using the template for calculation in Table 12.2.

Using the fourth and sixth column in Table 12.1, it follows that $w = 84$ so

$$s = 84 - \frac{13 \times 14}{4} = 84 - 45.5 = 38.5.$$

According to the output from SAS on page 513 the significance probability of H_0 is

$$p_{obs} = P(|S| \geq 38.5) = 0.0049$$

and the hypothesis is rejected, i.e., the distribution of the differences is not symmetric around 0, and, as above, we can conclude that the diet has influence on the quantity of calcium carbonate in the urine.

For the sake of illustration we calculate the U and T statistic even though N is less than 20. It is seen from Table 12.1 that there are tied observations since each of the values 0.1, 1.2, and 2.8 occur twice in the fifth column with the absolute values of the differences. Thus

$$\sum_j(a_j^3 - a_j) = 3(2^3 - 2) = 3 \times 6 = 18.$$

The formula (12.26) implies that

$$u(\mathbf{x}) = \frac{84 - \dfrac{13 \times 14}{4}}{\sqrt{\dfrac{13 \times 14 \times 27}{24} - \dfrac{18}{48}}} = 2.6931$$

and according to (12.19) the corresponding p-value is $2(1 - \Phi(2.6931)) = 0.0071$.

Similarly, from (12.27) and (12.21) we get

$$t(\mathbf{x}) = \frac{38.5}{\sqrt{\dfrac{13}{12}\left(\dfrac{13 \times 14 \times 27}{24} - \dfrac{18}{48} - \dfrac{38.5^2}{13}\right)}} = 3.8914,$$

which a significance probability equal to $2(1 - F_{t(12)}(3.8914)) = 0.0021$.

That the p-values for the u-test and the t-test deviate from the exact p-value $p_{obs} = 0.0049$ was to be expected since the criterion for applying these test ($N > 20$) is not satisfied. □

12.2.2 Two Samples

The discussion is based on the data in Example 12.2.

Example 12.2

We consider again the two samples in Example 2.5 consisting of the reaction times of flies exposed to a nerve poison where the flies in the first sample and the second sample have been exposed for 30 seconds and 60 seconds, respectively. The reaction times are seen in Table 12.4 together with the ranks of the observations in the *combined sample*.

As in Example 2.5 we are interested in investigating whether the contact time has an influence on the reaction times. □

As before we let x_{ij} denote the jth observation in ith sample, $j = 1, \ldots, n_i$, $i = 1, 2$. Here we consider a nonparametric model for two samples, i.e.,

$$M_0\colon X_{ij} \text{ has a continuous distribution } F_i\,,\ j = 1, \ldots, n_i,\ i = 1, 2,$$

where we implicitly assume that the X's are independent. We want to test the hypothesis that all the $n_{\cdot} = n_1 + n_2$ observations have the same distribution or that they constitute a sample from a common distribution F, i.e.,

$$H_0\colon F_i = F,\ i = 1, 2.$$

A nonparametric test of this hypothesis is the *Wilcoxon two-sample test*. The test is sometimes misleadingly referred to as the *Mann-Whitney test* which, however, is equivalent to the Wilcoxon test, see Remark 12.3 on page 504.

The Wilcoxon two-sample test is based on the ranks of the observations in the *combined sample*. Let R_{ij} denote the rank of X_{ij} in the combined sample and let

$$R_{1\cdot} = \sum_{j=1}^{n_1} R_{1j} \tag{12.28}$$

be the sum of the ranks in the first sample. The smallest value of $R_{1\cdot}$ occurs if the n_1 observations in the first sample all are less than the observations in the second sample and in that case $R_{1\cdot} = n_1(n_1 + 1)/2$, which is the sum of the numbers $1, 2, \ldots, n_1 - 1, n_1$ according to (12.5). Conversely, the largest value of $R_{1\cdot}$ occurs if all the observations in the first sample are larger that the observations in the second sample and in that case $R_{1\cdot} = (2n_{\cdot} - n_1 + 1)n_1/2$, which is the sum of the numbers $n_2 + 1$, $n_2 + 2, \ldots, n_2 + n_1 - 1, n_2 + n_1$.

Table 12.4 *The reaction times of two samples of flies exposed to a nerve poison together with the rank of the observations in the combined sample*

Contact time		*Rank*	
30 *seconds*	60 *seconds*	30 *seconds*	60 seconds
3	2	2	1
5	5	4.5	4.5
5	5	4.5	4.5
7	7	7.5	7.5
9	8	11	9
9	9	11	11
10	14	13	15
12	18	14	16
20	24	17	19
24	26	19	21.5
24	26	19	21.5
34	34	23.5	23.5
43	37	27	25
46	42	28	26
58	90	29	30
140		31	
	Sum	261	235

If $R_{1\cdot}$ is small, most of the observations in the first sample are less than the observations in the second sample, which is not in keeping with the hypothesis H_0 that the observations in the two samples have the same distribution. Similarly, a large value of $R_{1\cdot}$ indicates a deviation from H_0 since such a value only occurs if most of the observations in the first sample are larger than the observations in the second sample. To sum up, both small and large values of $R_{1\cdot}$ are critical for H_0.

It may be shown that under H_0 the mean of the discrete random variable $R_{1\cdot}$ is the average of the smallest possible value and the largest possible value of this variable, i.e.,

$$ER_{1\cdot} = \frac{n_1(n_\cdot + 1)}{2} \tag{12.29}$$

and, furthermore, that the variance is

$$VarR_{1\cdot} = \frac{n_1 n_2(n_\cdot + 1)}{12}. \tag{12.30}$$

The significance probability of the Wilcoxon two-sample test statistic of H_0 is

$$p_{obs} = P(|R_{1\cdot} - ER_{1\cdot}| \geq |r_{1\cdot} - ER_{1\cdot}|) \tag{12.31}$$

where $r_{1\cdot}$ is the observed value of $R_{1\cdot}$.

For small values of n_1 and n_2 the p-value of the Wilcoxon two-sample test may be found in tables or calculated using a statistical computer package. For large values of n_1 and n_2, it can be shown that the distribution of $R_{1\cdot}$ can be approximated by a normal distribution. Using (12.29) and (12.30) one has that

$$Z(\mathbf{X}) = \frac{R_{1\cdot} - \dfrac{n_1(n_\cdot + 1)}{2}}{\sqrt{\dfrac{n_1 n_2(n_\cdot + 1)}{12}}} \approx N(0,1). \tag{12.32}$$

The approximation may be used if $n_1 \geq 10$ and $n_2 \geq 10$ and if so, the significance probability in (12.31) may be calculated as

$$p_{obs}(\mathbf{x}) = 2(1 - \Phi(|z(\mathbf{x})|)), \tag{12.33}$$

where $z(\mathbf{x})$ is the observed value of $Z(\mathbf{X})$ and Φ is the distribution function of the $N(0,1)$ distribution.

The significance probability of H_0 is sometimes calculated using the t-distribution. To explain why, we first note that using (12.7) the variance of $R_{1\bullet}$ may be rewritten as

$$\frac{n_1 n_2 (n_\bullet + 1)}{12} = \frac{n_1 n_2}{n_\bullet} \frac{1}{n_\bullet - 1} \sum_{i=1}^{2} \sum_{j=1}^{n_i} (r_{ij} - \bar{r}_{\bullet\bullet})^2, \tag{12.34}$$

where

$$\bar{r}_{\bullet\bullet} = \frac{1}{n_\bullet} \sum_{i=1}^{2} \sum_{j=1}^{n_i} r_{ij} = \frac{1}{n_\bullet} \frac{n_\bullet (n_\bullet + 1)}{2} = \frac{n_\bullet + 1}{2}. \tag{12.35}$$

Using (12.34) and (12.35) formula (12.32) turns into

$$Z(\mathbf{X}) = \frac{\bar{R}_{1\bullet} - \bar{R}_{2\bullet}}{\sqrt{s_R^2 \left(\frac{1}{n_1} + \frac{1}{n_2} \right)}}, \tag{12.36}$$

where

$$\bar{R}_{i\bullet} = \frac{1}{n_i} \sum_{j=1}^{n_i} R_{ij}, \quad i = 1, 2,$$

and

$$s_R^2 = \frac{1}{n_\bullet - 1} \sum_{i=1}^{2} \sum_{j=1}^{n_i} (R_{ij} - \bar{R}_{\bullet\bullet})^2.$$

Note that the quantity in (12.36) is the t-test statistics in formula (3.30) on page 83 for the hypothesis that for the *ranks in the combined sample* the means of the ranks in the two samples are equal under the assumption that the variances of the ranks in the two samples are identical except for the modification that the estimate of the common variance, $s_1^2 = \frac{1}{n_\bullet - 2} \sum_{i=1}^{2} \sum_{j=1}^{n_i} (R_{ij} - \bar{R}_{i\bullet})^2$, in (3.30) has been replaced by s_R^2. Due to this interpretation the p-value of the test statistic $Z(\mathbf{X})$ is sometimes evaluated in a $t(n_\bullet - 1)$ distribution, i.e., as

$$p_{obs}(\mathbf{x}) = 2(1 - F_{t(n_\bullet - 1)}(|z(\mathbf{x})|)), \tag{12.37}$$

where $z(\mathbf{x})$ is the observed value of $Z(\mathbf{X})$ and $F_{t(n_\bullet - 1)}$ denotes the distribution function of the t-distribution with $n_\bullet - 1$ degrees of freedom.

Thus, again, the rank test may be interpreted as a well-known test by considering the ranks as being normal distributed.

Remark 12.2 In case of tied observations the expression for $Z(\mathbf{X})$ in (12.36) may still be used in connection with either (12.33) or (12.37). According to (12.11) the formula (12.34) in this situation takes the form,

$$\frac{n_1 n_2}{n_\bullet} \frac{1}{n_\bullet - 1} \sum_{i=1}^{2} \sum_{j=1}^{n_i} (R_{ij} - \bar{R}_{\bullet\bullet})^2 = \frac{n_1 n_2 (n_\bullet + 1)}{12} \left[1 - \frac{\sum_j (a_j^3 - a_j)}{(n_\bullet - 1) n_\bullet (n_\bullet + 1)} \right],$$

and the expression for $Z(\mathbf{X})$ in (12.32) is replaced by

$$Z(\mathbf{X}) = \frac{R_{1.} - \dfrac{n_1(n.+1)}{2}}{\sqrt{\dfrac{n_1 n_2(n.+1)}{12}\left[1 - \dfrac{\sum_j (a_j^3 - a_j)}{(n.-1)n.(n.+1)}\right]}}. \tag{12.38}$$

Note that the observed values y_j for which there are no tied observations, i.e., for which $a_j = 1$, do not contribute to the sum $\sum_j (a_j^3 - a_j)$ since for such values $a_j^3 - a_j = 1^3 - 1 = 0$. ▼

Example 12.2 (Continued)
From Table 12.4 on page 502, it is seen that in this situation we have tied observations, since the values 5, 7, 9, 24, 26, and 34 have been observed 4, 2, 3, 3, 2 and 2 times, respectively. Thus,

$$\sum_j (a_j^3 - a_j) = (4^3 - 4) + 2(3^3 - 3) + 3(2^3 - 2) = 126.$$

From the third column in Table 12.4, it is seen that $r_{1.} = 261$. Since $n_1 = 16$ and $n_2 = 15$ we have that $n. = 31$ and by means of (12.38) we find that

$$z(\mathbf{x}) = \frac{261 - \dfrac{16 \times 32}{2}}{\sqrt{\dfrac{16 \times 15 \times 32}{12}\left[1 - \dfrac{126}{30 \times 31 \times 32}\right]}} = 0.1981$$

and (12.33) implies that the p-value of the Wilcoxon two-sample test is

$$p_{obs}(\mathbf{x}) = 2(1 - \Phi(0.1981)) = 0.8430$$

or using (12.37)

$$p_{obs}(\mathbf{x}) = 2(1 - F_{t(30)}(0.1981)) = 0.8443.$$

The hypothesis H_0 is not rejected, which means that the contact time has no significant influence on the reaction times. □

Remark 12.3 The Mann-Whitney test is closely related to the Wilcoxon two-sample test. Mann-Whitney's idea was to consider all pairs of observations (x_{1j}, x_{2k}) that can be formed with one observation from each of the two samples and in case of no tied observations in the combined sample to calculate one of the two numbers,

$$mw_{1<2} = \#\{(x_{1j}, x_{2k})\,|\,x_{1j} < x_{2k},\ j = 1, \ldots, n_1,\ k = 1, \ldots, n_2\}, \tag{12.39}$$

or

$$mw_{1>2} = \#\{(x_{1j}, x_{2k})\,|\,x_{1j} > x_{2k},\ j = 1, \ldots, n_1,\ k = 1, \ldots, n_2\}, \tag{12.40}$$

i.e., the number of pairs such that the observation in the first sample is less than the observation in the second sample and the number of pairs such that the observation in the first sample is larger than the observation in the second sample, respectively. Small and large values of corresponding random variables, $MW_{1<2}$ and $MW_{1>2}$, are critical for the hypothesis that the two samples have the same distribution. Obviously, $MW_{1<2} + MW_{1>2} = n_1 n_2$ because $MW_{1<2} + MW_{1>2}$ is the number of pairs, so either $MW_{1<2}$ or $MW_{1>2}$ can be used as test statistics. It can be shown that

$$MW_{1<2} = n_1 n_2 + \frac{n_1(n_1+1)}{2} - R_{1.}, \tag{12.41}$$

so the Mann-Whitney test statistic is simply a translation of the Wilcoxon two-sample test statistic

Table 12.5 *The observations in Table 3.3 ordered according to size within each sample, together with the ranks of the observations in the combined sample*

Group 1		*Group 2*		*Group 3*		*Group 4*	
Area	*Rank*	*Area*	*Rank*	*Area*	*Rank*	*Area*	*Rank*
200	5	163	1	268	29.5	201	6
215	11.5	182	2	271	32	216	13
225	14	188	3	273	34	241	20.5
229	15	195	4	282	35.5	257	25
230	16.5	202	7	285	38	259	26
232	18	205	8	299	43	267	28
241	20.5	212	9	309	44	269	31
253	22	214	10	310	45	282	35.5
256	24	215	11.5	314	47	283	37
264	27	230	16.5	320	48	291	41.5
268	29.5	235	19	337	50	291	41.5
288	39.5	255	23	340	51	312	46
288	39.5	272	33	345	52	326	49
Sum	282	*Sum*	147	*Sum*	549	*Sum*	400

and with small and large values of both $MW_{1<2}$ and $R_{1.}$ being critical, the significance probabilities will be the same for the two tests.

If there are tied observations in the combined sample, the statistics in (12.39) and (12.40) are modified by adding the quantity $\#\{(x_{1j},x_{2k})\,|\,x_{1j}=x_{2k},\ j=1,\ldots,n_1,\ k=1,\ldots,n_2\}/2$. ▼

12.2.3 Several Samples

The discussion of the Kruskal-Wallis test, which is the nonparametric analogy to one-way analysis of variance, is based on Example 12.3.

Example 12.3

We consider again the data in Example 3.3 concerning the area of the leaves of 52 soya bean plants. The plants were divided into four groups with 13 plants in each group and the four groups were exposed to different treatments. The response, the total area of the leaves for each plant, was measured after 16 days. The observations, ordered according to size, are shown in Table 12.5, together with the ranks of the observations in combined sample calculated using Table 12.2 on page 497. As before we are interested in investigating whether the area depends on the treatments. □

Let x_{ij} denote the jth observation in the ith sample, $j=1,\ldots,n_i$, $i=1,\ldots,k$, and consider the nonparametric model for k samples, i.e.,

$$M_0\colon X_{ij} \text{ has a continuous distribution } F_i,\ j=1,\ldots,n_i,\ i=1,\ldots,k,$$

where all the X's are assumed to be independent. We want to test the hypothesis that all the $n_. = n_1+\cdots+n_k$ observations can be considered as a sample from a common distribution F, i.e.,

$$H_0\colon F_i = F,\ i=1,\ldots,k.$$

A nonparametric test of this hypothesis is the *Kruskal-Wallis test for k samples*, which is based on the ranks of the observations in the combined sample. Lad R_{ij} denote the rank of X_{ij} in the *combined sample* and let

$$R_{i}. = \sum_{j=1}^{n_i} R_{ij}$$

be the sum of the ranks of the observations in the ith sample. The smallest value of $R_{i}.$ occurs if the n_i observations in the ith sample all are less than the observations in the others and if so, $R_{i}. = n_i(n_i+1)/2$, which according to (12.5) is the sum of the numbers $1, 2, \ldots, n_i - 1, n_i$. Conversely, the largest value of $R_{i}.$ occurs if all the observations in the ith sample are larger than the observations in the others and in that case $R_{i}. = (2n. - n_i + 1)n_i/2$, which is the sum of the n_i numbers $n. - n_i + 1$, $n. - n_i + 2, \ldots, n. - 1, n.$. It may be shown that under H_0 the mean of $R_{i}.$ is the average of the smallest possible value and largest possible value of the variable, i.e.,

$$ER_{i}. = \frac{n_i(n. + 1)}{2},$$

and, consequently the average $\bar{R}_{i}. = R_{i}./n_i$ has mean,

$$E\bar{R}_{i}. = \frac{n. + 1}{2} = \bar{R}..,$$

which is the average of the $n.$ numbers $1, 2, \ldots, n. - 1, n.$. Under the hypothesis H_0, it must be expected that the k averages $\bar{R}_{i}.$ of the ranks within the samples vary randomly around $\bar{R}..$ and large values of the Kruskal-Wallis test statistic,

$$KW(\mathbf{X}) = \frac{12}{n.(n. + 1)} \sum_{i=1}^{k} n_i(\bar{R}_{i}. - \bar{R}..)^2, \tag{12.42}$$

are critical for H_0. Since the distribution of the test statistic may be approximated by a χ^2-distribution with $k-1$ degrees of freedom for moderate values of all the sample sizes n_i, the significance probability of the test is calculated as

$$p_{obs}(\mathbf{x}) = 1 - F_{\chi^2(k-1)}(KW(\mathbf{x})), \tag{12.43}$$

where $KW(\mathbf{x})$ is the observed value of $KW(\mathbf{X})$.

If $k = 2$, it may be shown that the Kruskal-Wallis test is equivalent to the Wilcoxon two-sample test because

$$KW(\mathbf{X}) = Z(\mathbf{X})^2 \tag{12.44}$$

and large numerical values of Z correspond to large values of KW and because the square of a $N(0,1)$-distributed random variable has a $\chi^2(1)$-distribution.

Using (12.7) the test statistic may be rewritten as

$$KW(\mathbf{X}) = \frac{\sum_{i=1}^{k} n_i(\bar{R}_{i}. - \bar{R}..)^2}{\frac{1}{n. - 1} \sum_{i=1}^{k} \sum_{j=1}^{n_i} (R_{ij} - \bar{R}..)^2} \tag{12.45}$$

Note that if the ranks in the combined sample are considered as being identical, normally distributed with variance σ^2, then $\sum_{i=1}^{k} n_i(\bar{R}_{i}. - \bar{R}..)^2 \sim \sigma^2\chi^2(k-1)$ and the test statistic in (12.45) appears by replacing σ^2 by $\frac{1}{n. - 1} \sum_{i=1}^{k} \sum_{j=1}^{n_i} (R_{ij} - \bar{R}..)^2$.

The expression in (12.42) is easily calculated on a pocket calculator when we know the sum and uncorrected sum of squares of the ranks in the k samples, since

$$\sum_{i=1}^{k} n_i(\bar{R}_{i}. - \bar{R}..)^2 = \sum_{i=1}^{k} \frac{R_{i}^2.}{n_i} - \frac{R_{..}^2}{n.}. \tag{12.46}$$

Remark 12.4 In case of tied observations, the expression for $KW(\mathbf{X})$ in (12.45) may still be used in connection with (12.43) and by means of (12.11) the formula in (12.42) becomes

$$KW(\mathbf{X}) = \frac{12 \sum_{i=1}^{k} n_i(\bar{R}_{i\cdot} - \bar{R}_{\cdot\cdot})^2}{n_\cdot(n_\cdot + 1)\left[1 - \dfrac{\sum_j(a_j^3 - a_j)}{(n_\cdot - 1)n_\cdot(n_\cdot + 1)}\right]} \tag{12.47}$$

Note that the observed values y_j for which there are no tied observations, i.e., for which $a_j = 1$, do not contribute to the sum $\sum_j(a_j^3 - a_j)$ since for such values $a_j^3 - a_j = 1^3 - 1 = 0$. ▼

Example 12.3 (Continued)
For the data in Table 12.5, we now calculate the Kruskal-Wallis test statistic in (12.47). From the table it is seen that there are tied observations, since the seven values 215, 230, 241, 268, 282, 288, and 291 all have been observed twice. Thus

$$\sum_j (a_j^3 - a_j) = 7(2^3 - 2) = 7 \times 6 = 42.$$

From Table 12.5 it follows that

i	n_i	$r_{i\cdot}$	$\dfrac{r_{i\cdot}^2}{n_i}$
1	13	282	6117.231
2	13	147	1662.231
3	13	549	23184.692
4	13	400	12307.692
sum	52	1378	43271.846

and using (12.46) and (12.47) we see that

$$KW(\mathbf{x}) = \frac{12\left(43271.846 - \dfrac{1378^2}{52}\right)}{52 \times 53\left[1 - \dfrac{42}{51 \times 52 \times 53}\right]} = \frac{81058.1520}{2755.1765} = 29.4203.$$

Formula (12.43) implies that the significance probability is

$$p_{obs}(\mathbf{x}) = 1 - F_{\chi^2(3)}(29.4203) = 0.000001827$$

and the hypothesis is strongly rejected. As in Example 3.3 the conclusion of the analysis is that the area of the leaves of the soya bean plants depends on the treatments. □

Example 12.2 (Continued)
From Table 12.4 on page 502, it is seen that the sample sizes and rank sums for the data are

i	n_i	$r_{i\cdot}$
1	16	261
2	15	235
sum	31	496

From page 504 we know that $\sum_j(a_j^3 - a_j) = 126$. The formulas (12.42), (12.46), and (12.47) imply that the Kruskal-Wallis test statistic is

$$KW(\mathbf{x}) = \frac{12\left(\frac{261^2}{16}+\frac{235^2}{15}-\frac{496^2}{31}\right)}{31\times 32\left[1-\frac{126}{30\times 31\times 32}\right]} = \frac{38.75}{987.80} = 0.0392.$$

From (12.43) the significance probability of H_0 is seen to be

$$p_{obs}(\mathbf{x}) = 1 - F_{\chi^2(1)}(0.0392) = 0.8430,$$

i.e., the same as was obtained by the Wilcoxon two-sample test. The reason for this is explained in (12.44) and the remark following (12.44) and here expressed by the fact that $\sqrt{0.0392} = 0.1981$. □

12.2.4 Spearman Rank Correlation Coefficient

In this section we discuss a nonparametric test of the hypothesis of independence of the two components of bivariate distribution on the basis of the sample $(x_1,y_1)^*,\ldots,(x_n,y_n)^*$ from the distribution. In the model,

$$M_0\colon (X_i,Y_i)^* \sim F_{(X,Y)},\ i=1,\ldots n,$$

for n independent observations from a bivariate distribution with distribution function $F_{(X,Y)}$ the hypothesis of independence can be formulated as

$$H_0\colon F_{(X,Y)}(x,y) = F_X(x)F_Y(y), \quad \text{for all } (x,y)\in\mathbb{R}^2, \tag{12.48}$$

where F_X and F_Y are the distribution function of the marginal distribution of X and Y, respectively.

Recall from page 268 that if $F_{(X,Y)}$ is the distribution function of the bivariate normal distribution, then the hypothesis H_0 may be expressed as

$$\rho = 0,$$

where ρ is the correlation coefficient,

$$\rho = \frac{Cov(X_i,Y_i)}{\sqrt{VarX_i VarY_i}} = \frac{E\{(X_i-EX_i)(Y_i-EY_i)\}}{\sqrt{E\{(X_i-EX_i)^2\}E\{(Y_i-EY_i)^2\}}}.$$

On page 270 it is shown that the maximum likelihood estimate of ρ calculated from the sample, $(x_1,y_1)^*,\ldots,(x_n,y_n)^*$, from a bivariate normal distribution is the empirical correlation coefficient (or the Pearson correlation coefficient),

$$r = \frac{SPD_{xy}}{\sqrt{SSD_x SSD_y}} = \frac{\sum_{i=1}^{n}(x_i-\bar{x}.)(y_i-\bar{y}.)}{\sqrt{\sum_{i=1}^{n}(x_i-\bar{x}.)^2\sum_{i=1}^{n}(y_i-\bar{y}.)^2}},$$

where $\bar{x}. = \frac{1}{n}\sum_{i=1}^{n}x_i$ and $\bar{y}. = \frac{1}{n}\sum_{i=1}^{n}y_i$. Furthermore, in Section 6.6.1 it is shown that the hypothesis $\rho = 0$ is tested by means of the statistic,

$$t = \sqrt{n-2}\frac{r}{\sqrt{1-r^2}} \sim\sim t(n-2). \tag{12.49}$$

The corresponding significance probability is

$$p_{obs} = 2(1 - F_{t(n-2)}(|t|)),$$

where $F_{t(n-2)}$ is the distribution function of the t-distribution with $n-2$ degrees of freedom.

A nonparametric version of the empirical correlation coefficient is obtained by replacing the observations with their ranks. Let $r_i(x)$ denote the rank of the observation x_i in the sample, $x_1,\ldots,x_n$, and, similarly, let $r_i(y)$ denote the rank of the observation y_i in the sample, $y_1,\ldots,y_n$. The *Spearman rank correlation coefficient* is then defined as

$$r_S = \frac{SPD_{r(x)r(y)}}{\sqrt{SSD_{r(x)}\,SSD_{r(y)}}} = \frac{\sum_{i=1}^{n}\left(r_i(x)-\bar{r}.(x)\right)\left(r_i(y)-\bar{r}.(y)\right)}{\sqrt{\sum_{i=1}^{n}\left(r_i(x)-\bar{r}.(x)\right)^2\sum_{i=1}^{n}\left(r_i(y)-\bar{r}.(y)\right)^2}}, \tag{12.50}$$

where $\bar{r}.(x) = \frac{1}{n}\sum_{i=1}^{n} r_i(x) = \frac{n+1}{2}$ and $\bar{r}.(y) = \frac{1}{n}\sum_{i=1}^{n} r_i(y) = \frac{n+1}{2}$ according to (12.5).

Using (12.7), formula (12.50) may be written as

$$r_S = \frac{\sum_{i=1}^{n} r_i(x)r_i(y) - \frac{n(n+1)^2}{4}}{\frac{(n-1)n(n+1)}{12}}. \tag{12.51}$$

Under the hypothesis of independence H_0, it may be shown that the corresponding discrete random R_S is symmetric around 0, which implies that

$$E R_S = 0, \tag{12.52}$$

and, furthermore, that

$$Var R_S = \frac{1}{n-1}. \tag{12.53}$$

The significance probability of H_0 is

$$p_{obs} = P(|R_S| \geq |r_S|). \tag{12.54}$$

For small values of n, this probability may be found in tables or using a statistical computer package.

For large values of n, the distribution of R_S can be approximated by a normal distribution. From (12.52) and (12.53) it follows that

$$Z(\mathbf{X},\mathbf{Y}) = \sqrt{n-1}R_S \approx N(0,1) \tag{12.55}$$

and the p-value may be calculated as

$$p_{obs} = 2(1-\Phi(|z(\mathbf{x},\mathbf{y})|)). \tag{12.56}$$

An alternative test of H_0 is obtained by replacing the original observations in the t-test statistic in (12.49) with the rank of the observations, i.e.,

$$t_S = \sqrt{n-2}\frac{r_S}{\sqrt{1-r_S^2}} \tag{12.57}$$

and calculate the significance probability as

$$p_{obs} = 2(1-F_{t(n-2)}(|t_S|)). \tag{12.58}$$

Remark 12.5 If there are $a_j(x)$ observations among the x's with rank $r_j(x)$ and $a_j(y)$ observations among the y's with rank $r_j(y)$, the expression for the rank correlation coefficient in (12.50) is still valid, but the formula (12.51) becomes

$$r_S = \frac{\sum_{i=1}^{n} r_i(x)r_i(y) - \frac{n(n+1)^2}{4}}{\frac{1}{12}\sqrt{\left((n-1)n(n+1)-\sum_j(a_j^3(x)-a_j(x))\right)\left((n-1)n(n+1)-\sum_j(a_j^3(y)-a_j(y))\right)}}. \tag{12.59}$$

The formulas (12.54), (12.56), and (12.58) are still valid in the case of tied observations. ▼

Example 6.1 (Continued)
From the output listing on page 517, it is seen that the Spearman correlation coefficient of the two samples, corresponding to the two dates, consisting of the logarithm of the length and the logarithm of the weight of the toad *Bufo bufo* are 0.969 and 0.923, respectively, and that the corresponding p-values are less that 0.0001. Thus, again a positive correlation between the two variables has been established. (Recall from page 289 that the Pearson correlation coefficients for the two dates were 0.978 and 0.942, respectively.) □

12.2.5 Two-Way Tables

Suppose that there exists a natural ordering of the s categories corresponding to the s columns in the $r \times s$ table below:

	1	...	j	...	s
1	x_{11}	...	x_{1j}	...	x_{1s}
⋮	⋮		⋮		⋮
i	x_{i1}	...	x_{ij}	...	x_{is}
⋮	⋮		⋮		⋮
r	x_{r1}	...	x_{rj}	...	x_{rs}

Often in examples the s categories are described in terms of s disjoint intervals and the observed numbers, $x_{i1}, \ldots, x_{ij}, \ldots, x_{is}$, in the ith row of the table correspond to grouping n_i observations from a continuous distribution into the s intervals.

In general, if a natural ordering of the columns in an $r \times s$ table exists and one is interested in testing whether the r distributions corresponding to the rows in the table are identical, a nonparametric alternative to the test of homogeneity of r multinomial distributions considered in Section 7.3.1 is the Wilcoxon two-sample test discussed in Section 12.2.2 if $r = 2$, and the Kruskal-Wallis test in Section 12.2.3 if $r \geq 3$.

If both the r categories corresponding to the r rows and the s categories corresponding to the s columns in an $r \times s$ table can be ordered in a natural way, a nonparametric test of independence of the two criteria corresponding to the rows and the columns is the Spearman rank correlation test in Section 12.2.4, which may be considered as a more powerful alternative to the test of independence in the multinomial distribution in Section 7.2.3 because it takes the ordering into account.

Remark 12.6 In a $2 \times s$ table the results of the Wilcoxon two-sample test and of the Spearman rank correlation test are identical since tedious calculations show that the statistic $z(\mathbf{x})$ in (12.38) is equal to minus the statistic $z(\mathbf{x},\mathbf{y})$ in (12.55). Thus, if $r = 2$ the test of homogeneity and the test of independence are identical from a mathematical point of view, although they are concerned with different hypotheses in different models. Recall from Section 7 that for a general $r \times s$ table a similar remark is true in models based on the multinomial distribution. ▼

Example 7.3 (Continued)
The data in this example is reproduced in the table below and clearly, there is a natural ordering of both the row and the columns. Both row and column variables can be considered to be grouped observations from continuous distributions due to very crude measurement techniques. In order to calculate the Spearman rank correlation r_S the table has been supplemented with row sums ($x_{i\cdot}$), the

column sums ($x_{\cdot j}$), and with the corresponding ranks calculated using (12.8).

		Breadth /Length				
		< 0.6	0.6–0.8	> 0.8	*Total*	r(row)
	< 25	6	31	11	48	24.5
Weight	25–50	8	29	19	56	76.5
	> 50	10	39	26	75	142.0
	Total	24	99	56	179	
	r(col)	12.5	74.0	151.5		

Since

$$(n-1)n(n+1) = 178 \times 179 \times 180 = 5735160,$$

$$\sum_{i=1}^{3}(x_{i\cdot}^3 - x_{i\cdot}) = 707904,$$

$$\sum_{j=1}^{3}(x_{\cdot j}^3 - x_{\cdot j}) = 1159560$$

and

$$\sum_{i=1}^{3}\sum_{j=1}^{3} x_{ij} r_i(\text{row}) r_j(\text{col}) = 1477794,$$

it follows from (12.59) that

$$r_S = \frac{1477794 - \dfrac{179 \times 180^2}{4}}{\dfrac{1}{12}\sqrt{(5735160 - 707904)(5735160 - 1159560)}} = 0.06979.$$

In the output from SAS on page 518, it is seen that the exact significance probability in (12.54) is 0.3520. Using (12.55) and (12.56) we find that the value $z = 0.93111$ of the approximately normal distributed test statistic gives the significance probability 0.3518, and finally, from (12.57) and (12.58), we get that $t = 0.93077$ with the significance probability 0.3532. Thus, the hypothesis H_0: $\rho = 0$ is not rejected, and there is no indication of a connection between the weight of the stones and their shape, expressed in terms of the ratio breadth/length. A similar conclusion was reached in Chapter 7 by means of a test of independence in the multinomial distribution. □

Example 7.4 (Continued)
The data in this example are reviewed in the table below. Obviously, the categories corresponding to the columns are ordered, and the data for each of the four municipalities may be considered as observations of the continuous variable concentration grouped into three intervals. For each of the two areas, test of the hypothesis that the two distributions within the area are identical are found on page 519.

		Concentration			
Area	*Municipality*	≤0.01	0.01–0.10	>0.10	*Total*
1	*Hadsten*	23	12	6	41
	Hammel	20	5	9	34
2	*Nørre Djurs*	37	6	4	47
	Randers	47	4	3	54

For area 1 the exact significance probability of the Wilcoxon two-sample test, or the Kruskal-Wallis test, is 0.7972 and the approximative test gives the p-value 0.8110. For area 2 the exact significance probability is 0.2865, which is approximated by the value 0.2742. Thus, for each of the two areas we do not reject the hypothesis of identity of the two distributions of the concentration corresponding to the two municipalities within the area.

In order to compare the two areas, we consider the table below, which contains sums of the distributions within the areas:

	Concentration			
Area	≤0.01	0.01–0.10	>0.10	*Total*
1	43	17	15	75
2	84	10	7	101

From the output from SAS on page 520, it is seen that with an accuracy of four decimal places the significance probability is 0.0002 for both the exact and approximative Wilcoxon two-sample test of the hypothesis that the concentration is the same in the two areas. As in Chapter 7, where the data were analyzed by means of homogeneity tests in the multinomial distribution, we can conclude that the distributions of the concentration are different in the two areas. □

Annex to Chapter 12

Calculations in SAS

The nonparametric tests in this chapter can be calculated in SAS using the procedures *PROC UNIVARIATE*, *PROC NPAR1WAY*, *PROC CORR*, and *PROC FREQ* as shown in the continuation of the examples below.

Example 12.1 (Continued)

The sign test and the Wilcoxon signed rank test can be found in the output listing from *PROC UNIVARIATE*. By default, the procedure considers a test of the hypothesis that the location of the observations is $\mu_0 = 0$. Other values of μ_0 may be considered using the `MU0=` option in the `PROC UNIVARIATE` statement.

```
TITLE1 'Example 12.1';
DATA calcium;
INPUT woman normal calcium@@;
difference=calcium-normal;
DATALINES;
 1 6.2 12.1    2 3.3  4.6    3 1.3 1.8    4 5.6 8.2
 5 3.8  3.9    6 5.1  7.9    7 4.1 4.0    8 1.9 4.6
 9 1.1  2.3   10 7.2 10.0   11 6.8 5.6   12 2.7 3.5
13 5.1  6.9
;
RUN;

PROC UNIVARIATE DATA=calcium;
VAR difference;
RUN;
```

The relevant part of the output is seen below in the table `Test for Location: Mu0=0`.

In the `Sign` test SAS considers the statistic $m = n^+ - (n^+ + n^-)/2 = (n^+ - n^-)/2$, i.e., the statistic n^+ minus its mean calculated in the binomial distribution with number of trials $n^+ + n^-$ and probability parameter 0.5, or half the difference between the number of positive and negative observations.

The p-value of the Wilcoxon `Signed Rank` test is calculated from the exact distribution of S if the number N of observations different from 0 (μ_0) is less or equal to 20. If $N > 20$, the p-value is calculated using the test statistic in (12.27) and the significance probability in (12.21) on page 499.

In the line `Student's t` the value and the significance level of the t-test statistic in formula (3.9) are given. In this example the t-test is actually the paired t-test in (5.11) and so the validity of this test has to be checked using the methods in Section 5.3 on page 252.

```
                          Example 12.1

                    The UNIVARIATE Procedure
                     Variable:  difference


                   Tests for Location: Mu0=0

Test             -Statistic-        -----p Value------

Student's t      t  3.274676        Pr > |t|      0.0066
Sign             M       4.5        Pr >= |M|     0.0225
Signed Rank      S      38.5        Pr >= |S|     0.0049
```

□

Example 12.2 (Continued)
For two samples the option WILCOXON in the call of *PROC NPAR1WAY* implies that the Wilcoxon two-sample test is calculated. SAS does not allways use the statistic $R_{1\cdot}$ in (12.28), but calculates the sum S of the ranks in the smaller of the two samples. As default the statistic Z in (12.38) on page 504 with $R_{1\cdot}$ replaced by S is calculated using continuity corrections, which we do not consider in this book. For this reason the option CORRECT=NO is used in the call of the procedure. The p-value calculated in the exact distribution of the statistic S in is obtained by means of the command EXACT. Finally, the option ANOVA implies calculation of the F-statistic of the hypothesis that the means in the two samples are equal under the assumption that the samples are normally distributed with a common variance.

```
TITLE1 'Example 12.2';
DATA reac;
INPUT group reactime@@;
DATALINES;
1   3 1   5 1   5 1   7 1   9 1   9 1  10 1  12
1  20 1  24 1  24 1  34 1  43 1  46 1  58 1 140
2   2 2   5 2   5 2   7 2   8 2   9 2  14 2  18
2  24 2  26 2  26 2  34 2  37 2  42 2  90
;
RUN;

PROC NPAR1WAY DATA=reac WILCOXON ANOVA CORRECT=NO;
CLASS group;
VAR reactime;
EXACT;
RUN;
```

In this example the two sample sizes are both larger than 10, so the two approximate tests based on the Z statistic are both valid; the EXACT statement is not necessary and is considered here only for the sake of illustration. SAS gives the following warning that the computation of the p-value in the exact distribution of S requires much time and memory.

```
/*WARNING: Computing exact p-values for this problem may require much time and
         memory. Press the system interrupt key to terminate exact
         computations.*/
```

The first part of the output, where caution should be exercized, is concerned with the analysis based on the normal distribution of the two samples. The F-test is based on the assumption that the samples are normally distributed, an assumption that should be checked using fractile diagrams if the sample sizes are large enough. In this example it is seen from Figure 2.16 on page 38 that the two samples of reaction times are **not** normally distributed. Consequently, the information about the F-test should not be used. The next part of the output gives information on the rank sums in the two samples. Under the heading, `Wilcoxon Two-Sample Test`, the observed values of the statistics S and Z are given together with the corresponding significance probabilities. The significance probability of the S statistic is found in the line, `Two-Sided Pr >= |S - Mean|`, under the heading `Exact Test`. The normal approximation (12.33) to the p-values of the Z statistic is given in the line `Two-Sided Pr > |Z|` under the heading `Normal Approximation` and the p-value in (12.37) calculated using the t-distribution is found under the heading `t Approximation` in the line `Two-Sided Pr > |Z|`. Finally, information concerning the statistic KW in (12.47) is given in the table `Kruskall-Wallis Test`.

```
                     The NPAR1WAY Procedure

         Analysis of Variance for Variable reactime
                Classified by Variable group

         group                  N                     Mean

         1                     16                28.062500
         2                     15                23.133333


Source     DF     Sum of Squares     Mean Square     F Value     Pr > F

Among       1         188.103360      188.103360      0.2217     0.6413
Within     29       24608.670833      848.574856


                Average scores were used for ties.


                     The NPAR1WAY Procedure

        Wilcoxon Scores (Rank Sums) for Variable reactime
                  Classified by Variable group

                          Sum of       Expected       Std Dev          Mean
group          N          Scores       Under H0       Under H0         Score

1             16           261.0          256.0      25.244610     16.312500
2             15           235.0          240.0      25.244610     15.666667

                Average scores were used for ties.


                   Wilcoxon Two-Sample Test

                Statistic (S)                  235.0000

                Normal Approximation
                Z                               -0.1981
                One-Sided Pr <  Z                0.4215
                Two-Sided Pr > |Z|               0.8430

                t Approximation
                One-Sided Pr <  Z                0.4222
                Two-Sided Pr > |Z|               0.8443

                Exact Test
                One-Sided Pr <=  S               0.4264
                Two-Sided Pr >= |S - Mean|       0.8529


                     Kruskal-Wallis Test

                     Chi-Square            0.0392
                     DF                         1
                     Pr > Chi-Square       0.8430
```

It is seen above that in this example the second sample is the smaller of the two samples. This explains why the value of z has the opposite sign of that calculated on page 504. □

Example 12.3 (Continued)

The program below creates the data set and *PROC NPAR1WAY* calculates the Kruskal-Wallis test and the F-test for the hypothesis that the distributions of the four groups in this example are identical.

```
TITLE1 'Example 3.3 nonparametric';
DATA area;
INPUT group area@@;
DATALINES;
1 200 1 215 1 225 1 229 1 230 1 232 1 241
```

```
1 253 1 256 1 264 1 268 1 288 1 288
2 235 2 188 2 195 2 205 2 212 2 214 2 182
2 215 2 272 2 163 2 230 2 255 2 202
3 314 3 320 3 310 3 340 3 299 3 268 3 345
3 271 3 285 3 309 3 337 3 282 3 273
4 283 4 312 4 291 4 259 4 216 4 201 4 267
4 326 4 241 4 291 4 269 4 282 4 257
;
RUN;

PROC NPAR1WAY DATA=area WILCOXON ANOVA;
CLASS group;
VAR area;
RUN;
```

The output listing from the program is given below. The Kruskal-Wallis test is calculated by means of the keyword `Wilcoxon` in the `PROC NPAR1WAY` statement. The `ANOVA` keyword request the analysis of variance F-test for identical mean assuming that the samples are normally distributed.

```
                     Example 3.3 nonparametric

                       The NPAR1WAY Procedure

              Analysis of Variance for Variable area
                  Classified by Variable group

              group               N              Mean

              1                  13        245.307692
              2                  13        212.923077
              3                  13        304.076923
              4                  13        268.846154

Source      DF    Sum of Squares    Mean Square    F Value    Pr > F

Among        3      57636.365385    19212.12179    21.4579    <.0001
Within      48      42976.307692      895.33974

                Average scores were used for ties.

                       The NPAR1WAY Procedure

          Wilcoxon Scores (Rank Sums) for Variable area
                  Classified by Variable group

                        Sum of     Expected      Std Dev          Mean
group         N         Scores     Under H0     Under H0         Score

1            13          282.0       344.50    47.313644     21.692308
2            13          147.0       344.50    47.313644     11.307692
3            13          549.0       344.50    47.313644     42.230769
4            13          400.0       344.50    47.313644     30.769231

                Average scores were used for ties.

                       Kruskal-Wallis Test

                    Chi-Square          29.4203
                    DF                        3
                    Pr > Chi-Square      <.0001
```

□

Example 6.1 (Continued)

The program for calculating the Spearman rank correlation test for the data is similar to that on page 288 except that the option SPEARMAN has been added in the call of *PROC CORR*. The program

```
TITLE 'Example 6.1 nonparametric';

PROC CORR DATA=bufobufo SPEARMAN;
VAR ln_weight;
WITH ln_length;
BY date;
RUN;
```

gives the output listing

```
                       Example 6.1 nonparametric

------------------------------- date=75-07-24 --------------------------------

                          The CORR Procedure

                      1 With Variables:    ln_length
                      1      Variables:    ln_weight


                          Simple Statistics

Variable          N        Mean     Std Dev      Median     Minimum     Maximum

ln_length        18     2.80167     0.09288     2.81500     2.67000     3.00000
ln_weight        18     5.77944     0.27541     5.82500     5.31000     6.27000


                 Spearman Correlation Coefficients, N = 18
                        Prob > |r| under H0: Rho=0

                                     ln_weight

                      ln_length        0.96881
                                        <.0001

                       Example 6.1 nonparametric

------------------------------- date=75-08-16 --------------------------------

                          The CORR Procedure

                      1 With Variables:    ln_length
                      1      Variables:    ln_weight


                          Simple Statistics

Variable          N        Mean     Std Dev      Median     Minimum     Maximum

ln_length        23     3.08652     0.07601     3.09000     2.97000     3.26000
ln_weight        23     6.57087     0.22573     6.59000     6.22000     7.03000


                 Spearman Correlation Coefficients, N = 23
                        Prob > |r| under H0: Rho=0

                                     ln_weight

                      ln_length        0.92297
                                        <.0001
```

□

Example 7.3 (Continued)

The Spearman rank correlation test in an $r \times s$ table may be performed by *PROC FREQ* using the statements `TABLES` and `EXACT SCORR`. The option `ORDER=data` in the call of the procedure is important since it ensures that the order of the rows and columns in the table is the same in the calculations as in the reading of the data.

```
TITLE 'Example 7.3 nonparametric';

DATA ex7_3;
INPUT weight$1-5 shape$7-14 count;
CARDS;
  <25     <0.6  6
  <25 0.6-0.8 31
  <25    >0.8 11
25-50     <0.6  8
25-50 0.6-0.8 29
25-50    >0.8 19
  >50     <0.6 10
  >50 0.6-0.8 39
  >50    >0.8 26
;
RUN;

PROC FREQ DATA=ex7_3 ORDER=data;
TABLES weight*shape/NOROW NOCOL NOPERCENT CHISQ EXPECTED;
WEIGHT count;
EXACT SCORR;
RUN;
```

In the output listing below it is remarkable that the statistic z, obtained by normalizing the rank correlation $r_S = 0.0698$ with its standard error, has the value 0.9607 with a p-value of 0.3367. On page 511 it is seen that in the calculation done by hand we have obtained the same value of r_S, but that the value of z there was 0.9311 and the corresponding p-value 0.3518, which is much closer to the exact p-value 0.3520 than that given by SAS. In the continuation of Example 7.4 below there is further indication that the normalization of the rank correlation coefficient that SAS considers differs from that in (12.55).

```
                  Example 7.3 nonparametric

                     The FREQ Procedure

            Statistics for Table of weight by shape

               Spearman Correlation Coefficient
               --------------------------------
               Correlation (r)           0.0698
               ASE                       0.0726
               95% Lower Conf Limit     -0.0725
               95% Upper Conf Limit      0.2121

                 Test of H0: Correlation = 0

               ASE under H0              0.0726
               Z                         0.9607
               One-sided Pr >  Z         0.1684
               Two-sided Pr > |Z|        0.3367

               Exact Test
               One-sided Pr >=  r        0.1766
               Two-sided Pr >= |r|       0.3520

                     Sample Size = 179
```

□

Example 7.4 (Continued)

PROC NPAR1WAY requires that the variable to be analyzed is continuous. In the program below we read the data the same way as in Example 7.3. Here the variable `concentration` is a text variable and in order to meet the demand above, we use this variable to define a new version of the data set `dichlor` with continuous variable `group` representing the three groups of the concentration. The statements `CLASS municipality`, `EXACT` and `BY area` imply that for each area we get the exact *p*-values of the Wilcoxon two-sample test for identity of the distributions corresponding to the two municipalities.

```
TITLE 'Example 7.4 nonparametric';
DATA dichlor;
INPUT area municipality$3-10 concentration$12-20 count;
DATALINES;
1 Hadsten    <=0.01 23
1 Hadsten  0.01-0.1 12
1 Hadsten      >0.1  6
1 Hammel     <=0.01 20
1 Hammel   0.01-0.1  5
1 Hammel       >0.1  9
2 Nr Djurs   <=0.01 37
2 Nr Djurs 0.01-0.1  6
2 Nr Djurs     >0.1  4
2 Randers    <=0.01 47
2 Randers  0.01-0.1  4
2 Randers      >0.1  3
;
RUN;

DATA dichlor;
SET dichlor;
IF concentration='<=0.01'   THEN group=1;
IF concentration='0.01-0.1' THEN group=2;
IF concentration='>0.1'    THEN group=3;
RUN;

PROC NPAR1WAY DATA=dichlor WILCOXON CORRECT=NO;
CLASS municipality;
VAR group;
FREQ count;
EXACT;
BY area;
RUN;
```

Part of the output listing from the program is:

```
----------------------------------- area=1 -----------------------------------

                      Wilcoxon Two-Sample Test

                 Statistic (S)                   1312.0000

                 Normal Approximation
                 Z                                  0.2392
                 One-Sided Pr >  Z                  0.4055
                 Two-Sided Pr > |Z|                 0.8110

                 t Approximation
                 One-Sided Pr >  Z                  0.4058
                 Two-Sided Pr > |Z|                 0.8116

                 Exact Test
                 One-Sided Pr >=  S                 0.3909
                 Two-Sided Pr >= |S - Mean|         0.7972
```

```
                    Kruskal-Wallis Test

               Chi-Square            0.0572
               DF                         1
               Pr > Chi-Square       0.8110

---------------------------------- area=2 ----------------------------------

                Wilcoxon Two-Sample Test

          Statistic (S)                      2501.5000

          Normal Approximation
          Z                                     1.0933
          One-Sided Pr >  Z                     0.1371
          Two-Sided Pr > |Z|                    0.2742

          t Approximation
          One-Sided Pr >  Z                     0.1384
          Two-Sided Pr > |Z|                    0.2769

          Exact Test
          One-Sided Pr >=  S                    0.1676
          Two-Sided Pr >= |S - Mean|            0.2865

                    Kruskal-Wallis Test

               Chi-Square            1.1954
               DF                         1
               Pr > Chi-Square       0.2742
```

The next program segment gives the tests of identity of the distributions in the two areas.

```
PROC NPAR1WAY DATA=dichlor WILCOXON CORRECT=NO;
CLASS area;
VAR group;
FREQ count;
EXACT;
RUN;
```

Output listing:

```
                Wilcoxon Two-Sample Test

          Statistic (S)                      7631.5000

          Normal Approximation
          Z                                     3.7805
          One-Sided Pr >  Z                     <.0001
          Two-Sided Pr > |Z|                    0.0002

          t Approximation
          One-Sided Pr >  Z                     0.0001
          Two-Sided Pr > |Z|                    0.0002

          Exact Test
          One-Sided Pr >=  S                 9.336E-05
          Two-Sided Pr >= |S - Mean|         1.526E-04

                    Kruskal-Wallis Test

               Chi-Square           14.2922
               DF                         1
               Pr > Chi-Square       0.0002
```

It follows from Remark 12.6 that for a $2 \times s$ table the normalized test statistics, both denoted by z, for the Wilcoxon two-sample test and for the Spearman rank correlation test are identical except for a sign. Here we have 2×3 tables and the following program calculates the Spearman rank correlation test for the data in area 2 in order to compare the results with those for the Wilcoxon two-sample test above.

```
PROC FREQ DATA=dichlor(where=(area=2)) ORDER=data;
TABLES municipality*concentration/NOROW NOCOL NOPERCENT EXPECTED
                                                 CHISQ;
WEIGHT count;
EXACT SCORR;
RUN;
```

Below it is seen that the value of the z statistic corresponding to the rank test is $z = -1.0909$. Since $r_S = -0.1093$ and $n = 101$, formula (12.55) gives the value -1.093. As seen from above, with an accuracy of three decimal places, this value is the same as the value of the z statistic of the Wilcoxon two-sample test. That the two tests are equivalent is seen from the exact p-value, which in both cases is 0.2865. This again indicates that the normalization of the z statistic that SAS considers in connection with the Spearman rank correlation in the `EXACT SCORR` statement in *PROC FREQ* differs from that in (12.55) resulting in the unfortunate fact that the p-values of two equivalent tests are different.

```
                    The FREQ Procedure

     Statistics for Table of municipality by concentration

              Spearman Correlation Coefficient
              --------------------------------
              Correlation (r)          -0.1093
              ASE                       0.0988
              95% Lower Conf Limit     -0.3030
              95% Upper Conf Limit      0.0844

                Test of H0: Correlation = 0

              ASE under H0              0.1002
              Z                        -1.0909
              One-sided Pr <  Z         0.1377
              Two-sided Pr > |Z|        0.2753

              Exact Test
              One-sided Pr <=  r        0.1676
              Two-sided Pr >= |r|       0.2865

                     Sample Size = 101
```

□

Main Points in Chapter 12

Sign Test

Test of the hypothesis,

$$H_0 : P(X_i > 0 \mid X_i \neq 0) = P(X_i < 0 \mid X_i \neq 0) = 0.5,$$

in the model for a sample, $x_1, \dots, x_n$,

$$M_0 : X_i \sim F, \quad i = 1, \dots, n.$$

Test statistic: Number of positive values in the sample, i.e.,

$$n^+ = \#\{ i \mid x_i > 0 \}.$$

p-value:

$$p_{obs} = 0.5^{(n^+ + n^- - 1)} \sum_{i=0}^{\min(n^+, n^-)} \binom{n^+ + n^-}{i},$$

where $n^- = \#\{ i \mid x_i < 0 \}$.

Rank Test: Calculation of Ranks

The ranks of the observations in a sample is calculated in the following way:
Let $y_1, y_2, \dots, y_m$ denote the m *different values* in the sample, $x_1, x_2, \dots, x_n$, in ascending order, i.e.,

$$y_1 < y_2 < \dots < y_m.$$

For $j = 1, 2, \dots, m$, let a_j denote the *number* of observations in the sample that are equal to y_j and let k_j denote the *cumulative number*, $k_j = a_1 + \dots + a_j$, and set $k_0 = 0$. The rank assigned to each of the a_j *tied* observations with value y_j is

$$k_{j-1} + \frac{a_j + 1}{2}.$$

Rank Test: One Sample

In the model for a sample from a continuous distribution F,

$$M_0 \colon X_i \sim F, \quad i = 1, \dots, n,$$

the hypothesis of symmetry of F around 0,

$$H_0 \colon F(-x) = 1 - F(x), \quad \text{for all } x > 0,$$

is considered.
Test statistic: The Wilcoxon signed rank test statistic is

$$S = \sum_{\{i: X_i > 0\}} R_i^+ - \frac{N(N+1)}{4},$$

where R_i^+ is the rank of $|X_i|$ in the sample consisting of the absolute values $|X_1|, \dots, |X_n|$ of the X's, and where $N = \#\{i \mid N_i \neq 0\}$ denotes the number of observations with value different from 0.
p-value:

$$p_{obs} = P(|S| \geq |s|).$$

Approximations ($N \geq 20$):

$$p_{obs}(\mathbf{x}) = 2(1 - \Phi(|u(\mathbf{x})|)),$$

where

$$U(\mathbf{X}) = \frac{S}{\sqrt{\frac{N(N+1)(2N+1)}{24} - \frac{1}{48}\sum_j (a_j^3 - a_j)}},$$

or

$$p_{obs}(\mathbf{x}) = 2(1 - F_{t(N-1)}(|t(\mathbf{x})|)),$$

where

$$T(\mathbf{X}) = \frac{S}{\sqrt{\frac{N}{N-1}\left(\frac{N(N+1)(2N+1)}{24} - \frac{1}{48}\sum_j (a_j^3 - a_j) - \frac{S^2}{N}\right)}}.$$

Rank Test: Two Samples

In the model for two samples,

$$M_0\colon X_{ij} \text{ has a continuous distribution } F_i\ ,\ j = 1, \ldots, n_i,\ i = 1, 2,$$

the hypothesis that all the $n. = n_1 + n_2$ observations have the same distribution is considered, i.e.,

$$H_0\colon F_i = F,\ i = 1, 2.$$

Test statistic: The Wilcoxon two-sample test statistic is

$$R_1. - ER_1. = R_1. - \frac{n_1(n. + 1)}{2}.$$

Here

$$R_1. = \sum_{j=1}^{n_1} R_{ij},$$

where R_{ij} denotes the rank of X_{ij} in the combined sample and $R_1.$ is the sum of the ranks in the first sample.
p-value:

$$p_{obs} = P(|R_1. - ER_1.| \geq |r_1. - ER_1.|)$$

Approximations: ($n_1 \geq 10$ and $n_2 \geq 10$)

$$p_{obs}(\mathbf{x}) = 2(1 - \Phi(|z(\mathbf{x})|))$$

or

$$p_{obs}(\mathbf{x}) = 2(1 - F_{t(n.-1)}(|z(\mathbf{x})|)),$$

where

$$Z(\mathbf{X}) = \frac{R_1. - \frac{n_1(n. + 1)}{2}}{\sqrt{\frac{n_1 n_2 (n. + 1)}{12}\left[1 - \frac{\sum_j (a_j^3 - a_j)}{(n. + 1)n.(n. - 1)}\right]}}.$$

The statistic $Z(\mathbf{X})$ may be written as

$$Z(\mathbf{X}) = \frac{\bar{R}_1. - \bar{R}_2.}{\sqrt{s_R^2\left(\frac{1}{n_1} + \frac{1}{n_2}\right)}},$$

where

$$s_R^2 = \frac{1}{n.-1}\sum_{i=1}^{2}\sum_{j=1}^{n_i}(R_{ij}-\bar{R}..)^2,$$

$$\bar{R}_i. = \frac{1}{n_i}\sum_{j=1}^{n_i} R_{ij}, \quad i=1,2$$

and

$$\bar{R}.. = \frac{1}{n.}\sum_{i=1}^{2}\sum_{j=1}^{n_i} R_{ij} = \frac{n.+1}{2}.$$

Rank Test: Several Samples

In the model for k (≥ 2) samples,

$$M_0\text{: } X_{ij} \text{ has a continuous distribution } F_i\text{, } j=1,\ldots,n_i,\ i=1,\ldots,k,$$

the hypothesis that all the $n. = n_1+\cdots+n_k$ observations have the same distribution is considered, i.e.,

$$H_0\text{: } F_i = F,\ i=1,\ldots,k.$$

Test statistic:

$$KW(\mathbf{X}) = \frac{12\sum_{i=1}^{k} n_i(\bar{R}_i. - \bar{R}..)^2}{n.(n.+1)\left[1 - \frac{\sum_j(a_j^3-a_j)}{(n.+1)n.(n.-1)}\right]},$$

where R_{ij} denotes the rank of X_{ij} in the combined sample and where

$$\bar{R}_i. = \frac{1}{n_i}\sum_{j=1}^{n_i} R_{ij}$$

and

$$\bar{R}.. = \frac{1}{n.}\sum_{i=1}^{k}\sum_{j=1}^{n_i} R_{ij}$$

denote the average of the ranks in the ith sample and in the combined sample, respectively.

An alternative expression for the statistic is

$$KW(\mathbf{X}) = \frac{\sum_{i=1}^{k} n_i(\bar{R}_i. - \bar{R}..)^2}{\frac{1}{n.-1}\sum_{i=1}^{k}\sum_{j=1}^{n_i}(R_{ij}-\bar{R}..)^2} = \frac{\sum_{i=1}^{k}\frac{R_i.^2}{n_i} - \frac{R..^2}{n.}}{\sum_{i=1}^{k}\sum_{j=1}^{n_i} R_{ij}^2 - \frac{R..^2}{n.}},$$

where $R_i.$ and $R..$ denote the sum of the ranks in the ith sample and in the combined sample, respectively.

p-value:

$$p_{obs}(\mathbf{x}) = 1 - F_{\chi^2(k-1)}(KW(\mathbf{x})).$$

Rank Test: Spearman correlation

In the model for n independent observations from a bivariate distribution with distribution function $F_{(X,Y)}$,

$$M_0\text{: } (X_i,Y_i)^* \sim F_{(X,Y)},\ i=1,\ldots n,$$

the Spearman rank correlation coefficient calculated on the basis of the sample, $(x_1, y_1)^*, \ldots, (x_n, y_n)^*$, is

$$r_S = \frac{SPD_{r(x)r(y)}}{\sqrt{SSD_{r(x)}SSD_{r(y)}}} = \frac{\sum_{i=1}^{n} \left(r_i(x) - \bar{r}.(x)\right)\left(r_i(y) - \bar{r}.(y)\right)}{\sqrt{\sum_{i=1}^{n} \left(r_i(x) - \bar{r}.(x)\right)^2 \sum_{i=1}^{n} \left(r_i(y) - \bar{r}.(y)\right)^2}}$$

Here $r_i(x)$ denotes the rank of the observation x_i in the sample, $x_1, \ldots, x_n$, and $r_i(y)$ denotes the rank of the observation y_i in the sample, $y_1, \ldots, y_n$.

If there are $a_j(x)$ observations among the x's with rank $r_j(x)$ and $a_j(y)$ observations among the y's with rank $r_j(y)$, the rank correlation coefficient may be calculated as

$$r_S = \frac{\sum_{i=1}^{n} r_i(x) r_i(y) - \dfrac{n(n+1)^2}{4}}{\dfrac{1}{12}\sqrt{\left((n-1)n(n+1) - \sum_j (a_j^3(x) - a_j(x))\right)\left((n-1)n(n+1) - \sum_j (a_j^3(y) - a_j(y))\right)}}.$$

The hypothesis of independence between the X's and the Y's, i.e.,

$$H_0\colon F_{(X,Y)}(x,y) = F_X(x)F_Y(y) \quad \text{for all } (x,y) \in \mathbb{R}^2,$$

is tested using the statistic R_S with significance value,

$$p_{obs} = P(|R_S| \geq |r_S|),$$

which for small values of n may be found in tables or using a statistical computer package.

Approximations of the p-value are obtained using either the statistic,

$$Z(\mathbf{X}, \mathbf{Y}) = \sqrt{n-1} R_S,$$

and the corresponding p-value,

$$p_{obs} = 2(1 - \Phi(|z(\mathbf{x}, \mathbf{y})|)),$$

where Φ is the distribution function of the standard normal distribution $N(0,1)$, or the statistic,

$$t_S = \sqrt{n-2}\frac{r_S}{\sqrt{1-r_S^2}},$$

and the corresponding p-value,

$$p_{obs} = 2(1 - F_{t(n-2)}(|t_S|)),$$

where $F_{t(n-2)}$ is the distribution function of the t-distribution with $n-2$ degrees of freedom.

APPENDIX A

Simulated Fractile Diagrams

In order to give the reader some experience in evaluating fractile diagrams, we show in this appendix fractile diagrams for different samples, $u_1, \ldots, u_n$, from the standard normal distribution $N(0,1)$. The samples have been generated by numerical simulation by means of the function `NORMAL` in SAS. Eight samples have been simulated for each of the sample sizes $n = 5$, 10, 15, 25, 50, 100, and 250. The corresponding fractile diagrams are shown on the following pages. The sample sizes appear from the heading of the diagrams.

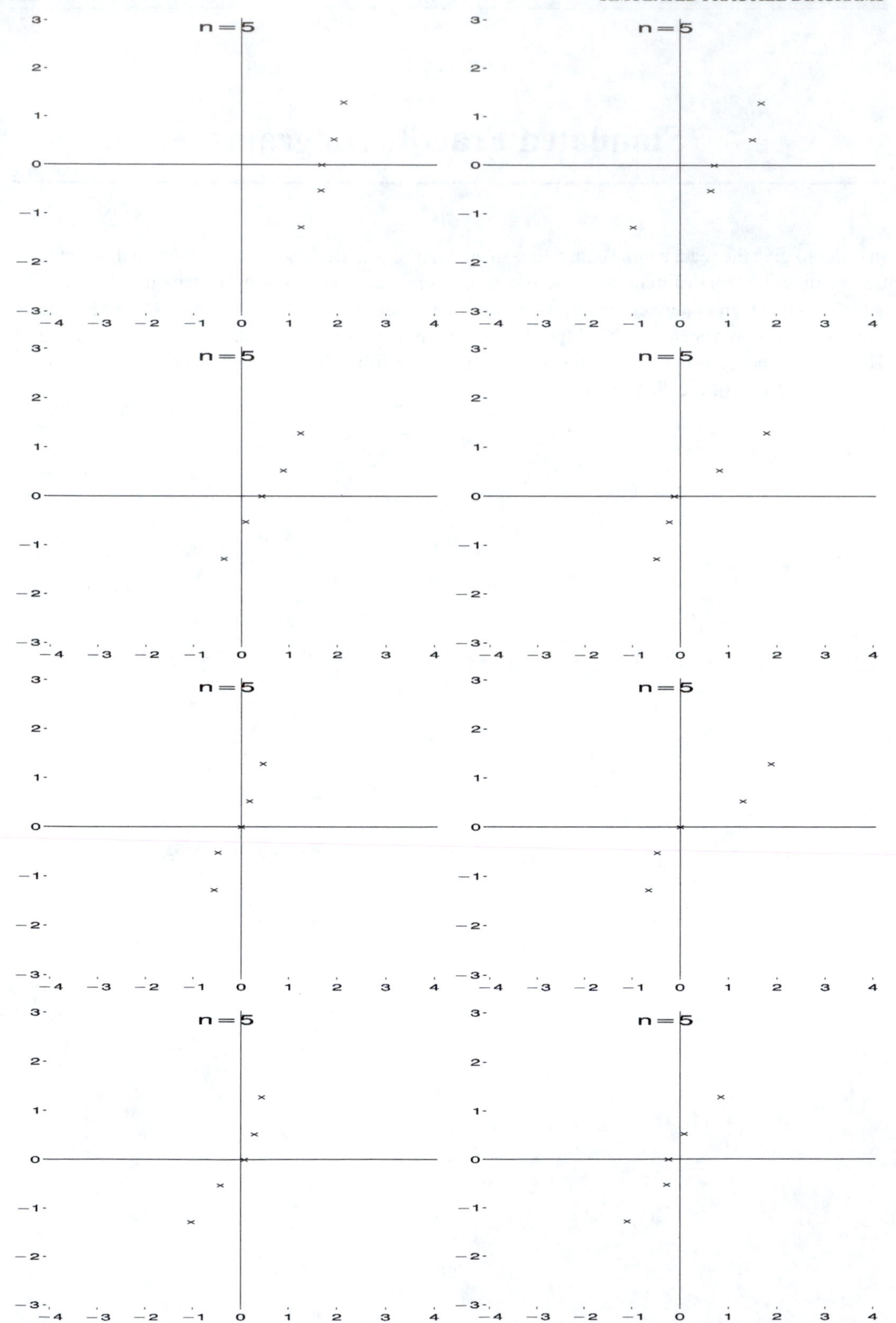
n = 5
n = 5
n = 5
n = 5
n = 5
n = 5
n = 5
n = 5

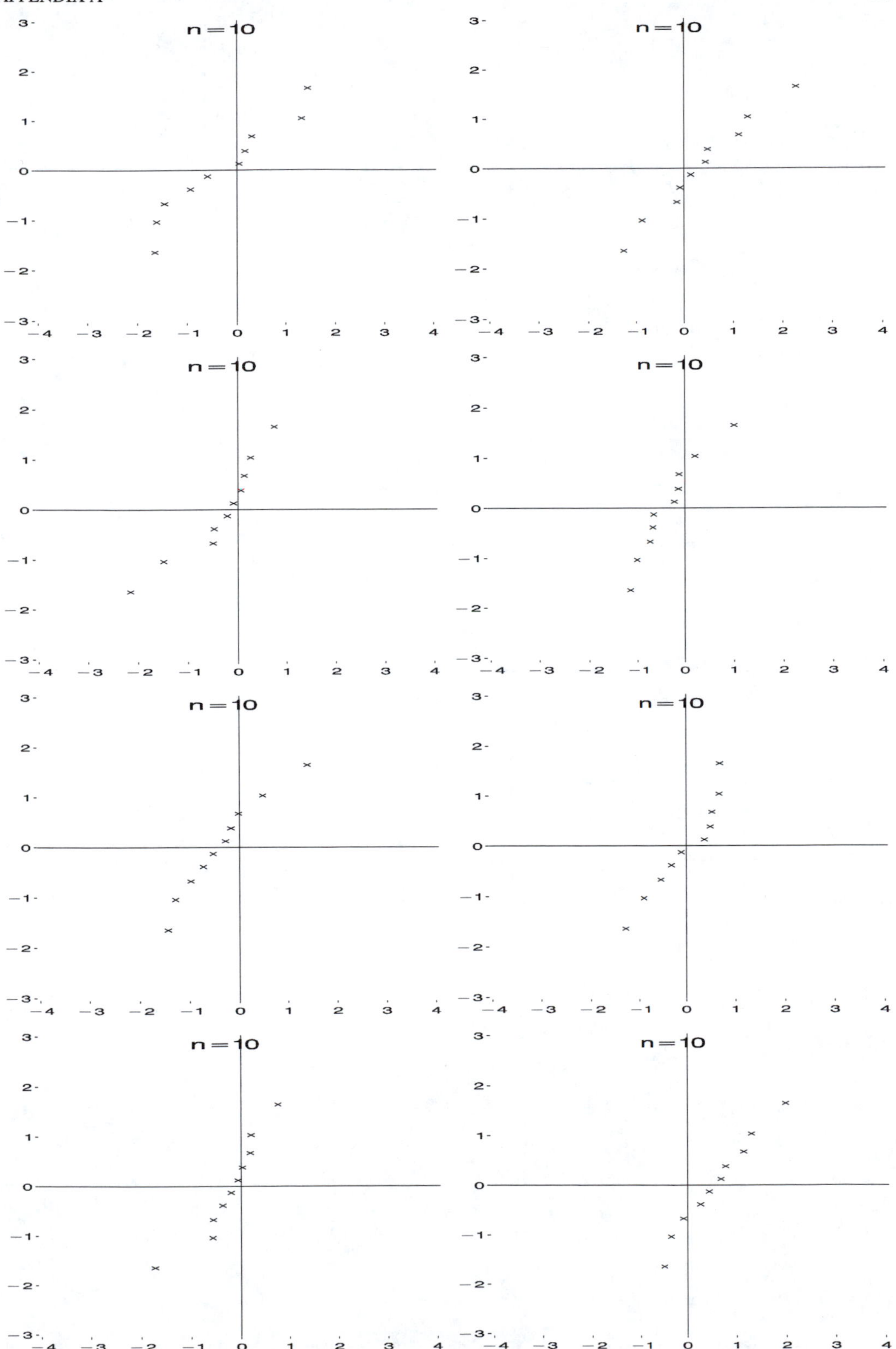
n = 10
n = 10
n = 10
n = 10
n = 10
n = 10
n = 10
n = 10
3
2
1
0
−1
−2
−3
−4 −3 −2 −1 0 1 2 3 4

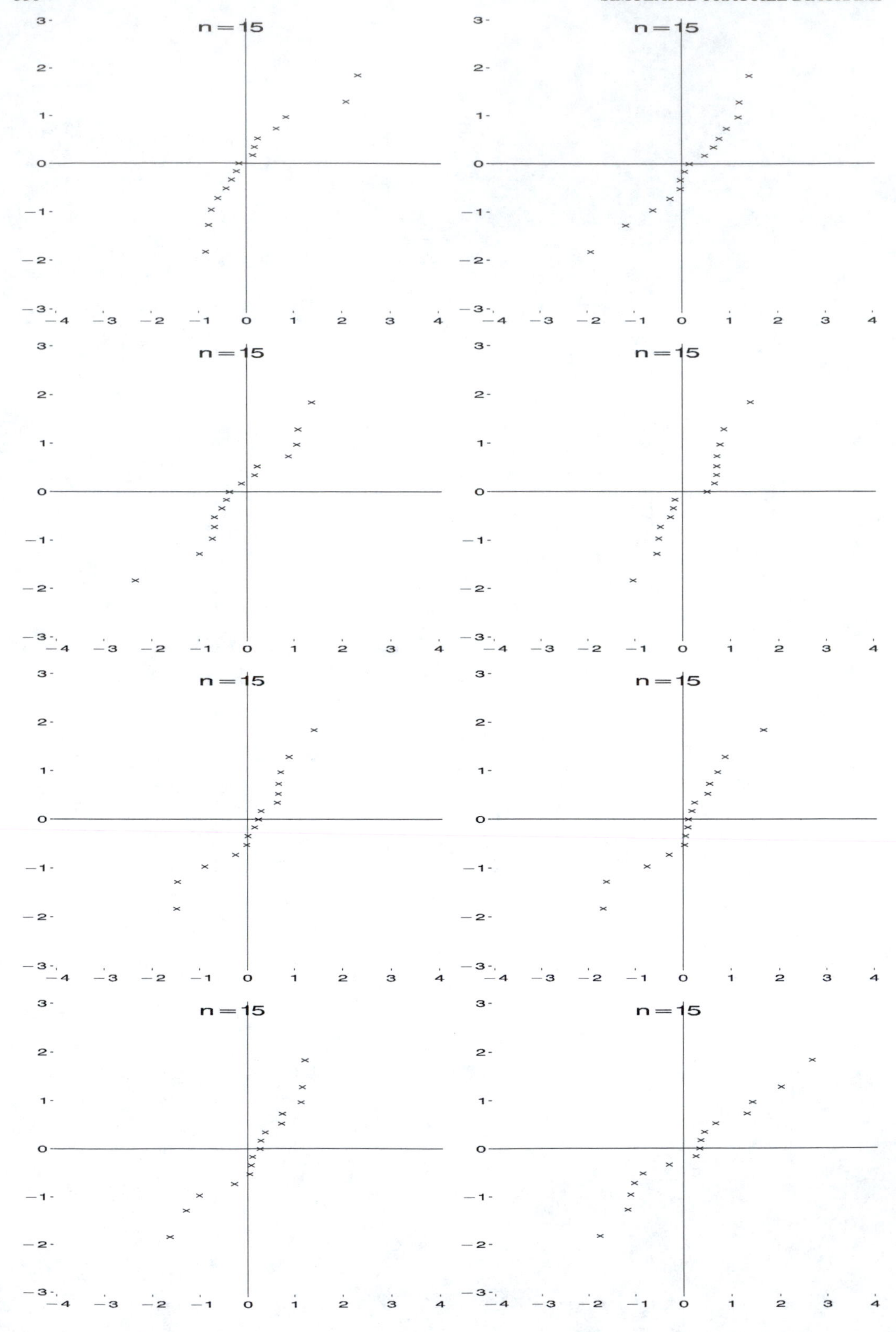
n = 15
n = 15
n = 15
n = 15
n = 15
n = 15
n = 15
n = 15

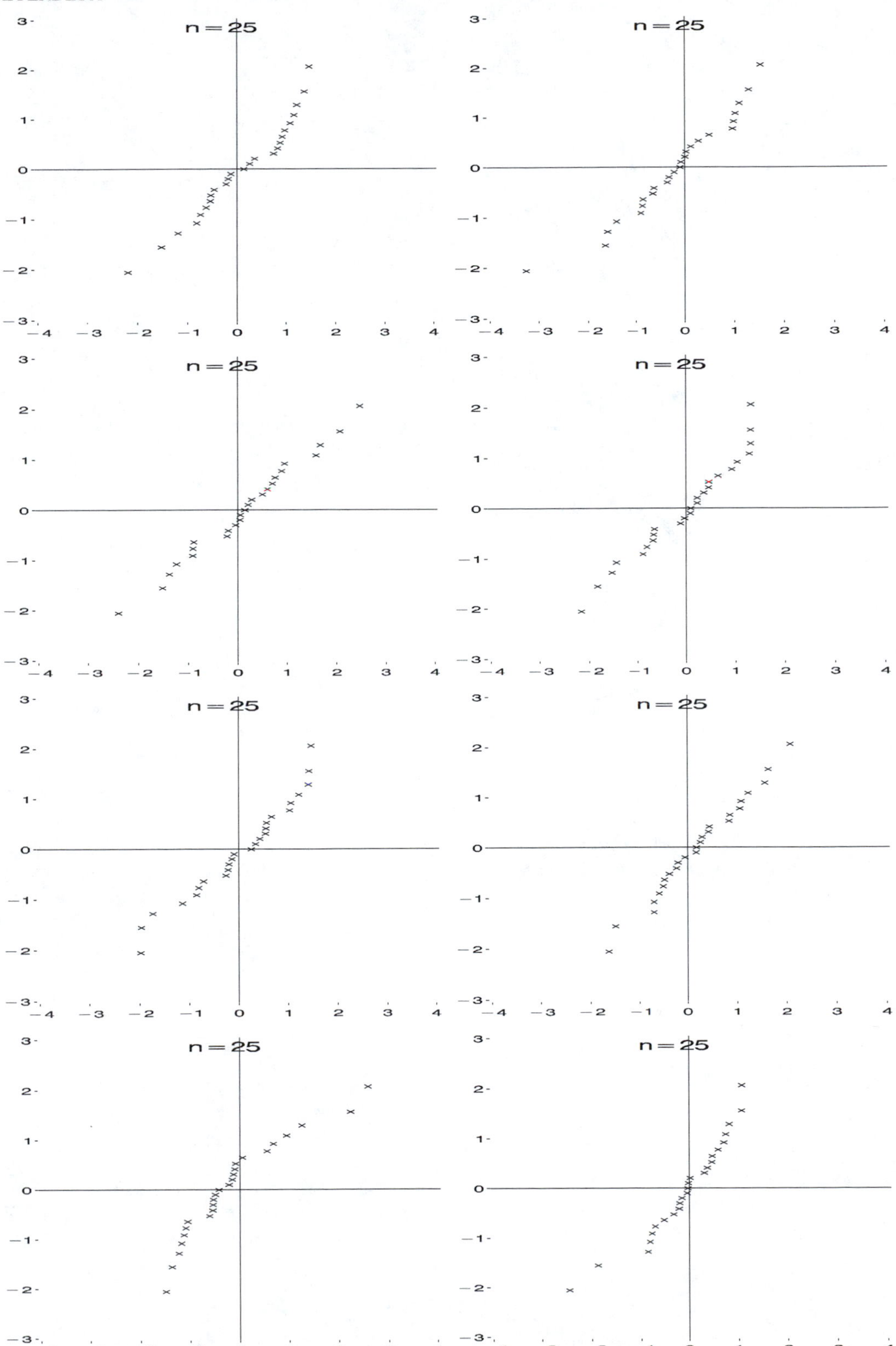
n = 25
n = 25
n = 25
n = 25
n = 25
n = 25
n = 25
n = 25

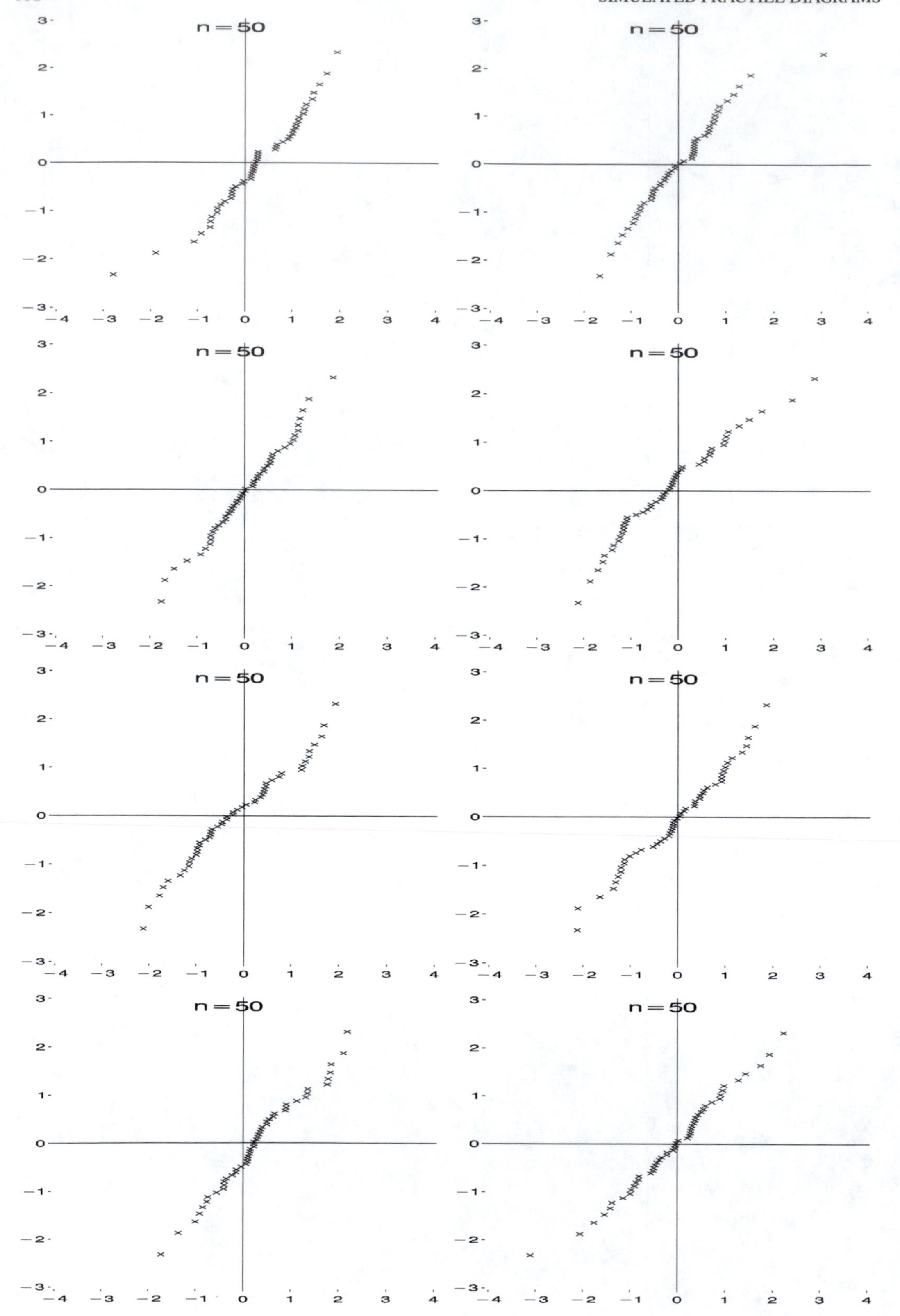
n = 50
n = 50
n = 50
n = 50
n = 50
n = 50
n = 50
n = 50

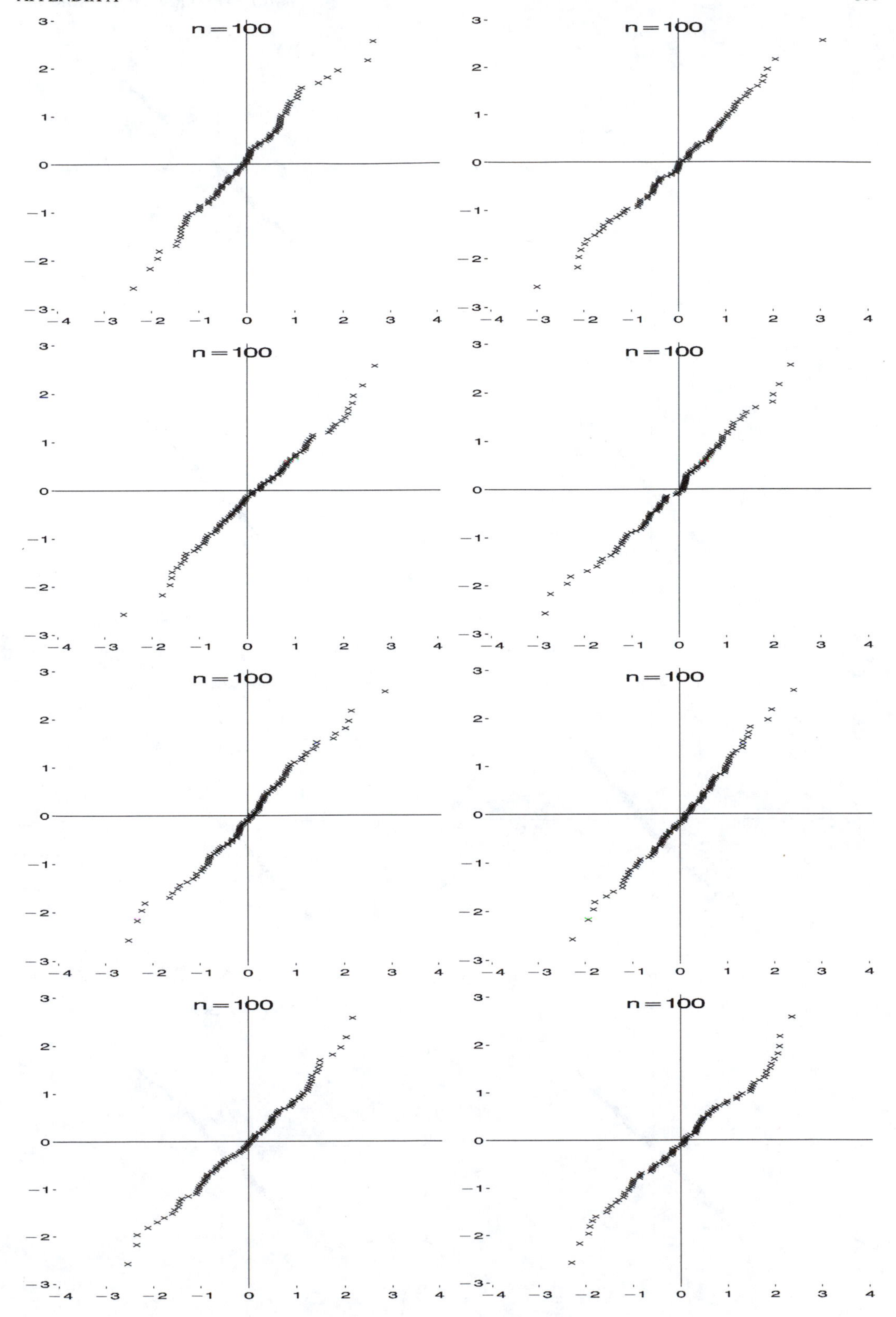
n = 100
n = 100
n = 100
n = 100
n = 100
n = 100
n = 100
n = 100

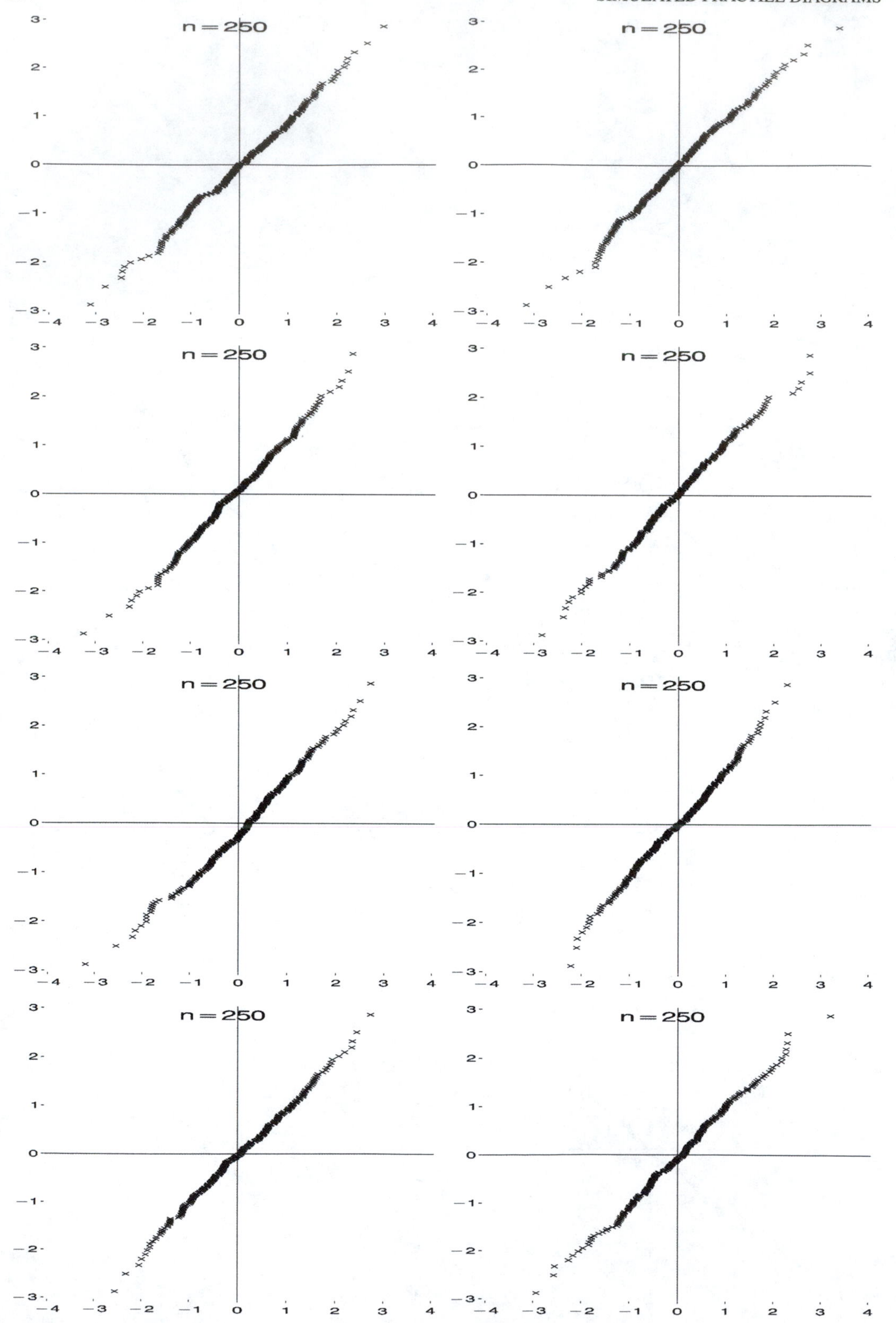
n = 250
n = 250
n = 250
n = 250
n = 250
n = 250
n = 250
n = 250

APPENDIX B

The Newton-Raphson Procedure

In Chapter 9 and Chapter 10 we have seen that it is sometimes necessary to use numerical methods in order to find the maximum likelihood estimate as a solution to the likelihood equation. Several numerical methods for solving equations exist. In this appendix we mention only one, namely, the *Newton-Raphson procedure*, which is an iterative method for solving equations of the form,

$$\mathbf{f}(\mathbf{x}) = \mathbf{0} \tag{B.1}$$

where $\mathbf{f}$ is a differentiable function from $\mathbb{R}^k$ into $\mathbb{R}^k$. We explain and illustrate the method in the case $k = 1$, where (B.1) takes the form,

$$f(x) = 0, \tag{B.2}$$

i.e., where f is a real-valued and differentiable function of a real-valued variable x.

Let x_0 denote a solution to (B.2). The Newton-Raphson procedure consists of finding a sequence of approximations, $x_m, m = 2, 3, \ldots$, to x_0 on the basis of an initial value x_1 by means of the formula,

$$x_{m+1} = x_m - \frac{f(x_m)}{f'(x_m)}, \tag{B.3}$$

which is the mth *iteration* of the procedure that specifies how to find x_{m+1} from x_m.

The idea behind the formula (B.3) is that the tangent to the curve $y = f(x)$ in the point $(x_m, f(x_m))$ has the equation,

$$y - f(x_m) = f'(x_m)(x - x_m),$$

and that the tangent therefore intersects the x-axis in the point $(x_{m+1}, 0)$, see Figure B.1.

If the initial value x_1 is sufficiently close to x_0, then under mild assumptions concerning f, the sequence $\{x_m\}$ will converge to x_0 as $m \to \infty$. Usually the methods converges quickly, but an unfortunate choice of the initial value x_1 may result in divergence of the method.

The iteration procedure ends when the distance between two consecutive x-values is sufficiently small and/or the value of $f(x_m)$ is sufficiently close to 0.

Example B.1

According to formula (10.14), the estimate $\hat{\kappa}$ of the precision κ in the Fisher distribution in Example 10.1 is found as the solution to the equation,

$$\coth(\kappa) - \kappa^{-1} = R/n,$$

where $R/n = 0.9747$, or, equivalently, as the solution to the equation,

$$f(x) = R/n - \coth(x) + x^{-1} = 0.$$

Since

$$f'(x) = -1 + (\coth(x))^2 - x^{-2},$$

formula (B.3) becomes

$$x_m = -\frac{R/n - \coth(x_m) + x_m^{-1}}{(\coth(x_m))^2 - 1 - x_m^{-2}}.$$

For the sake of illustration we set $x_1 = 1$ and find the following iterations:

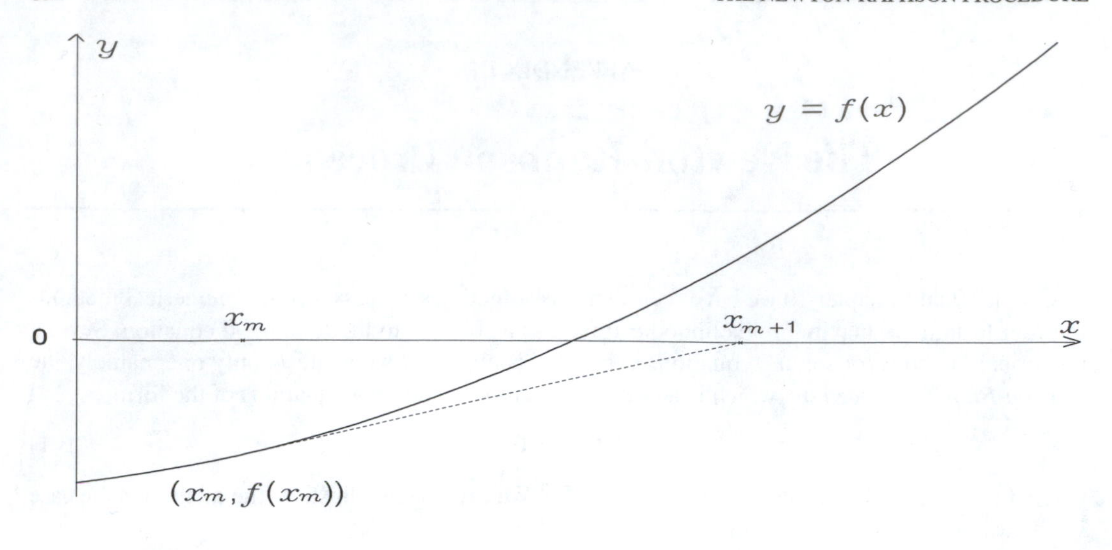

Figure B.1 *Illustration of the mth iteration in the Newton-Raphson procedure.*

m	x_m	$f(x_m)$
1	1.0000	0.6616
2	3.3978	0.2667
3	6.6453	0.1251
4	12.1737	0.0568
5	20.5935	0.0232
6	30.4448	0.0075
7	37.4117	0.0014
8	39.3708	7.0×10^{-5}
9	39.4788	1.9×10^{-7}
10	39.4791	$< 1.0\times10^{-10}$

Since the value of $f(x)$ in the 10th iteration is very close to 0, the iteration procedure is stopped and, consequently, we have found that

$$\hat{\kappa} = 39.4791.$$

The iteration procedure is illustrated in Figure B.2. (The initial value $x_1 = 1$ was chosen for the sake of illustration. According to Table 10.7, $x_1 = 40.0000$ is a better choice of initial value.) □

Example B.2

The likelihood equation for a one-dimensional parameter ω is

$$l'(\omega) = \frac{dl}{d\omega}(\omega) = 0,$$

i.e., it is an equation of the form (B.2) with $f = l'$. Since the observed information $j(\omega)$, according to (11.22), is

$$j(\omega) = -\frac{d^2l}{d\omega^2}(\omega) = -f'(\omega),$$

the formula (B.3) may in this case be rewritten as

$$\omega_{m+1} = \omega_m + l'(\omega_m)j(\omega_m)^{-1}. \tag{B.4}$$

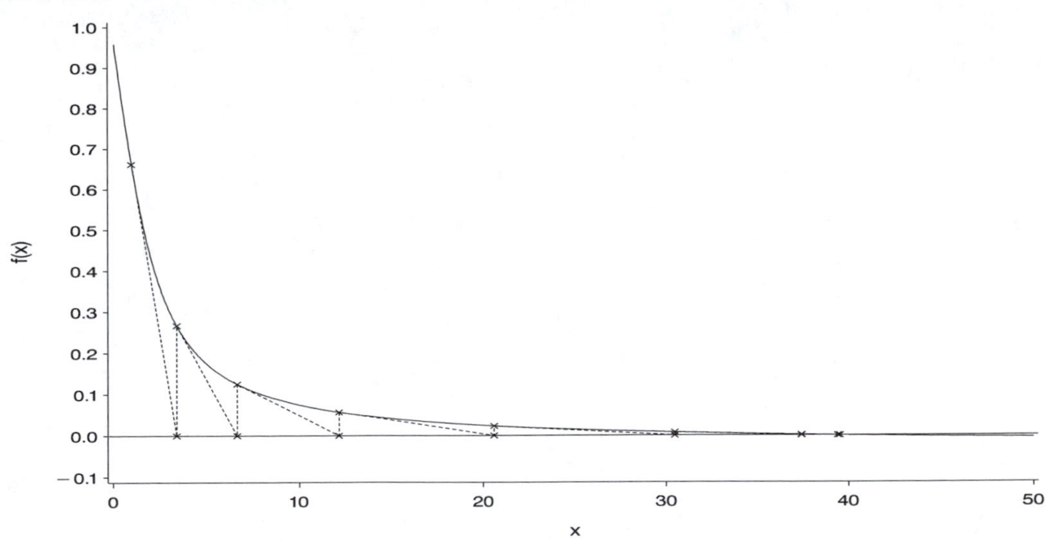

Figure B.2 *Solving the equation* $f(x) = R/n - coth(x) - x^{-1} = 0$ *by means of the Newton-Raphson procedure.*

As mentioned above the Newton-Raphson procedure may also be used in connection with functions of several variables. The equation (B.4) also determines the mth iteration if we want to solve the likelihood equation for a k-dimensional parameter, $\omega = (\omega_1, \dots, \omega_k)^*$. In that case we only have to interpret $l'(\omega)$ in (B.4) as the transpose of the vector of derivatives, i.e.,

$$l'(\omega) = (\frac{\partial l}{\partial \omega_1}(\omega), \dots, \frac{\partial l}{\partial \omega_k}(\omega)),$$

and $j(\omega)$ in (B.4) as the observed information matrix, i.e., as the $k \times k$ matrix with elements,

$$j(\omega)_{rs} = -\frac{\partial^2 l}{\partial \omega_r \partial \omega_s}$$

$r, s = 1, \dots, k.$ □

References

Andersen, A.H. and Keiding, N. (1983): *Anvendt Statistik*. Department of Theoretical Statistics, University of Aarhus, Denmark. [354, 397]

Anderson, E. (1935): The irises of the Gaspé Peninsula. *Bull. Amer. Iris. Soc.*, **59**, pp. 2–5. [27]

Anscombe, F.J. (1948): The transformation of Poisson, binomial, and negative binomial data. *Biometrika*, **35**, pp. 246–254. [381]

Anscombe, F.J. (1973): Graphs in statistical analysis. *American Statistician*, **27**, pp.17–21. [264]

Bartlett, M.S. (1936): The square root transformation in analysis of variance. *J. R. Statist. Soc. Suppl.*, **3**, pp. 68–78. [381]

Bartlett, M.S. (1937): Properties of sufficiency and statistical tests. *Proc. Roy. Soc., A,* **160**, pp. 268–282. [101]

Christiansen, F.B., Frydenberg, O., and Simonsen, V. (1973): Genetics of Zoarces populations. IV. Selection component analysis of an esterase polymorphism using population samples including mother-offspring combinations. *Hereditas*, **73**, pp. 291–304. [333]

Christiansen, F.B., Frydenberg, O., and Simonsen, V. (1984): Genetics of Zoarces populations. XII. Variation at the polymorphic loci PgmI, PgmII, HbI and EstIII in fjords, *Hereditas,* **101**, pp. 37–48. [302]

Duelli, P. and Wehner, R. (1973): The spectral sensitivity of polarized light orientation in *Cataglyhpis bicolor* (Formicidae, Hymenoptera). *Journal of Comparative Physiology*, **86**, pp. 37–53. [451]

Eanes, W.F., Gaffney, P.M., Koehn, R.K., and Simon, C.M. (1977): A study of sexual selection in natural populations of the milkweed beetle *Tetraopes tetraophthalmus*. In *Measuring Selection in Natural Populations*. Lecture Notes in Biomathematics, **19**. Springer-Verlag. Berlin. pp. 49–64. [347]

Fisher, R.A. (1915): Frequency distribution of the values of the correlation coefficient in samples from an indefinitely large population. *Biometrika*, **10**, pp. 507–521. [272]

Fisher, R.A. (1922): On the mathematical foundations of theoretical statistics. *Philos. Trans. Roy. Soc.*, **A222**, pp. 309–368. [481]

Fisher, R.A. (1925): Theory of statistical estimation, *Proc. Camb. Phil. Soc.*, **22**, pp. 700–725. [481]

Fisher, R.A. (1953): Dispersion on a sphere. *Proc. Roy. Soc. London Ser. A,* **217**, pp. 295–305. [449]

Fisher, R.A., Thornton, H.G., and Mackenzie, W.A. (1922): The accuracy of the plating method of estimating bacterial populations. *Ann. Appl. Biol.*, **9**, pp. 325–359. [363]

Gallucci, V.F. (1985): The Garrison Bay Project, Stock Assessment and Dynamics of the Littleneck Clam, Protothaca staminea. In Andrews, D.F. and Herzberg, A.M., *Data. A Collection of Problems from Many Fields for the Student and Research Worker*. Springer-Verlag, New York. [86]

Galton, F. (1885): Regression toward Mediocrity in Hereditary Stature. *Journal of the Anthropological Institute*, **15**, pp. 246–265. [269]

Hald, A. (1952): *Statistical Theory with Engineering Applications*. John Wiley & Sons, New York. [37]

Hotelling, H. (1931): The generalisation of Student's ratio. *Ann. Math. Stat.*, **2**, pp. 360–378. [285]

Keiding, N. (1976): Statistical analysis in the bivariate normal distribution. Department of Theoretical Statistics, University of Copenhagen. [266]

Kronmal, R.A. (1993): Spurious correlation and the fallacy of the ratio standard revisited. *J. R. Statist. Soc. A*, **156**, pp. 379–392. [299]

Li., J.C.R. (1964): *Statistical Inference*, v.1. Edwards, Ann Arbor, Mich. [211]

Mardia, K.V. and Jupp, P.E. (1999): *Directional Statistics*. John Wiley & Sons, New York. [447, 476]

Mendel, G. (1866): Versuche über Pflanzen-Hybriden, *Verhandlungen des naturforschenden Vereines in Brünn*, **4**, pp. 3–47. [302]

Neyman, J.R. (1952): *Lectures and conferences on mathematical statistics and probability*, 2nd ed, Washington, D.C.: U.S. Department of Agriculture. [282, 299]

Neyman, E. and Pearson, E.S. (1928): On the use and interpretation of certain test criteria for purposes of statistical inference. *Biometrika*, **20A**, pp. 175–204 and 263–294. [484]

Olsen, J.H., Nielsen, A., and Schulgen, G. (1993): Residence near high voltage facilities and risk of cancer in children. *British Medical Journal*, **307**, pp. 891–895. [326]

Poulsen, V., Rose-Hansen J., and Springer N. (1986): *Geostatistik*. Geologisk Centralinstitut, Københavns Universitet, Copenhagen, Denmark. [478]

Rutherford, E. and Geiger, M. (1910): The probability variations in the distribution of alpha-particles. *Philosophical Magazine, Series* 6, **20**, pp. 698–704. [393]

Samuels, M.L. and Witmer, J.A. (1999): *Statistics for the Life Sciences*. Prentice-Hall, Upper Saddle River, New Jersey. [96]

Saxov, S. (1978): *Noter til Geostatistik*. Laboratoriet for Geofysik, Aarhus Universitet, Denmark. [35]

Sick, K. (1965) Haemoglobin polomorphism of cod in the Baltic and in the Danish Belt Sea. *Hereditas*, **54**, pp. 19–48. [1, 353]

The Open University (1981) S237: *The Earth: Structure, Composition and Evolution*. [395]

Vistelius, A.B. (1966): *Structural Diagrams*. Pergamon, London. [477]

Watson, G.S. (1956): Analysis of dispersion on a sphere. *Monthly Notices Roy. Astr. Soc. Geophys. Suppl.*, **7**, pp. 153–159. [453]

Watson, G.S: (1970): Orientation statistics in earth sciences. *Bull. Geol. Inst. Univ. Uppsala*, **2**, pp. 73–89. [450]

Will, R.G., Ironside, J.W., Cousens, S.N., Estibeiro, K., Alperovitch, A., Poser, S., Pocchiari, M., Hoffman, A., and Smith, P.G.(1996): A new variant of Creutzfeldt-Jacob disease in the UK, *The Lancet* **347**, April 6. [399]

Williamson, E. and Bretherton, M.H. (1963): *Tables of the Negative Binomial Probability Distribution*. John Wiley & Sons, New York. [360]

Index

PRIFYSGOL CAERDYDD